# The Invention of the Silicon Engine

B. J. G. van der Kooij

In the Invention Series, the following studies have been published:

The Invention of the Steam Engine
The Invention of the Electromotive Engine
The Invention of the Communication Engine 'Telegraph'
The Invention of the Electric Light
The Invention of the Communication Engine 'Telephone'
The Invention of the Wireless Communication Engine
The Invention of the Internal Combustion Engine
The Invention of the Silicon Engine

In the Deep History Series, the following studies have been published:

Context for Innovation: British (R)evolutions in Perspective
Origins of Innovation: Ancient (R)evolutions in Perspective
Discovery of Innovation: Historic (R)evolutions in Perspective

This case study is part of the research work in preparation for a doctorate-dissertation to be obtained from the Erasmus University Rotterdam in the Netherlands. This publication of our research data is a of case studies about the Nature of Innovation in the *Invention*-series.

About the text: This is a scholarly case study describing the historic events and developments that resulted in the electronic engines called the Radio Tube, Transistor and Integrated Circuit. It is based on a large number of historic and contemporary sources. As we conducted hardly any research into primary sources, we made use of the efforts of numerous others by citing them to preserve the original character of their contributions. Where possible, we identified the individual authors of the citations. As some are not identifiable, we identified the source of the text. Facts and texts that are considered to be of a general character in the public domain (eg Wikimedia Commons licensed) are not cited but integrated in the text.

About the pictures: Many of the pictures used in this case study were found at websites accessed through the Internet. Where possible, they were traced to their origins, which, when found, were indicated as the source. For those that are not out of copyright, we feel that the fair use we make of the pictures to illustrate an aspect of the scholarly case is not an infringement of copyright.

Copyright © 2022 B.J.G. van der Kooij

Front Cover shows as illustrations in the top row prototypes of the Fleming Valve (1904, left) and the DeForest Vacuum Triode (1907, right). The middle row shows the first Computer of a Chip: a DIL packaged 4004 micro-processor (1971). Bottom row shows the die-shots of the biplanar transistor (Hoerni, 1962, left), the Integrated Circuit (Noyce, 1961, middle) and the Intel 4004 chip (1971, right).
Cover design by Maxime Delsaux (Crea-Mania).

Version 3.0 (June 2023). All rights reserved.

ISBN-13: 9798812779153

# Contents

**Contents** ........................................................................................................ iii
**Preface** ......................................................................................................... vii
   About the Invention Series .................................................................. ix
      The General Purpose Technology ................................................. xii
      Evolution and Revolution ............................................................... xiv
      About our Research ....................................................................... xv
   About Events, Evolutions and Revolutions ....................................... xx
   About Life Cycles ................................................................................ xxv
   About Organizations ......................................................................... xxix
   About the Context ............................................................................ xxxii
   About the Perspective ..................................................................... xxxvi
   About the Interaction ...................................................................... xxxix
   About the Acts of Novelty ................................................................. xli
   About the Metaphors used in this Case Study ................................ xlii
   About this Case Study ...................................................................... xliii
**Context for the Discoveries** .......................................................................... 1
   The Era of Mechanization and Mechanical Thinking ......................... 5
   Science and Engineering, Technical and Social Change .................. 10
      The Context of the Early Nineteenth Century (1800-1850) ............. 11
      The Context of the Late Nineteenth Century (1850-1900) .............. 12
      The Context of the Early Twentieth Century (1900-1920) ............... 37
      The Context of the First World War (1914-1918) ............................. 46
      The Context of the Interbellum (1920-1940) ................................... 55
      The Context of the Second World War (1940-1945) ....................... 87
      The Context of the Postbellum Period (1950-1975) ...................... 111
   The Context of the Industrial Revolutions ..................................... 150

**The Context of the Electric Era** ................................................................. **155**
   Scientists discovering the Nature of Lightning ...................................... 156
      The Broad Application of Electricity ................................................ 161
      A new Primary Mover for Electric Power ........................................ 169
**The Communication (R)evolution** ............................................................ **171**
   The Invention of the Communication Engines ...................................... 172
      Wireless Communication 0.0 (1860-1895) ........................................ 173
      Invention of the Wireless System ...................................................... 179
      Wireless Communication 1.0 (1895-1920) ........................................ 187
   The First Communication Revolution (1830-1920) ............................... 196
**The Calculating (R)evolution** .................................................................. **201**
   The Invention of the Calculating Machines .......................................... 202
      Calculating Machines 0.0 (1620-1800) .............................................. 205
      Calculating Machines 1.0 (1800-1900) .............................................. 212
      Calculating Machines 2.0 (1900-1920) .............................................. 230
      The First Calculating Revolution (1850-1930) .................................. 256
**The Computing (R)evolution** ................................................................... **275**
   The Invention of the Computing Machines .......................................... 276
      Computing Machines 0.0 (1780-1930) .............................................. 279
      Computing Machines 1.0 (1930-1950) .............................................. 281
      Computing Machines 2.0 (1950-1980) .............................................. 287
      The First Computing Revolution (1930-1980) .................................. 294
   The IC & IT-Revolution (1850-1950) ...................................................... 301
   The dawn of the Fourth Industrial Revolution ...................................... 303
**The Electronic (R)evolutions** .................................................................. **307**
   The First Electronic Revolution (1900-1950) ........................................ 308
      The Era of Classic Physics ................................................................ 309
      The Invention of the Radio Valve ..................................................... 315
      The Vacuum Tube in Communication Applications ......................... 327
      Early Development of Vacuum Tubes .............................................. 339
      Early Vacuum Tube Manufacturing Industry .................................... 341

    The Evolution of the Vacuum Tube ........................................................ 351
    Overview of the First Electronic Revolution (1900-1950) ................ 363
  The Second Electronic Revolution (1950-1980) ........................................ 367
    The Era of Solid-State Physics ................................................................ 367
    The Era of Modern Physics ..................................................................... 369
    The Invention of the Transistor .............................................................. 387
    The Transistor and its Manufacturers ................................................... 397
    The Evolution of the Transistor ............................................................. 406
    Industrial Dynamics in the Valley: from Start-up to Spin-off ........... 415
    The Invention of the Integrated Circuit. .............................................. 437
    The Invention of the Computer on a Chip: the Micro-computer ..... 457
    The Silicon Engine as a Technology Driving Force ........................... 471
    Overview of the Second Electronic Revolution .................................. 481
  Overview of the Electronic Revolutions .................................................... 486
    The Inventions in a Revolutionary Perspective ................................... 486

## The Radio Revolution ........................................................................... 489
  The Second Communication Revolution (1900-1950) ............................. 490
    Communication Systems ......................................................................... 490
    Radio Broadcasting 0.0. (1900-1920) ...................................................... 495
    Radio Broadcasting 1.0 (1920-1945) ....................................................... 516
    Radio Broadcasting 2.0 (1945-1970) ....................................................... 535
    Radio Broadcasting 3.0 (1970s-2000s) ................................................... 540
    The Radio Manufacturers (1900-1945) .................................................. 541
    Overview of the Early Days of Wireless Radio ................................... 573

## The Television Revolution .................................................................... 577
  The Era of Mass Entertainment .................................................................. 578
    Mass Entertainment ................................................................................. 578
    Television Broadcasting 0.0 (1900-1930) ............................................... 585
    Television Broadcasting 1.0 (1930-1950) ............................................... 601
    Television Broadcasting 2.0 (1950-1970) ............................................... 615
    Television Broadcasting 3.0 (1970-2000) ............................................... 629
    The Digital Television Revolution ......................................................... 632
    Overview of the Early Days of Television ........................................... 636

## The Mobile Telephone Revolution.................................................................643

Personal Communication...............................................................................644

Mobile Telephony 1.0 (1950-1975)..........................................................645
Mobile Telephony 2.0 (1975-2000)..........................................................649
The Mobile Networks and its Operators..................................................653
Development of the Cell Phones into Feature Phones..........................658
Personal Communication and Information Systems............................676

The Smartphone..............................................................................................688

Inventions contributing to the Smartphone...........................................689
Evolution of the Mobile Phone into Smart Phone................................695
The Mobile Telecom Industry..................................................................707

Overview of the Early Days of the Cell Phone..........................................723

The Impact on Mobile Telephony............................................................726

The ICT Revolution (1950-2000).................................................................731

The Rise of the ICT-Revolution................................................................732

## Conclusions.............................................................................................735

Change and Novelty of the ICT-Revolution..............................................735

The Historic Context for Change and Novelty........................................736

The Invention of Innovation..........................................................................740

The Different Dimensions of Novelty Creation......................................741
Acts of Organization..................................................................................748
The Overall Impact of the GPT-ICT.......................................................750

Epilogue............................................................................................................751

## References...............................................................................................753
## Acknowledgments...................................................................................756
## About the author....................................................................................756

# Preface

*When everything is said and done,*
*and all our breath is gone.*
*The only thing that stays,*
*Is history, to guide our future ways.*

My lifelong intellectual fascination with technical innovation within the context of society started in Delft, the Netherlands. In the 1970s I studied at the University of Technology, at both the electrical engineering school and the business school.[1] Having been educated as a technical student, I studied vacuum tubes, followed by transistors, and I found the novelty caused by the new technology of microelectronics to be mindboggling. Not only from a technical point of view but because of all the opportunities it created for new products, new markets, and new organizations.

During my studies at both the school of electric engineering and the school of business administration I was lucky enough to spend some time in Japan and California, where I noticed how cultures influence the context for technology-induced change and what is considered novel. In Japan, I explored the research environment. In the American Silicon Valley, meeting people like Ted Hoff —the inventor of the microcomputer— I saw the effects of the business environment. I observed the extremes: from the nuances in human interaction of the Japanese to the stimulating and raw capitalism of the United States. The impact of microelectronic technology, forecasted by my engineering thesis, made the coming 'technology push' a little clearer: the personal computer was on the horizon. The implementation of innovation in small and medium enterprises and the

---

[1] At the present time, it is the Delft University of Technology Electrical Engineering School and the Erasmus University Rotterdam School of International Business Administration.

subject of my management thesis left me with a lot of questions. Could something like a Digital Delta be created in the Netherlands?

During my life's journey, innovation has been the theme. After graduation in the mid-1970s, I joined a mature electric equipment company that manufactured electric motors, transformers, and switching equipment. Business development was one of my major responsibilities. How could we change a maturing/aging corporation by picking up new business opportunities? Japan and California were again on the agenda but now from a business point of view. I explored acquisition, cooperation, and subcontracting. Could we create business activity in personal computers? The answer was no.

I entered politics and became a member of the Dutch Parliament — quite an innovative move for an engineer— and innovation on the national level became my theme. How could we prepare a society by creating new firms and industries to meet the new challenges that were coming and that would threaten the existing industrial base? What innovation policies could be applied? In the early 1980s, my introduction of the first personal computer in Parliament caused me to be known as 'Mr. Innovation' within the small world of my fellow parliamentarians. Could we, as politicians, change Dutch society by picking up the new opportunities technology was offering? The answer was no.

The next phase on my journey brought me in touch with two extremes. A (part-time) professorship in the Management of Innovation program at the University of Technology in Eindhoven gave room for my scholarly interests. I was looking at innovation at the macro level of science. In addition, the starting of a venture company making application software for personal computers satisfied my entrepreneurial obsession. Now it was about the (nearly full-time) implementation of innovation on the microscale of a start-up company. With both my head in the scientific clouds and my feet in the organizational mud, my capabilities were stretched. At the end of the 1980s I had to choose, and entrepreneurship won for the next eighteen years. Could I start and do something innovative with personal computers myself? The answer was yes.

When I reached retirement in the 2010s and reflected on my past experiences and the changes in our world since the 1970s, I wondered what had made all this happen. Technological innovation was a phenomenon that had fascinated me along my entire life journey. What is the thing we call 'innovation'? In many phases of the journey of my life I tried to formulate an answer; with my first book, *Micro-computers, Innovation in Electronics* (1977, technology level), my second book, *The Management of Innovation* (1983, business level), and my third book, *Innovation, from Distress to*

*Guts* (1988, society level). In the 2010s I had time on my hands, so I decided to pick up where I had left off and start studying the subject of innovation again. As a guest of my alma mater, working on my dissertation, I tried to find an answer to the question 'What is the nature of innovation?'

My fascination with 'innovation' may have started in Delft. But seen from an intellectual point of view, *Corona volente*, it may end in Rotterdam.

## *About the Invention Series*

Our research into the phenomenon of innovation, focusing on technological innovation, covered quite a time span: from the late seventeenth century up to today. The case study of the steam engine marked the beginning of the series. That is not to say there was no technological innovation before that time. On the contrary, imitation, invention, and innovation have been with us for a much longer period of time and could have been investigated. However, we had to limit ourselves, as we wanted to look at those technological innovations that were the result of a 'General Purpose Technology' (GPT). An expression that is not a part of everyone's vocabulary. As, clearly, some clarification is needed here, we will start with some definitions of the major elements of our research: innovation, product, technology, GPT, and revolution.

We define *innovation* as the creation of something new and applicable. It is a process over time that results in a new combination: a new artefact, a new service, a new structure or method. Whereas *invention* is the conversion of a *discovery*[2] of a new phenomenon that does not need a practical implementation, innovation brings the initial idea to the marketplace, where it can be used. We follow Alois Schumpeter's definition: 'Innovation combines factors in a new way, or…it consists in carrying out New Combinations…' (Schumpeter, 1939, p. 84). We consider 'Innovation' being quite different from invention. Also, for Schumpeter:

> *Although most innovations can be traced to some conquest in the realm of either theoretical or practical knowledge, there are many which cannot. Innovation is possible without anything we should identify as invention, and invention does not necessarily induce innovation, but produces of itself…no economically relevant effect at all.* (Schumpeter, 1939, p. 80)

What about invention, then? We follow here Abott Usher's interpretation, where the creative act is the new combination of the 'act of skills' and the 'act of insight': "Invention finds its distinctive feature in the

---

[2] Discovery is the novel awareness of an existing but unknown phenomenon of facts; from geographical facts, astronomical facts to natural facts.

constructive assimilation of pre-existing elements into new syntheses, new patterns, or new configurations of behaviour" (Usher, 1929, p. 11). Again the element of a combination is recognizable. By the way, one has to realize that these definitions arose in the early twentieth century and their meaning has shifted over time.

As a great part of our research is related to *product innovation*, we define a product as an artefact (from the Latin 'arte' —by or using an 'art'— and 'factum' —something made) that, through its product-function, fulfils a need. As Herbert Simon stated:

> *An artefact can be thought as the meeting point—an interface in today's terms—between an 'inner' environment, the substance and organization of the artefact itself, and an 'outer environment', the surroundings in which it operates. If the inner environment is appropriate to the outer environment, or vice versa, the artefact will serve its purpose.* (Simon, 1996, p. 6)

Just imagine the product-function of timekeeping, realized by the 'inner' environment of the timepiece, that can be considered an answer to the need for timekeeping of the 'outer environment', the person who wants to know the time. Those environments, at a given moment in time, have to fit, as the example of the sundial illustrates; it is useless without the sun shining.

Those *needs* to be fulfilled by artefacts are ultimately related to human needs, from the basic need for shelter (the need for keeping warm creates the need for clothing) to derived needs (as in 'keeping the clothing closed') and aesthetical needs (as 'keeping the clothing elegantly closed'). There is a hierarchy in needs where the invention of the button certainly would fulfil a specific 'cloth-fixing need'. The concept of the product function thus can be quite abstract (as in the 'transportation' function) to quite detailed (as in the 'short-haul person and load transportation' function realized by a horse-powered cart). Basic needs are a constant, but derived needs come and go. So, over time, new product functions arise, as illustrated by the clock, which was an answer to a need when the agricultural society changed into an industrial society. The same clock —in the form of the marine chronometer— also played a vital role in navigation, being used to determine longitude by means of celestial navigation.

Innovation takes place in those product functions when the artefacts change. Take the timepiece, evolving over time from those early hourglasses and sundials into the pendulum clocks, marine chronometers, and pocket watches. It was the adaptation of the 'inner' environment' of the artefact to the requirements of the 'outer' environment. For example: It is the mechanical implementation as a wristwatch that realizes a timepiece with the function of 'easy portable timekeeping'. Thus, the product function 'time' — being a response to a universal need in our day, when the fourth dimension of time (ie in 'timing' work, travel) is dominating private and professional life— is implemented in many ways.

The realization of a certain implementation of the product-function is realized by people who know 'how to make it'. For the watch, there were people who mastered the 'fine mechanical watch technology', such as those nineteenth-century German and Swiss mechanics with their high-grade technological skills. This was soon followed by other skills when the digital timepieces were realized by people mastering microelectronic technologies. This leads us to the link between product innovation and technology.

We define *technology* as the 'knowhow' (based on knowledge) and the 'way' (based on skill) of making things. So, technology —ie 'knowing how to make things'— is part of the 'act of skills'. Technology is more than the 'technique' —ie a body of technical methods and procedures— from which it originates.

> *Technology is a recent human achievement that flourished conceptually in the 18th century, when technique was not more seen as skilled handwork, but has turned as the object of systematic human knowledge and a new 'Weltanschaung' (at that time purely mechanistic).* (Devezas, 2005, p. 1145)

> *Technique refers to any complex of standardized means for attaining a predetermined result. Thus, it converts spontaneous and unreflective behavior into behavior that is deliberate and rationalized.* (Ellul, Wilkinson, & Merton, 1964, p. vi)

We follow Anna Bergek and associates here:

> *The concept of technology incorporates (at least) two interrelated meanings. First, technology refers to material and immaterial objects—both hardware (e.g. products, tools and machines) and software (e.g. procedures/processes and digital protocols)—that can be used to solve real-world technical problems. Second, it refers to technical knowledge, either in general terms or in terms of knowledge embodied in the physical artefact.* (Bergek, Jacobsson, Carlsson, Lindmark, & Rickne, 2008, p. 407)

Technology can be simple. Just imagine the basic way of making coffee by pouring hot water over ground roasted coffee beans. Obviously, technology also can be quite complex. Just imagine the automatic expresso-machine grinding the beans, brewing the coffee and cleaning the machine. Machines, that in addition to their mechanical construction, depend on electronic technologies. And that electronic circuitry is depending on the semiconductor technology: knowing how to make those complex integrated circuits.[3] A technology that, on one hand, is based on fine-mechanical, optical, chemical, and electronic technologies creating the objects —with the production machines that work with astonishing accuracy— that developed over time as the result of impressive engineering efforts. On the other hand, it is based on the advanced knowledge of physical phenomena —the behaviour of electrons in semiconducting materials like silicon— that was acquired over decades by scientific efforts. So, as both science and engineering contribute, *technology* is 'knowledge' (originating from science), and 'knowhow' (originating from engineering) combined to fulfil a purpose.

However, technology is not only 'skills' (ie the ability to carry out a task) and 'insight' (ie the understanding of the act). As both science and engineering evolve over time, they create the infrastructure from earlier systems upon which the further developments build. The third element of technology, then, would be the 'accumulated experience' from earlier technical systems. Thus, as technologies are stacked on previous technologies, it is not hard to understand that the words 'electric' technologies and 'electronic' technologies represent, in fact, a collection of technologies: in way, 'electricity' and 'electronics' are *meta-technologies*.[4]

## The General Purpose Technology

A specific construct of a (meta-)technology is the concept of the *general purpose technology* (GPT). It can be defined in the following way:

> *A GPT is a single generic technology, recognizable as such over its whole lifetime, that initially has much scope for improvement and eventually becomes widely used, to have many uses, and to have many spill-over effects.* (Lipsey, Carlaw, & Bekar, 2005, p. 98)

This is a broad definition, hard to make operational or usable. Thus, in complement, we see a GPT as a cluster, or clusters, of innovations of which the fundamental new combinations, the basic innovations, have considerable impact on society (Scheme 1). We call these basic innovations the *general purpose engines* (GPEs). Henceforth, more narrowly, we define a general purpose technology as *the collection of 'general purpose engines' appearing in a range of interrelated clusters of innovations.*[5] In other words, a GPT is a cluster, or range of clusters, of innovations around the general purpose engines. And the 'engine' is the device that transforms, such as in the transformation

---

[3] An integrated circuit (IC) —popularly known as a 'chip'— is a miniaturized electronic circuit with a specific function (eg a memory IC or a central processing unit). It is at the core of modern electronic systems.
[4] We use the word 'meta' to indicate a higher level of abstraction.
[5] This definition is more precise than the one we used in the preceding case studies as the result of new insights developed in the micro-foundations of a GPT during those studies.

of heat into rotative power (ie the steam engine), and the transformation of electricity into (rotative and linear) motion (ie the electric dynamo/motor and relais), into light (ie the electric lamp) and sound (ie the loudspeaker). And, last but not least, the transformation of petrol power into rotative motion (ie the internal combustion engine).

One observes that a GPT has also been defined by its spill-over effects, the GPT being "the pervasive technologies that occasionally transform a society's entire set of economic, social and political structures" (Lipsey et al., 2005, p. 3). Lipsey also described a GPT as "a technology that initially has much scope for improvement and eventually to be widely used, to have many uses and to have many spill-over effects" (ibid, p. 133). Thus we refined Richard Lipsey's definition, by focussing on the general purpose engines being the micro-foundations of a GPT themselves. The 'spill-over effects' then are the (chains of) events (aka trajectories) that originate from the GPEs, giving the GPT their pervasive nature.[6]

> *In popular terms, a GPT is the meta-technology creating GPEs that results in techno-economic breakthroughs such as the Industrial Revolution, the Information Revolution, etc. It is the engine of economic growth but also the engine of technical, social, and political change—and it is the engine of creative destruction. The GPT is not a single-moment phenomenon; it develops over time: They often start off as something we would never call a GPT (e.g. Papin's steam engine) and develop into something that transforms an entire economy (e.g. Trevithick's high-pressure steam engine).* (ibid, p. 97).

The case studies in the Invention Series are about observing phenomena as they occur in the real world. For example, the development of the steam engine, from which one can conclude it was a GPT according to the definition. The observation of what caused the Second Industrial Revolution shows its complexity. Is 'electricity' in its totality the GPT, or are the electro-motor and the electric dynamo, engines with a complementary power-conversion function,[7] the GPT? Does the development of the electric motor and electric light, the telegraph, telephone, and wireless illustrate the pervasive nature of the GPT-Electricity? Or can it be that the resulting development trajectories of the telegraph, telephone, and wireless —engines with a communication function responding to the basic human need to communicate— are a GPT on their own? The interpretation becomes more complex, the opinions

---

[6] See: B.J.G. van der Kooij, *Lipseys Quest for the Micro-foundations of GPT—The General Purpose Engine*. Delft Repository: http://repository.tudelft.nl/islandora/object/ uuid%3A56fed0f9-8a38-487d-b93d-dd239c3e60c5?collection=research

[7] The electric motor uses electricity to create movement (rotational power). The electric dynamo uses rotational power to create electricity.

diffused, especially when one looks at the present time, for example, at the phenomenon of the Internet, part of the Information Revolution. As it is based on 'electricity', it could be considered a recent spill-over from the GPT-Electricity that started in the nineteenth century. Or it could be considered as a GPT on its own; the GPT-ICT. By restricting the GPT concept to (a limited collection of) GPEs with the same product function, the concept stays within limits.

## *Evolution and Revolution*

To conclude our definitions, a word about the use of the notion of *revolution*, as in 'Industrial Revolution' and 'French Revolution'. The word 'revolution' can be used to denote major social and political upheavals (eg the French Revolution) resulting in a major restructuring of a society (ie regime change) or the replacement of a former ruling elite with a new one (ie government change), often accompanied by a lot of violence and casualties (ie the Madness of the Times). In that sense, a *political revolution* is an internal war —in contrast to the external wars between nations— that attempts to alter state policy, its rulers, and institutions. *Societal revolutions* are the changes in the structure of society —often originating from the oppressed or neglected classes but also as a result of the Spirit of the Times— that are related to the concept of social change that we will go and explore. The companion concepts of scientific change, related to *scientific revolutions,* and technical change, related to *technological revolutions*, are discontinuities outside the political and societal spheres. In a technological revolution, the ruling meta-technology is replaced, or complemented, by another meta-technology: the new general purpose technology (eg steam technologies being replaced by electric technologies). As a consequence, the technological revolution restructures the material conditions of human existence and ultimately results in *socio-economic revolutions,* just as the preceding socio-economic revolutions did originally create the context for the following technological revolutions.

These drastic changes in the societal and social structures, caused by such major technological changes, are creating a broad spectrum of technical and organizational novelty. The socio-techno-economic disruptions are based on the technical and the economic dimension of the *industrial revolution*s. Although the violence aspect on the social level is not that obvious, like with the social revolution, the industrial revolutions also have 'victims.' The casualties of these socio-techno-economic revolutions —by unemployment or outdated technical knowledge and engineering practise— certainly can be identified as the victims. Schumpeter labelled this phenomenon as 'creative destruction'. New technologies created new jobs and destroyed old jobs. The lamplighter of the gaslights, the messenger boy for the telegrams, the male telegraph operator, and the female switchboard operator, they all faded away, to be replaced by totally different jobs in other technologies.

## *About our Research*

This exploration is the eighth manuscript in the *Invention Series*, a series of case studies on inventions that created the world we live in today. In the first case study, *The Invention of the Steam Engine*, we explored a methodology to observe and investigate the complex phenomena of technological innovation as part of a general purpose technology (GPT). In that case, it was about the steam technology that fuelled the First Industrial Revolution. One could consider that historical case study as a trial to see if our methodology could be applied. It looked promising enough to try again. The result was a case study on electro-motive engines and a case study on electric light. Followed by studies on the communication engines 'telegraph,' 'telephone', and on 'wireless communication.' [8] Now, in this case study, we focus on the events around the development of the 'Radio Tube', the 'Transistor' and the 'Integrated Circuit' (IC) that are the result of the GPT-Electronics. As well as the application trajectories that originated from, or were influenced by, these engines. So, let's start to describe the basic elements of our research approach into the electronic technologies.

Our *field of interest* in the GPT-Electronics is, next to its technological trajectory, the area of its application in mobility that was so important in the times of the Industrial Revolutions. To understand how this meta-technology could fuel the industrial revolution, we again applied the method of the case study. The case-study method offers room for context and content. The context is the real-life context: the scientific, social, economic, and political environment in which the observed phenomena occurred. The content is the description of the technical, economic, and human details of those phenomena. The reader will recognize this 'context and content' approach in the dualistic structure of the manuscript.

The case study is based on a specific scholarly framework to observe the phenomena as they occurred in the real world. Our *frame of analysis* is based on the construct of clusters of innovations/businesses, as identified by early twentieth-century scholars active in the domain of innovation research. Among those economists was Alois Schumpeter, who related the clusters of innovations to business cycles under the influence of creative destruction:

---

[8] These case studies are about electricity and its use as source of power and carrier of information. The case study 'The Invention of the Combustion Engine' —about power conversion of oil and gas by explosion— is an example of innovation in other realms.

*Because the new combinations are not, as one would expect according to general principles of probability, evenly distributed through time…but appear, if at all, discontinuously in groups or swarms.* (Schumpeter & Opie, 1934, p. 223)

Schumpeter continues: 'the business cycle is a direct consequence of the appearance of innovations' (Ibid, pp. 227–230). For Schumpeter, it was the entrepreneur who realized the innovation and, as imitators were soon following in the entrepreneurial act, thus created the business cycles nested within the economic waves. Later it was Gerhard Mensch and Jaap van Duijn who related the basic innovation within the clusters to the long waves in the economy with respect to industrial cycles. Mensch related the cyclic economic pattern to basic innovations: 'The changing tides, the ebb and flow of the stream of basic innovations explain economic change, that is, the difference in growth and stagnation periods' (Mensch, 1979, p. 135). Duijn referred to innovation cycles (Duijn, 1983). More recently it was scholars like Utterbach and Abernathy, Suarez, Dosi, Tushman, Anderson, and O'Reilly who developed and used, as part of their view on technological revolutions and technological trajectories, the construct of the 'dominant design' being the watershed in a technology cycle (Tushman, Anderson, & O'Reilly, 1997). This dominant design is the innovation that —at a given moment in time— has become the de facto industry standard. This dominant design we considered to be the basic innovation.

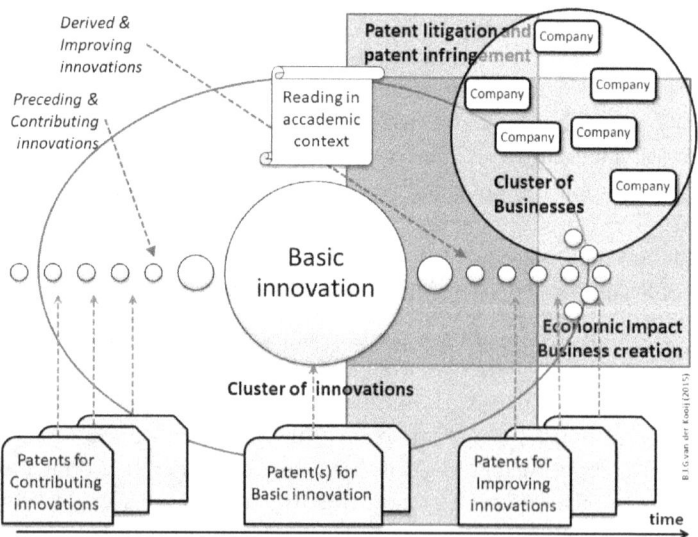

Scheme 1: The Construct of the Cluster of Innovations and the Cluster of Businesses.

Our *focus of analysis* is the cluster around the basic innovation with the preceding and derived innovations (Scheme 1). Our *units of analysis* are the contributions (ie events) made by individual people resulting in inventions and innovations. Then, for our domain of analysis, we first observed contributions in the GPT-Steam (a collection of many mechanical, hydraulic, thermic, and related technologies explored in the first study), followed by our observations of the electro-motive engines in the GPT-Electricity (second study) and electric light (third study). In the fourth, fifth, and sixth studies, we focus on the application area where communication technology based on electricity was applied. After the exploration of the Internal Combustion Engine in the seventh study, now in this eighth study, we focus on the electronic engines.

For our method of studying these complex phenomena, we chose the *embedded multiple case study*. The method is *multiple*, as we looked simultaneously at the scientific, technical, economic, and human aspects for a range of different individual cases. It is *embedded* because we looked at the individuals (the inventors, the entrepreneurs) within their organizations (their companies, the institutions) and societies, thus making the analysis multilevel and multidimensional. Our qualitative data originate from general, autobiographic, and scholarly literature (see references), creating a mix of sources that are quoted extensively. Our quantitative data were sampled from primary sources like the United States Patent Office (USPTO) and the European Espacenet), British and French equivalents.

Our *frame of perspective* was the identification of patterns that are related to the cluster concept. Can coherent clusters of innovation be identified within a specific General Purpose Technology? If so, how are they related, and how are the clusters put together? The first pilot case of the invention of the steam engine showed that it could be done. So in this case study, our objective was to identify the basic innovations that played a dominant role in the GPT-Electronics that extended the *Era of Electricity* in the Third and Fourth Industrial Revolution. As we used patents as innovation identifiers, and used patent wars (patent infringement and patent litigation) and economic booms (business creation, business and industry cycles) to identify the impact of basic innovations, this aspect is quite dominant in the study (Scheme 1).

Considering our *unit of analysis*, in view of the previously mentioned aspect of innovation being the result of a combination, we tried to refine the cluster concept by detailing the contributing innovations into specific technological development trajectories (see Scheme 2, left):

*Scientific contributions:* These include the trajectory of the 'scientific contributions' concerning the basic laws of nature the curious and ingenious people in the eighteenth and nineteenth century were inquiring into. We use the definition of *science* as:

> *The intellectual and practical activity encompassing the systematic study of the structure and behaviour of the physical and natural world through observation and experiment.* (Oxford Dictionary)

*Technology/Engineering contributions:* Next we distinguish the technological contributions and use —in addition to our previously mentioned definition— the definition of *technology* as 'The application of scientific knowledge for practical purposes' (Oxford Dictionary) and as the knowhow (knowledge) and way (skill) of making things. Or, as Giovani Dosi puts it:

> *[We] define technology as a set of pieces of knowledge, both directly 'practical' (related to concrete problems and devices) and 'theoretical' (but practically applicable although not necessarily already applied), know-how, methods, procedures, experience of successes and failures and also, of course, physical devices and equipment.* (Dosi, 1982, p. 151)

Practical knowhow is built up over time and transferred from generation to generation by the means of the apprenticeship. Thus, this incorporates the contributions of all those instrument makers

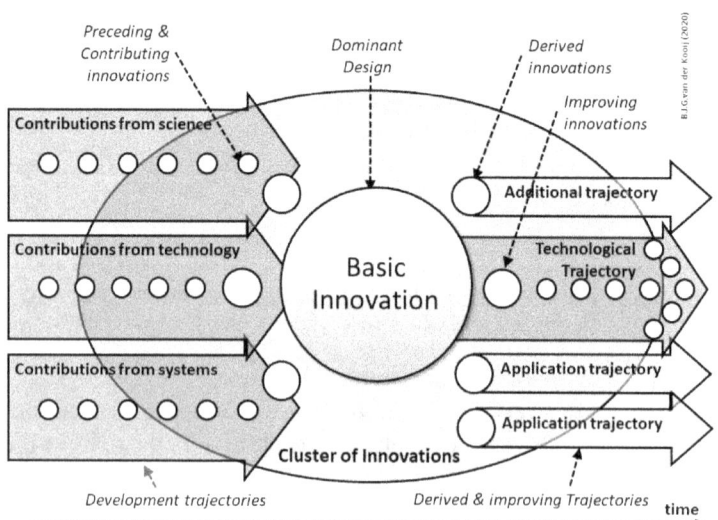

Scheme 2: The Construct of the Trajectories leading towards and from the Basic Innovation in a Cluster of Innovations.

using their fine mechanical skills to create engine components and subsystems, which were so essential to the creation of (electro)-mechanical machines. In today's parlance, their activity would be called electric and mechanical engineering that was transferred by imitation.

*System contributions:* A third development trajectory consists of the contributions that resulted in previously developed systems. The system concept being quite general, we will be using the definition of a system as 'A set of things working together as parts of a mechanism or an interconnecting network; a complex whole' (Oxford Dictionary). The keyword here is 'network', to which development so many creative minds contributed. However, these are contributions that are harder to classify. Let's, for example, consider our application area of communication (postal, optical, or electrical). Communication is always realized in a structure of several elements (parts, components) connected by a structure (network). For the classic postal system, it is the network of mail coaches, mail couriers, and the inns to change horses: the postal network. For optical communication, it is, as we have seen in an earlier case study, the network of semaphore relay towers and the organization of telegraphists that transmitted information in the semaphore code: the semaphore network. For electric telegraphy, it is similar. The electrical components like the transmitter, the cabling, and the receiver, the code used for the transmission, and the structure of the telegraph offices connected by copper cables created the network infrastructure for electric telegraphy: the telegraph network. People who contributed to that totality created the system contributions.

Given the genesis of the *basic innovation*, it will be followed over time by new contributions leading to other innovations (Scheme 2, right). Such as:

*Improvement contributions:* This includes contributions that enhance and improve upon the basic invention. The increasing knowhow of the ever-developing technology will add to the original invention step by step in in an incremental way. These improvement contributions create a technological trajectory of incremental innovations.

*Derived contributions:* In addition to the improvements, there will be contributions of another nature. In those cases, either to circumvent the patent protection or just by accident, the same functionality of the basic invention will be realized using a different concept, spinning off in a different technological trajectory.

The totality of these contributions creates trajectories; the *technological trajectory* of developments around the knowhow of the basic innovation, and the *application technologies* of the use of the engine as part of another system.

## About Events, Evolutions and Revolutions

In our historical analysis, we focus on discontinuities —the stepwise transition from one equilibrium to the next equilibrium, the shift from one steady state to another— in a specific development trajectory: the occurrence of specific *events* in the themes we observe. We distinguish between a process and an event. A process is continuous through time, whereas an event is an occurrence. In biology, speciation is a process, whereas a mutation is an event.[9] In technology, a rise in temperature is a process whereas a maximum temperature is a single occurrence. Events can differ in terms of their impact: from incremental change with hardly any impact, to disruptive change with massive impact. This gives both the incremental events labelled as 'evolutionary events', and the disruptive events labelled as 'revolutionary events.' Relating them to our earlier analysis of novelty (Scheme 1), we discriminate between the events with the character of a small step, and the events with the character of a large step. The totality of both the small and large steps constitutes the cluster of

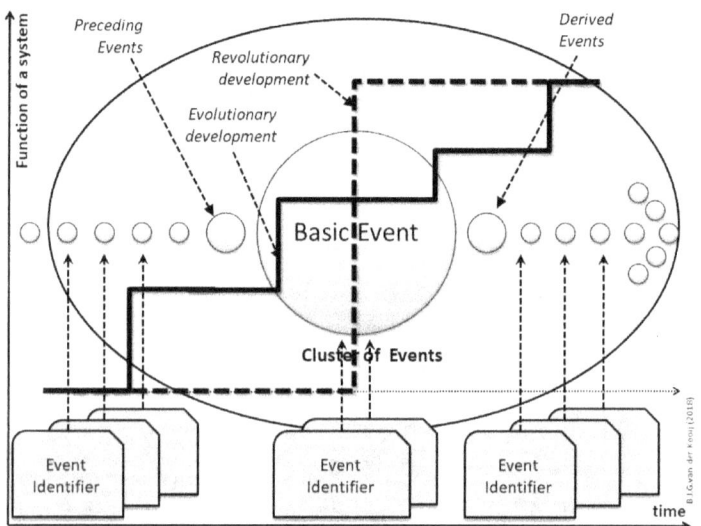

Scheme 3: Cluster of Events with Preceding Events and Derived Events.

---

[9] *Speciation* is the evolutionary process by which populations evolve to become distinct species. A *mutation* is the permanent alteration of the DNA; the nucleotide sequence of the genome, of an organism.

events (Scheme 3) in which the *basic event* is the core of the cluster. Seen in time, before the occurrence of the basic event there are the *preceding events*; following it, there are the *derived events*. Together with the basic event, they create a *cluster of events*.

We will define a basic event as a large, revolutionary change in a system with an impact outside the system itself. The event can be of a technical nature, as our previously mentioned technical innovations. However, it can also be of a political nature (eg the 'coup d'état), a social nature (eg a class conflict), or an economic nature (eg a conflict over resources). And we should not forget natural events (eg volcano eruptions), climatic events (eg the Middle Bronze Age Cold Epoch) or biologic events (eg pandemic viruses). This leads to the political event, the social event, the economic event, the military event, the technical event and the natural event we will observe in this analysis.

So, as the event itself is our *unit of observation,* it is relevant to define it. For us an event … 'is a specific change in a system over a certain timespan that has a considerable, measurable impact'. It can be a relatively small event occurring over a short period of time (eg a failed harvest in a region important to the local population); a moderate event with more drastic consequences over a short period of time (eg a volcanic eruption causing havoc in the region); or it can be a larger event over a longer period of time with large consequences (eg the event of the Fall of the Bastille during the French Revolution). The key to the identification of relevant events seems to be the cause, the timespan and the impact. Its subsequent interpretation is dependent on the point of view and the perspective of the observer.

As the event is our dominant phenomena identifier, how, then, do we identify events? The answer is simple: we don't, as many others did it for us. The body of knowledge concerning historic times is full of events; they are presented in the written, visual and graphical form (eg the more recent events), in archaeological form (eg the more ancient events) or in geological form (the prehistoric events). How do we select events? Using the themes, technology, economy, politics and society we select events described by scholars as revolutions, realising that semantics and academic bias[10] can play a confusing role.

And finally, how do we interpret these events? Again, we do not give our own interpretation of events. The events we present are observed, interpreted and qualified by others; from scholars active in a specific

---

[10] Academic bias (aka research bias), defined as any tendency which prevents unprejudiced consideration of a research question, occurs where an ideology influences the perspective of the observer.

discipline (eg chemical sciences) or scholars maintaining a broader perspective (eg historians of technology, economy, science). Product novelty claimed by inventors was —within the patent system— judged on by the peers of their time. They all studied the events, interpreted their findings from their particular point of view and concluded on their importance. The consequence is that we use secondary or tertiary resources, each with their own focus, perspective and bias. As scholars present their theory based on their perspectives, findings and interpretations, we are sometimes faced with different, complementary and conflicting views and theories. If possible, within limitations, we present them both.

Next about (r)evolutions. One could argue that 'an evolution equals to a revolution': many small steps over long periods of time with relatively small impact are equivalent to the big step within a short period of time with a large impact (Scheme 4). Or they can be considered as distinct phenomena, like the economic historian Abbott Usher hinted at when considering the differences in events that result in innovations and those that don't:

> *The structure of events in time reveals discontinuities of two distinct types. There are discontinuities between different systems of events that persist over long periods of time and, in many instances without any prospect of ultimate synthesis. There are other discontinuities that may be overcome, through some act of synthesis. <u>The establishment of new organic relations among ideas, or among material agents, or in patterns of behaviour is the essence of all invention and innovation</u>*. (Usher, 2013, p. 21) [underlining by me]

Historians use the notion of revolution also to identify non-social/non-political revolutions: they speak about an 'industrial revolution', a 'scientific revolution' and an 'information revolution.' What these types of revolutions have in common is the revolutionary result of evolutionary change creating novelty. They are processes that extend for more or less long time (ie the evolutions) or represent drastic short-timed changes for humanity (ie the revolutions). A much-debated construct as observed in the domain of evolutionary palaeontology.[11]

> *Large-scale evolutionary changes in continuous characters can result from two processes, punctuated equilibrium*[12] *(a series of steps/ saltations followed by stasis) or (phyletic) gradualism (whereby there is an accumulation of small incremental changes). If the tempo of evolution is short rapid changes followed by long periods*

---

[11] Palaeontology is the scientific study of life before the start of the Holocene Epoch. Palaeontology lies on the border between biology and geology, but differs from archaeology in that it excludes the study of anatomically modern humans.

[12] The concept of the *punctuated equilibrium* (developed by Elridge and Gould) considers long periods of little change, followed by a significant change in a short time. It created the notion that evolution is a stepwise process rather than a continuous process.

*of stasis, then it is likely that there are pulses of selective pressures associated with either external drivers or the appearance of novel phenotypes. Conversely, if evolutionary change is underpinned by a gradual and continuous process, then the selective pressures are likely to be either a ratchet process or driven by long-term low-level directional selection.* (Shultz, Nelson, & Dunbar, 2012, p. 2132)

Where this type of revolutions differ, is —next to their subject— the length of the timescale. The Neolithic Revolution covered millennia, the early Industrial Revolution covered centuries, the Information Revolution covers several decades, but the French Revolution itself covered just one decennium. For the people concerned, long timescale means a less dramatic impact for one specific generation, and societies can gradually adapt over the generations. Short time-scales, however, tend to impact individuals in a more prominent way.

To identify specific periods of time, historians use also other notions next to (r)evolution. Such as the notion of *Era* that is used to identify a period of time with particular events with a distinctive character (eg the Era of Steam). It is also used to identify periods of time characterised by a prominent feature (eg the Era of the Horse and Buggy). In addition, the notion of *Age* is often used to identify periods of time, characterized by person (eg the Victorian Age), a generation (eg the Age of the Baby Boomers), or a prominent phenomenon (the Stone Age). Both notions result from their specific perspective applied over a fixed period of time.

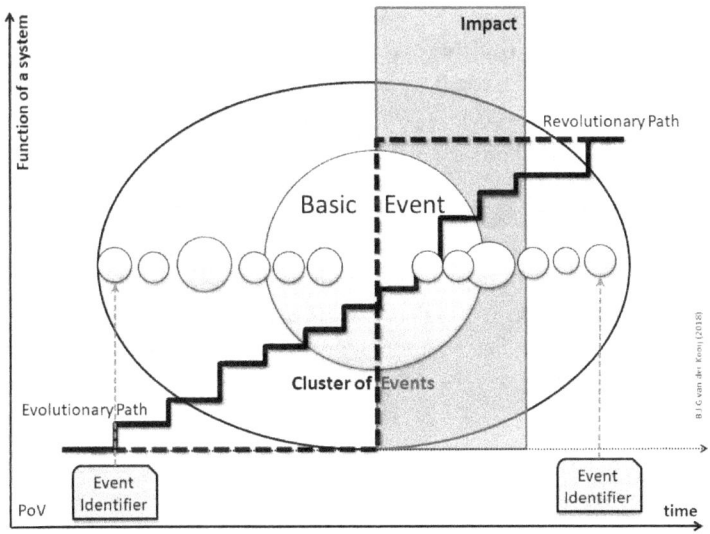

**Scheme 4: Difference between Evolution and Revolution.**
The figure illustrates the evolutionary path of small steps and the revolutionary path of the basic event.

However, there is no well-defined difference between the Era and Age. In our system of events an Era would encompass a range of clusters of events, and an Age would be the basic event with the subsequent derived events.

And there is the impact of the event; the social impact, the political impact, the economic impact. In a societal revolution the social order is changed when the ruling establishment is overthrown, and new social classes fight for a place in society (ie the social impact); a process of *social creative destruction*.[13] Also the institutions of the 'ancient regime' —such as the absolute monarchy— are overthrown, and new institutions —such as democratic parliaments with broader voting rights— are created (ie the political impact); the process of *political creative destruction*. But as the disturbances are more physically destructive (as in world wars, wars of religion), they have an increasing demographic and economic impact. For the technological revolution there are similarities recognisable when the 'old technological regime' is replaced with the 'new technological regime.' In the process destructing the old economy and creating the new economy in a process of *economic creative destruction*.

To conclude our description of the construct of Systems of Events with its clusters of events, a word about another use of the notion of *revolution*, as in 'Industrial Revolution' and 'French Revolution'. The word 'revolution' can be used to denote major social and political upheavals (eg the French Revolution) resulting in a major restructuring of the social order of dominant groups in a society (ie regime change) or the replacement of a former ruling elite with a new one (ie government change), often accompanied with by bloodshed, a lot of violence and casualties (ie the madness of the times). In that sense, a *political revolution* is an internal war (ie civil war) within a social system —in contrast to the external wars between social systems (ie nations)— that attempts to alter state policy, its rulers, and its institutions. *Societal revolutions* are the changes in the structure of a specific society —often originating from the unprivileged, oppressed or neglected classes but also as a result of the spirit/madness of the times— that are related to the concept of social change. The companion concepts of scientific change, occurring in *scientific revolutions,* and technical change, related to *technological revolutions*, are discontinuities outside the political and societal spheres in the realm of knowledge (ie Science) and knowhow (ie Technology).

In a Scientific Revolution the scientific thinking undergoes a fundamental change in its basic concepts and experimental practices

---

[13] The notion *Creative destruction*, coined by Schumpeter to describe an economic phenomenon related to product and process innovation, is the mechanism by which the 'old' (ie society, industry, business) is destructed and the 'new' is created by the event.

(the Kuhnian paradigm shifts). In a technological revolution, the ruling meta-technology is replaced, or complemented, by another meta-technology, such as the steam technologies being replaced over time by electric technologies and combustion technologies. As a consequence, the technological revolution restructures the material conditions of human existence and ultimately results in *socio-economic revolutions*, just as the preceding socio-economic revolutions originally created the context for the following technological revolutions. Thus, illustrating the *pendulum construct* of technology and society we study.[14]

Finally, a word about the use of the notions *invention* and *innovation* in the case study. We described before how we intend to define them, but in the explorations of the case study, we follow our sources. They use the words in the context of their time. A use that can be different from our time. For example: what would be called an invention in the early nineteenth century, that could be called an innovation today. There is quite a difference between the two, and even our present-day interpretations of both words show great variance, as we found in a survey of the word *innovation* as used by innovation scholars.[15]

## *About Life Cycles*

Change and Novelty evolves in a specific pattern that has similarities to the biological life cycle. That 'life cycle' represent the development of individual human life through pregnancy after fertilization, youth after birth, and growing up during adolescence into maturity (with its events like divorce, midlife crisis, empty nest syndrome and retirement gap). To end with decline and death after seniority. (Scheme 5, lower part). This lifecycle-concept can also be recognized in products, companies, industries and technologies (Scheme 5, upper part).

---

[14] The word 'society' relates to human societies that are characterized by patterns of relationships (social relations) between individuals who share a distinctive culture and institutions in a specific social environment. A 'society' is different from social groups (family, tribe, or clan), the *family* being a group of people affiliated by consanguinity (by recognized birth) and by affinity (by marriage), the *tribe* being a distinct people, dependent on their land for their livelihood, who are largely self-sufficient, and not integrated into the national society, and the *clan* being a group of people united by actual or perceived kinship and descent. The word society has different meanings such as *national societies*, countries/states with an economic, social, industrial, and/or cultural infrastructure, and *collective societies*, like companies with an economic, social, industrial, and/or cultural infrastructure.
[15] See: B.J.G. van der Kooij, 'Innovation Defined: a Survey'. Source: http://repository.tudelft.nl/view/ir/uuid%3A6a5624c9-e64e-4426-98e9-f239f8aaba18/

The analogy is quite basic. The creation of a new idea —similar to the moment of fertilization where sperm and egg create a new combination— needs a time to take shape; the incubation phase (aka pregnancy). The analogy goes even further as the development into a new life being gene-dominated, the development of the idea into a concept is meme[16]-dominated (with its cultural path-dependency).[17] The same goes for birth being the moment that the new biological life has to stand on its own, as does the 'proto-type' emerging from the original concept. From birth through puberty into becoming a grownup, is a process in which interaction with the environment shapes the new person. It has its equivalent in the mechanical world as market-testing, technical adaptation and improvement, shape the final product in its introduction phase.

Zooming in on technology —ie 'knowing how to make things'[18]—, one recognises that technology emerges (the birth of the technology) as a

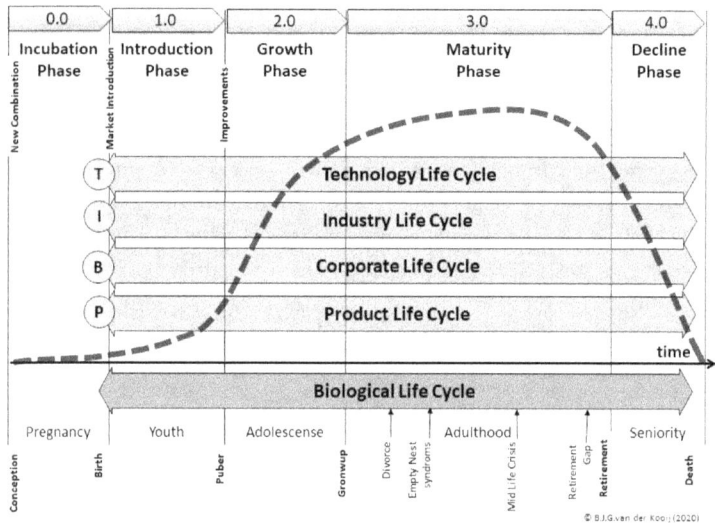

**Scheme 5: Life Cycle Concept and its Phases.**
Example: P 1.0 is the first phase of the Product Life Cycle; T 2.0 is the second phase of the Technology Life Cycle.

---

[16] Like the gene carrying biological information needed for reproduction, the *meme* is the carrier of cultural information that can be transferred.
[17] *Cultural path dependency* is about the fact that evolutionary activity shaping the future is based on, or even constrained by, the preceding historic cultural developments. It is similar to economic and technical path dependency.
[18] *Technology* is the sum of any techniques, skills, methods, and processes used in the production of goods or services.

discovery leads to an invention (the conception of the technology and the subsequent incubation period). If the technology survives, it develops in its growth phase (the dynamics of technological puberty) to reach the stability of technological maturity. Then the technology becomes an established 'best practice.' The technology has the potency to procreate as it spins-off other, related technologies for specifies application areas. Over time it can fade away, becoming obsolete as other improved technologies take over.[19]

We will apply this technology life-cycle model to explore the evolution of technology over time by using the life cycle periodization as follows:

*Period 0.0: Incubation Phase.* The time after the 'conjunction' up to the 'birth.' In the case of products, it is the time after conceptualization of the 'idea' (aka the Eureka moment) up to the realisation of the 'prototype.' In the case of a technology, it is the period between the first time that the practise/technique is observed (ie the sparking arc caused by an electric current), and the moment the tool is applied (eg the prototype of a rudimentary welding machine).

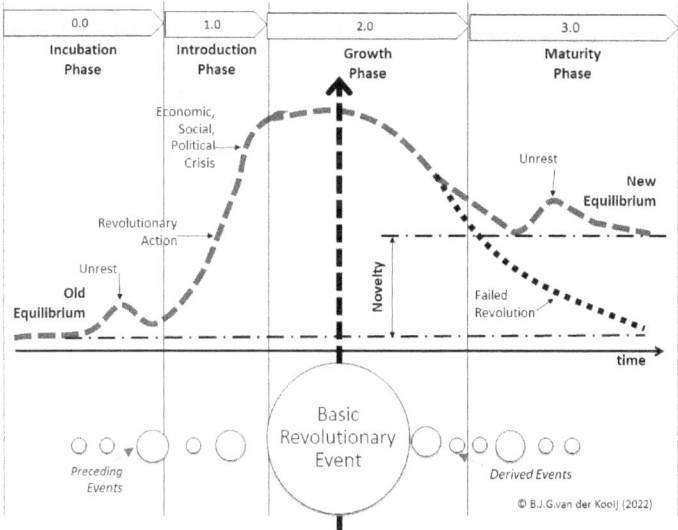

**Scheme 6: Relation Basic Revolutionary Event and the Dynamics of the Technology Cycle.**
The dotted line between the old equilibrium and new equilibrium, indicates the non-technological trajectories that create the business dynamics.

---

[19] A *best practice* is a method or technique that has been generally accepted as superior to any alternatives because it produces results that are superior to those achieved by other means or because it has become a standard way of doing things.

*Period 1.0: Introduction Phase.* This is the infancy period that follows the 'birth' and its early organization of realisation. In the case of products, it starts when they are introduced in the market place, and improved upon as technology develops. In the case of business, new entrants—ie start-up companies— launch their organizations, and start growing in size, capabilities and complexity [20] Some companies will not succeed in passing this phase and go down in the Valley of Death.

*Period 2.0: The Growth Phase.* This is the period after infancy up to the moment of maturity. In the case of products in this period, those products that survive the market introduction, are improved upon market responses. For companies that did survive the early introduction problems, competition and business development crisis, start growing in size, experience increasing sales volume, develop their business model and find their place in industry. In the case of technology, the tool/method is developing rapidly, offering better methods of realisation, cost reduction and quality improvement. And for industries, this is the period that they are getting shaped facing competition (when other business jump on the band wagon).

*Period 3.0: Maturity Phase.* The mature product has found its market(s) and market share(s), the business development created a stable company, and the companies created their industry. The firsts casualties of business life occur when companies fail. For example, during the competition shake-out, or as a result of internal factors (eg partner disagreements). In addition, depending on period of time and economic background, the process of vertical and horizontal integration takes place with mergers and acquisitions.

This periodization on the level of technology is the framework we will apply in the forthcoming analysis. Scheme 6 shows an overview of the different phases in relation to the Cluster of Innovation that surround the Basis Revolutionary Event. Phases that, when they succeed, create a new equilibrium after massive dynamics.

---

[20] The notion of a *business* entity refers to the individual company/corporation being an economic entity. But it can also refer to the collective of organisations (eg the manufacturing business). In addition, it can refer to the activity of organising the manufacturing, commercial and administrative activities of that economic entity: the 'doing business.'

## *About Organizations*

Just like humans live in social organisations —like the nuclear family, the extended family and the tribe/clan— they also take part in work organizations called 'firm' or 'company.' [21]

In early times the agricultural family was the basic work-unit. To be complemented with the individual artisan and/or travelling merchant. Each of them part of an early form of organization. It gave rise to the domestic system of the cottage-industry and artisanry. That artisan with his apprentices and apprentices was the *nuclear company* as basic unit of industrious organization. Later in time, the vehicle for realization of industrious activity during the industrial revolutions was basically the organizational form of the *extended company*. [22] An organisation that came in the form of the (industrial/merchant) entrepreneur creating an enterprise in combination with

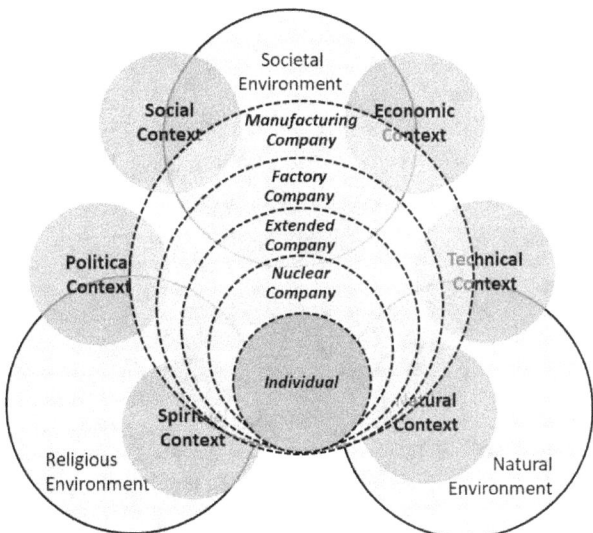

**Scheme 7: The Organizational Concept and its Environment.**

---

[21] Organization is essential for human activities that provide the basic physical needs of food, clothing, and shelter. Under conditions of chronic change, humanity has endeavoured to develop the most effective method of organizing these activities to compete for survival and meet the challenges of a particular era. The rise of organizations is marked by the constant adaption to changes in the technological, cultural, political, and economic environments.

[22] Companies can be classified in different ways. We draw in this classification of work organisations the analogy with the classification of social organisations: the nuclear family, the extended family and the subsequent forms of social organization.

the cottage-industry (ie the 'putting out system'). In the time of the First Industrial Revolution, the *factory company* (aka the organization implementing the Factory System[23]) emerged where the ownership of, and the working with the machinery as part of the division of labour, was separated. The executive powers to run the company, were the prerogative of the owner. And from there arose the *manufacturing company* where the division of labour was complemented by organisation; the organization of the work-flow, the organisation of producing the interchangeable parts (Scheme 7).

Concerning that ownership, originally the factory was owned privately by wealthy owners employing the workers. Over time they could become the family-businesses where ownership and executive power went from father to son. In addition, the factory company evolved into *stock companies* when that ownership became represented by stock certificates (documents of partial ownership) that could be traded on the stock market to non-participants (aka the general public of investors). The executive powers came in the hands of hired management as executive of the owners will. Next to the party that held the ownership and the party that supplied the labour, a third party had come on scene; the executive management. From this entity arose the *stock corporation* as an amalgamation of stock

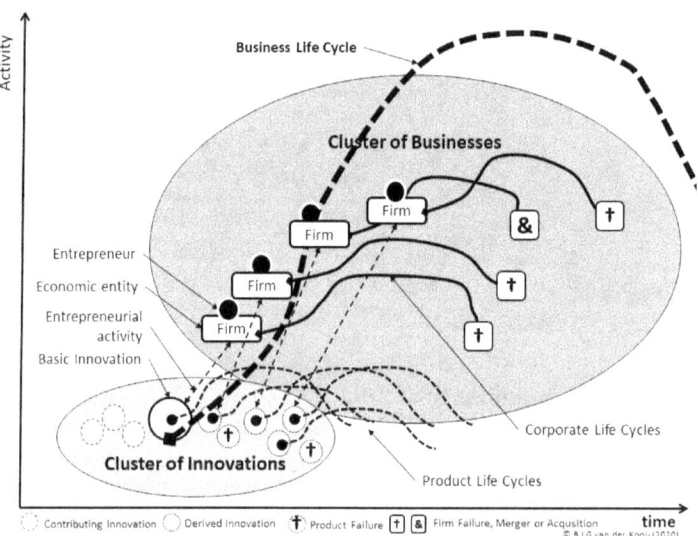

**Scheme 8: Relation Cluster of Innovations and Cluster of Business creating a Business Cycle.**

---

[23] The Factory system is a way of manufacturing using machinery and the division of labour.

companies. And, further down the road, came the appearance of the *mega-corporations*. Companies where, next to the (employed) executive(s), financial parties participated in the risks and profits of the enterprise (eg the colonial Indies Companies). And finally, there was the *corporate trust*; a corporation that administers financial assets on behalf of a group of business interests.

As the events we explore are related to individuals working in any form of organization, we will apply this distinction (eg the entrepreneur and his organization) when needed. Realizing that organizations operate in their specific environment (Scheme 7): the societal environment, the religious environment and the natural environment.[24] Environments creating the context for their existence and facilitating and restricting their space to manoeuvre.

Technology is practised by individual people organised in firms. Firms using a similar technology create a business (eg the motor car business) which in turn make up industries (eg the automobile industry). The creation of a(n) (basic) innovation within a Cluster of Innovations, can result in the creation of an economic entity (ie the firm) as the result of entrepreneurial

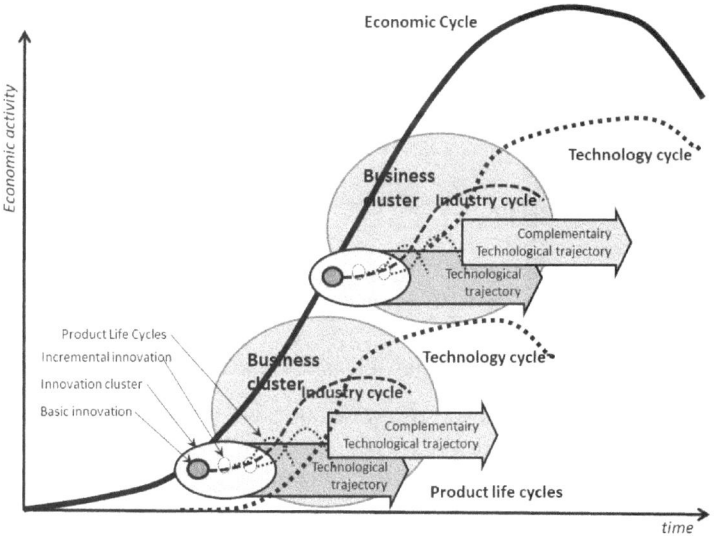

**Scheme 9: The Economic Life Cycle as the result of Industry and Technology Life Cycles.**

---

[24] The broader context of Change and Novelty is covered in separate case studies; the Deep History-series; Part 1: The Origins of Innovation (2018), Part 2: The Discovery of Innovation (2018), and Part 3: The Invention of Innovation (forthcoming).

activity. When other entrepreneurs, seeing the opportunities created by the new invention, are 'jumping on the bandwagon', they create a Cluster of Business (Scheme 8). A cluster of firms that has a dynamic of its own; among which the competition leading to the 'survival of the fittest' during its corporate life cycle.

The basic innovation, often protected by a patent, has its own product life cycle. Each of the following products resulting from the basic innovation (ie the incremental innovations derived from the basic innovation) has also a product life cycle of its own. They follow the original trajectory of development, but can also spawn of in a complementary trajectory. The same goes for the businesses that start, grow, and collapse (ie the corporate life cycle). Their totality creating an industry, that subsequently has a 'industry life cycle.'[25] Next to the original technological trajectory, with the rise of new technology cycles, additional technological trajectories emerge, each with their own business cluster (Scheme 9). All those businesses and industries are part of the economy; the system of production and consumption of goods and services. As a result of the business and industry life-cycles, also the economy has a life cycle; the 'economic life cycle' (Scheme 9). Such as the Long Waves and the Short Waves of economic development.

This classification and periodization —painted in rough brushstrokes on a general level— gives us a tool to analyse/classify over a period of time events and their development by subject showing the interrelation between product, technology, business, industry and economy.

## *About the Context*

As mentioned earlier, case studies are about content and context. Our specific case studies are about the *content* of technical change —they cover technological innovations— and we explore change from the perspective of the technological innovations themselves: the clusters of innovations.

These clusters are the result of contributions of many individual persons, individuals who lived within their specific 'Spirit of the Time', often even with its specific 'Madness of the Time'. People with personal hopes and fears, drives, ambitions, and limitations, honest people and cheating people, extraverted and introverted people, people who lived in — and whose behaviour was influenced by— times of war, physical destruction as well as psychological damage, and economic stagnation People who lived in times of peace, creation, and progress —and people

---

[25] This interpretation is based on Schumpeter's view as developed in his publication Business Cycles (1939).

# The Invention of the Silicon Engine

who lived in a specific society. Societies that could be under stress as the result of disruptions in their particular environment. Disruptions with a natural cause (eg pandemics, climate), or with a made-man cause (eg wars).

When observing the real world in all its complexity, one tries to create a mental model of that world. By definition, that mental model is simplified, limiting the complexity. The limiting is done by creating mental 'constructs.' For our contextual analysis, we will us the *construct of change*, eg the constructs called Social Change, Economic Change, Scientific Change, etc. Each of these change constructs covers a part of the total context. So — along with analysing the content, where we used the perspective of the clusters of innovations (Scheme 2) — now we analyse the *contexts* that influence the occurrence of those clusters by including the different change constructs to the cluster of innovations.

Those contexts are part of the environment in which the events take place. Spoken in general terms, we distinguish for the purpose of our exploration the following environments (Scheme 7):

*The Societal Environment:* Each event takes place in the societal environment as it existed at that moment in time. A complex structure of a society, its economy and its politics.

*The Religious Environment:* Each event takes place in a spiritual and political context, resulting from the human belief system giving him his belief system of religiosity as well as his ideologic/political worldviews.

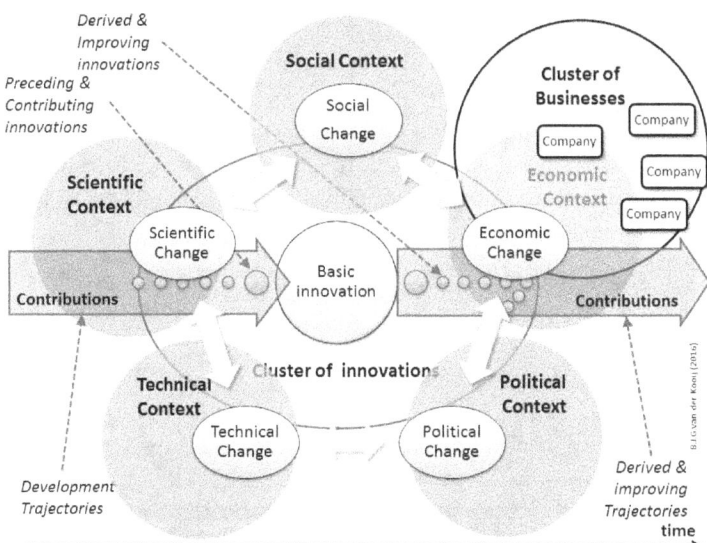

**Scheme 10: The Construct of Change in the different Contexts.**

*The Natural Environment:* These are the events that are outside man's direct influence: the volcanic eruptions, the extreme weather anomalies. But also the natural diseases spreading as pandemics.

These environments are creating the overall multi-dimensional context for the events. More specifically, they are a clustering of the following single-dimensional contexts (Scheme 10):

*Social Change:* The case observes the society defining the *social context* for the individual inventor and his inventions at that given period of time. A society that itself changed constantly. Hence, we speak about the autonomous change of social structures, social behaviour, and social relations in a society as being the result of social forces. When those changes are incremental, social change is incremental and evolutionary in nature. But sometimes the changes are discontinuous and disruptive, even revolutionary and transformative. Then we talk about revolutions such as the American, French, and Russian Revolutions. They are the drastic —sometimes even dramatic— forms of social change.

*Economic Change:* Part of the interaction in a society has to do with 'economic' activities like the production, distribution, trade, and consumption of goods and services. Together they create the *economic context* for innovation. Each of these activities has a dynamic of its own. Take 'trade' as an example, such as when the surplus of agrarian production was brought to the local market and traded for other surpluses, or when the local and regional trade grew into national trade and even colonial trade. These market-based local, regional, and national economies exchanged goods and services between participants by barter or a medium of exchange like currency. After the social system expanded into nations, the economies also evolved into larger structures, and each of the participating institutions developed on its own. Take the example of 'production': when the cottage industry developed into industrious mass production, creating clusters of businesses within specific areas of manufacturing.

When national economies develop, state policies—and their resulting laws—are needed to facilitate and control those economies. In creating these, the economic policies become part of the political structure, as they represent an (economic) interest. This process resulted in mercantile economic policies or—later in time—free trade policies. Economies have a dynamic of their own: they grow, and they contract, but normally, they change within limits, as the participants in the economies have an interest in maintaining a state of equilibrium. But sometimes the changes are discontinuous and disruptive, even revolutionary. Then we talk about revolutions: such as (the economic part of) the Industrial Revolution(s). It is the totality of these dynamics we consider to be economic change.

*Political Change:* In each society, be it small like a family, group, or organization, or be it big like a clan, tribe, or even nation, there is a power structure with centralization of authority. The power structure in a society is a means for the survival of the group. Politics is the dynamic interaction between the participants in that power structure.[26] Consequently, the participants of a group create the political system together as a system of entities: the participants, their institutions, and the relations between those entities. In short, there is the *political context* for innovation of those who rule and those who are ruled, and that context is the exercise of influence on someone's individual behaviour. This influence is based on personality, physical power, expertise, or just —historic— acceptance. That power structure is not fixed; temporary alliances between members of the group result in changes in the picking order. Hereditary power can be challenged, creating succession wars. The totality of these dynamics is considered to be political change.

*Technical Change*: Technology —in short, 'knowing how to make things'— is the collection of techniques, skills, methods, and processes. It is the basis of industrious human activity in the realization of goods and services. It is related to engineering, the process of designing and making tools and systems. On one hand, technology is based on the application of scientific knowledge, the understanding of the basic phenomena in nature. However, it is also based on engineering skills, acquired over time by 'doing things' and passed on by generations. Thus, technology is the combination of understanding (knowledge) and practise (knowhow). In their totality, they create the *technical context* for innovation. So, science, engineering, and technology are interrelated and undergoing changes, sometimes incremental or sometimes disruptive. We talk about the technical part of industrial revolutions as a drastic form of technical change. When technology changes, either evolutionary of revolutionary, we call that technical change.

---

[26] The word *politics* can have different meanings. Such as: *national politics,* the working out of forms of agreement or conflict solution between different groups of people and their specific interests; *parliamentary politics,* the formalized interaction in the structure of a democratic parliament; *party politics,* the collective views and wheeling and dealing of political parties to exercise political power; and *people's politics,* the representation of specific interests such as religious or economic interests. We will consider the political context to be the national execution (national politics) of the representation (parliamentary politics) of specific interests (party politics and people's politics). Related is the word 'policy': a policy is a statement of intent and is implemented as a procedure or protocol. Examples are: public policy, foreign policy, and economic policy.

*Scientific Change:* Science is about understanding the phenomena of the natural world we live in. That understanding is —in our present world— based on the scientific method, a disciplined way to study the natural world. It results in understanding the 'nature of matter', such as the 'nature of heat', 'nature of light', etc, subjects that have evolved in the scientific disciplines of chemistry, physics, etc, each discipline having its own knowledge base. This knowledge is represented in the form of concepts, theories, models, and laws, which create the *scientific context* for innovation. When scientific knowledge changes through evolution, we call that scientific change. And when it is based on a paradigm shift,[27] we call it a Scientific Revolution.

## *About the Perspective*

One has to be realistic and not try to cover the developments over centuries in their totality and all their complexity. At the time they occurred, the economic change and social change leading up to the Industrial Revolution received extensive scholarly attention, such as seen from a purely *economic perspective* (eg the Scot Adam Smith in the eighteenth century) or a *social perspective* (eg the German Karl Marx in the nineteenth century). Among all those manifest scholarly views, we will humbly limit ourselves to a *technological perspective* in line with the work of the economic historian Joel Mokyr. However, in doing this, we will first focus on the *context* for technical change, a context that itself was the result of a continuous process of change.[28]

Therefore, our particular perspective will be focused on the *socio-political environment* (Scheme 11) with the specific social context and political context, as the overall context that encapsulated the *techno-economic environment* for technical change and economic change. These social and political contexts can be described as follows:

*Social context:* The societal context is determined by the social world we are living in. It is our society that defines our existence, and that society is in a constant process of autonomous change. The societal context we are going to study is about (groups of) people, their (hierarchical) relations, and their (collective) social behavior. We look

---

[27] A paradigm shift is a fundamental change in the basic concepts and experimental practices of a scientific discipline. In the case of a paradigm shift, the prevailing framework of shared scientific views, theories, and models is replaced by a new framework.
[28] This general analysis draws heavily on the information as available in Wikipedia, and quite often partially edited text parts are used. As we consider this to be public knowledge, we do not quote individual pieces of text but incorporate them in our narrative. In addition, much of the details used were obtained from general sources. Finally, we quote from some general books about the Industrial Revolution.

at the contextual facts themselves and the way that context changes over time. Within that domain of social change, we see tensions build up between the participants of societies. Basically, the themes of the societal tensions and controversies in nineteenth century Italy[29] were of the same nature as those in eighteenth century France leading up to the French Revolution. Although the latter being about the legitimacy of the *absolute monarchy*, based on the 'divine right of kings,'[30] in relation to the evolution of *parliamentarian sovereignty*, based on the natural rights of man. It was also about the relation of Church and State, the place and role of religion in society based on the 'Divinity'. In addition, it was about 'court and country': the different groups of participants in society with the accompanying tensions between the societal classes as they developed over time.

*Political context:* Part of the social context is the interaction between social groups, where each group has its shared, specific interests. It created

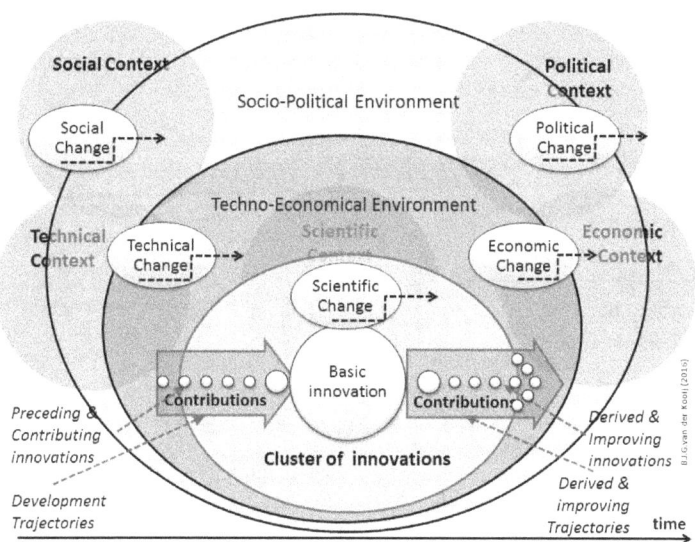

**Scheme 11: The specific Contexts for the Cluster of Innovations in its relevant Environment.**

The figure illustrates how the context of the techno-economic environment is shaped by the socio-political environment.

---

[29] The Risorgimento (1820-1870) was period of social unrest on the Italian Peninsula that resulted in the Unification of Italy under a monarchy.
[30] The doctrine of the *divine right of kings* —also called God's mandate— is about the royal legitimacy. It asserts that a monarch is subject to no earthly authority —only to the authority of God's representative on Earth, the pope— and that there is nothing to regulate the powers of the king, and he thus becomes an absolute power.

the 'politics' of that time, resulting in the (national) political context. The political context is about 'governance': both the *organizations* of governance and the *process* of governing.[31] Firstly, it is about the role of the state in governing the society: the executive form of government.[32] Secondly, it is about the relations between the 'governors' and 'those who were governed': the class structure of society and the interaction between the classes. The basic subject is *political power*: the ability to control the behavior and interests of others. Over the ages these political relations were based on feudal thinking.[33] However, in the period of time we are evaluating, it was the revolutionary views on the natural rights of man —as proclaimed by the Enlightenment philosophers— that were having an increasing influence on the relation between the 'rulers' and the 'ruled.'

In short, when we are looking at these social and political contexts, we look at the Constitution,[34] its form and institutions, and the holders of the related governing powers that came with it. Obviously these social and political contexts are not isolated issues. Therefore, we will —where appropriate— stray and immerse ourselves in neighbouring contexts, and we will describe some of the personalities that played an important role in those contexts.[35]

---

[31] *Governance* can be defined as all processes of governing, whether undertaken by a government, market, or network, whether over a family, tribe, formal/informal organization, or territory and whether through laws, norms, power, or language (Bevir, 2012, p. 1).

[32] *Government* is the political system by which a country or community (in general. a 'polity')is administered and regulated: ie how it is ruled. The political system is the set of institutions (eg the parliament, political parties) that controls (ie rules) the behavior of a 'State'. That includes the political organization of society, its law-defining and law enforcing institutions. We will use the legal institutions and their behavior, like monarchy and parliament, to describe the 'governmental context' (based on Encyclopedia Britannica definitions).

[33] *Feudalism*: The political structure in a social system in which the relations are derived from land ownership. It includes the concept of manorialism, where the landowning lords and the land working peasants are interrelated.

[34] *Constitution*: A body of fundamental principles or established precedents according to which a state or other organization is acknowledged to be governed (Oxford Dictionary).

[35] Our historic observations on what happened in society are by their nature based on the observations and interpretations of others. And each of these observers has his own point of view, his perspective, his focus and bias. The resulting interpretations they wrote down can be biased from a personal aspect (being a liberal thinker gives a different view than being a Marxist thinker) but also from a societal aspect (the nationalistic views). By using a wide variety of sources, both from different backgrounds and from different timeframes (contemporary authors and present-day others), we hope to eliminate this bias.

## *About the Interaction*

In our case studies, we primarily observe the phenomena related to technical innovation, phenomena that took place within the context of their times. In a way, it is a multi-layered context (Scheme 4). For one, there is the content shaped by the *individual environment*: the technical, economic, or scientific context on the individual level for the observed cluster of innovations. In addition to that direct context, there was the more indirect *techno-economic environment* with its technical change and its economic change. And finally, there is the context of the society in which 'it all happened', the *social-political environment* with its social change and political change. Seen in this way, it is a dynamic multi-layered environment that influences the phenomena we explore (Scheme 11). We look at each of the layers, and we look at the interaction between those layers.

We focus on *change*: such as at those artefacts that started to appear and that developed over time in an evolutionary process where some artefacts survived and others disappeared from the scene due to the properties of the environment. In analogy with biology, in the hot and dry desert, even the most potent plants struggle to survive, and only the best adapted ones survive. In the fertile environment of a tropical rainforest, plants in abundance compete with each other in a battle for survival. The same goes for the appearance of new artefacts (eg innovations): some thrive and prosper within their specific environments; others do not survive. It is —in Herbert Simon's concept— the properties of the 'inner environment' of the artefact that has to match the properties of the 'outer environment' if the artefact is to survive.

Therefore, trying to understand the dynamics of changes, we borrow from evolutionary biology the concept of Darwinian 'fitness for survival', which encompasses the fitness of the organism and the fitness of the environment. It is a concept that —in short— refers to the mutual relationship between organism and its relevant environment, between the properties of organisms to survive and the conditions of the environment in which the changes on a species level occur.

> *The fitness of the environment is one part of a reciprocal relationship of which the fitness of the organism is the other. This relationship is completely and perfectly reciprocal; the one fitness is not less important than the other, nor less invariably a constituent of a particular case of biological fitness.* (Henderson, 1914, p. 113)

In terms of technological innovation, this term refers to the fitness of a specific technology and its resulting artefacts in relation to the fitness of the environment in which it takes place. Clearly, some technologies 'make it and prosper'; other technologies proved to be 'dead ends.' They were not fit enough.[36] When the economic environment proves to be fertile —for example, in business terms— many technology-induced innovations and their artefacts will prosper.[37] That fitness of the environment may be the cause for the appearance of certain similar inventions at the same time, such as the parallel invention of the telegraph both in England and America.[38]

As this is not the place to dwell on evolutionary biology, we focus dominantly on the fitness of the relevant socio-political environment and techno-economic environment in relation to technological innovation itself (Scheme 11). As we analysed elsewhere,[39] the development of the GPT-child of the First Industrial Revolution, became one of the dominant catalysts of the Second Industrial Revolution. Both industrial revolutions took place within the techno-economic environment of their time. Hence, we talk about the *technical dimension* and the *economic dimension* of these revolutions. They will make up a large part of our analysis.

Also, as both the First and Second Industrial Revolution showed quite a bit of social dynamics —ie the American and French Revolutions, the 1848-Revolutions— obviously, there is a relationship between the phenomena and its socio-political environment, part of the described relationship between content and context. Hence, we talk about the *social dimension* and the *political dimension* of these revolutions, and we include in our analysis the social and political revolutions that took place when the foundations for these industrial revolutions were created. Our analysis for patterns of change in the different contexts is quite abstract; one could say we take a helicopter view. Not so much the larger and stationary 'satellite view', nor the more detailed 'birds-eye view', this helicopter view enables us to alter between pattern and detail by zooming in or out.

Borrowing from Shakespeare's play Julius Caesar the expression originated from Brutus and extending the metaphor of "There is an ebb and tide in the Affairs of Men", we refer to the periods in time where the

---

[36] An example would be the reciprocating electromotor of the early days of the electromotive engines. See: B.J.G. van der Kooij, *The Invention of the Electro-motive Engine* (2015), pp. 72–75. Another would be the arc light, which was replaced by the incandescent lamp. See: B.J.G. van der Kooij, *The Invention of the Electric Light* (2015), pp. 55–98.

[37] Here, the example is the availability of electricity when the electric dynamo came into existence and replaced the cumbersome voltaic batteries. Then the electric light, the telegraph, and telephone started to develop in force. See: B.J.G. van der Kooij, *The Invention of the Electro-motive Engine* (2015) pp.87–125.

[38] Both the Morse telegraph (based on the relay principle) and the original Cooke & Wheatstone telegraph (based on the galvanometer principle) appeared at the same time but in different environments (ie the American and British societies). The Morse concept survived. See: B.J.G. van der Kooij, *The Invention of the Communication Engine 'Telegraph'* (2015)

[39] See: B.J.G. van der Kooij, *The Invention of the Electro-motive Engine* (2015)

environmental context spurred a particular technological development.[40] Times of rising geo-political tensions reflected in the socio-political context of the concerned nations (eg the Madness of Times before both the World Wars). And that context would initiate increasing efforts to create novelty of a specific nature (eg military weaponry). Seen along that time-line of decades it creates an ebb and tide[41] in the *Affairs of Men*. The latter being the totality of human individual and collective behaviour and their social interaction (of which novelty creation is a part).[42]

## *About the Acts of Novelty*

The basic innovation at the core of the Cluster of Innovations, is the result of human induced intellectual acts of novelty creation. Thus, zooming in on the individual activities of the inventor —within the different acts of creation resulting in ideas, concepts and prototypes— could shed some light of the process that resulted in the creation of the (basic) innovation.

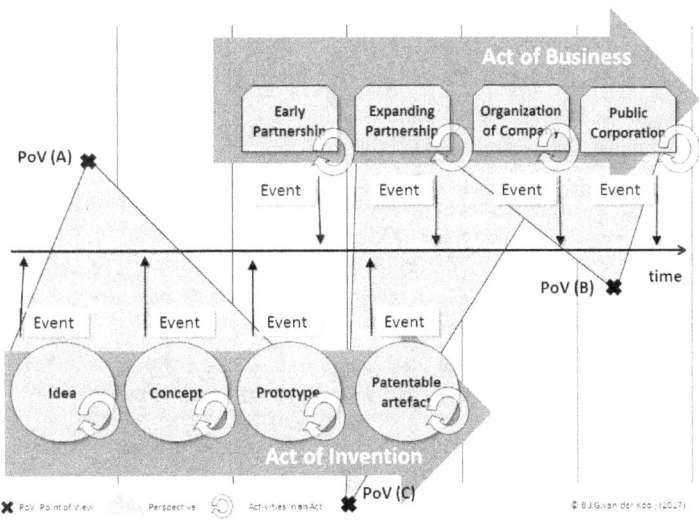

**Scheme 12: The Act of Invention and the Act of Business observed in different perspectives.**
PoV (A/B/C): Point of View

---

[40] Brutus: "There is a tide in the affairs of men. Which, taken at the flood, leads on to fortune; Omitted, all the voyage of their life, Is bound in shallows and in miseries. On such a full sea are we now afloat… (Julius Caesar, Act-IV, Scene-III, Lines 218-224)
[41] In the field of economics these variations are identified as the Long and Short Waves.
[42] The Construct of the Affairs of Men is used as generic term to identify the political, social and economic contexts.

Hence we use the construct of the *Act of Invention*[43] and the *Act of Business* (Scheme 12) to explore that aspect of novelty creation.

We define the Act of Invention as "the total process of inventive activity that starts with an idea and results in a (patented) artefact," and the Act of Business as "the total process of entrepreneurial activity that starts around a (patented) artefact and results in an organization." As identifier for those activities, we observed the related events.

The Act of Creation itself is the result of individual human behaviour in a specific context. Behaviour that is the outcome of one's personality, and personalities that are the consequence of Nature (ie biological characteristics based on genes) and Nurture (ie memes[44] based on upbringing). We limited us to the last and wondered why the inventors 'did what they did?' and look at factors resulting from their curiosity, ingenuity and creativity. But it is not only about the individual characteristics. Next to the individual Act of Creation, there is the collective Act of Creation, when a group in individuals is organised to create novelty (ie the R&D-department).

## *About the Metaphors used in this Case Study*

Throughout the Invention Series we have explored the events that resulted in the invention of 'engines.' This was the core event we labelled as basic innovations (Scheme 1) that had an impact on society and economy, and that that followed different application and technological trajectories (Scheme 2). Such as the steam *'engine'* (eg James Watt's invention) that became used in steam *'machines'* (eg the steam tractor, the steam locomotive) being the driving force of larger steam *'systems'* (eg steam boats, steam trains). Those systems were the driving force of the *'network infrstructures'* (eg the railway infrastructure) that had a massive impact on the Affairs of Man.

In a similar way we investigated in the other case studies the quartet 'engine-machine-system-network.' In the study Invention of the Silicon Engine, we apply the metaphor of the Silicon Engine. Being the development of a range of information processing engines (ie the GPE transistor, the single chip Integrated Circuit, the single chip microprocessor and the single chip microcomputer) that were the General Purpose Engine (GPE) of the Information Machines (eg Minicomputer and Personal Computer). Machines that were part of larger computing systems. (eg Time

---

[43] The word 'invention', used here for the individual contributions of the inventor-entrepreneur up to commercialization, would be replaced by the word 'innovation' in today's interpretation. For pragmatic reason we hold on to the historic use where appropriate.
[44] A meme is the cultural equivalent of genes (Dawkins, R.: The Selfish Gene. (1976).

sharing system, client-server system) empowering the information network (eg the computer networks).

## *About this Case Study*

This case study is another result of our quest to understand the Nature of Innovation. Where the other cases focused (1) on energy —the power of steam, the power of electricity and the power of explosions—, (2) the application of electricity —in light and rotative power applications—, and (3) the application of electricity in communication —telegraph, telephone and wireless—, in this case it is about (4) the application of electricity in information processing.

Of the dual roles of electricity —one offering means for transporting power and the other offering means for manipulating information— the latter's technological trajectory is explored. It is an exploration in two different trajectories (Scheme 2, right side). The first part of the *technological trajectory* starting with the invention of the electronic engine called 'Vacuum tube' (aka 'radio tube') in the early twentieth century. Followed by the invention of the electronic engine called 'transistor' in the middle of the twentieth century. The second part covers the *application trajectories*. The application of the engines that started the 'revolutions': the Radio Revolution, the Television Revolution, and the Mobile Revolution. All application trajectories where General Purpose Engines (GPEs) became the essential part of complex machines. Machines as part of technical systems, creating technical infrastructures and their institutions. As technologically-driven innovation is at the core of our explorations, we will focus on the basic technical development of those machines.

> To give an example. The vacuum tube (ie the Radio Tube, the Cathode Ray Tube) played a dominant role in the development of the hardware of radio television; from camera to receiver. Then came the service providers of the radio and television broadcasting; the organization (aka 'Stations') creating the broadcasted content. And finally, there were the regulatory institutions with their control over the airwaves: ie institutions like the General Post Office (GPO), the Federal Communication Commission (FCC), and the European Broadcasting Union (EBU).

The General Purpose Engines (GPE) of the vacuum tube and transistor were the building components for many larger devices, systems and infrastructures in specific application areas such as in communication and information processing. GPEs that created new technologies that became labelled as:

*The Communication Technology* (CT): This is the technology needed to realize communication system; from telegraph, telephone and wireless, radio and television broadcasting, to mobile telephony.

*The Information Technology* (IT): This is the technology needed to realize the calculating and computing systems; from the mechanical calculator to the electronic computer (mainframe, minicomputer and micro-computer).

In scholarly literature one notes a confusing use of the concept of ICT. In an effort to clarify (Scheme 13), one could say that the GPT-ICT has two contributing technologies; the Information Technologies (ie IT) and the Communication Technologies (ie CT). GPT's in which the General Purpose Engines (GPE) contributed to the development of the calculating and computing technologies as well as the cabled and wireless communication technologies. Each with a technological development trajectory where science and engineering were at the heart of economic progress as they ignited the individual application trajectories (Scheme 2). Although these technologies could be considered as a GPT of their own, we adhere to their combination as the GPT-ICT.

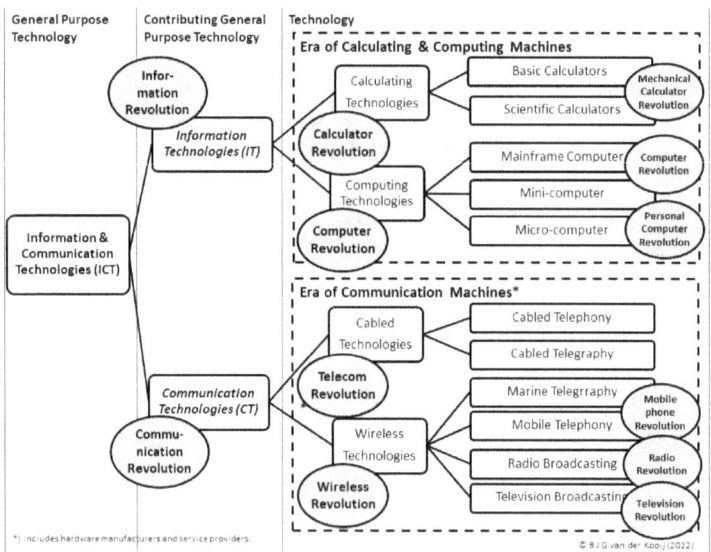

**Scheme 13: Overview Diagram of Technologies of the GPT-ICT.**
Diagram showing the breakdown of the GPT-ICT (left) into the contributing GPT's (middle) and the constituent Technologies (right). The Eras of development and the Revolutions are indicated separately.

A case study is about the events that define the Context and those that define the Content of specific events. In this case study that results in the following sections:

*The Context for the Discoveries:*[45] We will begin in the first section[46] with a thorough look at the events that created the general historical context. First, we explore the Industrial Revolutions in the nineteenth century after the French Revolution. A period that was dominated by the Mechanical Worldview and Mechanical Thinking. The century that saw major geo-political conflicts (eg Napoleonic Wars, 1848-European Wars) and where Science and Engineering were the walking companions of social, pollical an cultural change. In addition, we explore the contextual events in the twentieth century including both World Wars in the West and in the East.

Next, we explore in the second section the dawn of the Electric Era and the use of electricity as a source of power. We explore the major events of the *First Power Revolution* that emerged after Volta's invention of the electro-chemical battery creating DC-currents, as well as the *Second Power Revolution* creating also the AC-currents. We conclude this section with defining the *Third Power Revolution* where the internal combustion engine as a primary mover came on scene.

*The Content of the Inventions:* Then we move on to describe the events that put their mark on the use of electricity as a carrier of information.

*The Communication Revolution:* The third section of the case study is devoted to the content of the invention of the wireless communication engine itself. It covers the *First Communication Revolution* (1830-1920). We start with an analysis of the early days of wireless communication when the Hertzian waves became the focus of attention of scholars interested in the Nature of Lightning. Their work became part of the Wireless Mania in science. Then we describe the contribution of Guglielmo Marconi.

*The Calculating Revolution:* The fourth section of the case study is devoted to the content of the invention of calculating engine. It covers the *First Calculating Revolution* (1850-1930). We start with the ancient calculating tools and the early calculating machines. Next, we explore the development of the Stepped Drum Technology creating

---

[45] Our reader might wonder why such a long period is going to be observed. Basically, the answer would be that the foundations for our present-day technology-dominated, democratic society were created over that period of time.

[46] As both case studies share the same macro context, this reflects in the duplication in the first section, as we wanted that each case study could be read on its own.

calculating machines based on the Leibnitz wheel. Followed by the Pinwheel Technology used in calculating machines. For both we analyse the contextual events for contributors and their activities.

*The Computing Revolution:* The fifth section of the case study is devoted to the content of the invention of the computing engine. Starting with the Art of Mathematics, we explore the early endeavours from Babbage cs. creating their Logic Machines. Such as his ideas for a Differential Machine and Analytical Machine, to be realized in a mechanical technology. Those efforts failed, but the introduction of the electromechanical Relay Logic resulted in the early computing machines (eg the Zuse-machines). Next, we explore the first computers like the Colossus and the Eniac applying the vacuum tube technology. And we conclude with exploring the Era of Mainframes and the Era Minicomputers after the Second World War.

*The IC & IT-Revolutions:* Rounding of the described revolutions, we synthesize in the sixth section the analysed revolutions in the early construct of the Information Technology (IT) and Communication (CT) Technology creating the GPT-IC&IT. And we relate them to the Third Industrial Revolution (1920-1950).

Then we reach the core of our explorations into the content of the General Purpose Engines that fuelled the subsequent technological developments. Engines that we part of the new technology labelled as 'electronics.'

*The Electronic Revolutions:* In the seventh section we explore the General Purpose Engines (both the vacuum tube and the transistor, single chip Integrated Circuit and Micro-processor) that fuelled the technological development trajectories labelled as Electronic Revolutions. We explore the trajectory of the Radio Valve and the Cathode Ray Tube that created the *First Electronic Revolution* (1900-1950) in detail. Zooming in on the contributors like John Fleming (ie the Fleming Valve) and Lee DeForest (ie the Audion), we describe their endeavours, business practises and patent wars in detail. Next, we explore the invention of the Transistor and the de Integrated Circuit that created the *Second Electronic Revolution* (1950-1980) in detail. We analyse their contributors (resp. Bardeen, Brattain, Shockley and Kilby, Noyce) in the context of the development of the hi-tech region called Silicon Valley. And finally, we zoom in on the invention of the micro-processor; the micro-computer on a single chip and its early evolution.

These Electronic Revolutions would go and fuel the innovations in the subsequent application trajectories. Such as there are:

*The Radio Revolution:* In the eighth section of this case study, we explore the Era of Radio Broadcasting on the level of the machines (ie the radio receiver) and the level of broadcasting (ie the radio station). We trace the early efforts where wireless communication was used to transmit the 'spoken word,' as well as music (ie 'audio' creating the new phenomenon of early Home Entertainment. We zoom in exploring the Radio Broadcasting Boom with its development trajectory of the radio receiver and its trajectory of radio broadcasting.

*The Television Revolution:* In the nineth section of this case study, we explore the Era of Television Broadcasting, again on the level of the machine (ie the television receiver) as well as on the level of broadcasting (ie the television stations). We explore the contributions to transmit next to 'audio' signals, also 'video' signals (ie moving pictures) that made Home Entertainment an alternative to the earlier entertainment in Vaudeville theatres and the Cinema (ie the movie theatre) of that period in time. Zooming in on the Television Boom we explore how television broadcasting influenced the Affairs of Man drastically.

*The Mobile Telephone Revolution:* In the tenth section we explore the Era of Mobile Telephony, also on the level of the machines (ie the cell phones) and their infrastructure (ie the mobile phone networks) as well as the service industry around it (the mobile network providers). Starting to explore the subsequent generations of networks (ie 0G, 1G, 2G and 3G), we zoom in on the development of the accompanying cell phone, and its development trajectory into the Feature Phone. In addition, we describe the industrial development that saw the rise and fall of prominent companies (eg Nokia, Motorola, Blackberry). We conclude this exploration with an analysis of the invention of the smartphone.

*The ICT-Revolution:* Rounding of the described revolutions, we synthesize in the eleventh section the analysed revolutions in the concept of the GPT-ICT. And we relate them to the Fourth Industrial Revolution. Having explored the GPT-ICT is all its dimensions, we next try to synthesize our explorations.

*Conclusion:* In the last section we describe the themes that characterise the explored innovations. Such as the change from the sole inventor to the R&D-teams creating novelty. Or the Acts of Creation executed by individuals and teams. And the development of the organisable and manageable process of innovation.

This is the case study about the outgrowth of the *General Purpose Technology* of 'Electronics' with its 'Clusters of Innovations' and 'Clusters of Businesses' that created the Eras of Electronics and changed the world we live in. It will take you along a fascinating multitude of winding social, political, economic, technical and scientific roads into modern historic times and back to our present time.

B. J. G. van der Kooij

# The Invention of the Silicon Engine

## Context for the Discoveries

For some one living in the pre-electric era it was hard to image how the world in the twenty-first century would be dominated by a single, but basic phenomenon called 'electricity.' No woman, used to do the washing by hand and sharing the latest information about village affairs (Figure 1), could have imagined that one day there would be automatic washing machines that could be programmed to do the washing of clothes by themselves. Or could imagine that in their kitchen food could be stored in special cupboards that kept it cool, even in a frozen state, so it could be conserved over longer periods of time. And that the heat needed for cooking and baking did not require the maintenance of a wood of coal-fire, but could be acquired by turning a knob on the electric stove or oven.

**Figure 1: Chatting Women washing clothes by a Stream.**
Source: Daniel Ridgway Knight (ca 1898). Wikimedia Commons.

Nobody even was dreaming that one day there would exist rather simple communication engines, such as the telegraph, that would send the message at the speed of light to far away destinations all over the world. Neither would many people have imagined the communication engine of the telephone, which could be used to transmitting gossip as instant electric speech over long distances (Figure 2). Messages and speech transmitted around the globe by a network of copper wires

—spanning distances over land, crossing rivers, bridging the continents by undersea telegraph cables— may have seemed already magic for our ancestors, but there was more to come.

Likewise, it was hard for those people used to the limitations of the gas-, oil- or wax candle light, to imagine the abundance of electric powered lights in private and public places. Where the turning of a switch was sufficient to light a room, a restaurant or theatre and disperse of the darkness of the night. For the common man of the eighteenth century, it was impossible to imagine times to come where the pattern of daylight did not anymore control private, social and working life.

**Figure 2: *Party Wire* by Norman Rockwell (1919).**

Several subscribers connecting to the same 'party line' made any privacy hardly possible.

Source: http://www.best-norman-rockwell-art.com/norman-rockwell-leslies-cover-1919-03-22-the-party-wire.html. Courtesy Norman Rockwell Estate, www.img.com (Mike Mueller, July, 2018)

**Figure 3: *Wonders of Radio* by Normann Rockwell (1922).**

Source: Post Cover, May 20, 1922. Finch, Chr.: Norman Rockwell 322 Magazine Covers. P.116. . Courtesy Norman Rockwell Estate, www.img.com (Mike Mueller, July, 2018)

Where the traditional family life in the evening around the table, lighted by the oil lamp, would become dominated by boxes that produced music and speech (ie the radio phone, Figure 3).

In a similar way, for someone living in the pre-automobile era it was hard to image how the world in the twentieth century would be dominated by a single, but basic phenomenon called 'motor car.' In the time of our great-grand-parents, mobility was defined by the distance a person could walk (the common man), or by the distance he could ride on the back of a horse (the wealthier classes that could afford its ownership). The time when the horses powered the aristocratic carriage, the elegant buggy and the mail coach. Where the horse could be rented from the hackney men exploiting livery stables. Or where the horse pulled —next to the plough turning the fields— the barge, transporting the common people and goods through the canals.

Figure 4: Galloping Horseman.
Source. Frederick Remington. Wikimedia Commons.

Next came the days that the early strange contraptions of mobility showed up. Belching fumes and noises, these horse-less, steam-powered coaches would appear on the local roads (Figure 5). The horse-powered omnibus became replaced by the steam-powered omnibus. And soon, the steam-locomotive would pull trains of wagons running on tracks filled with people and goods over longer distances. The small world of the local market town and its surrounding became larger as people could travel over distance to other cities. Even more, when steam-power reached the factories, the industrious live changed as people went to work in the factory where the steam engine was powering the machines they operated.

Figure 5: Steam-powered Mobility.
Source: Wikimedia Commons. https://www.history.ac.uk/

May the penetration of the external combustion engine (aka steam engine) in their nineteenth-century lives have been a mental shock for our ancestors, it was the prelude of other machines that would disrupt the lives of our grandparents again a century later. By the twentieth century the carriages changed again. After the power of the clumsy steam engine had replaced the horse power, now the 'petrol motor' powered the coach and omnibus. And the horseless carriages that gave the privileged class their mobility. Even worse, their auto mania made the smoke-belching racing car appeared on the country scene (Figure 6), followed by the four-wheeled 'auto-mobile' and the two-wheeled 'motor-bike.' It took some decades before the exclusivity of the automobility reached the common people. But then the farmer's car lifted the rural isolation, and created the urban cities for the common man. Soon automobiles were cluttering the large cities in a similar way as the horse had done. That was the Automobile Revolution that gave them mobility they had been dreaming of.

**Figure 6: The Auto Mania of the Internal Combustion Engine.**

Source: Wikimedia Commons.

Like their dreams of flying in the skies like the birds, when the airgliders became powered and the airplane took off (Figure 7). Creating the Aviation Revolution where airlines connected 'airports' all over the world, like the motor-powered ships connected the 'seaports.' And all those vehicles were powered by the internal combustion engine running on petrol, belching smoke and noises again.

**Figure 7: The Airplane takes off.**

Source: Deanmosher.com.

Who could have foreseen in that period of time that one could travel by one's own people's car? Going to work or travel for leisure; holidaying and camping. Or where the longer distance mobility was not only by train or boat, but also by the seaplanes called 'Clippers.' But that was not all, as that Era of Mechanization not only brought our ancestors mobility, but also a wealth of new artifacts that changed the private and professional lives. From the time-devices used to keep track off their time, to the calculating machines used in their offices.

Trying to understand the Nature of Innovation —ie Change and Novelty that dominated the Affairs of Man in an historic perspective— one wants to know what is it origin.[47] Is 'innovation' a modern phenomenon pushed by ever progressing technology? Or is it an attribute of 'Human Life" that evolved over time?[48] And what were the underlaying evolutionary mechanisms that determined the subject of novelty in more recent times? So let's reflect for a moment on what happened in the first half of the twentieth century.

## *The Era of Mechanization and Mechanical Thinking*[49]

From its origin mankind depended on his muscular system to move around. His legs gave him mobility. A facility that was soon complement by the use of animal: the horse/donkey/mule that gave him extended mobility. The animal powered mobility was extended by mechanical contraptions; the buggy, the wagon, and the coach pulled by horses. Contraptions that became horseless carriages when the internal combustion engine was invented and used to replace the horse and extended again the range of mobility. Those horseless carriages developed over time into the automobile and motorbike. The Automobile Revolution was the result.[50] Even more, with the invention of the motor-powered glider (aka airplane), man entered the space before only accessible to the birds. The Aviation Revolution was the result.[51]

Also from its origin mankind depended on his vocal system to express himself. His voice gave him a means to communicate his thoughts and feelings with others. First with guttural sounds, later —when he got his language capability— with the spoken word. This form of communication had its limitations, though, as the distance he could cover with his voice was limited. So, he developed the coded word to be used with semaphores, that created the optical telegraphy system. And from there, the electric telegraph and telephone systems emerged, transmitting information over distance. In the process influencing the Affairs of Man.

---

[47] See: Van der Kooij, B.J.G.: *Origins of Innovation, Ancient (R)evolutions in Perspective*.
[48] See: Van der Kooij, B.J.G.: *Discovery of Innovation, Historic (R)evolutions in Perspective*.
[49] The English word *machine* comes through Middle French from the Latin *machina*, which in turn derives from the Greek (Doric: μαχανά *makhana*, Ionic: μηχανή *mekhane*) contrivance, machine, engine, a derivation from μῆχος *mekhos*' that means, expedient, remedy'
[50] See: Van der Kooij, B.J.G.: *The Invention of the Internal Combustion Engine*.
[51] See: Van der Kooij, B.J.G.: *The Invention of the Communication Engine 'Telegraph'*.

These are only some of the many inventions of a mechanical nature that influenced the Affairs of Man in a fundamental way up to the twentieth century. At their base lay the phenomenon of *mechanical thinking*. Where the rural and farm lifestyle —both dominated by the natural phenomena— had entertained *organic thinking*, after the Industrial Revolutions the processes of the mind had changed in character.

## Mechanistic Worldview

At basis of this shift was the collective worldview that emerged during the preceding period of the Scientific Revolution and Enlightenment. The first being the result of scholarly interest in physics (eg the Nature of Matter) and the latter the result of scholarly inquiries into metaphysics (eg the Nature of Being). Where the organic worldview had its base in human perception of his organic environment (aka biologic Nature) with the occurrences he understood (ie from the environment he lived in) being conceptualized as known entities and experiences. And where the phenomena he did not understand (the Powers of Nature) were conceptualized differently (eg the Mother Earth concept) that created belief in natural deities (eg Lightning being caused by the angry God Wodan throwing his hammer). Beliefs in the forces created by a supernatural entity outside nature itself; the gods. Beliefs that became the polytheistic and monotheistic religions that would rule the Affairs of Man. Ultimo culminating in the social and political and religious dominance of the western Roman Catholic Church. And where organic 'science' was the work of the alchemists (the Magicians of Nature) that looked at the material world as an organic entity, as a living body.

In contrast there was the abstract mechanical thinking where the mechanical worldview[52] resulted from investigations into the Nature of Matter and Motion creating the doctrine of 'Mechanisms.'[53] Such as the abstract time-concept that was materialized by clock-works. The cosmos was seen in mechanical terms; as the Clockwork Universe created by the (divine) Clockmaker. A universe ruled by laws, such as the Laws of Gravity defined by Isaac Newton. The Power of Heat was converted by a machine (ie the steam engine) in motion. Suddenly human power was extended, not by animal of water/wind power, but by the heat produced by a burning

---

[52] The Cartesian Worldview preceded the mechanical worldview and finds its origins in Descartes' dualistic thinking about mind and matter as expressed in "I think, therefore I am"(Cogito, ergo sum). He suggested the body works like a machine, and that mind (or soul) was non-material controlling the body, like the body influenced the mind.

[53] The doctrine of mechanism in philosophy comes in two different flavours. They are both doctrines of metaphysics, but they are different in scope and ambitions: the first (aka universal mechanism) is a global doctrine about nature; the second (aka anthropic mechanism) is a local doctrine about humans and their minds.

matter (eg coal, charcoal). It enhanced his mobility (with steam trains) and productivity (with steam-powered tools in the factory). As a side-effect the Mechanical Worldview implied the existence of a maker (a divine interpreter) and disarmed the objections of the Church, the representation of religion that grew out of the organic worldview.

The Mechanical Worldview, that was inspired by the mechanical artefacts (eg the tools) and constructs (eg the individual machines) as entities, was complemented by the thinking in systems. Both the Systems of Nature as well as the Systems of Men. The first paradigm considering nature as a complex biological system; from the molecular level to populations with interconnected subsystems and individual parts, existing in a biosphere (eg air, earth, oceans) and its dynamics (eg weather, climate). The second paradigm considering society from social and political systems (eg state formation) to economic systems (the 'economy') and socio-technical systems (the 'organization'). From the organic form of production by the single artisan, to the organized (equals 'programmed') system of production in the factory (ie the factory system). From the organic form of communication to the systems of tele-communication.

This evolution Mechanical Worldview into the *Mechanistic Worldview* was the result of *mechanical thinking* and *systems thinking*. The first as the form of cognitive thinking in terms of causes and their effects, of mechanical concepts and constructs. The second as form of cognitive thinking where interaction between the interrelated parts created its functions (both the primary, secondary functions and tertiary functions).[54] Combined with the growing knowledge base (the mechanical sciences) and knowhow of techniques (the mechanical technologies), it spurred of evolutionary developments of a specific nature.

*The Evolution in Mobility:* After our ancestors had tamed the donkey, horse and camel, this animal power had enabled them to travel longer distances and carry heavier loads. The more when he invented the carriage to transport people and freight. This mastering of 'animal powers' was the first step of the mastering of power that would follow; the next being the natural powers of water and wind. With the help of the 'waterwheel', he turned the streaming waters to his advantage. With the help of the 'windwheel' he turned the winds into a power source. The watermill and windmill were born to help him milling his grain, cutting planks, and pumping water.

---

[54] The automobile offers mobility (ie the primary function), the well-designed and well equipped automobile offers comforts to mobility (ie the secondary function), and the expensive and exclusive automobile creates social class effects like 'keeping up with the Jones' (ie the tertiary function).

In the evolutionary path of human development, a distinctive step took place when it was discovered that the Power of Heat[55] (ie fire) could be tamed by machines. It resulted in the creation of the steam engine (aka the external combustion engine), that lay at the foundations of the First Industrial Revolution. This was a next step in men's mastering of power. A step that brought him extended mobility as he could travel and trade over larger distances thanks to the steam locomotive and the steamboat.

The following step occurred when the Power of Explosion[56] (ie from black powder to dynamite) was mastered. Now power machinery (aka internal combustion engines) could be constructed creating the explosion engines in a broad variety. Then those engines were applied into movable carriages creating the horseless 'car,' and did extend his mobility again. Replacing the steam engine in sea ships, and powering the airships, it would become a major contributor to the Third Industrial Revolution. And now it extended his mobility again with the automobiles, motored boats and airplanes.

*The Evolution of Communication:* Communication is a basic function of biologic systems. The verbal and non-verbal exchange of information between members of biological systems, is part of its existential behavior. All over the Animal Kingdom, animals have their alarm-cries to warn for predators, do their courtship dances communicating their fertility, and growl to defend their territory. Both in an auditive form and a physical non-auditive form they try to get the message understood. Human beings developed next to the non-verbal communication (eg sign-language with hands and face), the languages of the spoken word resulting from the articulate sounds produced by the human vocal capabilities.

Communication fulfils a basic need for humans. People, belonging to families, clans and societies, always needed to communicate to survive. Not only as a practical tool to execute individual behaviour and express individual feelings, but also as a means to work together and to secure a place in the hierarchy of a group. If those collectives are small, limited to the family living in one shelter or the clan in one hamlet, communication is over limited distances as far as the voice carries; the *spoken word (*or *shouted word)*. When collectives expand into regional societies, another form of communication is needed; the *written word* that can be transported by courier —aka the post riders— on horseback.

---

[55] See: Van der Kooij, B.J.G.: *The Invention of the Steam Engine.* (2015)
[56] See: Van der Kooij, B.J.G.: *The Invention of the Internal Combustion Engine.* (2021)

So, the need for communication over distance increases when societies expand and interact over larger areas. But the means of communication were limiting its spreading over distance as a runner or horseman is limited in the distance he can cover. So, starting with smoke signals, man developed the mechanical transmission of information over distance; the optical telegraphy. With the *coded word* a new form of communication over longer distances —along a telegraphy chain of semaphore towers between which visual contact was possible— was realised.

*The Information Evolution:* Communication was based on the 'spoken word' and the 'coded word,' both containing information.[57] Information that was from ancient time restricted by the limitations of the human brain, both in terms of processing as well as transmission. It created the 'written word' of texts and the 'spoken word' of languages. In the process creating the *external memory*, the ability to store information over generations by extending the biological memory by artificial memory (eg tablets, books).

Processing information fulfils a basic need for humans. From the existential brain processing sensory information in the interaction of the individual with his environment, to the cognitive brain processing information on an abstract level according specific protocols (ie mathematics). Some of that information processed by the brain, could be processed by tools (ie the abacus) and later by machines (eg mechanical adder). First the mechanical technologies creating the calculating machines processing numerical data according certain algebraic protocols (ie addition, subtraction, multiplication, division). Next the electronic technologies creating the electronic calculators, the tabulating machines and the differential machines (later known as computers)[58] processing information more efficiently. Machines that over time complemented the biologic intelligence controlling human life, with machine-based intelligence controlling machines-based processes. And ultimo leading to Artificial Intelligence.[59]

These are some (technology-induced) evolutionary developments that influenced the Affairs of Man over time in response of some of his basic needs.

---

[57] Information is the collection of data; statements of facts about something or someone.
[58] A *computer* in modern parlour is an electronic device that manipulates data with its ability to store, retrieve and process data.
[59] *Artificial Intelligence* is the simulation of human intelligence processes by machines, especially computer systems. It covers both the theory and the development of computer systems able to perform tasks normally requiring human intelligence; eg visual perception and recognition, speech recognition, decision-making, and translation between languages.

# Science and Engineering, Technical and Social Change

The transition in the Western World from an agricultural society to an industrial society under the influence of technology —known as the times of the Industrial Revolution[60]— took place in some two centuries. In fact, it took place after the Renaissance Revolution in different but interlinked industrial transformations; the *First Industrial Revolution* that covers the period of 1760-1850, the *Second Industrial Revolution* that that covers from 1850-1920, and the *Third Industrial Revolution* that covers the period between 1920-1950 (Figure 8, upper part). Although their specific characteristics were quite different, these periods had the same mechanism in common: the control of mechanical power. As a result, societies were dominated by technological change and novelty on a scale never seen before in the history of mankind. Technological change that was the walking companion of Religious Change, Political Change, and Social Change.

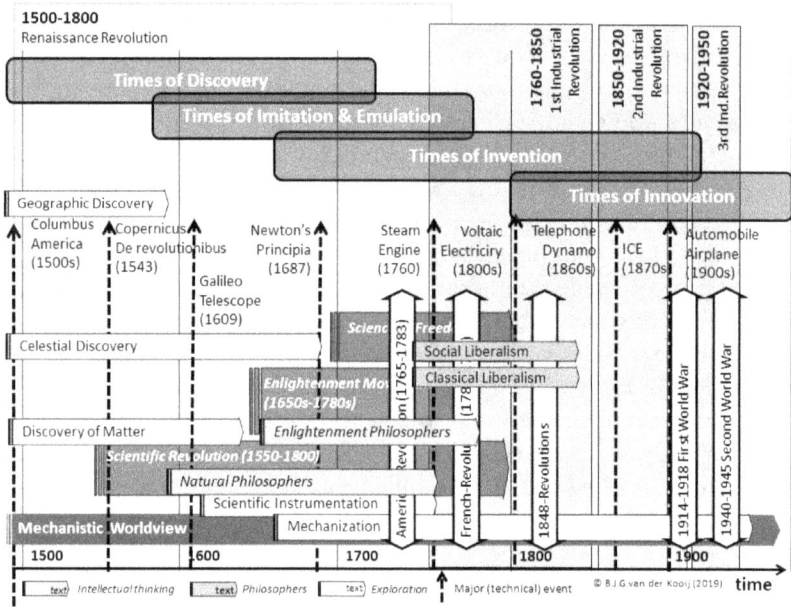

Figure 8: Overview of the Evolutionary and Revolutionary Context of the Industrial Revolutions.

ICE: Internal Combustion Engine.

---

[60] The concept of the Industrial Revolution is devised by historians to describe the totality of specific developments in that period of time. The concept was introduced by the English economic historian Arnold Toynbee (1852–83) to describe Britain's economic development from 1760 to 1840.

## *The Context of the Early Nineteenth Century (1800-1850)*

Technology created the 'engines' of change. From the steam engine, the electric engines to the internal combustion engine. Engines that were the offspring of the work of thinkers and tinkerers of the Scientific Revolution. The Scientific Revolution being the time when the thinkers reflected on the Nature of Matter. Such as the *Nature of Heat* and the Power of Steam, the *Nature of Lightning* and the Power of Electricity, and the *Nature of Chemical Substances* and the Power of Explosion. Each of these powers became subservient to man, when he invented their specific engines. Times that had seen the rise of the Mechanical Worldview creating the Age of Mechanization (Figure 8, lower part). Times when mechanical technologies would emerge in a broad range of applications; from the fine-mechanical technology of watch-, clock- and instrument-making, to the heavy mechanical technologies that created the power-based machines called locomotives and steam boats.

They all spawned into a wealth of application fields. A process to which many scientific and engineering minds contributed, creating all that novelty in a range of events during the *Times of Invention* and the *Times of Innovation* (Figure 8, top).

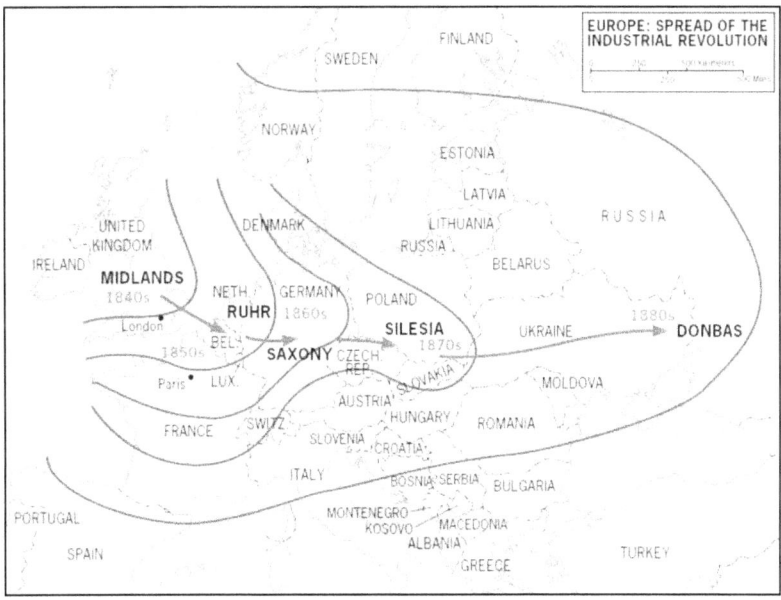

**Figure 9: Industrial Revolutions spreading over Europe.**
The regional industrial centres are indicated in bold.
Source: : Unknown. Wikimedia Commons.

Although the First Industrial Revolution sprouted in Britain becoming the Cradle of Industrialization, it was not confined to the British Islands. By the mid-nineteenth century industrialization spread over the Western World rapidly. It started by spreading on the European continent (Figure 9) and soon crossed the Atlantic Ocean to America.

## *The Context of the Late Nineteenth Century (1850-1900)*

The 1848-Revolutions were the culmination of the aftermath of the French Revolution. It was the end of a period of time that had seen on the European Continent the French Revolution (1789-1799), the French Revolutionary Wars (1792-1802), and the Napoleonic Wars (1803-1815) of the First French Empire. And it was the time that central European countries of the former Holy Empire (962-1806) laid the foundations for the German Unification, starting with the Confederation of the Rhine (1806-1813).[61] On the Northern Italian Peninsula the far-reaching Napoleonic reforms would ignite the freedom movements that became the catalyst of the Italian Unification; the Risorgimento (1848-1871).[62]

### **The Spring of Nations: the 1848-Revolutions**

All over Europe, but especially in Central/Eastern Europe, social turmoil erupted (Figure 10). The revolutionary wave began in France in February 1848 and immediately spread like a wildfire to most of Europe, affecting some fifty countries. The social uproar of the 1848-Revolutions shook the foundations of those European societies. Although royal dynasties dwindled, political reform took its impetus from the 1848-Revolutions, and nation-building got off the ground. Aided by the global effect of the Pax Britannica, the French Empire, the British Empire, and the Russian Empire all expanded and conflicts between the ruling European powers were amass.[63]

---

[61] The Confederation of the Rhine was a confederation of Napoleonic client states resulting from the conquests of Napoleon. It was a military alliance between of former lands of the Holy Empire that was dissolved in 1806. Although originally separated from the two largest states Prussia and Austria, in 1806 Prussia lost to Napoleon, and in 1810 large parts of what is now northwest Germany were quickly annexed to France.

[62] For more details on the Unification of Italy, see: Van der Kooij, B.J.G.: *The Invention of the Wireless Communication Engine* (2017).

[63] The Spanish Empire and Portuguese Empire by that time had collapsed under Napoleonic revolutionary pressure. Spain had lost its Western colonies in the Americas and renounced sovereignty over all of continental America in 1836. Portugal had lost most of their Eastern Colonies in the Indies. The British and Russian empires expanded significantly and became the world's leading powers. The Russian Empire expanded in central and far eastern Asia. The British Empire grew rapidly in the first half of the century, especially with the expansion of vast territories in Canada, Australia, South Africa, and heavily populated India, and in the

From the first technology-driven *Crimean War* (1853-1856)[64] between Britain, France, Turkey and Russia arose the *Wars of Italian Independence* (1848-1861) involving France, Austria and Piedmont-Sardinia; the *Wars of German Unification* in 1864 between Austria, Prussia and Denmark; Prussia's clash with Austria in 1866, and the *Franco-Prussian War* between the German states and France (1870–71). Then, the *Battle of Sedan* (1870) would mark a turning point on the European continent.[65]

The *Second French Empire*, the imperial Bonapartist regime of Napoleon III from 1852–1870, collapsed in 1870 and had given way to the *Third Republic* (1870-1940). This as the result of the disastrous outcome of the war with Prussia after the French lost the Battle of Sedan (September 1st, 1870). Then, at the end of the battle, Napoleon III surrendered his sword to Otto von Bismarck, the German chancellor who became the founding father op the 'Second Reich' (1871-1918).

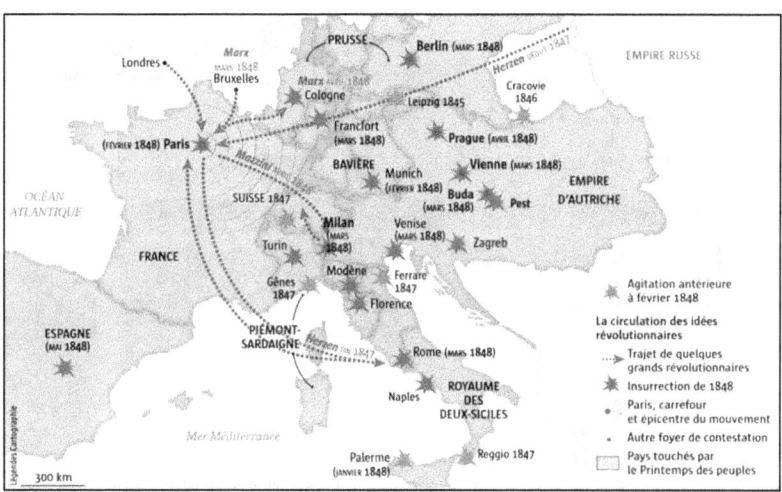

**Figure 10: The spreading of the 1848-Revolutions.**
Source: Unknown. Wikimedia Commons.

---

last two decades of the century in Africa. By the end of the century, the British Empire controlled a fifth of the world's land and one-quarter of the world's population.
[64] The Crimean War, aiming to stop Russia's expansion policy in the Near East into the crumbling Ottoman Empire, saw the use of the steam powered warship and steam locomotive for transport. In addition, telegraphy and photography was used for communication with distanced audiences.
[65] One factor that contributed to the German victory was rail transport. The Germans were capable of moving the military and their equipment fast over the extended railway infrastructure to the fronts.

Was the immediate aftermath of the 1848 Revolution during the Age of Capital a period of Restauration, during the next decades the geo-political situation would change dramatically (Figure 11). During the *Ages of Empires* (1875-1914), Europe at the end of the nineteenth century became the continent of the so-called imperial powers. (Hobsbawm, 2010a). The heritage of the former *First French Empire* under Emperor Napoleon Bonaparte (1804–1815), after becoming a republic, had exercised in the early nineteenth century a profound socio-political influence on civil societies in Europe. In the Second half of the nineteenth century, the *British Empire* (1815–1914) under the rule of Queen Victoria had become the most powerful colonial nation outside the European continent. The young *German Empire* (1871–1918) was in the process of being shaped by the members of the German Confederation and wanted a place between the Great Powers. The *Austro-Hungarian Empire* (1867–1918) had risen from the realms of the Habsburg dominions. And the *Russian Empire* (1721-1917) lay dormant in the East heading at a revolution of its own. All these empires, except Russia, were ruled by the remnants of absolute monarchies that had seen their 'divine rights' crumbled under rising parliamentary sovereignty during the preceding *Age of Revolutions* (1789-1848) (Hobsbawm, 2010b).

**Figure 11: Political Map of Europe (1867).**

The map shows the Kingdoms of Italy, Spain, France, Britain; the Russian, the Austrian-Hungarian and Ottoman Empire, and the fragmented remnants of the Holy Empire in Central Europe.

Source: Unknown. Wikimedia Commons.

## The European Empires

The second half of the nineteenth century —after the 1848-Revolutions— were for Europe, notwithstanding the wars and conflicts between nations, times of rapid population growth and economic progress (Figure 12). Britain had after centuries of expansion and imperialism developed into a world power. The (second) *British Empire* (1815-1914) — based on the exploration of the Far East creating the new eastern colonies in Australia and New Zealand— was reaching around the globe; 'Britannia ruled the waves.' The nineteenth century was the time of *Pax Britannica*, with the British Navy's 'showing of the flag' around the globe displaying her power. And they liked to show that techno-economic progress to the world during the *Great Exhibition of the Works of Industry of all Nations* in 1851.

> The *Victorian Era*, the period of Queen Victoria's reign (1837-1901), was a time of unprecedented population growth in Britain. The population rose from 13.9 million in 1831, to 32.5 million in 1901; with one-third of the population employed in manufacturing by 1870. That growth was founded after 1850 (Figure 12) by technological driven economic progress. A period of the rising middle class, attaining political and financial eminence, active as commercial or industrial entrepreneur. As capital and credit were easily available, many individuals and families became wealthy and successful, and had increasing time for leisure. The working class saw an increase in real wages, declining working hours, and yearly vacation. By the late-Victorian era the leisure industry had emerged and provided scheduled entertainment of suitable length at convenient locales at inexpensive prices. These included sporting events, music halls, and popular theatre.

All that acquired wealth from the colonies (ie the sugar and tobacco from Western Indies, teas from British India, wool from Australia/New Zealand), and the subsequent technological progress of the Industrial Revolution had made Britain into the 'Workshop of the World.'[66]

In the second half of the nineteenth century the British were going ahead, at full steam, dominating worldwide trade and commerce more and more. However, Britain, as the workshop of the world and the leading industrialized country, still had to find markets for all its mass-produced products. Not only home markets, but also international export markets, for which they needed the 'free trade practices.' Thus, next to British

---

[66] The 'workshop of the world' indicates that its finished goods were produced so efficiently and cheaply that they could often undersell comparable, locally manufactured goods in almost any other market.

products (eg locomotives, weaponry), also British ships, British capital and British financial institutions dominated world trading. Nevertheless, the Victorian boom came to an end, and was succeeded by the Long Depression (1873-1896). An economic recession that started with the financial British Panic of 1873 —originating from the US Banc Panic— followed by the financial failures of Vienna, industrial overproduction, and the bursting of the rail road bubble in America. All events inducing a twenty three-year period of worldwide growth and recession cycles which only ended in the late 1890s.

On the continent started the German uprisings in March 1848, such as the Prussian Revolution and Vienna Revolution, with the subsequent socio-political changes. Revolutions that took place during the Central-Europe's *Gründerzeit*, the economic phase in nineteenth-century in Germany and Austria before the great stock market crash of 1873 (Figure 12). It was the Golden Age for German speaking areas that competed on a world-class level in the domains of science, technology, industry and commerce. The industrialization had massive social repercussion as the middle-class participants (ie the Gründer or entrepreneurs) were committed to liberal principles, while the working class sought radical improvements to their working and living conditions.

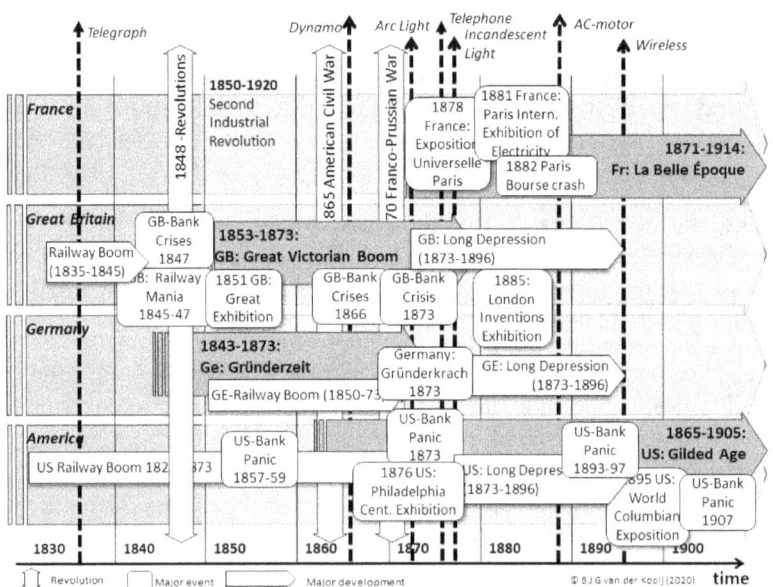

Figure 12: Overview of Economic Progress after the 1848-Revolutions.

From 1850 —after the failed March Revolution— Vienna, the capital and residence of Emperor Franz Joseph I, became the fifth largest metropolis in the world through the incorporation of the suburbs and the influx of hundreds of thousands of people, especially from Bohemia and Moravia, until 1910. Starting in the 1840s this industrialization had created a period of economic progress that paralleled the breakthrough of industrial capitalism in the German states. There, the national movements, frustrated in their effort to achieve civic reform, turned to the attainment of material progress. The victory of the reaction was followed by an economic expansion as the business community began to recover from its fear of mob violence and social upheaval.

Political tensions, however, between the Habsburg monarchy and the leading nationalist movements resulted in the Ausgleich ("Compromise") of 1867. Hungary was granted substantial autonomy, and separate parliaments, though based on limited suffrage, were established in Austria and Hungary creating the Austro-Hungarian Empire. At the same time the separatist movement to free Lombardy (Italy) from Habsburg colonialism and the *Third War of Independence* (1866) had resulted in the Kingdom of Italy. [67]

The geo-political tensions in Europe —especially in Central Europe under Habsburg rule— increased during the Russo-Turkish War (1877-1878) in which the Ottoman Empire as 'sick man of Europe' was defeated. The *Berlin Congress* of 1878 partially solved an international crisis between the European Empires, by granting varying degrees of independence to Romania, Serbia, Montenegro, Bosnia and Herzegovina. It fuelled, however the glowing embers of discontent and regional nationalism erupting in the Balkan Wars (1912, 1913) to come. The great loser at the Congress of Berlin was Russia, which left resentful that its enormous gains were nullified. The stage was set for intensification of European conflicts. The most obvious result of the Congress and of nationalist yearnings, juxtaposed with a more structured European map, was a new and general scramble for colonies in other parts of the world: the Scramble for Africa and the Scramble for the Pacific.

---

[67] After Napoleon's defeat in 1814-1815 the Kingdom of Lombardy-Venetia was brought under Austrian's Habsburg control. A large part of Austrian wealth came from taxing the occupied Northern Italy, specifically the wealthy Venetia region. However, after the Austrian 1848-Revolutions and the First Italian War of Independence (1849) had initiated separatist movements, in the Third Italian War of Independence (1866), the Austrians lost this territory to the newly formed Kingdom of Italy.

The Gründerzeit was not long-lived, as the US *Panic of 1873–1879*, which had serious worldwide effects, also ignited the crash of the Vienna Stock Exchange: the 'Gründerkrach' or 'Founders' Crisis' on May 9th, 1873. The Gründerkrach did fit in the European wide *Long Depression* (1873-1896) that resulted from the US Banc Panic of 1873 that occurred during the aftermath of America's Civil War (1861-1865).

After France lost the Battle of Sedan (1870) and the Siege of Paris (1870-1871), the fourth quarter of the nineteenth century was characterized by optimism, relative peace at home and in Europe, new technologies, and many scientific discoveries. An atmosphere shaped by a range of political, social, and technical changes that had emerged in the preceding decades. In France it was the time of the Belle Époque (1871-1914). As the political and social turmoil more or less ebbed away over the years —except for the Dreyfuss Affair and the odd assassination of President Carnot in 1894— the celebrations related to the 1889 *Exposition Universelle* (E: World Fair) also contributed to an atmosphere of optimism.

In America, the period after the Civil War (1863-1877), called the *Reconstruction Era* (1865–1877), was a period of political turmoil. President Lincoln was assassinated in 1865, followed in 1866 by a full-scale political war between Democrats and the radical Republicans. These were the times of racial segregation between the Caucasians (ie the white people) and the Afro-Americans (ie the black people). That Civil War was followed by the *Gilded Age* [68] (ca. 1870–1900), an era of enormous growth, especially in the north and west United States, attracting millions of émigrés from Europe.[69]

Both the Belle Époque and the Gilded Age heralded the fruits of the Second Industrial Revolution. Technological innovation blossomed into society; from electric light and telephone, to electricity powering machines in the manufacturing industries. But those times of progress, optimism and even gaiety were to change in the next century. It was —after the Victorian Boom and the Gründerzeit— the Gilded Age (America), and the Belle Époque (France) in the period up to the First World War, that were showing profound characteristics of transition, transformation, and progress creating their Golden Ages.

And all that time of social change on a large scale, also technological development had taken place on a scale not seen before. The 'electric' technology had spread —even more after the inventions of the electric AC-

---

[68] Mark Twain called the late nineteenth century the 'Gilded Age', meaning that the period was golden on the surface but underneath the thin veneer was a cesspool of greed and graft, shady business practices, scandal plagued politics and overt displays of upper-class consumerism and materialism.

[69] Between 1850-1900 the North-American population rose from 26 to 82 million people.

dynamo in the 1860s and the AC-motor in the 1880s— into homes, offices and factories. The electric light brightened the home, workshops, streets, public places. Electric machines powered the factories and public transportation. Technology was the walking companion of Social Change.

## Europe Empires intertwined with the Middle East.

The geopolitical development of Europe in the nineteenth century is closely related to the developments in the Middle East at that period of time. It was the time that saw confrontations, coalitions and warfarin between the French Empire, the British Empire and the Russian Empire, all trying to expand their spheres of influence. And it was the time of the Arab Ottoman Empire that had expanded into the Holy Roman Empire under the Habsburg monarchy up to Vienna. However, from the Battle of Vienna (1683) the Ottoman Empire began to lose its economic and military dominance in Europe, and over the next hundred years, the empire began to lose key regions of territory.

> Political intrigue within the Ottoman sultanate, strengthening of European powers, economic competition because of new trade routes, and the beginning of the Industrial Revolution all destabilized the once peerless empire. By the nineteenth century, the Ottoman Empire was derisively called the 'sick man of Europe' for its dwindling territory, economic decline, and increasing dependence on the rest of Europe.

*The Decline of the Ottoman Empire (1850-1914)*

The region of the Middle East, with its cultural inheritance running up from biblical times, was in the mid-nineteenth century under the rule of the Ottoman Empire (1299-1922). By the late eighteenth the Ottoman Empire had become instable —both in military terms as well in governmental terms—and gave rise to the *Eastern Question* in which the European powers engaged in a power struggle to safeguard their military, strategic and commercial interests in the Ottoman domains.

> By the mid-nineteenth century, the Ottoman-dominated region in Europe had undergone a range of separatist movements; the Greek War of Independence (1821-1829), the *Russio-Turkish War* (1828-1829), and the *Egyptian-Ottoman War* (1831-1833). Then the European 1848 Revolutions were keeping the European powers occupied, but by the time of the *Crimean War* (1853-1856) between France as guardian of the Roman Catholics, and Russia as the protector of the Orthodox Christians, the Eastern Question revived.

By then for the Ottoman economic problems started to become dominant. Such as the one started when they needed foreign loans (ie from French and British banks) to cover their financial deficits. Deficits caused by its lavish expenditures (by the imperial court), for entering the Industrial era (building railways and telegraph lines), and modernizing the army and building up a navy. Combined with a severe drought in Anatolia in 1873 and flooding in 1874 causing famine and widespread discontent in the heart of the empire. Subsequently, the *Great Eastern Crisis* (1875-1878) saw uprisings in the Ottoman territories on the Balkan Peninsula leading up to the *Russo-Turkish War* (1877-1878). In 1881, France occupied the Ottoman Beylik of Tunisia, A year later, in 1882, the British Empire occupied the Ottoman Khedivate of Egypt (Figure 13).

On the other hand, the rising state of Germany made overtures and closely allied with the Ottoman Empire. The German government took over the re-organisation of the Ottoman military and financial system; in return, it received several commercial concessions, including permission to build the Baghdad Railway (1899),[70] which secured for them access to

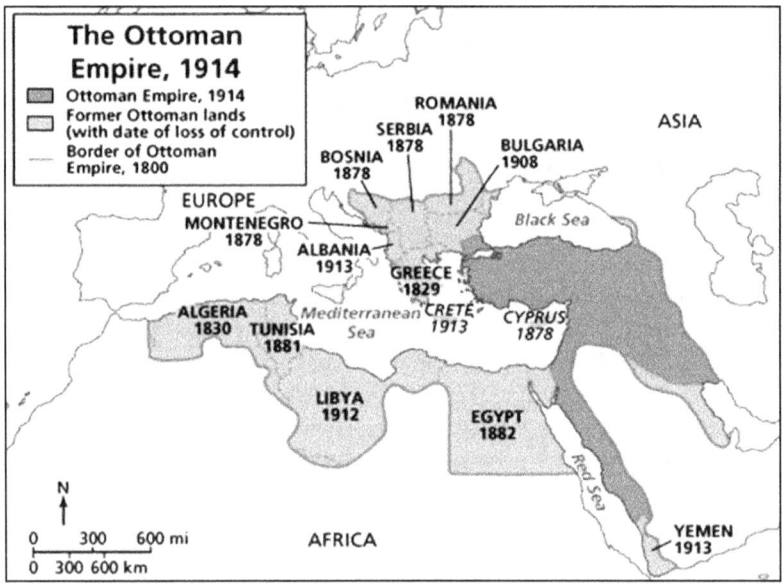

**Figure 13: The Territorial Decline of the Ottoman Empire up to 1914.**
Source: Unknown. Wikimedia Commons.

---

[70] The Berlin-Baghdad Railway was part of geopolitical views that political influence follows economic influence. For the Ottomans —ruling from Constantinople— it was about the control of the Arab Peninsula. For the Germans it was access to the Persian Gulf (where oil was suspected).

several important economic markets. It opened the potential for German entry into the Persian Gulf area and avoid the Suez Canal, by then controlled by Britain. German interest was driven not only by commercial interests, but also by a burgeoning geo-political rivalry with Britain and France. And Turkish interest was in countering Russian influence.

The need for reform within the Ottoman Empire had been obvious, and the *Tanzimat Reforms* (1839-1876)[71] re-invigorated Ottoman rule and were furthered by the Young Ottomans[72] in the late 1860s-1870s, leading to the *First Constitutional Era* (1876-1877) in the Empire that included the writing of the 1876 constitution and the establishment of the Ottoman Parliament. It was short-lived as the new sultan restored the absolute monarchy. It would take decades before the *Second Constitutional Era* (1908-1920), instigated by the Young Turk Revolution (1908), a coup d'état in the Balkan region of the Committee of Union and Progress (CUP), manned by exiled Turks.[73]

**Figure 14: Berlin-Bagdad Railway (1900-1910).**

Source: www.archaeoplan.com/Div01.htm

---

[71] The Tanzimat Reforms were a series of edicts reflecting western style modernization by guarantying life and property rights, instituting tax regulations, outlawing execution without trial, and other liberal reforms which recalled the French Declaration of the *Rights of Man and the Citizen* (1789). The Tanzimat reforms were directed at Europe to suggest that the Ottoman Empire belonged among the European nations as well as a commitment to transform the Empire based on European models.

[72] Young Ottomans, a secret society of Turkish intellectuals, sought to transform the Ottoman society by preserving the Empire and modernizing it along the European tradition of adopting a constitutional government.

[73] Young Turks were a heterodox group of secular/liberal intellectuals and revolutionaries, united by their opposition to the absolutist regime and desire to reinstate the constitution.

By 1875 the Ottomans were facing their *Great Eastern Crisis* and brought Europe on the brink of war. The regional uprising by the Serbs (in Herzegovina, 1875-1877), the April uprising of the Christian Bulgarians with the subsequent massacres (1876), the expulsion of Muslim Albanians (1877-1878), they all gave rise to a conflict of the Ottoman Empire with a coalition of the Eastern Orthodox Christians that were eager to get rid of Islam. After diplomatic manoeuvring had failed, the *Russo-Turkish War* (1877-1878) erupted. The Russian wanted to recover the territories and influence over the Black Sea they had lost after the Crimean War. And nationalist movements (led by leaders known as the 'apostles') in Romania, Serbia and Montenegro wanted their independence.

The Russian-led coalition won the war, pushing the Ottomans back all the way to the gates of Constantinople, leading to the intervention of the western European great powers in the *Constantinople Conference* (1877). In the subsequent *Treaty of San Stefano* (1878), the territories of the new nations were established. And the Ottoman were left with the Macedonian territories (Figure 15).

**Figure 15: Balkan Territories after the San Stefano Treaty (1878).**

Source: Foreign Exchanges (adapted).

After almost five centuries of Ottoman domination (1396–1878), an autonomous Bulgarian state emerged with the help and military intervention of Russia: the Ottoman vasal state *Principality of Bulgaria*, covering the land between the Danube River and the Balkan Mountains. By 1908 it became independent as the constitutional monarchy of the *Kingdom of Bulgaria*[74]

---

[74] The new kingdom was almost constantly at war throughout its existence, lending to its nickname as 'the Balkan Prussia'. For several years Bulgaria mobilized an army of more than

(Figure 13). In the meantime, the French had joined and occupied the Ottoman Beylik of Tunisia (1881), and the British had occupied the Ottoman Khedivate of Egypt (1882), gaining them access to the Middle East.

## The Far Eastern Empires

The Industrial Revolution had resulted in the modern industrialized societies in Western Europe and America, awakening geopolitical and social forces on a large scale. But they had also an effect on the other side of the world in the Pacific Region of the Far East. A region where the large European Empires had focussed their economics interests on in earlier times. Starting with the Portuguese traders exploring the new waters by the fifteenth and sixteenth century, followed by the British and Dutch 'East Indian Companies', and later by the French in Indochina and the Americans crossing the Pacific Ocean, the Pacific region had become a region of interest to western maritime expansion fuelled by western trade.[75] And with that came geopolitical events that shook the Affairs of (Asian) Men.

*The Fall of the Chinese Empire*

The Eastern part of the Asian continent, that became known as China (Figure 16) —the sino-centric world ruled by the Qing Dynasty (1644-1912) at the centre of interstate relations—, was the source of some of the world's most sought-after commodities; tea, porcelain, and silk. Western merchants, starting with the Portuguese, followed by the Dutch and Brits, had sought access to this highly lucrative trade since at least the seventeenth century.

By the nineteenth century, the activities of the merchants of the British East India Company, jumping in on the opium trade,[76] caused economic and social frictions. As the opium-fuelled China Trade increased in scope and value, the foreign presence in Canton and Macau grew in size and influence. The increasing addiction among Chinese made the Qing-officials to ban the opium trade in 1796. Next, the opium trade became a smuggling trade originating from Canton when private entrepreneurs joined the highly profitable opium business. By 1838, the British were selling roughly 1,400 tons of opium per year to China. Legalization of the opium trade was the

---

1 million people from its population of about 5 million, and in the 1910s, it engaged in three wars; the First and Second Balkan Wars, and the First World War.
[75] The Russian Empire focussed its imperialistic expansion in the 1850-1900 period on Siberia up to the Pacific Ocean, Central Asia (ie the Turkestan region), and South Asia (ie the Great Game including Afghanistan).
[76] To ease the trade deficit, opium —produced in the Indian colonies— was used as a barter payment replacing gold and silver.

subject of ongoing debate within the Chinese administration, but a proposal to legalise the narcotic was repeatedly rejected, and in 1838 the government began to actively sentence Chinese drug traffickers to death. After a range of conflicts, the escalation of tensions resulted in the seizure of British stock in Canton, the heart of the trade and the only Chinese city open to foreigners.

Following the Chinese crackdown on the opium trade, discussion arose as to how Britain would respond to the 'Chinese Question.' As advocates of free trade, opposing the Chinese restrictions of the Canton System,[77] they decided to apply gunboat diplomacy and send an expedition to China. Subsequently, British expeditionary forces ravaged the Chinese coast during the *First Opium War* (1839-1842). The subsequent 1842 Anglo-Chinese *Treaty of Nanjing* not only opened the way for further opium trade, but ceded the territory of Hong Kong, unilaterally fixed Chinese tariffs at a low rate, gave Britain the most-favoured nation status and permitted them diplomatic representation. The Treaty allowed British merchants, and eventually all foreign merchants, to deal with whomever they pleased in the newly-opened ports. The Americans joined in and negotiated in the Treaty of Wangxia (1844) similar conditions. The treaties —the first of the so-called

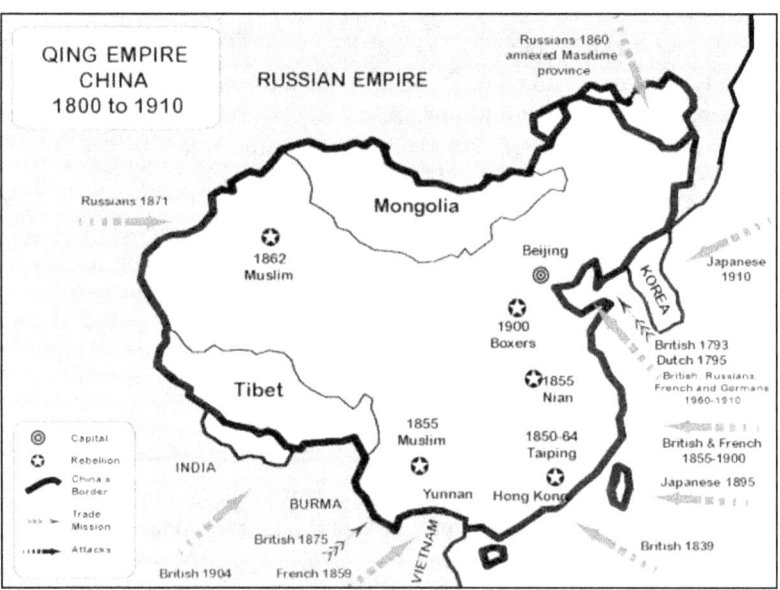

**Figure 16: The Chinese Empire under the Qing Dynasty (1800-1910).**
Source: https://www.china-mike.com/

---

[77] The Qing government had devised the Canton System in 1760 to control trade with the West, focussing all trade on the southern port of Canton.

'Unequal Treaties' between China and foreign imperialist powers— created five treaty ports open for Chinese-Western trade (Guangzhou, Xiamen, Fuzhou, Ningbo, and Shanghai) (Figure 17).

By 1850 the region known as Imperial China had become the 'sick man of Asia'. The rulers of the Qing Dynasty (1636-1912) ruled a multi-ethic collection of people, among which the minority of Manchu-Chinese and the majority of Han-Chinese. And that gave rise to rebellions.

> The *Taiping Rebellion* (1850-1864) —one of the bloodiest civil wars between ethnic groups, with 600 cities ruined and 20-30 million deaths— had started a process of massive social change as also other ethnic groups revolted; the Miao Rebellion (1854–1873) in Guizhou, the Panthay Rebellion (1856–1873) in Yunnan and the Dungan Revolt (1862–1877) in the northwest (Figure 17).

The rebellions brought western involvement as British and French troops with their modern weapons had come to the assistance of the outdated Qing Imperial Army. In addition, western Imperialism brought on the *Second Opium War* (1856-1860) —again about the economic interests of the opium trade— initiated by the Britain Empire and France Empire. The British and French forces attacked in range of battles Guangzou (1856) and Beijing (1860). Next, the French took further control over Indochina in the south (1844-1885), Russian Imperialism took chunks of the norther

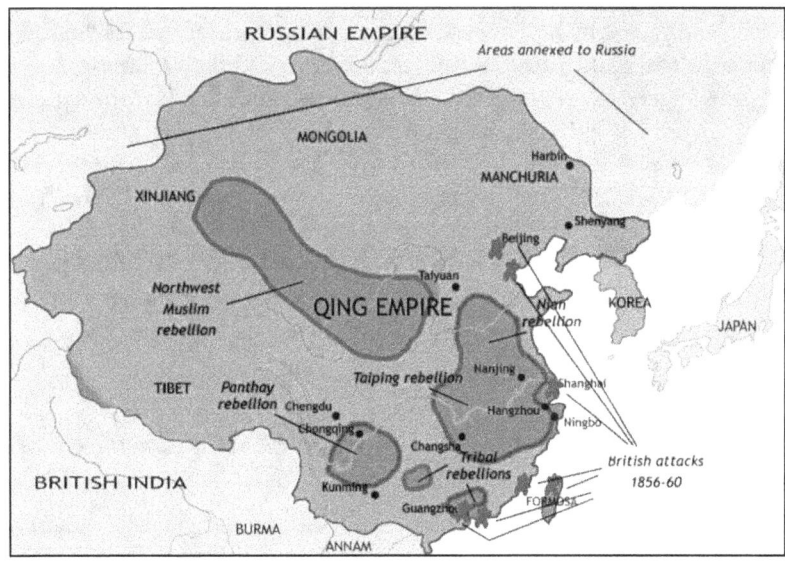

Figure 17: Taiping Rebellion and related Rebellions.
Source: Timemaps.com

territory (1860s). But that was not all as some decades later, Japanese Imperialism brought the *First Sino-Japanese War* (1894-1895), fought by the Beiyang Army. That war ultimo created Korean independence and saw Taiwan ceded to Japan.[78]

> One of the consequences of these wars —next to the loss of territory—, was that China lost effective control of her lucrative sea ports. Fifty of China's most prosperous ports were deemed 'treaty ports' which meant that they were open to foreign trade and residence. The western settlements in the treaty cities —under western governance (ie western law, policing)— became the base for diplomatic, mercantile and missionary activity. European nations divided China up into different spheres of influence.

> Another consequence was military reform. From the classic Chinese military emerged the Qing dynasty's Beiyang Army, a militia based on personal, rather than institutional, loyalties. They were the best military forces that the Qing dynasty could field at that time. After the defeat in the First-Sino-Japanese War it formed the base of the New Army, that became the dominant military force in China.

By 1900, China was a nation dominated by European nations, led by a conservative court resisting reforms. Natural disasters as the Yellow River Flooding and droughts—resulting in the Northern Chinese Famine of 1901—, fuelled the *Boxer Rebellion* (1899-1901) opposing foreign (ie Christian) influence in support of the Qing-rule. During the rebellion, Orthodox, Protestant, and Catholic missionaries and their Chinese parishioners were massacred throughout northern China. Europe's great powers reacted and quelled the revolt with the military force and brutality of the *Eight-Nation Alliance* during the *Siege of the International Legations* (1900). On August 14th, 1900, after fighting its way through northern China, an international force of approximately 20,000 troops from eight nations (Austria-Hungary, France, Germany, Italy, Japan, Russia, the United Kingdom and the United States) arrived to take Beijing and rescue the foreigners and Chinese Christians.

> Britain provided 10,000 troops from India and Australia, the Russian some 12,400 troops and 10 warships. Japan had —in close cooperation with the British— provided the largest contingent of

---

[78] British and French troops invading Beijing, ransacked the Old Summer Palace, and looted its treasures. After the looting, they set fire to the entire palace. The palace was so large – covering more than 3.5 square kilometres (860 acres) – that it took 4,000 men 3 days of burning to destroy it.

troops (20,300) and warships (18). Austria-Hungary, Germany, France, Italy and the US added marginally.

The subsequent *Boxer Protocol* of 1901 (ie the peace agreement), and the European 'scramble for concessions'[79] increasingly weakened the Qing Dynasty's control over China. The Western Europeans had learned a lesson, though, and gave up their aspirations to conquer China by military force. In the north the Russian troops, however, occupied Manchuria and aimed their attention at Korea that brought them in conflict with Japan; the result was the *Russo-Japanese War* (1904-1905).

And then in 1908 the deaths of Emperor Guangxu and his concubine Cixi —the real power behind the throne— gave the coup de grace to the Qing Dynasty. His successor, the two-year old nephew Puyi, became the Xuantong Emperor until the *Xinhai Revolution* (1911-1912) forced him to abdicate. That revolution culminated in a decade of agitation, revolts, and uprisings. Its success marked the collapse of the Chinese monarchy, and the beginning of China's early Republican Era.

*From Shogunate to Japanese Empire*

Not far from all those developments on the mainland of China, another development took place on the Japanese Islands. After centuries of feudal rule[80] under the shogunates[81] ruling from their castles (Figure 18) their isolation policies,[82] also in the 1850s, on the Japanese island Honshu the *Tokugawa Shogunate* (1603-1868) —the actual rulers of the Japanese Archipelago— had come under strain due to rising Western influence.

An influence that had already started in 1543 when the Portuguese explorers discovered the Japan Archipelago. A century later, after seizing the trade from the Portuguese, the 'eastindiamen' of the Dutch East India Company had acquired on the Island of Decima —an artificial island (120m

---

[79] Concessions in China were enclaves in key cities that became treaty ports. In these concessions, the citizens of each foreign power were given the right to freely inhabit, trade, do missionary reductions, and travel.

[80] Comparable to the European Medieval feudal clan-based society with the landowning knights serving his lords and monarch, in Japan the samurai warrior was the military backbone of the landowning daimyo serving the shogun.

[81] A Shogun was the military head over land-owing vassal lords (ie daimyo) controlling the warriors (ie samurai) of a clan. A Shogun of the Clan in power, a military and civil regent, served as the actual national leader, while the emperor— the Divine Son of Heaven who had descendent from the gods—was the symbolic head of the state and religion. A Shogunate was the government, office, or rule of a Shogun.

[82] Known as the Sakoku (lit. 'closed country'), the self-imposed isolationist policy of Japan under which no foreigner could enter nor could any Japanese leave the country on penalty of death.

x75m) in the Bay of Nagasaki—trading rights in 1634 (Figure 19).

*The Tokugawa shogunate had managed to attain great stature. It had drawn from the Emperor virtually all power, had moved the government from Kyoto to Edo, and had closely bounded each of the provincial rulers of the feudal system to the shogun. One of the primary reasons that this system, which had gone out of fashion in Europe almost a hundred years earlier, could remain working for such an extended period was a careful guarding by the shogunate of the principal values of the Japanese society. It allowed virtually no commercial contact with foreign countries (beyond the Netherlands and China, which were both allowed remote trading posts), and no contact whatsoever between the population and the foreigners, and had banned all voyages abroad.[83]*

The Dutch monopoly lasted till in 1854, when US Navy commodore Perry with a fleet of eight steam-powered warships executed gunboat diplomacy, forcing in the *Convention of Kanagawa* the opening of the ports to American vessels.

**Figure 18: Osaka Castle and Dejima.**

The stronghold of Osaka Castle (top), and the Nagasaki Island of Dejima (bottom).

Source: https://nagasakidejima.jp/, RIBApix.

Under dramatic climatic events[84] the Kanagawa Treaty was followed by similar agreements with the United Kingdom (Anglo-Japanese Friendship Treaty, October 1854), Russia (Treaty of

---

[83] Source: A Short History of Japan: from 1850 to the beginning of World War II. http://www.microworks.net/pacific/road_to_war/japan_1853-1941.htm (Accessed December 2021)

[84] The years 1854–1855 saw a dramatic series of earthquakes, known as the Ansei great earthquakes, with 120 major and minor temblors recorded over a less than two-year period including the 8.4 magnitude 1854 Tōkai earthquake on December 23th 1854, the 8.4 magnitude 1854 Nankai earthquake occurring the following day, and the 6.9 magnitude 1855 Edo earthquake, which struck what is today modern Tokyo, on November 11th 1855.

Shimoda, February 7th, 1855), the US (Treaty of Amity and Commerce in 1856), and France (Treaty of Amity and Commerce between France and Japan, October 9th, 1858). They became known as the Unfair Treaties. From 1859, the ports of Nagasaki, Hakodate and Yokohama. became open to foreign traders.

So, between 1853 and 1867, Japan ended its isolationist foreign policy and changed from a feudal Tokugawa shogunate[85] to the modern empire of the Meiji government.[86]

Swiftly, opposition against the foreigners erupted. Violence, hatred, open hostilities against the outsiders, led by young samurai who had the support of the powerless Emperor in Kyoto, flared up in the early

**Figure 19: Tokugawa Shogunate (1603-1868).**
Source: Digital Collection, BYU Library.

---

[85] A Shogunate (aka Bakufu, tent government), is a heredity military dictator. The Edo shogunate (aka the Tokugawa shogunate) came to power in 1603 and to its official end on November 9th, 1867.
[86] The Revolution, later called Restoration, led to enormous changes in Japan's political and social structure and spanned both the late Edo period (often called the Bakumatsu) and the beginning of the Meiji era.

1860s. Not surprisingly, this opening up of foreign trade also affected the social and economic stability. While some —those in power—prospered, many others went bankrupt. Unemployment rose, as well as inflation. Coincidentally, major famines also increased the price of food drastically. And, the foreigners brought cholera to Japan, leading to hundreds of thousands of deaths. All the elements of social disruption came together and exposed the ideological-political divide: the pro-imperial nationalists versus the shogunate-proponents.

The controversy between pro- and contra-factions of the 'sakoku' increased, after the initial diplomatic efforts by Japanese ambassadors to America (1860) and Europe (1863) to revise the Unfair Treaties had failed. Belligerent opposition to Western influence further erupted into open conflict when the Emperor Kōmei, breaking with centuries of imperial tradition, began to take an active role in matters of state and issued under the pressure of his court, on March 11th and April 11th, 1863, his "Order to Expel Barbarians." Under pressure from the Emperor, the Shogun was also forced to issue a declaration promulgating the end of foreign relations.

Military confrontations with the western powers erupted; such as the British bombardment of Kagoshima and the multi-national Shimonoseki campaign. The internal struggle for power initiated the civil *Boshin War* (January 1868-June 1869) between forces of the ruling Tokugawa shogunate and those seeking to return political power to the Imperial Court.

> Taking advantage of the disruption caused by these internal and external crises, in 1867 several powerful daimyo (ie regional warlords) banded together to overthrow Shōgun Yoshinobu, forcing him to resign authority. Marching into the imperial capital Kyoto, they 'restored' Prince Mutsuhito[87] —the second son of Emperor Kōmei born in 1852 —to power as Emperor Meiji and established the Meiji ('enlightened rule') Restauration (1868-1912).

Between May and July 1868, the capital Edo saw the fall of the Tokugawa Shogunate as in the end, the victorious imperial faction abandoned its objective of expelling foreigners from Japan and instead adopted a policy of continued modernization with an eye to eventual renegotiation of the Unfair Treaties with the Western powers. On October 26th, the Meiji era (1868-1912) started with the fifteen-year-old Emperor Meiji moving from Kyoto to Edo —renaming it Tokyo—and becoming the

---

[87] Emperor Kōmei had died suddenly on January 30th, 1867. Although rumours found his death quite convenient and suggested assassination, he could have the victim of a worldwide cholera pandemic. His second son Mutsuhito, though, became a clear victim of inbreeding due to the tradition of Consanguineous marriages. The hereditary diseases resulting from genetic defects were complemented with beriberi caused by malnutrition.

first monarch of the Empire of Japan. By the time of his death in 1912, Japan had undergone an extensive political, economic and social revolution, and emerged as one of the great geo-political powers on the world stage.

> *In the name of Emperor Meiji, numerous striking and far-reaching social, political, and economic changes are legislated through a series of edicts. Japan also opens its borders, sending several high-ranking expeditions abroad and inviting foreign advisors—including educators, engineers, architects, painters, and scientists—to assist the Japanese in rapidly absorbing modern technology and Western knowledge.* [88]

The first action, taken in 1868 while the country was still unsettled, was to relocate the imperial capital from Kyōto to the shogunal capital of Edo, which was renamed Tokyo ('Eastern Capital'). Dismantling the old regimes was not without internal resistance. By the 1870s disgruntled samurai participated in several rebellions against the government; such as the Satsuma Rebellion (1877) of unemployed samurai after military reforms.

During *Meiji Restauration Era* (1867-1912), the Frist Japanese Empire had undergone massive political, social, economic and technological change as a result of opening up to the Western influences. With the Industrial Revolution coming to Japan, it had seen the Freedom and People's Right movement during the 1880s introduce democratic institutions (ie the bicameral Imperial Parliament named Diet in 1890, a Prime Minister with a Cabinet), and the Emperor granted supreme control of the Army and Navy. The former independent feudal Daimyo domains were replaced with prefectures subordinate to the centralized power.

> This Meiji Constitution provided for a form of mixed constitutional and absolute monarchy. In theory, the Emperor of Japan was the supreme leader, and the Cabinet, whose Prime Minister would be elected by a Privy Council, were his followers; in practice, the emperor was head of state but the Prime Minister was the actual head of government.

## Japan's Industrial Revolution

After the American fleet under commander Perry had started his gunboat diplomacy in 1853, the Japanese —in order to avoid China's fate being colonised by Western powers— decided after a short civil revolt to apply 'defensive modernization.' However, the island country lacked many raw materials, including coal and iron.

---

[88] Source: Japan, 1800–1900 A.D. In: Heilbrunn Timeline of Art History. New York: The Metropolitan Museum of Art, 2000. https://www.metmuseum.org/toah/ht/10/eaj.html. (Accessed December 2021)

During the Meiji Restauration, Japan underwent an Industrial Revolution of its own. One early stage was to sweep away the feudal system of internal checkpoints, post stations, and merchant guilds as barriers to industrial development. In addition, the government did built railroads (eg Tokyo-Yokohama, 1869), improved roads, and inaugurated a land reform program to prepare the country for further development. It adapted a new Western-based education system for all young people, sent thousands of students to the United States and Europe, and hired more than 3,000 Westerners to teach modern science, mathematics, technology, and foreign languages in Japan.

> *The Meiji government concentrated its efforts on promoting industry and introducing modern forms of enterprise with the aim of fostering capitalism in Japan. [...] New infrastructure included the first telegraph line between Tokyo and Yokohama in 1869. Five years later, the telegraph network stretched from Nagasaki to Hokkaidō, while an undersea line further connected Nagasaki to Shanghai. In 1871, a modern postal service replaced the former courier system, and post offices were established around the country, selling stamps and postcards at set prices. In 1877, Japan joined the Universal Postal Union, linking its postal service to the world. It imported its first telephones the same year.*
>
> *A rail service started between Tokyo and Yokohama in 1872. This initial route relied greatly on British assistance, as the European power supplied financing, train cars, and even the chief civil engineer Edmund Morel. In 1874, a new line linked Kobe to Osaka, which was connected in turn to Kyoto in 1877. By the turn of the century, the network had spread across the whole of Japan. The government also invested in upgrading the country's major roads, enabling smoother transportation of goods by carts and other vehicles.*[89]

In addition, the state-led industrial development was stimulated by according special privileges to specific organizations and subsidizing the Zaibatsu.[90] These vertically integrated business conglomerates aided Japanese militarism and benefited from the conquest of East Asia by receiving lucrative contracts. This was one way the Meiji leaders aimed to

---

[89] Kawai Atsushi: *Japan's Industrial Revolution*. Source: https://www.nippon.com/en/japan-topics/b06904/japan%E2%80%99s-industrial-revolution.html (Accessed December 2021)
[90] The family-controlled financial cliques called Zaibatsu, such as the Mitsui Group and the Mitsubishi Group, were linked with either the Imperial Army of Navy. Financed by a bank, they covered specific sectors of the market through a changed of companies. The monopolistic business practices by the zaibatsu resulted in a closed circle of companies. In a way they were copies of the large western monopolistic companies like Standard Oil, Carnegie Steel Company, AT&T, General Electric, Western Union, Friedrich Krupp AG, Thyssen AG, Robert Bosch GmbH, Lloyd's of London, Reckitt and Sons, East India Company, and British Petroleum.

foster modern industry. But there was more.

> *The government also set up and operated many factories and establishments in fields like light industry and agriculture to boost the development of private industry. […] The Home Ministry organized its first Domestic Industrial Exposition in 1877 at Ueno Park, going on to hold five in total, with the last in 1903. These took inspiration from world's fair events in other countries and helped to promote modern industry and trade. The first displayed 84,000 products in six categories, including agriculture, horticulture, and machinery. It was highly successful, attracting 450,000 visitors over 102 days.*(ibidem)

## Japan-China Confrontation

Copying the Western industrial policies, one of those changes had been the expansion of Japan's sphere of influence outside the mainland (Figure 20). In 1878, it occupied the Ryukyu chain located at its south, with Okinawa as the new prefecture capital. Earlier, in 1875, the Kuriles located to the north had been occupied to provide a buffer against Russian aggression.

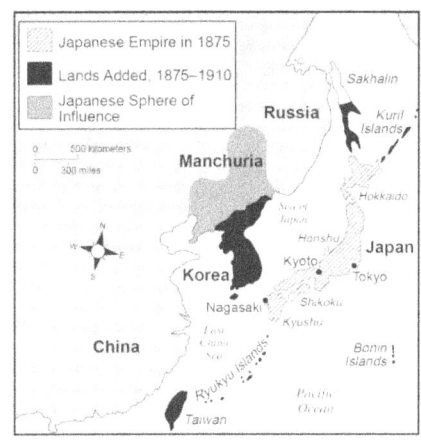

**Figure 20: Japanese Imperialism.**

Annexations (top, 1875-1910) and First Sino-Japanese War (bottom, 1894-1895)

Source: Henry Brun et al., Reviewing Global, /History and Geography, AMSCO, https://www.facinghistory.org (adapted).

Next it confronted the Korean Peninsula with gunboat diplomacy copied from the Western powers, and opened Korea to their trading and political influence after the Japan-Korea Treaty of 1876. It gave extraterritorial rights to Japanese citizens in Korea, and forced the Korean government to open 3 ports to Japan, specifically Busan, Incheon and Wonsan. With the signing of its first unequal treaty, Korea became vulnerable to the influence of imperialistic powers; and later the treaty led Korea to be annexed by Japan.

Then the Japanese Empire came into conflict with the Chinese Empire of the Qing Dynasty. Both wanted Korea under their own sphere of influence. The result was the Japanese victory in the *First Sino-Japanese War* (1894-1895). The Treaty of Shimonoseki (April 1895) ended the war by the Chinese recognizing independence of Korea and the transfer of Taiwan to Japan.[91] Just days after the signing of the treaty, three of the most active Western powers in China (Russia, France, and Germany) moved in quickly to establish their authority over certain ports. For the first time, regional dominance in East Asia shifted from China to Japan.

> Within China, the defeat was a catalyst for a series of political upheavals led by Sun Yat-sen[92] and Kang Youwei,[93] culminating in the 1911 *Xinhai Revolution* that ended China's last imperial dynasty and led to the establishment of the Republic of China (ROC). His political organization, the Revive China Society (Xingzhonghui) would later become the Revolutionary Alliance and would lay the foundation for China's Nationalist Party.

Within Japan —with copying western geo-political tactics had proven successful— the expansionist movements turned their eye to Taiwan. In a five-month invasion they controlled the Island of Taiwan and inaugurated five decades of Japanese rule in Taiwan (Figure 20).

---

[91] The Qing government also signed a commercial treaty permitting Japanese ships to operate on the Yangtze River, to operate manufacturing factories in treaty ports and to open four more ports to foreign trade. The Qing government also created the short-lived Republic of Formosa on the Island of Taiwan

[92] Sun Yat-sen was educated on the Kingdom of Hawaii, where he learned about the English and American ideals of a constitutional government, and the history of political struggles elsewhere. He had a close friend, the business man Charlie Jones Soong (1861-1918) who became part of Shanghai's anti-Qing resistance movement funding Yat-sen's travels. His descendants became —western-educated— some of the most prominent figures in the Republican China.

[93] Kang Youwei, political thinker, influenced many Chinese revolutionaries; among them Mao Zedong, a founding father of the People's Republic of China.

The Japanese Empire came also in conflict with Russia resulting in the *Russo-Japanese War* (1904-1905). The conflict was based on rival imperial ambitions in Manchuria and Korea. Seeing Russia as a rival, Japan offered to recognize Russian dominance in Manchuria in exchange for recognition of Korea being within the Japanese sphere of influence. Russia refused and demanded the establishment of a neutral buffer zone between Russia and Japan. Subsequent maritime confrontation led to a range of battles —eg the Battle of Port Arthur, the Battle of the Yellow Sea—, and the Japanese annihilated the Russian fleet in the *Battle of the Tsushima Straits*[94] on May 27th–28th, 1905. Imperial Russia's prestige was badly damaged and the defeat was a blow to the Romanov dynasty.

> The war had cost many casualties; the Japanese were estimated to have lost around 59,000 soldiers with around 27,000 additional casualties from disease. The Russians around 34,000 to around 53,000 men with a further 9,000–19,000 dying of disease and around 75,000 captured. China suffered 20,000 civilian deaths.

In the *Treaty of Portsmouth* (September 1905) — with US president Theodore Roosevelt being instrumental in the negotiations[95]—, Russia recognized Korea as part of the Japanese sphere of influence and agreed to evacuate Manchuria. Japan would annex Korea in the *Japan–Korea Treaty of 1910*, with scant protest from other powers. Russia also ceded the southern half of the Sakhalin Island to Japan.

> The Russo-Japanese War also had a profound cultural and political impact in the world. This was the first major military victory in the modern era of an Asian power over a European nation. The complete victory of the Japanese military surprised international observers and transformed the balance of power in both East Asia and Eastern Europe, resulting in Japan's emergence as a great power and Russia's decline in prestige and influence in Eastern Europe. It would also have an internal effect in Russia, as popular discontent in Russia after the war added more fuel to the already simmering Russian Revolution of 1905. Its events foreshadowed the 1917 subsequent Russian Revolution just twelve years later.

---

[94] This was naval history's first, and last, decisive sea battle fought by modern steel battleship (aka dreadnought) fleets, and the first naval battle in which wireless telegraphy (radio) played a critically important role. In October 1905 the British started the construction of HMS Dreadnought, which marked the beginning of a naval arms race between Britain and Germany in the years before 1914.

[95] In Japan, that considered both the Kuril Islands and the Sakhalin Island as an extension of their Hokkaido Island (Figure 20), public opinion was shocked by the very restrained peace terms which were negotiated at the war's end, and put a strain on the US-Japanese relations.

## Overview Europe, Middle East and Far East

From the preceding exploration emerges picture of the major geo-political shifts during the second half of the nineteenth century (Figure 21). A shift in the balance of powers between and within empires and nations. During a period in time when the technical progress of the second Industrial Revolution brought power to the people; from steam power to electric power to combustion power. Powering the worldwide maritime expansion, as well as enhancing the military's firepower, when imperialism scrambled for new colonies in Africa and the Far East. And powering the industrialization where technological innovation became a driving force of novelty. Creating new products spreading the artificial light (eg the incandescent lamp), and the spoken word (eg the telephone and the wireless) that would shape the private and business Affairs of Man.

A development in societies that originated from Britain as the *Craddle of Industrialization* in Western Europe with the tinkerers and thinkers discovering the Powers of Nature (eg the Power of Heat and the Power of Explosion), and inventing their engines (eg the steam engine and the internal combustion engine). A development picked up by the American entrepreneurial spirits developing technologies and applying those engines into remarkable innovations: such as the mobility machines from the motor

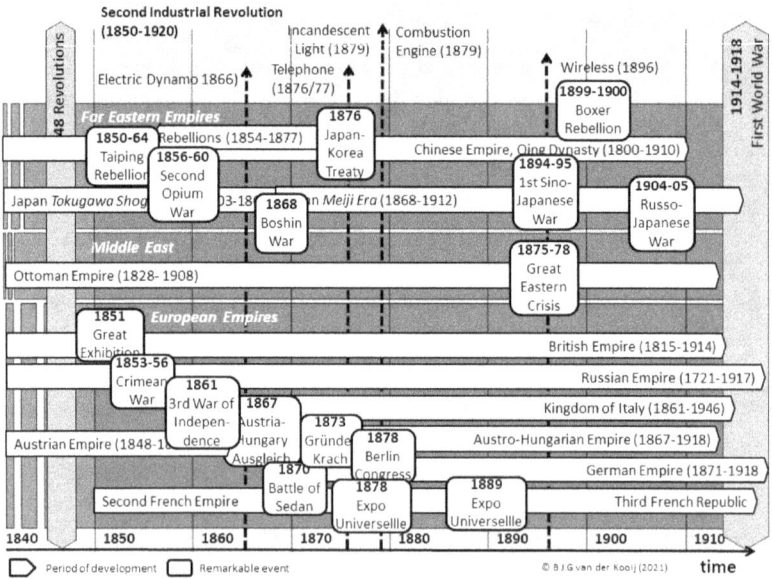

Figure 21: Overview Geo-political development of European and Middle Eastern and Far Eastern Empires (1850-1910).

bike and automobile to the airplanes concurring the skies. Next it was the Eastern industrious people who copied that industrialization within their own cultural setting.

## *The Context of the Early Twentieth Century (1900-1920)*

After the second half of the nineteenth century had seen the Second Industrial Revolution, technology continued to play an increasing dominant role in society. But, similar to the Scientific Revolution where the Natural Philosophers had experimented with the Nature of Matter (Figure 8) leading up to Newtonian Mechanics, now scholarly attention was drawn into the more fundamental questions of the properties of matter. With the discovery of electricity culminating in Volta's battery, they wondered what electricity was. They noticed the results (eg electro-magnetism), applied them in artifacts (ie the electromotor, the electromagnet), but they wanted to know the essentials of electricity.

### **The Rise of the Sciences**

After the technological-induced novelty of the second half of the nineteenth century (Figure 22), the beginning of the twentieth century was a period of great advances in scientific knowledge. [96] In addition to medicine and surgery, the physical sciences, chemistry, and mathematics were revolutionized. The new/modern physics[97] —relativity, quantum mechanics, particle physics— enabled physicists to probe to the very limits of physical reality. The revolution in physics spilled over into chemistry and biology. And the First World War played an important role.

> When the First World War broke out, disrupting Atlantic trade patterns, many countries were cut off from German products and know-how in the domains of optics, pharmaceuticals and chemical processes. The immediate national responses were to initiate their own research programs. In Britain the learned and professional societies —the Royal Society, the Chemical Society, the Institution of Electrical Engineers, etc. — formed comities to coordinate war work. As a result, in 1915 the Board of Inventions and Research

---

[96] Science is the pursuit and application of knowledge and understanding of the natural and social world following a systematic methodology based on evidence. By definition it covers a broad range of activities, uses the scientific method (developed by Newton), and has three major branches; natural sciences (eg biology, chemistry), social sciences(eg sociology), and formal sciences (eg mathematics).
[97] Physics can be periodized in Classic Physics (pre-1900) and Modern Physics (post-1900). Classic physics includes the study of mechanics, gravitation, heat, sound, light, electricity and magnetism. Modern physics includes the study of quantum mechanics, relativity, atoms, molecules, nuclei, elementary particles and condensed matter.

(BIR) was established for this purpose. In Canada, an 'Order in Council' in June 1916 created the Honorary Advisory Council for Scientific and Industrial Research, the basis for the later National Research Council (NRC) created in 1925. A similar development took place in the US.

The war affected scientific institutions and especially scientific education in all of the contending nations. Many scientists and engineers patriotically volunteered their professional services to their respective military authorities, some civilian scientific institutions took up military tasks, and for some purposes new military institutions or organizations were formed or old ones greatly expanded to make use of the professionals' services.

> The Germans created the first such organization in August 1914 with the *Kriegsrohstoffabteilung* (KRA), or War Raw Materials Office, located within the Prussian War Ministry. For the Germans, the British blockade made the effective development of substitute materials and alternative resources a critical priority, and it became a fruitful field of military-industrial cooperation.

*International Research Council:* As German defeat loomed in 1918, Allied scientists agreed to recreate the international system of science in a form that would carry on a cultural "war after the war." The result was the

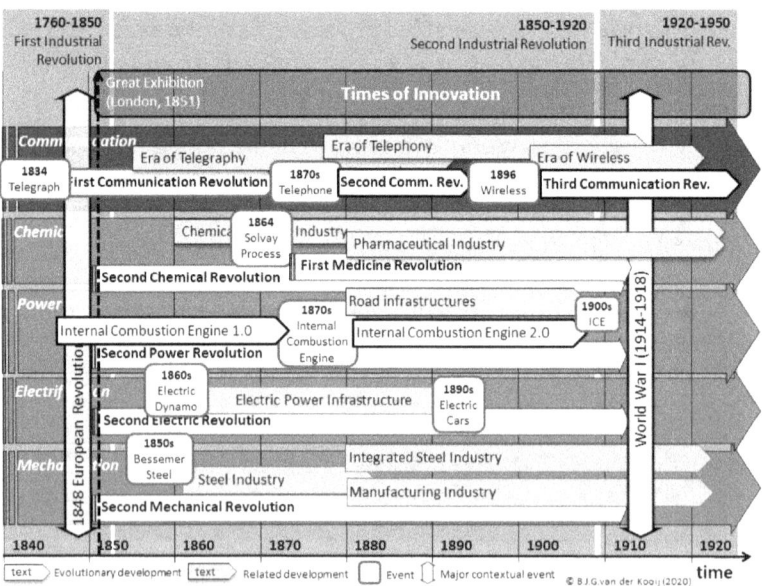

**Figure 22: Times of Innovation with its specific domains of Technology Induced Novelty.**

*International Research Council* (1919), to which the National Research Councils of the various Allied and neutral nations were admitted (but initially not the Germans, Austrians, or Russians). The resulting scientific associations —aka Unions— covered a broad field of sciences; from the Astronomical Union to the Mathematical Union, from the Chemical Union to the Union of Radio-Telegraphy, and from the Geographical Union to the Biological Union.

But there was more than these institutional developments around 'science.' The German theoretical physicist Albert Einstein revolutionized the way we looked at the Macro-Cosmos (aka the Universe) with his general theory of relativity. And the German physicist Max Planck the way we looked at the Micro Cosmos of the quantum mechanics of electrons, protons, neutrons, photons and positrons. The foundations of future technological developments were laid in this period of time.

## *Creating Novelty in an Organized Way*

From the days of the 'heroes of invention' —all those individual thinkers and tinkerers that had brought novelty—, the Second Industrial Revolution saw a massive creating of novelty. Under different names, from the Invention Factory (ie Edison's facilities in in Menlo Park where the incandescent lamp was developed) to the Model Room (Ford's secret facilities in Detroit where the Model T was designed), different disciplines of knowledge were brought together in the research facilities. The concept of the Research & Development (R&D) department was born.

The result was the systematic integration of scientific research into the process of industrial innovation in a host of areas, related to the *physical sciences* (chemicals, electric power and communications, metallurgy, and mechanical engineering including internal combustion engines for transportation).[98] With the war looming at the horizon, countries mobilized groups of scientifically trained professionals for the purpose of executing systematic research (ie applied research) for military purposes. And all that new *knowledge* acquired by the scientist was transferred to technological *knowhow* by engineers working in the R&D departments of big companies. The result was a tsunami of novelty in the areas of artillery and explosives, chemical warfare, tanks, communications, aviation, and submarines.

With the rise of organised collective R&D, scholarly interest into the phenomena of basic and applied research[99] started the exploration into the process of novelty creation. And in the field of economics the scholars of

---

[98] For a detailed exploration of these areas, see other case studies in the Invention Series.
[99] Basically, Basic Research is the search for knowledge and Applied Research is the search for solutions to practical problems using this knowledge.

the Austrian School of Economics contributed to economic thought on novelty that saw social phenomena result exclusively from the motivations and actions of individuals.[100]

## The Rise of the European Empires

The geo-political scene in Europe in the early 1900s was completely different from the Europe in Napoleonic times (1800-1815). Then it was the dawn of the rising evolutionary spirits, forcefully opposed by the establishment of that time (eg the European Ancient Regimes of the privileged class; Royalty with their 'divine rights', and the landowning, feudal Aristocracy). A century later, by 1900, it was the aftermath of the 'Springtime of Peoples,' that had started after the 1848-Revolutions. Now it were the new European Empires that were going to clash.

Similar to the early Portuguese, Spanish, Dutch and (first) British Empire erupting after the Age of Discovery (1400s-1520s) (Figure 8), these new empires were driven forth by a continuous population growth.[101] Next, the European continent had seen the rise of the empires with their colonial expansions; such as the (second) British Empire and the French Empire. All empires that were bringing colonial wealth back to their own societies. Wealth that largely financed the geo-political aspirations.

By the end of the nineteenth century, a new wave of Imperialism peaked during the *Scramble for Africa*. Sub-Saharan Africa, one of the last regions of the world largely untouched by 'informal' imperialism, was discovered by Europe's ruling elites for economic, political and social reasons. So, nearly all the great powers of the west, however with some lagging, turned their attention to

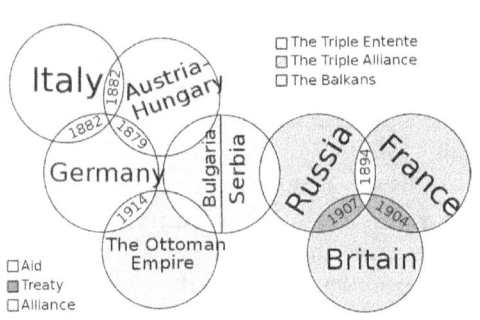

**Figure 23: European political Alliances before the First World War.**

Source: Wikimedia Commons.

---

[100] Such as the later contributions of Alois Schumpeter with his Theory of Economic Development (1934) and Abott Usher with his book 'A History of Mechanical Inventions.' (1929). For details: see 'The Intellectual History of Innovation.'
[101] By 1820, Western/Eastern Europe had some 133/91 million inhabitants, by 1870 this was grown to some 187/140 million, and by 1913 there were some 261/236 million.

the African Continent creating the *New Imperialism* (1881-1914). Next to valuable resources available throughout the continent, other factors like the quest for national prestige, tensions between pairs of European powers, and religious missionary zeal, played a role.

The efforts of unifications in Central Europe that had created the German Empire (1871) and the Kingdom of Italy (1870), and their subsequent expansions were now becoming intertwined with the interests of the older colonial empires. The result was —with Central-Europe as a boiling pot of nationalistic aspirations, and interstate tensions arising— the need to create alliances between states (Figure 24).

> France started diplomatic overtures with its former enemy Russia to win over the Russians to an alliance which could be used to further French purposes, defensive and possibly offensive as well. It resulted in a *Franco-Russian Alliance* (1894), a military treaty which remained in effect through the onset of the First World War (Figure 23). In Central-Europe, Germany on the other hand, also sought to create alliances and created the *Triple Alliance* (1882); an agreement between Germany, Austria-Hungary, and Italy. Facing that development of the German Empire, the former enemies France and Great Britain improved their relations and created the *Entente Cordiale* (1904). And, after the Anglo-Russian Convention (1907), in which Russia and Britain ended long-term disputes and conflicts (eg in the regions of Persia, Tibet, Afghanistan), it resulted in the *Triple Entente* of Russia, France and Britain.

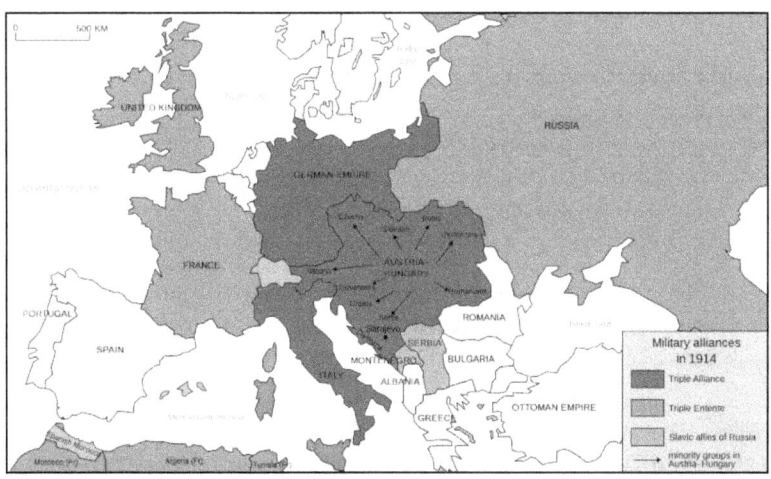

**Figure 24: The major Alliances in Europe (1914).**
Source: Unknown. Wikimedia Commons.

The period from the late nineteenth to the early twentieth century was a one of polarization. The political tension arose due to the social climate of imperialism, militarism, nationalism and patriotism that emerged on a broad front of states.[102]

> In many countries, militarism influenced the political scene, and in the nineteenth-century European mind, politics and military power became inseparable. The new German Empire dominated by Prussian Junker, eager to transform Germany into a global power through its Weltpolitik of aggressive diplomacy, colonial expansion, and the development of a large navy. Britain, with is massive naval force, responded. So, from 1897 to 1914, a Naval Arms Race between the United Kingdom and Germany took place.

Along with that imperialism and militarism, there was nationalism and patriotism. Most pre-war intellectual Europeans believed in the cultural, economic, and military supremacy of their nation. In concert with its brothers, imperialism and militarism, nationalism contributed to a mass delusion that made a European war seem both necessary and winnable.[103]

As the great powers beat their chests and filled their people with a sense of righteousness and superiority, another form of nationalism was on the rise in Southern parts of Central-Europe. This nationalism was not about supremacy or military power, but the right of ethnic groups to autonomy, independence, and self-government. With the world divided into large empires and spheres of influence, many different regions, races, and religious groups wanted freedom from their imperial masters. The regions of the Northern Italian Peninsula ever-exploited by the Austrian Empire wanted to control their own destiny, as did the peoples on the Balkan after being submitted for centuries to Habsburg rule followed by the rule of the Austro-Hungarian Empire.[104] This was the time in which the unified German States saw a massive economic development in which Germany became an industrial colossus. A colossus that wanted a 'place under the sun' and developed a foreign policy that troubled the other great powers of that time.

> The political context leading to the industrialization of the unified Germany, and its subsequent economic boom times, unsettled the balance of power in Europe. Also the 'New Course' —a period of

---

[102] Compare the developments of the late twentieth to the early twenty-first century that was also a period of polarization and geopolitical manoeuvring of individual states.
[103] Nationalism, patriotism: the social desire for national advancement or political independence combined with devotion and loyalty to one's own country.
[104] 'Nationalism as a Cause of World War I'. Source: http://alphahistory.com/worldwar1/nationalism/.

personal rule propagated by Emperor Wilhelm II in which chancellor Bismarck was dismissed in 1890— did not contribute to dampen the tensions of the approaching collision course of Germany's 'Weltmachtpolitik.' And with the Ottoman Empire —the 'sick man of Europe'— crumbling its domination in Central Europe (ie the Balkan), the great powers Russia, Britain and France were to embark on a road to destruction with Germany.

*The Mechanical Technologies of the Industrial Revolution*

After emerging in the cradle of industrialization, the industrial revolution spread over Europe. By that time Germany's Industrial Revolution had created industrial regions (eg the Ruhr Area, Berlin, Saxony, Silesia) connected by a vast railroad infrastructure as well as the industrial capacity for building steam locomotives (Figure 25).

Railroads helped Germany, like the United States, become a leading industrial power by the early twentieth century and complete its transformation from an agrarian to an industrial state. Regional trains emanating from Berlin and Cologne led to other German states as well as to European countries to the north, south, east, and west, connecting vast areas and markets. In the 1830s and 40s early railroad development in Prussia was not controlled by the state. Rather, German investors (including the Rothschilds and Mendelssohns), investment banking houses, and other European investors provided private capital.

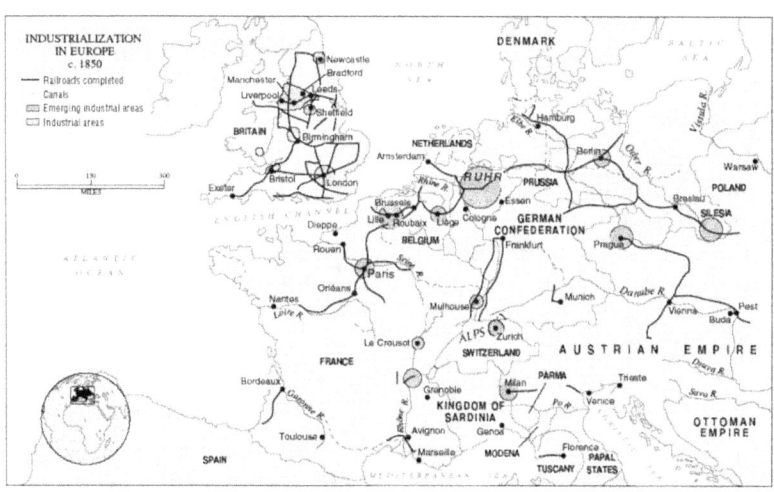

**Figure 25: Industrialization in Europe (c. 1850)**
Source: Unknown. Wikimedia Commons.

Railroads —both the infrastructure and the locomotives powered by the steam engines— became a 'facilitating' technology, and were key drivers of massive industrialization, urbanization, and economic growth. And they became instrumental in warfarin. Prussia's successful use of rail during its surprisingly quick victories in the wars of German Unification between 1864 and 1871 convinced other European countries, especially France and Russia, of the need to better utilize this technology to secure their own national security interests.

After 1871 the pace of rail construction intensified, and the technology of rail improved, all while European nations became increasingly cognizant of rail's potential military uses at the same time political tensions and geopolitical competition increased.

>Germany's construction of the Berlin-to-Baghdad railway (1900s), for example, was considered a blatant effort to exercise influence in the Near and Middle East and to threaten especially Russian and British interests. The German military planners (eg Field Marshall Alfred von Schlieffen), after the 1870 defeat of the Second French Empire (1852-1870), planned scenarios for different campaigns (eg an 'Exterminationskrieg' of France, a two-front war with France and Russia) that were based on military mobility. And the railroads played an essential role in the German War Plan (1906-1914) that foresaw for the 1914 occupation of France rapid troop movement by the German and French railroad infrastructure.

Nearly a century after Napoleon's defeat at Waterloo, in which the industrial revolutions had driven Britain to the peak of its worldwide (Navy-based) dominance, history repeated itself. But now with a new player that had developed into a nation possessing advanced industries based on scientific insights, an entrepreneurial spirit and an industrial skilled workforce. Again the industrial revolutions lay at its foundation, now creating a Railroad-based empire. And in this melting pot of geopolitical strive and nationalistic aspirations, with a dynamic techno-economic development under way, it only needed a spark to ignite…

*The Spark that ignited the Power Keg of Europe*

The Balkans was, a region full of imperialistic ambitions and growing nationalism. No nationalist movement had a greater impact in the outbreak of war than those of the Slavic groups in the Balkans. Pan-Slavism —the belief that the Slavic peoples of Eastern Europe should have their own nation— was a powerful force in the region. That Slavic nationalism was strongest in Serbia, where it had risen significantly in the late nineteenth and early twentieth centuries. It was particularly opposed to the Austro-

Hungarian Empire and its control and influence over the region. It was this imperialism, with the tensions between the Russian Empire, Ottoman Empire, and Austrian Empire, that brought the pre-war regional conflicts: the *Bosnian Crisis* (1908) and *Balkan Wars* (1912–1913).

The *Balkan Wars* (1912–1913), ending Ottoman presence on the Balkan and enabling the rise of individual Balkan states, resulted in the *Balkan Crisis of 1914*. The First Balkan War during October 1912/July 1913, was a war of aggression by the young nation states of Serbia, Montenegro, Bulgaria and Greece against the internationally isolated Ottoman Empire. In the Second Balkan War (June-August 1913), Bulgaria fought against all four original combatants of the first war, along with facing a surprise attack from Romania from the North.

And then came, in 1914, the ignition of the Powder Keg when Serbian irredentists assassinated Archduke Franz Ferdinand, heir of the Habsburg throne to the Austro-Hungarian Empire on June 28th, 1914 in Sarajevo by the 19-year-old Gavrilo Princip, a member of Young Bosnia.[105] His assassination resulted in a mechanized war the likes of which had never been seen before.[106]

Throughout July each of the major European powers engaged in intense deliberations, diplomacy, and signalling within their governments, among their allies, and with their adversaries. A complex web of alliances, coupled with miscalculations by many leaders that warfarin was in their best interests or that a general war would not occur, resulted in a general outbreak of hostilities among European nations in early August 1914. By May 1915 nearly every major European nation was involved and some 70 million military personnel was mobilized. A regional ethnic-based conflict had exploded into a First World War, in the context of a process of European polarization creating two sets of rival alliance systems. These two sets became, by August 1914, the Triple Alliance group of the *Central Powers* of Germany, Austria-Hungary, Bulgari and the Ottoman Empire on one side, and the Triple Entente group of the

---

[105] Young Bosnia was a separatist and revolutionary movement active in the Condominium of Bosnia and Herzegovina, and Austria-Hungary.
[106] Seen in a longer-term perspective, the Anglo-German naval race, the imperial competition, the rise of nationalism, the decline of the Ottoman and perhaps the Austria-Hungarian Empires, and the perceived increases in German and Russian economic and military capabilities all generated great instability, as did demographic pressures and ideological clashes.

*Allies* Russia, France, Great Britain, Italy, Belgium, Serbia, Montenegro and Romania on the other side.

## The Context of the First World War (1914-1918)

The First World War was the first geo-political conflict that spread out outside Europe over many regions in the rest of the world. Some of the main theatres of war were on the European continent, although other regions had their own specific conflict zones.

### The European Theatres of War

After the spark of the assassination, soon Europe was dotted with military conflicts in different theatres or war. The *Balkan Theatre* (1915-1916) saw the Austrian army invading Serbia, later followed by campaign to the east against Russia through Galacia. The *Italian Theatre* (1915-1918) saw Italy entering the war in order to annex the Austrian Littoral and northern Dalmatia, and the territories of present-day Trentino and South Tyrol. Soon, the *Eastern Front* stretched from the Baltic Sea in the north to the Black Sea in the south, involved most of Eastern Europe, and stretched deep into Central Europe as well (Figure 26).[107]

The *Eastern Front* saw the conflict

**Figure 26: The Eastern Front of World War I.**
Source: Encyclopedia Brittanica

---

[107] The Austrian Littoral was a crown land (Kronland) of the Austrian Empire, established in 1849 at the northern end of the Adriatic Sea. It consisted of three regions: the Istria Peninsula, Gorizia and Gradisca, and the Imperial Free City of Trieste. Trieste was the only sea-port of the Austrian Empire

explode between the Russian Empire and Romania, against the Austro-Hungarian Empire, Bulgaria, the Ottoman Empire and the German Empire. Military action saw the frontline move back and forth over Ukrainian Galicia. By 1917, the Russian Empire had collapsed, due to the October Revolution of 1917, and retracted from the war. In between the powers of the Russian Empire and the Austrian-Hungarian Empire, Ukrainian[108] nationalists, in an effort to declare independence from Russia, declared on November 20th, 1917 the Ukrainian People's Republic (UNR). After Vladimir Lenin and his radical Bolsheviks rose to power in November, Ukrainia —like its fellow former Russian property, Finland— took one step further, declaring its complete independence in January 1918. Between 1917 and 1919, several separate Ukrainian republics manifested independence. It was the start of a chaotic period of time for the Ukrainian Soviet Republic.[109]

More to the south, the Ottoman rule began to crumble, after in October 1908 Austria-Hungary seized the Ottoman province of Bosnia and Herzegovina, which it had occupied since 1878. Next in the Italo-Turkish

**Figure 27; Territorial Expansion of Ukraine.**
Source: Wikimedia Commons.

---

[108] One of pre-war Russia's most prosperous areas, the vast, flat Ukraine (the name can be translated as at the border or borderland) was one of the major wheat-producing regions of Europe as well as rich with mineral resources, including vast deposits of iron and coal.
[109] In 1922, Ukraine became one of the original constituent republics of the Union of Soviet Socialist Republics (USSR); it would not regain its independence until the USSR's collapse in 1991. Soviet collectivizing policies resulted in the Soviet Famine of 1932-1933 (aka Holodomor) that claimed ten million Ukrainian lives.

War (1911-1912), and in the Balkan Wars (1912-1913) the armies of the Ottoman Empire were defeated, that lost the bulk of its territory in Europe. The Balkan Wars brought to an end Ottoman rule of the Balkan Peninsula (except for Thrace and Constantinople) (Figure 28, top). On the *Southern Front*, the Italians took the Trentino region and the Austrian Littoral from Austrian rule (Figure 28, bottom).

The *Western Front* became the main theatre of war during the First World War. In August 1914 the German Army invaded in a rapid advance Belgium and Luxembourg on its way to France. Notwithstanding Belgium's fierce resistance, by early September the Germans reached the Marne River some twenty miles from Paris. However, the German advance stopped, both sides dug in along a meandering line of fortified trenches, stretching from the North Sea to the Swiss frontier with France, which changed little except during early 1917 and in 1918. It was the start of trench warfare where one million Allied troops lined up along a 100-mile stretch of the Marne River, facing 1.5 million German troops.

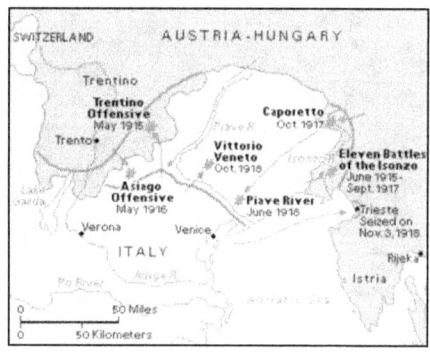

**Figure 28: South-Eastern Front.**
Ottoman Empire and the Balkan Wars (top), and Italian War (bottom).
Source: Brittanica.com

Between 1915 and 1917 there were several offensives along this front (Figure 29); from the second Battle of Ypres in April 15th, to the Battle of Verdun in February 1916, and the Somme Offensive during July 1st – November 18th, 1916. The attacks employed massive artillery bombardments and massed infantry advances. Entrancements, machine gun emplacements, barbed wire and artillery repeatedly inflicted severe casualties during attacks and counter-attacks and no significant advances were made. It became a chemical warfare, when after the first use of tear gas and poison gas, soon other more lethal chemicals such as chlorine, phosgene and mustard gas, were used.

America had maintained neutrality during the first years after 1914, but supplied the Allies with goods and loans. But military actions from the Germans against American activity increased as German policymakers argued that United States could no longer be considered a neutral party after supplying munitions and financial assistance to the Allies. Germany also believed that the United States had jeopardized its neutrality by acquiescing to the Allied blockade of Germany. They started an unrestricted submarine warfare, and throughout February and March 1917, German submarines targeted and sank several US ships, resulting in the deaths of numerous US seamen and citizens. That changed the political climate and the result was the direct involvement of America in the European warfare. On April 6th, 1917 Congress declared war on Germany. By June 1917,

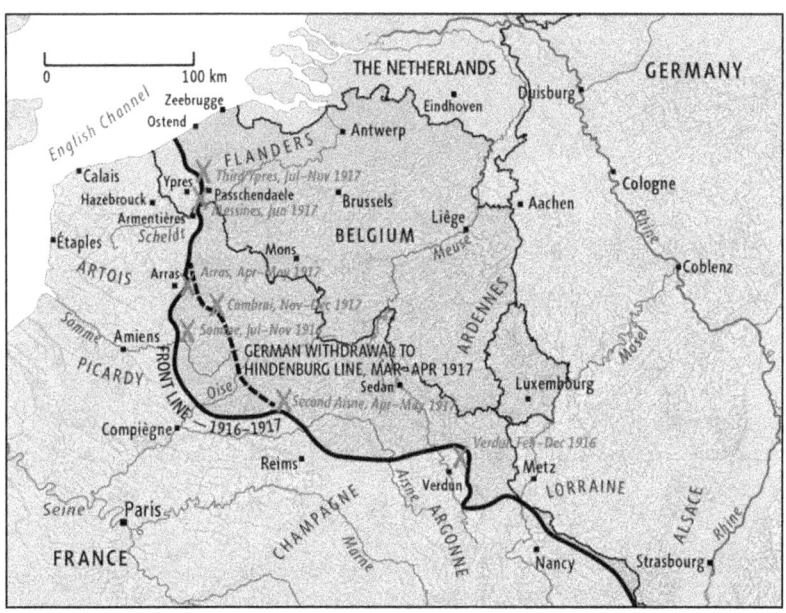

**Figure 29: Battles along the Western Front (1916-1917).**
Source: https://nzhistory.govt.nz/

14,000 American soldiers arrived in France, and the summer of 1918, about 2 million US soldiers of the *American Expeditionary Forces* were ready to combat. Overall, some 4,791,172 Americans would serve, some 2,084,000 would reach France, and 1,390,000 would see active combat.

By that time Germany was facing not only a new military force, but also internal problems. The eruption of the *German Revolution* (1918-1919), a reaction on the extreme burdens suffered by the population during the four years of war combined with social tensions between the aristocracy that lost the war and the general population that faced the consequences. It resulted in a defeat during the rapid series of Allied victories of the Hundred Days Offensive starting on August 8th, 1918. On November 11th, 1918 the armistice was signed. One of the deadliest conflicts in human history had ended.[110] And the geopolitical arena of Europe was reshaped during the Treaty of Versailles on June 28th, 1919 with Germany losing 13% of its territory (Figure 30).

**Figure 30: Political Map after Peace Treaty of 1919.**
Dark grey areas are the newly created states between Western Europe and Eastern Europe.
Source: Wikimedia Commons.

---

[110] The total number of military and civilian casualties (ie death and wounded) were about 40 million persons. Add to that epidemic of the Spanish Flu (1918), a virus that ultimo would kill somewhere between 50-100 million people.

In the west, Germany returned Alsace-Lorraine to France. It had been seized by Germany more than 40 years earlier. Further, Belgium received Eupen and Malmedy; the industrial Saar-region was placed under the administration of the League of Nations for 15 years; and Denmark received Northern Schleswig. Finally, the Rhineland was demilitarized; that is, no German military forces or fortifications were permitted there. In the east, Poland received parts of West Prussia and Silesia from Germany. In addition, Czechoslovakia received the Hultschin-district from Germany; the largely German city of Danzig became a free city under the protection of the League of Nations; and Memel, a small strip of territory in East Prussia along the Baltic Sea, was ultimately placed under Lithuanian control.

The end of First World War in Europe led to the emergence of the new 'minority problems' in the areas of collapsing German and Austro-Hungarian empires. Over 9 million ethnic Germans found themselves living in Poland, Czechoslovakia, Romania, and Yugoslavia. It would become a major cause of future problems. Outside Europe, the end of the war meant also the end of the German colonial empire. The war ended the Scramble for Africa, and in the post-war settlements, Germany's colonies were divided between Britain, Belgium, Portugal and South Africa. The German colonial possessions in Asia and Pacific fell without hardly any military bloodshed.

As result of the geo-political alignment in the Treaty of Versailles, four empires disappeared: the German Empire, the Austro-Hungarian Empire, the Ottoman Empire, and the Russian Empire. Including the dynasties of their rulers of the Romanovs (ie Nicolas II), the Hohenzollerns (ie Kaiser Wilhelm II), the House of Habsburgs-Lorraine (ie Emperor Charles I), and the Ottomans (ie Emperor Mehmed VI Vahideddin).

## The Middle Eastern Theatre of War

The Middle East —the lands running from Anatolia to Persia, and from the Caucasus to Egypt, encompassing its crossroads of civilization Mesopotamia and the Levant— were in the nineteenth century inhabited by a range of different native ethnic groups. They were ruled by the Ottoman dynasties, that from the sixteenth century on governed peoples of different religions (eg Islam, Christianity, Judaism) speaking a broad range of languages.[111] Although the nineteenth century had seen internal reform and modernization (1808-1839) coping with growing nationalism, during the Tanzimat-period (1839-1876) its geo-political dominance came under attack

---

[111] A detailed analysis of the history of this region aka the Cradle of Civilization can be found in the case studies of the Deep History series.

by the conflicts with the European Empires of Russia, Austria-Hungary and Britain (eg the Crimean War, 1853-1856), as well as the French and Italian expanding their influence in the North of Africa. By 1914 its territory was reduced to the Middle East (Figure 13).

From Napoleonic times on the British had a presence in Egypt and Turkey and confronted the Ottoman Empire in the Egypt-Syrian Campaign (1798-1801) and in the Anglo-Turkish War (1807-1809). They even had an increasing interest in the region after they initiated and financed the construction of Suez Canal (1859-1869), built to shorten the route to the East (ie British India), shifting the routes of world trade. Local unrest caused the British to invade Egypt in 1882 and take full control, although nominally Egypt remained part of the Ottoman Empire. By 1914, the Ottoman Empire joined the Entente of the Central Powers of Germany, Bulgaria and Austria-Hungary.

> By the end of the war, the Persian oilfields —explored by the Anglo-Persian Oil Company (est. 1909)—were becoming crucial for the British Royal Navy that was converting from coal to oil-powered warships. The Navy thus became the company's major customer, but the small royalties (16% of net sales) paid to the Arabs, created political discontent.

By the start of the First World War, the Middle Eastern theatre saw military campaigns between the Ottoman Empire (and allies) on one side, and the British, Russians and French (and local allies) on the other side. It started with the *Sinai and Palestine campaign* after the Ottoman raided the Suez Canal in January 1915 by invading the Sinai Peninsula. Next, in the Bosporus, the British initiated the *Gallipoli Campaign* (February 1915-January 1916) against the Ottoman Empire in order to get control over the Turkish Straits (aka Dardanelles) between the Aegean Sea and the Black Sea. And the Greeks armies were invading Western-Anatolia, Eastern Thrace and Constantinople that they were claiming on historical grounds.

> The idea behind the Gallipoli campaign was, when Turkey could be defeated, that both the Suez Canal and the access to the Black Sea would be safe. And that the British oil supply would not be interrupted.[112] However, the campaign failed and became a trench war, the warships and submarines on both sides crippling warfarin. At the end a stalemate situation arose, followed by the political war

---

[112] The British Royal Navy depended upon oil from the petroleum deposits in southern Persia, to which the British-controlled Anglo-Persian Oil Company (est 1908) had exclusive access. Also the lands around the Black Sea —eg the Donbass Region of Ukraine, Georgia) had early oilfields, next to its large coal reserves.

scare *Chanak Crisis*[113] that caused the British to evacuate their troops. The casualties on both sides mounted up to 500,000 people. Some 145,000 soldiers became sick due to insanitary conditions, especially from typhoid, dysentery and diarrhoea.

These campaigns were followed by the *Persian Campaign* (into modern Iran) between the Ottoman Empire and the British Empire cooperating with the Russian Empire, and the *Mesopotamian Campaign* (into modern Iraq) fought by British troops. The Persian Campaign proved to be a devastating experience for Western Anatolia and Persian Armenia. Over 2 million Persian civilians died in the conflict, mostly due to the Armenian genocide (1915-1917) by the Ottoman regime and the Persian famine of 1917–1919. In the Mesopotamian campaign, the Ottomans suffered 325,000 casualties, among which some 55,000 military. The British forces suffered some 85,000 casualties.

The Armenian Genocide was the systematic destruction and ethnic cleansing of the Armenian people living in Eastern Anatolia by the Turkish when the Ottoman Empire collapsed. These ethnic groups being adherents to Christian beliefs (eg the Apostolic Church, Hetanism, and the Catholic and Orthodox Church) became victim of Turkish nationalism that culminated in the Turkish War of

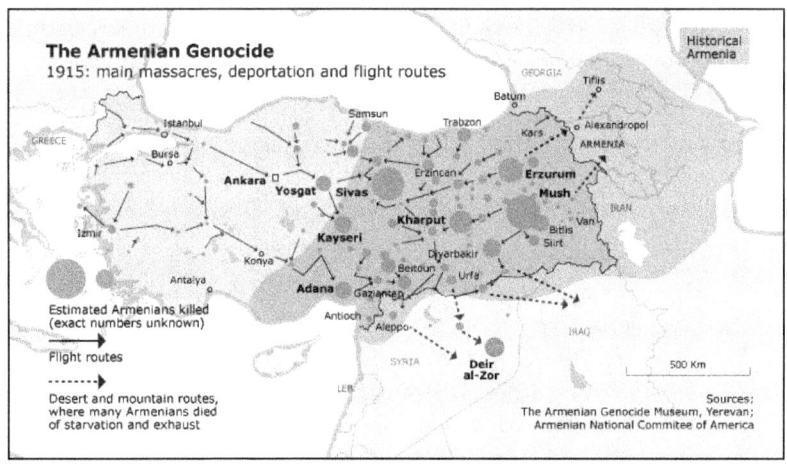

**Figure 31: The Armenian Genocide (1915).**
Source: https://www.vox.com/a/world-war-i-maps

---

[113] By August 1922, the Turkish army had reoccupied most of the Dardanelle regions. With the withdrawal of the French troop left the British with not enough troops in the region of Chanak, they requested additional troops from the dominions (eg Canada, New Zealand). Which was delayed by Canada, causing political tension that toppled the British government.

Independence (1914-1923). After earlier large-scale massacres in the 1895-1896 (est. 100,000 victims) and 1909 (est. 20,000-50,000 victims), in 1915 some 800,000-1,200,000 Armenians were sent on death marches to concentration camps in the Syrian Desert (Figure 31). In an effort to supress their Christian identity, Islamization of Armenians was carried out as a systematic state policy.

The Persian Famine of 1917-1919 after series of droughts after 1916 had created shortages of food, aggravated by the actions of the warring parties. Next to blockading road transport of grain, the requisitioning of pack animals, mules and camels for the oil industry in Khuzestan, and for the British and Russian armed forces, left the country's transport network in serious disarray, and disrupted the distribution of foodstuffs and other goods throughout the country, with disastrous consequences. Added by speculation and hoarding, the subsequent food revolts disrupted communities, and the shortages weakened the population making them receptive to diseases. Beyond nutrient-shortage starvation, epidemic and contagious diseases like influenza, cholera, typhoid and typhus claimed an estimated two million lives.

These events created a disrupted post-war Middle East where the colossal food crisis, plus large numbers of soldiers, refugees and destitute people constantly on the move in search of work and survival, created the unstable context for the disintegrating and collapsing Ottoman Empire. And it gave the impetus to the rising Turkish nationalism.[114]

The main character of the Middle Eastern Theatre of War was —like so many geo-political conflicts— one of shifting alliances, deception, irregular and guerrilla warfare, internal conflicts within the Ottoman Empire, religious confrontations as well as individual heroism.[115]

Another stalemate, the one in southern Palestine was broken by the *Battle of Beersheba* on October 31st, 1917. The British captured Palestine from the Ottomans in the *Third Battle of Gaza* (November 1st - 2nd, 1917). Next, on November 2nd, 1917, the British government issued the Balfour Declaration, favouring the establishment of a national home for the Jewish people in Palestine. The declaration had many long-lasting consequences. It greatly

---

[114] Turkish nationalism aimed at creating an independent Anatolia, felt threatened by the Greek claiming West-Anatolia with the large Greek populations, and the Italian invasions of West-Anatolia. In June 1920 the Greek army captured most of West-Anatolia around Smyrna, but by 1922 the Turkish victories ousted them out of Anatolia.
[115] Such as the actions of the British officer T.E. Lawrence (ak Lawrence of Arabia), whose daring raids with the Arab tribes against the Turkish Hejaz Railway made him a legend.

increased popular support for Zionism within Jewish communities worldwide, and became a core component of the British Mandate for Palestine in 1922 issued by the League of Nations.

The Middle Eastern warfarin had colossal human casualties. Between 1914 and 1918 one in six Ottoman civilians died, including 1.5 million Armenians in a state-led genocide, and hundreds of thousands were (re)moved. Also the food and health situation (eg cholera, plague, typhus, and influenza epidemics) remained precarious and caused the Persian Famine (1917-1919)[116] as a result of continued violence.

## *The Context of the Interbellum (1920-1940)*

By 1919, most of the military hostilities of the First World War may have been ended, the aftermath effects however, would shape the times to come. The war had killed more people —more than 9 million soldiers, sailors, and flyers and another 5 million civilians— involved more countries (28)—, and had cost more money —$186 billion in direct costs and another $151 billion in indirect costs— than any previous war in history. It had shaped the geo-political arena —with the downfall of four monarchies—, ignited the Bolshevik communism rise to power, and the nationalistic revolts against the colonial powers in the Middle East and Far East. Economically, the war severely disrupted the European economies and allowed the United States to become the world's leading creditor and industrial power. The war also caused vast social disruptions, including the mass murder of Armenians in Turkey and the Spanish influenza pandemic that killed over 25 million people worldwide.

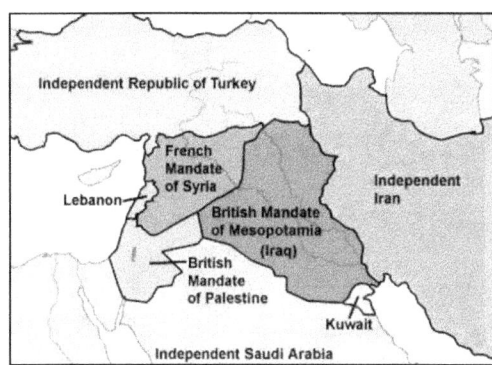

Figure 32: British and French Mandates in Middle East (1923).

Source: Unknown.Wikimedia Commons.

### Interbellum Period in the Middle East

The government of the Ottoman Empire had collapsed and its territory was partitioned in several new states between 1918 and 1922. It saw the creation of independent movements in the modern Arab world, the Republic of Turkey, as well as the

---

[116] Estimates of casualties of the Persian Famine range from 2-10 million people. The Spanish Flu Pandemic took the life of 17-50 million people worldwide

French and British Mandates (Figure 32). The mandates were thinly disguised licences for imperialism, issued under the policies of the League of Nations.

In the secret Sykes-Picot Agreement, already in 1916, the British, French and assenting Russians had decided to control the partitioning and keep a regional interest after the partitioning. However, the Russian Empire became engaged in the October Revolution and Russia's southward expansionism was (temporary) halted as the communist regime proclaimed itself an ally of the exploited masses.

France demanded a say in the Ottoman Empire's fate and compensation for its wartime losses in the form of southeast Anatolia and Iraq/Lebanon/Syria, claiming a cultural affinity from the Crusades (ie La Syrie Française). Britain wished to ward off the Bolsheviks and secure India by keeping the Suez Canal and Egypt, under British rule since 1882, and by protecting its economic interests from Palestine to Iraq (Figure 33). The British Mandate of Palestine would become roots of the state of Israel.[117]

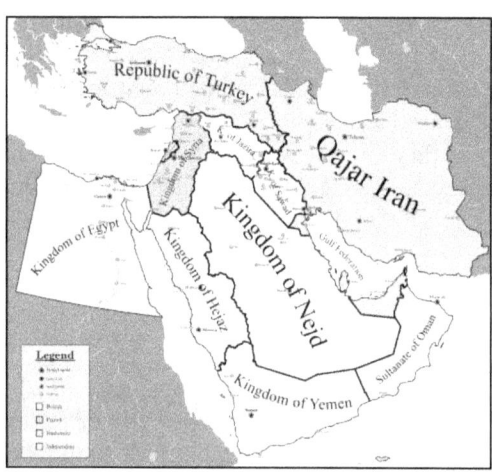

**Figure 33: Partitioning of the Ottoman Empire into New States (1923).**

Source: ocw.mit.edu

The result of the partitioning by the Treaty of Sevres (1920) was (1) the rise of *independent states*; like the Republic of Turkey, Iran (the Kingdom of Qajar) and the Saudi Arabian State (the Kingdoms of Njed and Hejaz). And (2) the establishment of *dependent states* in the form of protectorates: from the Kingdom of Syria to the Kingdoms of Jazira and Sawyad (Figure 33).

---

[117] After the October Revolution of 1917 —starting a week after the conception of the Balfour Declaration— some 1,200 Antisemitic pogroms started in the Ukraine massively killing Jews. An estimated 100,000 Jews were killed and 500,000 left homeless. Some 40,000 refugees arrived during the Third Aliyah (1919-1923) in Palestine.

The Turkish National Movement (TNM) rose to power during the *Turkish War of Independence* (1919-1923) and wanted revoking of the terms of the Treaty of Sevres. In November 1922 the TNM's leader, Mustafa Kemal Atatürk (1881–1938), strong-armed the Grand National Assembly to abolish the sultanate, forcing the Allies to accept the TNM as Turkey's only representative. The *Treaty of Lausanne* (1923) saw the recognition of the sovereignty of the newly formed 'Republic of Turkey' (Figure 34).

> Using nationalism, Atatürk encouraged people of the Muslim faith in Turkey to bond together and with his newly created forces defeated the Greek and Italian armies in West Anatolia and kept the Armenians in East Anatolia from taking over large amounts of land to create their own country (Figure 34).

In the Persian region, due to the ravages of the war waged by foreign belligerents on its soil from 1914 to 1919, the Kingdom of Qajar in 1921 was prostrate, ruined, and on the verge of disintegration. Then, the coup of 1921 brought the Shah Rheza Khan to power (1925-1941). Britain was welcoming the resultant decrease of the Bolshevik threat and accepting Iran's independence while preserving key interests, including the quasi-sovereign *Anglo-Persian Oil Company* (APOC) operating in southwestern Iran.[118] Britain also recognized Reza Shah when he established his own Pahlavi dynasty in 1925.

**Figure 34: Partitioning Treaty of Lausanne (1923).**
Source: Unknown. Wikipedia Commons

---

[118] The British *Anglo-Persian Oil Company* (est. 1909) was founded to extract oil from Iran. By 1927 the massive Baba Gurgur oilfield was discovered.

On the Arabic Peninsula —after Mohammed's Muslim era (ie Seventh Century) splitting up tribal groups into the Sunni Islam and Shia Islam religious factions— the century long ruling Hashemite dynasty had gotten control over a patchwork of tribal rulers[119] and arose as the rulers of the Emirate (ie Kingdom) of Hejaz. And the adjoining Sultanate of Njed of tribal people ruled by the Saud dynasty that had evolved in a range of phases; the First Saudi State (1744-1818), the Second Saudi State (1824-1891) and the Third Saudi State resulting from the Unification of Saudi Arabia (1902-1932).[120] It saw the Arab Revolt (1916-1918) and the participation of Lawrence of Arabia,[121] aiming to create a single unified and independent Arab state, establishing the Kingdom of Nejd and Hejaz. The rulers became entangled in tense negotiations with their former allies, the British and French, over the future of the region. In 1932 the territories were united in the Kingdom of Saudi Arabia ruled by the House of Saudi.[122]

With the progressing Interbellum and the rise of religion-based nationalism and independence, soon the Arab population in the British and French protectorates also wanted to loosen the imperialistic ties and started working towards independence along tribal and religious lines. It resulted in different geopolitical-based revolutions.

In Egypt, the status as protectorate was influenced by *Egypt's Revolution of 1919* against British occupation. The revolution led to the United Kingdom's later (partial) recognition of Egyptian independence in 1922 as the Kingdom of Egypt, and the implementation of a new constitution in 1923. But it was not until 1953 that Egypt obtained full independence.

In 1920 the *Great Iraqi Revolution* —of Sunni and Shia religious communities co-operating during the revolution as well as tribal communities, the urban masses, and many Iraqi officers in Syria— against British rule was

---

[119] The majority of the Arabs living in the Ottoman Empire were loyal primarily to their own families, clans, and tribes.
[120] The modern Kingdom of Saudi Arabia was founded in 1932 by Abdulaziz bin Abdul Rahman, known in the West as Ibn Saud. Abdulaziz united the four regions into a single state through a series of conquests.
[121] T.E. Lawrence was a British author, archaeologist, military officer, and diplomat. He was renowned for his liaison role during the Sinai and Palestine Campaign and the Arab Revolt against the Ottoman Empire during the First World War. He inspired the Bedouins to attack the Hejaz Railway.
[122] The discovery of oil in Saudi Arabia in 1938 gave impetus to the powers of the Saudi's. After the Second World War ended, the bilateral relation with US was established based on two objectives; together containing communism (ie security), and control over oil production (ie energy supply). Oil gave Saudi Arabia political leverage in the international community.

suppressed. However, the revolt, and its costs of 40 million pounds, caused British officials to drastically reconsider their strategy in Iraq. They created the Kingdom of Iraq, and although the monarch Faisal I of Iraq was legitimized and proclaimed King by a plebiscite in 1921, independence was not achieved before 1932, when the British Mandate officially ended.

In 1920 the short-lived Kingdom of Syria came to an end, and the region became a French Mandate (Figure 35, top). In 1923, that mandate over Syria (including Lebanon) came under stress when, in the aftermath of the Great Famine (1915-1918), Lebanese aspirations sought alterations of the demographic situation. The resulting Constitution of Lebanon (1926) created the republic state of Greater Lebanon. In parallel, the Syrian part of the mandate had resulted in the Syrian Federation (1922-1924), succeeded by the State of Syria (1925-1930) after the *Great Syrian Revolt* (1925-1927) opposing French rule.

> In Syria, in order to counter the Franco-British plans, the locals attempted to put a leader of their own choice, King Faisal, on the throne to try to escape colonial rule. The French responded in a similar fashion and crushed the rebels during the *Franco-Syrian War* (1920) at the Battle of Maysalun; once again thousands died and the French retained control. Faisal was forced into exile, and in 1921, the British placed him at the head of Iraq. Also the Great Syrian Revolt (1925-1927) suppressed the Syrian tribesmen, but also changed the French attitude toward imperialism in the Levant. As the French mandate lasted until 1943, when two independent countries emerged, Syria and Lebanon. French troops eventually left Syria and Lebanon in 1946.[123]

In July 1920, the British Mandate of Palestine (Figure 35, bottom) was created after the San Remo Conference. The British controlled Palestine for almost three decades, overseeing a succession of ethno-religious protests, riots and revolts between the Jewish and Palestinian Arab communities. The Mandate fell apart in two parts; Palestine and TransJordan. In Palestine the years 1923–29 was relatively quiet; Arab passivity was partly due to the drop in Jewish immigration in 1926–28. In 1927 the number of Jewish emigrants exceeded that of immigrants,

---

[123] Next to the Middle East, after the Scramble for Africa (1881-1914), France occupied territories along the North African coast (eg Morocco, Algeria), West and Central Africa. It also held territories in the Pacific (eg Pacific Islands, Indochina; Cambodia and Vietnam), and in the West Indies on the South-American continent. After its defeat in the Franco-Prussian War of 1870–1871 and the founding of the Third Republic (1871–1940) most of France's later colonial possessions were acquired.

and in 1928 there was a net Jewish immigration of only 10 persons. But tensions intensified after the Wailing Wall Incident (August, 1929).[124]

In 1923 the Emirate of Transjordan was recognized by the Brits. Transjordan became in 1928 nominally independent, and had its own Constitution, Parliament and Legislative Council that in a series of Anglo-Transjordanian treaties led to almost full independence for Transjordan.

During the Arab Revolt in Palestine (1936-1939), the Palestinian Arabs rose against the British Mandate in Palestine that had allowed the creation of the Jewish National Home[125] and the immigration of tens of thousands of Jews into region.[126] The Arabs, however, lost due to the military and financial support of the British. The British Peel Commission proposed the formation of a Jewish and Arab state. The Arabs opposed the partition plan and condemned it unanimously. It resulted in the *First Arab-Israel War* (1947-1949)

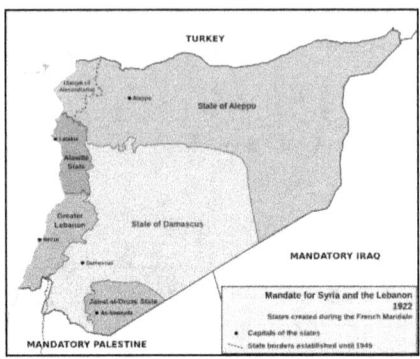

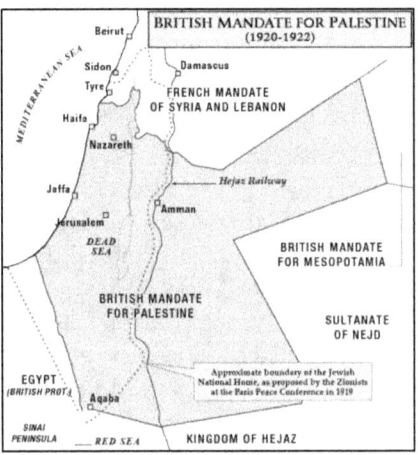

**Figure 35: The French and British Mandate (1920-1923) in the Levant.**

Dotted line (.....) indicates boundary of Jewish Home Proposal.

Source: Wikimedia Commons, Edmaps.com

---

[124] This incident was about was access to and custody over the Western Wall of the Temple Mount (al-Haram al-Sharif, the Noble Sanctuary) in Jerusalem.
[125] The European powers mandated the creation of a Jewish homeland at the San Remo conference of April 19th–26th, 1920
[126] The Jewish population had grown under British auspices from 57,000 to 320,000 in 1935. In the year 1936 some 60,000 Jews arrived.

when military forces of Egypt, Transjordan, Syria, and expeditionary forces from Iraq entered Palestine. The Israeli's proclaimed in the Declaration of Independence (1948) their state of Israel.[127]

This brief exploration into the events of the Near East Interbellum shows how French and British imperialism still exercised geopolitical influence after the end of the First World War. But in a complicated mix of social, religious and political changes, the multitude of tribal states of the Modern Middle East got their independence, and the British and French imperialism more or less collapsed.

*The Middle East Oil Corridor*

It was also during the Interbellum, that large deposits of fossil oil were discovered in the Middle East; in the region called the 'Oil Corridor' (Figure 36) Starting with exploration of Texan and Californian oil deposits, as well as the Caspian Sea, Azerbaijan and Armenian deposits at the end of the nineteenth century, in the early twentieth century large deposits were discovered in Persia (later Iran) and around the Perian Gulf Basin. It started the race for the Middle East oil, as the Third Industrial Revolution was powered by the oil-fuelled internal combustion engine. With the advent of mechanized warfare on land, the increasing use of aircraft, and the

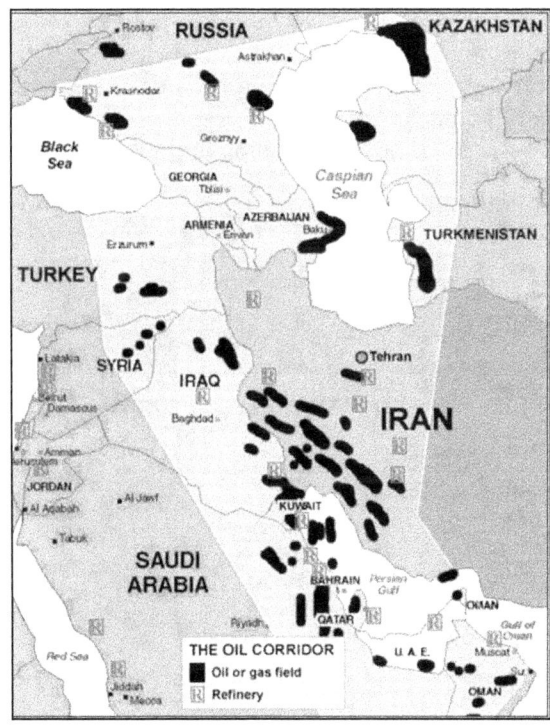

**Figure 36: The Middle East Oil Corridor.**
Source: people.wou.edu

---

[127] Around 700,000 Palestinian Arabs fled or were expelled from their homes in the area that became Israel, and they became Palestinian refugees in what they refer to as the Nakba ('the catastrophe'). In the three years following the war, about 700,000 Jews emigrated to Israel.

transition of naval propulsion from coal to oil, petroleum became a vital strategic commodity.[128]

Since that time, the discovery and exploitation of oil in the Middle East has had a profound influence on modern society and their foreign politics. Oil created vast fortunes and industrial empires (eg the 'Seven Sisters'), directed war campaigns, promoted the widespread use of petroleum, gave birth to OPEC (1960), realigned twentieth-century geo-politics, and started oil crises (eg 1973, 1979).

*Overview Interbellum Middle East*

From the preceding exploration emerges picture of a major geo-political shift in the Middle East during the Intebellum. A shift in the balance of powers between and within empires and nations that arose during the Third Industrial Revolution after the First World War (Figure 37). It was the Interbellum period that became the dawn of the next conflict on a world scale.

Had the First World War seen the end of the European Empires (ie the German Empire, the Austro-Hungarian Empire, the Ottoman Empire, and the Russian Empire), it was also the time of awakening of the Middle East.

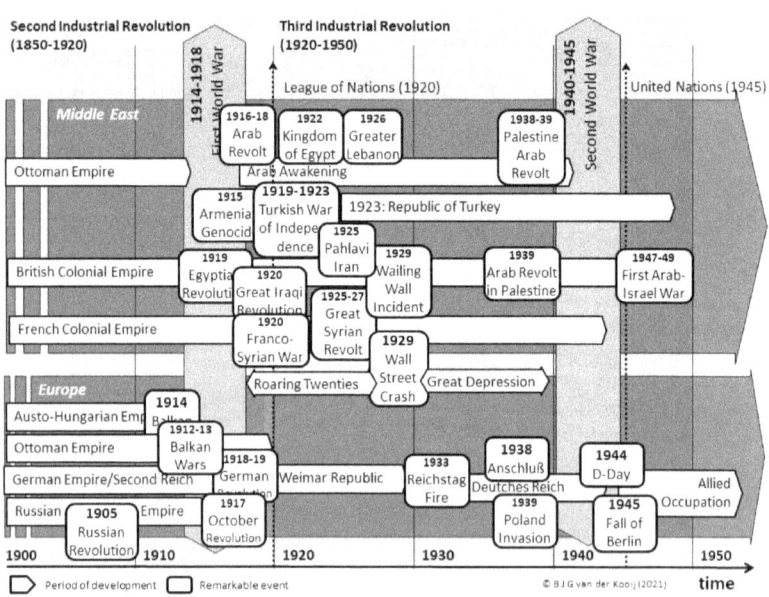

**Figure 37: Overview Geo-political development of European and Middle Eastern Empires (1900-1950).**

---

[128] See: Van der Kooij, B.J.G.: *The Invention of the Internal Combustion Engine* (2021)

The collapse of the Ottoman left a range of independent states (eg Saudi Arabia, Turkey, Iran), as well as the mandates from the European colonial empires Brittain and France over the emirates ruling a multitude of nomadic and settled tribes. The mandates that kept the region in the grip of the 'sphere of influence' of Britain and France, both with their own (geo-politcal and geo-economic) interests.

## The Rise of Colonial Empires in the Pacific

The Far Eastern region saw in the late nineteenth century the *Scramble for the Pacific* as Western powers rushed to defend (eg British, French) and expand (eg Germany, America) their sphere of influence in the Far East. After the earlier days where explorers and whalers, settlers and missionaries had travelled the Pacific Ocean, now —similar to the Scramble of Africa (1881-1914)—, the western powers became anxious to expand their sphere of influence in the Pacific (Figure 38).

At the *Berlin Conference* (1884-1885) the European diplomats decided on a division of both Africa and Oceania. In Oceania it concerned the island clusters of Micronesia (eg the Carolinians, Marshall Island), Melanesia (eg Figi, New Guinea, Solomon Islands) and Polynesia (eg Hawaii Samoa, Cook Islands) (Figure 38). Pacific islands were seen by the US as valuable since they were refuelling and supply stations for trade routes with China. The islands also were valuable for guano, nickel, sugar, vanilla and fruit. So Britain, France, Germany and the US began a rush of annexations.

*The British Colonial Empire in the Pacific:* By the start of the nineteenth century the heritage from the first British Empire (1479-1763) was already enormous, and in the 1815–1914 period some 10 million square miles of territory and 400 million people were added. So that the British Empire became to cover most continents (eg Africa, India, Australia). The Victory over Napoleon (1815) had left Britain without any serious international rivals (except Russia) in the Far East, and the Brits expanded again their sphere of influence. First by the Pacific Islands creating the *British Western Pacific Territories* (1877) where the Royal Navy ruled the waves, next by extending their influence by sending in the military to prevent French and German expanding territory in the *Scramble of the Pacific.*

*The French Colonial Empire:* The French, next to their colonial aspirations in Africa, had their eyes on the Far East and became active in Polynesia. They expanded their activities into the rest of the Pacific and Indochina (1858-1870) by creating dependencies in French Polynesia. Such as the Society Islands with Tahiti (1842), New Caledonia (1854), and by Vietnam (1864) and Cambodia (1867). After the Sino-French War

(1884-1885), the French created French Indochina from their protectorates Vietnam, Laos, and Cambodia.

*The American Colonial Empire:* The Great Leap Westward over the American continent did not stop at the beaches of California. The subsequent maritime expansion was the result of a naval expansionist strategy, but also by the promising commerce with China. After commercial exploration (eg whaling), by 1868-1874 the US Navy had established its presence on the Hawaii islands with the Reciprocity Treaty (1875) embracing free trade and the leasing of Pearl Harbour (1887). After several conspiracies and rebellions, the Hawaiian Kingdom was overthrown in a coupe d'état (1893). Next, Hawaii was annexed by America in 1898, and the US created a naval base in Pearl Harbour. More to West, the island of Guam (part of the Mariana Islands), was occupied, followed by the control over the more than 7000 islands of the Philippine Archipelago. Both territories were acquired in 1898 after the defeat of the Spanish in the Spanish-American War. Next it took the *Philipino-America War* (1899-1902) to cement that control. Together with the island of Guam, the Philippine archipelago were depicted as stepping stones to the riches of China. By 1900, the United States was a recognized world power with substantial commercial, political, and military interests and territorial holdings throughout the Pacific region.

*The German Colonial Empire:* As the German Empire was late in rising to a

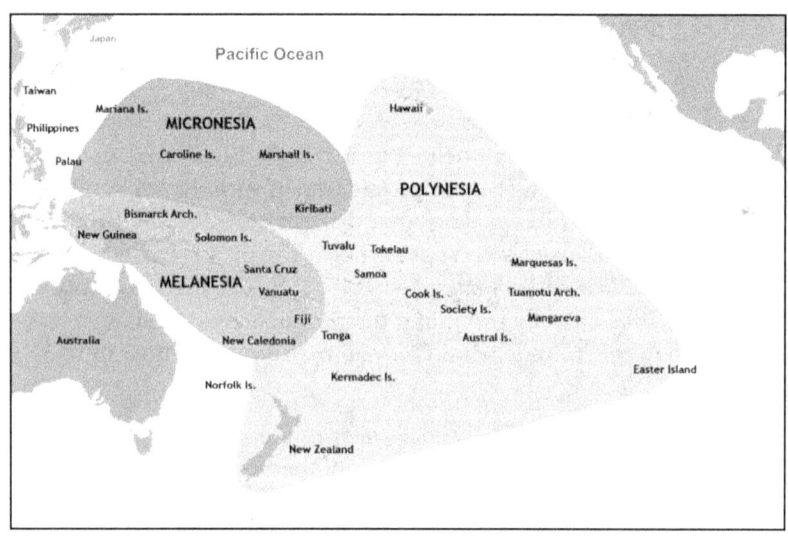

**Figure 38: The Specific Clusters of the Pacific Islands.**
Source: Wikimedia Commons.

world power (1870s), it was also late in the Scramble for Africa (1884) and the Scramble for the Pacific. Nevertheless, they became the third-largest colonial empire after the Brits and French. In the Pacific they focussed on the Spanish occupied Micronesian Islands (eg Marshall Islands, Caroline Islands). However, the German Empire's presence in the Pacific came to a sudden end following the outbreak of the First World War.

The extension of European control over Asia added a further dimension to the rivalry and mutual suspicion which characterized international diplomacy in the decades preceding the First World War.

## The Rise of the Chinese Republic and the Japanese Empire

Simultaneously with the rising geo-political tensions in the run-up to the First World War in Europe and the rise of the Soviet Communist ideology, in Asia there had been geo-political developments of its own. On the one hand the fall of the Chinese Empire in 1912, on the other hand the rise of the Japanese Empire after the Meiji-Restauration Era.

### China's Civil Revolution

Just before the First World War was to break out in Europe, China's *Xinhai Revolution* (1912) was ending the rule of the Qing dynasty that had seen both internal rebellion and foreign imperialism, and the start of the Republic of China (ROC). The internal rebellion, in which different political factions/cliques competed for hegemony, led to a fracturing of the country as different cliques in the Beiyang Army[129] claimed individual autonomy and clashed with each other.

**Figure 39: The Chinese Warlords of the different factions during the Warlord Era (1925).**

Names of dominant warlords heading their specific cliques. The KMT ruled the lower regions without a name.

Source: Wikimedia Commons.

Sun Yat-sen—a proponent of Chinese nationalism and democracy— as the

---

[129] The Beiyang Army was the western-style army of the imperial Qing Dynasty.

leader of the Kuomintang (KMT) who had played an instrumental role in the run-up to the Revolution, became shortly the first provisional president of the Republic. [130] However, the Beiyang Army split into several mutually hostile factions that battled for supremacy over the following years. In the preceding 1911-Revolution, General Yuan Shikai —who had held the different factions of the Beiyang Army together— became his successor, and proclaimed himself emperor in 1915. But that was to no avail as he had to back down due to massive opposition followed by his death in 1916.

The following decade, known as the Warlord Era (1916-1928), was characterized by constant civil war between different factions/cliques headed by warlords large and small. Such as the warlord Wu Peify heading the Zhili Clique (Figure 39), and his protegee the Nanking warlord Sun Chuanfang. And these warlords were in turn supported by the foreign nations like Japan that already had formulated its 'Twenty-One Demands on China' in 1915. The result was shifting alliances and territories ruled by cliques losing and winning internal battles. Such as in 1920 when the discredited Anhui clique was overthrown in a war with the Western-backed Zhili clique (Figure 40).

In 1919 the *May Fourth Movement*, idealising the Bolshevik Revolution and Marxist theory in Russia, contributed in the founding of the Chinese Communist Party (CCP) in 1921. In 1923, the Kuomintang (KMT) and its

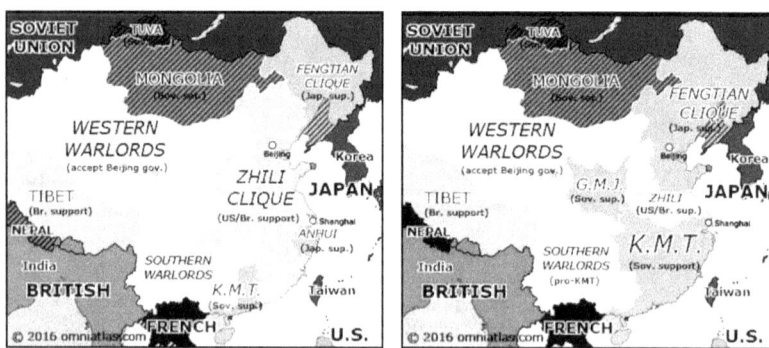

Figure 40: Foreign Support to shifting cliques during the Warlord Era (1923, left and 1929, right).
Source: Omniatlas.com

---

[130] The Chinese Nationalism contrast with the ethnic based Han Nationalism of the Qing dynasty, as it considers that the Chinese people are a nation and promotes the cultural and national unity of all Chinese people.

Canton government accepted aid from the Soviet Union after being denied recognition by the western powers. Soviet advisers —the most prominent of whom was Mikhail Borodin, an agent of the Comintern— arrived in China in 1923 to aid in the reorganization and consolidation of the KMT along the lines of the Communist Party of the Soviet Union.

By 1925, the CCP failing to dominate the KMT, a new political leader general Chian Kai-shek succeeded founder Sun-Yat-sen after his death. When Chiang gradually gained the support of Western countries, the conflict between him and the communists became more and more intense. In April 12th, 927, Chiang turned on the communists, and marched to communist controlled Shanghai and massacred some 5,000-10,000 communist militias during the *Shanghai Massacre*. That May, tens of thousands of communists and their sympathizers were killed by nationalist troops, with the CCP losing approximately 15,000 of its 25,000 members. Next, Chiang defeated the communist in the *Nanchang Uprising* on August 1st, 1927.

The Warlord Era reach an end in 1928 when the Kuomintang under General Chiang Kai-shek officially unified China through the *Northern Expedition* of the National Revolutionary Army fighting the Beiyang armies and their warlords (Figure 41). After the *Central Plains War* (1929-1930), which involved more than one million soldiers, Chiang was ultimately victorious which ensured his status as the singular leader of all China.

However, although regionalism and war-lordism would continue in weakening the

**Figure 41: Routes of the Northern Expedition (1926-1928).**
Starting in the South in Guangzhou, the KMT-armies moved up to Beijing.
Source: Wikimedia Commons.

country, the victory started the reunification of China marking the beginning of the *Nanking Decade* (1927-1939). And it saw the rise of the CCP's military wing —aka the Red Army— when communist elements from the National Revolutionary Army defected in the Nanchang Uprising (1927). It was the start of the first phase of the *Chinese Civil War* (1927-1937) when the Kuomintang (KMT) purged its Communist members in Shanghai and other cities (ie Hunan, Jiangxi).

By 1930, the Communist Red Army[131] had established the *Chinese Soviet Republic* (CSR) in the provinces of Jiangxi and Fujian. In 1930, Mao claimed a need to eliminate alleged KMT-spies and Anti-Bolsheviks operating inside the Jiangxi Soviet and began an ideological campaign featuring torture and guilt by association, in order to eliminate his enemies. The campaign continued until the end of 1931, killing approximately 70,000 people and reducing the size of the Red Army from 40,000 to less than 10,000. In order to eliminate the threat of the CSR, in 1933, Chiang's National Revolutionary Army besieged Jiangxi with a range of encirclements.

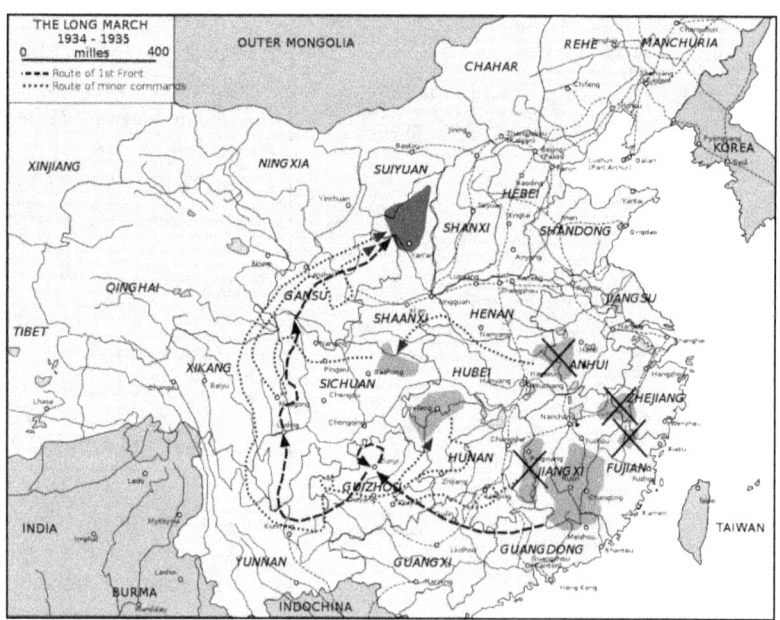

Figure 42: The different Marches constituting the Long March of the Communists (1934-1935).
Source: Unknown. Wikimedia Commons.

---

[131] The People's Liberation Army (1928-1937) traces its roots to the 1927 Nanchang Uprising of the communists against the Nationalists. Initially called the Red Army, it grew under Mao Zedong and Zhu De from 5,000 troops in 1929 to 200,000 in 1933.

To escape the pursuit of the KMT-Army, the Red Army of the CCP retreated in a range of marches aka as the Long March (October 1934-October 1935) (Figure 42). The Long March began the ascent to power of Mao Zedong, whose leadership during the retreat gained him the support of the members of the party.

The Nanking Decade was marked by both progress and frustration. Chiang was elected President of the National Government by the KMT central executive committee in October 1928. His Nationalist government suppressed the CCP, but dissent, corruption and nepotism were rampant and revolts broke out in several provinces; internal conflicts also perpetuated within the government. The CCP-adhering warlords moved in and the decade came to an end with the political crisis of the X'ian Incident in 1936, in which Chiang was arrested in order to force the KMT to enter into a truce with the insurgent CCP and form a united front against Japan. The incident was solved after negotiations about ending the civil war and resist the Japanese instead together.

The Xi'an Incident was a turning point for the CCP. Chiang's leadership over political and military affairs in China was affirmed, while the CCP was able to expand its own strength under the new united front, which later played a factor in the Chinese Communist Revolution that erupted after the war and created the second phase of the *Chinese Civil War* (1945-1949). And it was the start of *Second Sino-Japanese War* (1937-1945).

## The Rise of the Japanese Empire

During the Industrial Revolution of the Meiji Restauration era, Japan had opened up to the West. With the start of the First World War in 1914, Japan practically sided with the Allies by taking German territories in the Far-East; the Mariana, Caroline, and Marshall Islands (aka Micronesia). Followed by seizing German Holdings in China's coastal region of the Shandong Peninsula. Just as Australia did with the other scattered German colonies in Oceania.

Politically, the Japanese Empire seized the opportunity to expand its sphere of influence in China, and to gain recognition as a great power in post-war geopolitics. So, in 1915, Japan presented the Chinese their 'Twenty-One Demands,'[132] that would reduce China to a Japanese Protectorate. After a lot of diplomatic stalling, a reduced set of 'Thirteen Demands' —without the disputed section 5— was accepted. With this

---

[132] The demands of Section 1 concerned Japan's recent seizure of German ports and operations in Shandong Province. The demands of Section 2 expanded Japan's sphere of influence in southern Manchuria and eastern Inner Mongolia. The most extreme demands (in section 5) would give Japan a decisive voice in finance, policing, and government affairs.

action, the Japanese Empire —now under Emperor Taishō (1912-1926)— gained more mistrust than goodwill among the allies.

Toward the end of the war, Japan increasingly filled orders of war material for its European allies. The wartime boom helped to diversify the country's industry, increase its exports, and transform Japan from a debtor to a creditor nation for the first time. The 1920s were the most liberal phase yet in Japanese history.

The nation, pretty much settled in its enlarged empire, still harboured imperialist sentiments, but for the moment, sensible leaders grasped that the time for expansion abroad had come to an end. However, neither parliament nor government, to an extend not even the emperor, controlled the military clique. On the contrary, it was the military that controlled the civilian government. The Japanese constitution, more to the point several later amendments, explicitly required the ministers of the Navy and Army, respectively, to be serving officers of their services.

> The military had already a strong influence on Japanese society from the Meiji Restoration. Almost all leaders in Japanese society during the Meiji period (whether in the military, politics or business) were ex-*samurai* or descendants of *samurai*, and shared the same set of values and outlooks.

During the Taishō period (1912-1926), Japan had seen a short period of democratic rule. But the victory of the Bolsheviks in Russia in 1922 and their hopes for a world revolution, threatened Japan's societal structure and imperial aspirations.[133] The Japanese Communist Part, originally a underground political association, came to the surface in 1922 for a short time as it was outlawed in 1925. So, during the reign of Emperor Shōwa (Hirohito) in the Shōwa Period (1926-1945) up to the Second World War, right-wing policies became dominant in opposing socialism and communism.

In the western societies the Interbellum had seen both the Roaring Twenties as well as the Great Depression. The latter threatening the collapse of the economic world order. That Great Depression hit the Japanese economy hard, as its traditional customers did impose customs barriers on their products.

> The Japanese military, realising the lack of raw materials (ie coal, oil, minerals and metals) on their island creating a dependence on foreign imports, became concerned. So, with the *Invasion of Manchuria* in 1931,

---

[133] The announced goals of the Japanese Communist Party in 1923 included the unification of the working class as well as farmers, recognition of the Soviet Union, and withdrawal of Japanese troops from Siberia, Sakhalin, China, Korea, and Taiwan.

The Invention of the Silicon Engine

the military without informing the government, annexed he mineral rich Chinese Manchuko region, securing its industrial base in the form of a puppet state; the Republic of Manchuko (Figure 43).[134]

The Western states —organized in the League of Nations to solve international conflicts— were not pleased with the invasion, and voices called for an economic boycott. After an investigation of a League commission, it was concluded that Manchuria should be returned to Chinese sovereignty. The Japanese government rejected the Commission's

Figure 43: Japanese Annexation of Manchuria (1931).
Source: Wikimedia Commons.

---

[134] Manchuria was an important region due to its rich natural resources including coal, fertile soil, and various minerals. For Japan, Manchuria was an essential source of raw materials. Under Japanese control Manchuria was one of the most brutally run regions in the world, with a systematic campaign of terror and intimidation against the local Russian and Chinese populations including arrests, organised riots and other forms of subjugation.

findings and withdrew from the League in March 1933. Just as Germany had done after the League had refused to accept Germany's demands to increase its military. Despite China's appeals, it was decided not to extend sanctions on Japan.

The 1930s were a decade of fear in Japan, characterized by the resurgence of right-wing patriotism, the weakening of democratic forces, domestic terrorist violence (including assassinations),[135] and stepped-up military aggression abroad. With the introduction of mass education, conscription, industrialization, centralization, and successful foreign wars, Japanese nationalism began to foment as a powerful force in society.

> The rise of Japanese nationalism paralleled the growth of nationalism within West-Europe. Both Germany and Japan, as nation-state rising from the Holy Roman Empire, respectively the Shogunates, had in common a politically weak position of the emperor, powerful authoritarian oligarchies of military rulers (knight and daimyo), the conversion from agrarian societies to industrialized societies, and the military supporting imperialism in order to get access to natural resources.

And with the Japanese nationalism came the rise of militarism. From 1932 to 1936, the country was governed by generals and admirals. Mounting nationalist sympathies led to chronic instability in government. Japan's invasion of China in 1937 started the *Second Sino-Japanese War*. Following the Marco Polo Bridge Incident about a missing Japanese soldier, the Japanese scored major victories, capturing Beijing, Shanghai and Nanjing in 1937.

> In the Rape of Nanjing, starting December 13th, 1937 and lasting six weeks, at least 200,000 Chinese perished in the mass-scale random murder, wartime rape, looting and arson committed by the Imperial Japanese Army against the residents of Nanjing, the capital of the Republic of China (ROC). The captured Chines troops were machine gunned to death along the Yangtze River. At least 80,000 Chinese soldiers and POWs, or possibly over 100,000, were massacred by the Japanese.

These were the political developments among the populations of the Far East. A region after the expeditions in earlier times by western explorers discovering the riches of the East, where the Western Empires rushed to

---

[135] During the May 15 Incident of 1932, a coup d'état was attempted, launched by reactionary elements in the Imperial Japanese Navy and Army. The Prime-Minister Tsuyoshi was assassinated. Their light punishment further eroded the rule of law and the power of the democratic government tin Japan to confront the military.

extend their sphere of influence beyond trade and Christianisation. A region that by the twentieth century —especially during the Interbellum— had seen a development on its own.

## Overview Interbellum Far East

From the preceding exploration emerges picture of a major geo-political shifts in the Far East and Pacific. A shift in the balance of powers between and within empires and nations that arose during the Third Industrial Revolution after the First World War hit Europe and the Middle East (Figure 37). It was the Interbellum period that became the dawn of the next conflict on a world scale as the world war hit next to Europe and the Middle-East, also the Far-East (Figure 44).

The Chinese Empire of the Qing Dybasty collapsed in the early twentieth century after the Xinhai Revolution (1912), and became the Republic of China in two distinct periods; the Warlord Era and the Nanking Decade. The rise of communist influence resulting in the Chinese Communist Party (1921) created political turbulence and disruptions between warlords and their cliques. In the same Interbellum period the Japanese Empire, after the Meiji Restauration, rose to power. Competing with the Western Powers, the nationalist military fractions expanded their

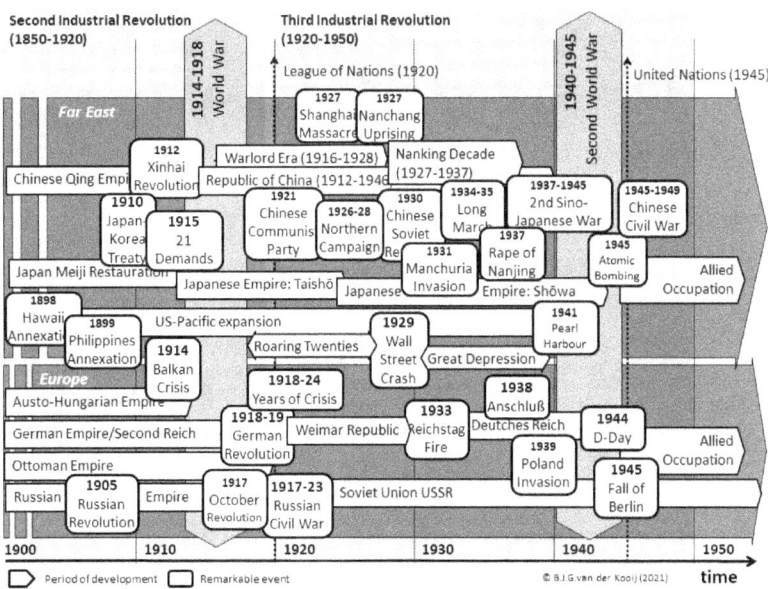

Figure 44: Overview Geo-political development of European and Far Eastern Empires (1900-1950).

sphere of influence into the Chines territory by occupying Korea Peninsula (1910) as well the Manchuria region (1931).

Already before the First World War, America had expanded its sphere of influence into the Pacific by annexing Hawaii (1898) and creating a naval base in Pearl Harbour and acquiring the Philippines from Spain (1899). Faced with the Japanese-Chinese conflict, in the first two decades of the twentieth century, the relationship between the United States and Japan was marked by increasing tension and corresponding attempts to use diplomacy to reduce the threat of conflict.

## The Interbellum in Europe and the US

After the end of the First World War in Europe, the subsequent Interbellum was, notwithstanding being a short period of time with its two decades, a period that saw many fundamental changes in the Affairs of Man. Technological leadership —based on a broad range of technological driven novelty resulting from increasing collective organised Research and Development (R&D) activities— had passed from Britain and the European nations to the United States in the course of the First World War.

### The Aftermath in the US

By then America was a country where on the background was simmering the mental aftermath of the war.

> *The nation was spiritually tired. Wearied by the excitements of the war and the nervous tension of the Big Red Scare, they hoped for quiet and healing. Sick of Wilson and his talk of America's duty to humanity, callous to political idealism, they hoped for a chance to pursue their private affairs without governmental interference and to forget about public affairs. There might be no such word in the dictionary as normalcy, but normalcy was what they wanted.* [136]

In that Spirit of Time, petroleum-based energy production and the associated mechanisation (eg in the automotive industry) expanded mobility dramatically, and started to influence the geo-political power relations. The early commercial airplanes had taken off and started conquering the skies along (inter)-continental sky ways served by clippers following 'airlines.' The radio squirmed the ether, opening up the world even to most-isolated citizen in the remote corners of the country.[137] Culturally, it gave rise to the movements of the *Roaring Twenties*, a period of economic prosperity and growth for the middle class in North America, Europe, Asia, and many

---

[136] Source: Allen, F.L.: Only Yesterday, an Informal History of the 1920s. (1931) https://gutenberg.net.au/ebooks05/0500831h.html
[137] In 1922 the US sales of radio sets, parts, and accessories amounted to $60 million. By 1925 that had become $506 million, and 1929 it his nearly $850 million.

other parts of the world. Automobiles, electric lighting, electric house hold appliances, and radio broadcastings became all commonplace among populations in the developed world.

But the accompanying economic progress came to an end as rampant speculation in stock led to the *Wall Street Stock Market Collapse* in 1929. The indulgences of this era were followed by the *Great Depression*, an unprecedented worldwide economic downturn which severely damaged many of the world's largest economies. Pollical it was the time in which soviet-communism got a dominant position in Russia after the October Revolution of 1917. And it was also the time of rising racial-based and nation-based nationalism, that would dominate the governmental attitude of the Central-European states. In other words, the Interbellum saw in two decennia the world changing dramatically.

## *The Aftermath in Europe*

The military confrontations on the European continent may have been ended, and the peace-treaty being signed, by 1918 the destabilising effects of the political collapse of the Russian, German, Austro-Hungarian, and the Ottoman Empires (Figure 30) were largely present. The German *Years of Crisis* (1918–1924) were marked by socio-economic turmoil as roving 'Freikorps' units[138] brutalized German political life. In the East, the Russian *Civil War* (1917-1923) raged on in Eastern Europe, and Central Europe struggled to recover from the material devastation of the First World War.

Next to the four empires that collapsed due to the war, old countries were abolished, new ones were formed, boundaries were redrawn, international organizations were established, and —competing with the old ideologies— many new ideologies took a firm hold in people's minds. Minds that carried the mental scars of wartime-experiences. Minds that became indoctrinated with nationalistic dreams. Again, it was the time of revolutionary events and pandemics, as well as massive social turmoil.

But there was also the less visible side of destruction, as people carried with them their atrocious experiences from the wartime having an impact on their mental health. Not only of the military returning maimed, disabled and shellshocked from the theatres of war, but also of the civilians having been confronted with the collateral effects (from rape to terror and random killing). As so many social fabrics (in family-ties and social bonds) had been

---

[138] Freikorps were irregular German and other European military volunteer units, or paramilitary. They consisted out of war veterans and were the early organizations that developed into the German Braun shirts (ie Sturmabteilung SA), and the Italian Black shirts.

disrupted, and traditional mechanisms of coping with stress (ie fight, flight or freeze) proved inadequate, other social mechanisms kicked in. Suspicion and mistrust ruled social interaction.

The vacuum left behind by the dissolution of the German, Russian, Austro-Hungarian and Ottoman Empires was filled with a number of new countries in Central Europe and the Middle East (Figure 30, Figure 33). With the imperial centres gone irretrievably, national groups —both of Slavic and of German ethnic origin— began to assert themselves and make bids for independence, and demanded the right to create their own states. Political ideologies like 'communism' rose to power, creating the federation of soviet republics —after the Russian Soviet Republic was founded as result of the October Revolution (1917)— and became known as the Soviet Union (USSR, 1922).

> Germany came out of the war with most of its industrial infrastructure intact, leaving it in a better position to become the dominant economic force on the European continent after the First World War. But that development was faced by problems of its own. After the *German Revolution* (1918-1919) following the revolt of the Kiel Mutiny, the *Weimar Republic* (1919-1933) was proclaimed. It started the times of the hyperinflation[139] due to the German response to the war-reparations, the aftermath of the French Revanchism claiming the regions of Elzas-Loraine, and the occupation of the Ruhr-and Saar-region by French and Belgium troops. The revolution was not only a transformation that ended a monarchy and the rise of a parliamentary democracy, it was also a socialistic revolution as programs of *progressive social reform* was introduced.

### The Rise of European Dictatorships

Dictatorships[140] in all variations are of all times; from the Roman dictators/Emperors to the Japanese Shogun-dictators and their powerless Emperors. With the advent of the nineteenth and twentieth centuries, dictatorships and constitutional democracies emerged as the world's two major forms of government, gradually eliminating monarchies as the most widespread form of government in the pre-industrial era.

*Nazi Germany under Adolf Hitler:* The German Years of Crises (1918-1924) proved to be a fertile ground for the rise of early nationalism during the

---

[139] The social and political cost of the hyperinflation was high. Savings evaporated, a lifetime of savings would no longer buy a subway ticket. Pensions planned for a lifetime were wiped out completely. Not surprisingly, politically, the hyperinflation fuelled radicalism on both the left and the right.

[140] Dictatorship is a government form dominated by one person (ie the dictator) backed up by military power and/or political power.

Weimar Republic (1919-1933). It started with Kaiser Wilhelm II's abdication after the German Revolution (1918–1919) —modelled after the Soviets of the Russian Revolution of 1917— eroded his powerbase. His exile to a mansion in Doorn, the Netherlands ended the House of Hohenzollern's 300-year reign in Prussia and 500-year reign in Brandenburg. The former general of the Imperial Army Paul Hindenburg initiated the Weimar Constitution creating a republic under a parliamentary republic system with the Reichstag elected by proportional representation. The new republic held its first election on January 19th, 1919. However, it did not take long for the rising star of the NDSAP leader *Adolf Hitler*, to grab power in the Beer Putsch of Munich, Bavaria, on November 8-9th, 1923. It was the start of is rise to power. By 1932 Hitler, after becoming Chancellor, clamped down on the opposition. Meetings of the left-wing parties were banned and some of the moderate parties found their members threatened and assaulted. Even the centre of German democracy, the Parliament buildings in Berlin called 'Reichstag,' came under fire (1933). By 1938, the Anschluss' meant the annexation of Austria.

*Fascist Italy under Benito Mussolini.* Although the Kingdom of Italy had been one of the Central Powers of the First World War, Italy was the only victorious country unable to exploit the positive effects —in the political, economic and social sphere— of the military victory. The internal political arena in the post-war period, unable to solve economic afterwar-difficulties (eg rising inflation, high cost of living, and unemployment) and social conflicts (ie mass demobilization, industrial and rural strikes) during the *Biennio Rosso* ('Two Red Years') ended with a political and parliamentary crisis. Failing to liberal reform, fascism took over and Italy became governed by the National Fascist Party from 1922 to 1943 with *Benito Mussolini* as prime minister. The Italian Fascists imposed authoritarian rule and crushed political and intellectual opposition, while promoting economic modernization, traditional social values and a rapprochement with the Roman Catholic Church.

*Spain under Francisco Franco:* Spain was neutral during the First World War; however, it felt its effects economically and politically as the opposing groups of 'Francophiles' and 'Germanophiles' divided society. In 1923 General Miguel Primo de Rivera seized power with the approval of the king. He dissolved Parliament, imprisoned democratic leaders, suspended trial by jury, censored the press, and placed the country under martial law. In 1931 a republic was proclaimed, headed by a provisional government of republicans and socialists. In 1936 a group of army officers led by General *Francisco Franco* staged a fascist revolt that result in the Spanish Civil War (1936-1939).

*The Spirit of Time: Roaring Twenties*

Clearly, the First World War had cost Western civilization its self-congratulatory pre-war optimism accumulated in the US Gilded Age and the French Belle Époque (Figure 12). The wartime experiences created the feeding ground for the subsequent rise of nationalist movements (eg Germany, Italy, Spain) and their dictatorships (ie Hitler, Mussolini and Franco). It initiated a period of time in which the Interbellum generation[141] experienced socio-economic ups and downs: the Roaring Twenties as well as the Great Depression. Times that saw —next to the anti-war sentiment showing up in literature and movies— revolutions in Arts and Technology.

On the other hand, the Interbellum was also the time of the combined maturing and spreading of earlier technology-induced novelty. It was the period of time that saw the *Era of Electric Light* brightening up societies, the *Era of Tele-Communication* connecting peoples and businesses, as well as the *Era of Radio Broadcasting* spreading information and entertainment. In addition, the *Era of Mass-Production*[142] and the *Era of Auto-Mobility* had come together with the rising 'leisure time' (ie free time not spent working) for the masses of the working class.

> Leisure time that facilitated a lifestyle where 'work' to earn an income became complemented by recreational activities; from social to family leisure. Such as the outdoor-recreation of traveling (by automobiles), of sports (like the team sports of rugby, soccer and football) and events (such as festivals, cycling and motor races). Or indoor activities of live performances in theatres (music, spectacles), films in movie-theatres. And last-but not least with the rising part of free-available income, the new art of shopping creating the Commercial Revolution of the 1920s. It was the time of massive cultural novelty; from the *Jazz Age* and the *Era of Art Deco*.

Next to the revolutionary events that took place on the European continent, other events took place that would mark the Spirit of Time. The

---

[141] This is the generation of people born in the first decade of the twentieth century. The name comes from the fact that those born during this time were too young to have served in the military during World War I, and were generally too old to serve as enlisted personnel in World War II.

[142] The production of automobiles, petroleum, steel, and chemicals skyrocketed during this period. This was largely due to the adoption by industry of the technique of mass production, the system under which identical products were churned out quickly and inexpensively using assembly lines. The changeover to mass production drove down prices for objects that were previously made in much more individual, time-consuming methods and subsequently enabled an increase in new, affordable technology. A middle class of Americans emerged in the post-war period with surplus money and a desire to spend more, spurring the demand for consumer goods, especially the automobile.

devastation of the First World War had not only caused demographic, economic and socio-political havoc, it had also inflicted psychological and physical damage on a scale hardly seen before. For example, among the military that were confronted with the new technology-induced warfare.

> Obviously, war is about defeating the enemy. Often that means killing their soldiers. During the First World War the German killing power —of their artillery and machine guns, their U-boats and mines, and their poison gas— had increased massively. The new weapons generated new, horrible injuries that took life and limb in a flash or festered into gangrenous wounds that could further maim and kill. The carnage traumatized some men into shellshock, and poison gases like mustard gas burned and suffocated others so horribly that nurses dreaded caring for them because they could provide little comfort. War diseases —notably the soldiers' nemeses diarrhoea, dysentery, and typhus— flourished, and the trenches offered new maladies such as the 'trench foot,' an infection caused by wearing sodden boots and standing in water and mud for days on end, and the 'trench fever,' a debilitating fever transmitted by body lice. Add to that the horrors of the influenza epidemic, the infectious disease that costed the lives of a staggering 8.7 million enlisted men. [143] All in all, more men died to the influenza virus than in military action. Not surprisingly, the 'shell shock'[144] of artillery fire, the terror of chemical warfare, and the diseases of the trench war had —next to the 'normal' experiences of warfare— left deep mental impressions, mental scars, mutilations and psychoses.

The utter carnage and uncertain outcome of the war was disillusioning, and many began to question the values and assumptions of Western civilization and of Christian religion.[145] The war had opened eyes to unimagined horrors and when it was over people simply wanted to live again. The result was observable in some specific cultural changes:

*The Lost Generation:* The unprecedented carnage and destruction of the war stripped the 'lost generation'[146] of their illusions about democracy, peace, and religion. The old norms and values that gave them their ethical and moral reference in life, were doubted. Their feelings were

---

[143] Source: Jones, M.M.: The American Red Cross and Local Response to the 1918 Influenza Pandemic: A Four City Case Study. Public Health Reports, Volume 125 (2010).: p.83
[144] The claustrophobic environment of trenches was conducive to nervous and mental breakdowns amongst soldiers due to unrelenting shellfire from the enemy.
[145] Parallels can be found in the period of time after the Black Death Pandemic (c. 1348) where the dominant Catholic Church failed to answer the question "Why…".
[146] The people born between 1883-1900, in France known as the Génération du feu, the '(gun)fire generation'.

represented by some of the most famous American writers of that time; F. Scott Fitzgerald, Gertrude Stein, T.S. Eliot, Ernest Hemingway, John Dos Passos, and John Steinbeck.

F. Scott Fitzgerald in his 1925-novel *The Great Gatsby*, explored themes of decadence, idealism, resistance to change, social upheaval and excess, creating a vivid portrait of the Roaring Twenties. It was made into several movies (1926, 1949, 1974, 2013) with same title.

These shifting cultural norms and values were expressed in the attitudes of a hard-drinking, fast-living set of disillusioned young people; it created the American Spirit of Time.

*The Jazz Age:* In America a new style of music making developed in the speakeasies; places (often owned by organized criminals) where customers could drink alcohol and relax, or speak easy. The competition for patrons in speakeasies created a demand

**Figure 45: Some Impressions of the US-Roaring Twenties.**

From partying in a Speakeasy (top, 1920s), a Flapper dancing at a Party (middle), to performances of Duke Ellington's Band (bottom, 1930s).

Source: History Collection (top) https://historycollection.com; Penn State (middle); Chicago Herald photo (https://www.chicagotribune.com/).

for live entertainment. The already-popular jazz music, and the dances it inspired in speakeasies and clubs, fit into the era's raucous, party mood (Figure 45). With thousands of underground clubs, and the prevalence of jazz bands, liquor-infused partying grew exponentially. The sheer number of speakeasies that popped up over time in big cities —New York had some 32,000— indicates the growth of jazz popularity. Jazz legends like the black Louis Armstrong and Duke Ellington were both invited to play at white only clubs though they were both African American. In addition, the radio and the record player brought the music to attention of a wider public.

Figure 46: Political Poster for Prohibition (1919)

Source: https://prohibition.osu.edu/gallery?field_image_terms_1_tid%5B%5D=61

*The Prohibition Era:* Originating from the American Progressive Era, where the negative social effects of rapid industrialization and urbanization had become on the political agenda, a ban on the sale and import of alcoholic beverages was already lobbied for before the First World War (Figure 46). Over time calls for a ban had become more strident, Saloons and the heavy drinking culture they fostered were associated with immigrants and members of the working class, and were seen as detrimental to the values of a Christian society. The Anti-Saloon League, with strong support from Protestants and other Christian denominations, spearheaded the drive for nationwide prohibition. In fact, the Anti-Saloon League was the most powerful political pressure group in US history; no other organization had ever managed to alter the nation's Constitution. All these socio-political efforts resulted in the Eighteenth Amendment that reflected the Progressives' faith in the federal government's ability to fix social problems. Meant to reduce criminal activity, it had the opposite effect.

*Crime and Decadence:* Between 1920 and 1933 organized crime penetrated the business of the speak-easies. The gangsters relied on their music and the musicians relied on the gangsters for income. Such as the famous gangster Al Capone earning an estimated $100 million per year from his rackets. The Prohibition gave rise to a criminal class and the mob-wars, the corruption of public officials, and a widespread disrespect for the rule of law.

Not only America had its roaring Spirit of Time, also on the continent in the big cities such as Berlin, the First World War left its heritage. There the Roaring Twenties showed similar patterns of the 'Zeitgeist': a time for hedonism, sexual liberties and artistic experimentation. Following the afterwar turmoil and hyperinflation, the period 1923-1929 created in Germany the Goldene Zwanziger ('Golden Twenties'); a period with a cultural renaissance and an explosion of intellectual productivity.

> *Artist Experimentation with Design:* After the First World War had ended, during the Weimar Republic, a renewed liberal spirit allowed an upsurge of radical experimentation in all the arts, which had been suppressed by the old regime. In the visual arts and architectural construction new movements developed, such as in architecture the Bauhaus's schools that reflected the new ideas fusing arts, crafts and technology. Industrial design and art integrated in the Arts Décoratives (ie Art Deco) and resulted in novel product designs, and architectural design that expressed functionality. In liberal arts, the movements known as expressionism, Dada, and surrealism all played major roles in reconfiguring focus and perception. The Dada movement consisted of artists who —in reaction to technology's dominance— rejected the logic, reason, and aestheticism of modern capitalist society, instead expressing nonsense, irrationality, and anti-bourgeois protest in their works. The Surrealism was a social and cultural revolutionary movement in reaction to horrors of technology-powered warfarin.

> *Hedonism and Sexual Liberation:* Weimar Berlin was known to be a place that broke the social conventions of its epoch. Following American trends, the cabaret scene and jazz band became very popular. Modern young women were Americanized, wearing makeup, short hair, smoking and breaking with traditional mores. The euphoria surrounding Josephine Baker, where she was declared an 'erotic goddess' and in many ways admired and respected, kindled further 'ultramodern' sensations in the minds of the German public. Many women questioned gender roles and some of them defied patriarchal traditions by becoming economically independent from men. The cabaret environment also created room for sexual minorities to express themselves in a relatively freer way. Many gay and lesbian-targeted establishments opened and survived during those years, even though sexual intercourse between males was criminalized.

> *Crime and Prostitution:* In the areas of Europe left ravaged by the First World War, without jobs or any form of social relief, women having lost their partner or unable to find a partner among the decimated male population, lived in dear circumstances. As a result, prostitution rose in cities such as in Berlin. Crime in general developed in parallel with prostitution in the city, beginning as petty thefts and other

crimes linked to the need to survive in the war's aftermath. Berlin eventually acquired a reputation as a hub of drug dealing (cocaine, heroin, tranquilizers) and the black market. The police identified 62 organized criminal gangs in Berlin, called *Ringvereine*.

## Prelude to the Second World War

The Great Depression that spread from America after the Wall Street Crash in 1929, not only affected the US, but also European countries. In addition, in Germany the Weimar Republic had collapsed after the Nazi-Party had gained dominance in the national elections of 1932. In the context of the two competing political ideologies —communism and nationalism— the latter would see the rise of dictatorship.

By 1932 the leader of the Nazi Party, Adolf Hitler, after becoming Chancellor, clamped down on the opposition. Meetings of the left-wing parties were banned and some of the moderate parties found their members threatened and assaulted. Even the centre of German democracy, the Parliament buildings in Berlin called 'Reichstag,' came under fire.

The Reichstag Fire on February 27th, 1933 (Figure 47) was blamed by Hitler's government on the Communists. Hitler used the ensuing state of emergency to obtain the presidential assent of president Hindenburg to issue the *Reichstag Fire Decree* the following day. The decree invoked Article 48 of the Weimar Constitution and 'indefinitely suspended' a number of constitutional protections of civil liberties, allowing the Nazi government to take swift action against political meetings, arresting and killing the Communists.

**Figure 47: Reichstag Fire (Berlin, 1933)**
Source: https://www.wbez.org/

With the Nazi-Party becoming the dominant political power in 1933-1934, Hitler abolished democracy, espousing a radical, racially motivated revision of the world order (the 'Neuordnung'),[147] and soon

---

[147] The *Neuordnung* entailed the creation of a pan-German racial state, structured according to Nazi ideology, to ensure the existence of an Aryan-Nordic master race, consolidate a massive

began a massive rearmament campaign. Combined with governmental programs to stimulate the economy —eg the creation of the Autobahn-network and reorganizing the automobile industry—, the repudiation of the Treaty of Versailles —taking the Saar-territory back under German control—, and the coalition with Italy in the Rome-Berlin Axis, Hitler embarked on this path to create Gross Deutschland for a nation that needed 'Lebensraum'.

At the same time, elsewhere in Europe other dictatorial regimes had gained political dominance. In Italy it was the Fascist movement of *Benito Mussolini*, whose imperialist aspiration brought on the invasion of the Ethiopian Empire. And in Spain General *Francisco Franco* came to power during the Spanish Civil War (1936-1939). After a difficult start of the nationalist rebellion in Spanish Morocco, Franco asked for international help. Hitler and Mussolini responded and their intervention —with the bombardment by the Germans of

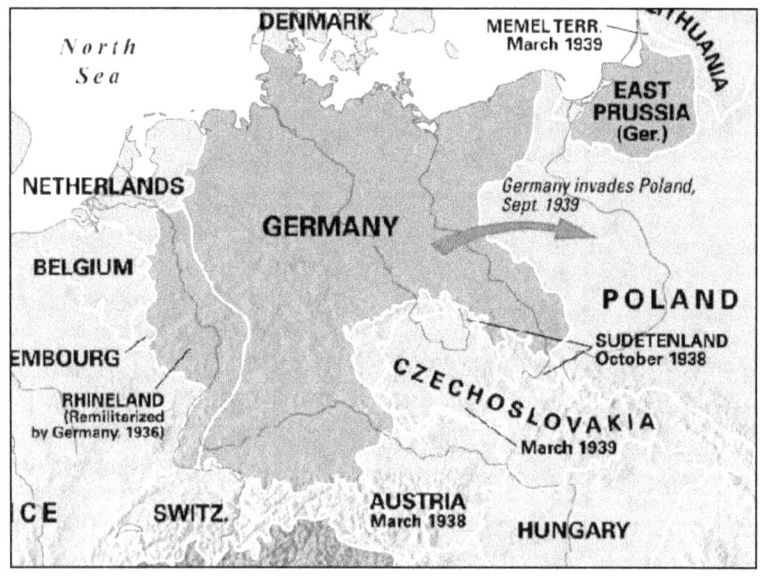

Figure 48: Germany's Anschluss of Austria and Annexation of Sudetenland (1938).
Source: Unknown. Wikimedia Commons.

---

territorial expansion into Central and Eastern Europe through colonization by German settlers, achieve the physical annihilation of Jews, Slavs (especially Poles and Russians), Roma ('gypsies',) and others considered to be "unworthy of life", as well as the extermination, expulsion or enslavement of most of the Slavic peoples and others regarded as "racially inferior".

the Bask village of Guernica (1937), and the 78,000 Italian Corps of Volunteers supporting nationalist troops— this became the 'dress rehearsal' for next World War.

In the immediate aftermath of the dissolution of the Habsburg Monarchy —with Austria left as a broken remnant, deprived of most of the territories it ruled for centuries and undergoing a severe economic crisis— the idea of unity with Germany seemed attractive also to many citizens of the political left and centre. It fitted the German 'Heim ins Reich' doctrine and in 1938, after much Austrian Nazi-political manoeuvring cultivating pro-unification tendencies, Austria was annexed into Germany —the 'Anschluß'— creating Greater Germany.

Also that year, Hitler pressed for the Sudetenland's[148] incorporation into Germany and the Czech part of Czechoslovakia was subsequently invaded by Germany in March 1939 (Figure 48).[149]

The annexation of the Sudetenland by Germany was, to a large degree, prepared by the Sudeten Germans. During the Great Depression these mainly mountainous regions populated by ethnic Germans were hurt by the economic crisis more than the interior parts of Czechoslovakia. High unemployment made the people more

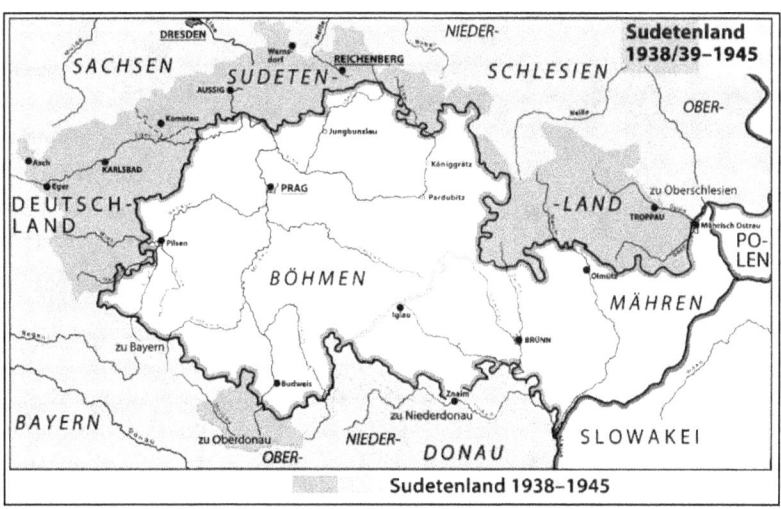

**Figure 49: German Occupation of Sudetenland (1938).**
Source: Unknown. Wikimedia Commons.

---

[148] Sudetenland are the native German-speaking regions of Czecho-Slovakia in the 1930s.
[149] On February 24th, 2022 Russia invaded the Ukraine region. The situation in and around Ukraine was increasingly reminiscent of the Sudetenland Crisis in 1938.

open to populist and extremist movements, and the parties of the German nationalists gained immense popularity among the ethnic Germans within Czechoslovakia. And their propaganda became anti-Semitic and anti-Czech.

In the parliamentary elections of May 1935, the Sudeten German Party (ie Nazi-party) received almost two-thirds of the Sudeten German vote and sent the second largest bloc of representatives to the Czechoslovak Parliament. Afterward, the Sudeten Nazis increased their activities, which were basically aimed at uniting the Sudetenland with Germany and included hostile outbreaks and provocative incidents. The Sudetenland affair would be an excuse for Hitler to the takeover of the weakened Czechoslovakia after the invasion of Poland.

**Figure 50: The Free City of Dantzig (1938).**

Source: Unknown. Wikimedia Commons.

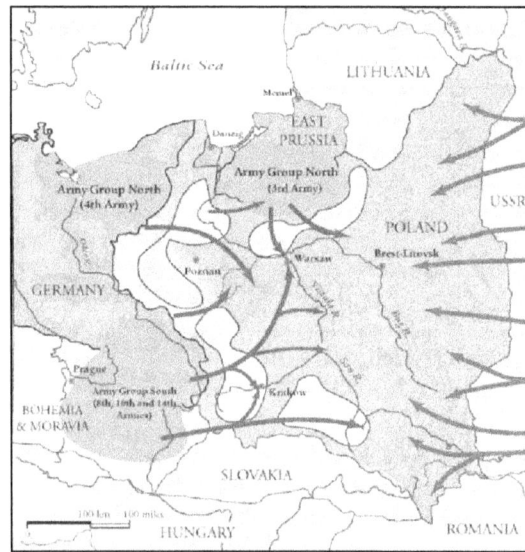

**Figure 51: German and Russian Invasion of Poland (September 1939).**

Source: Unknown. Wikimedia Commons.

The next natural target for German expansion was Poland, so Hitler claimed the free city of Danzig, (Figure 50) and Germany invaded Poland on September 1st, 1939. On September 2nd, 1939 Germany officially annexed the Free City of Danzig (Figure 51). [150] Subsequently,

---

[150] The Baltic Sea coast that had been part of Germany from the early 1800's, however after the First World War the free city of Dantzig (aka Gdansk) was created by the Allied powers in the Treaty of Versailles. The Polish Corridor separated both Dantzig and East Prussia

France and Britain declared war on September 3rd, 1939. The Second World War had started.

By September 8th the Germans had already reached the suburbs of Warsaw. As also the Russian Soviets invaded Eastern Poland on September 19th, Poland was defeated. Germany annexed the western and occupied the central part of Poland, and the Soviet Union annexed its eastern part. Thus, began for Poland a nightmarish occupation that would last more than five years and determined by the Nazi-policy that classified the Polish as racially inferior. Their occupation was disastrous for the Polish population as about 5.6 million Polish citizens died as a result of the German occupation and about 150,000 people died as a result of the Soviet occupation.

## *The Context of the Second World War (1940-1945)*

During the Second World War in a quest for race and space, Germany fought battles on different fronts; the Western Font, the Eastern front and

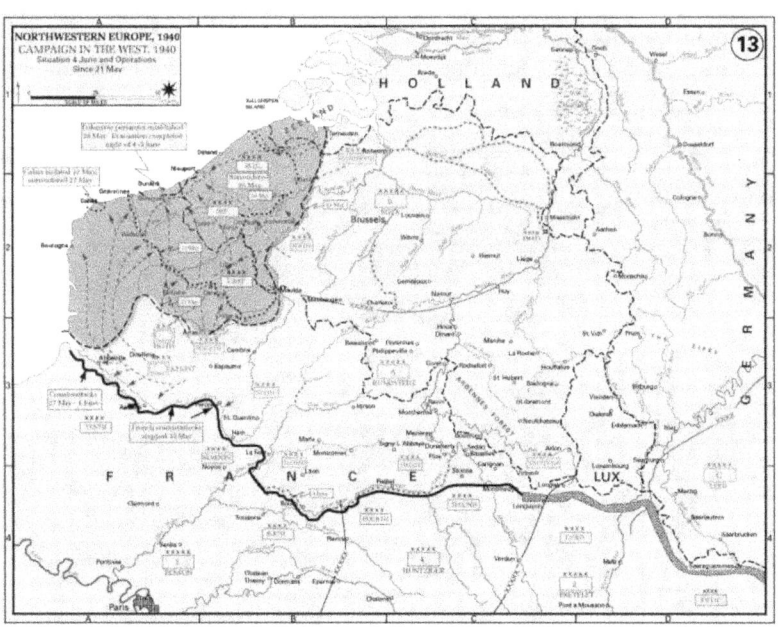

**Figure 52: The Situation on June 4th during the Blitzkrieg on the Western Front (1940).**
Source: Unknown. Wikimedia Commons.

---

from Germany. It was populated for ninety percent by Germans and a Polish minority. The German incorporation of Danzig was a territorial claim that every government of the Weimar Republic put on its agenda.

the South-Eastern Front. Although they had destruction in common, each had its own timeframe and characteristics.

## The Opening Gambit of the Second World War (1940-1941)

The *Western Front* (Figure 52) saw a *Blitzkrieg* when in April 1940, Germany invaded Denmark and Norway. In May it launched an offensive against France, circumventing the Maginot Line of fortresses (Figure 61) by attacking the neutral nations of Belgium, the Netherlands, and Luxembourg. After the Luftwaffe bombed the harbour city Rotterdam to destruction on May 10th, the Dutch surrendered on May 14th and by May 17th the whole country was occupied. Next, in the *Battle of France*, on May 10th, Belgium and Luxemburg were invaded. The Wehrmacht rapidly advanced to the English Channel through the Ardennes (ie the 'Ardennes Offensive'), and cut off the Allied forces in Belgium, trapping the Allied armies in a cauldron on the Franco-Belgian border near Lille during the German Race to the Sea. By May 28th, 1940 the Belgium Army surrendered.

Belgian, British and French troops were trapped and surrounded by German troops after the six-week in the Battle of France (May-June 1940). They were pushed back to the sea by the German armies.

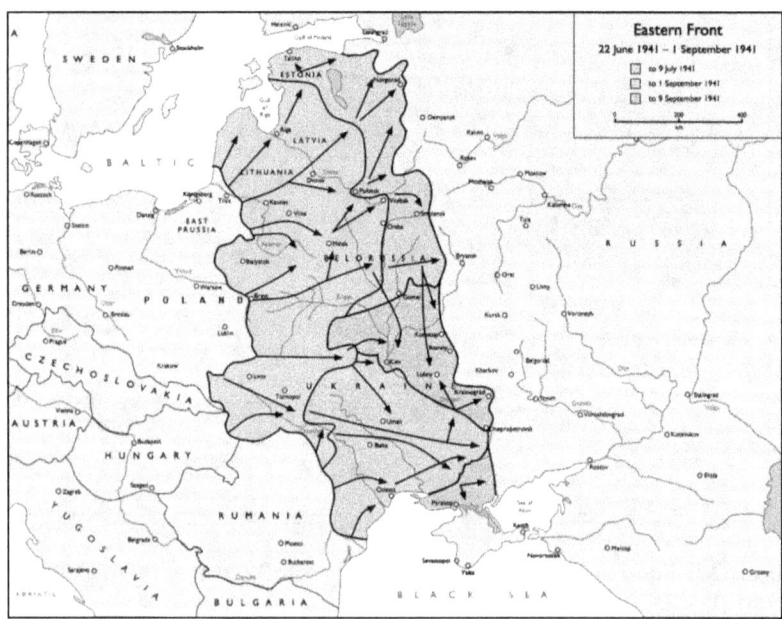

Figure 53: The Eastern Font (June-September 1941) at the beginning of Operation Barbarossa.

Source: Unknown. Wikimedia Commons.

Then during 'Operation Dynamo,' the British Expeditionary Force, French and Belgian troops were evacuated from the beaches of Dunkirk by a fleet of 800 vessels. Among the many vessels of Navy and Merchant Navy ships (destroyers, hospitals ships, ferries, etc.), many fishing boats, pleasure craft, yachts, and lifeboats participated. Some 340,000 man were saved.

Next the Germans turned south against the weakened French army, and Paris fell to them on June 14th. Nearly forty years after their victory over Paris in 1871, they marched again over the Champs Elysée along the Arc de Triomphe. Eight days later France signed an armistice with Germany in Compiegne. It was the start of the military German occupation of France. And the Germans prepared to cross the Channel and invade Britain — 'Operation Sea Lion'— by air raids of the Luftwaffe. The ensuing Battle of Britain (July-October 1940) saw massive destruction London and other cities, but ended as the first defeat for the German military forces.

The *Eastern Front* (Figure 53) saw the largest military confrontation in history in which the arch enemies Germany and Russia fought each other. The battles on the Eastern Front were characterised by unprecedented ferocity, wholesale destruction, mass deportations, and immense loss of life due to combat, starvation, exposure, disease, and massacres. It started when, following the German-Soviet non-aggression pact, Poland was invaded by both the Nazi Germany on September 1st 1939 and by the Soviet Union on September 17th, 1939. The campaigns ended in early October with Germany and the Soviet Union —as a result of the Ribbentrop-Molotov Pact, the 1939 non-aggression treaty between Nazi Germany and the Soviet Union— dividing and annexing Poland.

However, the cooperation between Russia and Germany did not last, as by early 1941 the Germans turned their war effort to South-East Europe where Italy had already started to expand its sphere of influence. And they started their 'Operation Barbarossa' —the military plans for the invasion of Russia— with the 'Operation Typhoon.' It resulted in the Siege of Leningrad in September 1941, and continued with the Battle for Moscow in October 1941. The Siege of Leningrad became one of the longest and most destructive sieges in history.

The Battle of Moscow would become the first time since June 1941, that Soviet forces, supported by the fortification of the Stalin Line near Kiev (Ukrainia) had stopped the Germans and driven them back. The Blitzkrieg stalled, even more when the Russian bombed the oil refineries in the area, disrupting petrol supply of the German Army. Hitler decided not to proceed to the nearby Moscow but focus his armies on the South. in the Battle of Sevastopol on the Crimean Island (October 1941-July 1942).

The *South-Eastern Front* (Figure 54) focussed on the Mediterranean and Middle East theatre. It was also the area where the Italians operated. The fighting in this theatre lasted from June 10th 1940, when Italy entered the war on the side of Germany, until May 2nd, 1945 when all Axis forces in Italy surrendered.

The Italian involvement started when Italy tried to invade Southern France during the *Battle of the Alps* (June 10th-25th, 1940) while the German were busy invading France in the North. Here the Alpine line of

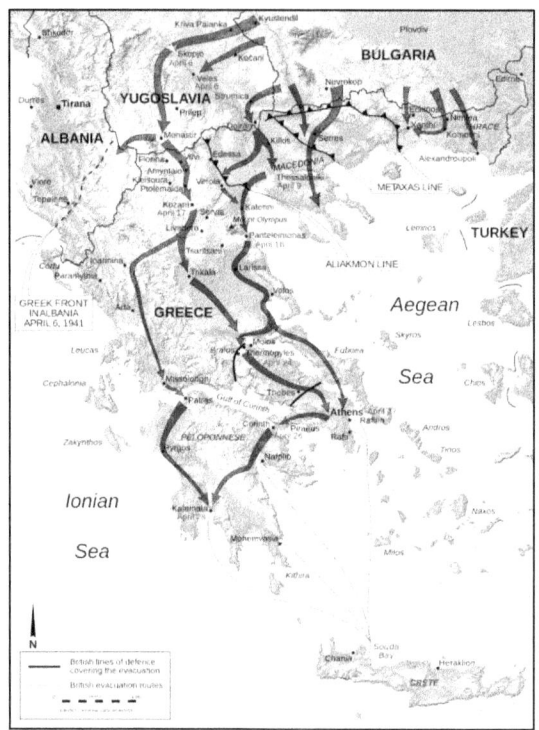

**Figure 54: The South-Eastern Front.**
Source: Wikimedia Commons.

defensive installation —similar to the Maginot Line in the North and known as the Little Maginot Line— proved (together with the Alp Mountains) a too formidable barrier for the Italians. The Franco-Italian Armistice of June 25th ended the brief invasion.

Next Italy invaded Greece from Albania on October 28th, 1940 in the Greco-Italian Campaign (1940-1941) starting the *Battle of Greece*. It was a fiasco as the badly organized Italian Army was defeated by the Greek. It took the help by German forces in the Marita-Campaign starting from Bulgaria, to conquer Greece in April 1941 (Figure 54). With confidence high from early gains, German forces planned elaborate attacks to be launched to capture the Middle East and then to possibly attack the southern border of the Soviet Union.

Italia's expansionist aspirations engaged in a colonial war by 1935, occupied Ethiopia and the East African region of Somaliland and Eritrea in 1936. But that did not last long as by 1941 the British East African Campaign (June 1940-November 1941) brought the region under Allied control. Italian defeats prompted Germany to deploy an expeditionary force to North Africa and at the end of March 1941, and General Rommel's Afrika Korps launched an offensive which drove back the Commonwealth forces. In under a month, Axis forces advanced to western Egypt and besieged the port of Tobruk for eight months. By June 21st, 1942, Rommel captured it. For the British Empire, the surrender after the Fall of Tobruk equalled the Fall of Singapore, and started the crisis in the Mediterranean.

This was the Axis opening gambit of the Second World War that set Europe on fire, the German armies fighting on three fronts and Italy fighting in the South. Not much later, the European War would become a World War that also involved non-Western states such as Japan.

For the British it was more than defending the home (is)land. From September 1939 to mid-1942, the UK led Allied efforts in multiple global military theatres. Commonwealth, Colonial and Imperial Indian forces, totalling close to 15 million serving men and women[151], fought the German, Italian, Japanese and other Axis armies, air-forces and navies across Europe, Africa, Asia, and in the Mediterranean Sea and the Atlantic, Indian, Pacific and Arctic Oceans. Britain became the nucleus of the Allied war-effort in Western Europe, while the US —although they (financially) sympathised to the British cause— kept aloof till they were provoked by the Japanese attack of Pearl Harbour.

## The Opening Gambit of the Pacific War

Also in the Far-East of Asia the military in the Japanese Empire had converted their expansionist aspirations in military actions after its invasion of China in 1937, the occupation of northern French Indochina by 1940 and the invasion of Thailand in 1941. Then, the Japanese surprise attack on Pearl Harbour on December 7th, 1941 (Figure 55).[152] brought the US into

---

[151] The colonial contribution in terms of soldiers numbered 2.5 million men from India, over 1 million from Canada, just under 1 million from Australia, 410,000 from South Africa, and 215,000 from New Zealand.
[152] The attack was part of a range of surprise attacks; on the Philippines, Guam, and Wake Island and on the British Empire in Malaya, Singapore, and Hong Kong. For Pearl Harbour, 6 regular aircraft carriers (all that the Japanese Navy then had), 2 battleships, 3 cruisers, and 11 destroyers were allocated. Since surprise was of the essence, a Sunday, December 7, was chosen as the date for the attack. The attack started with two waves of 360 aircraft, targeting the eight US Navy battleships present, all were damaged, with four sunk. Also three cruisers,

the war with a Declaration of War the next day. But it took a while before the US could initiate its Pacific Campaigns as Europe became the first priority of the American war effort. In the meantime, the Japanese continued their expansion with the rapid occupation of the Malay Peninsula, Singapore and the oil-rich British colonies in February 1942 and the Dutch colonies in March 1942.

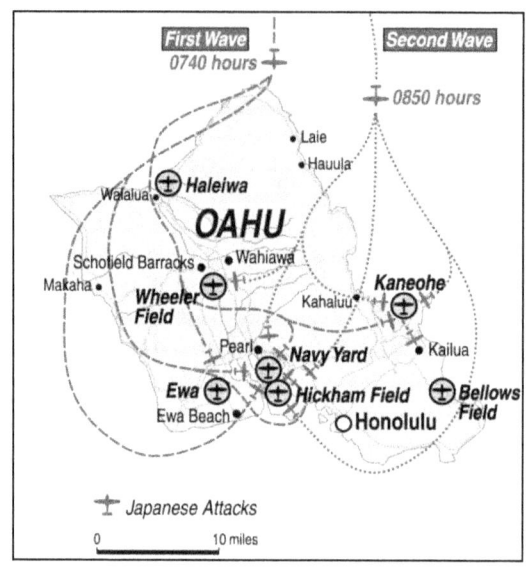

**Figure 55: The Japanese Attack on Pearl Harbour.**
Source: Warfare History Network

Japan's main confrontation was with China. The Second Sino-Japanese War (1937-1945) between the Empire of Japan and the Republic of China had been in progress since July 7th, 1937. Japan's control of Manchuria was the start, followed by further expansion to the North. However, in the *Battle of Khalkin Gol* (June, 1939) they were halted at China's border with the Soviet territory in the Far East (Figure 56). With the escalation of the European Eastern Front in 1941, Japan and the Soviet Union signed a Neutrality Pact. The Japanese *Hokushin Doctrine* of expansion into the North being halted by the Soviet, Japan started their Southward expansion under the *Nanshin-ron* doctrine (the Pacific Islands and South East Asia were considered to be a historic part of their sphere of interest).[153]

Japan had joined the Axis Powers of Germany and Italy, by signing the *Tripartite Pact* on September 27th, 1940. Together with the *Anti-Comintern Pact* (November 1936) between Japan and Germany, and the *Pact of Steel*

---

three destroyers and 188 aircraft were destroyed. In total, 2,403 Americans were killed, and 1,143 were wounded,
[153] Resource-rich areas of Southeast Asia were earmarked to provide raw materials for Japan's industry, and the Pacific Ocean was to become a 'Japanese lake.'

(May 1939) between Germany and Italy, it governed their wartime relation. Japan has chosen the Axis side and was by now in conflict with the US and Great Britain, and the Imperial Japanese Navy and Army started surprise attacks on American territories (Hawaii, the Philippines, Guam) and British territories (Malaya, Singapore, Hong Kong). Their opening gambit was the attack on Pearl Harbour on December 7th, 1941. It was the start of the Japanese Navy's 'Blitzkrieg' across the Pacific.

After Pearl Harbour the Japanese campaigns in South East Asia progressed rapidly,[154] and in January, Japan invaded British Burma, British Borneo, the Dutch East Indies, New Guinea, the Solomon Islands and captured Manila, Kuala Lumpur and Rabaul.

Figure 56: Japanese Control over China.
Source: Encyclopaedia Brittanica.

---

[154] Panic, hysteria, and fear, swept across America as the war continued to expand. Roosevelt's War Relocation Authority was established to administer the evacuation and internment of over 100,000 Japanese emigrants and American citizens of Japanese heritage.

The Attack on Singapore (February 8-15th, 1942) was the result of their next wave of attacks that occurred virtually simultaneous with the Japanese attack on Pearl Harbour. With a land force of 30,000 troops fighting down the Malayan Peninsula attacking Singapore the landside, the air campaign of long-range Japanese bombers, and the Japanese Imperial Navy engaging the British fleet form the seaside, they overwhelmed the British. By February 15th, about a million civilians in the city had fled and that afternoon the Brits capitulated. About 80,000 British, Indian and Australian troops in Singapore became prisoners of war. The capture of Singapore resulted in the largest surrender in British history.

With the fall of Singapore, the islands of the Dutch East Indies were the next target of the Japanese military campaigns. Their natural resources from rich rubber plantations and oil wells being the focus of Japanese interests. Starting with a campaign at Borneo on December 17th, followed by the *Battle of Balikpakan* (January 23-25th, 1942), they captured Celebes, Amboina and Timor, and by February 1942 they occupied most of the Dutch East Indies islands (Figure 57).

The *Battle of Bataan* (January 7th-April 9th, 1942) resulted in the largest military surrender the US had experienced since the American Civil War. Next, the *Battle of the Java Sea* (February 27th, 1942) where

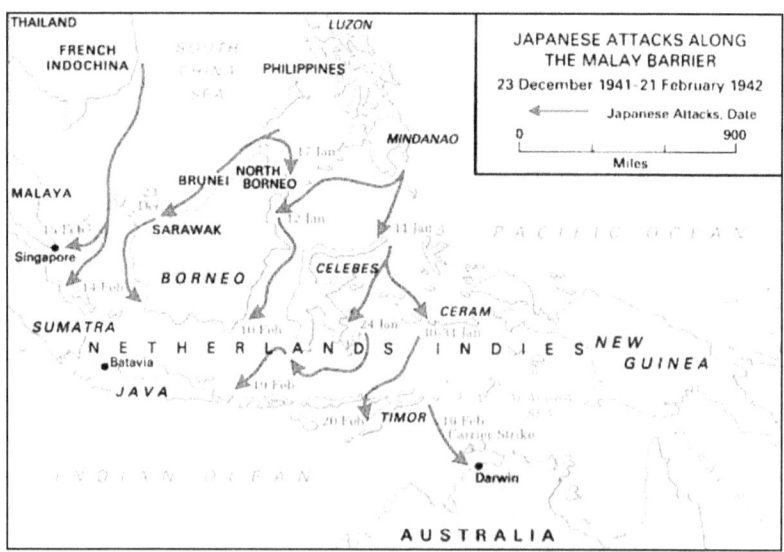

Figure 57: Japanese Invasion of British and Dutch Colonies.
Source: Unknown. Wikimedia Commons.

the Japanese defeated the Allied naval forces under the Dutch admiral Kael Doorman, cemented the Japanese victory. By May 1942 they also had invaded the Philippine Islands and, after conquering Australia Territory of Papua, threating to invade Australia itself.

However, the Sea *Battle at the Coral Sea* (May 4-8th, 1942) and the *Battle of Midway* (June 4-7th, 1942) between the Imperial Japanese Navy and the naval and air forces of the US and Australia, checked their progress and became a turning point in the Pacific War. Especially after the Japanes defeat during the clash of the carriers at the Battle at Midway,[155] the Allies switched to the strategic initiative[156] paving the way for the landings on Guadalcanal and the prolonged attrition of the Solomon Islands campaign. The Battle marked the height of the Japanese expansion as they were prompted to cancel their plans to invade New Caledonia, Fiji, and Samoa and lost all but the last vestiges of their earlier strategic initiative (Figure 58).

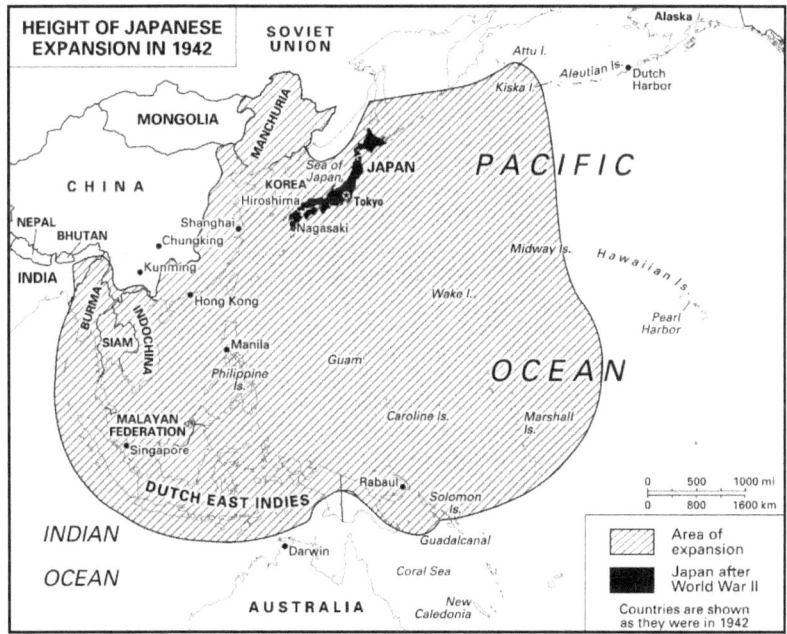

Figure 58: Japanese South-Eastern Expansion (1942).
Source: Encyclopaedia Brittanica.

---

[155] The Japanese lost four large fleet carriers (the Akagi, Kaga, and Soryu in battle and the Hiruya some days later) and two heavy cruisers. The US lost the carrier Yorktown.
[156] This was the dawn of a years-long series of island-hopping invasions and several even larger naval battles.

## Times of German Occupation of Europe (1941-1944)

After the Blitzkrieg, the Germans occupied most of Europe and established civil and military administration. The Schutzstaffel (aka SS), the elite corps of the Nazi Party, possessed exceptional powers throughout German-dominated Europe and in the course of time came to perform more and more executive functions, even in those countries under military administration. Compulsory enrolment of foreign workers into the German armaments industry were soon applied to German-dominated Europe and ultimately turned 7,500,000 people into forced or slave labourers.

In addition, upon the German conquest of Poland in 1939, Hitler ordered the SS to kill a large proportion of the Polish intelligentsia. A reign of terror against the nationalistic-minded Polish ruling classes began, and by the war's end 3,000,000 Poles (in addition to 3,000,000 Polish Jews) had been killed.

The SS started to implement their program of the *Final Solution* with the start of the invasion of the Soviet Union in 1941, and special mobile killing squads began systematically killing the Jewish population on conquered Soviet territory in the rear of the advancing German armies. In a few months, up to the end of 1941, they killed about 1,400,000 people. In the

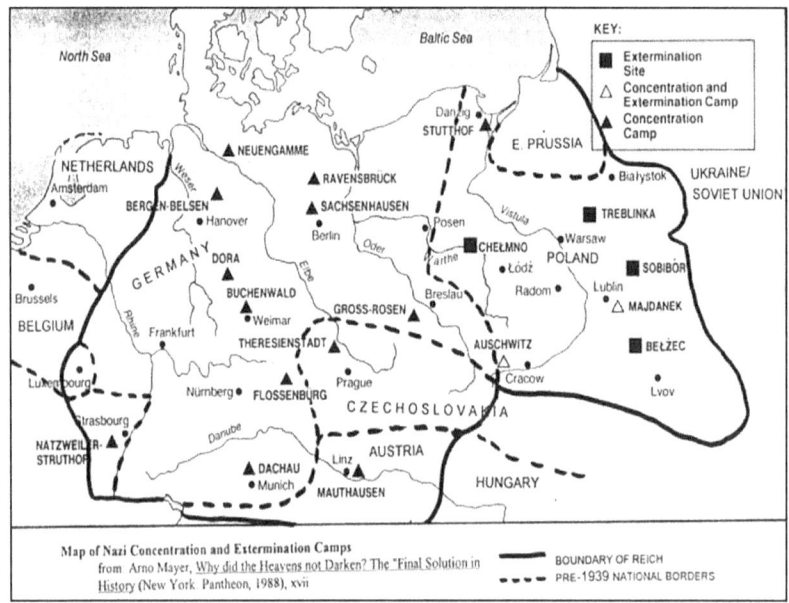

**Figure 59: Concentration Camps in Germany.**
Source: https://marcuse.faculty.history.ucsb.edu/classes/33d/33d05/33d05L05Camps.htm

period from May 1942 to September 1944 more than 4,200,000 Jews were killed in extermination camps such as Auschwitz, Treblinka, Belzec, Chełmno, Majdanek, and Sobibor (Figure 59). About 5,700,000 Jews died in the course of the Final Solution. [157]

These inhumane occupation policies were practiced to a greater or lesser extent in all the countries occupied by the Germans, and the result was the beginning in 1940–41 of armed, underground resistance movements in those countries. The German occupation authorities' attempts to eradicate the Resistance in most cases merely fanned the flames, due to the Germans' use of indiscriminate reprisals against civilians.

Also in other ways the occupied countries were disrupted under German occupation. Such as the many soldiers kept as prisoners of war (POW) and used in forced labour. Other civilians were obliged to work in German agriculture and industry. And there was the collaboration of their own people with the Axis powers; by joining the German army and by the local pro-Nazi political parties taking part in ruling the occupied country, and by people joining the German army units as volunteers.

> About 45,000 Norwegian collaborators joined the fascist party Nasjonal Samling (National Union). The Dutch Nationaal-Socialistische Beweging (National Socialist Movement) grew to 100,000 members, and played a role in lower government. In France, one of the conditions of the armistice was that the French pay for their own occupation by a 300,000 men army (ie a re-enactment of Napoleon's occupation of Germany a century before). With an overpriced artificial exchange rate, the 400 million French francs payable per day, created a system of organized plunder. Add to that the more than 600,000 men sent to Germany as part of the 'Service du Travail Obligatoire' (ie forced labour), and the more than two million prisoners of war that were sent to Germany.

Less visible was the organized economic plunder —such as the confiscation of Jewish possessions in the early war— and the systematic organized art plunder on a broad scale.[158]

---

[157] The "Final Solution to the Jewish question" was the official code name for the murder of all Jews within reach, which was not restricted to the European continent. This policy of deliberate and systematic genocide starting across German-occupied Europe was formulated in procedural and geopolitical terms by Nazi leadership in January 1942 at the Wannsee Conference held near Berlin, and culminated in the Holocaust, which saw the killing of 90% of Polish Jews, and two-thirds of the Jewish population of Europe.
[158] Plundering occurred from 1933, beginning with the seizure of property of German Jews, until the end of the war, particularly by military units known as the Kunstschutz, although most plunder was acquired during the war. In addition to gold, silver and currency, cultural

## Times of Japanese Occupation in the Pacific (1942-1944)

The expansion doctrines of Japan resulted in the Japanese Empire at its apex in 1943 being one of the largest empires in history. It accounted for more than 20% of the world's population at the time with 463 million people in its occupied regions and territories. This was the outcome of a long procession of colonisations —enhanced by the Second World War—, of the Japanese Empire that originated after the Meiji Restauration in 1868.

The Island of Taiwan, formerly known as Formosa, —already under Japanese rule after the First Sino-Japanese War (1894-1895) and the first step in implementing their 'Southern Expansion Doctrine' of the late nineteenth century— in order to become a model colony, saw the modernization of the island's economy, infrastructure, industry, public works, and cultural Japanization (ie assimilation of its population with Japanese culture). With the Colonial Government as the primary driving force, as well as new immigrants from the Japanese Home Islands, Taiwanese society was sharply divided between the rulers and the ruled. By the 1920s modern infrastructure and amenities had become widespread, although they remained under strict government control. Then the rise of militarism and the Kempeity[159] —sometimes compared to the rise of European fascism— started the Second Sino-Japanese War (1937-1945), Japanese rule in Taiwan was reinvigorated and Japan sought to use resources and material from Taiwan for their war effort.

Korea, already under Japanese rule since 1910, changed from an agricultural society into a more industrialized country during the 35 years of Japanese colonial rule. Rapid urban growth, the expansion of commerce, and forms of mass culture such as radio and cinema, which became widespread for the first time. Industrial development also took place, partly encouraged by the Japanese colonial state, although primarily for the purposes of enriching Japan and fighting the wars in China and the Pacific rather than to benefit the Koreans themselves. By the time of the Japanese surrender in August 1945, Korea was the second-most industrialized nation in Asia after Japan itself.

---

items of great significance were looted, including paintings, ceramics, books and religious treasures. Their destination was Germany, in particular the Nazi-kingpins and the Führermuseum planned in the city of Linz. The Allied recovery team of the Monuments Men found a number of hiding places for looted art, including the famous salt mine in Altaussee, in the Austrian Alps. It contained some 12,000 stolen artworks.

[159] The Kempeitai was the military police arm of the Imperial Japanese Army from 1881 to 1945. It was both conventional military police and a secret police force. In Japanese-occupied territories, the Kempeitai also arrested or killed those who were suspected of being anti-Japanese.

After the Fall of Singapore February 15th, 1942, the 'Strait Settlements' British Malaya and Borneo, and the colonies of the Dutch East Indies, were swiftly overrun and occupied. Once the Japanese had taken Malaya and Singapore, their attention turned to consolidating their position. Under a military administration the Japanese brought in Japanese companies that got monopolies for goods essential to Japan; food priority was given to the Japanese troops. Together with the rest of Southeast Asia, Japan exploited Borneo as a source of raw materials and natural resources. And the large part of the population with a Chinese background were treated harsh. The members of the Chinese population perceived as threats were rounded up and executed during the *Sook Ching Massacres* (January - February 1942). Malay and Chinese girls and women were taken to serve as comfort women in brothels. And the Chinese Community was forced to raise $50 million as atonement for its support of the Japanese war effort.[160]

The occupied Dutch East Indies colonies underwent a similar fate, welcomed by a local population sympathetic to the Japanese liberation from their Dutch colonial masters. Together with the destruction of the Dutch colonial regime, the Japanese facilitated Indonesian nationalism during their efforts to create a puppet state. The Indonesian ruling class (composed of local officials and politicians who had formerly worked for the Dutch colonial government) co-operated with the Japanese military authorities, who in turn helped to keep the local political elites in power and employ them to supply newly arrived Japanese industrial concerns, businesses and the armed forces.

> The Japanese had erected major prisoner of war (POW) camps at the regions they occupied, some of which held both military and civilian prisoners. For detaining the Dutch governmental and corporate administrators they build detention camps and Dutch nationals were rounded up by Japanese soldiers and put in internment camps.[161] In addition to the 100,000 European (and some Chinese) civilians interned, 80,000 Dutch, British, Australian, and US Allied troops went to prisoner-of-war camps where the death rates were between 13 and 30 percent.

---

[160] Chinese sources indicated that some 97,000 suspected anti-Japanese Chinese had been imprisoned or killed by the Japanese in Singapore and Malaya. The number of comfort women—sex slaves— in the territory occupied by the Japanese numbered 360,000 to 410,000, among whom the Chinese were the largest group, about 200,000.
[161] One a personal note, the parents and ten-year old sister of your author were detained in Japanese detention camps. They survived the camps and were repatriated to Europe.

The local sentiment changed, however, as between 4 and 10 million Indonesians were recruited as forced labourers (aka 'Romusha') working for economic development and defence projects in Java. Between 200,000 and half a million were sent away from Java to the outer islands, and as far as Burma and Siam to build railways. In addition, to regulate raping by the Japanese soldiers, many of the occupied female population had to serve in brothels, as 'comfort women' (est. 50,000-200,000) Among which many Dutch women and girls.

## Turning the Tables in Europe (1943-1945)

By 1943 the Axis-force in Europe were facing a different situation. Germany and Italy were defeated in North Africa and then, decisively, the Germans were stopped at Stalingrad in the Soviet Union. Key setbacks in 1943 —which included a series of German defeats on the Eastern Front, the Allied invasions of Sicily and Italy, and Allied victories in the Pacific— cost the Axis its initiative and forced it into strategic retreat on all fronts. In addition, by June 1943 the British and Americans began a strategic bombing campaign against Germany with a goal to disrupt the war economy, reduce morale, and 'de-house' the civilian population.

On the western front, on June 6th, 1944 (known as D-Day), the Western Allies invaded the coast of Normandy in northern France during the 'Operation Overlord' (Figure 60). After a massive preparation —1.5 million soldiers and their material were brought to Britain, including the deception of Germans of the exact landing area— nearly 160,000 troops crossed the English Channel on June 6th, to land on the beaches of Normandy. In the subsequent Normandy Campaign, they fought their way from the beaches, and more than two million Allied troops were in France by the end of August. By August 25th, 1944, Paris was liberated and the Allies continued to push back German forces in

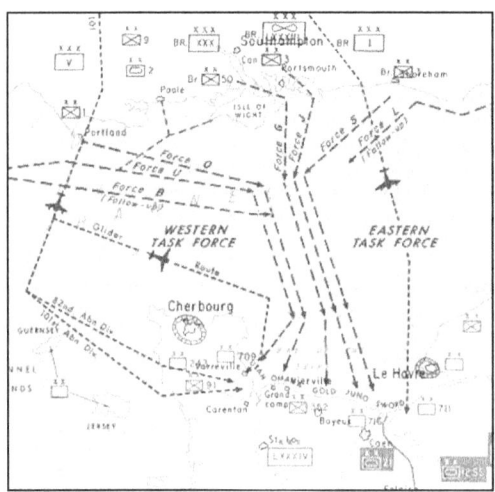

**Figure 60: Operation Overlord on D-Day June 6th, 1944.**

Source: Unknown. Wikimedia Commons.

western Europe during the latter part of the year.

During the Siegfried Line Campaign, the allied forces moved up to the natural barrier of the River Rhine, combined with a defensive line of some 18,000 bunkers, tunnels and tank traps stretching over 630 km built during 1938-1940 facing the French Maginot Line; the Siegfried Line (aka Westwall) (Figure 61).

In August 1944 the first clashes took place along the northern part of the Siegfried line. On December 16th, 1944, Germany made a last attempt on the Western Front by using most of its remaining reserves to launch a massive counter-offensive in the Ardennes in the Battle of the Bulge, and the 'Operation Nordwind' (December 1944-January 1945) in Rhineland-Palatinate, Alsace and Lorraine. To no avail, as both operations failed and the Western Allies crossed the River Rhine in March 1945. Once the Allies had crossed the Rhine, the British fanned out northeast towards Hamburg crossing the river Elbe and on towards Denmark and the Baltic. The Americans and went south, fanning out into Czechoslovakia, Bavaria and Austria. By May 1945 Allied forces were over-running all of western Germany from the Baltic in the north to Austria in the south. It did not take long before the Germans surrendered on May 8th, 1945.

**Figure 61: The Siegfried Line.**
Source: Wikimedia Commons.

- On the eastern front the failure of 'Operation Barbarossa' reversed the fortunes of the Third Reich. After the Germans had lost by February 1943 the Battle of Stalingrad —under the appalling weather conditions of the Russian winter— the tides turned for the Germans in Eastern-

Europe. On January 27th, 1944, Soviet troops launched a major offensive that expelled German forces from the Leningrad region. By late May 1944, the Soviets had liberated Crimea, largely expelled Axis forces from Ukraine, and made incursions into Romania. On June 22nd, the Soviets launched a strategic offensive in Belarus ('Operation Bagration') that destroyed the German Army almost completely. In September 1944, Soviet troops advanced into Luthania, Poland, Romania and Yugoslavia. By early August the Russians reached the Polish capital of Warsaw. And, in a race with the western allies, in April 1945 they reached Berlin (Figure 62).

So, the war in Europe concluded with an invasion of Germany by the Western Allies and the Soviet Union, culminating in the capture of Berlin by Soviet troops, the suicide of Adolf Hitler, and the German unconditional surrender on May 8th, 1945. Germany came out of the war in ruins after the massive bombing raids targeting industrial areas, but in the process also wiping complete cities for the face of the Earth; such as the fire bombings that had been carried out against Hamburg (1943) and Dresden (1945). The German Empire was reduced to ruins, as were many nations on the European continent. And Germany became occupied by the Allied forces ruling in different zones till 1949.

**Figure 62: The Eastern Front with the Race for Berlin (1945).**
Source: Wikimedia Commons.

## Turning the Tables in the Pacific (1943-1945)

In 1943 the war in the Pacific was coming to a climax. The Japanese were at the apex of their expansion and the Allies were turning their force to the Pacific. However, the 'Europe First' doctrine of the Allies,[162] brought limited, defence-oriented actions in the Pacific. The turning point in the Pacific war came with the Allied[163] victories at the *Battle at the Coral Sea* (May 4th-8th, 1942) and the *Battle of Midway* (June 4th-7th, 1942) (Figure 63).

Midway proved to be the last great naval battle for two years. The US used the ensuing period to turn its vast industrial potential into increased numbers of ships, planes, and trained aircrew.

**Figure 63: Allied Pacific Campaigns (1943-1945).**
Source: NPS.gov.

---

[162] With much of the US fleet destroyed and a nation unprepared for war, America and its allies decided they needed to save Great Britain and defeat Germany first.
[163] The major Allied participants were the US, the British Empire and China. The British supplied the colonial troops and troops from Canada, Australia and New Zealand).

Subsequent fighting in the Pacific theatre consisted of some of the largest naval campaigns in history, and incredibly fierce battles and war crimes across Asia and the Pacific Islands, resulting in immense loss of human life.[164] In strategic terms the Allies began a long movement across the Pacific, seizing one island base after another.

Two strategic targets were chosen to cut off Japan's supply lines to Southeast Asia; the Philippines and Formosa. It resulted in two major campaign routes (Figure 63).

The eastward drive up to Formosa started in June 1943 with Operation Cartwheel to recapture New Guinea. Next the Gilberts and Marshall Island were recaptured in a combined air-, sea- and amphibious-warfare. With the subsequent victories in the *Marianas Campaign* (on the islands of Saipan, on Guam, and on Tinian, during June and July 1944), American forces were getting close to Japan itself. From the Marianas, the very long-range B-29 Superfortress heavy bombers started to bomb the Japanese home islands from well-supplied air bases; the ones with direct access to supplies via cargo ships and tankers. On March 29th, 1945 the B-29 fire-bomb attack on Tokyo leaved much of the city in ashes, killed an estimated 100,000 people, and inaugurated a series of incendiary strikes against other Japanese cities. By mid-June, Japan's six largest cities had been devastated, as well as a hundred smaller towns and cities.

In the Battle of the *Iwo Jima Bay* (February 19th–March 26th, 1945) the five-week battle saw some of the fiercest and bloodiest fighting of the Pacific War. From the strongly fortified positions, the 21,000 Japanese solders nearly all fought to death. Then came the *Battle of Okinawa* (April 1st–June 22nd, 1945) with a large amphibious assault. After a long campaign of island hopping, the Allies were planning to use Kadena Air Base on the large island of Okinawa as a base for Operation Downfall, the planned invasion of the Japanese home islands, 340 mi (550 km) away. It was supposed to be starting with the invasion of the Southern Island of Kyūshū on X-Day planned to be on November 1st, 1945.

Already having experienced the strategic bombing raids on the homelands since 1944,[165] for their defence the mainland the Japanese organized their last resources; from kamikaze pilots and suicide divers, to volunteer conscripts. The deeply ingrained Japanese code of Bushido (ie the

---

[164] Estimates for the Allied casualties mount up to 4 million+ military and 26 million+ civilians, and for the Axis 2,5+ million military and 1 million+ civilian deaths.

[165] During the air raids some estimated that 160 tons were dropped on the large cities like Tokyo, Osaka, Kobe killing between 241,000 and 900,000 people, wounding between 213,000 and 1.3 million, and rendering some 8 million homeless. The most commonly cited estimate of Japanese casualties is 333,000 killed and 473,000 wounded.

way of the warrior) would prohibit surrender and made them willing to fight to death.

The westward drive started in June 1943 with Operation Cartwheel to recapture New Guinea and the islands up to the Philippines (Figure 63). After the Philippines were liberated, they would form the main base for the final assault against the Japanese Homeland. So, after the recapturing of the Green Islands, Solomon Islands and the Admiralty Islands the Allied force steamed up to the Leyte Gulf. There they engaged in the *Battle of the Leyte Gulf* from October 23rd-26th, 1944; the largest naval battle of the Second World War. [166] This was the first battle in which Japanese aircraft carried out organized kamikaze attacks, and the last naval battle between battleships in history.

It was complemented by a range of battles —such as the *Battle of Ormoc* (November 11th-December 21th, 1944)— on several islands where a vast armada of battleships, aircraft carriers,[167] cruisers, and destroyers pounded the area with shells and bombs before the landing and during its early stages. The Japanese suffered irreparable defeat in the Leyte campaign with the Japanese army losing four divisions and several separate combat units, while the Japanese navy lost 26 major warships and 46 large transports and hundreds of merchant ships. The United States lost one light carrier, two escort carriers, and several other vessels. The campaign for Leyte proved the first and most decisive operation in the American reconquest of the Philippines. The stage was set and the Allies were ready. In the Philippines, the people were eager to assist in the long-awaited battle for their liberation after the harsh Japanese occupation.

The next strategic objective was the main island of the Philippines. By mid-1944, American forces were only 300 nautical miles (560 km) southeast of Mindana —the largest island in the southern Philippines— and able to bomb Japanese positions using long-range bombers. The campaign for Mindanao started on March 10th, 1945 and on August 15th, 1945 the island was recaptured and the route to Manila opened up. The *Battle of Manila* (February 3rd–March 3rd, 1945) resulting in the death of over 100,000 civilians and the complete devastation of the city, was the scene of the worst urban fighting in the Pacific theatre. Japanese forces

---

[166] Among which the Musashi, a battle ship of the Yamato-class, the heaviest and most powerfully armed battleships ever constructed. During the Battle of Leyte Gulf, the Musashi was sunk by an estimated 19 torpedo and 17 bomb hits from American carrier-based aircraft on 24 October 24th, 1944.
[167] The American fleet aircraft carriers played a major role in winning decisive naval battles. These carriers, typically with thirty to ninety aircraft, tended to form the core around which naval striking task forces were assembled.

committed mass murder against Filipino civilians during the battle. The battle left 1,010 U.S. soldiers dead and 5,565 wounded. An estimated 100,000 to 240,000 Filipinos civilians were killed.

As the Allies advanced towards Japan, conditions became steadily worse for the Japanese people at home. Japan's merchant fleet declined from 5,250,000 gross tons in 1941 to 1,560,000 tons in March 1945, and 557,000 tons in August 1945. Lack of raw materials forced the Japanese war economy into a steep decline after the middle of 1944. The civilian economy, which had slowly deteriorated throughout the war, reached disastrous levels by the middle of 1945.

Then, the Allied prepared to invade the Japanese mainland with the Operation Downfall, the final assault on Japan's mainland. It would become a major operation involving 6 million Allied troops against some 4 million Japanese soldiers and 31 million civilian conscripts defending their homeland. The Japanese planned an all-out defence of Kyūshū, with little left in reserve for any subsequent defence operations. But before 'X-Day' was to happen, atomic bombs were dropped on Hiroshima (August 6th, 1945) and Nagasaki (August 9th, 1945) (Figure 64).

The US had developed a new devasting weapon in the Manhattan project; the nuclear weapons known as atomic bomb. Two of those bombs were used as the Airforce deployed on August 6th the B-29 bomber *Enola Gray* from the Island of Tinian to Hiroshima, followed by the B-29 bomber *Bockscar* on August 9th to Nagasaki. The result was devastating. In Hiroshima some 70,000–80,000 people —around 30 percent of the population of Hiroshima at the time— were killed by the blast and resultant firestorm. The same fate killed 20,000 soldiers, and another 70,000

**Figure 64: The Atomic Bomb Missions (August 1945).**

Source: Unnown. Wikimedia Commons.

civilians were injured. In Nagasaki at least 35,000–40,000 people were killed and 60,000 others injured. The radius of total destruction was about 1 mi (1.6 km).

It took the Japanese rulers —the military and emperor— some days of intense negotiations, an attempted coup d'état, and a massive additional bombing raid of 400 B-29 bombers with conventional bombs, but on August 15th, the Emperor announced his decision to surrender that was broadcasted to the public. The formal surrender of Japan ceremony took place aboard the battleship USS Missouri in Tokyo Bay on September 2nd, 1945.

After the war, Japan lost all rights and titles to its former possessions in Asia and the Pacific, and its sovereignty was limited to the four main home islands and other minor islands as determined by the Allies. The Japanese Empire was reduced to ruins, as were many nations in the Pacific. And it was the start of the decolonization of Asia.

*Demobilisation and Displaced persons*

The end of the war became a time with massive movement of people: former prisoner of war (POW), displaced persons, refugees and demobilized forces. One of them being the Allied military not immediately engaged in occupation or stabilization duties in Japan, but send home or to other post-war conflict areas (eg Korea).

The demobilization of troops was one of the first great afterwar challenges. The British demobilized a million servicemen, among those the Commonwealth troops in the Pacific originating from their colonies (the Indian, Australian and New Zealand troops). The homeward bound US troops were brought back from the different theatres of war by Operation Magic Carpet.[168]

Similar to the situation in Europe with millions of displaced persons and refugees on the move, directly after the Japanese capitulation the Allies were faced the problem to return some six million Japanese back to Japan, and some 1.7 million aliens from Japan. The resulting mass repatriation came on top of the disarmament of the Japanese scattered over the Pacific Island, as well as the return of the Prisoners of War originating from the different pacific regions (Malaya, Singapore) where they had been forced to work for the Japanese (eg on the River Kwai/Burma Railway).

---

[168] In total, ten aircraft carriers, 26 cruisers, and six battleships were converted into troopships to bring soldiers home from Europe and the Pacific. By December 1945 some 700,000 service men were returned from the Pacific.

## De-Colonization of the East

With the Japanese capitulation, the Allies (ie Britain, the US and Australia) were faced by a post-war power vacuum in the Pacific. China had its Communist Revolution (1945-1949), but also elsewhere the communist ideology spread its wings where nationalist movements attempted their decolonization efforts.

**Figure 65: First Indochina War (1945-1954).**

A CIA map from 1950 showing areas of communist activity.

Source: https://alphahistory.com/ vietnamwar/first-indochina-war

*First Indochina War:* In French occupied Indochina in 1945 the Viêt Minh —a national independence movement created by the Indochinese Communist Party (ICP) that sought independence from the French Empire— organized the August Revolution. Their leader, Hồ Chí Minh declared Vietnam's independence by proclaiming the establishment of the Democratic Republic of Vietnam on September 2nd, 1945. Also in September 1945, a force of 200,000 Chinese National Revolutionary Army troops arrived in Hanoi to accept the surrender of the Japanese occupiers in northern Indochina. It was the start of diplomatic negotiating to get Vietnam into the French Union, its newly created colonial empire. However, in 1946, the states of French Indochina (Cambodja, Laos, Vietnam) withdrew from the French Union, leading to the First *Indochina War* (1946–1954) (Figure 65).

By late 1946, the French had mobilised 50,000 troops in Vietnam and regained control of Saigon. From 1946 to 1954, the French opposed independence, and Ho Chi Minh led guerrilla warfare against them. The first two years of the war (1947-48) was marked by sporadic fighting. In early 1949, the French, frustrated by a lack of progress in the war, changed tack. Paris began looking for a political solution rather than a military victory. The new government instigated by the French government, however, failed. Notwithstanding the United States direct political, economic, and military support to French forces in Vietnam.

> It was the start of the US involvement as part of their foreign policy during the Cold War opposing communist expansion. And the Chinese Communists provided the Viet Minh with considerable military assistance throughout the remainder of the Indochina War.

With the help of Chinese weapons and other equipment, the Viet Minh expanded quickly into a significant military force. The year 1952 saw some of the most bitter fighting of the war, as the Viet Minh launched a series of advances in the north, to restore their supply lines and expel the French. By the end of 1952, French casualties (dead, wounded, and missing) had reached 80,000 to 90,000, and the Communists showed no signs of letting up. Then came the confrontation at Dien Bien Phu, a military post in North-western Vietnam. For months the Viet Minh forces laid siege on the garrison, and on May 7th, 1954 it was overrun. The victory triggered a tremendous shakeup in the French government and created an outcry against the war throughout France that could not be ignored. On July 20th, 1954, France agreed to permanently withdraw from Vietnam under terms of an agreement known as the Geneva Accords. The country was divided in North Vietnam supported by the USSR and Chine, and South Vietnam. And there the US became heavily involved leading to the Second Indochina War (1955-1975) aka the Vietnam War.

*The Indonesian War of Independence:* When Japan surrendered on August 15th, 1945, leaving a power vacuum in the Indonesian Archipelago, the Indonesian leaders Sukarno and Mohammed Hatta, proclaimed the independence of Indonesia. It started the *Indonesian National Revolution* with its guerrilla warfare and internal Indonesian political and communal upheavals. The 'Pemuda' —the youth wing of the Communist Party of Indonesia— created the violent Bersiap-period (1945-1946) that started immediately after the proclamation of Indonesian independence and before the arrival of the first British and Dutch armed forces. Chinese houses and shops were looted and their families killed, as well as thousands of Eurasian people (eg Moluccans, Ambones). In addition, they attacked sultans and other members of the Indonesian elite, retaliated violently against those village heads who had assisted Japanese oppression of Indonesian peasants, and they attacked other alleged 'traitors' and fought for turf and weapons.

As the Dutch military were crippled during the German occupation in Europe and the Japanese occupation in the Pacific, the British were charged with restoring order and civilian government in Java. After arriving in September 1945, they evacuated the 10,000 Indo-Europeans and European internees from the volatile Central Java interior. Having colonial problems of their own, and eager to pull its soldiers out of Indonesia, the British allowed for large-scale infusion of Dutch forces into the country throughout 1946. Together the Dutch and British troops confronted the newly formed regional Indonesian defence forces. However, by November 1946, all British soldiers had been withdrawn from Indonesia. They were replaced with more than 150,000 Dutch soldiers. As a consequence of civil turmoil created by the Pemuda some 100,000 Dutch and Indonesian-Dutch were repatriated to the Netherlands. Among them the survivors of the Japanese internment camps.[169] Some 30,000 of them returned to later Indonesia again.

By August 1946 a Republican government had been established in Jakarta. It would become a four-year struggle for freedom with sporadic but bloody armed conflicts, internal Indonesian political and communal upheavals, and two major international diplomatic interventions.

After failed negotiations to create the United States of Indonesia, the Dutch became entangled in 'politionele acties' (ie Police actions) to restore law and order by July 1947. International pressure led to a ceasefire, and what was originally an internal Dutch/Indonesian affair took on an international dimension resulting in a loss of the (financial) US-support and the (diplomatic) UN-support.

After the Indonesian Republic under Sukarno successfully suppressed a large-scale communist revolt (ie the Madiun Rebellion in October 1948), the United States realized that it needed the nationalist government as an ally in the Cold War. Dutch possession was an obstacle to American Cold War goals, so Washington forced the Dutch to grant full independence on December 27th, 1949. On August 17th, 1950, the fifth anniversary of his declaration of Indonesian independence, Sukarno proclaimed the Republic of Indonesia as a unitary state. A few years later, Sukarno seized all Dutch properties and expelled all ethnic Dutch —over 300,000— as well as several hundred thousand ethnic Indonesians who supported the Dutch cause.

---

[169] On a personal note; being survivors of the Japanese internment camps, the parents and sister of your author were among the people repatriated after the war to the Netherlands.

The Second World War had nearly bankrupted many European countries. As a result, the cost of maintaining thousands of soldiers and administrators in colonies halfway around the world became increasingly unfeasible. Particularly for the United Kingdom, which emerged from the conflict with crippling debt. In addition, long-standing independence movements —based on the notion of self-rule— gained momentum and placed increased pressure on colonial powers. The result was the withdrawal of European colonialism. [170]

## *The Context of the Postbellum Period (1950-1975)*

The Second World War had been fought in the European, the Middle Eastern and Pacific theatres of war. The Fall of Berlin in April 24th, 1945 heralded the end in Europe, the dropping of the atomic bomb on August 6th and 9th 1945 did the same in Asia. With the massive destruction of the 'old', it was the dawn of changing times; with new institutions, new social movements, new leadership,[171] and with new technologies that would affect radically the Affairs of Man. The Second World War changed the geo-political alignment and power structures of the globe with rise of the super powers; the United States and the Soviet Union. And it laid the foundations for the geo-economic powers of the European Union and the Chinese Republic to come. The global economy suffered heavily from the war, although nations were affected differently.

> The United States emerged much richer than any other nation, experiencing a baby boom[172] that fuelled a period of deep political instability. In contrast, the United Kingdom was in a state of economic ruin and continued in relative economic decline for decades. The burgeoning political ideologies of socialism and communism were confronting the capitalist ideologies in a way never seen been before. Known as the Cold War it would affect the Affairs of Man on a worldwide scale for the next decades.

---

[170] On a personal note; your author —born in 1947 while his parents were repatriated to the Netherlands— returned that year with his Dutch family to Bogor, Indonesia. In 1953 they were among the returnees that travelled on the SS Oranje to Amsterdam in November 1953. Source; https://igv.nl/wp-content/uploads/2018/03/Oranje_1945-1962.pdf. (Accessed January 2022)

[171] Several changes in leadership occurred during 1945. On April 12th President Roosevelt died and was succeeded by Harry S. Truman. Benito Mussolini was killed by Italian partisans on April 28th. Two days later, Hitler committed suicidein besieged Berlin.

[172] The post-war Baby Boom created the *Baby Boomer Generation* (born from 1946 to 1964), next to the children of the *Silent Generation* (born from 1928 to 1945) or the *Greatest Generation* (born from 1901 to1927). In the West, boomers' childhoods in the 1950s and 1960s experienced the ideological confrontation during the Cold War, creating the foundations for the counterculture of the 1960s.

## Postbellum in Europe

That German surrender in early 1945 may have been the military surrender, but the war had not ended on the civilian side, as there were numerous scores to settle. For example, by the those who had been the focus of the German extermination program for Slavic people; Hitler's Master Plan East of his 'Lebensraum' policy, and his scorched earth tactics. And, not to forget, the organized extinction —aka Holocaust— of more than six million Jews. As a result, between 1944 and 1948, millions of displaced people, including ethnic Germans (Volksdeutsche) and German citizens (Reichs-Deutsche), were permanently or temporarily on the move in Central and Eastern Europe.

**Figure 66: Material Destruction from bombing during the war.**

Ruins of Rotterdam (top, 1940) Monte Casino (middle, 1944) and Dresden (bottom, 1945).

Source: Stadsarchief Rotterdam, Defensie.nl. Wikimedia Commons.

But there was more, as the wartime-destruction of Europe was massive, apart from some 60 million lives that were lost. Europe was in ruins as the result of 'strategic bombing' from both sides. From the German side by the blitz-bombings of Rotterdam (1940) and London (1940-41), and later raids during the war. From the allied side by the bombing of Rome (1943) by 110,000 sorties —including the destroyed Italian Monastery of Monte Casino (1944)— to the sustained aerial bombardments by the Allied that had erased most German cities (eg Cologne, Hamburg, Dresden) (Figure 66). In the West of Europe, bombing had destroyed 5 million houses and apartments. Industrial facilities were hard-hit. Especially damaged was the transportation infrastructure, as railways, bridges, and docks had been specifically targeted by airstrikes, while much merchant shipping had been sunk. The region's trade flows had been thoroughly disrupted; millions of dispersed persons were collected in refugee camps, and some 12 million refugees from the east had crowded in.

Had the war been a period of destruction, the period direct after the war was a period of clearing and reconstruction, both in materialistic terms as well as in psychologic terms. The war had destroyed communities and families, and disrupted the social and economic fabric of nations. Societies were ruined by the war casualties of the military and civil population, and large parts of its people were mentally damaged by the war-experiences. Families were torn apart as their members failed to return from the theatres of war. And those that returned were scarred for life.

*The European Post-war Resurrection*

Europe may have been in ruins —material and financially as well as mentally— the Spirit of Time was based on the feelings of 'Never Again' that had emerged after the First World War. Although the League of Nations (est. 1920) as a forum for resolving international disputes had failed, now similar ideas arose again and shaped the United Nations. It created the new *Zeitgeist of (social and economic) Resurrection*.

As a result, the years 1948 to 1952 saw the fastest period of recovery and growth in European history. During the Post-war economic boom (1945-1973) industrial production increased by 35%. Agricultural production substantially surpassed pre-war levels. The poverty and starvation of the immediate post-war years disappeared, and Western Europe embarked upon an unprecedented two decades of economic growth that saw standards of living increase dramatically.

After the war had ended, there was the issue of the German war reparations,[173] the contribution payments to the cost of the occupation forces, the physical plunder of industry and rail roads by the Soviet[174] and the intellectual plunder by the US.[175] Ideas of further suppressing Germany in order to prevent any future repeats,

---

[173] War Reparations was already an issue in the Napoleonic Times, the 1870-War and the First World War. Then it had crippled the Weimar Republic. After the Second World War, according to the Potsdam conference held between July 17th and August 2nd, 1945, Germany was to pay the Allies US$23 billion mainly in machinery and manufacturing plants.
[174] The Soviet dismantled complete industries and the railway infrastructure to haul it off to Russia. The Soviet Union also relocated more than 2,200 German specialists—a total of more than 6,000 people including family members—with Operation Osoaviakhim during one night on October 22nd, 1946.
[175] The Allies confiscated intellectual property of great value, all German patents, both in Germany and abroad, and used them to strengthen their own industrial competitiveness by licensing them to Allied companies. In addition, Operation Paperclip had brought more than 1,600 German scientists, engineers, and technicians to the US for government employment.

[176] however, politically faded away and became replaced by the Marshall Plan (1948-1951). This US recovery plan for Europe with a budget of $15 billion, was crafted as a four-year plan to reconstruct cities, industries and infrastructures heavily damaged during the war and to remove trade barriers between European neighbours.

That first decade of the postbellum period was marked by a range of economic and institutional events, that shaped the Resurrection.

Figure 67: Division of Germany in East-Germany and West-Germany (1949).

Source: https://www.militaryhistories.co.uk/berlin/germany

*German Post-war Division:* The Yalta Conference (February 4-11th, 1945), showing the ideologic differences between Russia (aka the East) and the Western countries (aka the West), resulted in dividing their spheres of influence. From the Occupation Zones arose in 1949 the Soviet-controlled *German Democratic Republic* (GDR, East Germany), and the Allied controlled *Federal Republic of Germany* (FDR, West-Germany) (Figure 67).

East Germany came under communist rule that nationalized infrastructures and industrial plants, and implemented a centrally planned economy. Backed by Russian occupation forces, the Soviet created in the Eastern Bloc ruled by the Socialist Unity Party of Germany (SED), a command economy and an agrarian reform of land ownership. Next to land confiscated from the Junker estates, State

---

[176] The Morgenthau Plan was a proposal to eliminate Germany's ability to wage war by eliminating its arms industry and removing or destroying other key industries basic to military strength.

farms were also set up, called Volkseigenes Gut (State-owned Property). And the newly formed State Security Service (Staats Sicherheitsdienst, SSD), commonly known as the Stasi, became one of the most effective repressive and intelligence agencies. The German war reparations owed to the Soviets impoverished the Soviet Zone of Occupation and severely weakened the East German economy. The poverty induced or deepened by these war reparations, provoked the massive Republikflucht ('Desertion from the republic') to West Germany, further weakening the East German economy. [177]

West-Germany, also faced with war reparations, instituted a 'social-free market economy' with reforms regarding social welfare policies. Such as social security, social relief with health care and state pensions, subsidized education and housing programs. As a result, it would from the 1950s onwards come to enjoy prolonged economic growth. The 1960s was a decade of extraordinarily high and sustained rates of economic growth in Germany.

The creation of the Berlin Wall and its extension into the Iron Curtain separated the two ideologies of Soviet communism and American capitalism, started the Cold War and fuelled the American foreign polities opposing communism worldwide.

*The Iron Curtain:* The postbellum time, notwithstanding the Zeitgeist of Resurrection, was the period dominated by the confrontation of the Eastern communist ideology, and the Western capitalist-democratic ideology.[178] It was the start of the Iron Curtain that divided the European continent in East and West Europe, each with its own socio-political doctrines. It became the definite political, military, and ideological barrier separating East-Germany and West-Germany. And it was the time in which the Soviet-Union annexed the Baltic States, created the satellite states (eg the German Democratic Republic, Poland, Hungary) and extended the *Union of Soviet Socialist Republics* (USSR) with the Eastern Block as part of the Warsaw Pact in 1955.

---

[177] Between October 1945 and June 1946, 1.6 million Germans left the Soviet zone. Next, despite these increased security measures of the Iron Curtain, 675,000 people fled to West Germany between 1949 and 1952.

[178] In short, the communist ideology has a totalitarian system governance of a single-party state, state owned property, no free enterprise, state supplied education and health care, state-controlled press and aims at a classless society. The capitalist-democratic ideology has a democratic system governance of multiple parties, privately owned property, driven by free enterprise, education and health care provided by private entities, freedom of the press and results in class distinctions.

On June 26th, 1963, after visiting the just erected Berlin Wall, President John Kennedy, held a famous anti-communist speech to an audience of 120,000 people in which he declared 'Ich bin ein Berliner.' Not much later, on November 22nd, 1963, he was assassinated in Dallas.

*The West-German Economic Miracle:* Germany might have been in ruins, its industrial output in 1947 being only one-third its 1938-level, and a large percentage of Germany's working-age men dead or disabled, it had a skilled workforce and a high technological level by the end of the war, partly as a result of the deportations and migrations which affected up to 16.5 million Germans. However, they were still restrained by the Allied continuation of the Nazi-policies of price-control and food rationing. But when those policies were abandoned in 1948, as a result of currency reform, decontrol of prices, cutting taxes and abolishing food rationing, and aided by the economic support of the Marshall Plan, that changed. Together with the breaking down of old trade barriers and traditional practices, and the opening of the global market, it made Germany and Austria recover rapidly. Based on the liberal-conservative traditions of continental Europe, it added to the social market economy its attendant 'Wirthschaftwunder.' Aided by the temporary workforce of the many 'Gastarbeiter' of unskilled workers, its economy recovered rapidly.

Economic output continued to grow by leaps and bounds after 1948. By 1958 industrial production was more than four times its annual rate for the six months in 1948 preceding currency reform. Industrial production per capita was more than three times as high.

*The Dawn of the United Nations:* As the western world was trying to avoid to repeat the horrors of the first half of the twentieth century, the failed League of Nations was transformed in the *United Nations* by 1945. In 1948 it adopted the Universal Declaration of Human Rights as a common standard for all member nations.

The United Nations became an institution with a range of specific organisations. From the UN *Security Council*, the UN *Economic and Social Council*, the UN *International Court of Justice*. the *International Labour Organisation* (ILO), the *Food and Agriculture Organization* of the United Nations (FAO), the *United Nations Educational, Scientific and Cultural Organization* (UNESCO), and the *World Health Organization* (WHO). And, as the tension between East and West flared up, the defence organization of the *North Atlantic Treaty Organization* (NATO).

*The Dawn of the European Union:* In Europe the first not too successful steps towards political and economic integration of Europe into an 'European community' were taken. In addition to the creation of *Council of Europe* (1949) with human rights and common legislation as its objective, in 1951 the creation of the supranational organisation of the *European Coal and Steel Community* (ECSC) focussed on the economic subjects. In 1957, an effort to create a more extensive common market in Europe without trade barriers, the *European Economic Community* (EEC), was more successful. The EEC created a common market that featured the elimination of most barriers to the movement of goods, services, capital, and labour. Something changed during the Zeitgeist of Resurrection. It was the 'Idea of a United Europe.'

*European 1968-Protests heralding the end of the Postbellum miracle.*

As part of the escalation of social conflicts, predominantly characterized by popular rebellions against state militaries and their bureaucracies, Europe saw protest movements erupt in 1968. In France the May student protests led to a general strike involving some 10 million workers, and French society did undergo a sea of socio-political change in the May Revolt's aftermath. In Italy, in May 1968 all universities were occupied. The following Hot Autumn saw massive general strikes, and during the following 'Years of Lead' more violent groups such as the *Red Brigade*, emerged. In West-Germany, the 1968 student protest movement of the maturing baby boomers, spread into opposing the old social institutions (eg the conservative media) and authoritarian policies (eg by former Nazis in government). And the rise of radical movements in 1970 resulted in the violent *Red Army Faction* (Rote Armee Fraktion, RAF) and the *Baader-Meinhoff Group* that engaged in bombings, kidnappings and murder.

In Czechoslovakia, Poland, and Northern Ireland similar student protest had erupted, but with a different societal impact. The Prague Spring reforms saw the creation of a dual federation of the Czech Socialist Republic and Slovak Socialist Republic, but those were suppressed by a Soviet initiated Warsaw Pact intervention.[179] In Poland the March-protests of a dissident movement against the communist regime were suppressed, and an anti-Jewish campaign resulted in a mass exodus. In Ireland it was the time of the Civil Rights Campaign starting the ethnic and sectarian based 'Troubles' (ie the Northern Ireland Conflict). In the Netherlands the actions of the 'Dolle Mina' (Mad Mina) group, created a second wave of feminism.

---

[179] The Warsaw Pact was a collective defence treaty signed in Warsaw, Poland between the Soviet Union and seven other Eastern Bloc socialist republics of Central and Eastern Europe. It was a reaction to the western NATO created in 1949.

The major European economies were (directly and indirectly) affected by this 1968 wave of social protest of the maturing baby boomers. It shook in societies the educational system, the job markets, the balance of political powers, the issue of civil rights. Inflation rose, full employment came to an end, machines replaced more and more costly human labour. The more visible end of the European post-war economic boom came with the Oil Crisis in 1973. Had the oil price-level been relatively low during the preceding economic boom, then it rose dramatically due to the Arab-policies of oil embargos for those countries that supported Israel during the Yom Kippur War. Europe's economy tumbled into a recession.

## Postbellum in America

The United States had emerged from the Second World War as one of the foremost economic, political, and military powers in the world. Wartime production had pulled the economy out of its earlier depression and propelled it to great profits. In order of avoiding another global war —next to President Wilson's initiative to create the United Nations— for the first time the United States began to use economic assistance as a strategic element of its foreign policy, and offered significant assistance to countries in Europe and Asia struggling to rebuild their shattered economies.

*The Rise of Pax Americana*

As a consequence of its rise to economic superpower, its technological dominance in wartime, and the aspirations of its foreign policies (eg Never Again, fighting communism), America as a capitalistic, constitutional democratic society, obtained a new geo-political role: the *Pax America*. It became the global policeman, after the demise of the earlier Pax Britanica.

*The Marshall Plan:* Fanned by the fear of Communist expansion and the rapid deterioration of European economies in the winter of 1946–1947, Congress passed the Economic Cooperation Act in March 1948 and approved funding that would eventually rise to over $12 billion for the rebuilding of Western Europe.

*The Truman Doctrine:* After the war and the US involvement in nearly all its theatres of war (Europe/Mediterranean, Middle East, Pacific), the US reoriented its foreign policy in what became known as the Truman Doctrine. A policy that stated that the United States would provide political, military and economic assistance to all democratic nations under threat from external or internal authoritarian forces.

*Red Scarce and McCarthyism:* The rise of the communist ideology after the Russian Revolution of 1917, and the visible postbellum confrontations of the communist's ideology and the capitalist-democratic ideology, had increasingly infected the popular opinion about US national security. The Un-American Activities Committee of the House (HUAC) — created already in 1938 to investigate alleged disloyalty and subversive activities on the part of private citizens, public employees, and those organizations suspected of having either fascist or communist ties— revealed after the war Russian atomic espionage. It had given way to another form of Red Scarce; McCarthyism that was characterized by heightened political repression and persecution of left-wing individuals, and a campaign spreading fear of alleged communist and socialist influence on American institutions and of espionage by Soviet agents.[180]

By 1952, several US states had enacted statutes against criminal anarchy, criminal syndicalism, and sedition; banned from public employment or even from receiving public aid, the communists and 'subversives'; asked for loyalty oaths from public servants, and severely restricted or even banned the Communist Party. On the federal level, the *Communist Control Act* of 1954 was passed with overwhelming support in both houses of Congress after very little debate.

The primary targets of McCarthyist[181] persecution were government employees, prominent figures in the entertainment industry, academics, left-wing politicians, and labour union activists. Many people suffered loss of employment and the destruction of their careers and livelihoods as a result of the crackdowns on suspected communists, and some were outright imprisoned. Such as the people on the Hollywood Blacklist,[182] as well as the 140 leaders of the Communist Party.

---

[180] The Atomic spies, active in the US, Britain and Canada, conducting nuclear espionage and revealing it to the Soviets. Notable was the Klaus Fuchs British spy ring penetrating the Manhattan Project. Due to Fuchs's position in the atomic program, he had access to most, if not all, of the material Moscow desired. Also the US network of Julius and Ethel Rosenberg, obtained classified information on the Manhattan project. Convicted of espionage in 1951, they were executed by the federal government of the United States in 1953.
[181] McCarthyism, named after US Senator Joseph McCarthy, is the practice of making accusations of subversion and treason, especially when related to communism and socialism.
[182] The blacklist involved the practice of denying employment to entertainment industry professionals believed to be or to have been Communists or sympathizers. Though many of the entries on the blacklist were the result of rumours, the hint of suspicion was enough to end a career and ruin the lives of many in the entertainment industry. As part of the wider sweep of anti-Communist activities of the post-war period, the Hollywood blacklist brought media workers into the web of suspicion and fear that characterized the era.

*The Technological Powerhouse*

Developments in military technology paralleling the civil applications had made a dramatic difference in supporting the new geo-political role of America. The mechanization of war —from tanks and long-range air bombers to battleships and aircraft carriers— was fuelled by a massive scientific activity in military-financed R&D institutions; those in corporations, universities and the independent institutions. From the nuclear research leading to the atom bomb, to the modern communications and information technologies (ICT). As a result, America's military power ruled the waves where it showed its presence.

*The Nuclear Arms Race:* The Nuclear Arms Race had its origins in the ballistic missile-based nuclear arms race between the two superpowers following the Second World War. It had emerged from the US military industrial complex that developed, tested and manufactured the missiles originating from the work of the scientists behind the German V2 program,[183] that had been taken to America as part of the operation Paperclip.[184] A similar action had started Russian efforts into missile building, and the Sputnik launch proved that their rockets ICBMs were capable of delivering nuclear warheads at any point of the world.

The aerospace defence contractors, part of the US Military-Industrial complex —already designing and building the long-range strategic bombers— jumped on the bandwagon of the Medium Range, Long Range and Intercontinental ballistic missile programs. Others companies developed the warheads and navigation systems. Among others, the Aerospace Corporation, Space Technology Laboratories of TRW Inc., and Lockheed Missiles & Space were three firms that proclaimed proprietary expertise. Missile guidance systems were manufactured by a range of defence contractors: from Raytheon, McDonald Douglas to General Dynamics and Boeing.

Next to the military use of atomic power with its controlled unleashing of energy instantly (ie the atomic bomb), there was also another development taking place. That was the controlled unleashing of atomic power in a controlled way over longer time; nuclear reactions created

---

[183] The development of the *Vergeltungswaffe* (Vengence Weapon) was Germany's effort for aerial bombing of cities: London, Antwerp and Liège were the main targets. The V-weapons (eg V-1, V-2) were unmanned long-range bombing vehicles that were used as retribution against the allied mass bombing of German cities. At the height of the campaign against Britain, some 20,000 houses/day were damaged by the V-weapons.

[184] Operation Paperclip identified the scientists that had worked in German Army Research Center in Peenemunde, developing the V-2 rocket. By 1947 this evacuation operation had netted an estimated 1,800 technicians and scientists, along with 3,700 family members.

heat/steam that could be used to create electricity. This civilian use of nuclear energy would become the start of the Era of Nuclear Power.

*The Space Race:* After the Soviet Union had announced on August 2nd, 1955 its space program, the Sputnik Crisis[185] was the result of the launch of a tiny satellite 'Sputnik' on October 4th, 1957, followed by Luna program of robotic spacecraft missions to the moon. Sputnik 2 was the second aircraft launched into Earth orbit on November 3rd, 1957 with a dog named Laika on board. As part of the next Vostok program, on April 12th, 1961 the Soviet Union launched the first man Yuri Gagarin in Earth's orbit. This all heralded the start of the Space Race as on September 12th, 1962, President John Kennedy announced the Nations Space Effort. The plan was to land a man on the Moon before 1970. It would be realized with the Apollo 11 mission in 1969.

As a result of the Sputnik Crises, the perceived threat to America's national security had created political consensus among the interest groups. The military at the Pentagon wanted to match the Soviet military achievements, corporate America was looking for new business, and the universities were eager to commit (and fund) their basic research. The result was the creation of a special organization on July 29th,1958; the *National Aeronautics and Space Administration* (NASA). While this new federal agency would conduct all non-military space activity,[186] the *Advanced Research Projects Agency* (ARPA) was created in February 1958 to develop space technology for military applications. Initiatives and programs of the ARPA that would have an impact on the Internet.

> *Project Mercury* was the first human spaceflight program of the United States, running from 1958 through 1963. It was followed by the *Project Gemini* (1961-1966) and the *Project Apollo* (1960-1972). As the satellites needed power to reach their orbits, it needed rockets. So from the ballistic missiles (ie the intercontinental ballistic missiles (ICBM) that had grown out of the guided-missile program, the powerful space boosters were developed.

*Atom Bomb Craze:* With the proven military powers, after the Japanese surrender, America realised that its atomic monopoly could support its geo-political foreign policies. The monopoly did not last long, however, as the Russians developed also nuclear technology and exploded their first atomic bomb in 1949. Followed by others countries with their own

---

[185] The crisis was the perceived technological gap between the US and the Soviet Union, creating a period of public fear and anxiety.
[186] A significant contributor to NASA's entry into the Space Race was the technology from the German V-2 rocket program led by Wernher von Braun who had come to the US as part of the Operation Paperclip.

military nuclear programs; the United Kingdom in 1952, France in 1960 and the People's Republic of China in 1964. So, the ideologic conflict between Communism and Capitalism had become a conflict between two superpowers —and their allies— armed with atomic power that could destruct the world. Superpowers that held each other in a fragile power balance, because neither could make the first strike without the threat of a counterstrike. So, the benefits of using nuclear weapons in a conflict —even in a proxy war[187]— were greatly diminished. And in between the two forces was Europe, a potential battle field for a nuclear war. America extended the 'atomic umbrella' over Europe and became member of the NATO (the North Atlantic Treaty Organization).

*The Unavoidable Conflicts of Policing*

On the one hand, the economic dominance and geopolitical role as guardian of the Pax Americana had lured the US into its role as the policeman of the world defending the capitalist-democratic ideology. On the other hand, its technology-driven power had made it a well-armed policeman. A combination of factors that led to America's increasingly complex involvement in geo-political conflicts all over the world.

*The Berlin Blockade:* The postbellum period after the Second World War, notwithstanding the original Zeitgeist of Resurrection, was the period dominated by the confrontation of the Eastern communist ideology, and the Western capitalist-democratic ideology. An early confrontation resulted in the *Berlin Blockade* (June 1948-May 1949) when the Soviet Union led by Joseph Stalin blocked the Western Allies' railway, road, and canal access to the sectors of Berlin under Western control. Suddenly, some 2.5 million civilians had no access to food, medicine, fuel, electricity and other basic goods. The Allies reacted with the *Berlin Airlift*,[188] transporting supplies to the people of West Berlin by air. This Soviet hostility was the first major conflict of the Cold War.

*The Pacific Turmoil:* As an afterwar effect of the military contributions in the Pacific, the US became heavily involved in the geo-politics of the Pacific Region. The confrontation with both the old communism of Russia after the Communist Revolution (1917-1924), and the new communism of China after the Chinese Revolution (1945-1949), would lead to US military involvement on a large scale. An involvement that ran from the

---

[187] A proxy war means that there is no direct involvement of the conflicting nations as the opposing powers in a proxy war use others to fight on their behalf.
[188] American and British air forces flew over Berlin more than 278,000 times carrying in total 2,334,374 tons of cargo. With the American C-47 and C-54 transport airplanes, together, flying over 92,000,000 miles (148,000,000 km). At the height of the Airlift, one plane reached West Berlin every thirty seconds.

Korean War (1950-1953) to the Vietnam War (1955-1975).

*The Cuba Missile Crisis:* While the Soviet had supported the pro-communist Cuban Revolution by solidifying their relations economically, the failed *Bay of Pigs Invasion* of April 1961 had deterred US-Cuban relations and increased the US-Soviet tensions. At the same time US were deploying Jupiter ballistic missiles in Turkey and Italy that covered Soviet territory. The Soviet grabbed the opportunity to match that by shipping missiles to Cuba for their deployment close to the US territory (ie 140 km from Florida). Under the argument of deterring the US from invading Cuba again, they prepared a range of missile launch facilities on Cuba.

When a U-2 spy plane produced photographic evidence, president Kennedy had to act. Instead of an invasion proposed by the military — that de facto was a declaration of war—, he decided for a maritime blockade (and calling it a naval quarantine). After some tense events of Soviet ships loaded with missiles challenging the blockade, in which the world was holding its breath, first-secretary Khrushchev backed off and the US ships transporting missiles were called back. A Third World War had been avoided and was replaced by the fragile balance of the *Mutual Assured Destruction* (MAD) doctrine.[189]

*The Problems at Home*

The manpower for the Pax Americana originated from the US population. Although reduced in size after the Second World War with the 12 million military personnel to some 1.5 million in 1947, it still was a considerably army by the 1950s with total numbers floating between 3,3 (1954), 2.4 (1960) and 3.5 (1968) million military personnel.

*The Year 1968:* It is not surprising that the Cold War, the Nuclear War Race, and the Space Race —in combination with the US-military presence in Asia (after the Korean War becoming involved in the Vietnam War)— had its effect of American society itself. Paralleling the European student protest, all over the US the summer of 1968 heralded a period of nationwide turmoil, with political assassinations, anti-war protests, racial unrest and highly publicized clashes with police. It was the dawn of the anti-establishment counterculture to come.

The *Civil Rights Movement* (1954-1968) had given birth to the *Black Power Movement* and the *Black Panther Movement*. The assassination of Martin Luther King on April 4th, 1968 inflamed racial tensions. In addition, the escalating military presence in Vietnam —and the

---

[189] As part of a national security policy the full-scale use of nuclear weapons by two or more opposing sides would cause the complete annihilation of both the attacker and the defender.

massive drafting of new recruits[190]— had led to nationwide Anti-Vietnam protests polarizing the country, and even disrupted the 1968 Democratic Convention in Chicago. The murder of senator Robert Kennedy on June 5th, 1968 already had derailed the convention, but the presence of 23.000 Chicago police officers and National Guardsmen to suppress any protest, infuriated the peace-protesters outside the convention into violence. It resulted in the Battle of the Michigan Avenue that was widely covered by the media. The convention itself soon became a battleground between wing of anti-war supporters and the different nominees of a divided Democratic Party. Hubert Humphrey eventually won the nomination, but it wasn't much of a victory. Richard Nixon prevailed in the general election ushering in a Republican wave that prevailed for almost thirty years.

The described events would determine the Zeitgeist in the Western World to come. The time where the beeps of a small satellite (aka the Sputnik) launched into orbit in 1957 would ignite a Nuclear Arms War that sprouted the Space Race. An arms war with long range ballistic missiles as delivery vehicles of nuclear warheads capable of destructing each of the superpowers. The time where the geo-political confrontations between East with its communist ideologies, and the West with its capitalist ideologies, would be influencing the Affairs of Man again. The time where the Red Scarce paranoia would cloud American society poisoned with the fear of communism fuelled ad absurdum by people like McCarthy. Where the early Cold War (1947-1962) would increase the tensions between nations culminating in the Cuba Missile Crisis (1962) that brought the world on the brink of the next world war. But also the time where the Sino-Soviet Split (1956-1966), followed by President Nixon's visit to China in 1972 changing the global balance of power, had its effect on cooling down the Cold War.

The subsequent *Détente* had both strategic and economic benefits for both sides of the Cold War. Negotiations about arms reduction between the US and the USSR, had resulted in the SALT I-agreement (1972) and —during the Helsinki Process— the SALT-II (1972-1979) agreements.

This was the period of the US Postbellum that would create the foundations for the context of the Late Twentieth Century.

---

[190] In 1967, the US military was drafting as many of 15,000 man/month for the first half year, mainly from minorities and lower/middle class whites. Notwithstanding the many exemptions, as many as 100,000 draft eligible baby boomers fled the country.

## Postbellum in the Middle East

With the Decline of the Ottoman Empire (Figure 13), after the First World War, the partition of the Ottoman Empire in the Treaty of Sevres (1920) had created the new independent states (Figure 33) and the mandates for the Middle Eastern British, French and Italian Protectorates. During the interwar-period of the Interbellum (1919-1939) the Middle-East had seen some fundamental geo-political dynamics (Figure 37). After the Second War had ended the post-war period revived this all. Some even more prominent as a result of what had happened during the war.

> One of the consequences of the Second World War was that the British were unable to finance any longer the cost of their worldwide presence. Faced with a post war financial crisis bringing it on the brink of bankruptcy and disrupting politic consensus, notwithstanding US-loans and Marshall plan funding, they shed traditional overseas military roles as fast as possible. [191]

*Pan-Arab Movement and Pan-Islamism*

The remnants of the collapsed Ottoman Empire had, despite their sectarian and religious difference, a common ideology; pan-Islamism. It was the concept of the Ummah:[192] 'One family, one cause.' Building forth on the creation in 1945 of the *Arab League* —a political effort to forward the Arab economic and social welfare— during the 1950s and 1960s the *Pan-Arab Movement* aiming to unify Arabian states reached its height.

*Egypt:* It was in Egypt that Pan-Arabism took hold after a military coup of the Free Officers movement in 1952 led by Gamal Abdel Nasser. Embracing the communist ideology of a state-controlled economy, the Egyptian Republic with a one-party political system was created. The subsequent Nasser Era witnessed a rapid increase in living standards. The national economy grew significantly through Arab Socialism with agrarian reform breaking up the large private estates, major modernisation projects —such as the Helwan Steel Works (1957), and the Aswan High Dam (1954-1960)[193] — and the nationalisation of key

---

[191] The post-war military cost £200 million a year, to put 1.3 million men (and a few thousand women) in uniform, keep operational combat fleets Stationed in the Atlantic, the Mediterranean, and the Indian Ocean as well as Hong Kong, fund bases across the globe, as well as 120 full RAF squadrons.
[192] Ummah means 'community' and is a synonym for Ummat al-Islām, 'the Islamic community.' In the context of pan-Islamism and politics, the word ummah can be used to mean the concept of a Commonwealth of the Muslim Believers.
[193] The Helwan site was containing both the basis steel factories (steel mills, aluminium plant) as well as the automobile factories. Other industries include flour milling, food

parts of the economy, notably the Suez Canal Company (1956). The latter resulted in the *Suez Crisis* in which the Israeli invaded the Egyptian Sinai during the Second-Arab-Israeli War.[194] In a complex diplomatic game, Nasser played Western Powers as well as the Soviet Union off against each other; the US fear of communism, the British fear of losing their Middle-Eastern oil interest, and France's Algerian problem.[195]

The Suez Crisis of 1956 made Nasser a heroic figure in pan-Arab states. In an effort to counter Russian action after the Suez Crisis, Nasser's first attempt to unification was the *United Arab Republic* (UAR) of Egypt and Syria in 1958, but that already failed by 1961. Instead of a federation of two Arab peoples, as many Syrians had imagined, the UAR turned into a state dominated by the Egyptians.

Next to Nasser's dream of unifying the Arabs, there were other initiatives. The *United Arab Republic* (ie a political union between Egypt and Syria) was a short-lived state from 1958-1961/71. The *Federation of Arab Republics* (ie a political union between Libya, Egypt and Syria) lasted from 1972-1977. And the *Arab Islamic Republic* (ie a political union between Libya and Tunisia), although proposed in 1974, was never implemented.

*Arab Cold War* (1950s-1970s): Although Pan-Arab movement failed to unify Arab states, it had found a common opponent in the Zionism[196] ideology. This was a period of political rivalries of varying degrees of ferocity in the Arab world. Various sectarian and social differences within the Arab societies was another fuelling factor for pan-Arabism's decline. A great obstacle to Arab unity was the religious divide between the Sunni Islam and the Shia Islam; a schism originating after the death of the prophet Muhammed in 632 AD. This schism deepened after the First World War mandates were cutting through centuries-old religious and ethnic communities. The Sunni-Shia divisions would fuel a long-running civil war in Syria, fighting in Lebanon, Iran, Iraq, Yemen and elsewhere, and terrorist violence on both sides. And that religious

---

processing, and textile printing and dyeing. It was electrically powered by the generators in the Aswan High Dam.

[194] In a wave of minority expulsion of western foreigners as retaliation of the Suez Crisis, in 1956–1957 some 25,000 Jews —almost half of the Jewish population of Egypt— were expelled from the country.

[195] France was facing an increasingly serious rebellion in their colony Algeria, where the Algerian National Liberation Front (FLN) rebels were being verbally supported by Egypt via emissions of the Voice of the Arabs radio, financially supported with Suez Canal revenuehttps://en.wikipedia.org/wiki/Suez_Crisis - cite_note-77 and clandestinely owned Egyptian ships were shipping arms to the FLN.

[196] Zionism was initially based on the idea that Jews would thrive if they could form a separate community of their own, but it developed into a complex plan for building first a community, then a state.

involvement of the Islam became complemented with Judaism (ie the Jews) returning to the Middle East. And last but not least, there was the oil-factor creating wealth and power, both for the nationalists as well as for the Western powers.

## Iran from Crisis (1946) to Revolution (1979)

Another partition of the collapsed Ottoman Empire had been the independent and short-lived Kingdom of Qajar (Figure 33). After the coup of 1921 had brought the prime minister Rheza Khan Pahlavi to power (1925-1941) in the *Imperial State of Iran* (1925-1935). After his coronation as Shah (ie King) in April 1926, he continued the reform policies democratizing the country and freeing it from foreign interference, renaming the state 'Iran' in 1935. By 1941, however Russia and Great Britain —in order to prevent cooperation with Nazi Germany— occupied the country and forced his abdication. His son Mohammed Reza Shah Pahlavi became Shah (King) in 1941.

After the war ended, the British and American withdrew their forces, but the Soviet Union set up two pro-soviet republics; the Azerbaijan People's Republic and the Kurdish Republic of Mahabad (Figure 68). Military confrontations resulted in the *Iran Crisis* (1946) that ended when the Soviet Union withdrew its forces and the Iranian army —with American support— restored control over the area. In the next years (1947-1951), initiatives to create a constitutional monarchy failed, and the political arena became quite unstable with so many foreign powers involved.

**Figure 68: The Iran Crisis of Soviet Republics in Iran (1946).**
Source: Wikimedia Commons

Then, the *Abadan Crisis* (1951-1954) —nationalizing the Iranian assets of British Petroleum controlled Anglo-Iranian Oil Company (AIOC)— challenged British presence in Iran. The British Navy reacted with an economic blockade. In a coup d'état in 1953, Mohammad Reza Pahlavi's rule was restored and became firmly aligned with the Western Bloc. The US/UK-sponsored coup d'état — planned by their secret service CIA and the British MI6— brought Reza Shah Pahlavi to absolute power as a vasal state of the Kennedy Administration.

The 1953-conflict —in this oil rich region with a large ethnic and religious diversity (Figure 69)— was the first in a range of Cold War conflicts to come and was a decisive turning point in Iranian history. For Iran it had broader consequences as oil and religion started to dominate state affairs. After the coup, the tangible gain for the United States in overthrowing the elected government of Iran consisted of a large share of Iranian oil wealth.[197] For Iran it meant repression between 1953-1958 as — supported by the Shah' SAVAK (ie secret police)— a regime of political suppression ruled.

**Figure 69: Iran's ethnic and religious diversity (2004).**
Source: https://maps.lib.utexas.edu/maps/thematic.html

---

[197] Already in 1951 nationalisation of the oil industry —ie the Anglo-Iranian Oil Company (AIOC) owned for 50% by the British government— was on the political agenda. By that time, the nationalization of the oil industry —by creating the National Iran Oil Company (NIOC)— was nominally maintained, although in 1954 Iran entered into an agreement to split revenues between NIOC and a newly formed international consortium called the 'Seven Sisters' that was responsible for managing production and exploration.

Notwithstanding the land reforms, [198] nationalisation and civil reforms while preserving traditional power patterns —instigated by the Shah during the *White Revolution* (1963-1978)— it fuelled the resentment of the intelligentsia as well as the urban working class.[199] Also the power of (Shia) scholarly clergy was curtailed, creating a deep rift with their leaders that resulted in the arrest of Ayatollah Ruhollah Khomeini on January 22$^{nd}$, 1963. In 1964, Khomeini was sent into exile to France where he remained for 15 years until the next revolution brought him back. It would take another two decades before the *Iran Revolution* (1978-1979) ended the Pahlavi dictatorship, and brought the Shia clergy to power creating Islamic Republic with Khomeini as the first 'supreme ruler.'

## *Iraq from British Mandate to Tribal Clique.*

The cradle of civilization in ancient times in the Euphrates-Tigris region of Mesopotamia known today as Iraq and Kuwait —a battle zone of rival regional empires and tribal alliances over the ages— had also become part of the Ottoman Empire in the sixteenth century. The umbrella of the Ottoman law held supreme sway over this complex interwoven web of alliances and ethnicities, creating provinces around Mosul, Bagdad and Basra. In each of these provinces, one of the three ethnic groups (Shia's, Sunnis & Kurds) held the reigns to governance (Figure 70).

After the collapse of the Ottoman Empire in 1922 the Hashemite Kingdom of Jazira and Sawad (Figure 33) had been governed under a British mandate of Mesopotamia.[200] This delicate Ottoman system was undone by the British, and a geographical entity called Iraq was carved out making a poorly woven patchwork of the three provinces.

By 1933 it had become the Hashemite Kingdom of Iraq, ending British mandatory role, and starting a turbulent political period. Establishment of Sunni religious domination in Iraq was followed by the Assyrian, Yazidi and Shi'a tribal unrests and revolts, which were all brutally suppressed. Multiple coups followed in a period of political instability, peaking in the *Anglo-Iraqi War* (1941) that restored British rule after an anti-British coup d'état.

---

[198] Khomeini had denounced the Shah and was arrested. After his arrest three days of riots broke out, some hundred thousand protesters marching past the Shah's palace shouting "Death to the Dictator …," completely surprising the government. In June the masses revolted again, and their demonstrations were crushed again.
[199] The land reform, instead of allying the peasants with the government, produced large numbers of independent farmers and landless laborers who became loose political cannons, with no feeling of loyalty to the Shah.
[200] The Brits obtained exclusive right to Persian Oil Reserves. The Anglo-Persian Oil Company struck oil in 1927 in the Baba Gurgur oilfield.

At the end of the Second World War the Iraqi Kurds revolted, and —after being suppressed by the Iraqi in 1945— retreated into Iranian Kurdistan joining the local Kurdish elements in establishing the Republic of Mahabad (Figure 68).

In 1945, Iraq joined the United Nations and became a founding member of the Arab League. Soon, massive civil unrest in Bagdad erupted against renewing British political and military influence in the Anglo-Iraqi Treaty (1948) and culminated in the Iraqi Intifada (1952). Next, in July 1958, the Hashemite Kingdom of Iraq was overthrown in the *14 July Revolution*, with the king, crown prince and prime minister all killed by the

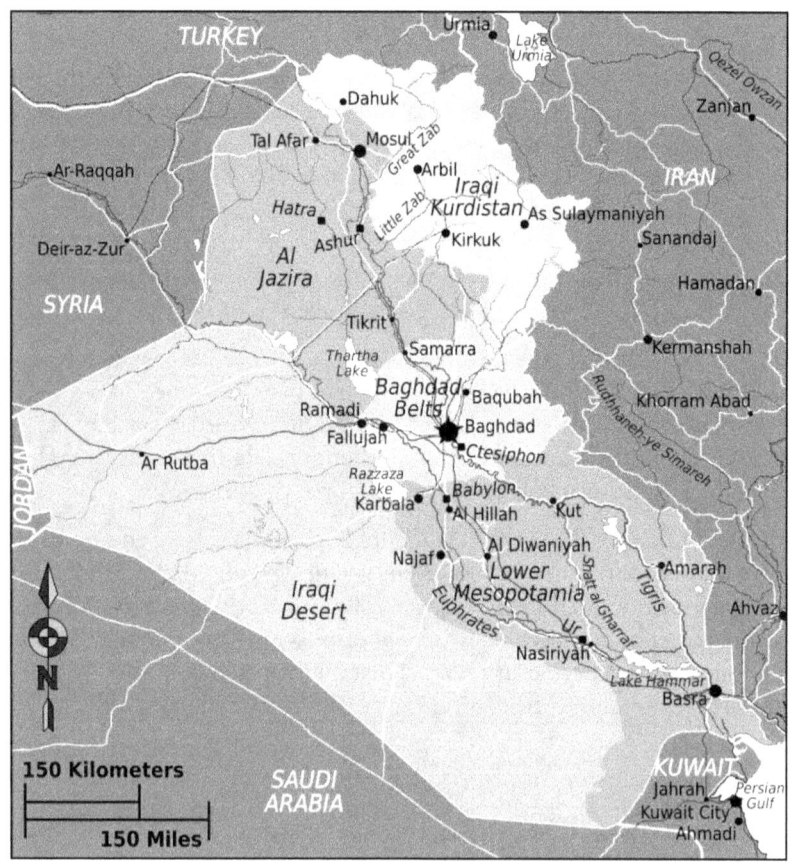

**Figure 70: Ethno-religious regions in Iraq**

Region of Iraqi Kurdistan with Sunni-Kurd population, Al Jazira region with Sunni-Arab/Sunni-Kurd population, Bagdad with Shia-Arab/Suni-Arab population and Shia-Arab Basra.

Source: Wikimedia Commons.

nationalist revolutionaries. Iraq's monarchy was also replaced by the *Iraqi Republic* (1958-1968) with an Arab orientation. Iraq became controlled by the Arab Socialist Ba'ath Party, under the leadership of Saddam Hussein and his ruling tribal clique of Tikritis-Sunni Arabs. [201]

Between 1968 and 1973, through a series of sham trials, executions, assassinations, and intimidations, the party ruthlessly eliminated any group or person suspected of challenging Ba'ath rule.

> In the early 1970s, Saddam nationalised the Iraq Petroleum Company and independent banks, and created an 8000-man strong security force (ie the Mukhabarat). The wealth from oil after the 1973-Oil crisis enabled him to modernize Iraq; eg improving infrastructures, industrial expansion, mechanizing agriculture and expanding public health care.

## Saudi Arabia from Tribal Society to Oil Empire

A third player in the Middle East scene after the Ottoman Collapse was Saudi-Arabia that emerged after the unification of the Kingdom of Nejd with the Kingdom of Hejaz (Figure 33). In 1932, the two kingdoms of the Hejaz and Najd were united as the 'Kingdom of Saudi Arabia,' and ruled by the House of Saud.

> Already in the eighteenth-century Muhammad Ibn Saud (reign 1744-1765), chief of an Arabian village that had never fallen under control of the Ottoman Empire, rose to power together within the Wahhābī religious movement. He and his son Abd al-Azīz I (reigned 1765–1803) conquered much of Arabia; and Saud I (reigned 1803–1814) conquered the holy cities of Mecca and Medina in the early years of his rule. After the Second Saudi State (1824-1891), in the Third Saudi State (1902—today) by 1932 Ibn Saud established the Kingdom of Saudi Arabia by royal decree. A number of his sons later ruled the country: Saud II (reigned 1953–64), and Faisal (reigned 1964–75). The success of the Saud family was in no small part due to the motivating ideology of Wahhābism, an austere form of Islam that was embraced by early family leaders and that became the state creed.

Though far smaller in population than Egypt, the Kingdom of Saudi Arabia would have, next to the oil wealth discovered in the Al-Has region along Persian Gulf cost in 1938, prestige as the land of the holy cities Mecca and Medina, the two holiest cities of Islam that attract a constant flow of pilgrims in the haji and umrah-pilgrimages.

---

[201] Most the governmental positions were occupied by people united by close family and tribal ties. Increasingly, the most sensitive military posts were going to the Tikritis.

In the 1940s and 1950s, the Al-Ghawār oilfield near Dahnam (1948), and the Saffāniyyah offshore field in the Persian Gulf (1951) (Figure 71) —the largest source of petroleum in the world— put Saudi Arabia on the world map of oil producing countries. The lucrative petroleum trade fostered sophisticated diplomatic relations between Saudi Arabia and the West, as well as Japan, China, and Southeast Asia. During the Second World War, the foundation was laid for Saudi's relation with the US. The US guaranteed they would protect the Saudi regime, and Saudi Arabi would support the US national security geo-policies and subsidize its oil exports to the US.

The wealth generated by the oil production, together with tribalism and religious conservatism, fuelled the country's subsequent development. The large influx of foreigners fuelled the latent xenophobia from tribal times; non-essential foreigners were kept out as much as possible. And the increasing wealth did not immediately filter down to the masses, but financed the lavish lifestyle of the scions of the ruling family that expanded in a huge family of princes and princesses.

Another factor was the socio-economic and domestic reform instigated by absolute monarchy of the former tribal society. Coming from an economy based on conventional agriculture and herding, the oil revenues financed the construction of roads, hospitals, airports, and schools. It changed the two-class-society from the 1950s by adding a growing middle class.

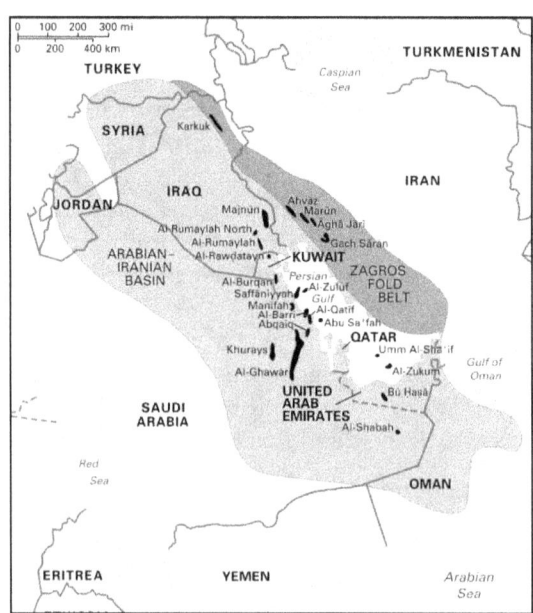

Figure 71: Oilfields around the Persian Gulf Basin.

Source: Encyclopaedia Britannica.

Notwithstanding the US-Saudi relation, during the 1973 Arab-Israeli war, Saudi Arabia participated in the Arab oil boycott of the United States and other Western allies of Israel. A founding member of

OPEC, Saudi Arabia voted in favour of the group's decision to moderate oil price increases beginning in 1971. After the 1973 war, the price of oil rose substantially, increasing Saudi Arabia's wealth and political influence.

## The Birth of OPEC

With rise of the internal combustion engine as power of mobility during the Third Industrial Revolution (1929-1950), the need for fossil fuels had increased rapidly.[202] Starting with exploration of the oilfields in Azerbaijan by some 270 oil producing companies, in 1920 these were nationalized by the Russian soviet government. The Second World War had shown how important oil was, and on August 8th, 1944, the *Anglo-American Petroleum Agreement* was signed, dividing Middle Eastern oil between the US and the UK. In addition, by 1949 some of the world's largest oil fields were just entering production in the Middle East. By then, the international oil market was dominated by the 'Seven Sisters';[203] a cartel of multinational companies that largely controlled the Middle East's oil production after Second World War.

> The political power the Seven Sisters exercised —controlling prices, production and royalty's— angered the oil producing countries, among which Iran. Beginning in 1950, also the Saudi Arabian government began trying to increase government shares from oil revenues. In addition, the Suez Crisis of 1956, blocking the oil transport vital to the West, made the British aware of their strategic vulnerability on the Middle Easter oil production.

As Western nations' dependency on oil increased, Middle Eastern nations realised the strength of their bargaining position and renegotiated existing deals with Western oil companies for a greater share of the oil profits. To create more political momentum, the *Organization of the Petroleum Exporting Countries* (OPEC) was created at the Baghdad Conference on September 10–14th 1960, by Iran, Iraq, Kuwait, Saudi Arabia and Venezuela. During the 1970s the primary goal of OPEC members was to secure complete sovereignty over their petroleum resources. The formation of OPEC marked a turning point toward national sovereignty over natural resources, and OPEC decisions came to play a prominent role in the global oil market and international relations.

---

[202] See: Van der Kooij, B.J.G.: *The Invention of the Internal Combustion Engine* (2021).
[203] The Seven Sisters consisted of three companies formed by the breakup by the US Government of Standard Oil, along with four other major oil companies (eg Royal Dutch Shell, British Petroleum). With their dominance of oil production, refinement and distribution, they were able to take advantage of the rapidly increasing demand for oil and turn immense profits.

*The Birth of the State of Israel (1947-1949)*

Already before the Second World War, an increasing number of Jews had sought refuge from Nazi Germany in Palestine, claiming historical rights from Babylonian and Roman times. The ensuing clandestine immigration was executed by ship, supported by Jewish Agency[204] and the paramilitary organization Haganah, and brought some 150,000 people in 141 voyages to Palestine. And, after the war 250,000 Jewish refugees were stranded in displaced persons (DP) camps in Europe, the problem became manifest and was brought to the agenda of the UN Special Committee on Palestine (UNSCOP) in 1947.

After the British had in 1920 obtained their Mandate for Palestine, and with their commitment in the Balfour Declaration (1917) to create a Jewish National Home, the United Nations General Assembly (UNGA) proposed in their *1947 Partitioning Plan* the forming of a Jewish State and an Arab state (Figure 72). It was accepted by the Jews but rejected by the Palestinian Arab leadership.

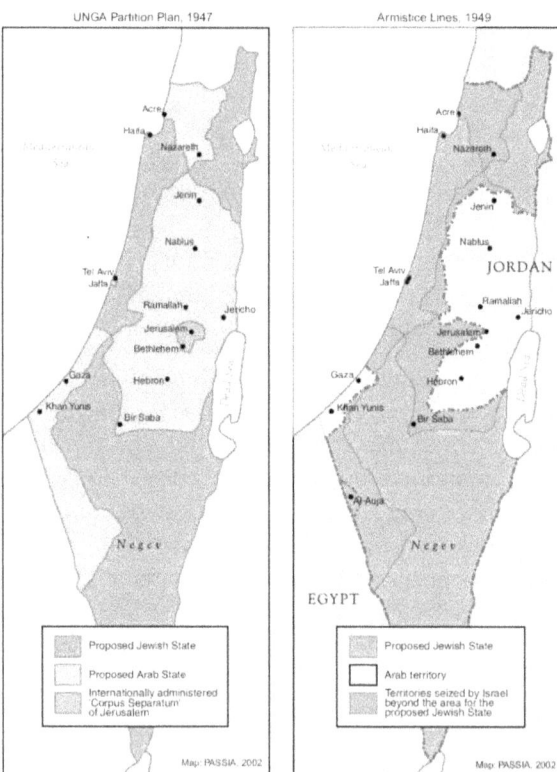

**Figure 72: The Birth of the State of Israel.**
Source: https://worldhistorycommons.org/map-partition-israel-and-palestine

It resulted in the 1947-1948 Civil War in Mandatory Palestine. Anti-British Jewish militancy increased and the situation required the presence of over 100,000 British troops in the

---

[204] Zionists formed their own government system called the Jewish Agency.

country. So, the Brits being caught between their commitment to a Jewish Home and their mandatory obligations, ended their Mandate for Palestine in 1948.[205]

However, the original inhabitants —the Arab-Palestine— also claimed rights over the former Ottoman territories and sought to prevent Jewish migration into Palestine, leading to growing Arab–Jewish tensions. Then, after the Israelian *Declaration of Independence* (1948) created the State of Israel, in the subsequent First *Arab-Israeli War* (1948-1949) a military coalition of Egypt, Transjordan, Syria, and expeditionary forces from Iraq entered Palestine. Next, the Jews facing the formidable armies of neighbouring Arab countries, in three phases a range of battles and truce periods took place. The 1949-Armistic Agreements stopped military hostilities and created the Jewish State, but did not solve the basic conflict as it saw an exodus of Arabs from Israel (est. 700,000)[206] and Jews into Israel (est. 230,000).

Within three years (1948-1951), immigration doubled the Jewish population of Israel and left an indelible imprint on Israeli society. Some 300,000 arrived from Asian and North African nations, more than 270,000 came from Eastern Europe. After 1949, many of the newcomers of Russian origin were settled in agricultural communities (aka Kibbutz), others had been creating 480 new towns and villages.

The sheer volume of immigrants created internal and external tensions. Such as the mounting tension between the new State of Israel and its neighbouring Arab states. In 1956, in the *Second Arab-Israeli War*, Israel invaded Egypt, triggering the Suez Crisis; among Israel's rationale for the invasion was its goal of forcing a reopening of the Straits of Tiran, which had been closed by Egypt for all Israeli shipping since 1950. That closing became again an issue in 1967 when the Egyptian announced that the Straits would be closed to Israeli vessels (Figure 73). It was the start of the *Third Arab–Israeli War* (1967) when Syria, Egypt and Jordan amassed troops along the Israeli borders and the Israeli launched pre-emptive airstrikes, followed by a ground offensive, conquering the entire Sinai Peninsula in six days. In other campaigns, at the cessation of hostilities, Israel had seized the Golan Heights from Syria, the West Bank (including East Jerusalem) from Jordan, and the Gaza Strip. The war saw short-lived but over 20,000 Arab troops killed while Israel lost fewer than 1,000 of its own.

---

[205] The negative publicity resulting from the situation in Palestine caused the Mandate to become widely unpopular in Britain, and caused the United States Congress to delay granting the British vital loans for reconstruction.
[206] The Palestinians forced from their homes and villages became refugees in camps in Egypt, Jordan, Lebanon, and Syria.

Because a key part of their national identity was built around destroying Israel, Palestinians received little support from many Western countries in the 1960s and 1970s. So, the Palestinian conflict continued till it came to another war in 1973, triggered by the 1972 Munich Olympics massacre; a Palestinian hostage action directed at Israeli athletes. The *Yom Kippur War* (October 6-25th, 1973) started with an attack from Syrian and Egypt forces. The Israel Defence Forces' response was swift in both the Golan Hights, the West-Bank and the Sinai Peninsula (where they crossed the Suez Canal). With international pressure mounting, the hostilities finally ceased on October 26th. Israel signed a formal cease-fire agreement with Egypt on November 11th and with Syria on May 31st, 1974.

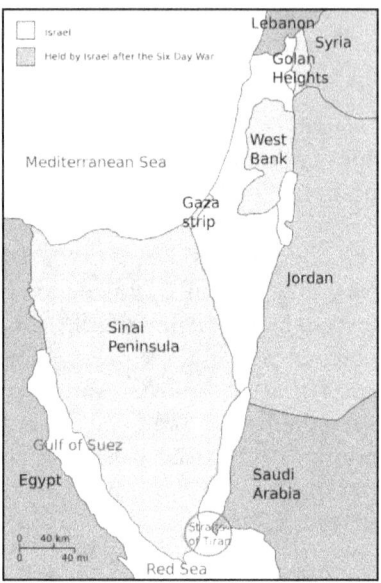

**Figure 73: Arab Israeli War (1967).**
Source: Wikimedia Commons.

The Yom Kippur War had far-reaching implications; the Arab world had experienced humiliation in the collapsed of the Egyptian–Syrian–Jordanian alliance in 1967, but felt psychologically vindicated by early successes in the 1973 conflict. The Israelis recognized that, despite impressive operational and tactical achievements on the battlefield, there was no guarantee that they would always dominate the Arab states militarily. And the world experienced the 1973 Oil crisis as the Arab oil producing states created an Arab oil embargo and trade boycott against Israel's supporters.

*The Oil Crisis of 1973*

After the tension between the new State of Israel and the surrounding Arab States had culminated in the 1973 Yom Kippur War —basically a conflict between Egypt, Syria and their backers versus Israel— the conflict expanded. The conflict itself lasted only from October 6th to October 25th, but the subsequent OPEC oil embargo did shake western economies when the Arab countries participated in an oil boycott against Israel's allies; Canada, Japan, the Netherlands, United Kingdom, and United States.

By the end of the embargo in March 1974, the price of oil had risen nearly 300%. As a result, the oil-exporting nations began to accumulate vast wealth. Much went for arms purchases that exacerbated political tensions, particularly in the Middle East. Although the higher oil price had considerable economic effects on western (and Japanese) industrial production, the embargo-affected countries did not undertake dramatic foreign policy changes. However, the crisis had a major impact on international relations and created a rift within NATO with policies shifting from pro-Israel to pro-Arab over time.

The Western countries instigated reduction of public oil consumption by rationing policies and maximum speed limits, fuel saving design of automobiles, house isolation programs, and the search for alternative energy sources.[207] Regions outside the OPEC-control increased exploration and production (ie North Sea, Alaska, Caspian Sea) creating the 'oil glut' of the 1980s.[208] Next to the effect of oil-prices on national economies, the religious tension was influencing the political situation in the Middle-East.

> In 1978 Iran saw a strike consisting of 37,000 workers at Iran's nationalized oil refineries reduce production from 6 million barrels (950,000 m$^3$) per day to about 1.5 million barrels (240,000 m$^3$). It was the start of the Iran Revolution that toppled the pro-Western monarchy of the Shah and replaced it by the Islamic Republic under the rule of Ayatollah Khomeni. From early 1979 to 1983 Iran was in a 'revolutionary crisis mode.'

Next the invasion of Iran by Saddam Hussein's Iraq in 1980 started an eight-year period of turmoil; the *Iran-Iraq War* (1980-1988). Now it was caused by the religious rivalry between the Sunni and Shia factions, and the fear that the Iran Revolution would be exported to Iraq. The turmoil ended in a stalemate with no territorial changes. Both the superpowers Russia and America became entangled in the conflict, and many European countries sided with Iraq.

> The number of casualties was extreme and estimates reach over a million military and civilian deaths. The Iran–Iraq War was the deadliest conventional war ever fought between regular armies of developing countries.

---

[207] The American automobile industry, used to design large, heavy and powerful (12 cylinder, 15 miles/gallon) cars, saw the crisis reduce the demand for large cars. Japanese and European economy/compact cars became successful and forced American manufacturers to adapt their designs.
[208] Later in time, the decline in Russian oil production (30% between 1988 and 1992), effected economies of the Soviet Republics that created their collapse in that period.

At the end of the war, the conflict spilled over to the Kurdistan region. Using 60,000 troops along with helicopter gunships, chemical weapons (poison gas), and mass executions, Iraq hit 15 villages, killing rebels and civilians, and forced tens of thousands of Kurds to relocate to settlements. Many Kurdish civilians fled to Iran. By September 3rd, 1988, the anti-Kurd campaign ended, and all resistance had been crushed; 400 Iraqi soldiers and 50,000–100,000 Kurdish civilians and soldiers had been killed.

*Overview Postbellum Middle East*

The preceding exploration paints in large brush strokes a sketch of the Middle-East. A region that, from old times a tribal-oriented range of societies, as the cradle of the Islam, and part of the Ottoman Empire, had seen the spread of the Islam accompanying the spread of the Arabian culture, language and customs. However, by the twentieth century frustrated Arab nationalist ambitions and socialist and fascist ideologies had given rise to several movements and political parties (eg the Pan-Arab movement and the Ba'ath party). And during the Second World War, the Middle East changed from a strategic geo-political backwater to a new front of the Axis powers. The US realised the importance of the region when they became involved in the war after the Japanese attack on Pearl Harbour.

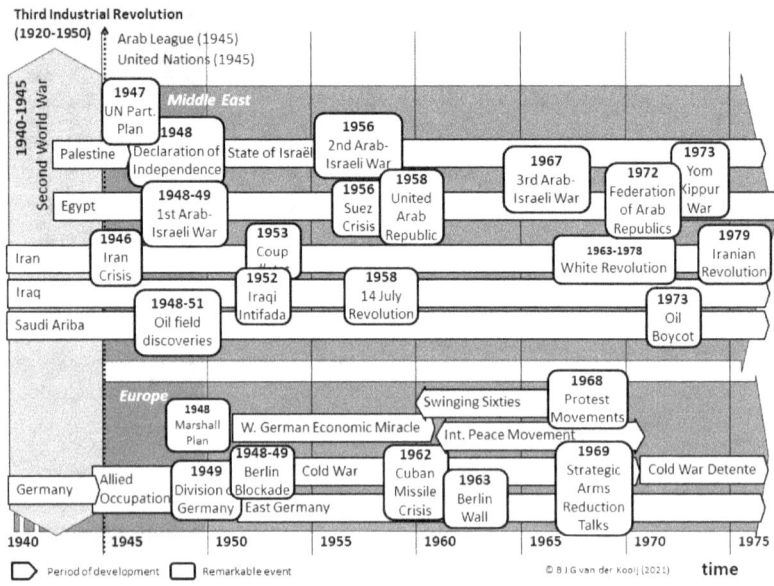

Figure 74: Overview European and Middle East Postbellum (1950-1975).

Britain's influence at the close of the Second World War had dramatically increased. It had troops in almost every Middle Eastern country, and was facing the conversion of its sphere of influence in peacetime over the access to the Far East (ie the Suez Canal), the oil flow from the Middle East, and safeguard it from the communist domination. However, after World II, the geo-political power balance had changed dramatically and this had a profound effect on the Arab and Muslim world. Britain and France were no longer the geo-political powers that they once were, and this provided opportunities for the emerging Arab states to gain independence. The US under the Truman Doctrine stepped in, took over much of the British positions in the Middle East by increasing its military presence. The United States and the Soviet Union had become superpowers and the Middle East became a major theatre of Cold War conflicts.[209]

> Major events in the Postbellum period were the creation of the State of Israel, and the discoveries of massive oil reserves. The first gave rise to the Arab-Israel conflict with a range of wars, and unsuccessful attempt to unify the Arabs states. The second made the Middle-East a political powder keg as the oil-dependence of the West made them mingle in Arab politics (Figure 74).

## Postbellum in the Pacific

In an effort to contain the worldwide spreading of Russian Communism, America's foreign policy in the post-war geo-political situation also focussed on the Far-East. It saw the occupation of Japan (1945-1952), drew America in the 'limited war' of the Korean War (1950-1953),[210] and involved it in the Vietnam War (1955-1975). During that war, America utilized a diplomatic opportunity in the deteriorating relation between the USSR and China. Nixon's visit to China in 1972 —the first US president to visit mainland China while in office— had restored diplomatic relations with the People's Republic of China, and resulted in a significant shift in the Cold War balance.

### *The Chinese Communist Revolution*

The Eastern-confrontation between the Soviet-originated communism and the American imperialistic capitalism had a deeper background. The conflict was between the Nationalists and Communists.

---

[209] The end of Cold War (1989) and Soviet support was the final nail on the coffin for communism and socialism and with them Arab nationalism, in the Middle East.
[210] The war cost the lives of millions of Koreans and Chinese, as well as over 50,000 Americans. It had been a frustrating war for Americans, who were used to forcing the unconditional surrender of their enemies.

On the Chinese part of the Asian continent, since 1912 the political revolts had ended 2000 years of imperial rule, the period of 1927-1924 saw the nationalist Kuomintang (KMT) party under Chiang Kai-shek, and the Chinese Communist Party (CCP), founded in 1921 with help of the Communist Party of the Soviet Union, engaged in the *Chinese Civil War* (1927-1949). After Japan had invaded Manchuria in 1931, the *Second Sino-Japanese War* (1937-1945), in which Japan full-scale invaded China, caused a temporary pause in the civil conflict for political dominance. China's communists and nationalists fought Japan during the Second United Front with aid from both the Soviet Union and the United States, and —after a range of counter offensives in the China theatre of war— on September 2nd, 1945 China regained all territories lost to Japan.

Immediately after the war ended by the surrender of Japan, the Civil War erupted again with the communist-ruled and Soviet-backed Chinese Communist People's Liberation Army (under Mao Zedong) conducting a guerrilla warfare against the Nationalists government (under Chiang Kai-shek). In the North, the Russians made the Japanese settlers leave, dismantled the industrial base in Manchuria build up by the Japanese, and withdrew in March of 1946. In the East, the Japanese confiscated island of Taiwan was placed under the governance of the People's Republic of China (ROC). But that was soon to change with the eruption of the *Chinese Communist Revolution* (1945-1949).

The fight against the Japanese had weakened the nationalists (both military and politically) and strengthened the communist as the confrontation in the *Battle of Liaohsi* (1948) made clear. The nationalists saw their military defeated with heavy losses, and lost their Manchurian territorial base. The victorious communists moved down to the south where they defeated the nationalist forces in the *Battle of Pingjin* and the *Battle of Hsupeng* (November 1948-Janary 1949). In the three confrontations the nationalist lost 2.5 million troops, as well as the support of the west. The communist's leader Mao Zedong proclaimed in October 1st, 1949 the *People's Republic of China* (PRC) on the mainland. Chiang Kai-shek, 600,000 Nationalist troops and about two million Nationalist-sympathizer refugees retreated to the island of Taiwan. It became the start of the Authoritarian Era (1949-1991) in which the People's Republic of China (PRC) and the Republic of China (ROC) on the island of Taiwan continued a state of war until 1979.

Under Mao's rule, China went through a socialist transformation from a traditional peasant society, leaning towards heavy industries under planned economy, while campaigns such as the *Great Leap Forward* (1958-1962) and the *Cultural Revolution* (1966-1976) wreaked havoc on the entire country.

It started with *Hundred Flowers Campaign* (1956-1957) where, to reduce tensions in the communist party, citizens were stimulated to openly express their opinions of the Communist Party. The campaign failed and CCP Chairman Mao Zedong conducted an ideological crackdown on those who had criticized the party. Citizens were rounded up in waves by the hundreds of thousands, publicly criticized, and condemned to prison camps for re-education through labour, or even execution. By 1958 private ownership was abolished and all households were forced into state-operated communes. Besides these economic changes the Party implemented major social changes in the countryside including the banishing of all religious and mystic institutions and ceremonies, replacing them with political meetings and communist propaganda sessions.

> In 1958, after China's first *Five-Year Plan*, Mao called for 'grassroots socialism' in order to accelerate his plans for turning China into a modern industrialized state. Mao launched the *Great Leap Forward*, established People's Communes in the countryside, and began the mass mobilization of the people into collectives.

Nevertheless, these economic policy reforms contributed to the *Great Chinese Famine* (1959-1961) with an estimated death toll due to starvation that ranged in the tens of millions (est. 15-55 million). After a period of less radical leadership to recover from the failures of the Great Leap Forward, in order to regain political dominance, Mao charged that bourgeois elements had infiltrated the government and society with the aim of restoring capitalism. He launched a movement to preserve Chinese Communism by purging remnants of capitalist and traditional elements from Chinese society, and to re-impose Mao Zedong Thought (known outside China as Maoism) as the dominant ideology in the PRC.

> Using the *Red Guards* —a para-military organization originating from militant students— he went after the enemies of state (aka the Five Black Categories[211]) and the old establishment (the 'Four Olds')[212] with the massacres of the 1966 'Red August.' Ransacking many famous temples, shrines, museum and other heritage sites in Beijing old books and art were destroyed. By early 1967 Red Guard units were overthrowing existing party authorities in towns, cities, and entire provinces. It was the origin of the Red Terror that was

---

[211] The landlords, rich farmers, counter-revolutionaries, bad influencers and Rightist.
[212] The Chinese society's old customs, old culture, old habits, and old ideas.

spread over China as the *Cultural Revolution*. The resulting anarchy, terror, and paralysis completely disrupted the urban economy. [213]

*The Rising Suns in the East: Japan, Korea and Taiwan*

While China converted more and more to their own version of the communist ideology, the Pacific region saw some remarkable developments of countries rising from the wartime destruction while adhering to the capitalist-democratic ideology.

## Japan's Recuperation

The war with Japan ended with the dropping of the atomic bombs on Hiroshima and Nagasaki on August 6th and 9th, 1945, respectively. As a result, on August 12th the Emperor stepped down, and on September 2nd, 1945 formally signed Japan's surrender. The military confrontations of the Second World War in the Pacific had come to an end and the Asian Postbellum started.

> In Japan, with US air raids taken off from the Marinas Island and from carriers, conventional bombing and firebombing had resulted in destruction of 67 Japanese cities, as many as 500,000 Japanese deaths and some 5 million more made homeless. Among those cities Tokyo (some 100,000 deaths), Osaka, Nagoya, Kobe, Yokohama and Kawasaki. And the atomic bombs on Hiroshima and Nagasaki had not only caused some 129,000-226,000 direct and indirect deaths, the cities were destroyed by firestorms.

Subsequently, America occupied Japan —an occupation that involved a total of nearly 1 million Allied soldiers— and established the Supreme Commander for the Allied Powers (SCAP, 1946-1952) to suppress Japanese militaristic nationalism, and to restructure the traditional Japanese society into a Western democratic society.[214]

> The first challenge was to feed the starving population and its returning colonists (est. 5 million up to 1946, 1 million in 1947). Next came the disbarment and demobilization of the Japanese soldiers, the release of political prisoners and the purge of wartime

---

[213] The Cultural Revolution damaged China's economy and traditional culture, with an estimated death toll ranging from hundreds of thousands to 20 million.

[214] The new Japanese Constitution of May 3rd, 1947, replacing the Meiji Constitution of 1890, changed Japan's previous totalitarian system of quasi-absolute monarchy and stratocracy with a form of constitutional monarchy and parliamentary democracy seen in Western societies. It featured prominently the 'rights and duties of the people'; from the liberal freedoms (freedom of assembly, speech, writing), to the human rights (eg property rights). Including the individual rights of the people "to be respected as individuals" and, subject to "the public welfare", to "life, liberty, and the pursuit of happiness".

public officials. Then came the creation and implementation of a new Constitution. The emperor became a ceremonial state symbol rather than a divine ruler with power. Similar to the Nuremburg Trials (1945-1946) against German war crime, the Tokyo War Crimes Tribunal (1946-1947) prosecuted and convicted Japanese war crimes.

With rising Cold War tension, the US changed their policy in 1947 and shifted from the demilitarization and democratization of Japan to the economic reconstruction and remilitarization of Japan; the Reverse Course Policies. In many ways, the Reverse Course resembled the Marshall Plan in Europe, especially in that it prioritized economic reconstruction while attempting to limit the influence of communism in the region.[215] In 1952 the US-occupation of Japan ended.

*Japan's Post-war Economic Miracle:* The recovery stage of US occupation of Japan started with agricultural land reform,[216] followed by industrial reform. Actions that incorporated the reforming of the Zaibatsu that had worked so closely with the Japanese military. Next, the resulting economic recovery of Japan became known as the *Japanese Post-war Economic Miracle* (mid-1950s to early-1970s). Under the industrial policies executed by the Ministry of International Trade and Industry (MITI), Japan's industrial base strengthened. The policies to promote domestic industry and to protect it from international competition were strongest in the 1950s and 1960s. First focussing on the shipbuilding industry, then the steel and automobile industries, and finally the electric and electronic industries.[217] As a result, just like the German post-war economy, the devastated Japanese economy rose quickly from the ashes left after the Second World War. Japan rapidly became the world's second largest economy (after the US) and by the 1980s the Japanese economy had become one of the world's largest and most sophisticated.

Japan underwent not only an industrial and economic revolution of its own into 'Japan Incorporated', also society changed. The massive urbanization caused housing problems (eg the single room rabbit

---

[215] Japan's post-occupation relationship with the United States was founded on the understanding that Japan would have access to the US market in exchange for the continued presence of American bases on Japanese soil.
[216] Japan had also its equivalent of the British Lords of the Manor; the absentee landlords who owned agricultural land but did not farm it themselves, and tenant farmers who rented the land in exchange for giving the landlord a high proportion of the crop. Under Japan's 1946 land reform, landlords who owned more than the permitted amount had to sell the excess land to the government at a fixed price. The government then sold it at the same price, giving first preference to any tenant who had been farming the land.
[217] On a personal note, your author stayed in in 1971 as an exchange student three months in Japan at Matsushita Denki's Laboratories near Osaka.

apartments), and mobility problems (eg overcrowded networks of trains), but also created department stores, large shopping areas, movie houses, coffee shops, arcades, bars, nightclubs and restaurants. As the national centre for government, finance, business, industry, education, and the arts, Tokyo became a magnet for many Japanese and the quintessential expression of Japanese urban life. The Japanese society Americanized as the impact of American culture was everywhere. Young urbanites, in particular, took with gusto to jazz and rock music, pinball machines, American soft drinks and fast foods, baseball, and the freer social relations that typified American dating patterns.

## The Division of Korea

Notwithstanding Russia had entered the war against Japan in a late stadium (ie August 8th, 1945) with the invasion of the Japanese puppet state of Manchuria, it extended its sphere of influence on the Chinese mainland to the Korean Peninsula. After the end of war in 1945, the Allied divided the country into a northern area (protected by the Soviets) and a southern area (protected primarily by the United States).

The Japanese occupation of the Korean Peninsula ended with the landing of American troops, and subsequent repatriation of Japanese; over 500,000 by December 1945 and 786,000 by August 1946. Local left-wing uprising and rebellions were suppressed, killing 30,000-100,000 people. Although the Yalta Conference had planned for a four-power multinational trusteeship over the Korean Peninsula, it turned in a Soviet-American rule temporarily dividing the country in two occupation zones (ie comparable with the division of Germany) along the 38th parallel. However, during the early confrontations between the communist and capitalist ideologies resulting in the Cold War, the division became permanent with the establishment of the American supported *Republic of Korea* (ROK) in the south of the Korean Peninsula on August 15th, 1948, promptly followed by the establishment of the Russian supported *Democratic People's Republic of Korea* (DPRK) in North Korea on September 9th, 1948 (Figure 75).

In 1950, after years of mutual hostilities and bloody clashes, North Korea invaded South Korea —with Russia's blessing— in an attempt to re-unify the Peninsula under its communist rule. The subsequent *Korean War* (1950-1953), saw the North (with Russian support) invade the South conquering Seoul, and the South (with American support)[218] counter invade the North. The campaigns ended by 1951 with a stalemate as the front

---

[218] By 1950, under the flag of the United Nations, and under the command of the US, an international military force of twenty-one nations (with the United States providing with over 326,000 soldiers around 90% of the military personnel), became involved.

stabilized, and the last two years were a war of attrition that saw many flee from the North.[219] It was a war in which almost every substantial building and much of the infrastructure in North and South Korea was destroyed.[220] In 1953 the Armistice Agreement left Korea divided by the Korean Demilitarized Zone (DMZ). Tensions between North and South escalated in the late 1960s with a series of low-level armed clashes known as the Korean DMZ Conflict.

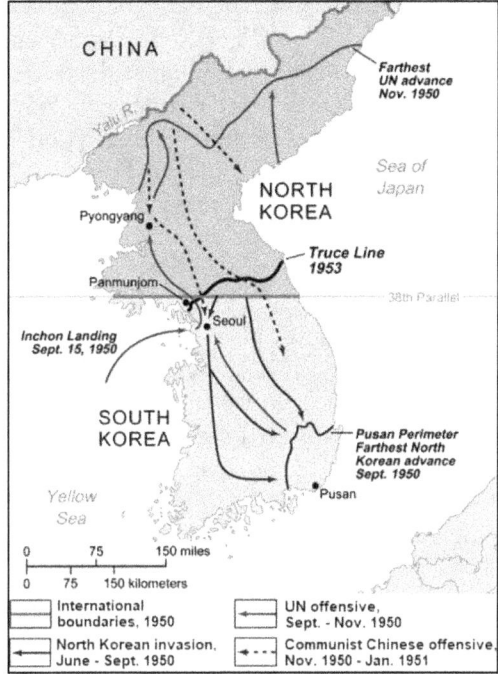

Figure 75: Korean War Campaigns (1950-1953).
Source: https://dpaa-mil.sites.crmforce.mil/

The aftermath of the Korean War set the tone for Cold War tension between all the superpowers. The Korean War was a turning point in America's current foreign policy of active intervention across the globe to stem the expansion of Soviet influence in the defence of US interests. Part of America's foreign policy were concerned with the economic development of South-Korea by creating an independent and self-supporting economy.

The result of all this was the split up of the Korean Peninsula into two different states, each with an opposing ideological background that would paint international relations in the region.

---

[219] Attrition warfare represents an attempt to grind down an opponent's ability to make war by destroying their military resources by any means including guerrilla warfare, people's war, scorched earth. Several million North Koreans are estimated to have fled North Korea,
[220] The Korean War was among the most destructive conflicts of the Cold War era, with approximately 3 million war fatalities and a larger proportional civilian death toll than Second World War or the Vietnam War.

*North Korea:* The Democratic People's Republic of Korea was proclaimed on September 9th, 1948, with the former guerrilla-leader and Communist puppet Kim (aka Kim Il-Sung) as the Soviet-designated premier. By 1949, North Korea was a full-fledged Communist state (with the Marxist-Leninist concepts of collectivized land, etc.) mixed with the Juche (ie self-reliance) ideology. However, North-Korea's recovery from the war was slowed by a massive famine in 1954–55. After the central government took its share, starvation threatened many peasants; about 800,000 died. The first Five Year Plan (1957-1961) consolidated the collectivization of agriculture and initiated mass mobilization campaigns as part of a centrally planned command economy. Industry was fully nationalized by 1959. The 1960s saw a massive state investment in heavy industry, state infrastructure and military strength. In the 1970s, expansion of North Korea's economy, with the accompanying rise in living standards, came to an end as foreign aid declined (eg due to the Chinese Cultural Revolution, and the death of Stalin in 1953 that heralded the ideology-based Sino-Soviet Split (1956-1966).

> Some decades later, after the collapse of the Soviet Union and the Eastern Bloc in 1991, North Korea experienced severe economic hardships. Combined with the mid-1990s natural disasters, including floods and drought, poor agricultural policies, economic mismanagement and rising black markets, the North-Koreans had to liberalize the state-run economy. After his father's heart attack, Kim Jong-Il became in 1994 president of a military dominated junta under the policy "Songun Chong'chi," or 'Military First.' The subsequent international isolation, enhanced by the North Korea's quest for nuclear weapons and intercontinental ballistic missiles (ICBMs), started a period of geo-political isolation in the next millennium.

*South Korea:* After the Japanese defeat in 1945, the transfer of American military governance in 1948 created the *First Republic* (1948-1960) with a presidential system of government. The rule of its first president Syngman Rhee —an American educated Christian— was characterized by authoritarianism and massive corruption. The main policy of South Korea was anti-communism and 'unification by expanding northward'. When the communist army attacked from the North in June 1950, in the *Bodo League Massacre*, some 60,000-200,000 people suspected of communist affiliation or sympathy were killed by the retreating South-Korean forces. In the subsequent years South Korea experienced political turmoil under years of autocratic leadership. The administration became increasingly repressive while dominating the political arena. However, the *April Revolution* (1960) of student protest made president Rhee resign (and flee into exile to Hawaii). The subsequent *Second*

*Republic* (1960-1963) adapted a parliamentary system to remove power from the office of the president.

South Korea up to well into the 1960s truly represented a backward, desolate economy based on subsistence agriculture with all the difficulties of a developing country. Under the earlier Korean administrations, South Korea's economy had grown at a painfully slow rate. Notwithstanding South Korea had received a $3.1 billion from the United States, a very high figure for the time, a privilege for being on the hottest frontier of the Cold War.

The Second Republic was short-lived as a military junta (in the *May-16-Coup* of 1961) dissolved the National Assembly and declared their 'Pledges of the Revolution': anti-communism; strengthened relations with the United States; an end to government corruption termed "fresh and clean morality"; a self-reliant economy; working towards reunification; and a return to democratic civilian government. By 1963, the junta formally stepped back, and the National Assembly created the *Third Republic* (1963-1972) under the presidency of general Park Chung-hee. With military now in the Presidency, it became a dictatorship that prioritized economic development and anti-communism. Park began a series of economic policies that brought rapid economic growth and industrialization to the nation that eventually became known as the Miracle on the Han River.[221]

*South Korean Economic Miracle*: The watershed in Korea's economic history was in the 1960s when, after the military coup of 1961, a protectionist economic policy began, pushing a bourgeoisie that developed in the shadow of the State to reactivate the internal market. The First Five-Year Plan (1962-1966) started and aimed at the development of basic industries (chemicals, iron and steel, shipping and manufacturing) as well as the promotion of science and technology in its consumer electronics industry. Crucially, strong support for R&D was central to the plan and resulted in the establishment of the Korea Institute of Science & Technology (KIST) in 1966, and the Ministry of Science & Technology in 1967.

# The Rise of Taiwan

During the five decades of Japanese rule the 'model colony' of the Taiwan had seen improvements in the island's economy, public works, industry, and a cultural Japanization. After the war had ended, a period of turmoil characterized the political development of Taiwan. The

---

[221] Under his government, South Korea saw the development of chaebol, family companies supported by the state similar to the Japanese zaibatsu.

Kuomintang (KMT) administration on Taiwan was repressive and extremely corrupt compared with the previous Japanese rule, leading to local discontent. It started with the expulsion of the Japanese, some 300,000 people, back to Japan. Then, after the *February-28-Incident* (1947), tens of thousands of people were killed or arrested (est. 18,000-28,000 deaths). It was the start of the period of *White Terror* (1949-1992); political repression in which the KMT persecuted perceived political dissidents, imprisoning some 140,000 people and executing some 3,000-4,000 people. Following the anti-government uprising —or '228' Incident— on February 28th, the KMT retreated from mainland China to Taiwan during the closing stages of the Chinese Civil War in 1949.[222]

**Figure 76: Taiwan and the Taiwan Strait.**
Source: Wikimedia Commons.

Up and until 1958, small-scale military campaigns between the ROC forces and the People's Liberation Army (PLA) were carried out across the strategic islands in the Taiwan Strait (Figure 76). Such as the First Strait Incident (1955) and Second Strait Crisis (1958), in which the United States pledged to protect the island from the mainland.

Chiang Kai-shek developed a top-secret military policy of return to the mainland; the project National Glory (1961). As the catastrophic Great Leap Forward hit the Chinese mainland, he saw a

---

[222] Approximately 2 million ROC troops took part in the retreat, in addition to many civilians and refugees, fleeing the advance of the People's Liberation Army of the Chinese Communist Party (CCP).

crisis-opportunity to launch an attack to reclaim mainland China. By 1965 the plans were completed, but the plan was abandoned by 1972 due to implementation problems and America's opposition.

*The Taiwan Economic Miracle:* After a period of land reform, in 1959, a 19-point program of Economic and Financial Reform, liberalized market controls, stimulated exports and designed a strategy to attract foreign companies and foreign capitals. At first, the Japanese companies moved in, reaping the benefits of low salaries, the lack of environmental laws and controls, a well-educated and capable workforce, and the support of the government. Soon other nations followed. The governmental policies attracted foreign companies in order to obtain more capital and to get access to foreign markets. The big foreign companies got contracts with the huge network of small sized, familiar and national companies, which were an important percentage of the industrial output.

*Overview Far East Postbellum*

While the Second World War had been a global disruption in the Affairs of Man of a magnitude never seen before, its Postbellum period was also unprecedented (Figure 77). Next to being a Reconstruction Era in which material, social and economic recovery from the war took place, both Europe and the Pacific had seen the economic miracles, as well as the

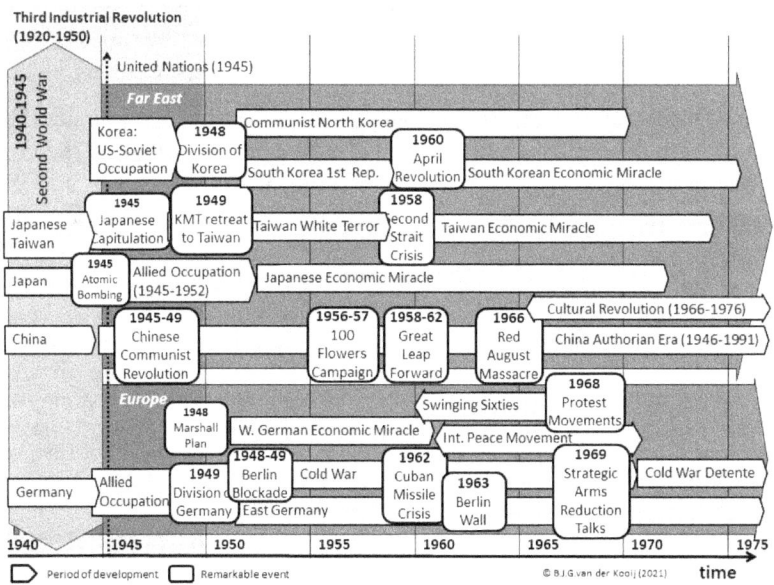

Figure 77: Overview European and Far East Postbellum (1950-1975).

constitutional revolutions where the communist ideology and capitalist-democratic ideology interacted on a large scale in national political affairs.

This all in the context of the Cold War between two emerging superpowers that paralleled societal tensions resulting in massive geo-political change. One of them being the rise of a united Europe, paralleled with the rise of Pax Americana. And it saw the rate of decolonization increase as the former British, French and Dutch colonies freed themselves from colonial governance.

## *The Context of the Industrial Revolutions*

From the Middle Ages on, the agricultural and feudal society[223] had undergone gradual change in which technical change and social change, political change and religious change had been walking companions (Figure 8). The Scientific Revolution had explored the physics of the *Nature of Matter*, the Enlightenment movement had explored the meta-physics of the *Nature of Being*. Had the investigations of the natural philosophers into the Power of Heat and its utilization brought the steam engine, the intellectual endeavours of the enlightenment philosophers had brought the Freedoms of the Body, Soul and Spirit.[224] And the Renaissance Revolution had it all brought together. The physical and mental bondage to Land and Lord was replaced by freedom to move physically and professionally. The bondage to the Sacraments of the Church loosened, giving freedom of religious beliefs. And the mental bondage to Church and State, both Absolute Monarchies, loosened to give way to the liberal freedoms of thinking and living.[225]

These were the times where the Mechanical Worldview dominated the affairs of Man. Times that saw next to 'Change and Destruction,' the times of 'Change and Novelty.' The massive disruptions of Revolutions and Wars, creating the specific Zeitgeists of Warfarin, preceded by the Madness of

---

[223] In the feudal society the landowners provided land to tenants in exchange for their loyalty and service. With the absolute monarch as holder of all lands on religious grounds, it became a system of reciprocal legal and military obligations which existed among the warrior nobility and revolved around the three key concepts of lords, vassals, and fiefs. The peasantry as lowest class were bound to the landowner —the Lord of the Manor and his domain (aka demesme)—in a system of *manorialism*. This system bound the agrarian labourer (aka serf) to the manor and its seigneur. It was a system of limitations, duties (household assistance, land labour, military service), and obligations (demesme labour, paying for the manor's common services, and the Droit the Seigneur before marriage).
[224] What modern-day democratic societies consider as normal and call the Human Rights. Among which the Freedom of Thought, Opinion and Expression, encompassing the Freedoms of Religion and Speech; the right to equal protection under law; and the right to organize and participate fully in political, economic, and cultural life of society.
[225] See: Van der Kooij, B.J.G.: *The Discovery of Innovation: Historic (R)evolutions in Perspective.* (2018)

Times, and succeeded by the Spirit of Times.

The social revolutions, toppling the ancient regimes and replacing them with a new regime (eg the French Revolution), as well as the major geopolitical wars (eg the First and Second World War) saw massive destruction. Material destruction as well as human destruction (eg the physical casualties of revolution and warfarin) and mental destruction (eg the posttraumatic stress disorder, PTSD) that disrupted societies on a massive scale.

The First Industrial Revolution, originating from the Cradle of Industrialisation of the British Isles, had brought novelty to society in an unprecedented way. Steam power affected the Affairs of Man in many ways (eg factorization, mechanized mobility, urbanization). Even though social institutions had the tendency to repeat itself —"The windmill gives you society with the feudal lord; the steam mill, society with the industrial capitalist." (Karl Marx, The Poverty of Philosophy)— societal structures changed as the burgeoning new middle class of the bourgeoisie emerged. New riches not originating from landownership but from colonizing and worldwide trading, gave rise to new societal powers (eg the British colonial nabobs, the Dutch Merchant class). Also, the emerging citizens class wanted a say in their society; its ruling and its taxation. Colonial societies

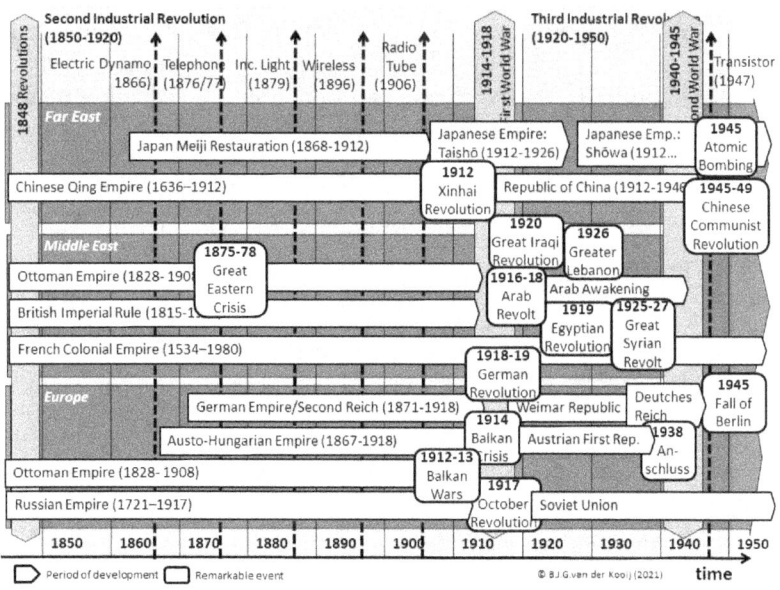

Figure 78: Contextual Overview of major events during the Second and Third Industrial Revolution.

revolted against their overlords (eg the motto of the American Revolution: "No Taxation without Representation").

The *First Industrial Revolution* (1760-1850) brought steam power to the people, and it was the French Revolution (1789-1799) that brought democratic power to the people. It took some decades, but then the seeds of the revolutionary ideas Liberty, Equality and Fraternity developed into social pastures, troubling the Ancient Regimes —ie the absolute monarchies and their aristocracies— desperately clinging to their old powers and their acquired rights (ie the Manorial rights, the Royal prerogatives) of earlier times. Over time the former absolute monarchies were replaced by the constitutional democracies[226] —based on the principles of the Trias Politca[227]— with their classical and social liberalism in *Times of Innovation* of the nineteenth century (Figure 8).

The *Second Industrial Revolution* (1850-1920) (Figure 78) —characterised by the evolution of electricity as source of power (van der Kooij, 2016), and as carrier of information (van der Kooij, 2017)— had brought a new wave of industrialization next to the maturing systems of the Era of Steam (eg national railroad networks, steam shipping lines traveling the globe). Fuelled by the development of electro-mechanical technologies, technological communication systems (eg telegraphy, telephony) spanned the world. And the dark times of candle and gaslight faded away when the electric lamp brightened up the industrial, business and private living environments. The workshop factories of the industrial society started with organizing mass production of mechanical machines (from the automobile to the typewriter). Followed by the assembly line, the factory saw increasing productivity with the new methods of production. The socio-economic impact was enormous, and the period from 1870 to 1890 saw the greatest increase in economic growth in such a short period as ever in previous history. Local markets became (inter-) national markets. The division of labour made both unskilled and skilled labour more productive, and led to a rapid growth of population in regional industrial centres (Figure 9).

---

[226] Democracy rests upon the principles of majority rule and minority rights. Majority rule must be coupled with guarantees of individual human rights that, in turn, serve to protect the rights of minorities and dissenters: whether ethnic, religious, or simply the losers in political debate.
[227] The Trias Politica concept concerns —in order to prevent the concentration of power— the separation of state power into three branches: a legislature, an executive, and a judiciary branch. It is a system of checks and balances.

Next it was the *Third Industrial Revolution* (1920-1950),[228] bounded by the First and Second World War, that saw a new wave of industrialization. The development of the Internal Combustion Engine (ICE) as a source of rotary power contributed, together with other technologies like chemicals (eg rubber, plastics) to the Automobile Revolution as well as the Aviation Revolution (van der Kooij, 2021b). The mechanical arithmetic and computational machines, and the analytical machines fuelled science and business. In addition, the development trajectories of the radio tube and transistor started the 'electronic' technologies replacing mechanical technologies, and their subsequent revolutions of computing systems (eg the Revolutions of Mainframes and Mini-computers) and communication systems (eg Mobile Phone, Radio and Television Broadcasting Revolutions).

In the period after the 1950s, that we labelled as the *Fourth Industrial Revolution* (1950-2000), the invention of the transistor (1947) gave society the Power of Computing. The new development in miniaturization of electronic circuits had created the Integrated Circuit (1960). And originating from the electromechanical computing machines, the Computer on a Chip had sprouted (1971). Subsequently the progress of semiconductor technology and its offspring had fuelled technological evolution on a massive scale.

## Changing Perspective

The preceding exploring into the context of Change and Novelty shows the contextual picture —painted in rough brushstrokes— of a societal development in the Western World (ie North America and Continental Europe) over the period from the 1800s up to the 2000s. A period of time in which Technical Change had become the walking companion of Social Change. Technical Change fuelled by Science and Engineering creating technological-induced novelty on a scale not seen before. Where Social Change was influenced by the times of collapsing Imperialism and Colonialism, as well as the horrors of worldwide warfarin, creating the context for what was happening with the Affairs of Man.

One the one hand there were the *evolutionary developments* in the societal institutions such as the geo-political power structures. On the other hand, there were the *revolutionary developments* that took place in societies. Such as socio-political revolution like the American, French and Russian Revolution. And the technical revolutions like the Industrial Revolutions. It

---

[228] In contrast to the historian's prevailing interpretation of the Third Industrial Revolution (IR3) as the Digital Revolution, we define this part of the Third Industrial over a shorter time frame focussing on the maturing mechanical artifacts and their power conversion: IR1 is steam-based power, IR2 is electricity-based power, IR3 is solid-fuel based power.

was the mechanisms of Social Change, Political Change, and Economic Change that fuelled a process of constant change shaping societies.

It is time to zoom in in on the technological developments that accompanied these changes. So, next we will start and explore the events of that technological induced novelty by focussing on the developments within the Electric Revolution; the period of time where Man mastered the new General Purpose Technology of Electricity that proved to be a fertile source of novelty.

# The Context of the Electric Era

For somebody living in the pre-electric times, it would have been impossible to imagine our present society. Living in the times of the candlelight and gaslight, horse power and windmill, the farmers following the rhythm of nature, the citizen the dirty, overcrowded and noisy urban lifestyle. With only one mobile power source at their disposal; the horse polluting the towns with their manure and the clatter of their horseshoes on the cobbles as the horse carriages of the wealthy passed by. How could some-one living in that context have imagined our present times dominated by electricity powering the machines?

In the preceding exploration the dynamic *context* of the Western Word in the 1800-1970 period was painted in large brushstrokes. Creating a picture of a context where western societies underwent —in different degrees and at different times— a massive evolution from an agrarian society to an industrial society. An evolution dotted with civil turmoil and uproar, by limited regional warfarin as well as by large geopolitical wars. An evolution where times of economic progress were interspersed by times of economic depression determining

**Figure 79: Benjamin Franklin and his kite experiment.**

Source: Wikimedia Commons.

the Affairs of Man. And an evolution where 'technology' as the result of science and engineering —the totality of the contributions of all the thinkers and tinkerers— initiated, paralleled and shaped those developments. It was the Era of Mechanization where mechanical technologies offered the foundation for the more specific technologies; from the chemical technologies to the electric technologies.

## *Scientists discovering the Nature of Lightning*

The Electric Revolution found its roots when, by the end of the seventeenth century. the Scientific Revolution had resulted in a new view of nature. A development in which 'science' over time became an autonomous discipline, distinct from philosophy and technology. This Scientific Revolution was the time where natural philosophers all over Europe focussed on specific natural phenomena —both in the Macro-Cosmos (explored by the telescope) and the Micro-Cosmos (explored by the microscope)— to investigate the Nature of Matter.

The work of these *natural philosophers* resulted in the progress of 'natural philosophy' that grew apart in different directions, later to be called scientific disciplines. Scientific thinking was about the basic constituents of the natural word: matter and energy as studied by the classical physics. One

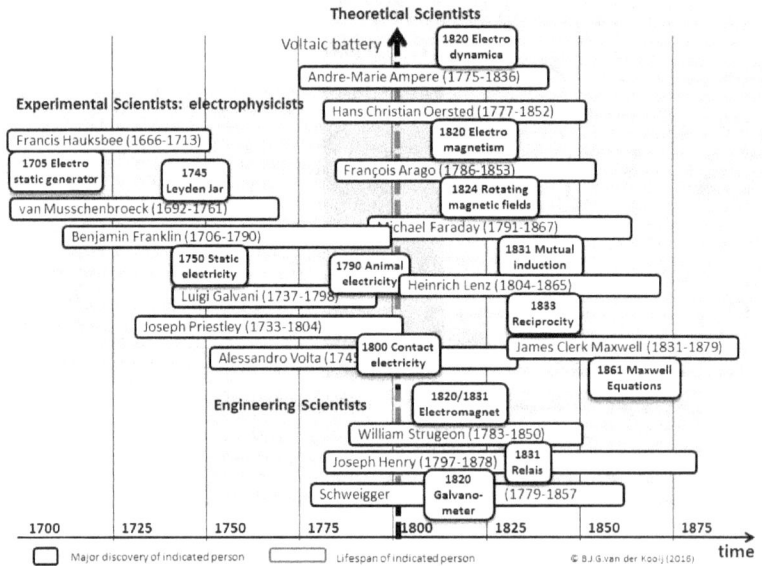

Figure 80: Contributors to the Discovery-trajectory of the GPT-Electricity.

of those constituents was 'electrical' energy that could be found in the electric lightning —aka 'electric fire'— that had frightened people from old times on. From the eighteenth century on, that new phenomenon of 'electricity' came under the scrutiny of many curious and inventive people.

Already by the early eighteenth-century people had been observing 'electric fish'; different species of fish that produced with their electric organs shocks when touched; from the electric eel to the electric ray. They could produce electric shocks up to 200 volts that were strong enough to fell a human adult. It gave rise to theoretical views on *animal electricity*. The same effect of electric shocks was discovered when two suitable materials (such as glass and a cat's fur or a woollen cloth) were rubbed against each other. This gave rise to the theoretical views of *frictional electricity* that created the electrostatic machine. Others curious people —such as Benjamin Franklin (Figure 79)— studied the dangerous lightning accompanying thunderstorms and were trying to unravel its secrets: that form of electricity became known as *static electricity*. Thus the *General Purpose Technology* (GPT) of 'Electricity' came into existence when the secrets of the 'electricity' were slowly unravelled by different kinds of experimental scientists (Figure 80, left); people called the 'electro-physicists', each with different theories.

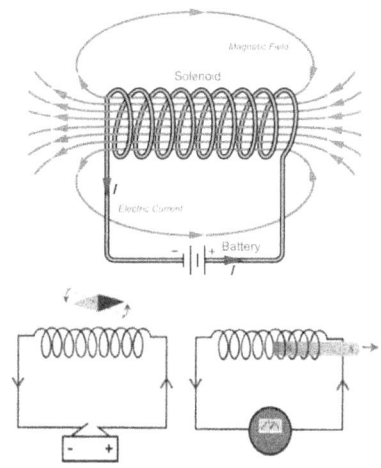

**Figure 81: Principles of Electro-magnetism.**

An electric current from a battery creates an electromagnetic field moving (top), for example moving a compass needle (lower left). And a permanent magnet moved in a coil creates an electric current (lower right).

But that confusion changed around the turn of the nineteenth century when Alessandro Volta explored 'electro-chemistry' and developed the electro-chemical cell.[229] His discovery of *contact electricity* created by the electro-chemical 'Voltaic' battery, was a breakthrough that resulted in a scientific frenzy of the *Battery Mania*: the experimenting with the new source of energy. All that interest soon resulted in the first novel applications; the electromagnet, the 'direct current' electric rotative motor, the glowing wire, and the early spark lights. It was a trajectory of

---

[229] In the electro-chemical cell, a chemical reaction produces a current of electrons. When those electrons flow in one direction, it is called Direct Current (DC).

constant discovery by the engineering scientists (ie Schweigger, Henry, Sturgeon) (Figure 80, bottom). In parallel, major properties of electricity — such as electromagnetism and magnetic fields (Figure 81)— were unravelled by the theoretical scientists like Oersted, Ampere, Arago and Faraday (Figure 80, top).
230

## The Discovery of Basic Properties of Electricity

Much exploration in the new field of electricity was about the phenomenon of the magnetic field created by a solenoid (Figure 81, top); such as the experiments by Michael Faraday. Experimenting with a voltaic battery, the Danish physicist and chemist Hans Christian Oersted (1777-1851) discovered that the electric current, when running through a coil (aka solenoid) could move a nearby compass needle (Figure 81, bottom left).

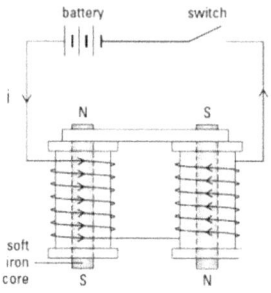

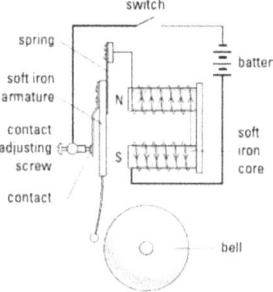

**Figure 82: Principle of the Electromagnet and the Relay.**

The electromagnet (top) being activated attracts other iron pieces by its South and North pole. Such as a soft-iron armature (bottom). That force could be used, with an interrupter contact, to ring a bell (bottom).

Soon others engineering scientists showed that a wire coiled in a spiral acted just like a natural magnet. The iron rod moved in or out the coil depending of the polarity of the battery. Son the reciprocal effect was discovered, as moving the iron core in and out the coil even created an electric current in the circuit (Figure 81, bottom right). Central to these phenomena were the *magnetic fields* created by the electric current in the coil; the property of electro-magnetism was discovered (Figure 81, top).

This property of electro-magnetism was used to create artifacts such as the electromagnet. Adding a soft iron armature that was attracted by that electro-magnetism of the (non-moveable/fixed) electro-magnets, created movement of that armature. For example, to ring a bell. So, electro-magnetism was used to create artifacts that could convert electrical energy

---

[230] See for more details: B.J.G. van der Kooij. *The Invention of the Electromotive Engine* (2015).

in mechanical work (ie the property of linear movement).

In addition to the development of the relay, the same property of electro-magnetism creating movement was used in another device. The device was called the (mirror)-galvanometer. DC-electricity made the coil of the (mirror)-galvanometer to move when a (feeble) current would pass through the coil placed around a stationary iron core. A

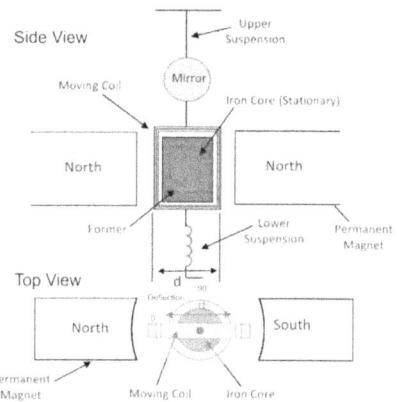

**Figure 83: Principle of the Mirror Galvanometer.**

An electric current running through the Moving Coil placed between the two magnet poles, caused a movement of that coil.

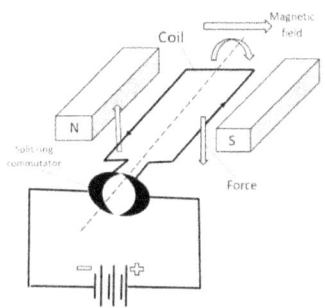

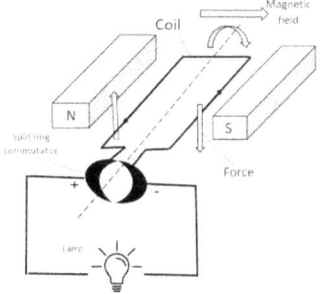

**Figure 84: Principle of the Electric Motor and the Electric Dynamo.**

The Electromotor: a battery supplies an electric current to the coil that rotates in the magnetic field (top). The Dynamo: rotating the coil in a magnetic field creates an electric current that lights a lamp (bottom).

displacement in proportion to its strengths over an angle of max. 90° (Figure 83). The result was a measuring instrument called 'galvanometer' (using a mirror connected to the coil), as well as the needle telegraphy (using a needle connected to the coil).

This principle started the initial rotation (aka deflection) of a coil (max 90 degrees), but to rotate a full circle of 360 degrees a change of the DC polarity was needed. The following step was to add a switch to the coil called 'commutator.' This switch would reverse the polarity of the DC-electricity when the coil reached its maximal deflection. The combination of coil and commutator made the coil rotate

360° in a magnetic field (ie the property of rotation). So it was discovered that a moveable coil in a magnetic field created rotation, and adding a switch to change polarity to the coil, the continuous

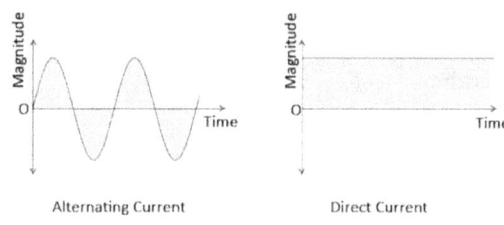

**Figure 85: AC electricity versus DC electricity.**

rotary movement became possible. The electromotor was born (Figure 84, top).

Next the reciprocal version of the electric motor was discovered, when a coil rotating in a magnetic field —caused by an external force called the prime mover[231]— created electricity. Electricality that could be used, for example, to light an electric lamp. The electric dynamo was born (Figure 84, bottom).

The result of the development of the electric dynamo was the creation of a new form of electricity (Figure 85). Had the battery supplied a direct current (DC), now the rotating dynamo supplied an alternating current (AC). This would have massive implications we will explore further on.

These were the basic properties of electricity that would create the artifacts like the electro-magnetic relays and the electric motor/dynamo that would have a major impact of the further development of the electric technologies. Next to the relay, galvanometer and electric motor/ dynamo, the electromagnetic properties of a coil were also applied in the sound-wave transducers. These artifacts made the 'acoustic telegraph" (ie telephone) possible (Figure 86, top) as well as later acoustic communication systems (eg radio) with their microphone and the loudspeaker (Figure 86, bottom). The mechanism was simple; the moving coil microphone converting the air movement from acoustic waves (ie speech) on a membrane into an electric frequency (ie audio signal), and the moving coil loudspeaker converting an electric frequency into an air movement (ie sound).

In addition to the 'active' elements, the 'passive' elements came on scene. Next to the already earlier developed concept of 'resistance' (Ohm's

---

[231] The prime mover can be from a natural source like wind, water. But also from a external combustion engine (aka the steam engine) or internal combustion engine (eg the Diesel engine).

Law as created by Georg Simon Ohm), also the Inductor (L) was applied in circuits. The same goes for the Leyden Jar that evolved into the Capacitor (C). Together with their derivatives —the variable resistance (Rheostat) and variable capacitor— the resistor, the inductor and the capacitor became the passive building blocks of electric circuits in telegraphy, telephony and wireless transmission (Figure 97, Figure 101, Figure 104). In addition, there emerged the passive components based on a specific effect of materials. From isolators to conductors and semi-conductors. Or materials like the quartz crystals that oscillated due to the piezo-electric effect. In different combinations these passive components were used in electric circuits. Both active and passive elements became the building blocks of telegraphic circuits, followed by telephonic circuits and in wireless circuits, and in logic circuits. Electric circuits that defined the functionality of the device; from amplifier and oscillator to modulator/demodulator.[232]

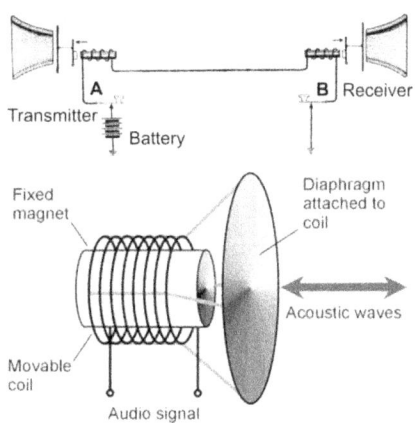

**Figure 86: Principle of Sound Conversion.**
Principle of the Acoustic Telegraph with transmitter and receiver (top), and the principle of a Wave Transducer (bottom).

## *The Broad Application of Electricity*

The preceding description of some basic artifacts of electricity, illustrates the versatility of the new phenomenon. A versatility that created a range of application trajectories that became technologies of their own.

### **Electricity as Creator of Movement**

Further experimenting led to a range of specific technical-trajectories. The electromotive force could be used to create movement, and that resulted in engineering scientists creating the early electric motors. In reciprocity, the mechanical movement of a coil rotating in a magnetic field

---

[232] A modulator is an electronic circuit that superimposes a low-frequency (information) signal onto a high-frequency (carrier) signal for the purpose of wireless transmission. The demodulator does the reverse.

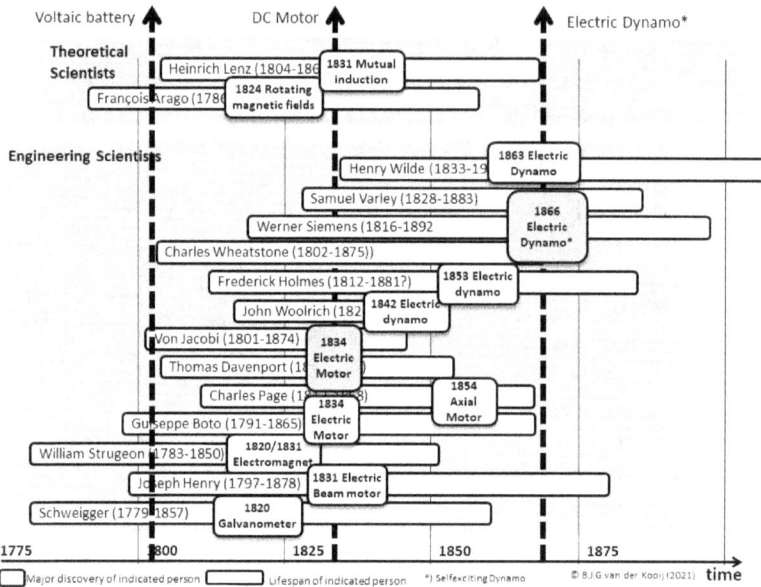

Figure 87: **Overview Contributors to the Development of the Electric Motor and Dynamo.**

created electricity; the early dynamos. It was the discovery of the fundamental relation between magnetism, electricity and movement.

Movement that was created first in the form of the *reciprocal electrical motor* where (two) magnets created the action, just as the cylinder in a steam machine created rotation. But that proved to be a dead-end technology. However, when the properties of electromagnetic fields were further explored, it became clear that rotative motion could be realised easier with a different construct: the movable coil between the poles of two magnets.

Thus, the DC-electric motor was born in the 1830s-1840s. From the early crude artefacts slowly engineering efforts developed a range of electric motors that were powered by direct current (DC) from batteries (eg Froment, Page, Davenport, Jacobi) (Figure 88). However, this development of the linear electromotor proved to be a dead-end technology. Notwithstanding the Battery Craze among scientist, the electro-chemical process that created the battery was a quite cumbersome affair that hindered the further applications considerably. But that was to change, although it took some time.

*Electric Motor and Electric Dynamo:* After it had become clear that creating movement with electricity was possible, soon the reciprocal version was discovered (ie by Lenz): now electricity could also be created from rotating actions. The electric dynamo was born. A range of engineering scientist jumped on the bandwagon and started experimenting (Figure 87). During the Battery Craze they focussed om improving the Voltaic Cell. But soon further experimenting resulted in a range of DC-dynamo's that could replace the battery as source of electricity. The next step was the development of the more powerful self-exciting dynamo in the 1860s. Electricity, both Alternating Current (AC) and Direct Current (DC) became available in abundance, creating the spring board for further application of electricity.

**Figure 88: Early Reciprocal Electric Motor.**
Replica of Gustav Froment's reciprocal motor imitating a steam engine (1840s).
Source: Wikimedia Commons.

## Electricity as Carrier of Information

In parallel to the development of the electric motor and dynamo — where electricity was used a source of energy— many curious people investigated another trajectory. Now electricity was no so much used to transport energy, as well as the use of electricity to transport information. It was the trajectory of the communication engines that used electricity to transmit messages (later called telegrams) over a network of cables: the telegraph that could write at a distance was born. It was the parallel development that took place in both Great-Britain and in America that resulted in the mid-1830s in distant writing with lightning speed: Cooke & Wheatstone's telegraph, and Samuel Morse's telegraph (Figure 89). [233]

Soon, the further development of the telegraph focussed on two development trajectories: the technical trajectory (with technical improvements such as communication speed, longer distances and automatic connection), and the techno-economic trajectory. The later trajectory was dominated by the push to find a solution for when the need for more capacity arose and the high cost of the long copper telegraph lines was hindering further applications of the new communication medium. It was the search for the 'harmonic telegraphy' where one telegraph line could transmit several individual telegraph signals at the same time.

---

[233] See for more details: B.J.G. van der Kooij. *The Invention of the Communication Engine 'Telegraph'* (2015).

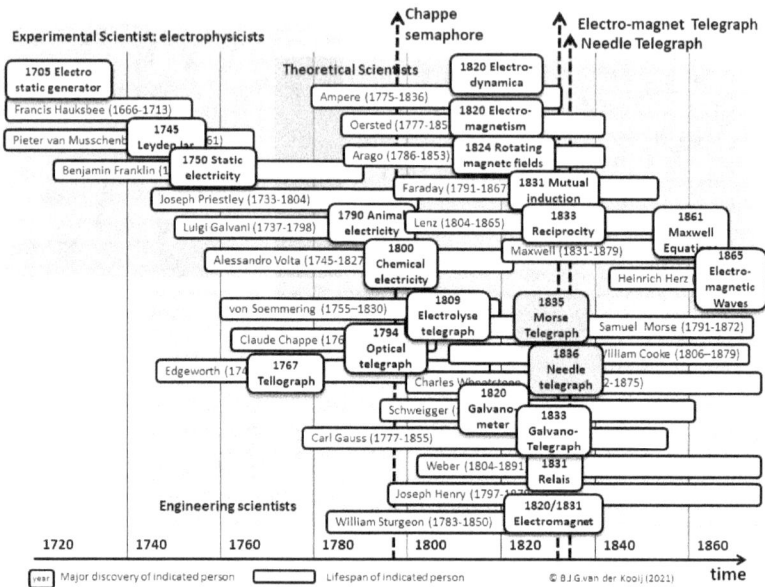

Figure 89: Contributors to the Development of Cabled Telegraphy.

In a totally different field of physics —the field that came to be known as 'acoustics'— the experimenting of many curious engineering scientists into the producing (eg Helmholtz) and the recording of sound (eg Edison) had created artefacts such as the 'Helmholtz Resonator' and the 'Phonograph' (lit. 'sound writer'). In addition, the work of the more theoretical scientists had resulted in a better insight in the Nature of Sound and had created theories of sound (eg by John Strutt). Then, it was the fusion of the build-up knowhow of 'telegraphy' and the build-up knowledge the field of 'acoustics' (Figure 90) that created something unheard of: 'Electric Speech' [234]. It was Alexander Graham Bell —the teacher of deaf fascinated by telegraphic communication over distance— who, after much experimenting with the 'harmonic' telegraph, transmitted audible sounds over a considerable distance in the 1880s. The new phenomenon of 'distant speaking' complemented 'distant writing'. Communication over distance had found a new dimension. But there was more to come.

## Electricity as Creator of Light

Light had fascinated scientist already for a long time. Several 'Nature of Light'-scientists (eg Herschel, Ritter, Fresnel and Fraunhofer) had discovered the properties of natural light (Figure 91). That electricity could

---

[234] See for more details: B.J.G. van der Kooij. *The Invention of the Communication Engine 'Telephone'* (2015).

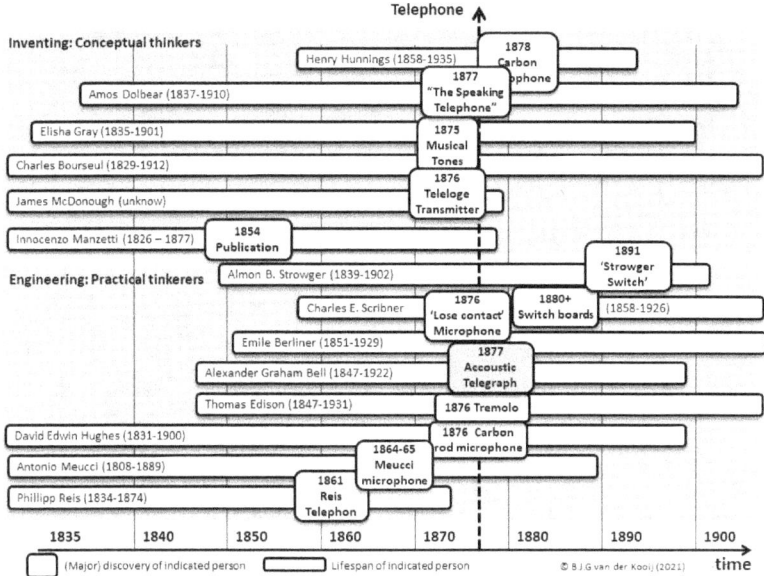

Figure 90: Overview Contributors to the Development of Telephony.

create light became obvious when the battery experimenters (eg Humphry Davy) discovered that shortcutting a battery created a spark, or that a thin wire would start to glow when the current that passed it was large enough. It resulted in two different developments. [235] One were the sparks that created the 'arc light,' similar to the light one observes in a welding process today. It gave rise to the development of the Arc Lamp; a bright source of white light uncomfortable to the eye. The other was the incandescent light that was emitted by a glowing wire encapsulated in a glass vessel (aka bulb). Among the many experimenters in this trajectory, it was Edison who created in the 1870s an incandescent lamp that had a considerable life-span. For more than a century in a range of incremental improvements and product variations —till the rise of the LED-light in the twenty-first century— it would be an important source for electric illumination that brightened up the Affairs of Man.

## Electricity as Creator of Electromagnetic Waves

The development of telegraphy was hindered by the cables that were needed to transmit the electric signal over distance. It was a fixed point-to-point system that did not allow for mobile application. Then, a young Italian called Guglielmo Marconi realized a system for wireless

---

[235] See for more details: B.J.G. van der Kooij. *The Invention of the Electric Light* (2015).

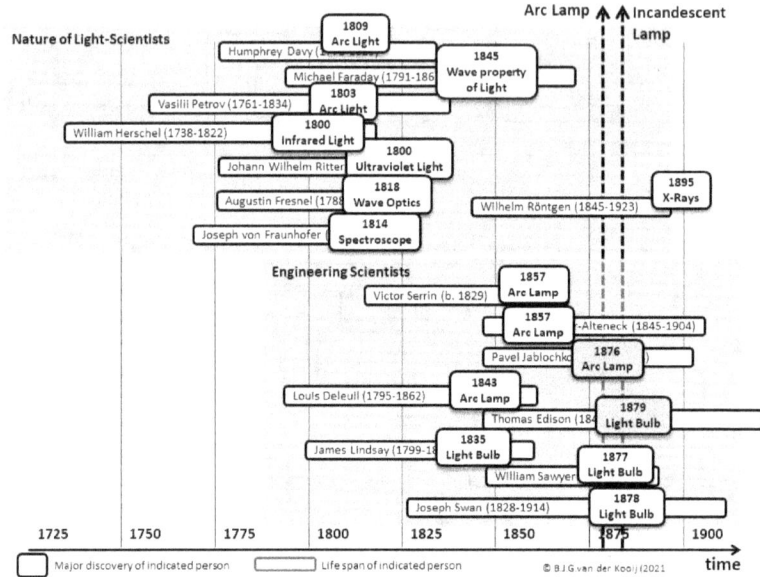

Figure 91: **Overview Contributors to the Development of Electric Light.**

communication, using the Hertzian waves —the electro-magnetic waves discovered by Heinrich Hertz— to transmit the information through the air.[236] Soon, after he took his invention to England, experimenting and improving his system, he covered increasingly large distances. Within a period of two decades his invention would change how the world communicated as result of the Wireless Mania among scientists and engineers (Figure 92). By the start of the First World War it had become clear to many national governments —including the military—that wireless communication would be the next part of the Communication Revolution.

So, originating from the basic innovation of Volta's battery, the GPT-Electricity had followed a technical trajectory as well as several application trajectories (Scheme 2) that would influence the Affairs of Man in a revolutionary way.

---

[236] See for more details: B.J.G. van der Kooij. *The Invention of the Wireless Communication Engine* (2015).

## The GPT-Electricity: Evolution and Revolution

This in a nutshell described the development of the General Purpose Technology Electricity (GPT-E) over a period of time (Figure 93) along different application trajectories. A development that started with the DC-battery that had practical limitations of its own. However, after electricity could be created by dynamos in abundance (by the mid-century), the distribution of electricity created a network of local, regional and national electric powerlines: the infrastructure of electricity distribution known as the 'electric grid.' Now electricity came available to power a myriad of electric motors used in streetcars, elevators, ventilators and manufacturing equipment. It also powered electric light in houses, factories, and public places (Figure 93, top).

In addition, the rise of the communication engines telegraph and telephone resulted in a worldwide cabled-communication infrastructure that —over decades— evolved from the early point-to-point telegraph lines. First covering the high-populated regions, then followed by regional and continental coverage, and finally by Trans-Atlantic submarine cables connecting the continents. Cables spanned the globe, transmitting messages that crossed the continents (eg the British All Red Line, 1902). Cables that offered the solution for distant communication, but also cables that were

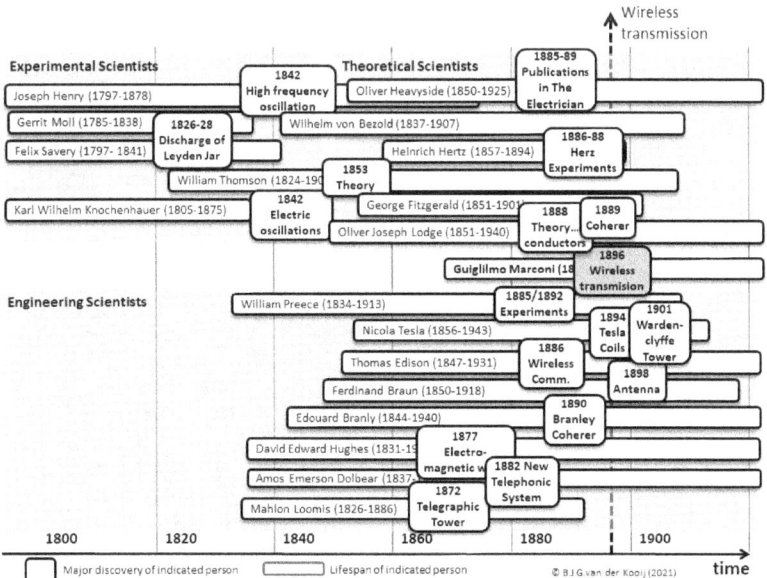

Figure 92: Contributors to the Discovery trajectory of Wireless Transmission.

the problem, both in a technical and in an economic sense. But the development of wireless telegraphy —soon followed by (early) wireless telephony— would solve that problem.

From the early nineteenth century up to the early twentieth century, a range of discoveries, inventions and innovations in the field of electricity had changed society: it had been an *Electric Revolution*. Starting with the *First Electric Revolution* (1800-1860) that emerged after Volta's invention of the electro-chemical battery creating DC-currents, it was followed by the *Second Electric Revolution* (1860-1900) after the AC electric dynamo and AC electro-motor came into maturity. The abundance of electricity changed the way people lived and worked, freed from the limitations of daylight by the electric light available at the click of a switch. And freed by electric power replacing human and animal muscle power, as well as the natural powers of wind and water. In addition, the new means of electricity-based communications not only enabled new means of communication, it also had massive side-consequences.

Such as enabling the rise of standard time that would go and rule private and professional live. It also ended the marine incommunicado-problem when ships left ports, and would facilitate transparency in trade and commerce. All those technical changes made possible by electricity had, next to the civil freedoms created by

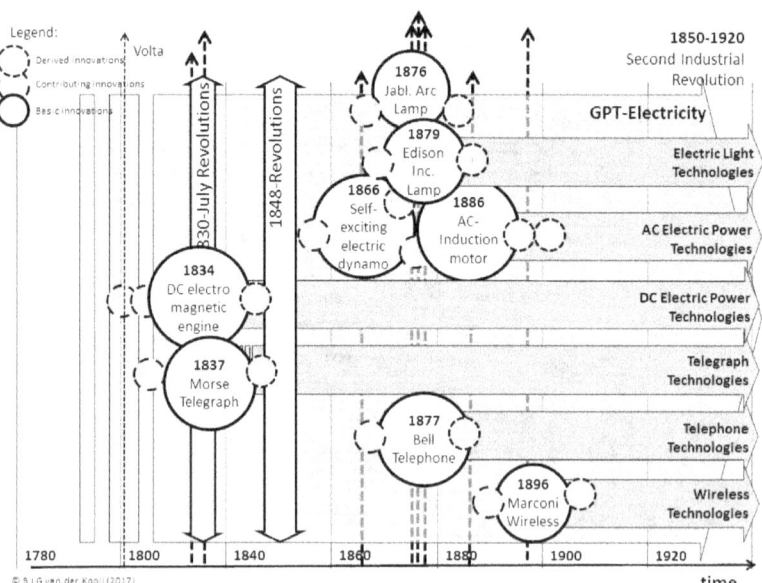

Figure 93: Overview of **GPT-Electricity** with its Power Applications and Communication Applications.

the Enlightenment, given humankind additional *technology-based freedoms*. At the end of the nineteenth century and the beginning of the twentieth century, the world was ready for the next phase in the development of the GPT-Electricity. It was to be found in electricity's capability to carry information.

## A new Primary Mover for Electric Power

Those were the applications of electricity, but there was also another side to its existence. Electricity after its electro-chemical (ie battery) phase needed a primarily mover for its creation. From the wind and water turbine propelling the electric dynamos, to the external combustion engine (ie the coal and oil fuelled steam machine) and the internal combustion engine (ie the diesel-fuelled combustion motor) driving massive electricity generators. In their totality the GPT-Electricity had being fundamental in the *First Power Revolution* (the Power of Heat from wood, coal and oil creating Steam), the *Second Power Revolution* (in which chemical reactions created DC-electricity, and steam power AC-electricity), and the *Third Power Revolution* (the Power of Explosion derived from oil and gas creating the combustion-

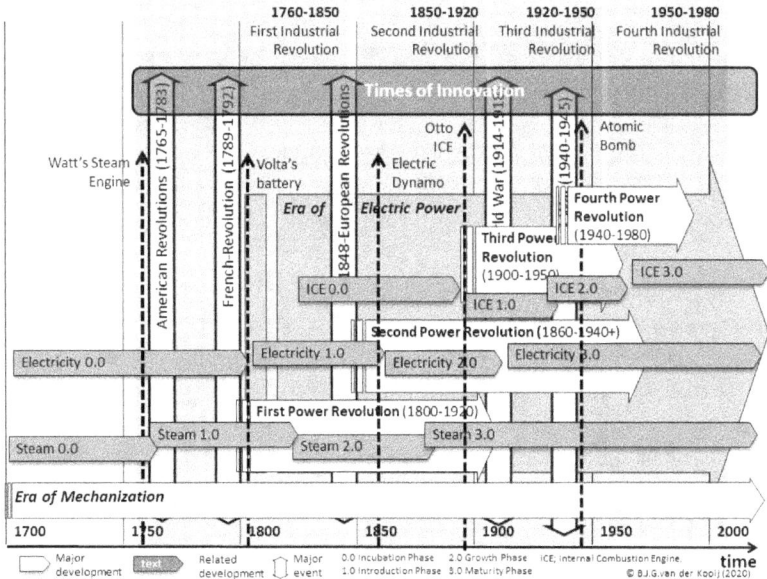

**Figure 94: Overview Era of Electric Power.**
The different Power Revolutions related to the life cycles of the contributing technologies marking the subsequent Industrial Revolutions.

power).[237] With the rise of nuclear power reaching civil applications, a new prime mover emerged that could heat water into steam. Steam that could power the electricity generators. It would create the *Fourth Power Revolution* (the Power of Nuclear Fusion creating steam) (Figure 94).

Once the basics of nuclear *fusion* —ie the combining of two atomic nuclei into one atom nuclei— and nuclear *fission* —in which the nucleus of an atom splits into two or more smaller, lighter nuclei— were understood, next to the application in military weaponry, other civil applications were explored. Such as nuclear medicine where the nuclear medical applications saw the rises of therapeutic and diagnostic procedures. In the field of power generation, it became the nuclear powerplant used for naval propulsion (eg in submarines, aircraft carriers), but also as part of electricity production for civil use, replacing coal and oil. There was a general feeling that everything would use a nuclear power source of some sort, in a productive way, from irradiating food to preserve it, to the development of nuclear medicine.

However, in a relatively short period, from the rise of the Atomic Age to its demise, in a period of some thirty years, nuclear power became controversial. The more when it became a major component of the Cold War threatening human's existence. Soon the early dramatic rise of nuclear power went into equally meteoric reverse. Not only the nuclear incidents, but also the nuclear waste problem, and the relation to nuclear weapons, as well as the Zeitgeist of the Nuclear Holocaust had shaped the public opinion. Rallies against the use of nuclear power, combined with the demonstrations of the peace movements, flared up at late 1970s, early 1980s. Influencing the political climate drastically.

## Changing Perspective

In the preceding exploration we focussed on the evolution of the early electric power technologies and electric communication technologies, characterized by events that were of an electro-mechanical nature. Events that occurred up into the twentieth century. Now, changing our perspective, we will zoom in on the specific events that resulted in the development of the Communication Technologies (CT).

---

[237] The Power Revolutions became historian's prime indicator for the concept of the Industrial Revolutions (IR); steam for IR1, electricity for IR2 and explosion for IR3. Continuing in that perspective, the Fourth Power Revolution should be heralding IR4.

# The Communication (R)evolution

From the early development of Life, the exchange of information is part of biologic lifeforms. Take the animal kingdom, where verbal and non-verbal communication support the physical, safety and social needs. Communication in the form of the cry of a bird to warn for a predator, the pheromones of ants marking a food trail, the courtship dance as part of mating rituals.

Communication, here observed as the exchange of information between a 'sender' and a 'receiver' by technical means, is fundamental in the Affairs of Man. Not only does it enable social entities like the family, clans and nations to function by the spoken word, it also spreads and transfers their culture by the written word; from the handwritten postal mail to the monks creating the early handwritten books. The development of the technical means —aka technology— influenced that communication process in different ways. One was the covering of distances.

Over time communication of the coded word and the spoken word started covering

**Figure 95: Postal Mail distributed by the riders of the Pony Express pass the newly erected telegraph poles.**
Source: George Ottinger (artist). Wikimedia Commons.

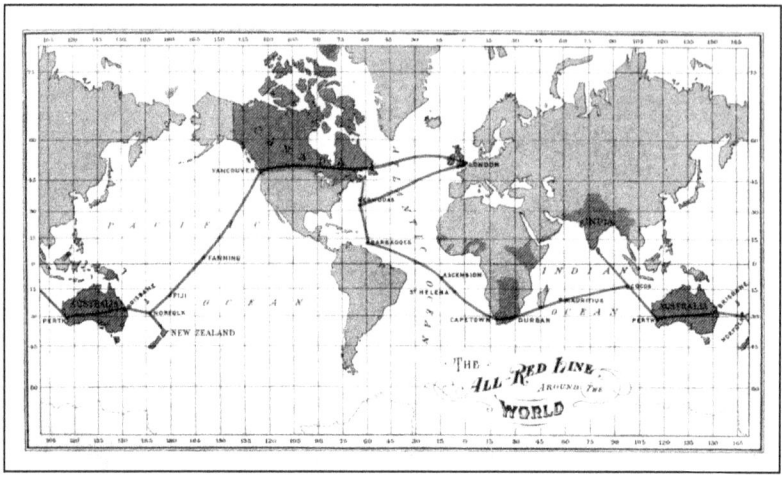

**Figure 96: Cabled Telegraphy spanning the World.**
Source: Wikimedia Commons.

large distances replacing earlier forms of distant communication; distant writing (ie tele-graphy) was replacing postal mail (Figure 95), and distant speaking (ie tele-phony) enabled the spoken word to cover large distances.

This early development of tele-communication would be the dawn of the Era of Tele-Communication where technical means would affect the Affairs of Man again both in his private life and business life. A period of time that would create the cabled telegraphy infrastructures spanning the world. Such as the 'All Red Line' system of telegraphs that linked the British Empire by a network of sub-marine cables (Figure 96).

## *The Invention of the Communication Engines*

From the preceding analysis, it becomes clear that electricity had two major fields of application. Next to the Power applications (ie electric light, electric heating and electric motors), electricity found a new application area in the transportation of information creating the Communication Engines (Figure 93). It were those communication engines that exploded into the massive development of different forms of communication; from early telegraphy in the 1830s, the telephony of the second half of the nineteenth century, and to radio broadcasting systems nearly a century later. A period of time that spans several fundamental technological developments in creating the communication engines.

The development of communication over long distances started with Optical Telegraphy —ie the semaphore system of Claude Chappe used during Napoleonic times— being used for communication over distance at the beginning of the nineteenth century. His mechanical system of moving panels, was followed by Electric Telegraphy when electricity became available in the form of *Galvano-electric telegraphy* (ie the needle telegraph of Cook & Wheatstone) and *Electromagnetic telegraphy* (ie the relay telegraph of Samuel Morse)[238] in the 1840s. It took some more decades but by the 1870s the coded signal of dots and dashes was replaced by electric speech (ie the human voice) by Alexander Graham Bell: *Acoustic Telegraphy* (aka telephony) was born[239]. By that time communication over long distances along copper wires —aka cabled tele-communication— had changed society, and electricity had played a crucial role in its development. One could say that the Era of Telegraphy and the Era of Telephony had ignited a new *Communication Revolution*.

## *Wireless Communication 0.0 (1860-1895)*

Then came the *Era of Wireless Telegraphy*. The early development of the wireless telegraph is the story about electromagnetic waves; how to create them, how to use them, and how to receive them. It is a story that found its origins[240] in the work of Hans Oersted in 1820 when he published about his experiments with an electric current and a compass needle[241]. It was a sensation in the scientific world of that time; from Paris to Berlin and St. Petersburg, from the London scholars to the Geneva- and Turino scholars. His original experiments were soon to be followed by the experiments of other curious scientists all over Europe. Among them the Englishman Michael Faraday (1791-1867) who by then was working on the phenomenon of magnetic induction. He concluded in the 1830s that a force between magnetism and electricity existed.[242] He observed the changes in that force which, taking some time to interact, resulted in wave motions like 'the vibrations upon the surface of disturbed water'[243]. Faraday had created

---

[238] Basically, the galvanometer is a movable coil that moves in a stationary field between to magnets. And a relay is a magnetised iron object that moves under the influence of a magnetic field created by a stationary coil. See: B.J.G. van der Kooij: *The invention of the Communication Engine 'Telegraph'*. (2015)

[239] See: B.J.G. van der Kooij: *The invention of the Communication Engine 'Telegraph'*. (2015)

[240] See: B.J.G. van der Kooij: *The invention of the Electro-motive Engine*. (2015)

[241] H.C.Oersted, *Experimenta circa effectum Conflictus Electrici in Acum Magneticam* (Experiments in the Effect of a Current of Electricity on the Magnetic Needle), Copenhagen 1820.

[242] M. Faraday, *On the Induction of electrical currents and of the Evolution of Electricity from Magnetism*. (Faraday, 1832)

[243] For reasons of his own, on March 12, 1832 Faraday had deposited his observations from his 'Experimental Researches in Electricity', due to a controversy about priority with Humphry Davy, in a sealed document to be guarded in the strongbox of the Royal Society.

the 'lines of force' with a circular motion: the magnetic curves that were moving in an unknown medium called the 'ether.'

That was quite in contradiction to the then popular Newtonian thinking of forces acting in a straight line (the 'action at a distance' concept with its universal gravitation). In the 1850s Faraday's work, published in his 'Experimental Researches in Electricity,' inspired the Scot James Clerk Maxwell (1831-1979). He transformed the 'lines of force' into a mathematical approach that explained how they propagated.[244] But in what medium were the waves propagated? Were it the classical 'fluids', or was it the problematic medium called 'ether'? And, what was the nature of the medium in which those waves propagated? Maxwell developed from his mathematical explanations a theory published in a paper called 'A Dynamical Theory of the Electro-magnetic Field' (1864) linking the phenomena of light with those of electricity and magnetism.[245] This was the 'space that contains and surrounds bodies in electric or magnetic conditions'. His work became the foundation of electromagnetic theory. And it was later published in a classic textbook 'A Treatise on Electricity and Magnetism' in 1873.

His theory excited many scholars that became known as the 'Maxwellians' (Figure 92). Although his mathematical approach was for many scientists hard to grasp, some more experimental British scientists like David Edward Hughes (1831-1900) came close to creating electromagnetic waves. As did the more theoretical scientists like George Francis Fitzgerald (1851-1901)[246], Oliver Lodge (1851-1940)[247], and Oliver Heaviside (1850-1925)[248], and the Dutchman Hendrik A. Lorentz (1853-1928). Although

---

When the document was opened more than hundred years later —on June 20, 1937— this text was part of his original views expressed.

[244] J.C. Maxwell, *On Faraday's Lines of Force* (read 1855/1856). (Clerk Maxwell, 1864). Part of 'On Physical Lines of Force'.

[245] With the publication of *A Dynamical Theory of the Electromagnetic Field* in 1865, Maxwell demonstrated that electric and magnetic fields travel through space as waves moving at the speed of light. Maxwell proposed that light is an undulation in the same medium that is the cause of electric and magnetic phenomena. The unification of light and electrical phenomena led to the prediction of the existence of radio waves.

[246] G.F. FitzGerald, *On the Electro-Magnetic theory of Reflection and Refraction of light*. (Fitzgerald, 1880).

[247] As early as 1879 Lodge became interested in generating (and detecting) electromagnetic waves, something Maxwell had never considered. Quite some time later he published a series of papers proving the existence of electromagnetic waves and their propagation in free space. On June 1, 1894, at a meeting of the British Association for the Advancement of Science at Oxford University, Lodge gave a memorial lecture on the work of Hertz who had recently died.

[248] Oliver Heavyside published about his electromagnetic theories from 1882-1902 numerus articles in The Electrician. Such as in 1885, 1886, and 1887 about "Electromagnetic

they all expanded on Maxwell's theories, it was not till some years later that these theories were solidly confirmed. That was done by the German Heinrich Hertz (1857-1894) who, in the late 1880s, after studying Maxwell's theory, showed that electromagnetic waves could be created and detected. Waves that were the result of electro-magnetic forces. In parallel to the proof of the experimental work of Heinrich Hertz —cut short by his premature death at the age of thirty-six in 1894— it was Oliver Lodge who played a vital role in the practical radio communication that followed. (Garratt, 1994)

## Early Days of Wireless Communication

In its essence, the development of the wireless telegraph is a story about the replacement of the transmission carrier of the wired telegraphy: the electric copper cables that by then spanned the globe and oceans, enabling communication around the world. It is also a story of how wireless transmission complemented the cabled telegraphy. From the origins of the electric bell that could ring at a short distance, to the telegraphic writing at a large distance, it was the cable that conducted the information carried by electricity. Next to the *transmitter* at the beginning point, and the *receiver* at the end point, the cable was the connecting medium in point-to-point communication.

Except the fundamental contributions of a few individuals —such as Samuel Morse and Cooke & Wheatstone[249]—, the telegraph's development over time, although rather important, was not a development heralded in the scientific annals of invention. Its overall development was more the result of the practical works of the engineers whose thinking and tinkering solved the problems at hand.

So, basically the telegraph is a device for the transmission of information from one point to another point —at a long distance— by the means of a wire. It is a so-called *wired point-to-point communication* device. And those 'points' show its limitations as they are both fixed in their position. That also goes for the cable that connects the points; it is fixed (often on poles, Figure 95). Cabled telegraph thus is a static system of communication that is limited to land application and that is not usable in the mobile situations. So, ships being mobile 'points' that cannot be connected by a cable, are excluded from cabled telegraphy.

---

induction and its propagation" and in 1888, 1889 "Electromagnetic waves, the propagation of potential, and the electromagnetic effects of a moving charge".
[249] See: B.J.G. van der Kooij: *The Invention of the Communication Engine 'Telegraph'*. (2015)

## Early Wired Broadcasting

It is not hard to image that, after the first cabled transmissions between two points, soon appeared the next concept of sending messages from one point to many points. The *wired (one-way) point-to-many communications*, also called 'broadcasting', where information is distributed from one point to many recipients. Take for example the 'financial' information of stock brokers and the stock markets that were distributed using special systems (such as the *ticker telegraph* developed by Thomas Edison for the Gold and Stock Telegraph Company in 1867). Or imagine the distribution of news items by telegraph from one central source to many editorial offices of local newspapers, by the news-company Associated Press (est. 1864).

It was not only the telegraph that saw this development of 'broad casting', also the telephone that came on stage some decades later, underwent the change from the point-to-point concept to the point-to-many concept. It became the 'telephone newspaper' where a range of 'programs' was 'broadcasted' by telephone to a large audience of subscribers. Such as the telephone newspaper of the Hungarian *Telephon Hirmondo* news service as developed by Tividar Puskas, and the French *Theatrophone* created by Clément Adler, both demonstrated at the Electrical Exhibition in Paris in 1881[250]. It was not only the spoken word, also music was transmitted by wire. Already in the fall of 1876 experimental 'concerts' were transmitted over wire line by Bell from Paris, France to Brantford, Ontario, using a triple mouthpiece telephone transmitter to accommodate several soloists. From 1876 through 1880 a variety of cabled broadcasting transmissions were conducted, both in America and in Europe.

Surprisingly enough, this concept of broadcasting was not the focus of early wireless communication. When the use of electric waves came to its full existence in the late 1800s/early 1900s, it was seen as a replacement for and/or addition to wired telegraphy. A replacement that came about due to a range of inherent properties; such as the fact that wireless telegraphy could service places unserviceable by cabled systems. Among them the mobile maritime applications with the ship-to-shore and ship-to-ship communication. And those communications could be long-distance *point-to-point* transmissions in the case of the massive steam ships crossing the Atlantic Ocean. Shipping that till then always been 'incommunicado' once they left port. Ultimo, after wireless communication earned its place as an addition for cabled communication, it would encompass broadcasting. But before that was going to happen, quite some other events happened.

---

[250] See: B.J.G. van der Kooij: *The Invention of the Communication Engine 'Telephone'*. (2015) pp 283-287.

## Experimenting to Prove a Theory

The preceding examples illuminate the development of the early 'wired' communication systems in different *application trajectories* and *technological trajectories*. They have one thing in common, they all were based on the wired transmission concepts that originated from the early days of the telegraph system and developed into the telephone system. As one realizes that a transmission system using wires (both in plain air and underground/underwater) in its essence is limited in application to fixed points, expensive to install and to maintain, limited in their capacity, and vulnerable to disruptions caused by vandalism or weather conditions, then it is not hard to understand that inventive minds tried to circumvent the wire for communication purposes (Figure 98, bottom). They were looking for the ways and means of *wire-less communication*.

## The Hunt for the Electric Waves

After it had become clear that 'electric sparks' —the flashes of intense light created by shortcutting a battery— created more than just the waves of visible light, but also electric waves of different frequencies outside the visible spectrum, the hunt was on. Soon those invisible 'electro-magnetic waves' were explored by a range of theoretical and practical scientists (Figure 98). They followed different approaches; from the more theoretical approach of the Maxwellian scholars, to the practical approaches realising components for the creation and detection of the electric waves. Trajectories that paralleled the further development of the telegraphic and telephonic systems in the late nineteenth century.

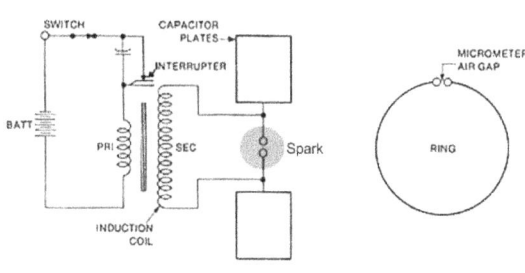

**Figure 97: Herz's Experiments.**

Principle of wireless transmission as discovered in the Hertz experiments. The electromagnetic waves generated by the spark (the transmitter, left), are received by the 'ring' (right).

Source: http://people.seas.harvard.edu/ ~jones/cscie129/ nu_lectures/lecture6/hertz/Hertz_exp.html

One of those experimentalists was Heinrich Herz (1857-1894) who proved the existence of electro-magnetic waves as previously theorized by James Clerk Maxwell and published in his 'Treatise on Electricity and Magnetism' in 1873. Between 1886 and 1889, being fully conversant by now with Maxwell's theory and dedicated to applying it, Hertz would conduct a series of experiments that would prove that the effects he was observing were results of Maxwell's predicted electromagnetic waves. In 1887 Hertz tested Maxwell's hypothesis. He used an oscillator[251] made of polished brass knobs, each connected to an induction coil and separated by a tiny gap over which sparks could leap (Figure 97). And the aerial antenna that he had created, registered those sparks as a distant action. The spark-induced electro-magnetic waves were proved to exist and they could be created by mechanical devices.

This confirmation of electromagnetic waves in 1888 started a surge of interest in the scientific community interest equivalent to Oersted's discovery of electromagnetism in 1820. It became the scholarly *Hunt for the Electric Waves*. Thinkers and tinkerers all over Europe started experimenting with electromagnetic waves within their specific domains of interests; some from a scientific point of view, others with a more practical approach (Figure 98). Expanding on Heinrich Herz's experiments with 'Herzian waves', they discovered that 'electric waves' created by electric sparks, could be used as a carrier for coded information.

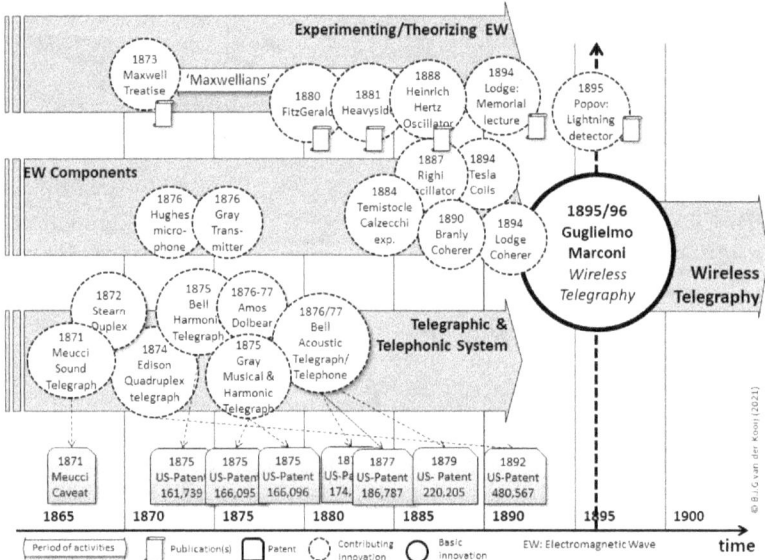

Figure 98: Overview early Contributions to the Wireless Transmission.

---

[251] These devices —called spark-oscillators— created in the 'ether' a burst of untuned electric waves that could be detected over increasing distances. Waves detected by other devices related to the conductivity of electricity: devices later called the 'coherer.'

## Invention of the Wireless System

Remarkably enough, in all the years before and after Hertz's contribution in 1888, with all those scientists contributing to the basic understanding and all those engineers tinkering with parts of the *concept* of the wireless communication (Figure 92), no effective *system* of wireless communication had been developed yet by the mid-1890s. Then, a young Italian with British roots was about to change that.

*Guglielmo Marconi* (1874-1937), fascinated by Herz' concept of electric waves, experimented at the attic of his parental home Villa Griffone by meticulously copying the experiments of Hertz with coils, batteries, and wires. Soon he set up experiments in the garden with his brother Alfonso, who assisted him with bridging increasingly longer distances between transmitter and receiver. Step by step, his experiments improved the total system. As a result, the transmitter gave of a better signal due to the improved antenna, the receiver became more sensitive due to his coherer, the Morse telegraph functioned better due to the additional relay and battery, etc. By the 1895s, it was obvious that Guglielmo's experimenting had created something that could be of value. It was also clear that for converting the early prototypes into a saleable system, much more experimenting was needed. Thus, to expand his experimenting opportunities and find financial support to pay for it, he had to be in England. So, in 1886 he went to England with his Black Box (Figure 99), where he found support by his mother's family. [252]

**Figure 99: Marconi's Black Box (1896).**
Source: Museum of the History of Science, Oxford. Press pack https://www.mhs.ox.ac.uk/marconi/presspack/images/receiver.jpg

The young man who disembarked in early 1896 on the English shore brought with him the crude artefact of a system for wireless telegraphy. Despite its simplicity, in those days it

---

[252] The Marconis, originally mountain men ('montanari'), by then were landed gentry, owning land in the Pontecchio region near Bologna. His mother was the daughter of John Jameson, a Scot who had gone to Ireland and into the business of distilling whiskey. (Jameson & Sons) in the late 1780s. Jameson exported their distillation products all over the world (incl. Italy), and much of the business was based in London.

could be called a high-tech invention. The artefact in the black box did mysterious things that were pushing the frontiers of existing knowledge and knowhow. It made communication at a distance possible through the dark space of the invisible ether. It was built with that new magical technology able to control lightning and with the maturing electro-mechanical techniques of that time. It was operated on by the technical priests of that time. That all made wireless communication magical and fascinating for the general public.

After patenting his invention —GB Patent № 12,039 granted on June 2nd, 1897— he came in contact with William Preece, Chief Engineer to the British Government Telegraph Service of the Post Office. It was the beginning of a close cooperation of experimenting, and the start of a conflictual relation with the British scientific community.[253]

**Figure 100: Marconi's Radio for Wireless Telegraph Communication (1912).**

Titanic's Radio Room (top, 1912), and Valve Tuner with radio tubes (bottom, 1912).

Source: Wikimedia Commons.

The following period of 1896–1899 had been a time of constant experimenting and testing of Marconi's equipment for wireless communication (Figure 100, bottom) in and around Britain. In 1899 America had also become an interesting testing ground, and the Atlantic crossing in 1901 —testing the long-distance capabilities of his system—cemented Marconi's public image. The main theme had been increasing the distances to be bridged by the Hertzian waves. And that created a massive amount of publicity for the young Marconi.

Marconi became known as 'Mister Wireless' among the general public. But the British establishment had ideas of their own, and soon the Post Office withdrew its cooperation, and legislation started frustrating his

---

[253] For details, see: Van der Kooij, B.J.G.: *The Invention of the Wireless Engine.* (2017)

entrepreneurial activities. Marconi would soon become entangled in a web of conflicting scientific aspirations, engineering frustrations, political and economic interests.

Facing the state opposition for his entrepreneurial activities in England, as well as the opposition of the British telegraph companies, Marconi abandoned focusing on land-based telegraphy in Britain and focused on marine applications. He went international, gave demonstrations to the Russian Tsar and Italian King as well as the American Navy. He demonstrated the possibility of long-distance wireless communication with his transatlantic transmissions. Marconi, not without a keen sense for publicity, saw his experiments heralded in the press; the more after the *Titanic* hit an iceberg in the early morning of April 15th, 1912 on her maiden voyage, sending emergence message by is wireless telegraph (Figure 100, top). His fame and business acumen would, after additional technical improvements in the following years, result in the worldwide wireless monopoly of the Marconi System[254] in maritime applications.

In the meantime, he continuously improved on his system tackling the unavoidable 'childhood' defects one by one (eg the 'interference problem').

> In the preparation before setting up the Atlantic connection experiment, Marconi travelled to New York by steamer. His arrival was a wild event, as he was met at the quayside and his hotel by dozens of reporters. He also was approached by a young graduate from Yale Sheffield Scientific School, who, having been working on his dissertation about Hertzian waves and looking for a way to fulfil his dreams, wrote Marconi a letter on September 22nd, 1899:
>
> *I have been exceedingly fascinated by the subject and hope for this opportunity of following work on that line. Knowing that you are about to conduct experiments for the US Government in the wireless telegraphy, I write you begging to be allowed to work under you. It may be that some assistants well versed in the theory of Hertz waves will be desired in that work, if so may I not be given the chance? … It has been my greatest ambition since first working with electric waves to make a life work of that study. If you can, signor, aid me in fulfilling is desire, you will win the lasting indebtedness, as you have already the admiration of, Your obedient servant, Lee De Forest.* (Jolly, 1972, p. 74)

The eloquent young author of this letter, being one year older than Marconi, was Lee DeForest (1873–1961), who would become

---

[254] The Marconi System was a system of wireless communication for coastal application (shore-to ship) and maritime applications (ship-to-ship) that was dominated by Marconi companies. It also would be the basis for long distance communication, where the colonial empires benefitted.

known later for his work related to the invention of the vacuum tube known as Radio Tube, an invention that would change the world once again. But that did not happen as Marconi's employee. Even worse, the different companies he was involved in would later compete with Marconi's activities, and they would become entangled in costly patent litigation.

The invention of Wireless Communication sparked an explosion of entrepreneurial activity. Some entrepreneurs took licenses from Marconi, some pirated and copied his systems bluntly, and other countries (eg Germany) had limited patent systems, making Marconi's patenting efforts fruitless. The result was a cluster of businesses that erupted in America, Britain and Germany in the next decade. A cluster that saw mergers and acquisitions that would who cast their shadow on the next decades. Decades that would see the rise of new electronic technologies and the application of radio broadcasting.

## Wireless Mania in Science and Engineering

At the end of the nineteenth century, there was a massive quest in the scientific community to understand the phenomenon of wireless transmission made possible by the electromagnetic waves (aka Herzian Waves) created by sparks. In England, France, Italy, Germany as well as in America the *Hunt for the Electric Waves* involved many people; from scientists to experimentalists and scientific electricians. Their focus was on wireless telegraphy and its application, but in the process, they also explored the more fundamental phenomena.

The scientific community worldwide jumped on the bandwagon of Marconi's concept of spark-tuned electric waves. Even more with the First World War (1914-1918) looming at the horizon, wireless telegraphy would become important in military applications. And every state would try to get control of the new technology for communication over distance.

The German scientists like Ferdinand Braun (1850–1918), Adolf Karl Heinrich Slaby (1849–1913) and Count Georg von Arco (1869–1940) — working withing the context of their time where the Germany Empire wanted a place under the sun— were stimulated to explore wireless communication. The more because the worldwide cabled telegraphy infrastructure was under British control. Already in the early 1870s, Braun had started experimenting with crystals as part of his work in wireless telegraphy. This in order to improve upon the weak element in wireless transmissions; the sensitivity of the receiving unit (aka coherer). His efforts —and those of many others—resulted in a range of different solutions, among which the crystal *diode rectifier* (aka cat's whisker' diode).

Braun introduced the closed circuit of oscillation (Figure 101) into wireless telegraphy and was one of the first to send narrow frequency electric waves in definite directions. It was his 'closed circuit tuned oscillator'-concept in the electric wave part of the transmitter, and its separation from the antenna (aka 'uncoupling), that would enhance the quality of long-distance wireless telegraphy greatly. With generous support from Emperor Wilhelm II, and with the practical assistance of naval units, Slaby and his assistant Arco started further experiments in the summer of 1897. As soon as it was evident that they could supply useful arrangements, they got the order to develop wireless equipment. And in doing so to circumvent Marconi's patenting efforts that were not granted in Germany at that time. [255]

Also in other countries Marconi's wireless experiments had sparked the attention of governments and scientists.

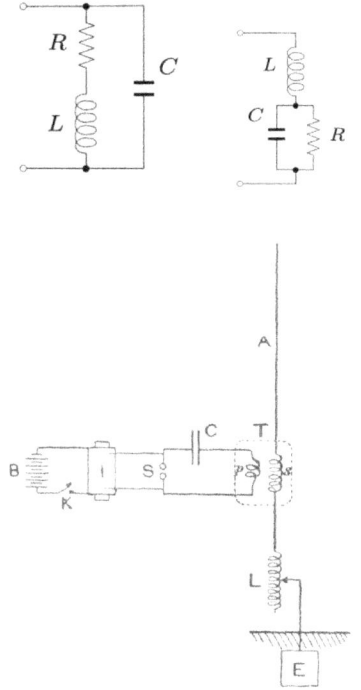

**Figure 101: Electric Oscillating Circuits.**

Oscillating LCR circuits composed out of passive components (top), and the circuit of Marconi-telegraphy sender (bottom).

B: Battery, S: Spark generation, T: transformer, K: Key, L: Coil. A; Antennae, E: Earth connection.

In England it was the Englishman Oliver Joseph Lodge (1851–1940) who would contribute to the theoretical understanding of the new phenomenon of electromagnetic waves. Another scientist was George Francis Fitzgerald (1851–1901), professor in natural and experimental philosophy at the School of Engineering, Trinity College in Dublin. Together with Oliver Heaviside (1850–1925) they became known as the 'Maxwellians' who revised, extended, clarified, and confirmed James

---

[255] See: Van der Kooij, B.J.G.: *The Invention of the Wireless Communication Engine* (2017).

Clerk Maxwell's mathematical theories of the electromagnetic field during the late 1870s and the 1880s.[256] The physicist William Crookes (1832–1919), later a pioneer of vacuum tubes, published in the *Fortnightly Review* an article entitled 'Some Possibilities of Electricity' (1892). In this look into the future, he clearly discerned the coming of a new form of wireless telegraphy based on the application of Hertzian waves.

It took a while after the early controversy[257] with Marconi, but by 1897 Lodge, a busy man active in many scientific fields, picked up his experiments with wireless transmission again due to Marconi's growing popularity. His experimenting resulted in August 1898 in US-Patent № 609,154, titled 'Electric Telegraphy'. His novelty was that he utilized the concept of 'syntonic' tuning: it applied in a closed-circuit both capacitors and a variable inductance, both in the receiver and transmitter. In January 1898, Lodge had made his invention public in a paper read before the Physical Society. In his lecture, he claimed to have created the means for electrical oscillations at a *particular frequency of oscillation*. Thus, individual messages could now be transmitted to individual stations without disturbing the receiving appliances of other stations, which were tuned at a different frequency. Already fascinated by the application of Hertzian waves, after this presentation, the scientific community became even more interested in 'syntony'. Among them was John Ambrose Fleming, who would be working later as scientific advisor in close cooperation with Marconi.

In America, another development had taken place in the 1880s when Thomas Edison was busy developing the incandescent lamp; in its essence a light-emitting filament within a vacuumized tube (Figure 182). But it had also another property as the heated filament did emit, next to light in the visual spectrum, also electrons; a property that became known as *thermionic emission*.

In addition, as wired electric telegraphy was in full development, many American scientists and engineers were aware of Hertz's work. Next to Nicolas Tesla and Thomas Edison, experimenters like Amos Dolbear, John Stone Stone, Reginald Fessenden, and Lee DeForest were, each in their own way, involved in spark-generated wire telegraphy. Many of the work was done on the early detectors that had emerged from the

---

[256] In 1859, James Clerk Maxwell discovered the distribution law of molecular velocities. Maxwell showed that electric and magnetic fields are propagated outward from their source at a speed equal to that of light and that light is just one form of electromagnetic radiation.
[257] The scientist of those time both in America as well as in Britain considered Marconi's invention —similar to that of Morse— as an application of scientific principles that were discovered by them. It gave rise to confrontations in the patent wars.

lightning experimenting. Amos Dolbear (1837-1910), for example, was granted US-Patent № 350,299 on October 5th, 1886, for a 'mode of electric communication'. Lee DeForest (1873-1961) worked on improving the early receivers and wave detectors. His friend, the American mathematician John Stone Stone (1869–1943), also became interested. Stone would be the one who made the link between resonant circuits and the problems of wireless telegraphy, creating *selective wireless telegraph*.

John Stone Stone contributed much to 'Space Telegraphy' and 'Radio Telephony,' both fundamental to later radio-development. He would be granted some seventy wireless patents over time. Among which seven in December 2nd, 1902 (eg US-Patent № 714,756 and US-Patent № 714,831 for 'selective electrical signalling'). His most remarkable publication was "The Practical Aspects of the Propagation of High Frequency Electric Waves along Wires" (1912) hailing high frequency wire telephony.

In Italy, scientific interest by the end of the nineteenth century encompassed also the electromagnetic waves. Professor Augusto Righi (1850-1920) had, already in 1887, experimented with a spark-gap sphere oscillator (aka the 'Righi oscillator'), creating shorter electric waves, and published results of his work in a treatise, *Optice Elettrica*, in 1897.

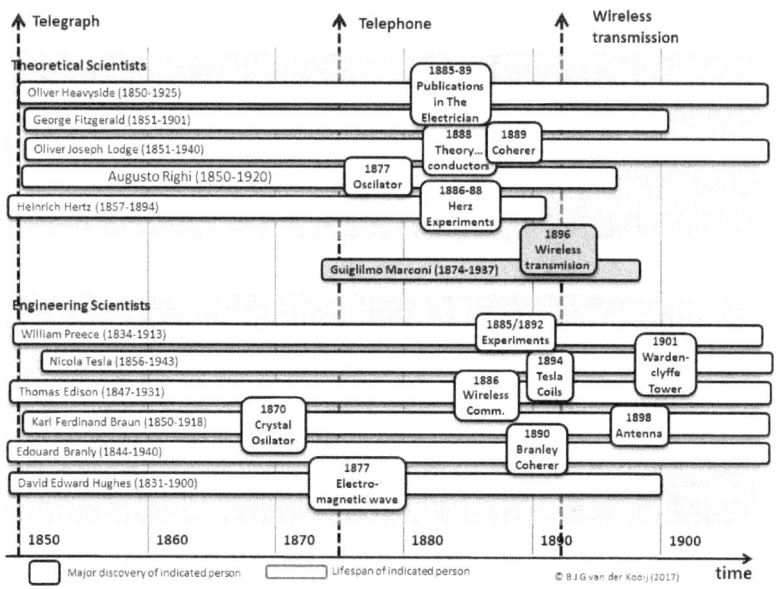

Figure 102: Overview Contributors of Electromagnetic Waves.

In France the physicist Edouard Branly (1844–1940) published in 1890 'On the Changes in Resistance of Bodies under Different Electrical Conditions' in a French Journal, where he described his investigation of the effect of minute electrical charges on metal and many types of metal filings. It would result in his famous detector (aka Branly Coherer) used in spark-induced telegraphy.

There was also the New Zealander Ernest Rutherford (1871–1937), later the First Baron Rutherford of Nelson, graduated from Canterbury College in Christchurch (NZ) in 1893. Being knowledgeable about Hertzian waves, he experimented with the magnetization of iron by the discharge of a Leyden jar, thus, by a spark with high-frequency emissions of electric waves. He found that when he changed the magnetic intensity (the cause), the changes of the magnetic induction (the effect) were lagging: the hysteresis effect. From these experiments, he developed a device for the detection of electromagnetic waves — using the hysteresis property of iron— that would soon replace the unreliable coherer (Rutherford, 1897).

The combined efforts of these theoretical scientists and the engineering scientist (Figure 102) resulted in the application of electromagnetic waves in wireless telegraphy. And it also created the devices (eg the coherer detector) that made the practical systems possible. In addition, the scientific community became curious about the nature of the electro-magnetic waves. As a result, the properties of invisible particles of the electromagnetic waves had the full attention of scientists. And with that interest came the scientific exploration of the properties of the fundamental materials. However, by the turn of the century, it was Marconi's contribution, to combine the fruits of all these different development trajectories into a communication system that would rapidly conquer the ether.

As a consequence of the Wireless Mania in science as well as in engineering, wireless technology was rapidly improving on component- and on system level.[258] It was still based on the spark-induced electric waves that could be used as a crude carrier of information. Information transmitted from a 'sender' creating the electric waves, to a 'receiver' receiving the electric waves that travelled the through the space called 'ether.'

> In addition to the more theoretical approaches of some theoretical scientist, there had been the engineering scientists who implemented the ideas and concepts in practical applications. Such as William Preece (1834-1913), heading the English Post Office. It was

---

[258] For details, see: Van der Kooij, B.J.G.: *The Invention of the Wireless Communication* Engine. (2017)

the British/American David Edward Hughes (1831-1900) who in 1877 experimented with electromagnetic waves —before Herz's experiments— discovering electromagnetic induction. And Thomas Edison (1847-1931) explored wireless transmission (aka grasshopper telegraphy')[259] as did Nicola Tesla (1856-1943) focussing on his system for the worldwide transmission of electrical energy. It was Guglielmo Marconi (1874-1937) who created the wireless telegraphic system by combining all the system components others had invented.

## *Wireless Communication 1.0 (1895-1920)*

Marconi's contribution heralded a rapid growth of the technical development trajectory. Wireless communication came on the agenda of many engineering scientists, as well as government officials who became aware of the potential of wireless communication in the affairs of governance. Marconi, in the meantime, continued improving his system. As a result, in a relatively short period of time (Figure 103) both on component and on system level, the wireless technology developed rapidly.

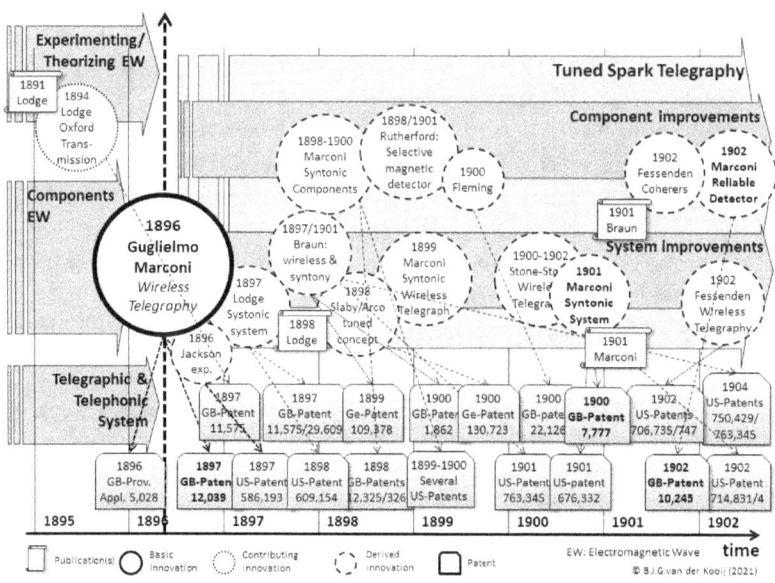

Figure 103: Overview of Improvement in Wireless Telegraphy after Marconi's Invention.

---

[259] This was an electromagnetic induction system which allowed telegraphic signals to jump the short distance between a running train and telegraph wires running parallel to the tracks.

## From Sparked-based to Single-Wave Transmission

The original wireless telegraph was spark-based (Figure 97). Hitting the telegraph key, creating long and short sparks —with DC electricity from a battery— resulted in a burst of electric waves of different frequencies in the radio spectrum of frequencies. And was thus transmitting the Morse-coded information. A rather crude method that had many drawbacks, both technically and application wise. Despite early improvements that made the spark-based telegraphy quite usable, the system had its limitations as it used a lot of bandwidths. And that was something that was going to become scarce with the rising number of telegraph stations. The need was thus rather clear; a continuous 'carrier' wave that could be controlled in its transmission frequency, would have to be found.

> Over time the spark-based system was constantly improved upon and, after the application of condensators (C), inductive coils (L) and resistances (R), a more efficient single frequency wireless transmitter (aka the *arc oscillator* and *arc convertor*) could be created. It were the LCR-oscillator circuits (Figure 101) that operated at their resonant frequency in a tuned circuit. But the spark-generated electric waves were by the nature of their creation multifrequency bursts.

The next step in its development was related to refining that burst of electric waves into a single wave of a specific frequency. And emitting that single frequency —ie switching it ON and OFF— would the indicate a specific telegraphic signal. The creation of single frequency waves was made possible by using the mechanical 'alternators' —a variation of electric dynamos— producing an electric wave of a single high frequency. That single wave, being switched On and Off, created improved wireless communication. The question was how to create those single waves. It was Reginald Fessenden —looking at electric waves from a different background— who created his *continuous wave transmitter* based on an electro-mechanical dynamo (aka generator, alternator).

## The Invention of the Modulated Single frequency 'Radio waves'

The Canadian Reginald Fessenden (1866-1932), after moving from Canada to New York, started to work for Thomas Edison as an assistant-tester for underground electrical mains.[260] He soon became as a junior technician in Edison's laboratory. When Edison, running into financial problems, had to let go of most his laboratory employees, Fessenden took other jobs and ended up as professor at the Electrical Department of the Purdue University. After helping George Westinghouse to install the

---

[260] See: B.J.G. van der Kooij: *The Invention of the Electric Light*. (2015) pp 164-166

lighting at the Columbian Exposition in Chicago (1893), he became professor at the University of Pittsburgh where he started experimenting with wireless communication in 1898.

In 1900 he left the university and started working for the US Weather Bureau, and experimented with a network of coastal wireless radio stations to transmit weather information, thus avoiding the need to use existing telegraph lines. His early experiments with a rotary spark-transmitters resulted in the transmission over a distance of 1.6 km on December 23rd, 1900. Then he started experimenting with higher-frequency 'alternator'[261] transmitters and came up with the idea of using two Alexanderson alternators operating at closely spaced frequencies to broadcast two signals, instead of one. The receiver would then receive both signals, and as part of the detection process, only the 'beat frequency'[262] representing the Morse code would exit the receiver. The *heterodyne system* —meaning 'generated by a difference (in frequency)— was born.

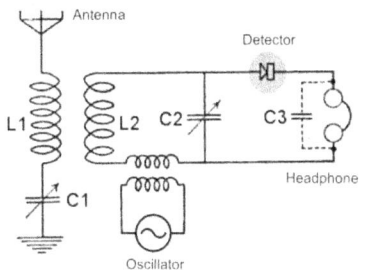

**Figure 104: The Principle of the Heterodyne Circuit.**
Source: Wikimedia Commons.

In addition, he developed the idea of superimposing an electric signal, oscillating at the frequencies of sound waves, upon a radio carrier wave of constant frequency. The sound signal would modulate the amplitude of the radio wave into the shape of the sound wave: the principle that became known as Amplitude Modulation (Figure 326). The receiver of this combined wave would separate the modulating signal from the carrier wave and reproduce the sound for the listener (Figure 104). But it was more than just creating a single 'carrier wave' on which the telegraphic signal was modulated. He also found a way to put on that carrier another multi-frequency signal known as speech. It was the birth of a new way of mixing (aka heterodyning) a *sound signal* with a carrier signal. It would become the transmission system of Amplitude Modulation (AM).[263]

---

[261] In its essence, an alternator is a dynamo creating an AC-current. The frequency of the AC-current is depending on the rotation speed, the gear box, and the arrangement of the coils. The higher the rotation speed, the more revolutions, and the higher the frequency.
[262] These beat frequencies were a low-frequency 'pure sine waves' (up to 10 kHz).
[263] A heterodyne is a signal frequency that is created by combining or mixing two other frequencies using a signal processing technique called heterodyning. It involves the processes of modulation (ie combining the signals) and demodulation (de-combining the signals).

*On December 23, 1900 Fessenden said into his microphone, "One, two, three, four. Is It snowing where you are Mr. Thiessen? If so telegraph back and let me know." Thiessen replied by telegraph in Morse code that it was indeed snowing. In great excitement Fessenden wrote at his desk, "This afternoon here at Cobb Island, intelligible speech by electromagnetic waves has for the first time in World's History been transmitted." This was almost a year before Marconi's transmission in Morse code from England to Signal Hill in Newfoundland, on December 12, 1901.*[264]

In his US-Patents № 706,737 through № 706-747, all granted on August 12th, 1902 he protected his concept (Table 1). Over time, Fessenden not only worked on transmitting electric waves, he also conducted many experiments on detecting electric waves; for example creating an electrolytic detector (his so-called 'Barretter', resulting later in US-Patent s № 727,331 of May 5th, 1903 and № 793,684 of July 8th, 1905)[265]. Followed by a thermal detector: the hot-wire barrater. (Sarkar & all, 2006, pp. 369-371)

After leaving the US Weather Bureau as the result of a conflict about his rights of the inventions he had made, with the financial support of two Pittsburgh millionaires, Hay Walker and Thomas H. Given, he had started in November 1902 the *National Electric Signalling Company* (NESCO).

Table 1: Patents granted to Reginald Fessenden on August 12th, 1902.

| Patent № | Granted | Description |
|---|---|---|
| US 706,735 | August 12, 1902 | Wireless Telegraphy (filed December 15, 1899). |
| US 706,736 | August 12, 1902 | Apparatus for Wireless Telegraphy (filed May 17, 1900). |
| US 706,737 | August 12, 1902 | Wireless Telegraphy (filed May 29, 1901). |
| US 706,738 | August 12, 1902 | Wireless Telegraphy (filed May 29, 1901). |
| US 706,739 | August 12, 1902 | Conductor for Wireless Telegraphy (May 29, 1901). |
| US 706,740 | August 12, 1902 | Wireless Signalling (filed September 28, 1901). |
| US 706,741 | August 12, 1902 | Apparatus for Wireless Telegraphy (filed Nov. 5, 1901) |
| US 706,742 | August 12, 1902 | Wireless Signalling (filed June 6, 1902). |
| US 706,743 | August 12, 1902 | Wireless Signalling (filed June 26, 1902) |
| US 706,744 | August 12, 1902 | Current Actuated Wave Responsive Device (filed July 1, 1902). |
| US 706,745 | August 12, 1902 | Signaling by electromagnetic Waves (filed July 1, 1902). |
| US 706,746 | August 12, 1902 | Signalling by electromagnetic Waves (filed July 1, 1902) |
| US 706,747 | August 12, 1902 | Apparatus for Signalling by electromagnetic Waves (filed July 22, 1902). |

Source: USPTO

---

[264] Source: An Unsung hero: Reginald Fessenden, the Canadian inventor of radio telephony. http://www.ewh.ieee.org/reg/7/millennium/radio/radio_unsung.html

[265] Fessenden would later contribute in solving the basic problem of dampening in spark-generated electric waves. He also developed the heterodyne principle, used to combine two frequencies. He was granted US-Patent 706,740, filed on September 28th, 1901 and granted August 12th, 1902.

Fessenden contributed his patents, the millionaires the funding (for $330.000 they obtained a 55% share in the company). The company started trying to create business, build some experimental wireless stations with 400-foot antenna towers at Brant Rock, Massachusetts. As a result of their excellent performance, three more stations were built in New York, Philadelphia and Washington. However, the company, trying to compete with Marconi, was not too successful and the relation between Fessenden and his financial backers soured.

> *Believing that he had already made major compromises to suit his backers, compromises that interfered with his experimentation and that required him to fill too many roles at once, Fessenden became increasingly uncompromising and abrasive. He came to see every negotiation over every detail as a battle over preserving the autonomy and discretion he had left. His backers, who by 1905 had already invested half a million dollars in Fessenden's visions, stoked the embers of their own resentment, which Fessenden fanned with each new demand. The increased tensions within the company, which were exacerbated by external events such as the panic, left both sides feeling beleaguered and frustrated. Fessenden, Given, and Walker had never been able to agree on and pursue long-term business strategies, and the erosion of their superficial alliance during these years precluded the discovery of a remedy for the situation. They continued to pursue short-term projects that were sustained only through the first intoxicating flush of enthusiasm. When the endurance and determination necessary to sustain a strategy over years rather than months was not summoned, one short-term plan replaced another. A productive alliance can provide the sustenance a company needs, but such an alliance did not exist at NESCO, and this deficiency had major repercussions not only on the company, but also on how and by whom radio would be developed.* (Douglas, 1989, pp. 151-152)

Without informing his partners, in 1906 he created the *Fessenden Wireless Company of Canada* for the long-distance transmission between Canada and Scotland. However, he was confronted with technical and climate problems as a gale broke down his transmission tower on December 6th, 1906. As the relation between the partners went for worse, Fessenden was dismissed in 1911. A decade after its start, the Nesco company went bankrupt to re-emerge in 1917 as the *International Radio Telegraph Company*.

His scientific activities, however, were more successful than his business activities. They covered fields such as the incandescent lamp (US-Patent № 452,494), and the wireless telegraphy (US-Patents № 706,735-706,747) granted on August 12th, 1902 (Table 1). He would later continue with the development of radio broadcasting of human speech and sound, was an early contribution to an alternative way of wireless communication that would replace the 'whiplash' effect of the spark-based wireless telegraphy. Fessenden eventually became the holder of more than 500 patents. His US-Patent № 706,735 with a priority date of December 15th, 1899 would later cause problems for Marconi. In the 1930s he would become interested in mechanical television and design a rotating mirror-based 'television system' to transmit moving pictures (US-Patents № 2,059,221/222) granted on November 3rd, 1936.

## The Rise of Electro-Mechanical Alternators

Next to the described electromechanical development, there were efforts in the field of the generating the transmission carrier waves. In the same year 1902 the Swedish-American engineer Ernst Alexanderson (1878-1975) emigrated to America and started working for General Electric. That company had received an order from Reginald Fessenden to build an alternator able to produce a high frequency. So, in 1904 he designed, continuing on Fessenden's ideas, an alternator generating continuous waves in the frequency range of 50-100 kHz: the longwave alternator. In 1906 the machine was installed in Fessenden's radio station in Brant Rock, Massachusetts. By fall its output had been improved to 500 watts and 75 kHz. On Christmas Eve, 1906, Fessenden made an experimental broadcast of music, including him playing the violin, which was heard by Navy ships and shore stations down the East Coast as far as Arlington.

> Alexanderson was granted US-Patent № 1,008,577 on November 14th, 1911 for his high frequency 'Alexanderson Alternator.' Anderson received in total some 322 patents for his work and his alternator' would be widely employed in high-power/low-frequency wireless stations.

Improving on his system, he developed 50kW machines that were massive in size and weight and were only suitable for land use. Next to maritime communication, they were used for commercial wireless telegraphy to transmit over intercontinental distances. Some were placed in Marconi's transoceanic station in New Brunswick, N.J. As the First World War (1914-1918) had interrupted the development of wired based telegraphy, and the military needed transoceanic communications with the military operations in France, in 1918 the installation was commandeered for official transoceanic service by the US Navy.

But there were more developments around the electro-mechanical creation of electric waves. In 1908 the Westinghouse engineer, professor Rudolph Goldschmidt (1876-1950) from German origin, devised an intricate method to enable an alternator to generate high frequency without requiring excessive rotation speeds. His contributions were used in high power longwave telegraphic transmission stations between the US and Germany, inaugurated on June 19th, 1914. In the same time as Alexanderson developed his mechanical wave generator, the Danish electrical engineer Valdemar Poulsen (1869-1942) followed another mechanical trajectory to create sparks of a more single frequency. He created the 'arc convector' that produced an undampened wave as he converted the direct current into an alternating current of a quite high frequency (2-20kHz). The arc was created within a glass vessel that contained hydrogen between a revolving carbon cathode and a copper anode. He combined it with a tuned circuit, added a traverse magnetic field, and the result was a continuous wave.

> He applied for a patent on June 19th, 1903 and was granted US-Patent № 789,449 in May 9th, 1905 for a 'Method of Producing Alternating Currents with a High Number of Vibrations."

Next to the further developments in Britain and Germany, the system became used in America. Poulsen's patents were acquired by the Stanford-educated Cyrill F. Elwell and the *Poulsen Wireless Telephone & Telegraph Company* was created in 1909. In 1910 the company was renamed in the *Federal Telegraph Company* with help of the capitalist Beach Thompson. This company would employ a remarkable man: Lee DeForest —fleeing from the East Coast to escape his problems there— to develop practical receiver for the Poulsen wireless system. Which he did using Fleming's valves and his own experimental development of the vacuum tube called Audion.

Poulsen's technology made wireless transmitters possible that could create high power continues waves (although still using a lot of bandwidth) It was used till the First World War when many battleships were equipped with Poulsen transmitters. However, after the development of the new 'electronic' technologies —with the electronic devices like vacuum tubes developed by Fleming and DeForest— these technologies would make this development trajectory obsolete by the end of the war. Later in time, other systems for creating and detecting electromagnetic waves became available, but their origin was simple. Just as the incandescent lamp —ie an electric-powered red-hot wire enclosed in a glass vessel emitting radiation in the visible frequencies— succeeded the spark light, an electric single-wave generator circuit based on a glass-enclosed vessel (the radio tube) would succeed the spark-based electric wave generator.

These were some of the major contributions along the 'electro-mechanical' trajectory of the electric alternators. The creation of modulated single-frequency electric waves had been a major step forward, but it was the merger of the *single-frequency wireless telegraphy technology* with the *vacuum tube technology* that would create a breakthrough for wireless telegraphy.

## Overview of Early Development of Wireless Technology

The preceding analysis shows how the development of wireless technology took place in distinct sub-technologies. From the spark-based systems to the continuous wave systems.

*Spark-gap based systems:* The contribution of Guiglielmo Marconi (Figure 103) in controlling Herz's electric waves was fundamental. [266] Although his spark-induced electric waves were a rather crude means of transmission, it showed a novel way of communication. His contribution spawned the development of a range of improvement trajectories on component level (eg the coherer improvements) and on system level (eg the tuning technology), resulting in the system of *tuned-spark telegraphy*. Starting with the marine applications that solved the century old incommunicado-problem of seafaring, the long wave communication system was soon spanning the globe (Figure 105, bottom).

*Continuous Wave:* In addition, there were other thinkers and tinkerers who improved on his concept in other ways. Leaving the spark-based technology they created the continuous wave, single-frequency systems by electro-mechanical means known as 'arc convertors' (aka alternators (Figure 105, top). A single-wave system that was capable of creating a single frequency carrier wave that could transmitted the telegraphic code. It still was a crude system, but the heterodyne system would change that.

These were the developments of a technical nature. In addition, there were developments of a more economic nature as all those companies that were created soon became an important industry.

> Next to the British and American Marconi Companies who were licensed to use Marconi's numerous Patents, also other companies entered the wireless market. Like the German company Telefunken, which had its equipment manufactured by its parent companies AEG and Siemens & Halske. The British Marconi companies and the German Telefunken would become fierce competitors.

As the telegraph and telephone manufacturing industry[267] was well developed by that time, many of the new wireless companies depended on their collective mechanical skills and technology to engineer their own systems. For example, it was the company General Electric —already used to building electric generators and electric motors— who did build the high

---

[266] See: Van der Kooij, B.J.G.: *The Invention of the Wireless Communication Engine* (2017)
[267] See: B.J.G. van der Kooij: *The Invention of the Communication Engine 'Telephone'*. (2015) pp. 302-310.

frequency Anderson-alternators. And, as this all happened during the prelude to the First World War, the military of many imperial nations became highly interested as they realized the wireless communication was a ground-breaking alternative to the wired communication they used.

By the early 1900s, the Marconi companies —not in the least as a consequence of its patent position making it the de facto owner of the emerging wireless technology— had created quite dominance in the marine market: the Marconi Monopoly.[268] In the coming decade much would influence the survival of that position. Many of them were on a scale out of the sphere of influence or the control of the British and American Marconi management. While for the moment the prospects for the American business looked good, the first clouds of the different times to come were appearing at the horizon.

The wireless technology would become applied after the First World War also in non-military applications. It would be the fundament for the broadcasting systems in which the 'radio receiver' (ie the wireless receiving unit) would play a dominant role in the Affairs of Man.

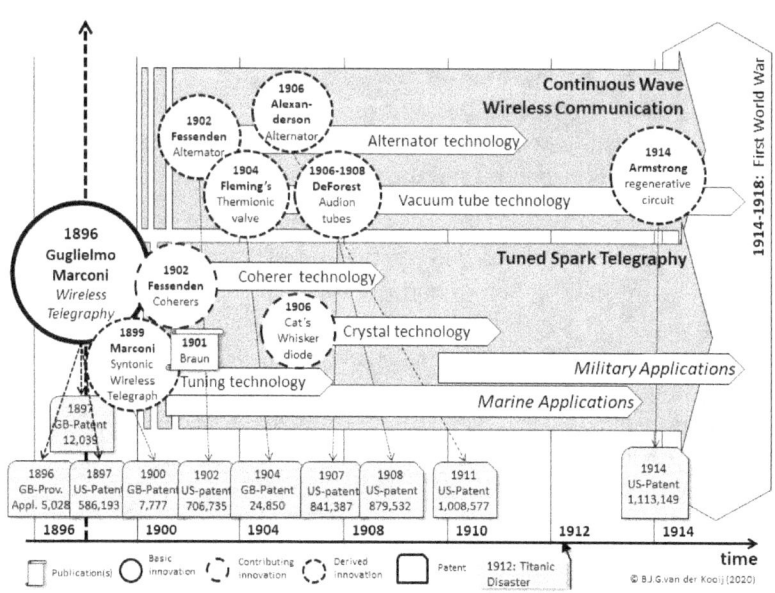

Figure 105: Overview Development Trajectory of Wireless Telegraphy.

---

[268] The first company created by Marconi was The Wireless Telegraph & Signal Company (1897), renamed in 1900 as Marcon's Wireless Telegraph Company. He also created subsidiaries such as the Marconi Wireless Telegraph Company of America (idem Canada).

## The First Communication Revolution (1830-1920)

In the Preface of this case study, reflecting on the context for innovation, we mentioned the basic human needs. From the existential needs for food and safety, to the social human needs. Among them we find the communication needs. From early times when man developed his languages (ie the spoken word), to the Ancient Times when man developed his scripts (ie the written word).[269]

Jumping ahead in time, as a result of the Electric Revolution that made electricity available, its property of carrying 'information' would start a revolution in communication. It was the inventions of Morse and Cooke & Wheatstone, Alexander Bell and Guillermo Marconi that created the communication engines of the telegraph, telephone and wireless. Their contributions were instrumental to the *First Communication Revolution* that shaped the Affairs of Men (Figure 106). The period of time where technology revolutionized human communication over distances.

### Periodizing the First Communication Revolution

This First Communication Revolution, was in its core based on simple devices. It was the electro-magnet —creating linear movement used in the relay—, and the electromagnetic coil (Figure 86) —creating multi-frequency currents in the microphone/loudspeaker— that became the components to transmit information over long distance through a cabled infrastructure. They created the electro-mechanical 'general purpose engines' that became the essential parts of telegraphy systems. Systems that realised the function of communication over distance.

The Era of Communication with its Communication Technologies (CT) had three major technologies as it contributors. Technologies that created the Telegraphic, the Telephonic and the Wireless communication, each with their own technological development trajectories (Figure 106) creating their General Purpose Engines (GPE). In their totality these technologies created a revolutionary development that can be periodized as follows along their subsequent General Purpose Engines: the GPE-Telegraphy, the GPE-Telephony and the GPE-Wireless.[270]

---

[269] See: Van der Kooij, B.J.G.: *The Origins of Innovation, Ancient (R)evolutions in Perspective.* (2018).

[270] Periodization is the process of categorizing the past into discrete, quantified, named blocks of time in order to facilitate the study and analysis of history. This results in descriptive abstractions that provide convenient terms for periods of time with relatively stable characteristics.

*Periodization of Telegraphy*

In its essence the *communication-technology* that realised the function of transmission of the coded word over distance (aka telegraphy), and later the spoken word (aka telephony), also over distance, is the collection of many mechanical and electric technologies. Seen in the perspective of the life cycle (Scheme 5), the communication technology is marked by specific events creating the basic innovations.

*Telegraph 0.0* (1780s-1830s): The evolution of technology behind telegraphy started somewhere in the late eighteenth century (Figure 89) when the first forms of communication over distance by semaphore were developed (eg Chappe's Optical Telegraph).

*Telegraph 1.0* (1830s-1850s): The breakthrough for telegraphy came with the parallel inventions of Cooke & Wheatstone in Britain and Samuel Morse in the USA during the 1830s that both applied electricity as carrier for the coded word (Figure 89). It gave rise to the first experimental local networks. And a telegraph industry that started manufacturing telegraph equipment in volume.

*Telegraph 2.0* (1850s-1900s): After the first experimental and private networks, by the mid-nineteenth century telegraphy boomed, with

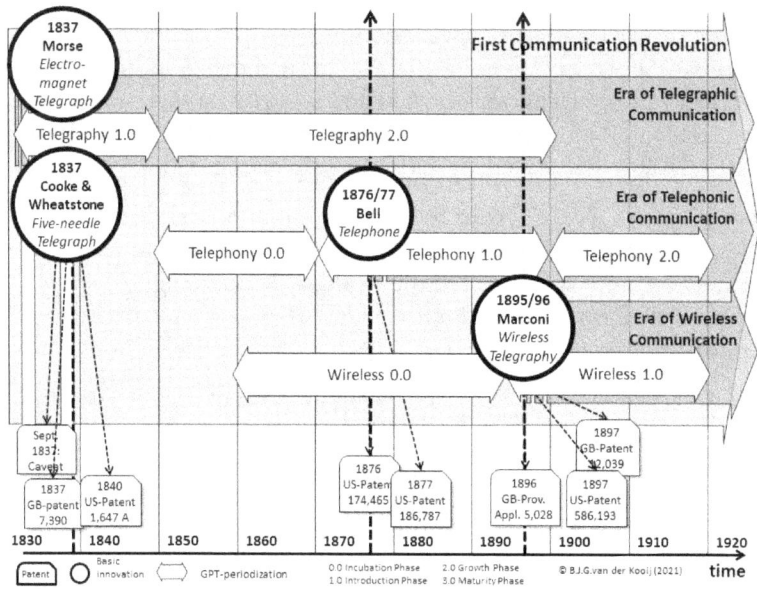

Figure 106: Timeline of the Major Contributions to the First Communication Revolution (1830-1920).

networks spreading over regions, countries and continents. With rise of new entrants, regional monopolies erupted, and long-distance communication monopolised in the US by Western Union. With the British 'All Red Line' connecting the continents by the early 1900s linking the colonies of the British Empire.

*Telegraph 3.0* (1900s-1950s): These were the consolidation times of the large vertical and horizontal integrated telegraph companies that were monopolizing the industry. But also the times of increased government regulation, new entrants into the field of telegraphy, and new competition from the cabled telephony.

*Periodization of Telephony*

Next it was the application of Telephony[271] that made transmission of the 'spoken word' (aka sound) possible using electricity. It was a spin-off of the harmonic telegraphy where frequencies were used to separate specific telegraph-connections using the same copper cable.

*Telephone 0.0* (1850s-1870s): Early experimenters with sound had developed both the components to convert sound into an electric signal as well as converting that electric signal into sound (Figure 90).

*Telephone 1.0* (1870s-1900s): But it was Bell's invention of the acoustic telegraph (aka telephone), that created the telephone system (Figure 90). First the simple local 'party lines', soon followed by the local telephone networks where telephone switchboards created the connection between two parties. Soon telephony boomed and regional and national networks emerged along the telegraphy networks. Network that became part of the Bell System that was the result of Bell's patent monopoly.

*Telephone 2.0* (1900s-1920s): After Bell's patent monopoly ended an explosion of independent telephone network providers and telephone equipment manufactures served the world-wide booming telephone markets. Telephone systems improved, networks expanded. The fixed telephone landline system penetrated every corner of the globe.

*Telephone 3.0* (1920s-1980s): The fixed telephone with its massive cabled infrastructure and large telecommunication companies, being the dominant means to transmit the spoken word in 'tele-communication,' stayed on scene till the rise of the mobile phone. The AT&T monopoly came under the influence of antitrust and deregulation policies, resulting in the Bell breakup creating room for the small start-ups.[272]

---

[271] Van der Kooij, B.J.G.: *The Invention of the Communication Engine 'Telephone'* (2016).
[272] AT&T was founded as Bell Telephone Company by Alexander Graham Bell, Thomas Watson and Gardiner Greene Hubbard after Bell's patenting of the telephone in 1875.

*Periodization of the Wireless*

Over time the cabled telephony was complemented after the wireless telephony matured during the Wireless Revolution. The new engines of 'Wireless Communication' got rid of the cabled infrastructure both for the coded word and spoken word by applying the electro-magnetic waves.[273]

*Wireless 0.0* (1860s-1895s): Early experimenters with the Herzian waves (aka electro-magnetic waves) had developed both the components to convert a signal into a pulsed electromagnetic wave (ie the arc transmitter) as well as converting that electromagnetic wave back to the signal (ie the receiver aka coherer) (Figure 92).

*Wireless 1.0* (1895s-1920s): Marconi's experimenting with a wireless system resulted in the Marconi System of wireless communication protected by the Marconi Monopoly (1905-1914). First applied in marine communication (ship-ship and ship-shore), further improvements in wireless telegraphy—during the 'Wireless Hype'—a wireless tsunami created wireless networks that soon spanned continents.

*Wireless 2.0* (1920s-1940s): After Maroni's patent protection had ended by 1915, government involvement by the British Post Office and the American government, controlled the further business development of wireless service providers. Next to telegraphy also telephony used the wireless means of transmission.

After the wireless technology had been used dominantly for telegraphy, the use of wireless in telephony and broadcasting and its subsequent development sparked a new technology cycle; Mobile telephony technologies, Radio/Television technologies and Broadcasting technologies.

*The 'C' of Communication in ICT*

This periodization illustrates the evolutionary development of the three major General Purpose Engines (GPEs) of the *Communication Technologies* in the application trajectories of the GPT-Electricity (Figure 107). It was the Telegraph Technology that realised the cabled telegraphs, and the Telephone Technology that realised the cabled telephones. And it was the Wireless Technology that created the cable-less communication along the air waves. A development that expanded into the technologies that made the broadcasting systems of the radio and television possible. Together with the Information Technologies, it would become part of the General Purpose Technology of the Information and Communication Technology (GPT-ICT) that dominated the Twentieth Century.

---

[273] Van der Kooij, B.J.G.: *The Invention of the Wireless Communication Engine* (2017).

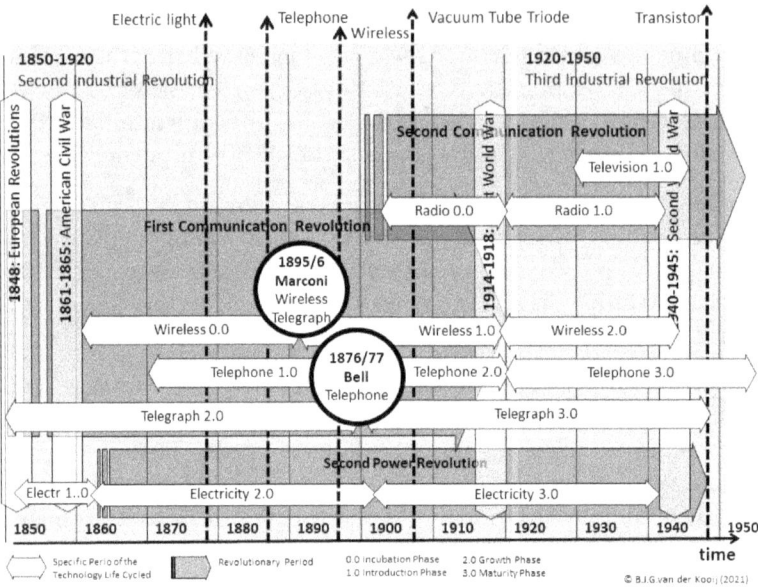

**Figure 107: Overview of the Periodization of the Communication Technologies.**

## Changing Perspective

In the preceding exploration we focussed on the events in the communication technologies that were of an electro-mechanical nature. Events that occurred up to the early twentieth century. Now, changing our perspective, we will focus and explore the events that created another result of the mechanical technologies; the machines that would arise in the application field of calculation. Their totality would become known as the 'Information Technologies' (IT).

# The Calculating (R)evolution

When one considers the 'mechanical technology' as knowing how to make artefacts of metals (eg Iron, Copper), one has to go back in time to find its roots (ie the Bronze Age, the Iron Age). When one narrows 'mechanical technology' down to the art of making machinery,[274] one reaches the more recent times; the medieval times of the (fine-)mechanical technologies creating the time-machines like the clocks and watches. And when one considered 'mechanical technology' the fruit of mechanical engineering' one touches on the times of the First Industrial Revolution. The time where the power-machines like the steam engine were conceptualized, designed, manufactured and applied as source of rotary power.

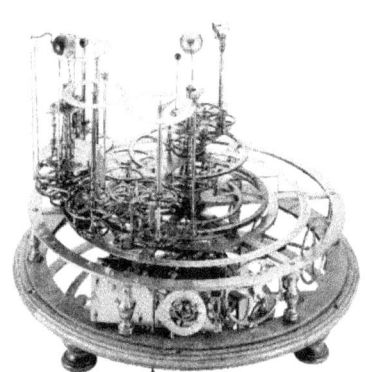

**Figure 108: The Universe Clockwork.**
Source: Unknown. Wikimedia Commons.

The Scientific Revolution saw the rise of the Mechanical Worldview; from the Universe as a Clockwork to the Newtonian Mechanics underlaying the view that all physical phenomena would be understood in mechanical terms. And where all of the wonders of nature worked like a well-oiled set of cogs; the clockwork was the grand machine of the Universe (Figure 108). The conception of the universe as machine amenable to holy rules was the seed of the Enlightenment. That was the

---

[274] The word *Machinery* covers a broad range of constructs. A machine is any physical system with ordered structural and functional properties. As a mechanical system it manages power to accomplish a task that involves forces and movement.

time of *Physics* (eg the sciences of the Nature of Matter), and of *Metaphysics* (the sciences looking at the Nature of Being/Existence and the Nature of Society/Government). From the classical physics of the Cosmos, to the classical physics concerned with bodies acted on by forces and bodies in motion. With the resulting mechanical thinking stimulating the creative contributions of the 'thinkers and tinkerers' to create machinery. And with the rise of the mechanical technologies themselves came the professionals labelled as engineers (eg the mechanical engineers).

That evolutionary path emerged over the centuries in the time of the Industrial Revolutions (Figure 8) with their mechanization in which man extended his powers and capabilities with tools and machinery. A period that saw the development of the Power Machines; both the external combustion engine (aka steam-engine), as well as the internal combustion engine (aka petrol-motor). Machines that brought humanity 'individual mobility' and that powered the Automotive Revolution and the Aviation Revolution.[275] In a similar way during the Second Industrial Revolution, mechanization enabled the creation of the 'Machines of the Mind'; from calculating machines to computing machines.

## *The Invention of the Calculating Machines*

The roots of the Computing Engine—that became later known as the 'computer'—are to be found in the time when the mechanical calculators were invented. That were the machines that could execute arithmetic operations, ie they could 'calculate', a capability till then exclusive belonging to the human mind. The mechanical Art of Calculation that started in the Renaissance times when the desire to economize time and mental effort in arithmetical computations—and to eliminate human liability to error—created the first machines of calculation.

But the origin of calculation is much older.[276] In pre-modern times, counting sticks, knots, pebbles and tally sticks —with values denoted by specific notches— were common forms of counting and numerical record-keeping throughout the world. Advances in the numeral system[277] and

---

[275] See: Van der Kooij, B.J.G.: *The Invention of the Internal Combustion Engine.* (2021)
[276] Tracking and recording the motion of the sun, the moon, and the planets as they paraded across the desert sky, ancient Babylonian astronomers used simple arithmetic to predict the positions of celestial bodies. The Egyptian priests, with their need for calculation of areas (after the Nile flooding) and volumes (their pyramids), developed geometry. The Greek mathematicians—eg Pythagoras, Archimedes—expanded with the calculating of the surface of a circle, the volume of a conic sections.
[277] The Sumerian and Babylonian numerals were a combination of the decimal (ie 10-base) and the sexagesimal system (ie 60-base). The Egyptian, Greek, Phoenicians and Roman

mathematical notation eventually led to the development of mathematical operations such as addition, subtraction, multiplication, division. These numeral systems, along with the use of Roman numerals (I, II, III, IV...), persisted through the Renaissance, as many were hesitant to adopt the Hindu-Arabic numerals (1, 2, 3, 4....) used today out of concern for accuracy and the potential for forgery.

*Numeral System:* Counting with the fingers might have been the origin of the numeral system: the two arms of a person (base-2 system), the fingers of one hand (base-5 system), the fingers of both hands (base-10 system), or the total of all a person's fingers and toes (base-20 system). In addition, the position of the numeral in the sequence became significant. Had the Romans used a special indicator (L=50, C=100, D=500, M=1000), the Arabic system used the position to indicate the unit, tens, hundreds and thousands (1, 10, 100, 1000). Or, in another notation, the position of the unit $1=10^0$, the tens $10=10^1$, and the hundreds $100=10^2$, etc.

**Table 2: Example Decimal System**

| Decimal number | Structure with base number 10 |
|---|---|
| 10 | $1 \times 10^1 + 0 \times 10^0$ |
| 11 | $1 \times 10^1 + 1 \times 10^0$ |
| 111 | $1 \times 10^2 + 1 \times 10^1 + 1 \times 10^0$ |
| 101 | $1 \times 10^2 + 1 \times 10^1 + 1 \times 10^0$ |

**Table 3: Example Binary System**

| Binairy Number | Structure with base number 2 |
|---|---|
| 001 | $0 \times 2^2 + 0 \times 2^1 + 1 \times 2^0$ |
| 011 | $0 \times 2^2 + 1 \times 2^1 + 1 \times 2^0$ |
| 101 | $1 \times 2^2 + 0 \times 2^1 + 1 \times 2^0$ |
| 111 | $1 \times 2^2 + 1 \times 2^1 + 1 \times 2^0$ |

**Table 4: Comparison Decimal-Binary**

| Decimal Value | Binary equivalent | Decimal Value | Hexa-decimal equivalent |
|---|---|---|---|
| 0 | 0000 | 1-9 | 1-9 |
| 1 | 0001 | 10 | A |
| 2 | 0010 | 11 | B |
| 3 | 0011 | 12 | C |
| 4 | 0100 | 13 | D |
| 5 | 0101 | 14 | E |
| 6 | 0110 | 15 | F |
| 7 | 0111 | 16-25 | 10-19 |
| 8 | 1000 | 26 | 1A |
| 9 | 1001 | 27 | 1B |

Jumping to present times, the hexadecimal (base-16) and binary system (base-2 system) became important. The hexa-decimal system uses the ten numbers of the decimal system, but add letters A-F for the values of 10 to 15. Instead of the zero and the nine digits in the decimal system, the binary system only has the zero ('0') and the one ('1'), and the location $2^0$, $2^1$, $2^2$, etc.

Table 2 illustrates the decimal system built up of units $1 \times 10^0$), tens ($1 \times 10^1$), and hundreds ($1 \times 10^2$). Table 3 shows the binary system built up of the values '0' and '1'. Table 4 shows the conversion from the decimal to the binary system.

---

numeral are decimal-based. Present day French counting is a combination of the 60-base (76=soixante-dix-neuf), 20-base (80-quatre vingt), and the 10-base notation.

*Operations:* Calculations are the algebraic operations executed on *operands*;[278] the elements of a numeral system (eg the decimal system, the binary system) and the mathematical symbolic notation (eg Arabic numeral 1,2,3, etc. and Roman numerals, I, II, III, etc.). Next to the basic algebraic *operations* of addition (the '+' sign) and subtraction (the '-' sign), other operations such as multiplication (the 'x' sign of repeated additions) and division (the ':' sign) developed.

Together with the development of the numeral systems and the calculating operations,[279] came the development of the calculating tools that replaced the fingers. And with the tools came the development of specialized devices; such as the tally sticks,[280] the notches in the tally sticks, the knotted strings, and the marked pebbles.

## Early Days of Calculating Tools

To support the human mind in his Art of Mathematics,[281] mechanical tools for 'counting' came on scene. The abacus is perhaps the most well-known pre-modern calculating tool, and is often associated with the wire-and-bead devices that originated in the Middle East. While its true origins remain debatable, the word abacus would have referred to an ancient practice of moving pebbles ('calculi') along lines written in sand. That inspired the Roman hand-hold abacus as well as the Chinese Suanpan. It was followed by the Japanese Soroban in use since the sixteenth century, and the Russian 'Schoty' in the seventeenth century (Figure 109).

During the so-called Dark Ages in Western Europe the art of counting with abacus was more or less forgotten. One of the first scientists, who not only popularized the Hindu-Arabic digits, but also reintroduced the abacus, was Gerbert d'Aurillac (c. 946-1003) aka Pope Sylvester II. His manual tool 'reckoner' became widely used in Western Europe once again during the eleventh century.

---

[278] The arithmetic expression '3+6=9' shows the operands '3' and '6' and the addition operator '+'. An operand, then, is also referred to as 'one of the inputs (quantities) for an operation.'

[279] The word 'calculate' comes from the Latin word 'calculus" (ie 'small pebbles).

[280] The tally stick was an ancient memory aid device used to record and document numbers, quantities, or even messages. The single tally with the tally marks was used for calculation (5-base system), the split tally was used as proof of a loan, and in tax collection (up to Napoleonic times). If someone lend the bank money, the amount was cut widely on a stick, and the stick was cut in two; one part (aka the foil) kept by bank, the other part (aka stock) kept by the lender. The holder of the stock, became the stockholder owning 'bank stock.' And, as the stock could be used for payment, it became a form of money.

[281] Pure mathematics has its own history alongside that of counting. We will meet that subject later on.

The Invention of the Silicon Engine

Other devices emerged from the seventeenth century on, when the slide rule became the most commonly used calculating device for nearly three hundred years.

The slide rule was invented around 1620–1630, shortly after John Napier's publication of the concept of the logarithm.[282] In 1620 *Edmund Gunter* (1581-1626) of Oxford developed a calculating device with a single logarithmic scale; with additional measuring tools it could be used to multiply and divide. In c. 1622, *William Oughtred* (1574-1660) of Cambridge combined two handheld Gunter rules to make a device that is recognizably the modern slide rule. He also developed a circular slide rule in 1632. In 1675 Isaac Newton solved cubic equations using three parallel logarithmic scales. The device became known as the Newton Calculator.

## *Calculating Machines 0.0 (1620-1800)*

The idea of mechanizing arithmetic calculation dates back to the seventeenth century.[283] The time where the role of the timing-device known as 'clockwork' was highly

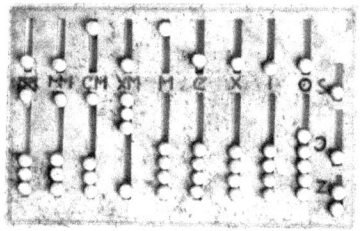

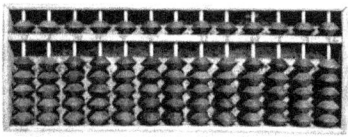

**Figure 109: The Abacus.**

The Roman handheld abacus (top, 2$^{nd}$-5$^{th}$ century AD), Chinese Suanpan (upper middle), Japanese Soroban (lower middle) and Russian Schoty (bottom,1600s).

Source: Wikimedia Commons, Maas Collection.

---

[282] In mathematics, the *logarithm*—log (x)—is the inverse function of the *exponentiation*—$x^n$—as shown in the example: $1000 = 10 \times 10 \times 10 = 10^3$, the "logarithm base 10" of 1000 is 3, or $\log_{10}(1000) = 3$.
[283] The natural philosophers of the Scientific Revolution—Hobbes, Descartes, Beekman, Newton—created the mechanical philosophy (aka Mechanism).

celebrated in scientific thinking (eg comparing the Universe with a clockwork). Times where clockmaking blossomed[284] and the first mechanical calculating machines appeared. Such as the calculating clock 'Rechenuhr' designed by the German physicist and astronomer Wilhelm Schickard (1592-1635) in 1623/24, the 'Pascaline' developed in 1642 by Blaise Pascal (1623-1662), and the 'Stepped Reckoner' created in 1694 by Gottfried Von Leibnitz (1646-1716) that added the functions of multiplication and division to the basic functions of addition and subtraction (Figure 110).

## Gears, Wheels and Buttons

This was only the beginning of the development of calculating machines,[285] as other thinkers added their own ideas, concepts and versions. Machines with a dedicated function that were realized in the fine-mechanical technologies of that time by the artisan instrument/clock makers. The Era of Mechanization was the time of fine-mechanical parts of levers, gears, and wheels combined in machines to process data. And the combination of the parts into a hardwired system[286] defined their function.

*The Calculating Clock*: Astronomy needed a lot of manual calculations (eg the orbits of planets, meteors). While toiling over the many tedious calculations necessary in astronomy work, Wilhelm Schickard's thoughts turned to the notion of mechanically performing mathematical calculations. In 1623, he seemed to have succeeded in designing (not building) an incomplete mechanical device which could perform additions and subtractions. As could be concluded from a letter he wrote on September 20th, 1623, to his fellow astronomer Johannes Kepler describing the calculating clock. It was to be built by a professional, a clockmaker named Johann Pfister. However, the prototype was destroyed in a fire while still incomplete.

At the same time others like the Italian scholar Tito Livio Burattine (1659), the English academic Samuel Morland (1666) and the French clockmaker Rene Grillet (1673), worked on devices that could perform calculations.

---

[284] These were the days of the Pendulum clock (Christiaan Huygens, 1656), the escapement (William Clement, 1671), and the repeating clock (Daniel Quare, 1675).
[285] Calculators perform arithmetic operations. Algebraic equations of the type $y=a + b$, $y=a - b$, $y=a \times b$ and $y=c : d$ contain respectively addition, subtraction, multiplication and division as basic operators,
[286] A clock is a device with a time-keeping function. It has all the parts combined into a system with a single function: to tell the 'time of the day.'

*The Pascaline:* Blaise Pascal's father Étienne Pascal was a tax collector in France, faced with the time-consuming calculation of the ever-increasing taxes in millions of deniers, sols and livres. His son Blaise designed a machine for him to do those calculations, a clockmaker in Rouen made it. The machine used input dials to enter data, and presentation windows with output drums to show the result. Soon it was noted by others, and on May 22$^{nd}$, 1649, by royal decree signed by Louis XIV of France, Pascal received a patent (or privilege as it then was called) for his arithmetic machine. It stated...

**Figure 110: Early Mechanical Calculators (17$^{th}$ Century).**
Replica Schickard calculator (top, 1623), Blaise Pascal's Pascaline (middle, 1645), and Leibnitz's Stepped Reckoner (bottom, 1673).
Source: Wikimedia Commons, History-computer.com

"[...] *arithmetical machine, according to which the main invention and movement is this, that every wheel and axis, moving to the 10 digits, will force the next to move to 1 digit and it is prohibited to make copies not only of the machine of Pascal, but also of any other calculating machine, without permission of Pascal. It is prohibited for foreigners to sell such machines in France, even if they are manufactured abroad. The violators of the privilege will have to pay penalty of 3 thousand livres.*" (text Privilige)

## The Invention of the Stepped Drum Technology

The polymath Gottfried Leibniz, working during his lifetime for three rulers of the House of Brunswick in Hannover, Germany, focussing on logic thought, dreamed about a logical (thinking) device. These thoughts he published in treatises like 'Dissertatio De Arte Combinatoria' (On the Art of Combinations, 1666/1690), 'De Progressione Dyadica' (On the Dyadic Progression, March, 1679), and 'Explication de l'Arithmetique Binaire' (Explication of Binary Arithmetic, 1703). Leibniz, trying to improve upon

Blaise Pascal's calculator, described a machine for solving algebraic equations. Comparing logical reasoning to a mechanism, he reflected on the goal of reducing reasoning to a kind of calculation and of ultimately building a machine, capable of performing such calculations. By 1673, Leibniz had a concept of the calculator, sketched out design possibilities and began working on a model with unnamed Parisian craftsmen.

> *Who exactly did work on the machine after 1679 is unclear, [...] Only beginning in 1693 does the picture become clearer. The artisan Georg Heinrich Kölbing was employed. He and the butler Balthasar Ernst Reimers were busying themselves with the machine as Leibniz wanted to present the invention to his patron, Ernst August von Hannover, which he finally managed to do in 1695.[...] After 1710, Leibniz took back to Hanover the almost finished, but still not quite working, second machine and the finished pieces of the newly commissioned third machine. [...] Leibniz died in 1716 and the project of building the third model of the calculator was never completed.* (Morar, 2014, pp. 6-7)

It took some years, but —based on the *Leibnitz Wheel*, a cylinder having nine teeth of increasing length (Figure 111)— he had conceptualized and prototyped the pinwheel calculator that became known as the 'Stepped Reckoner.'[287] A number of such machines were made (but not completed?) during his years in Hanover by craftsmen working under his supervision.

> *Conceptually, the Stepped Reckoner was a remarkable machine whose operating principles eventually led to the development of the first successful mechanical calculator. The key to the device was a special gear, devised by Leibniz and now known as the Leibniz wheel, that acted as a mechanical multiplier. The gear was really a metal cylinder with nine horizontal rows of teeth; the first row ran one-tenth the length of the cylinder, the second two tenths, the third three-tenths, and so on until the nine-tenths length of the ninth row. The Reckoner had eight of these stepped wheels, all linked to a central shaft, and a single turn of the shaft rotated all the cylinders, which in turn rotated the wheels that displayed the answers.*[288]

Leibnitz's Stepped Reckoner was the start of a specific range of calculating machines along a technical trajectory of improvements. Although conceptualized and designed by the polymaths of those days,

---

[287] If the machines were completely realized by the artisans he hired and actually functioned is unclear. There seems to have been a large distance between Leibnitz' rhetoric and the actual machine. He described it in the 'Brevis descriptio machinae arithmeticae' in the first issue of the Miscellanea Berolinensia, a self-promoting account.

[288] Source: Leibnitz and the Stepped Reckoner. http://ds-wordpress.haverford.edu/bitbybit/bit-by-bit-contents/chapter-one/1-8-leibniz-and-the-stepped-reckoner/ (Accessed June 2010).

these machines were the fruits of the mechanical technologies available at that time; the artisans working metals, building clocks and instruments. And with the calculating machines, it was all about gears, wheels and pins (Figure 111).

*The Calculation Craze*

By the start of the eighteenth-century, all-over Europe the creative minds picked up the concept of the new machines, copied them, or developed their own versions. The century became a festival of beautiful mechanical machines implementing calculating functions. Especially in the social circles that could afford the patronage of scientific thinking; the aristocratic and monarchical societies. As they had done in the preceding centuries characterized as the Scientific Revolution.

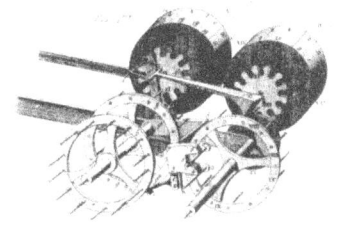

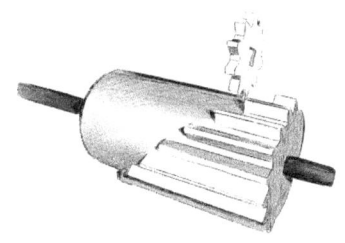

**Figure 111: Gears, Wheels and Pins.**

Interior of Pascal's Pascaline (top), interior of Leibnitz Stepped Recknoner (replica, middle), and the Leibnitz Wheel (bottom).

Source: Wikimedia Commons. https://www.uni-hannover.de/en/universitaet/profil/leibniz/leibnizausstellung/

*During the sixteenth and seventeenth centuries, savants and literati with mixed mathematical inclinations were keenly interested in simplifying and automating calculation. This goal was something that unified the interests of two very different groups: mathematicians or astronomers engaged in the social-credit business of the commercium literarium on the one hand, and shopkeepers, apothecaries and accountants preoccupied with the real-money business of burgeoning global capitalism on the other.* (Morar, 2014, p. 1)

By the dawn of the eighteenth century, a range of events took place that left their traces in history.

The Italian aristocrat Giovanni Poleni (1683-1761), born in the Republic of Venice as son of wealthy marquis, at the age of twenty-four heard news of Pascal's and Newton's calculators. He decided to build his own, and had created in 1709 the Poleni calculator in the shape of a grandfather' clock (Figure 112, top). Differently from Leibniz' 'stepped drum', the marquis conceived the 'pinwheel' as input device; a wheel with an adjustable number of teeth (0-9) that could be raised or lowered to set the figure to be operated upon.

There was the German mathematician and instrument maker Jacob Leupold (1674-1727). He designed a calculator with the mechanism of the 'Schaltkinke' (aka switching latch, intermittent contact, adjustable pawl, and selectable ratchet), but his death prevented its realization. However, the Leupold mechanism became quite popular at the end of the XIX century in several calculating machines. He wrote the encyclopedia 'Theatrum Machinarum Generale ('The General Theory of Machines') (1727), a nine-volume series on machine design and technology, published between 1724-1739. decades before. It included a description of the Leibnitz' machine.

The German mechanic, constructor and optician Anton Braun (1686-1728) from Möhringen (Baden-Württemberg, Germany), was appointed in 1724 as a mechanician and optician of the imperial court in Vienna, Austria. He started to design a calculating machine for the purposes of the court that he finished in 1727,

**Figure 112: Early Mechanical Calculators (18th Century).**

Replicas of Giovanni Poleni's calculator (top 1709), and Braun's Sprout Wheel calculator (middle, 1727). Lord Stanhope's Machine (1777, lower middle), and the Hahn Calculator Machine, Schuster replica (bottom, 1783).

Source: Arithmeum, Bonn. Museo Nazionale della Scienza e della Tecnologia "Leonardo da Vinci", Milan. https://history-computer.com/

producing a calculating machine using pinwheels in beautiful workmanship (Figure 112). When in 1727 he presented the machine to the Holy Roman Emperor Karl VI, he got into favour of the emperor and was appointed as imperial instrument maker. In addition, he was rewarded a 12-diamond chain, occupied with the portrait of the emperor and a huge sum of money; 10,000 guilders.[289]

All over Europe there were many others from different backgrounds who contributed in different degrees and different capacities to the development of the mechanical calculator: Hillerin de Boistissandau (France) in 1730, C.L. Gersten (Germany) in 1735, Jacob Isaac Pereire (France) in 1750, Phillip Mathieus Hahn (Switzerland) in 1773. Charles Stanhope, the 3rd Earl of Stanhope (England) in 1775; Johan Helfreich Müller (Germany) in 1783, Jacob Auch (Germany) in 1790, and Reichhold in 1792. (Figure 112).

The German vicar Philip Matthaus Hahn (1739-1790), born in a wealthy family, was known in church circles through his theological writings, with princes and nobles by his large astronomical clocks and machines, and in the upper middle class through its pocket watches, scales and his calculating machine. Around the summer of 1770 he started to design a calculating machine using Leibnitz' stepped drum concept. The first working copy of the device was ready in 1773, but was demonstrated as late as in 1778, because Hahn has difficulties with the reliability of the tens carry mechanism. Until 1779 four machines were made, till the end of his life, Hahn manufactured about 5-6 devices.

The British aristocrat, polymath and politician Charles Stanhope (1753-1816), member of the Royal Society, was devoting a large part of his wealth to experiments in science and philosophy. In the 1770s he was designing calculating machines, using both an adapted stepped drum technology as well as the pinwheel-technology. Stanhope worked on his logic machines for some 30 years, desining several versions. The calculating machines itself were made by the skilful mechanic James Bullock, the first machine in 1775, the second in 1777, and the last in 1780.

Notwithstanding all these early experimentations with clockwork machines that performed actions originally reserved before to the human mind, by the end of the eighteenth century, calculating machines were still curiosities used for display purposes, rather than for actual use. The

---

[289] In those days the royal courts had their own scientists working for them; astronomers, mathematicians, and instrument makers. Working the game of patronage, it was important to be well-connected to the courts, so they dedicated their publications and artefacts to those aristocratic/royal patrons.

limitations imposed by the mechanical technology made it impossible to meet Pascal's dream of making them a practical calculation device.

## Calculating Machines 1.0 (1800-1900)[290]

It was not until the beginning of the nineteenth century that the calculator became a more popular device outside the specific social circles of the well-to-do aristocracy employing scientists, instrument makers and master-artisans. And that shift in exclusivity was caused by the later improvements in the mechanical calculator as conceptualized by Leibnitz. It was the construct that became known as the 'Arithmometer.'

### The Arithmometer

The calculator called Arithmometer was created in 1820 by the French engineer Charles Xavier Thomas de Colmar (1785-1870). After his —non technical— studies in Napoleonic times, he had entered the tax administration, which in fact governs the collection of taxes. After occupations in the Napoleonic army in Spain he was responsible for the Army's food supply, next —after working for the Duke of Angoulême in France— he went to England to study the insurance business. In 1819, he founded in France with Jacob Dupan, a wealthy Swiss businessman, the insurance company 'Compagnie du Phénix.' Followed by the creation in 1829 of the insurance company against fire, called 'Le Soleil.'

In all these occupations he must have been faced with incessant accounting calculations.[291] So he became interested in the idea of a calculating machine and started —assisted by the clockmaker Devrine— working on a prototype based on Leibnitz' concept. In 1822, Devrine builds for him the first recorded arithmometer in history (Figure 113). A reckoning device that consisted out of three parts: one part concerned with the setting of the input numbers, a second part concerned with the counting operation, and a third pard recording (and presenting) the result.

> He obtained the 5-year French Patent № 1,420, on November 18th, 1820, for his calculating machine based on Leibnitz's design. It was described in the November, 1822 issue of the 'Bulletin de la Société d'Encoragement pour l'Industrie Nationale'.

---

[290] This chapter is based on numerous sources in the public domain. Much data (without citing them, sorry) are obtained from websites like www.arithmometre.org; https://history-computer.com/ inventions/; https://www.officemuseum.com/calculating_machines_adders.htm; https://www.whipplemuseum.cam.ac.uk/; https://www.xnumber.com/
[291] Insurance is a business of gambling, but also a business of risk-taking, preferable 'calculated risks.' And for that calculation, something other than the capabilities of the human mind could be useful.

However, in the following years between 1822 and 1844, not too much work was done to proceed with the machine concept,[292] as he spent all of his time and energy on his insurance business after he had founded the Compagnie du Soleil in 1829. The 1840s had been the period when Thomas de Colmar began to really develop his insurance business. In 1843 he founded a second insurance company, the Compagnie de L'Aigle. And at the end of 1854 it absorbed the portfolios of the companies Le Globe, La Lyonnaise, La Palladium and the mutual society La Nivernaise. The insurance business was successful, and being by now quite a rich man, this is also the time —after this hiatus of some thirty year— when Thomas decided to invest in the further development of his arithmometer. By July 1848 he had his first prototype being built by a watchmaker.

He obtained a range of related patents; the French FR-Patent № 8,282 granted on April 25th, 1849, and the British GB-Patent № 13,504 in 1851, and the Belgium Patent № 5,618 granted on March 24th, 1851.

He demonstrated his machine successfully at the Great Exhibition of London in 1851, and the Universal Exhibition of Paris in 1855. To promote his machine, Thomas also offered the crowned heads of Europe special editions of his machines. In return he received numerous distinctions from those recipients.

*Between 1851 and 1854, he will receive in return numerous honorary distinctions: He receives from his Highness The Bey of Tunis the Nicham in*

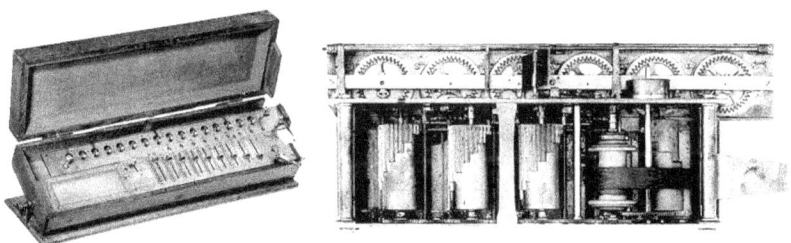

**Figure 113: The Arithmometer.**
Arithmometer by Thomas de Colmar (left, 1820), and the interior mechanism with the Leibnitz Wheels (right, 1822).

Source: Wikimedia Commons. http://www.mhs.ox.ac.uk /staff/saj/arithmometer
http://www.arithmometre.org

---

[292] Production figures differ from source to source. Some indicate that in total, Thomas' workshop completed five hundred machines from 1821(1845) to 1865, three hundred machines from 1865 to 1870, four hundred machines from 1871 to 1875, and three hundred machines from 1876 to 1878. Others claim that more than 5000 specimens of the Calculating machine were manufactured during the 90 years.

*diamond (rank of commander). In 1852, the prince president (future Napoleon III) offered him a gold snuffbox with his number. He was appointed Commander of the Order of Saint Gregory the Great by His Holiness Pope Pius IX in December 1852. The list is long: The King of Greece, the Grand Duke of Tuscany, the King of Sardinia, the King of Portugal, the king of the two Sicilies and so on offer him distinctions to make a gurnard mullet turn pale!*
*In 1855, he presented at the Universal Exhibition in Paris a gigantic machine of 30 figures that looked like a piano. [...] Serial production of his arithmometer begins. We see the first serial numbers appear, the first user manuals. During this period, Thomas had around 200 machines made, including 150 10-digit models and around 50 16-digit models. [...] The second peak corresponds to the period of 1860-1863, when Thomas de Colmar put on the market a new much more efficient arithmometer.* [293]

By the 1860s his arithmometers were being produced in a form which remained relatively stable[294] until the First World War. At the time of his death in 1870,[295] around 900 Arithmometers had left his workshops, making it the first mass produced mechanical calculator in the world, and at the time, the only mechanical calculator reliable and dependable enough to be used in places like government agencies, banks, insurance companies and observatories.

After his death, his son and grandson continued the business, untill his engineer Louis Payen bought in 1887 the business. And when Payen died in 1902, his widow took over and created the Veuve Payen's models. In March 1915, the widow Payen sold her business to Alphonse Darras, who had to liquidate the company as the war effort had made copper and brass unavailable. In total about 1,500 machines were made between 1820 and 1930.

## Cloning the Arithmometer

Colmar's design was so successful, that about twenty clones of the Arithmometer were manufactured in. Starting with Burkhardt in 1878, followed by Layton, Saxonia, Gräber, Peerless, Mercedes-Euklid, XxX, Archimedes, TIM, Bunzel, Austria, Tate, Mada etc. (Figure 114). These clones, often more sophisticated than the original arithmometer, were built until the beginning of the Second World War.

---

[293] Source: The Thomas de Colmar Arithmometer: a historical perspective.
http://www.arithmometre.org/Biographie/PageBiographie.html. (Accessed July 2010)
[294] In modern parlance we call this a dominant design.
[295] The War of Sedan and the Shelling of Paris (1870-1871) in which Germany started their conquest of France, saw the Alsace/Lorraine-region occupied by Germany. In 1871 the German government suddenly prohibited French insurance companies from exercising their industry in Alsace and Lorraine.

Arthur Burkhardt (1857-1918), using the classical stepped-drum system of Leibniz and Colmar, created a version of the stepped-drum calculator: aka the Burkhard Arithmometer. The production of first model of the machine (the so-called Model A or Model I, Figure 114) was started in 1880. It was not a success as Arthur Burkhardt did little to make his "arithmometer" known, and after a slack in sales, he started again in 1885 and by1892 the number of sold machines of Burkhardt reached 500. Burkhardt obviously did too little to catch up with the up-and-coming German competition: Saxonia (Glashütte), Peerless (in the Black Forest), XxX (Dresden), Archimedes (Glashütte) and TIM (Berlin) passed Burkhardt in a few years. In 1905 the 'Glashütte calculating machine factory Arthur Burkhardt' was created. Over the following years it introduced improved models (eg Model D, E and G, Figure 115), and, although the Burkhardt Arithmometer obtained medals on national, international and world fairs, it hardly found any buyers. In 1929 the factory in Glashütte was closed.

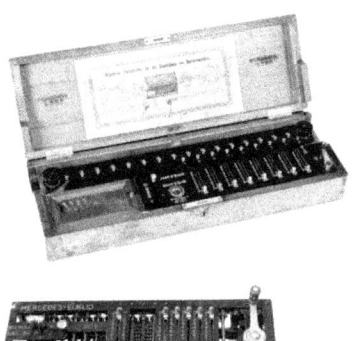

**Figure 114: Clones of the Arithmometer.**

Burckhardt's Model A Arithmometer (top, 1873), Mercedes-Euclid Model I (1906, middle).

Source: Wikimedia Commons, History-computer.com http://modell-sammlung.uni-goettingen.de/

One of those competitors was the Archimedes calculator. Its director Reinhold Pöthig had worked in Glashütte as an apprentice for Arthur Burkhardt, and had ideas for improvements on the classic Arithmometer design He developed his own version and began making them in 1906, and named the machines Archimedes. By 1912 the company was only making calculators, having abandoned all other activities, so it was renamed to *Glashütter Rechenmaschinen-Fabrik Archimedes, Reinhold Pöthig.*

The first English stepped drum machine (Thomas de Colmar-type arithmometer) was devised in early 1880s by Samuel Tate (1840-1917), an iron worker and mechanical engineer from Sedgley, Staffordshire. Tate made some improvements in the construction of the Colmar's device and patented his improvements (UK-Patent № 65, granted January 1st, 1884). In 1883 the brothers Charles and Edwin James Layton, booksellers and authors of insurance books, exhibited the first arithmometer as the agents of Tate and soon afterwards acquired the patents, arranging a manufacturing workshop on Farrington Road in London.

In 1895 three of the mechanics, who worked in the factory of Arthur Burkhardt in Glashütte, Germany decided to leave and to found in 1902 a new factory for calculating machines; the *Glashütter Rechenmaschinen-Fabrik Saxonia*. In this factory was produced the successful calculating machine brand Saxonia (Figure 116). Over 12,000 Saxonia machines were manufactured and sold from 1895 to 1914. First models of the machine certainly are almost identical to the Burkhardt's Arithmometer, but later models had many improvements. In 1920 the factory Saxonia merged with the factory of Burkhardt and the joined company *United Glashütte Rechenmaschinenfabriken* was opened. It would continue production up to 1929, when it bankrupted.

During these early years, the name 'Arithmometer' became synonym with the mechanical calculator. For forty years, from 1851 to 1890, the Colmar arithmometer was the only type of

**Figure 115: Evolution of Burkhardt Arithmometer Machines.**

From the wooden Model 514 (top, 1900), to the metal Model E (bottom, 1925).

Source: http://modellsammlung.uni-goettingen.de/

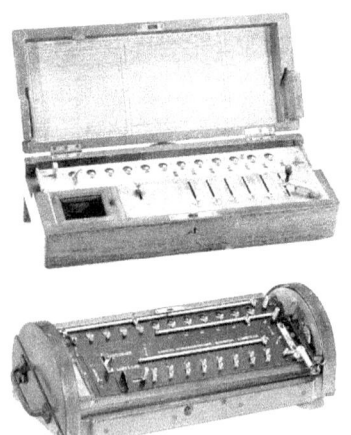

**Figure 116: Saxonia Arithmometer Calculators.**

Burkhardt's Aritmometer clone (top, unknown dat), and Improved Arithmotere (bottom, 1895).

Source: https://history-computer.com/

mechanical calculator in commercial production, and it was sold all over the world. However, with rise of the Comptometer around 1890, a new generation of mechanical calculators emerged. It was a trajectory that started with the development of an easier ways to enter the numbers: the numerical keyboard.

## The Quest for Keyboard and Printing Facilities

Operating the Arithmometer was a complex affair using the input wheels. So the quest was on for an easier input device. Between 1850 and 1887, many attempts were made to develop a calculating machine that would use buttons (aka 'keys') as means to enter the input data.

In 1844 the Frenchman Jean-Baptiste Schwilgué (1776-1856), a watch and clockmaker from Strasbourg, together with his son Charles, patented a key-driven calculating machine (French Brevet d'invention № 624, and Brevet d'invention № 623, granted December 24th, 1844), which seems to be the third key-driven machine in the world, after those of James White and Luigi Torchi. From this development the rather simple 'adding' machines (aka Adder) appeared on scene (Figure 118).

In Europe, key-driven adding machines were developed by Victor Schilt (1851), F. Arzberger (1866), A. Stettner (1882), Bagge (1882), d'Azevedo (1884), Petetin (1885) and Maq Meyer (1886). In the United States key-driven machines were developed by: D.D. Parmelee (1850), Thomas Hill (1857), G.W. Chapin (1870), W. Robjohn (1872), D. Carrol (1876), Borland & Hoffman (1878), M. Bouchet (1883), C.G. Spalding (1884), A. Stark (1884), L.W. Swem (1885), P.T. Lindholm (1886) and B.F. Smith (1887).

Engineers explored the calculating machines in different functional directions using different concepts. Among them the 'Additionneur mécanique' from Alexis Petetin (1826-1890) from Besançon (Doubs, France); the adder developed and perfected by Cyrus G. Spalding (1835-1910) by 1884, and the adder made by Peter. L. Lindholm (1850-1890), who emigrated from Sweden becoming professor of mathematics at Bethany College in Virginia USA.

Alexis Petetin, was granted two French patents (№ 110,349, November 27th, 1875 and № 163,925, granted April 27th, 1884) and one German patent (№ 321,48, granted January 24th, 1885) for adding machines. Spalding was granted US-Patent № 146,407 on January 13th, 1874, and US-Patent № 293,809 on February 19th, 1884 for an 'adding machine.' P.T Lindholm was granted US-Patent № 343,770 on June 15th, 1886. He also received British Patent № 7874 of June 15th, 1886 (Figure 117).

Also *Thomas Hill* (1818-1891), a Harvard-educated mathematician, obtained a patent for a calculating machine; his improved arithmometer.

> *The key-driven calculating machines were already built in Europe (...the machines of Torchi, Schwilgué, and Schilt), and when Hill heard about them, he decided he would make his own. Hill applied for patent and sent a wooden model to the US patent office (up to 1880, the Patent Office required inventors to submit a model with their patent application).²⁹⁶ Inventors placed great importance on their models and viewed a well-executed model as the key element in obtaining a patent). The US patent № 18,692 for an arithmometer was granted on November, 24th, 1857. [...] Other key-driven machines were introduced in the following years, in particular by the Americans Leonard Nutz (in 1858) and Joseph Alexander (in 1864), Austrian Friedrich Arzberger (1866), Gilbert Chapin from New York (1870) and David Carroll from Pennsylvania (1876) among others. Changes in these machines led to the first practical key-driven machine—famous Comptometer by Dorr E. Felt, introduced in 1885.²⁹⁷*

**Figure 117: Early Adding Machines.**

Spalding Adder (top, 1884), Lindholm Adder (bottom, 1884)

Source: Wikimedia Commns, https://history-computer.com/

---

²⁹⁶ A typical patent application has two main parts. The first is a specification of the invention, which is written like a brief science or engineering article describing the problem the inventor faced and the steps she took to solve it. It also provides a precise characterization of the "best mode" of solving the problem. The second part of the patent application is a set of claims, which usually encompass more than the material set out in the specification. Claims define what the inventor considers to be the scope of her invention, the technological territory she claims is hers to control by suing for infringement.

²⁹⁷ Source: Tomas Hill. https://history-computer.com/people/thomas-hill/ (Accessed June 2021).

None of them would equal the success of the next dominant design of the mechanical calculator. The rather simple adding/multiplication machines (Figure 118) however, paralleled the development of the pinwheel calculator. The Comptometer used a 'multiple order keyboard' —also called full keyboard— which consisted of a matrix with 9 rows of keys, one for each digit (1 to 9). The number was entered by pressing one digit in each column.

Next to the improvement on the input side of the calculating machines, efforts were taken to improve the output side of the machine. Efforts focussing on making a physical copy of the calculation result with a 'printing facility.'

Already in 1853, the Swedish engineer *Edvard Scheutz* (1821-1881), designed and constructed a machine to compute tabular differences and print the results. The prototype was completed in 1853, patented in Great Britain in 1854, and shown at the Paris Universal Exhibition in 1855. However, it would take some decades before the printing facility became incorporated in calculating machine. That moment came when Dorr Felt developed his Comptograph in 1886.

## The Invention of the Pinwheel Technology.

Even though numerous calculating machines had been designed till the start of the nineteenth century, only the arithmometer of Charles Xavier de Thomas patented in 1820 had some practical importance. It took half

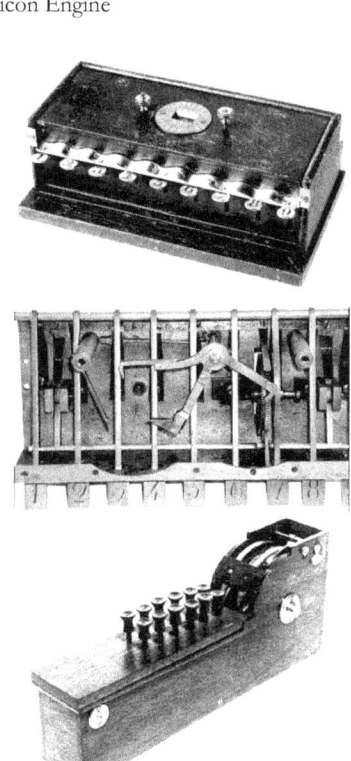

**Figure 118: Keyboard-driven Adding Machines.**

Schwilgué's calculating machine (top, 1844). Victor Schilt's keyboard Adder (upper middle, 1851). Thomas Hill Adder (lower middle, 1857). Madden/Bouchet Adding Machine (bottom, 1884)

Source: Swiss Federal Institute of Technology Historycomputer.org, Officemuseum.com.
https://americanhistory.si.edu/

century and a new development, however, for the calculating machine to create any impact.

> Some early development around 1840 preceded the new concept of calculation. Such as the pin-wheel based calculator developed by the watchmaker Izrael Staffel (1814-1884). After a decade of development of mechanical calculating machines, in 1845 he exhibited his calculating machine that was able to add and subtract. In 1851 he showed his improved calculator that could also divide, multiply and obtain a square root, at the Great Exhibition in London. He was awarded a gold medal for the best machine of this kind. But is seemed not to have been commercialized.

In America, Frank Stephen Baldwin (1838-1925), a machine designer by profession, had seen a Thomas-arithmometer in the office of a life insurance company at St. Louis, which, at the time, was the only mechanical desktop calculator in commercial production. He already had invented and patented a 'Recording Lumber Measure', and started thinking about calculating machines. Soon, he had realized a major change in von Leibnitz's principle in 1875; the later called 'Baldwin Principle' used a 'pin wheel': a cylinder with movable pins (Figure 121,

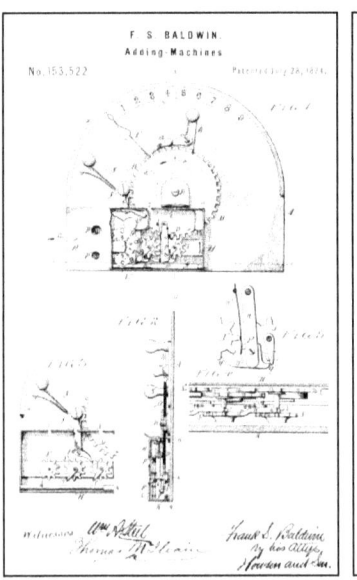 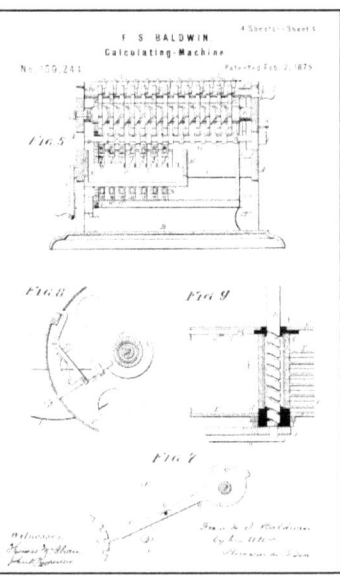

**Figure 119: Baldwin's Pinwheel Patents.**

Baldwin's US-Patent № 153,522 for his Adding Machine (left), and US-Patent № 159,244 for his Calculating Machine (right).

Source: USPTO.

left). Based on this principle,[298] Baldwin made several models between 1873 and 1912. Starting with the Baldwin Adding Machine up to the Baldwin Calculating Machine (Figure 120).

For his work, he was granted several patents. From US-Patent № 153,522 filed in 1873 and granted on July 28th, 1874 for an adding machine, US-Patent № 159,244 granted on February 2nd, 1875 for a calculating machine; up to US-Patent № 641,065 granted on January 9th, 1900 also for a calculating machine (Figure 119).

Baldwin placed both machines on an exhibition at the Franklin Institute. He sold his calculator to various insurance companies in New York, and to various government departments in Washington. Later on, Baldwin designed a listing machine in 1905 with only ten keys and a spacer. After the award of a further Patent in 1908 for the Baldwin Recording Calculator, which combined the listing facilities of the 1905 machine with a calculator, he collaborated with Jay Randolph Monroe (1883-1937), of the Western Electric Company in New York City, from 1911 onwards to commercialize his calculating machines.

**Figure 120: Baldwin's Machines.**

Baldwin's Adding Machine (top), his Calculating Machine (1875, middle), and the improved Baldwin Calculating Engine (bottom, 1905).

Source: Wikimedia Commons, https://americanhistory.si.edu/collections, History-computer.com.

In Russia the Swedish engineer Willgodt Odhner (1845-1905), working for Ludvic Nobel, was asked to repair a Thomas arithmometer somewhere between 1871-1874. Stimulated, he made a first prototype of a pinwheel calculator in 1875 of his own ideas for a calculating that paralleled

---

[298] In the *Baldwin Principle* the nine stepped drums are replaced by a pinwheel. This is a variable-toothed cylinder with sets of nine radial pegs that protrude or retract by the action of setting levers through slots located in the opposite side of the cylinder.

Baldwins ideas (Figure 121). Odhner—in Russia—seems to have been unaware of Baldwin's work—in America—on a calculator because it was patented in February 1875. Next, Nobel made a deal with Odhner for producing 14 calculating machines at his factory. He also presented his arithmometer at the Imperial Russian Technical Society in 1878.[299]

For his invention, he obtained DE Patent № 7393, granted on November 19th, 1878. He received a similar US-Patent № 209,416 on October 29th, 1878 (Figure 122). In addition, he filed his patent application in other countries. The corresponding Swedish patent is № 123 of 1879. The Russian patent № 148/1879 for three years was filed on February 14th, 1879 and granted December 31st, 1879.

In 1877, Nobel became occupied by the war between Russia and Turkey, and lost interest in the project about calculators. By that time Odhner met many difficulties, personal and professional, as his relations with the directors of the Nobel-factory were quite bad. Odhner had to make debt to finance his living. From 1878 to 1892 he worked for a state-owned Russian printing company called the Expedition.

In the meantime, he sought to commercialize his patent stimulated by Karl Königsberger, a serious businessman who was not interested in production. They created a partnership, and the first selling efforts after patenting the invention were directed to United States, where Karl had a business partner, but these efforts were not successful. However, a representative of Königsberger succeeded in

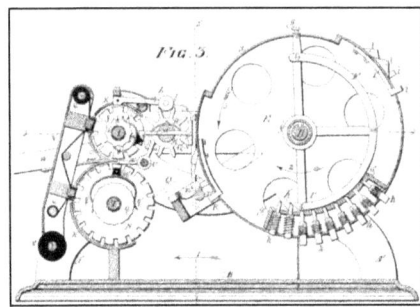 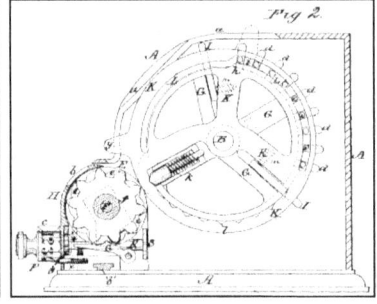

**Figure 121: Baldwin's and Odhner's Pinwheel.**
Detail from Baldwin's US-Patent № 159,244, Figure 3 (left, 1875) and Odhner's US-Patent № 209,416 Figure 2 (right, 1878).
Source: USPTO

---

[299] Source: http://w1.131.telia.com/~u13101111/odhner.html; http://odhner.com/kevin/wtodhner/calcs.html; http://www.rechenmaschinen-illustrated.com/timo%20leipala_Odhner_Part%202.pdf

selling the licence[300] in 1892 to German sewing machine company *Grimme, Natalis & Co* (GNC) to produce Odhner calculators, who soon after started production in Brunswick, Germany. They sold their machines under the Brunsviga brand name. From the 500 machines sold by the end of 1892, by 1912 they built the 20,000th machine.

In 1887 Odhner was granted an official permission, next to his work in the printing facility of the Expedition, to open his own workshop, which later on became the 'W. T. Odhner' factory in St. Petersburg. This was in 1890, and Odnher got his rights for the calculating machine back from Königsberger. He improved on his calculating machine (Figure 123), patented its again and was granted a Russian Patent for an improved calculating machine.

The patents were also registered in France (№ 261806, 1890), Luxemburg (1890), Belgium (№ 91812, 1890), Sweden (№ 3264, 1890), Norway (№ 2117, 1890), Austria-Hungary (№ 45538, 1890), England (№ 13700, 1891), Germany (№ 64925, 1891) and Switzerland (№ 4578, 1892), USA (№ 514725, 1894).

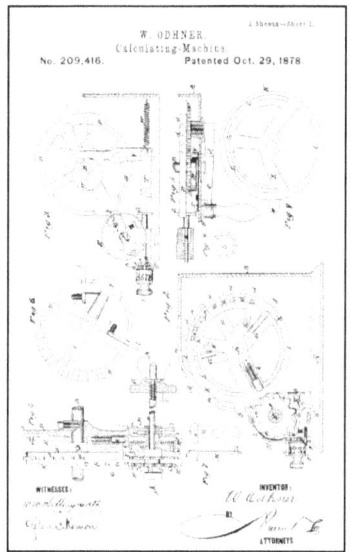

**Figure 122: Odhner's First Calculator.**

Odhner Patent № 209,416 (top, 1878), and the Odhner Airthmetic Calcuator (bottom, 1877). The name "Arithmometer" is written across the top and above it is written W. Odhner. In the upper left is written "L Nobel", and on the upper right "St. Petersburg".

Source: USPTO, National Museum of American History, Smithsonian Institution.
http://odhner.com/kevin/wtodhner/calcs.html

---

[300] GNC paid 100,000 Reichsmarks for the patent and 10 marks for each calculating machine sold.

In the same year 1890 Odhner started —next to the production of the Orlov printing press and his calculating machines— a powerful publicity campaign for his new calculator and in 1893 the arithmometer was exhibited with success at the World's Columbian Exposition in Chicago. In the meantime, because Odhner did not have money, he took an English investor, Frank Hill, as his partner and they founded in 1892 the *Mechanical Factory of Odhner & Hill*. The company expanded swiftly, but their cooperation ended soon and in 1895 and the company was renamed *Maschinenfabrik & Metallgiesserei W.T. Odhner*. [301]

By 1897 some 5,000 machines were manufactured. However, the Russian production only lasted some thirty years until the Russian factory was nationalised and closed down during the Russian Revolution of 1917.[302] By then some 29,000 calculators were made.

Odhner's sons Alexander and Georg had fled to Sweden, and there they continued to produce the machines by the 'AB Original

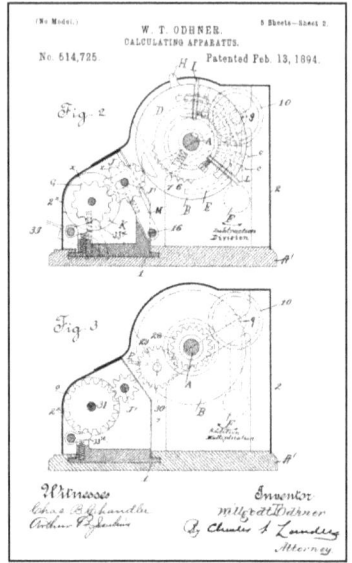

**Figure 123: Odhner's Improved Calculator.**

Odhner US-Patent № 514,725 (top, 1894), and Improved Calculator (bottom, 1890).

Source: USPTO, Swedish National Museum of Science and Technology

---

[301] The production palette of Odhner's factory varied. Of the various products, one can mention cigarette machines designed by Odhner capable of producing 4000 cigarettes in an hour, Orlov printing presses, small mechanical precision instruments and castings of brass, aluminium and cast iron. The bestseller of the Odhner's factory certainly was not the arithmometer, but the Orlov printing press.

[302] As in every other factory in Petrograd—the new name for St.Petersburg—at the revolutionary times, much time at the Odhner factory went to meetings of the workers. As the only Russian calculating machine factory the company was very important to soviet economy and was nationalized in September 1918.

**Figure 124: Clones of Odhner's Calculator.**
The 'Iron Felix' mechanical calculator (left, 1920s). Thales Mez, Germany (middle, 1930). Multo Addo, Olympia (Sweden, right, date unknown).
Source: https://www.nzeldes.com /HOC/IronFelix.htm; http://public.beuth-hochschule.de

Odhner' company in Gothenburg under the product name Arithmos. In 1924, the Russian government moved the old production facility to Moscow and cloned the Odhner calculator under the name Felix Arithmometer (aka Iron Felix, Figure 124, left),[303] which went on well into the 1970s. Between 1929 and 1978 several million were manufactured, in more than 20 modifications.

Odhner's arithmometer was copied, manufactured and sold by many other companies all over the world. In France there was the brand name of Dactyle. In Germany there was Thales, Triumphator, Walther and Brunsviga. In England there was Britannic and Muldivo. In Sweden Multo, in Russia Felix and in Japan Tiger and Busicom (Figure 124). [304]

*The Key-driven Calculating Machine*

All of the early calculating machines operated on different principles, testimony both to the ingenuity of their inventors and to the anarchy of the state of the art of mechanical calculation. Although Baldwin's and Steiger's calculators could perform all four arithmetic functions, they failed to crack the mass market, and the reason was chiefly technical. Along with all their rivals, these calculators had at least two major limitations. First, they lacked a truly convenient method for entering numbers, which made them somewhat awkward to use. Second, they lacked a printer for recording results, which meant that you had to keep track of your results and write the answers down on a sheet of paper, creating many opportunities for error.

---

[303] Felix Dzerzhinsky founded the Soviet calculator factory in Moscow in 1924, and was also the founder of the famous Cheka (Bolshevik secret police). He was sadly notorious for his hardness and was awarded the nickname 'Iron Felix', just like the calculator.
[304] Busicom would later come to play an important role in development of the first computer on a chip.

Then came the American Dorr Eugen Felt (1862-1930) who, after working in various machine shops, had become a proficient mechanical draftsman. At the age of 23, during the winter of 1884-85, he constructed in a macaroni box his first prototype of a new type of calculator he called 'Comptometer' (Figure 125). The machine used, next to a keyboard, a different mechanism from the arithmometer's stepped drums: the pinwheel.[305]

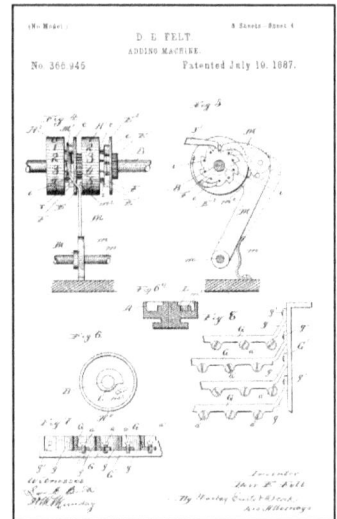

The original comptometer design was patented by Felt. He was granted US-Patent № 366,945 on July 19th, 1887, followed by US-Patent № 371,496 on October 11th, 1887 (Figure 127) and US-Patent № 405,024 on June 11th, 1889.

Two years later, on June 11th, 1889, he was granted US-Patent № 441,232 for the type writing machine called Comptograph, a Comptometer with a printing facility. This patent was followed by many others labelled as 'tabulating machines.' In total he was granted some forty patents between 1886 and 1914.

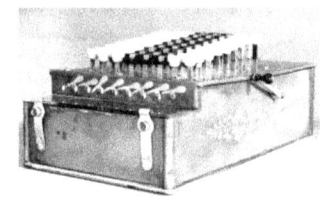

**Figure 125: Felt's Pinwheel Calculator 'Comptometer.'**

Felt's Patent № 366, 945 (top, 1887), the prototype in Macaroni box (middle, 1885), and the wooden box model A (bottom 1905).

Source: Wikimedia Commons. USPTO http://www.vintagecalculators.com/html/comptometer.html

Financed by his cousin Chauncy Foster, and facilitated by his employer, the Chicago businessman Robert Tarrant, he further improved his machine. Over the next two years, the design was refined into a metal mechanism while retaining its

---

[305] Others had already explored the Poleni pinwheel concept of 1709. Such as Izrael Staffel, a polish clockmaker introduced his pinwheel machine in 1845 at an industrial exposition in Warsaw, Poland and won a gold medal in 1851 at The Great Exhibition in London.

wooden case (about 6,500 machines were built in between 1887 and 1903) (Figure 125).

By the end of 1887 four of the eight machines built that year were installed in the US Treasury offices. The remaining four machines were then bought by Chicago businesses. The eager acceptance of his machines by progressively-minded business men encouraged him to enlarge his manufacturing facilities. So Felt and Tarrant signed a partnership contract on November 28th, 1887, and incorporated the *Felt & Tarrant Manufacturing Company* on January 25th, 1889. By 1930, when Felt died, the firm had $3.1 million in sales and 850 employees.

Several of Felt's Comptograph patents were licensed to others (including Burroughs), and eventually formed the basis of the American 'accounting machine' industry. Also known as 'adding typewriters', these machines could post combinations of two or more records and furnish a proof-journal on accounts payable, payroll, general ledger, and other records. The mechanization of bookkeeping was the result of the conjunction of the typewriter and the calculating machines that evolved in the both the accounting machines and the tabulating machines.

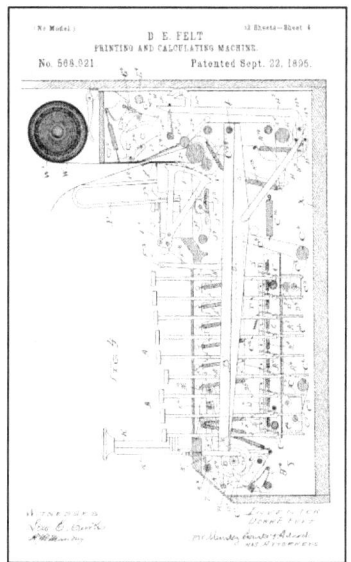

**Figure 126: Felt's Pinwheel Calculator 'Comptograph.'**

Felt's Patent № 568,021 (top, 1896), the Comptograph printing and calculating machine (bottom, 1900).

Source: Wikimedia Commons. USPTO http://www.vintagecalculators.com/html/comptometer.html

William Seward Burroughs (1857-1898), working as a clerk at a bank, was bored with the monotonous work of calculation. At this time Burroughs —gifted with a natural affinity and talent for mechanics— became interested in solving the problem of creating an adding machine. He moved to St. Louis, Missouri where he rented space in the Boyer

Machine Shop. There —after meeting many inventors including Baldwin— he created a calculating machine that combined many of the earlier calculation concepts. His machines borrowed Pascal's carry mechanism, Leibnitz' stepped cylinder, Babbage's sequence of calculations, and the Colmar Arithmometer concept.

And he obtained US-Patent № 388,116 filed in January 1885, granted on August 21st, 1888 (Figure 127). He also filed other patent applications and obtained US-Patent № 388,117, US-Patent № 388,118, and US-Patent № 388,119, all granted on August 21st, 1888.

After he had developed and sold early versions of his adding machines and the 'Burroughs Registering Accountant', the machines had to be manufactured. He organized the *American Arithmometer Company* (AAC)[306] in 1886 and a contract was entered into with the Boyer Machine Company for the manufacturing of the device, the selling operations were established and from time-to-time different models were put out, the beginning of the long line of models. As William Burroughs died already in 1888, he would not participate in the success of his machine. But his company continued to expand.

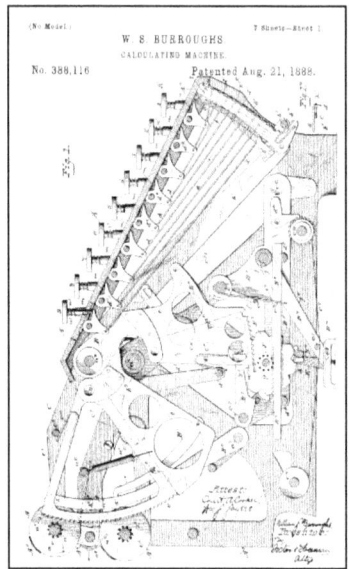

**Figure 127: Burroughs Adding Machine.**

Burroughs Patent № 388,116 (top, 1888) and the Adding Machine (bottom, 1892).

Source: USPTO, https://history-computer.com/

---

[306] The word 'Arithmometer' had become a generic name for a calculator, and was not tealted any more to the stepped-drum technology.

American Arithmometer experienced its first major expansion in 1898, when Sir John Turney of Nottingham acquired the rights to manufacture Burroughs machines for all the countries and continents of the Eastern hemisphere, including Britain and Europe.

As the machine was designed to ease the monotony of clerical work, by 1890, the machines were well known in the banking industry, and adoption was soon spreading. Between 1895 and 1900, business really took off with sales jumping to 972 machines, and the machine won gold medal at the Paris Exposition. Sales progressed from 2,000 units (1901), 3000 (1902) to 4,500 (1903). In 1905, the American Arithmometer Company was reorganized into the *Burroughs Adding Machine Company* and relocated in Detroit, Michigan. Burroughs manufactured its 50,000th machine in 1907.

By 1910 Burroughs offered 74 models with between 6 and 17 columns of keys and began advertising some of its models as bookkeeping machines. In 1911 there were 78 Burroughs models ranging in price from $175 to $850 and Burroughs introduced its Burroughs Class 3 visible adding machines based on the Pike design. Also in 1911, Burroughs introduced a key-driven calculator that looked very much like a Felt & Tarrant Comptometer. In 1912, the

**Figure 128: Corporate Advertisement.**
Russian advertisement for a 13-digit Odhner-Hill arithmometer (left.1895). Corporate promotion for the Burkhardt Arithmometer (middle, 1908). Burroughs promotion as the Nation's Business (right, 1919).

Source: National Library of Finland. https://mycompanies.fandom.com/

Burroughs Calculator was $150. By 1913 Burroughs had a 60% US market share.[307]

The Thomas-calculator concept of the Arithmometer and Felt's concept of the Comptometer were cloned by many others creating a new industry. Like Arthur Burkhardt's Arithmometer developed from 1873 on in a range of variants. His company, the *Erste Glashütter Rechenmaschinenfabrik* located in Glashütte —a small town in Sächsische Schweiz-Osterzgebirge, which was the birthplace of the German watchmaking industry— became the heart of the German calculator industry. Or the company of the Swede Willgodt Odhner who carried out his activities in St. Petersburg, Russia, creating the foundation for the Russian calculator industry. And in America William Seward Burroughs established the American Arithmometer Company in 1886 that would start the American calculator industry. A development they promoted by Corporate Advertisement (Figure 128).

**Figure 129: Early Mechanical Calculators (1870s-1920).**

Brunsviga Model B, (top, 1898), and the Burroughs Adding Machine Class 1, Model 9 (1910, bottom).

Source: Wikimedia Commons. John Wolff's Web Museum. https://www.nzeldes.com/HOC/IronFelix.htm

## *Calculating Machines 2.0 (1900-1920)*

By 1900, four machines —the Comptometer and Burroughs' adding machine in the US, Odhner's Arithmometer in Russia, and the Brunsviga calculator in Germany— were the dominant calculating machines in the marketplace (Figure 129). A range of companies were manufacturing these mechanical calculators in volume. Improvements soon followed by other clone makers and their commercialization created the mechanical calculator industry that would appear in different manufacturing regions.

---

[307] Burroughs championed the principle of complete interchangeability. The parts making up the infrastructure of his machine were fully interchangeable, which aided mechanics making repairs.

In America, the market for calculating machines grew quickly (despite protests of office clerks that feared loss of their jobs). So, next to the before mentioned companies of Burroughs and Felt & Tarrant, other companies popped up. During the first decade of the twentieth century, Burroughs faced competition from both key-driven calculators and a number of rival adding-listing machines, including Dalton, Pike, Standard, Universal, and Wales. Machines manufactured by the *Standard Adding Machine Company* (est. 1892s), the *Addograph Manufacturing Company* (est. 1902), the *Pike Adding Machine Company* (est. 1904), the *Universal Adding Machine Company*, the *Adder Machine Company* and a range of other companies.

This period so also massive Merger & Acquisition (M&A) activity. As part of a monopolizing business strategy, Burroughs bought many of the before mentioned companies; and become a de facto monopolist. That gave reason to the American government to file in 1913 a law suit against Burroughs. It became the lawsuit of the *US Government vs Burroughs Adding Company* that stated:

> *III. Conspiracy, Attempt to Monopolize and Monopoly. [...] Certain of the individual defendants as managers, officers, agents, or employees of the defendant company, and its predecessor, have been and are now attempting to monopolize such trade and commerce and are monopolizing the same, and they and their predecessors in office ever since 1901 have conspired and they are now conspiring to restrain such trade and commerce in the commodities aforesaid, and have restrained and are now greatly restraining the same. Some of the overt acts committed in pursuance of said conspiracies, and the methods by which said restraints have been accomplished, are as follows. [...]*[308]

The further ruling dwelled on the way each company was acquired. Although the defendants all denied acquisitions by unfair practises, the judge's rule forbade any more acquisitions and ordered the company to instruct its personal company to change their business practises.

> *[...] to absolutely desist and refrain from interfering with or directing, or permitting others under their control or under the control of either of them, to unlawfully interfere with the business, machines, or appliances of competitors engaged in the manufacture, sale, and shipment, or in the sales or shipment in interstate and foreign commerce of adding machines or appliances, by inducing or trying to induce such purchasers to cancel their contracts with competitors and to return to such competitors the adding machines or appliances so purchased, or by wrongfully obtaining information respecting the business, sales, or shipments of such competitors, or by fraudulent or illegal means of inducing the employees of*

---

[308] Source: 1913 Legal Challenge. http://www.burroughsinfo.com/1913-challange.html. (Accessed June 2021)

*said competitors to give them such information, or permitting agents or employees of the defendant company, or of either of the individual defendants, to seek or to induce others to seek employment of said competitors for the purpose and with the intent thereby of wrongfully securing information as to the business of said competitors, or by any other method specific in said subdivision "e" of the IV paragraph of the petition, or by any other similar wrongful and unlawful means acquiring such information as to the business of a competitor, and such servants and agents and other person or persons connected with the defendant company are hereby perpetually enjoined from violating such instructions.* (ibidem)

## Improved Calculating Machines

All these calculators were in fact simple adding machines, as they implemented multiplication by multiple addition under operator control. But that was about to change as the multiplication operation became a direct-commanded facility.

Léon Bollée (1870-1913): Next to the passion to the for automobiles that ran in the family, the young Frenchman Léon Bollée was obsessed by another passion; calculating machines and devices. In 1889 he invented a machine that required only one turn of the crank handle to multiply the number entered on the sliders by a multiplier number. The Bollée-Multiplier was a large manually operated non-printing direct multiplication calculating machine. Three versions of the large multiplier and several smaller machines were developed by Bollée. But as the manufacturing cost were exorbitant, and sales did not pick up.

The devices were patented in France (FR-Patent № 201,033. 1889), Belgium, Germany (DE-Patent № 88,936 and DE-Patent № 82,963), Austria-Hungary, England (GB-Patent № 16,677) and the USA (US-Patent № 556,720 granted March 17th, 1896).

A few years after Bollée introduced his machine, the Swiss inventor Otto Steiger (1858-1923) patented a direct multiplication calculating machine that would sell widely as the 'Millionaire' (Figure 131). He obtained in Germany DE-Patent № 72,870 granted on December 23rd, 1892. Manufactured by his company H.W.Egli AG, it was the first

**Figure 130: The Multiplying Calculator.**
The Bollee multiplier (top, 1889) and

Source: https://www.histoire-informatique.org/musee/1_2_16.

commercially-successful direct multiplication calculating machine. Made of brass and steel, the Millionaire weighed about 100 to 120 pounds and had the dimension 15 (h) x 55 (l) x 27 (w). Its enormous weight made it quite uncomfortable to use. It was made initially for businesses, but scientists found it to be quite beneficial. It eventually became a popular demand with the governments as well. It was in production from 1893 until 1935, with a total of about five thousand machines manufactured.

The brothers William W. Hopkins (1850-1916) and Hubert Hopkins (1859-1920), worked on a functionally advanced adding machine and had developed their early adding machines by 1870.

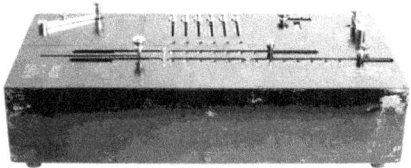

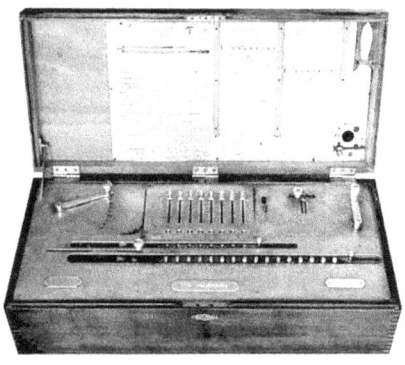

**Figure 131: The Millionaire.**
Prototype Steiger's Millionaire (top, 1895). Steiger's Millionaire case (middle) an interior (bottom, 1893).

Source; Wikimedia Commons,
https://www.arithmeum.uni-bonn.de/;
http://www.johnwolff.id.au/calculators/Egli/Egli

Minister William Hopkins obtained his first patent for a chain driven calculator (US-Patent № 203,151 granted April 30th, 1878), and in 1892 he filed his patent application for a 10-key-driven adding, subtracting and recording machine (and was granted US-Patent № 517,383 for it on March 27nd, 1894) (Figure 132).

Sometime in the early 1890s William Hopkins founded *Standard Adding Machine Company*. By 1905 over 3400 machines were sold. However, when in 1916 William Hopkins died, Standard Adding Machine Co. began to decline, to be closed in 1921.

The brothers Hopkins, in need of money to set up production, had found James Lewis Dalton (1866-1926) and his associates interested. So, in December 1902 the *Addograph Manufacturing Company* was founded, half of shares owned by the Hopkins brothers, the other half owned by Dalton. Dalton became president, and Hubert Hopkins became director.

> *However, in 1903 Hubert Hopkins began an unfair game and secretly sold his stock in the Addograph Co. to American Arithmometer Co., the manufacturer of the famous Burroughs Adding Machine, and with additional shares it had bought, gave it control of the Addograph Co […] As the patent application for the machine, filled in January 1903 (granted as US patent No. 1,039,130), was assigned to Addograph Co., this action threatened the investments of Dalton. To regain control, Dalton paid to American Arithmometer $40000 for the stock (Hopkins had sold this stock for only 5000$, and this prove to be a very bad decision, because using his patent Dalton will become a millionaire later). Thus Dalton was granted the exclusive right to make and sell the machines, what he did, founding in July 1903 a new company—the* Adding Typewriter Company *in Poplar Bluff, Missouri. The company later changed its name to Dalton Adding Machine Company.*[309]

In 1902, the 'Dalton,' an adding/ printing machine designed by Hubert Hopkins was introduced by the Addograph Manufacturing Company,

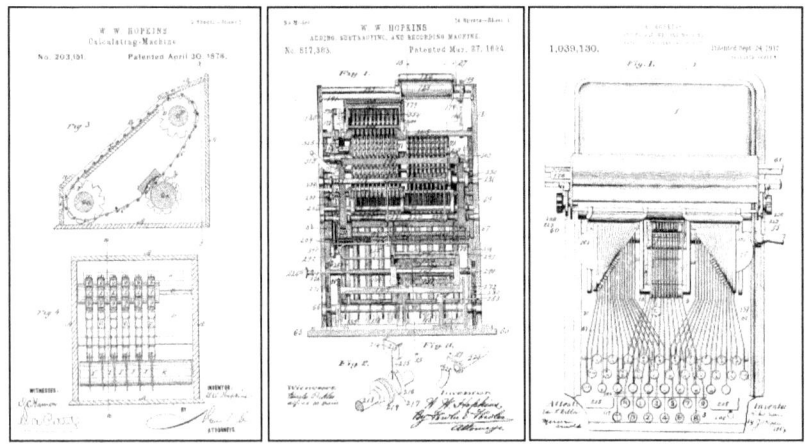

**Figure 132: Hopkins Brothers' Patents.**
William Hopkins US-Patent № 203,151 (left, 1878) and US-Patent № 517,383 (middle, 1894). Hubert Hopkins US-Patent № 1,039,130 (right, 1912),

Source: USPTO, Smithsonian.

---

[309] Source: Dalton Adding Machine: https://history-computer.com/inventions/dalton-adding-machine-complete-history-of-the-dalton-adding-machine/ (Accessed June 2021)

and in 1909 the company relocated to a purpose-built factory building. The machine began to sell on the market and 200 sales offices were ultimately opened up in different parts of the world, and sales ran up to $1,000,000 worth a month.[310] By the 1920's over 150 models of Dalton Adding Machines had been designed and over 60,000 machines/year were sold. After the death of James Dalton of acute appendicitis on January 11th, 1926, the Dalton Adding Machine Co. merged with other companies to become Remington Rand in 1927.

Other entrepreneurs of those days, noting the work of the engineering inventors, grabbed the opportunity and jumped on the bandwagon of the calculating machines.

It was Jay Randolph Monroe (1883-1937), born in a family of bankers who worked for the Western Electric Company in a clerical and legal capacity. After becoming acquainted with Frank Stephen Baldwin, owner of patents for his calculating machine, they collaborated in 1911. Baldwin had an interesting calculation machine, and Monroe saw a need for a simple and portable calculator for business applications. A machine that could be used with minimal training, and which would perform all four arithmetic functions with equal ease. So, they adapted the Baldwin calculating machine for commercial production and created 1912 the *Monroe Calculating Machine Company*.

**Figure 133: Standard and Dalton Adding Machine.**

The Standard Adding Machine Model B (top, 1907), and the Dalton Model 181-4 Adding Machine (bottom, 1925)

Source: USPTO, Smithsonian.

His first calculating machine Model A, designed with the user's needs for simplicity in operation and simplicity in construction as a design rule, was soon followed by others. Larger scale production started with the

---

[310] That real price of that sales would be equivalent to $12,9 million in 2019. Source: https://www.measuringworth.com/

model D in 1914. This was followed by the models E, F, and G over the next few years. The Monroe product range covered an enormous variety of styles and features, from basic hand-cranked units to high-speed motor-driven machines with multiple registers and fully-automatic multiplication and division. By 1925 an electromotor-driven machine was introduced.

## Merging Functionality: Accounting Machines

The development of the typewriting machine and the mechanical calculating machine ran in parallel. Each had its own trajectory of continuous improvements and functional enhancement, but their functionality also merged over time into a special type of machine; the special purpose typewriter. From 1904 through the 1920s, a number of companies sold combination typewriting-adding machines for clerical use in accounting departments. These machines were variously called 'writing -adding,' 'typewriter-adding,' 'typewriter billing,' and 'typewriter bookkeeping' machines, as well as 'adding and subtracting typewriters.'

**Figure 134: Monroe Calculating Machines (1920s).**
Monroe Model K20 (top, 1921) and Monroe electric Model KA-160 (bottom, 1925).

Source: National Museum of American History, Smithsonian

In the early 1900s machinist and inventor William Hopkins (1850-1906) — with his brother Hubert Hopkins— founded the *Standard Adding Machine Company*. In 1904 he applied for a patent for a 'multiplying and typewriting machine.' (Figure 135) This was the first of several related patents for similar machines.

> William was granted US-Patent № 844,519 (filed on May 12th, 1904) on February 19th, 1907 for a multiplying and writing machine; and US-Patent № 1,133,029 filed on August 28th, 1905 and granted on March 23rd, 1915 for a computing and writing machine. Hubert Hopkins filed on July 31st, 1905 an application for an adding and writing machine that was granted as US-Patent № 1,049,093 on December 31st, 1912.

With the backing of local businessman John C. Moon, Hubert organized the *Moon-Hopkins Billing Machine Company* and had a commercial machine manufactured and out on trial by 1908. Its business success proved elusive the following years, and after extensive negotiations, the Burroughs Adding Machine Company purchased rights to the machine in 1921.[311]

A range of other companies also manufactured a multitude of these combination machines by 1910. Such as the *Elliott-Fisher Co.*, the *Ellis Adding-Typewriter Co.*, the *Remington Typewriter Co.*, the *Underwood Typewriter Co.*, and the *Howieson Calculating Machine Co.* (Figure 136).

This was the period of time when the mechanical calculating machines, their extended and improved version as well as the combined functionality of the accounting and billing machines, conquered the business world with their offices. Often the machines were designed by individuals —their 'inventors'— who soon sought the cooperation of investors/ businessmen to realise the manufacturing and marketing of their machines. That cooperation would take different forms. Sometimes it was purely financial: from financial loans to early venture capital. Or it was of a technical nature; offering prototyping of production facilities to the inventor.

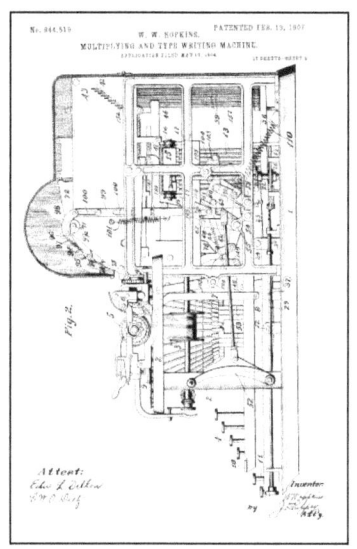

**Figure 135: William Hopkins' Billing Machine.**

US-Patent № 844,519 (left, 1904), and Burroughs Moon-Hopkins Style 7205 Bookkeeping Machine on Stand (right, 1904).

Source: USPTO, https://womenshistory.si.edu/

---

[311] In 1911, electric motors were added to the Moon-Hopkins, and models were priced at $500-$750. One model was $750 in 1916. In 1924, different models were priced at $650 to $1000. NB. $1 in 1921 had a real price worth of $14,50 in 2020 (based on Consumer Price Index) (source: Measuringworth.com)

The result was an industry of calculating machines manufacturers that served a range of office-markets. After the early adapters of large companies, technology penetrated with a range of machines the clerical workplace. But also other businesses discovered that they could profit from the new type of machines. Sometimes embracing the standard adding machines, sometimes the special machines like the cash register.

## Specialized Calculation Machines

During this development of merging functionality, also dedicated calculating machines appeared on scene. Such as the cash registers, accounting machines, and tabulation machines.

**Figure 136: Machines with Combined Functionality.**
Howieson Calculating Machine with Underwood Standard Typewriter (top, 1911). Ellis Adding Typewriter (bottom, 1913)

Source: https://womenshistory.si.edu/, Officemuseum.com

*The Cash Register*

The *cash register* is basically an adder with a printing facility to register sales transactions. Those original machines —the dial registers (Figure 137)— were nothing but simple adding machines, originally to prevent misconduct by the salesperson. James J. Ritty, a pub owner from Dayton, Ohio faced with less than honest behaviour, enlisted the mechanical skills of his brother John to design and build what they dubbed a cash register. The first prototype was patented in 1879 as 'Ritty's Incorruptible Cashier.'

>Ritty filled for a patent on March 26[nd],1879, that was granted as US-Patent № 221,360 on November 4th, 1879. Followed by Patent № 271,363 granted on January 30th, 1883, and US-Patent № 318,506 granted on May 26th, 1885.

In the small manufacturing company, he set up to develop the machine largely constructed of wood. Most of the employees were carpenters and cabinet makers. Ritty himself was, next to organizing the start-up, still running the pub. As sales did not pickup, after nearly two years of struggle, James and John Ritty sold their entire cash register business, including patents, to Jacob H. Eckert for $1,000. Subsequently Eckert set up the *National Manufacturing Company* of Dayton, Ohio, in 1881 to manufacture and sell the mechanical cash register. One of his customers was John Patterson, owner of a successful business selling coal and miner's supplies, who solved with the machine his problem of dishonest clerks.

In 1884, John Henry Patterson and his brother Frank Jefferson Patterson, bought out Eckert and his investors for $6500, got full control over the company and renamed it as the *National Cash Register Company (NCR)*. Patterson set up an 'inventions department' to create bigger, better and more thief-proof registers. In addition, he organized a sales force and began a training program for his sales people. The first factory building was constructed on the Patterson farm in 1888. It was the start of a company that would dominate the cash-register business.

*Technical development:* Improvements in the cash register followed rapidly with different models with functions like multiple drawers and transaction display. The first generation cast metal cabinet registers —the cast brass-encased registers known as the Brass Cash machines— were made

**Figure 137: Cash Registers.**
Ritty Dial Cash Register (top, 1879). National Cash Register Co. Model #5 (middle, 1903), and Model class 500 (bottom, 1911)

Source: Wikimedia Commons, NCR.

in the early 1890's and continued through 1908. Over the years it grew into a dedicated machine, even more when it became powered in 1906 by an electromotor.

*Commercial development:* A sales network was created all over the Eastern US with a highly motivated sales force to place the product in stores. He paid his salesmen generous commissions and introduced the idea of exclusive territory for each salesman. To allay customer fears of maintaining such complex machinery, he established a force of repairmen to service the products after the sale. As part of his marketing techniques, Patterson was using every conceivable means to display his products to the public. Such as promotional brochures (eg the Hustler Series), advertisement and direct mail campaigns. By 1894 a half million Hustlers and other circulars were mailed and an extra man was added to the Dayton post office just to handle the company mailings. His campaign, "Get a Receipt," convinced the world on the need to get a receipt with every cash transaction.

The turnover was slow to pick up, and in its first decade of operation, NCR sold some 16.000 machines. But that changed and the company built more than a million machines between the years 1908 and 1918. In the decade of the 1920's, they built another million plus machines.[312]

NCR's success inspired also other manufacturers to jump on the bandwagon. During the period 1888 to 1915 the cash register, clothed in fancy cast-metal cases, spread into nearly every retail establishment. Between 1884 when the first register became popular, and 1916 (in the middle of the First World War), more than 1.5 million were sold. The mechanical cash registers improved steadily over the years, with the addition of features like an adding mechanism for individual sales (the first cash registers only recorded total amounts received); printed receipts; machines with several drawers, one for each clerk in a department store; and, in a few high-priced machines, automatic change making. With the cash register business highly successful, eighty-four companies sold cash registers between 1888 and 1895.

Not many of these early companies survived as NCR used every means of legal protection and every competitive technique to weaken the competition or to put them out of business altogether. One of those was acquisition, the other was claiming patent infringements. NCR continued to buy out the competition or to sue them for twenty more years, creating de facto a monopoly in the cash register business. Many of those fraudulent practises (see citation) were executed by a protégé of Patterson: Thomas Watson. A practise that held till 1913 when NCR was sued by the Federal government under the Sherman Antitrust Laws and was found guilty.

> *While at NCR, in 1903 he [Thomas Watson] was selected to head a new secret NCR subsidiary, the purpose of which was to gain control of the second-hand cash register industry. The Watson Cash Register and Second Hand Exchange had as its purposes to buy, secretly, the second hand stores for National –*

---

[312] Source: https://www.ncr.org.uk/jacob_h._eckert.html ; https://www.ncr.org.uk/jhp_-_p2.html

*especially the ones adjacent to competitors' store fronts. He then used NCR's vast stock of traded-in competitor machines that were "weakened" to create havoc among targeted competitors. He would undercut the adjacent store no matter how low they went until he drove them to selling their business or going bankrupt. At times, Watson would acquire the competitive company and the target never knew they were actually selling to NCR until it was too late. In 1906 NCR repurchased all the Watson stores, their job apparently having been completed. NCR rose to a 95% market share!*[313]

That was the way American business operated.

*The Accounting Machine*

The Accounting machine was a mechanical calculator especially designed for accounting tasks (ie billing, payroll, ledger) in business use. The Billing machines were used by department stores, savings institutions, and hotel front offices as billing machines and for writing checks.

**Figure 138: National Accounting Machines.**

NCR Class 2000 Bookkeeping and Distributing Machine, (top, 1921) and Class 3000 Billing Machine (bottom, 1930s).

Source: https://americanhistory.si.edu/, Officemuseum.

The *Inventions Department* at National Cash Register Company developed the firm's first bookkeeping machine, the NCR Class 2000. It went on the market in the early 1920s and it was an adding machine with a twelve-column keyboard and six separate registers designed for accumulating totals. While some customers used it, the machine lacked a typewriting facility that could describe transactions in detail. But, after in 1928 NCR purchased rights to the Ellis adding typewriter, they could add the typewriter function to the accounting

---

[313] Source: Crandall, R.L.; Robbins, S.: The Incorruptible Cashier (1988/1990). https://www.rickcrandall.net/cash-register-history/

machine. As a result, they introduced in 1929 the successful Class 3000 bookkeeping machine.

*The Tabulating Machine*

The typewriter became the base for added functionality. Such as the Gorin Tabulator (1898), that could write columns of figures anywhere on a page. Next the adding typewriter came on scene; such as the Elliot-Fisher flatbed adding typewriter (Figure 139, top). The introduction of the ledger-posting machines gave a great advance to the calculation and printing of mathematical tables as they could be made into 'difference' engines.

A rather special calculating machine was the Burroughs Machine for Vote Calculation (Figure 139, bottom). The machine was developed in 1936 for use in the United States House of Representatives to keep record of the voting of 435 individual members on roll call votes for publication in the Congressional Records. Recording the yeas and nays, absent and present, paired for and paired against votes of each individual member, the machine which is similar to an adding machine, did the job that costed month to do manually, in less than two weeks.

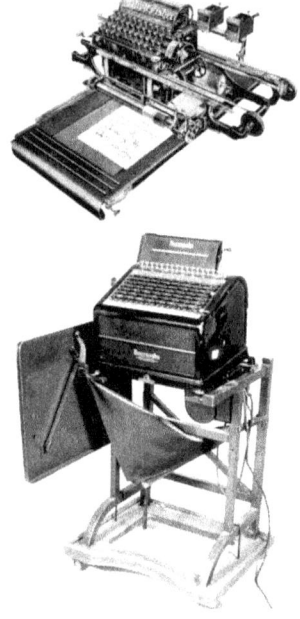

**Figure 139: Tabulating Machines.**

Elliot-Fisher flatbed adding typewriter (top, 1903). Burroughs Machine for Vote Calculation (bottom, 1936)

Source: Smithsonian, https://americanhistory.si.edu/

*The Census had a Problem.*

Already earlier, the need for tabulation machines had become evident. The more when the amount of data to be processed manually became a problem. As was the case with the Census Bureau.

> The US Census was —from 1790 onwards— a ten-year collection of statistic demographic data of the US population. From the quite basic decennial censuses, by 1880 the Census —and the population— had expanded and the collected data on was enormous. It concerned tallies on the population by sex, race, and birthplace; in

other tallies, it collated these statistics with literacy, occupation, and other characteristics. Although the 1880 census itself had taken only a few months, the work of tabulating and analysing the data promised to drag on for years. But that was about to change.[314]

Herman Hollerith (1860-1929) —born in a family of German immigrants that came to the US after the 1848-Revolution— after a complicated youth studied engineering. In 1879 the bright and unmanageable teenager Herman Hollerith graduated with distinction at School of Mines at Columbia University in New York. After graduating Hollerith became an assistant to Professor Trowbridge, who was so impressed of his mind that he asked him to become his assistant. William Petit Trowbridge (1828–1892) was a mechanical engineer, military officer, and naturalist, who in 1879 was appointed Chief Special Agent to the Census Bureau, and took with him Hollerith, as a statistician. In 1883 Hollerith obtained a post (as assistant patent examiner) in the US Patent Office in Washington, D.C. In 1884, he resigned from the Patent Office, became an independent patent agent, and embarked on his main career as an independent inventor and entrepreneur.

In 1882, the Census Bureau was facing the processing of the data of the 1880-Census that would take a decade to

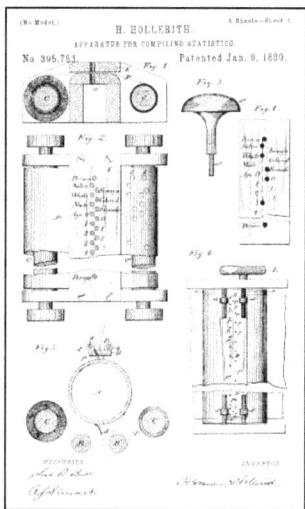

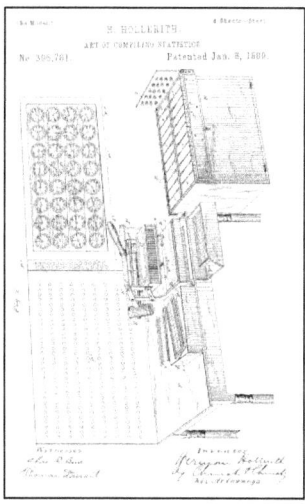

**Figure 140: Hollerith Patents.**

Hollerith's US-Patent № 395,783 for a punched paper roll (top, applied 1884), and US-Patent № 395,781 (bottom, applied 1887).

Source USPTO.

---

[314] Source: https://history-computer.com/inventions/tabulating-machine-history-of-the-hollerith-tabulating-machine/; http://ed-thelen.org/comp-hist/CBC-Ch-04.pdf; https://www.immigrantentrepreneurship.org/entries/herman-hollerith/; https://www.computerhistory.org/

tabulate and publish. Hollerith became involved and he designed a tabulating machine with a punched paper tape. Soon he improved from the punched tape to punched cards to contain the primary data.

The decision to use cards led Hollerith to redesign his initial system. The resulting Hollerith Electric Tabulating System consisted of several subsystems. He designed a *special puncher* (a pantograph punch consisting of a template and two connected punches); when the operator punched the template, the second puncher perforated the card. The *card reader* was a small press made up of an overhead array of pins and an underlying bed of tiny cups of mercury; when the operator slipped a card into the press and pulled down on the handle, the pins passed through the holes into the mercury, closing electrical circuits that advanced the counters (each completed circuit caused an electromagnet to advance a counting dial by one number), 40 simple clock-like dials set into a wooden table. When the bell signalled the card had been read, the operator recorded the data on the dials, opened the card reader, removed the punch cards, and reset the dials. And next there was the tabulating, sorting of data, and the accounting of data by an accounting machine.

The Census Office was impressed with Hollerith's work, but it decided to conduct an official test of the system before making a commitment. In a trial with different machines designed by others, Hollerith's system smashed the rivals. Pleased with the results, the Census ordered 56 tabulators and sorters. Hollerith was in (big) business (with one customer) starting in Georgetown, Washington his workshop in 1892. After expansion, the business was in 1896 incorporated as the *Tabulating Machine Company*. Hollerith sold his patent rights and business assets to the company and gained a controlling 50.2 percent of the shares. Ten of Hollerith's friends bought shares, and the company raised approximately $12,000 in cash.

In 1884 he applied for a patent application, and in 1889 received his first patent for a tabulator with US-Patent № 395,783 issued on January 8th, 1889, for a tabulator with a punched paper tape. On the same date he was granted US-Patent № 395,781 (applied on June 8th, 1887) his first patent for the "Art of Compiling Statistics' with a tabulator using punched cards. (Figure 140). During his lifetime he was granted 30 US-Patents.

At that time Hollerith had one customer; the Census bureau. Hollerith's machines went to work in July 1890, shortly after the completion of the head count of the 1890-census. Compared to the 1880 census, which had taken nine years and cost $5.8 million, the 1890 count was completed in fewer than seven years, but it had cost $11.5 million, twice as much.

## Expanding Hollerith's Tabulating Machine Company

Hollerith's punch card system received a great deal of attention in the popular and scientific press in the USA and abroad. As a result, Hollerith's system was promptly adopted all over the world. In late 1890, Austria, then Russia, Canada and France ordered several tabulators and sorters for their census. From 1901 to 1903 he worked hard to attract private industry, and — after some initial resistance— also private industry began using his tabulation machines too.

**Figure 141: Hollerith's Tabulating Machine.**

Hollerith tabulator (left) and sorting table (right) as used in the 1900 US Census.

Source: US Census Bureau, Science Photo Library.

Swamped with paperwork, large companies like the Chicago department stores, the New York Central Railroad Company, the Pennsylvania Steel Company, and the Insurance companies like Continental Casualty moved the tabulation equipment into their accounting and inventory departments.

As he was leasing the machines to the Census Bureau, and needed customers in the non-census years, he was constantly facing financing problems. So, he needed to expand his sales volume. The more when the Census Bureau cancelled its leasing contract for 1904-1905.

> Hollerith went to the UK to explore business opportunities. As a result, *TMC Ltd.* was formed in 1902 as a private company by a London-based syndicate selling Hollerith's machines. In 1907, they reformed it as the *British Tabulating Machine Company Ltd* (BTM) to sell Hollerith machines —made in the US— in UK and the British Empire. BTM had to pay a royalty of 25% on their revenues to the American company TMC for every Hollerith machine they sold. [315]
>
> They soon realised that they had come off extremely poorly in the bargain, and for a number of years made very little profit as most of

---

[315] During the Second World War, BTM were helped enormously by the British Government, who employed them to build the Turing Bombe machines at Bletchley Park to crack the Enigma codes.

their money was taken up by buying machines from TMC at dictated prices. Soon the relation deteriorated, and by 1920 they started to build their own punch card machines.

In the meantime, to solve the financing problem for Hollerith in America, the best solution seemed to be a merger, and in 1911 Tabulating Machine Co. joined with three other outfits —the International Time Recording Co, Bundy Manufacturing, and Computing Scale of America— to become the *Computing Tabulating Recording Company* which eventually in 1924 became renamed as the *International Business Machines Corporation* (IBM).

> The financier Charles R.Flint, experienced in creating holding companies like US Rubber, was the organizer of the merger that brought Hollerith and his partners $ 2,3 million.[316] Hollerith himself earned $1,210,500 out of the deal, and after selling TMC he engaged in a life of leisure, which did not include operating in a large organization or making any substantial attempt to start a new business. He died in 1929.

The relation between the American IBM holding the rights of TMC, and British BTM worsened. The more when BTM, after it started to make computers, merged into the British giant of computers ICL

> In Britain, during the dawn of the Second World War, BTM was called upon to design and manufacture a machine to assist breaking the German Enigma machine ciphers. This machine, known as a Bombe, was initially conceived by Alan Turing, but the actual machine was designed by BTM chief engineer Harold 'Doc' Keen, who had led the company's engineering department throughout the 1930s. The project was codenamed "CANTAB". By the end of the war, over two hundred bombes had been built and installed.

Expanded on the acquired knowledge BTM built a valve-based computer called the HEC (Hollerith Electronic Computer). The first model (HEC 1) was built in 1951 from double triode and diode valves. It was Britain's first mass-produced business computer. The next development was a machine that was the HEC 2 with a number of enhancements specifically designed for a commercial workload, followed by the HEC 4. When in 1959 the *International Computers and Tabulators (ICT)* was formed by a merger of the BTM and Powers-Samas, the HEC 4 became renamed as the ICT 1201 (1200 series). Some 100 of these machines were sold.

---

[316] That real price of that commodity would be $63,8 million in 2019. Source: https://www.measuringworth.com/

## The Quest for the Differential Machines

The calculation machines emerged in the time when more complex mathematical operations (eg logarithms, trigonometric functions, polynomial functions) were performed by hand, and presented on paper as *mathematical tables*. The use of tables was not confined to research scientists and mathematicians. There were tables for astronomy, navigation, physics, engineering, statistics, trade and finance, in the army and in many other areas. Engineers, architects, mathematicians, astronomers, bankers, actuaries, journeymen, insurance brokers, statisticians, navigators —anyone with a need for calculation— relied on printed numerical tables for anything more than trivial calculations (Figure 142). And in the process of making the books, containing those tables, they were calculated, copied, checked and typeset by hand. A laborious process prone to error.

Among the publications of the Swiss-French polymath Johann Lambert (1728-1777) was his 'Zusätze ... Tabellen' (Additions to Logarithmic and Trigonometric Tables) of 1770. These logarithmic and trigonometric tables were the most accurate and complete of their time. By the end of the nineteenth century the mechanical calculators were used to produce a wide range of tables. Machine manufacturers supplied their customers with manual containing instructions for printed tables to assist in routine calculations. These often involved reducing non-metric measurements to decimal portions of a given unit (eg currency). And, up to present time,

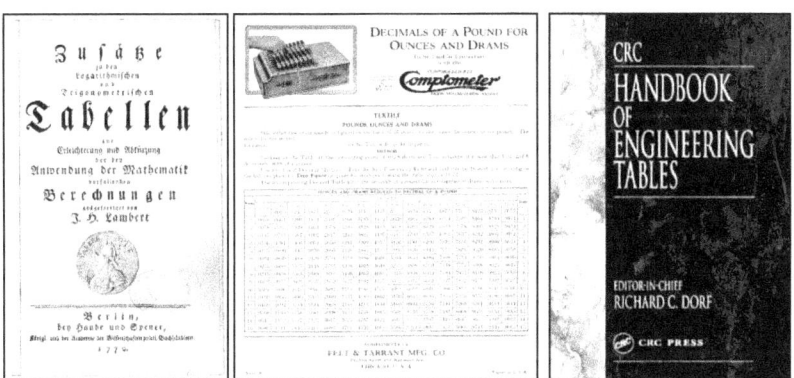

Figure 142: Mathematical Tables.
Titel page 'Zusätze ... Tabellen' (left, 1770), Manual for the use of the Comptometer by Felt & Tarrant (middle, 1913-1925), CRC-Mathematical Tables (right, 2003)

Source: Wikimedia Commons, Mathematical Association of America.
https://americanhistory.si.edu/

Mathematical Tables (ie the CRC Mathematical Tables) are still used in engineering.

The publication of such tables, however, required a lot of manually calculating work and the final product was full of errors. Scientist were looking to develop machines —just like the calculator did with the arithmetic function— that could reduce the labour of calculation and perform these complex calculations 'error free'. Their solutions became known as the 'difference engines' that could handle a specific set of calculation known as Differential Calculus. [317]

> Although the mathematical art of Differential Calculus finds its roots in Greek geometers (eg Euclid, Archimedes) and the medieval Islamic astronomers and mathematicians (eg Aryabhata, Sharaf al-Dīn al-Tūsī , Alhazen), the modern version is attributed to Isaac Newton (1643–1727) and Gottfried Wilhelm Leibniz (1646–1716), who provided independent and unified approaches to differentiation and derivatives.[318]

The Difference Engine was in fact a mechanical calculator that could calculate the first, second, third derivates of a mathematical function and print it on paper. The resulting book with mathematical tables was a collection of those prints. Seen in the light of the potential use, the issue of creating a difference engine was occupying many scholars (Williams, 1976).

In 1819/1820 the English polymath and mathematician Charles Babbage (1791-1871) —born into a wealthy family[319]—, had a passion for accuracy and became distressed with the errors found in mathematical tables. Being responsible for one of the earliest sets of accurate logarithm tables —published in 1827—, and aware of the problem with printed mathematical tables, he started conceptualized a mechanical machine that could perform the complex mathematical operations. The result was the concept for his *Difference Engine Nr. 1*. Next he started working on the prototype, after he had —with support of the Royal

---

[317] The word 'difference' relates to the mathematical construct of Differential Calculus, complement of Integral Calculus (*differentiation* being the reverse process of *integration*). Differentiation is a method of finding the derivative of a function. So a difference machine is a machine that could perform the calculus of differentiation.

[318] *Derivatives:* For non-mathematical schooled people, the real world of movement illustrates the use of derivatives. For a moving object (eg a car) from point A to point B (a distance), 'velocity' (eg km/hour) is the first derivative (with respect to time) of the object's displacement. The object's 'acceleration' (ie $m/s^2$, the rate of change of the velocity of an object with respect to time) is the first derivative (with respect to time) of an object's velocity, and thus the second derivative of the objects' displacement.

[319] On his father's death in 1827, Babbage inherited a large estate (value around £100,000, equivalent to £8.51 million or $11.7 million in 2020), making him independently wealthy.

Society— received £1,500 from the Chancellor of the Exchequer and put up between £3,000-5,000 of his own money.

This was the beginning of a machine made by the instrument maker Joseph Clement that would contain some 25,000 parts, weighting tons. However, the project was never completed, as it was plagued by a range of troubles and interruptions. Financial troubles and personality clashes stopped the work on the first Difference Engine on April 10th, 1833 due to a conflict between Babbage and Clement.

> *Babbage failed to build a complete machine despite independent wealth, social position, government funding, a decade of design and development, and the best of British engineering. The reasons are still debated and the cocktail of considerations is a rich one. Babbage was a prickly character, highly principled, easily offended and given to virulent public criticism of those he took to be his enemies. Runaway costs, high precision, a disastrous dispute with his engineer, fitful financing, political instability, accusations of personal vendettas, delays, failing credibility and the cultural divide between pure and applied science, were all factors.*[320]

**Figure 143: Babbage's Machines.**

Part of the Differential Machine Nr. 1 (top, 1834), and the Analytical Machine, (bottom, 1871).

Source: Wikimedia Commons, Science Museum Group.

For the British Government that had bankrolled the venture, the project was a costly failure. When the final bills were paid the Treasury had spent £17,500 and Babbage privately some £6,000.[321]

---

[320] Source: https://www.computerhistory.org/babbage/history/ (Accessed August 2020).
[321] In 2020, this would be £1,945,000 on the basis of a real cost calculation. Source: https://www.measuringworth.com/calculators/ukcompare/relativevalue.php.

In 1834, with the Difference Engine project stalled, Babbage designed some improvements to his first computer, and the result was the *Analytical Engine* which basic design was completed by 1837. Despite its failure, it had created the foundation for the concept of 'computers' as the machine had an input-device, an output device, a memory and a central processing unit.

After finishing of the work on the design of the Analytical Engine in 1847, Babbage turned to the design of a Difference Engine №2, exploiting the improved and simplified arithmetic mechanisms developed for the Analytical Engine. The new design was elegant and efficient requiring one third the number of parts of Difference Engine No. 1 for greater computing power (Figure 143). With 8,000 parts, the Engine would weigh five tons and measure eleven feet long and seven feet high. The machine was designed to consist of four components: the mill (ie the calculating unit), the store (ie the data memory), the reader (ie input device), and the printer (ie output device). However, Babbage made no attempt to construct the machine, and his attempt to gain broad exposure and honour for his machine designs at the Great Exhibition in London's Crystal Palace in 1851, failed.

The Difference Engines were more than a simple calculator, however. It mechanized not just a single calculation but an operation of a whole series of calculations on a number of variables to solve a complex problem. And it could change its operation by changing the instructions contained on a punched card (so it became programmable).

> The Analytical Engine, as Ada Lovelace argued in 1843, went beyond the bounds of simple arithmetic as it established *"a link ... between the operations of matter and the abstract mental processes of the most abstract branch of mathematical science."* It was a physical device that was capable of operating in the human realm of abstract thought called mathematics.

It went far beyond mechanical calculators in other ways as well.[322] These concepts of analytical machines were among the first mechanical 'automatic computing machines' (later known as computers). However, the nineteenth century movement to automate *mechanical computation* failed and the movement largely died with Babbage in 1871. It would take nearly half a century before the concept of the Difference Engine resurfaced. In the meantime, some other interesting events took place.

---

[322] Like modern computers, the Difference Engine had storage—that is, a place where data could be held temporarily for later processing—and it was designed to stamp its output into soft metal, which could later be used to produce a printing plate.

The Swede Per Georg Scheutz (1785-1873) —a technical editor, printer, and publisher in Stockholm, Sweden— became aware of Babbage's work. With his son Edvard Scheutz (1821-1881), an engineering student, he started to work on a machine that could create and print logarithmic tables. The subsequent model of the Scheutzian calculation engine (aka as Tabulation Machine), conceptualized in 1837, was finalized in 1843.

In 1851 they obtained funds from his government to build an improved model; the Difference Machine Nr 1. Funds that became only available after the King intervened, and fifteen friends guaranteed it in case of non-compliance. By February 1st, 1852, the first working drawings of the ultimate design were finished, and the machine (roughly the size of a piano) was completed in October 1853 (Figure 144). It was patented in Great Britain (GB-Patent № 2216, of October 17th, 1854) and subsequently demonstrated at the World's Fair in Paris, 1855.

> *In 1854, the Scheutzes brought their invention to London, where they demonstrated it before the Royal Society of London. Babbage, ever the gentleman, welcomed his fellow inventors with open arms. The following year, the Swedes entered the machine in the Great Exhibition in Paris and it won a gold medal – thanks, in part, to Babbage, who was a highly respected member of the Institute of France and who had lobbied on their behalf.*[323]

The following years it was used to produce and print a table of a polynomial of the third degree, which he published in 1849 titled 'Nytt och enkelt sätt att lösa nummereqvationer af hogre och lägre grader efter gardhska teorien.' ('New and simple way to solve number equations of higher and lower degrees ….'). The machine also was evaluated by astronomers in Sweden, Paris and England and finally sold to the Dudley Observatory in Albany, US. In 1854 a second machine — the Difference Engine No. 2, aka the Scheutz-Donkin machine— was developed, exposed on the 1855-Exposition Universelle in Paris, and sold to America in 1857. However, the British industrialist Bryan Donkin and Edvard Scheutz failed in their efforts to find further buyers for difference machines, which, after taking a loss on the construction of the No.2 machine, they offered to produce at wholesale discounts for £ 2000 apiece. Nevertheless, and despite their limited accessibility, the two Scheutz machines received considerable publicity.[324]

---

[323] Source: http://ds-wordpress.haverford.edu/bitbybit/bit-by-bit-contents/chapter-two/2-5-the-scheutzes-tabulating-machine/
[324] Source: Merzbach, U.C.: Georg Scheutz and the First Printing Calculator https://repository.si.edu/bitstream/handle/10088/2435/SSHT-0036_Lo_res.pdf?sequence=2&isAllowed=y (Accessed July 2021)

The Scheutz' calculator inspired the Swedish prolific inventor Martin Wiberg (1826-1905) to develop a difference engine (ca. 1860, Figure 144). His calculator became especially well known as it produced a set of logarithm tables, which included logarithms of the trigonometric functions, and appeared in 1875.

The American mechanical engineer George Barnard Grant (1849-1917), while a student at Harvard, worked on the problem of inventing a mechanical calculator improving on the work of Babbage and Scheutz. Grant described his design in the August 1871 issue of the magazine American Journal of Science, entitled "On a New Difference Engine." Following the publication, he created two machines. By 1876 he could display his Difference Engine at the Philadelphia Centennial Exposition (Figure 144). The machine was approximately 2.5 meters long and 1.5 meters high, weight about 900 kg. It consists of up to 15,000 pieces and is worth about $10,000. While the difference machine soon faded into relative obscurity, Grant continued working on mechanical calculators, and invented a second small all-purpose calculators; such as the 'Grants Barrel' in 1877 (Figure 145).

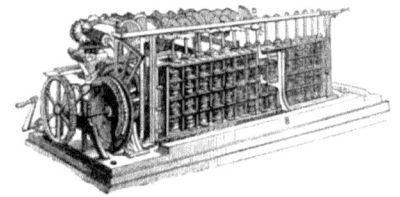

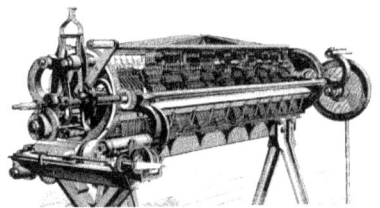

**Figure 144: Early Differential Machines.**

Scheutz' Difference engine (top, 1853), (part of) Wiberg Differential Machine (upper middle, 1860), Grant's Difference Engine lower middle, 1876) and Torres Differential Machine (bottom, 1893).

Source: https://history-computer.com

The Invention of the Silicon Engine

Grant was granted a range of patents for his calculators; US-Patent № 129,355 granted on July 16th, 1872; US-Patent № 138,245 on April 29th, 1873, US-Patent № 368,528 on August 16th, 1887, and US-Patent № 605,288 granted on June 7th, 1898.

In 1893 the Spanish mathematician Leonardo Torres y Quevedo (1852-1936) presented his first paper to the Spanish Royal Academy of Sciences. It was devoted to an algebraic machine, able to calculate the roots of an any-grade equation (Figure 144). Torres' major written work on the subject of automatics was his fascinating *Essays on Automatics*, published in 1913, in which he devised the term Automatics. The paper showed the main link between Torres and Babbage. And he demonstrated that all of the cogwheel functions of a calculating machine like that of Babbage could be implemented using electromechanical parts.

His 1914 analytical machine used a small memory built with electromagnets; his 1920 machine, built to celebrate the 100th anniversary of the invention of the arithmo-meter, used a typewriter to receive its commands and print its results.

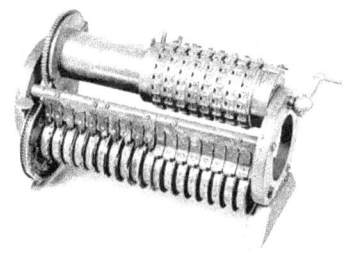

**Figure 145: Grant Barrel Calculator (1877).**
US-Patent № 368,528 (top, 1887) and Barell Calculator (bottom, 1877)
Source: USPTO, History-computer.com.

Clearly, all these contributors struggled with converting their ideas and concepts into reliable machines that could be manufactured and commercialized. However, the fine-mechanical technology, and the

253

craftsmanship that was needed, made them massive and costly to manufacture. The funding of their work, eg to pay for the work of the artisans, often was problematic. Even more when governmental institutions became involved. Nevertheless, their models and machines were the foundation for broadening the range of applications of the calculating machines. And the later difference machines that emerged from an unexpected development: the accounting machines produced in a large variety by the calculator industry.

## Functional Merging Two Development-trajectories

After the rise of the typewriting machine —that was capable of handling texts—, the emergence of the calculating machine —capable of arithmetic calculations— and the accounting machine —that could do arithmetic's as well as printing the results in a large variety— spurred the scientific community that started to realize that these machines maybe could be the modern version of Babbage's Difference engine.

> On the one hand, there hand been the development trajectory of the maturing calculating machines as started with the Arithmometer and the Comptometer. Basically, machines that could perform a single calculation with basic arithmetic operations of addition, subtraction, multiplying and division. Machines that found their way into a range of mass-markets as well as in a myriad of market niches. With calculators modified for special purpose applications, the manufacturers developed their calculator markets step by step, offering a range of machines over the years (Figure 146).

Figure 146: Advertainments for Calculating Machines (1920s).
Dalton's Simple Calculator (left), NCR product range (middle) and Burroughs Typewriter-Billing machine (right).
Source: http://www.vintageadbrowser.com/

On the other hand. there was the development trajectory of the Differential machines as started with Babbage's contribution. Machine that could executed more complex mathematical operations —a series of single calculations using a data storage-facility— like the function of 'differential calculus.'

In the 1920s the scientific world as well as the business world was experimenting and designing mathematical calculating machines of more extended capacity. The great mover was the need of accurate and reliable mathematical tables that were fundamental to a broad range of professions.

**Figure 147: Specialized Calculating Machines.**

Brunsviga Dupla (top, 1928) and Brunsviga Twin Model D18R (bottom, 1955)

Source: Arithmeum, Science Museum.

In 1928 the Brunsviga Company produced the crank operated Brunsviga Dupla (Figure 147, top) that could also be used to calculate 'differences' as it had two registers. This spurred the development of tandem machines in the 1930s. Such as the Brunsviga Twin (Figure 147, bottom), the Marchant Twin, the Britannic Duo. The NCR Accounting Machine Class 2000 (Figure 138) was one of those calculating machines that could calculate difference as far as the fifth difference or to integrate from sixth differences.

It was the astronomer Leslie J. Comrie (1893-1950) —since 1926 deputy-director of the Nautical Almanac Office— who explored the use of these type of machines designed for business use, to the preparation of (astronomical) mathematical tables.

> *Comrie was early concerned with the search for a 'difference' engine which could difference a numerical table automatically. He soon found that two commercial machines-the Brunsviga-Dupla and the Burroughs Class II- could be used for the inverse process of building up functions from their second differences. This was followed by the remarkable discovery of the possibilities of the six-register*

> *National Machine as a difference engine. Comrie showed that this machine could be used to difference automatically as far as the fifth difference or to integrate from sixth differences. The speed of sub- tabulation which can be achieved in this way is very great and the whole problem of interpolation was revolutionized.* (Massey, 1952, p. 102)

After initiating the foundation of the Scientific Computing Service (1936) —a company that specialised in scientific calculations generally, and particularly those where mechanical computation and mass production methods were employed— he focussed his attention to the creation of mathematical tables. Work that became even more relevant with the approaching Second World War when academic research (eg nuclear physics) became involved in military affairs, and the war-effort needed 'Ballistic Tables' predicting the trajectory of a projectile while aiming their canons. His focus on existing commercial engines to scientific use, instead of specially designed devices, paid off. It might not have been envisioned by the manufacturers of commercial machines, but it led to a new application. The scientific world embarked on a new trajectory of intelligent machine using relay-based logic; the electro-mechanical computer.

## *The First Calculating Revolution (1850-1930)*

In the preceding analysis we explored the multitude of events that were part of the development of the mechanical 'calculation' machines. We observed the rise of the different machine-concepts that took over a part of the mental 'arithmetic' activity; addition, subtraction and (later) division and multiplication. We noted the events of the merging functionality (eg the typewriting function merging with the calculating function). And we observed the events that contributed to the specialized calculating machines (eg the cash register), and the more complicated accounting machines. In its totality these events created the First Calculator Revolution.

### Patents for Calculating Machines

The First Calculating Revolution had several milestones. When one looks at the patents issued in the nineteenth century for mechanical calculators (Table 5), it shows that, after the first patents of Thomas de Colmar by the mid-nineteenth century, the inventors of calculating machines started increasingly patenting their inventions. Especially the US saw a rise in patenting activity during 1840-1900 for improvements of calculating machines (ie adding machines, arithmometer-type). But also European countries —depending on their law situation— saw increased patent-activity in the 2$^{nd}$ half of the nineteenth century (Figure 148).

Patenting, in earlier times starting as an economic privilege given by royalty, was —under the influence of the Enlightenment philosophers— seen as the right of intellectual property. But that right was not implemented in the same way and at the same time in different countries. Also the conditions could for issuing a patent be different; in France for example a patent was only issued when the invention was also made in France.

In America, the first Patent Act of the U.S. Congress was passed on April 10th, 1790, and required a model of the invention till 1793. In 1836 and 1870 US-Patent Law was revised. The British Patent system —originating from the Royal Monopolies aka Letters Patent— became institutionalized with the Patent Law Amendment of 1852. Then the separate patents for each of the participating countries in the United Kingdom were replaced with the issuing of a single UK patent. The modern French patent system for "brevets d'invention" was created by the revolutionaries during the French Revolution in the 1790s. Patents were granted for 5, 10 or 15 years, the patent tax prohibitive high. The French patent law was adapted in 1844 and focussed on protecting the French industry.[325] And the German Imperial Patent system, successor to the Patent systems of

| Year | Patent for Calculating Machine | Inventor |
|---|---|---|
| 1820 | Arithmometer (FR Patent № 1420) | Thomas de Colmar |
| 1844 | Key-driven calculating machine | Jean-Baptiste Schwilgué |
| 1849 | Arithmometer (FR-Patent № 8,282) | Thomas de Colmar |
| 1872 | Calculating machine (US-Patent № 129,335) | George Barnard Grant |
| 1878 | Calculating machine (US-Patent № 209,416) | Willigot T. Odhner |
| 1885 | Calculating machine (US-Patent № 3 88,116) | William S. Burroughs |
| 1887 | Full keyboard (US-Patent № 371,496) | Dorr E. Felt |
| 1888 | Full keyboard Calculator (US-Patent № 388,116) | William S. Burroughs |
| 1889 | Comptograph (US-Patent № 405,024) | Dorr E. Felt |
| 1889 | Multiplier (FR-Patent № 201,033. 1889 | Léon Bollée |
| 1890 | Calculating machine (US-Patent № 420,667) | Edward Selling |
| 1892 | Millionaire Calculator (DE-Patent № 72870) | Otto Steiger |
| 1894 | Adding, Subtracting, and Recording Machine (US-Patent № 517,383) | W.W.Hopkins |
| 1900 | Calculating Machine (US-Patent № 641,065) | Frank Stephen Baldwin |
| 1905 | Multiplying and Type Writing Machine (US-Patent № 844,519) | W.W.Hopkins |

Table 5: Overview of Important Patents for Mechanical Calculating Machines.

Source: Espacenet, USPTO

---

[325] Source: Source: Galvez-Behar, G.: *The Patent System during the French Industrial Revolution: Institutional Change and Economic Effects.*

the preceding states, was founded during the creation of the First German Empire (1871-1886) with the Patent Act of 1877. In 1817, the first Patent Act had already come into force in the United Kingdom of the Netherlands, but in 1869 the Act was repealed by the Patent Abolition Act and the Netherlands became a more free-spirited nation. Till the new 1910 Patent Act, patents were notrecognized in the Netherlands.

*The Patent Explosion (1900-1920)*

The total number of patents issued for 'calculating machine(s)' and 'adding machines' from 1840 into the 1960s is comprehensive (Figure 149). The result can be divided into two distinct periods.

*1840-1900:*[326] This was the dawn of the calculating machine development, and of related patents (Figure 148). Headed by the US, the European countries followed suit, although lower in numbers.

*1900-1960:* In this period, where the calculating machine found their dominant designs, many patents were granted for improving the

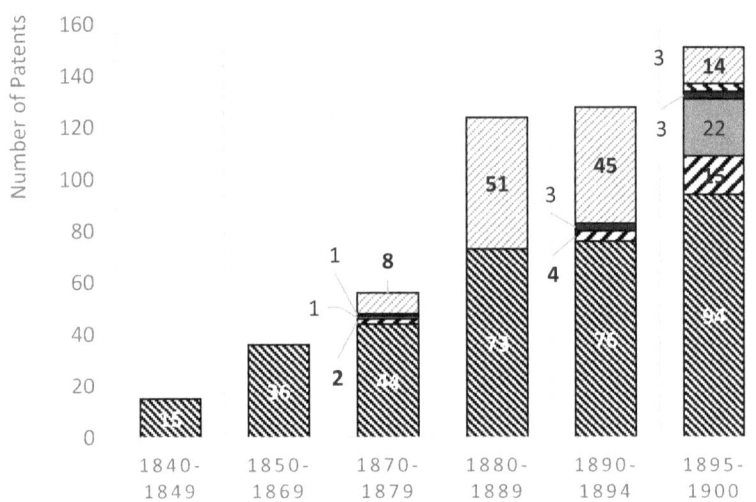

**Figure 148: Number of Worldwide Patents for Mechanical Calculating Devices (1840-1900).**

Source: http://www.ami19.org/BrevetsEtrangers/PatentsList1800-1849.html

---

[326] For a detailed overview of the patents related to mechanical calculators issued in this period 1840-1900, see: https://www.aconit.org/histoire/calcul_mecanique/documents/Brevets_machines_de_calcul.pdf (Accessed August 2021).

calculating machines. Especially the period 1900-1920 saw a Patent-explosion issued for calculating machines (Figure 149, top).

This explosion was also notable for other inventions of the late nineteenth century (Figure 149, bottom). These were the Times of Innovation (Figure 8) when in many domains, inventors created a massive novelty from the basic innovations (eg the electric light, the Internal Combustion Engine) that characterized the Second Industrial Revolution (1850-1920).

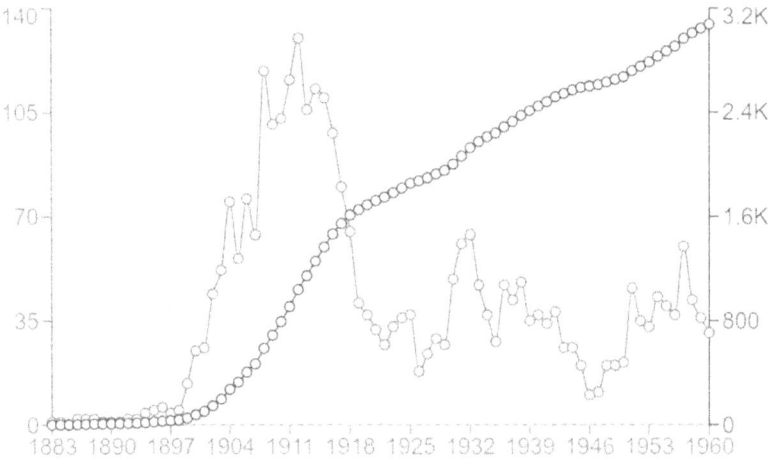

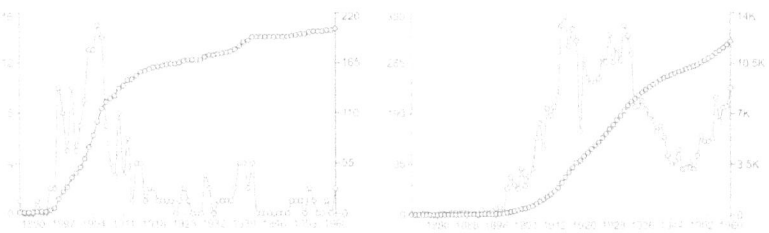

**Figure 149: Overview Patent Explosion; the Calculating machines (top), the Electric Light (bottom left), and the Internal Combustion Engine.**

Tables show yearly and accumulated number of patents. Left axis is the number of patents in a given year, Right axis the accumulated volume of patents.

Query (top): (ti all "Calculating Machine" OR ti all "Adding Machine") AND pd within "1840-1960". Query (bottom left): (ti all "Incandescent Lamp" OR ti all "Electric Light") AND pd within "1840-1960". Query (bottom right): (ti all "Gas Engine" OR ti all "Internal Combustion Engine") AND pd within "1840-1960".

Source: Espacenet

The patents for calculating machines were assigned dominantly to the manufacturing companies —and their employees in the R&D departments— and paralleled the explosion in the machines manufactured by the mechanical calculator industry. Some of the more remarkable inventors were:

The American engineer Joseph A.V. Turck (1870-1956) was a famous figure in the world of mechanical calculators. He was the author of about 40 US-patents in this area; the first received in 1899 (US-Patent № 622,091), and the last in 1957. In 1898 Turck designed a key-driven calculating machine, which he patented in 1898 (US-Patent № 631,345), in 1901 (US-Patent № 679,348) and in 1903 (US-Patent № 720,086). Later on, this machine will be manufactured under the name Mechanical Accountant. After 1910, Turck worked as a chief designer of the calculator firm Felt & Tarant company for over 20 years.

The German mechanical engineer Franz Trinks (1852-1931) joined in 1883 the company Grimme, Natalis & Co (GNC). He started working on the printing calculator —the Trinks Arithmometer— in 1892. As director of "Grimme, Natalis & Co", he paved the way for the 'Brunsviga' calculating machine acquiring the Odhner patents for 10,000.00 RM plus a 10.00 RM license for each machine sold. Through clever advertising and commercial strategies, he did build up a worldwide sales network. Trinks developed twenty-two Brunsviga models in more than 30 years: Models A, B, C, D, F, G, H, K, J, JA, arithmotype (1908 - first 4-species printing machine in the world), M, MA, MR, M24, MD, Trinks-Triplex, MDIIR, MH , MJ, MJR and N. Trink's inventiveness is demonstrated in more than 250 patents (in Germany, Austria, Switzerland, Great Britain, USA, Denmark, France). He is the author of about 37 US-Patents in this area, the first received in 1906 (US-Patent № 832,375), and the last in 1923.

The inventive activity of these two men —as well as that of many others that are hidden in the fogs of history— was concerned with a new phenomenon called calculus and its arithmetic activity. As 'arithmetics' are a mental activity, the question was if such an invention was patentable. Was a 'calculation' an abstract idea where systems took over part of human intelligence, and thus excluded from patentability? Or was it a 'thing' (ie product) that performed the arithmetic operation?[327]

---

[327] The Patent Subject Eligibility test relates if the claimed invention is applicable in processes, machines, manufactures and compositions of matter. They should not belong to the exceptions of abstract ideas, laws of nature and natural phenomena. Together with criteria such as novelty, inventive step or non-obviousness, utility, and industrial

> [...] *That such devices were patent-eligible subject matter seemed beyond dispute, and there are no federal cases in which claims to such devices or their methods of operation were held to be unpatentable subject matter. Calculating machines also perform simple arithmetic that a human could easily do by "head and hand", but that does not disqualify them as patentable subject matter. This is because the mathematical operations had been mechanized into physical elements: the "locus of the operation" was in the mechanical or electrical elements of the machine.* [...][328]

With the increasing number of patents applied for by different inventors, some failed the eligibility test (eg W.Burroughs' US-Patent № 388,116). And for those that were granted, their owners were prone to litigate over patent violations. Debates over priority raged; such as the lawsuits between Felt &Tarrant and the American Arithmetic Corporation (later Burroughs).

*Pioneering Patents and Basic Innovations*

Some patents were more important than others, due to their nature and the impact their invention had. Its pioneering nature could be in a new field (eg electricity). Combining the patent's nature with the impact of the invention (Scheme 1) —both the technical impact and the economic impact— is another indication of the importance of the invention. The technical impact can be derived from the industrial activity that emerged by the increasing number of manufacturers. The economic impact becomes clear when one looks how the calculating machine did have a massive impact on clerical work done in offices. And last but not least, the impact of an invention can be noted from the patent-disputes that emerged when patent-rights were infringed upon.

> Such as the disputes between Felt and Burroughs who became engaged in lengthy and costly legal disputes about the various patents. A similar situation was the court case Dalton Adding Mach. Co. et al. vs. Moon-Hopkins Billing Mach. Co. et al (1915).

So, the inventions that were covered by a basic patent, as recognized by the scientific community, are labelled in our exploration as basic innovations.

---

applicability, which differ from country to country, the question of whether a particular subject matter is patentable is one of the substantive requirements for patentability.
[328] Sachs, Robert R.: The Mind as Computer Metaphor. (p.15) https://assets.fenwick.com/bilski-documents/sachs_aipla_2016_spring_paper-22.pdf. (Accessed August 2021)

## The Machines that changed the Office.

Next to the Invention Rooms —the later R&D departments where the engineers worked—, the Office Rooms were the places where the clerks worked. Their work was of an administrative nature, supporting the business operation (eg the trading activities of the East India Companies of the seventeenth century's British Empire). Over the course of the nineteenth and twentieth centuries, increasingly specialised office designs —from the office towers of Chicago and New York to the post-war suburban corporate campuses— reinforced a distinction between the workplace and home.

Over time that workplace changed as work became supported by new tools of a technological nature. Morse's telegraph, Bell's telephone and Edison's dictating machine, Remington's typewriter and Burroughs accounting machine (Figure 150), they all revolutionised both concepts of work and office design. Telecommunications meant offices could be separate from factories and warehouses, separating white- and blue-collar workers. Office machines increased productivity of the clerk/white-collar

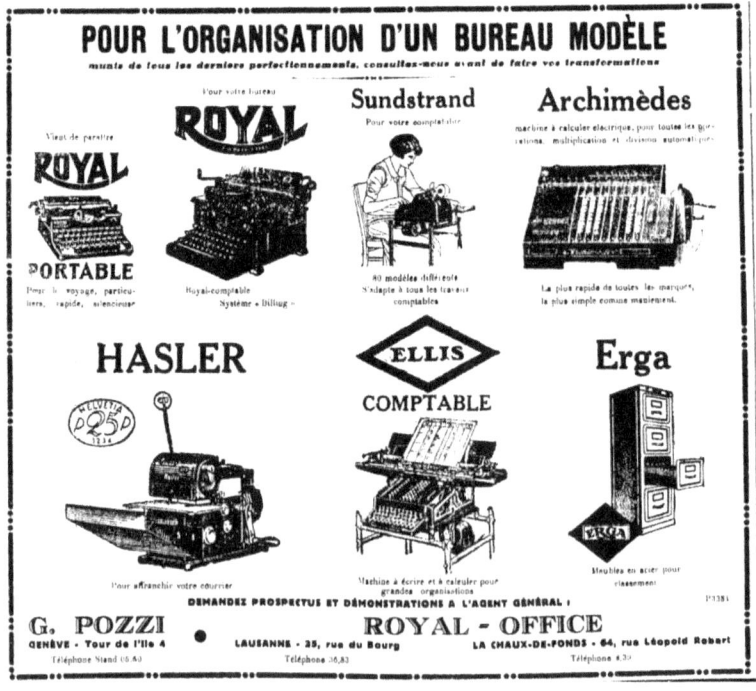

Figure 150: Advertisement for Office Equipment.
Source: https://www.jaapsch.net/mechcalc/archimedes.htm

office worker; the typewriter filled the pre-printed forms with financial information. Calculating machines became a standard desktop tool. As a result, job contents changed; the original recording bookkeeping 'accountant' became the manager supervising the bookkeeping department.

The first clerical workers were men. Early clerical work was seen as a craft, developed to help business owners keep current records of their enterprises and to maintain relations with the outside world. The clerical workplace was his desk, his tools were pen & paper, his arts were writing and calculating. That changed when technology made the type-writer (1870s), the shorthand machine (aka stenotype machine (1880s), the calculating machine (1890s), and the accounting machine (1910s) possible. And for each of them there was the natural resistance to change. At first many people were sceptical of 'mechanical writing'; handwritten documents and correspondence were the norm, even in business. But after Remington started mass manufacturing typewriters in the 1870s, the machine found a new home: the Office. The same took place with 'mechanical calculating' as mental calculations were the norm. After Felt introduced his Comptometer —the multiple-order-key-driven-calculating-machines— Burroughs' and Monroe's machines changed that norm in the early twentieth century. Places with large admirative needs (eg banks, railroad companies, insurance companies) were the first to adapt the new machine.

And with mechanization of clerical work came another change. Women were brought into offices to fill these new 'clerical' positions, with firms taking advantage of the supply of middle-class, high school or college-educated women who would work for lower

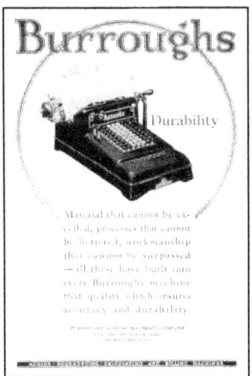  

**Figure 151: From Adding Machine to Bookkeeping Machine, the whole product range offered by Burroughs.**
Source: Vintage Ads, Wikimedia Commons.

wages than men of comparable education. And the calculator manufacturers focussed their advertisement not only on their product and product range, but also on the female user (Figure 151).

Just like the manufacturing technology had changed the factory with its assembly lines, the blossoming office machine technology had changed the office. As the nature of work progressed, so did the machines in couple of decades. From the machines for 'calculating' arose those that assisted in 'book keeping and billing,' and from book keeping came accounting machines. And the size of the markets seemed endless.

## Manufacturing Mechanical Calculators; the Calculator Industry

At the end of the nineteenth century, with the Era of Mechanization in full swing and the demand for mechanical calculators sharply rising, the machines of writing and calculating —later known as the 'office machines' in a large diversity— grew up together as their common knowledge base was the fine-mechanical technology. By 1900 many companies in the US, and Europe (eg Germany and Britain) manufactured mechanical calculators.

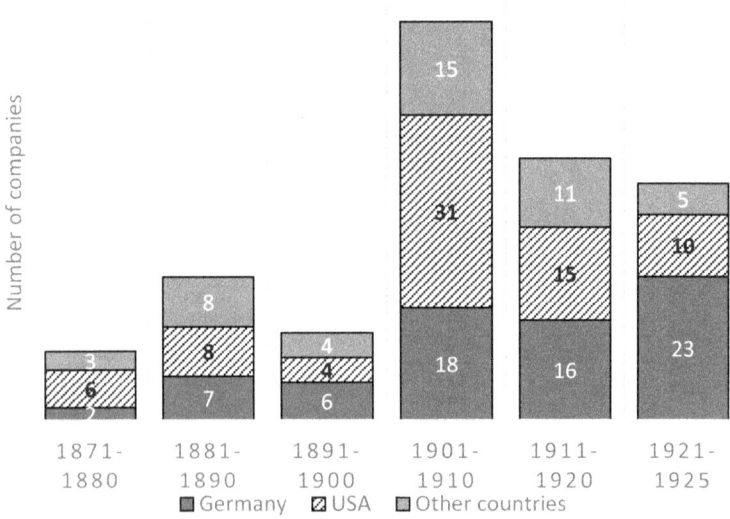

Figure 152: New entrants (ie Companies and Inventors) of Manufacturing Mechanical Calculators.

Rough Indication of new entrant for Germany, USA and other countries.

Source: http://www.rechenmaschinen-illustrated.com/Martins_book/Ernst%20Martin%20-%20Rechen%20Machinen%20OCR%204.pdf

The rise of the mechanical calculator paralleled the rise of the big corporations: until the Financial Panic of 1893 the inventors struggled to find a market, but towards the end of the decade their revenues skyrocketed.

The period after 1900 saw a sharp increase in manufacturing companies. Dominant were the new entrants of the German and US manufacturers (Figure 152). During those years the manufacturing of the mechanical calculators changed. Had earlier machines been made individually by watch-, clock- and instrument makers, later followed by the specialized workshop where artisan labour produced the machines —aka craft production—, now new production techniques evolved. The manufacturing of mechanical calculators paralleled the development in the clockmaking industry, the typewriter industry, the cash register industry. All companies that applied the technology of interchangeable parts that was fundamental to the mass-manufacturing techniques.[329]

In Europe the regions that had already developed a clock and watchmaking industry, soon embraced the new opportunities the mechanical calculator was offering. In Germany, since 1890, the office machine industry developed in small villages with loyal skilled workers in remote mountain valleys of Thuringia and the Erzgebirge. Such as the villages of Sömmerda and Zella-Mehlis in Thuringia and Glashütte in the Erzgebirge mountains. In Britain, Samuel Tate had developed and his stepped-drum calculator. In 1883 the brothers Charles and Edwin James Layton, booksellers and authors of insurance books, exhibited the first arithmometer as agents of Tate and soon afterwards acquired the patents, arranging a manufacturing workshop on Farrington Road in London.

In many countries in Europe, entrepreneurs picked up the development of the calculating machines, either by taking a licence or just by cloning the popular machines. Many companies, already active in the field of fine-mechanical production (eg sewing machines), saw an opportunity to expand their existing business of typewriters, cash registers, tabulating machines, but other countries took longer to enter the Era of the Mechanical Calculating Machines.

> *In spite of Italian inventiveness and of growing national market, no calculator industry was created before 1930. It was only with the autarky enforced by the fascist regime — protectionism due to the economic crisis of 1929 and to the embargo imposed by the League of Nations as a consequence of the Italian-*

---

[329] This technology originated from the fire-arm industry mass producing firearms with standardized part that were assembled on 'assembly lines' as part of the factory system. It was also the technique applied in the mechanical industries (eg the automotive industry).

*Abyssinian war of 1935 — that an effort to create national industries to minimize import from abroad was carried on. As a matter of facts, both causes were not so severe — the embargo lasted only a few months and was strictly observed by Great Britain only — but fascist propaganda exploited autarky as a stimulus for national pride and popular consensus. Thus, after 1930 a bunch of Italian manufacturers came up, producing several models of full-keyboard (Comptometer's style) and reduced-keyboard adders (printing and non-printing) and also four-operations machines (Brunsviga's style).*

*Only one Italian company was able to reach a world-wide market: Olivetti. Founded by Camillo Olivetti in 1908 for the production of typewriters, with the headquarters in Ivrea, near Turin. [...] In 1935 Adriano Olivetti realized that, although successful in the typewriter market, his company was missing the opportunity of other office appliances, like desk top calculators. The first mechanical calculator was produced in 1940 (the MC 4 Summa).*[330]

**Figure 153: Manufacturing Industries.**
NCR Manufacturing cash registers in Dayton, Ohio, USA. Daylight Room (top, 1893), and Final Assembly Room (bottom, 1894).
Source: NCR

After the European success of the Arithmometer in the 1850s, the novelties in the mechanical calculator had transferred across the Atlantic to the USA ready to be implemented. But it was not until after the American Civil War (1861-1866) —when new forms of (steam-powered) manufacturing allowed the American industry to grow and spread across the nation—, that a vibrant office equipment industry emerged. An industry with calculator

---

[330] Source: Silvio Hénin: Early Italian Computing Machines and Their Inventors. https://hal.inria.fr/hal-01526799/document. (Accessed June 2021)

manufacturing centres in cities such as Chicago, Detroit, Dayton, St. Louis, and Philadelphia. Like the typewriter, the mechanical calculator was constructed out of interchangeable part that were assembled at working stations (Figure 153).

This was the prelude to the wave of the mechanical calculating machine that were developed up to the end of the nineteenth century. It would start the Golden Age of the 'hand cranked' calculating machines (1887-1915). From the turn of the century, electrical motors were increasingly used to power the calculating machines. It gave rise to the electromechanical calculators. Also, other functions were added; such as the printing function.

## The Golden Age of Calculating Machines

Before 1900, makers of calculating machines had found a regular though small clientele in science and technology with limited demand. Calculating machines had been used alongside other aids in particular for complex calculation in geodesy and astronomy, as well as in insurance matters. Then, a rapid increase in demand from commerce, industry and government started in about 1900. A calculator manufacturing industry with over 100 companies throughout the world appeared in the next two decades (Figure 152). The main manufacturing countries being the US and Germany.

Europe saw the rise of specific regions of manufacturing. After the earlier production of around 1850 with the Thomas stepped drum machines in Paris, and the Burkhardt calculator in Glasshutte (Saxony) in 1878, the mass production of the Odhner pinwheel machines started in the Russian town of St.Petersburg in 1878. Manufacturers of early mechanical machines (eg sewing machines, typewriters), observing the increasing demand for calculators in new markets (eg office work), grabbed the opportunity. Such as Grimme, Natalis & Co., which was based in Brunswick and originally manufactured sewing machines, who purchased in 1892 the rights to manufacture pinwheel calculating machines of Odhner's design and sold them under the name Brunsviga. Other companies were started by their inventor's. Such as the Felt & Tarrant Manufacturing Company, located in Chicago USA, that was founded to manufacture the Comptometer developed by Felt.

> A similar concentration of manufacturing was seen in the US. Manufacturers in the East-USA primarily made adding machines.[331] Until 1900, Felt & Tarrant was the market leader with its Comptometer, but soon had to concede its leading position to the Burroughs Adding Machine Company promoting their 'Registering

---

[331] The city of St. Louis, by 1880 the fifth largest city in the US, had some thirteen manufacturers of calculating machines in that period of time.

Accountant.' he USA became the largest exporter of calculating machines by the First World War.

> *America entered the age of mechanical calculators in late 19th century, much later than Europe. When major European countries were undergoing extensive industrialization at the time of the appearance of Thomas calculators, the United States was still primarily involved in agriculture while Canada was not even on the map as a country. [...] It was not until after the Civil War when new forms of manufacturing (steam powered) allowed the American industry to grow and spread across the nation. It was at that time, when a vibrant office equipment industry was created with calculator manufacturing centers in cities such as Chicago, Detroit, St. Louis, and Philadelphia.* [332]

Over the first decades of the twentieth century, the manufacturing of the mechanical calculators exploded, due to mass-production techniques that were applied.[333] However, the calculating machines, used everywhere where basic 'calculation' was needed, reached the limits of their application.[334]

> *By the 1930s, mechanical and electromechanical calculators from Burroughs, Felt & Tarrant, Marchant, Monroe, Remington, Victor, and other manufacturers had penetrated all aspects of a modern office operations. It seemed that the mathematical tables could handle the rest in science and engineering. However, certain branches of science and engineering reached a calculating barrier that prevented further progress unless an automated method of dealing with complex operations, such as solving linear equalities, differential equations, etc. could be found. What was needed was a new type of a calculating machine that could perform large-scale error-free calculations by following, in a mechanical way, a predefined sequence of operations, that is, by following a program.* (Ibidem)

But those more advanced machines were only on the brink of the horizon. Although the majority of the key developments of the mechanical calculator took place before 1900, the first decades of the next century saw the market accept and embrace the new type of machines. Then came the expansion of the early twentieth century that were the result of two key-innovations in the technology;

*Stepped-drum technology:* The first key-innovation found its roots in the development of the Thomas-calculator concept—adopted by

---

[332] Source: Stachniak, Z. The World of Calculators: from office equipment to pocket gadgets. (2011-2014 http://www.cse.yorku.ca/~zbigniew/nats1700/lecture4.pdf
[333] Remember, this was the time of Taylor's Scientific Management theory (aka Taylorism) that saw productivity increase through new ways of organizing labour.
[334] The end of the Era of Mechanical Calculating Machines was around 1968-1973 when the electronic pocket calculator came into existence.

Burroughs—into the *Arithmometer* that became known as the stepped-drum technology.

*Pinwheel-technology:* The second key-innovation was based on Baldwin's contribution to the pinwheel-technology, and the keyboard for data entry as developed by Dorr Felt. Their work gave the foundation for the key-innovation of the *Comptometer*.

The multitude of calculating machines that were realized with these technologies fulfilled a limited need for calculation devices in the marketplace. During the early twentieth century that market for mechanical calculators developed in several product categories.

The first were the adding machines used for basic calculations of addition and subtraction (ie single functionality). The second was the four-function calculators that added the multiplication and division functions to addition and subtraction. The third category were the accounting machines that performed 'accounting operations'[335] that evolved in the bookkeeping machines (ie a combined functionality) (Figure 154).

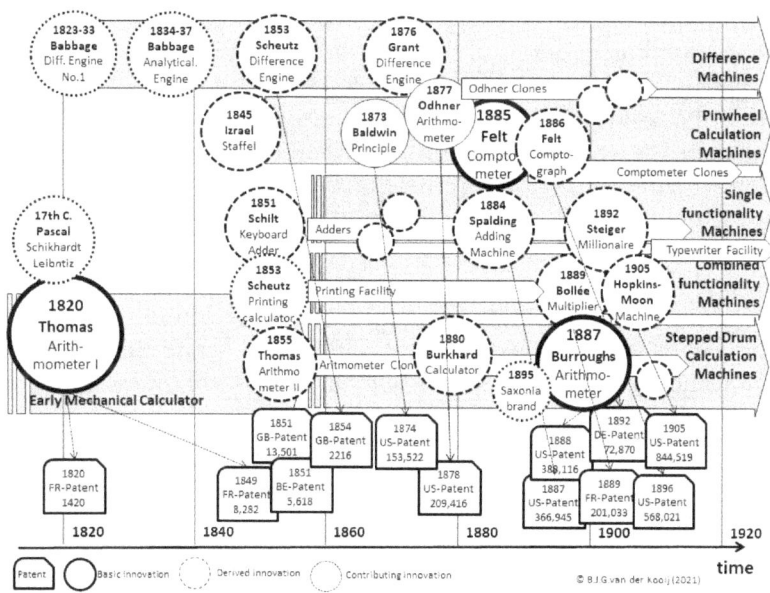

Figure 154: Overview Contributors to the Evolution of Calculating Machines (1850-1920).

---

[335] Among those operations we find tabulation; the systematic representation of data in rows and columns.

However, the calculating machines only focused on the simple basic arithmetic calculations. But there was —especially in the more scientific circles— a latent need for machines that could perform more complex mathematical operations. Those machines became known as the Differential Machines (aka Difference Machines) (Figure 154, top).

The Golden Age of Mechanical Calculating Machines, followed the times of rapid growth of the industry that manufactured the machines. And with maturing industry came the survivors—often after a range of merging activities (Figure 156)—, and the losers that stopped production, went bankrupt or were acquired by the survivors. For a technology start-up it was not the right time, as the large companies having the resources and market share, were not waiting for a new competitor. They might have been interested in new technological developments, but certainly weren't eager to have additional competition arising from those new developments.[336]

## Periodizing the First Calculating Revolution

This total development of calculation machines —labelled as the First Calculator Revolution—can be periodized (Scheme 5) in three different periods.

*Mechanical Calculation 0.0* (1700s-1800s): A period that covers the early conceptual contributions of the polymaths like Wilhelm Schickard (1592-1635), Blaise Pascal (1623-1662), and Gottfried Von Leibnitz (1646-1716), as well as the Calculator Craze in which mechanical solutions for realizing the arithmetic function by machines (Figure 155) were sought by many curious minds. Concepts that with a range of contributions created the foundations for the early stepped-drum calculating machine and the early pinwheel calculating machines developed by the more practical oriented inventors of the nineteenth century. This was the period where concepts, developed by polymaths and natural scientists, were realised by the master-craftsman of the fine-mechanical technologies. Resulting in the single-item (or limited items) that could be paraded in aristocratic circles.

*Mechanical Calculation 1.0* (1800s-1900s): We also observed how the practical implementation of the concept led to the early 'calculating machines' in two dominant technologies; the stepped-drum technology and the pinwheel technology. Both technologies, realized by artisan craftsman in the fine-mechanical technologies of those days, rose to prominence after the 1850s (Figure 155, right).

---

[336] This phenomenon of the industry life cycle (Scheme 5) consist of several periods/phases; startup, growth, shakeout, maturity, and decline.

These were the days of the First Industrial Revolution. In terms of inventions in the field of calculation, their seemed not much progress in the first decades of the nineteenth century. Maybe because it were also times of social and geo-political unrest. From the French Revolution, the Napoleonic Wars up to the 1848-Revolutions. The second halve of the nineteenth century showed a different picture.

The first calculating machines that were introduced at Expositions, Shows and Fairs —eg the World Fairs like the Great Exhibition in London (1851), the Expositions Universelle in Paris (1855)— attracted attention of the scholars and aristocrats, even were honoured with awards. For some it were 'intelligent' machines as they took over activities formerly belonging to the human brain. Soon the new concept of mechanical calculating machines called 'arithmometers' blossomed in the next decades. Next the inventor-entrepreneurs brought the fruits of their engineering activity to the latent markets. Or —together with investors/businessmen— they organised for its manufacturing. From the difficult to operate-stepped drum machines to the simple adders and calculating machines and to the rather complex accounting machines.

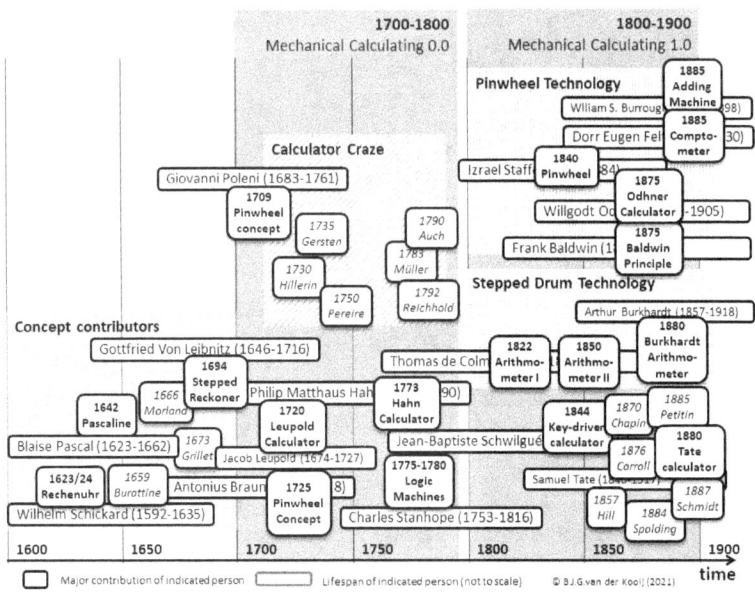

Figure 155: Overview of Early Contributors to Calculating Technologies (1600-1900).

Machines that were bought by the early adapters; ie the banks, the insurance companies. [337]

*Mechanical Calculation 2.0* (1900s-1930s): In this period the calculation machines got their extended arithmetic functionality, and even were combined with the functionality of type-writers. As they fulfilled a latent need, their market expanded swiftly. By 1920, calculating machines — from the simple adder to the complex bookkeeping machines— were not unusual wonders any more. Entire industries in Europe and the United States manufactured them (Figure 156). Manufacturers that cooperated with the inventors, of just cloned their machines, developed over the years their improved models, adding functionality. The widely used machines served as common tools of scientists, engineers, statisticians, actuaries, government officials, accountants and payroll clerks. Also the retail business adapted to the cash register machines en masse. Using a broad range of models offered by the manufacturers. Originally the calculating machines had found their standard design

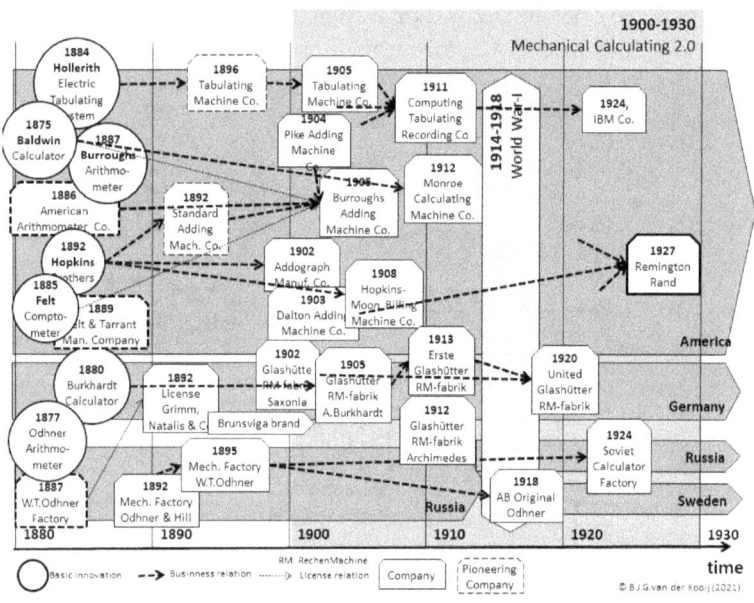

Figure 156: Overview of Major Manufacturing Companies during the Era of Mechanical Calculation 2.0 (1900-1920) of the First Calculator Revolution.

---

[337] An accounting machines generally is a calculator and a printer combination tailored for a specific bookkeeping function. These bookkeeping and billing machines were the combination of a typewriter and a recording adder.

either as the box-based Arithmometer, later in the desktop-models of the Comptometer. Later they would develop their own characteristic design and functionality.

In the period of some eight decades, the *First Calculating Revolution* (1850-1930) had taken place. Based on the fine-mechanical technologies available in that period of time, early calculating machines were handcrafted, but by the end of the century, series and mass production techniques became applied. A development that parallels other developments at that time (eg typewriter, sewing machine). However, the fine-mechanical technology was not suitable for implementing more complex logic functions.

That First Calculating Revolution was to be followed by the *Second Calculator Revolution* when the new 'electronic' technologies developed. A development that started with the vacuum-tube technology being applied in calculating machines. Soon followed by the transistor technology and the Integrated Circuit-technology. They subsequently fuelled a range of development trajectories: the trajectory of the electronic desktop and pocket calculator, and the trajectory of the scientific and programmable calculators. Trajectories paralleled by the Analytical Machines spinning off in the application trajectory of 'computing.'

## Changing Perspective

We explored how the mechanical calculating machines developed over a period of decades from the basic arithmetic calculating machines to rather complex bookkeeping machines. A development trajectory that was highlighted by remarkable events in which basic innovations contributed to its growth, implementing the fine mechanical technology of that time to create their machines. A technology that both facilitated and limited the early design at the same time.

Now we will go and change our perspective to the next technology that came available with the rise of the electro-mechanical relay. But before exploring the events of that trajectory, we will zoom in on the development of 'mathematics' that were to play a prominent role in de new concepts of machines that would later be labelled as 'computers.' A development that started with the 'Art of Calculus.'

# The Computing (R)evolution

The dynamic inter-war period called Interbellum (1920-1940) saw massive novelty; from social innovation and cultural innovation to economic and technical innovation. Culturally the 'Roaring Twenties' experienced social, artistic and cultural dynamism. Breaking away from traditionalism, a range of social and culture movements (from the rise of Jazz Age to the Art Deco, from the Prohibition to Women's Suffrage) changed society profoundly.

**Figure 157: Artist Impression of The Thought Machine.**
This cartoon by Frank R. Paul from 1927, depicts the spirit of time where thinking machines would arise. The bottom right part could be inspired by the 'Millionaire' calculating machine.
Source: https://www.collectorshowcase.fr/

As part of the Machine Age (1750-1950), the technological novelty had created a broad range of mechanical machines. From the machines that could write at a distance at lightning speed (aka the telegraph) and the talking machine (aka telephone), to the machine that created individual mobility (ie the motor bike and the automobile) and gave man access to the skies (ie airplanes). And amongst all that novelty arose the *Machines of Thought* (Figure 157) that were taking over a part of the work done by human machine of thought; the Brains.

## *The Invention of the Computing Machines*

Already in ancient times, the Art of Calculus was developed by the Mesopotamian scholars studying the skies, creating Babylonian Calculus. Also the Egyptian mathematicians used a numeral system for geometric calculations after Nile flooding. And the Greek mathematicians like Plato, learning from the Egyptians, founded the Platonic Academy in Athens that became the mathematical centre of the world in the fourth century BC.

That Art of Abstract Thinking called 'mathematical thinking' has a long history. The scholarly interest in 'logic' [338] —often associated with 'correct thinking'— was already practised by the old Greek philosophers.[339] Such as the mathematical logic —aka the Philosophy of Numbers— practised by Pythagoras (569-475 BC). The Pythagoreans studied and practised 'arithmetic'; creating their cult of Music and Arithmetic. Next, the philosophy of 'correct reasoning' culminated in ancient Greece with Aristotle's (384-322 BC) teachings of logic; from categorical forms, propositions (being True or False) and syllogisms.[340] Then — jumping ahead in time— quite some centuries later we see the lost work of the Greek philosophers being rediscovered by the Arab scholars. Aristotle's views rose to prominence by the efforts of the Islamic Aristotelians (800-1000) working in the House of Wisdom in Bagdad. By the late nineth century, the mathematical School of Baghdad was the focus of logic studies in the Arab world.

In Europe, after the contributions to mathematical thinking of the 'Church Scientists' (1200s-1450s) —eg Albert of Saxony publishing his 'Per Utilis Logica'— in the fourteenth century, arose in England the Neo-mathematical thinking as practised by the 'Oxford calculators,' who took a strikingly logico-mathematical approach to philosophical problems.[341] Using Aristotelian logic and physics, they studied and attempted to quantify every physical and observable characteristic, like heat, force, colour, density, and light. Although during the Scientific Renaissance (1450s-1550s), mathematical concepts were further explored (eg the real number system,

---

[338] *Logic* is defined as that field of inquiry which investigates how we reason correctly (and, by extension, how we reason incorrectly). The aim of logic is the elaboration of a coherent system that allows to investigate, classify, and evaluate good and bad forms of reasoning.
[339] For a more extended analysis on this topic, see the Chapter: *The Genesis of Scientific Thinking* in: Van der Kooij. B.J.G.; The Origins of Innovation (2018).
[340] Syllogisms are structures of sentences each of which can meaningfully be called true or false: the 'assertions.' Aristotle defined a syllogism as a "discourse in which, certain things being stated something other than what is stated follows of necessity from their being so." But in practice he coined the term to arguments containing two premises and a conclusion.
[341] Such as the Oxford natural philosopher John Dumbleton (c. 1310-1349) who wrote his 'Summa Logicae et Philosophiae Naturalis' (Summary of Logic and Natural Philosophy).

trigonometry, arithmetic, inductive reasoning), the scholastic attention to logic dwindled in the context of the religious upheaval caused by the Reformation. The Aristotelian Worldview became rejected and was to be replaced by the Natural Worldview.

Then came the Scientific Revolution of the sixteenth/seventeenth century, and the *Natural philosophers* (eg Boyle, Hauksbee, Franklin) explored the Nature of Matter, such as the Nature of Lightning. In addition, *Enlightenment philosophers* (eg Hobbes, Locke, Hume) were exploring the Nature of Being/Existence and the Nature of Society/Government. And the *Mathematical philosophers* (aka the mathematicians) went on a more abstract (ie mathematical) trajectory of thinking (eg Descartes, Pascal, Leibnitz, Newton, Euler); they started the new art of *metaphysic thinking* (Figure 158).[342]

Within the progressing nineteenth century, the phenomena of electricity —with the discoveries of Luigi Galvani and Alessandro Volta opening whole new areas of investigation— became the focus of their attention of the *Electrical mathematicians* (eg Ampere, Arago,

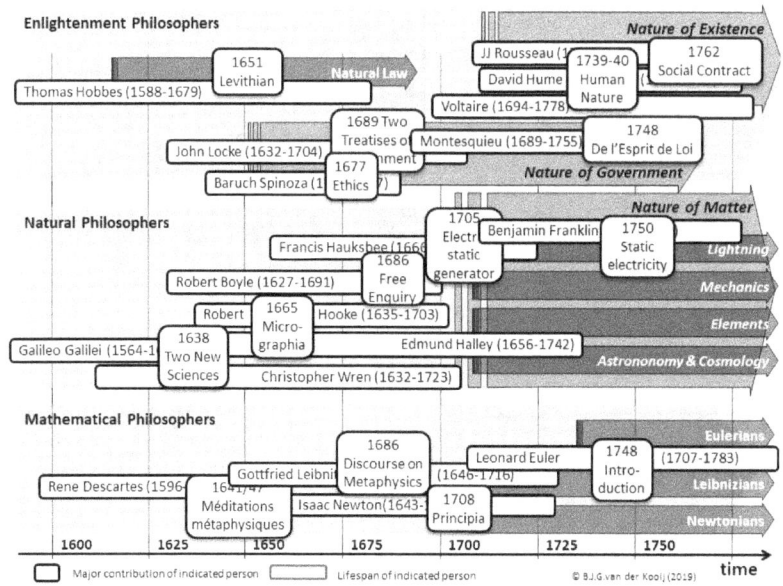

Figure 158: Overview Natural, Enlightenment and Mathematical Philosophers during the Scientific Revolution.

---

[342] For a more extended analysis on this topic, see the Chapter: *The Genesis of Scientific Thinking* in: Van der Kooij. B.J.G.; The Discovery of Innovation (2018).

Maxwell).[343] And with the rise of economic thinking, the *Economic mathematicians* (eg Cournot, Walras, Jevon) focussed on the mathematical aspects economic phenomena (Figure 159).

With the seventeenth century came increasing interest in symbolizing logic. The logical work of the German mathematician, philosopher, and diplomat Gottfried Wilhelm Leibniz (1646-1716) was one of the great triumphs, as well as tragedies, in the history of logic.[344] He created in the 1680s a symbolic logic that is remarkably similar to George Boole's system of 1847. Although Leibniz might seem to deserve to be credited with great originality in his symbolic logic —especially in his equational, algebraic logic— it turns out that such insights were relatively common to mathematicians of the seventeenth and eighteenth centuries who had a knowledge of traditional syllogistic logic.[345]

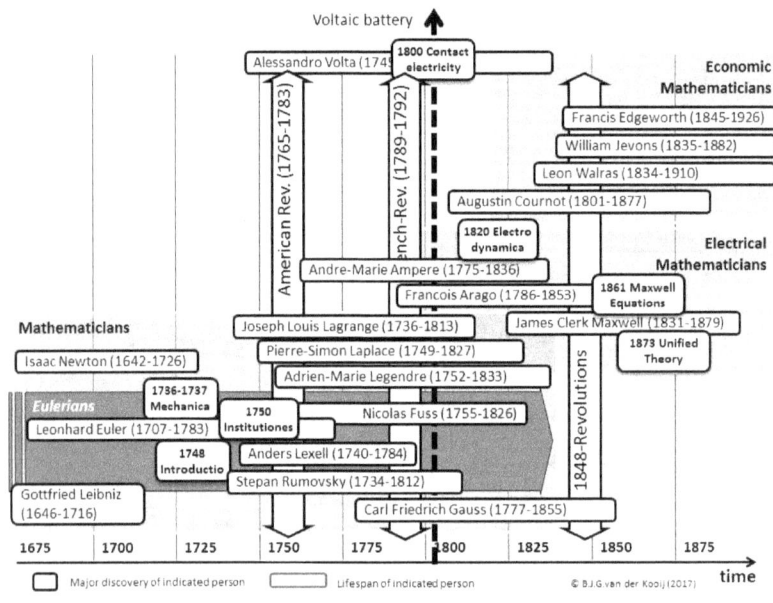

Figure 159: Overview Mathematicians after the Scientific Revolution.

---

[343] By 1827, André-Marie Ampère had published a series of mathematical and experimental memoirs on his electrodynamic theory that not only rendered electromagnetism comprehensible but also ordinary magnetism, identifying both as the result of electrical currents. The Scottish physicist James Clerk Maxwell developed his profound mathematical electromagnetic theory from 1855 onward.
[344] The Leibnitz-Newton Calculus controversy, where both Newton and Leibniz both claimed to be the founder of differential and integral calculus. The prevailing opinion was against Leibniz (in Britain obviously, not in the German-speaking world).
[345] Syllogistic logic is the formal analysis of logical terms and operators and the structures that make it possible to infer true conclusions from given premises.

So, over time mathematics as a philosophy, turned in mathematics as a way of understanding Nature; from the physical nature to the metaphysical nature. Mathematics became the language in which the knowledge of the sciences became written. Mathematics as an art that expressed order in the form of axioms, theorems and logical or numerical relations. And it was mathematics that covered both the Macro-Cosmos (ie Einstein's Theory of Relativity) and the Micro-Cosmos (ie the Quantum Mechanics).[346] Mathematics in the form of algebras —dealing with symbols and the rules for manipulating those symbols— focussing on specific areas; from elementary algebra and the algebra of logic, to the algebra of sets.

## Computing Machines 0.0 (1780-1930)

As described before, the Art of Mathematics is a mental activity of abstract thinking and reasoning that covers a broad range of applications. Over time in the specific fields —such as the algebras— specific techniques developed; the arithmetic techniques of addition, subtraction, multiplying and division. And from the ancient tools that assisted the arithmetic operations of the mind, grew —as we have been exploring before— the dedicated calculating machines that could perform these arithmetic operations. The next step was the development trajectory of machines that could perform more complex mathematical calculations: the computing machines. A trajectory that started with the Logic Machines.

### The Mechanical Logic Machines

Mathematics is an activity of the mind, sometime a repetitive activity, especially when it involves calculus. However, the human mind, being a creative and complex 'machine', is not too good at repetitive tasks. So, the aspiration of developing tools to assist in that repetitive activity called calculus, became apparent and the calculating machines came on scene. Soon that aspiration extended in the logic machines, that could do more than basic calculus. The more when the English Lord Charles Stanhope (1753-1816), after constructing several mechanical calculators, built his logic machine he named the *Stanhope Demonstrator*. The machine proved that problems of logic could be solved by mechanical means. His Demonstrator became the catalyst for the further development of logic thinking that saw many contributors.

The English logician William Stanley Jevons (1835–1882) was interested in economics and started applying mathematical methods in economic

---

[346] For a more extensive exploration, see: Suarez, Juan Luis Vazquez: The Importance of Mathematics in the development of Science and Technology.
https://sema.org.es/images/site/boletines/boleti19.pdf#page=69 (accessed May 2021).

thought with his publication 'A General Mathematical Theory of Political Economy' (1862). He perfected Boole's system of logic, and published in 1864 'Pure Logic; or, the Logic of Quality apart from Quantity,' based on the Boolean system of logic.[347] His publication 'Elementary Lessons on Logic' (1870) soon became the most widely read elementary textbook on logic in the English language.

>He also designed in 1869 a logic machine capable of solving a complicated problem faster than a human could solve it without the aid of a machine; the 'Logic Piano.' The machine was built for him by a clockmaker at Salford in 1869 and first demonstrated by Jevons in 1870 at a meeting of the Royal Society of London.

Building forth on the logical piano, his student the American historian Allan Marquand (1853-1924) built first a mechanical, and later an electro-mechanical logical machine in the 1885s. It is possible even the whole electromagnetic design of the machine to have been made by Peirce, because Marquand after 1886 wrote no more on logical machines, while Peirce wrote in 1887 an amazing article titled 'Logical Machines.'

Also others contributed to the development of these mechanical machines. Such as a third machine of the Jevons type invented in 1910 by Charles P. R. Macaulay, an Englishman living in Chicago. It is a compact, ingenious boxlike device with interior rods operated by tilting the box a certain way while pins on the side are pressed to put statements into the machine. And a curious contrivance for evaluating the 256 combinations of syllogistic premises and conclusions was constructed in 1903 by Annibale Pastore, a philosopher at the University of Genoa. Their contributions became the dawn of the machines based on Relay Logic.

## Relay Logic

These calculation machines were realized in the mechanical technologies available at that time. However, the electromechanical technologies used in telegraphy and telephony spurred the development and use of 'Relay-logic.' Where the 'logic' was the Boolean Logic; a form of algebra that used Boolean operators. And where 'relay' was the electromagnetic device known as the electromagnetic relay.

During the late 1930s several researchers realized that Boolean operations could be given physical form as arrangements of switches; a

---

[347] The mathematician George Boole (1815-1864) developed in the 1840s the algebraic logic, a branch in Algebra (the Art of the operators of addition and subtraction, multiplication and division) that integrated logic thinking with operators conjunction (AND), disjunction (OR) and negation (NOT). It is thus a formalism for describing logical operations, in the same way that elementary algebra describes numerical operations.

switch being a two-state device, it is 'ON' or 'OFF'. And, the relay being equivalent to such a switch, it created the circuits of Relay Logic. In the 1930s, while studying switching circuits, the American mathematician Claude E. Shannon (1916-2001) observed that one could also apply the rules of Boole's algebra in this setting, and he introduced *switching algebra* as a way to analyse and design circuits by algebraic means in terms of logic gates. It was a transformative work, turning circuit design from an art into a science.

> In 1937, he wrote his master's degree thesis, 'A Symbolic Analysis of Relay and Switching Circuits.' A ground-breaking paper on the application of symbolic logic to relay circuits was published in 1938 from this thesis. His work became the foundation of digital circuit design.

Shannon, both scientist and engineer, would later become known for his work in the general *theory of communication*, that became as fundamental as the physics of nature.

> Shannon published his masterpiece in 1948: 'A Mathematical Theory of Communication.' The heart of his theory was a simple but very general model of communication: A transmitter encodes information into a signal, which is corrupted by noise and then decoded by the receiver. Despite its simplicity, Shannon's model incorporates two key insights: isolating the information and noise sources from the communication system to be designed, and modelling both of these sources probabilistically.

The next contribution was to *information theory*. In it, Shannon defined the units of information, the smallest possible chunks that cannot be divided any further, into what he called 'bit' (short for binary digit), strings of which can be used to encode any message. For example in a 'byte,' a string of of eight bits. The most widely used digital code in modern electronics is based around bits that can each have only one of two values: 0 or 1 (aka binary system).

## *Computing Machines 1.0 (1930-1950)*

In the preceding exploration we identified some important events in the development of computing machines that were originally based on the electro-mechanical technology of relays. However, that technology had its limitations; the computing machines were huge in volume, maintenance intensive, and consumed much electricity. They were also expensive to build, although that was less of a problem where governmental (eg military) and scientific institutions were involved.

But in the meantime, a new technology had emerged that was replacing the relay as logic device; the vacuum tube we will explore further on in more detail. Although having similar limitations of cost, size and reliability, those 'valves' would become the building blocks of the next generation of computers, that were developed before, during and after the Second World War. Computing machines that emerged from Von Neuman' proposal 'First Draft of a Report on the EDVAC'[348] that described a design architecture for an electronic digital computer. After the war, Von Neuman published in 1946 with Arthur Burk and Herman Goldstine, the paper 'Preliminary Discussion of the Logical Design of an Electronic Computing Instrument' that became the birth certificate of computer science. His design for a stored-program computer was an advancement over the hardwired program-controlled computers of the 1940s, such as the Colossus (Figure 160) and the ENIAC (Figure 163).

*The Colossus Computer in Britain*

The British engineer Thomas Flowers (1905-1998), working for the Research Branch of the Post Office, was engaged in switching electronics for telephone exchanges. From his wartime experiences with code-breaking machines (ie the bombes), he proposed a more sophisticated alternative based on vacuum tube technology. The resulting machine became known as the Colossus.

This Colossus Mark I (Figure 160) contained 1500 vacuum tubes, was used breaking the cipher codes of the Germans, and came into operation in 1943. Soon the Mark II followed —using 2,400 vacuum tubes— in 1944. The machine immediately produced vital information for the imminent D-Day landings planned for Monday June 5th, 1944.

**Figure 160: The Colossus (1943).**
Source: Techrepublic.com

Other projects at British Universities laid the foundation for the further developments. One of those research teams from the University of Manchester created the computing machines that became known as the *Manchester Computers*. The first of these universal (as opposed to special-purpose) computers to come into operation was the *Small-Scale*

---

[348] EdVac stands for Electronic Discrete Variable Automatic Computer.

*Experimental Machine* (SSEM) —known as the Manchester Baby (Figure 161, top)— developed at the computer laboratory of the Manchester University, which first ran a program on June 21th, 1948. By 1949 the Baby extended in to a full-size computer, the *Manchester Mark I* (Figure 161, upper middle), a stored program computer that contained 4,050 vacuum tubes.

The British National Physical Laboratory (NPL) started the development of the *Pilot ACE* based on Turing's design of the Automatic Computing Engine (ACE). The Pilot ACE went into service in late 1951. Lacking hardwired capabilities for multiplication and division, Pilot ACE started out using fixed-point multiplication and division implemented as subroutines aka 'software.'

That fixed-point arithmetic, however, proved problematic was soon replaced by floating-point arithmetic.

Also the Cambridge University by the 1930s had seen their scientist pay attention to developments in the world of mathematical computing. Resulting in designs such as the *Mallock Machine* (1933) (Figure 161, lower middle). Next, inspired by Von Neuman's reports on

**Figure 161: Early British Computing Machines.**

Replica of the Small Scale Experimental Machine aka Manchester Baby (?) (top, 1948). Manchester Mark I (upper middle, 1949); Cambridge's Mallock Machine (lower middle, 1933), Cambridge's Electronic Delay Storage Automatic Calculator (EDSAC) (bottom, 1948)

Source: Wikimedia Commons; https://history-computer.com/; http://curation.cs.manchester.ac.uk/; SSPL.

computer architecture as well as the US developments around the ENIAC, at the Cambridge Mathematical laboratory the *Electronic Delay Storage Automatic Calculator* (EDSAC) was proposed in 1947. And subsequently built with vacuum tubes in 1947-1949 (Figure 161, bottom).

### The ENIAC in America.

Between 1927 and 1943, the mathematician and scientist Vannevar E. Bush (1890-1974), working at the Massachusetts Institute of Technology (MIT), had developed a series of electromechanical analogue computers which greatly facilitated the solution of complex mathematical problems. In 1931 he created the mechanical *Differential Analyzer* (Figure 163), a machine that was built from electrical components (2000 vacuum tubes, thousands of relays) and mechanical components (including

**Figure 162: The IAS Computers.**
The IAS computer (top, 1951), and the IBM 701 development version (bottom, 1952).
Source: IAS Edu, IBM.

150 motors, shafts, and gears) that filled a large room. An even more successful machine, the so-called *Rockefeller Differential Analyzer* (DA) — funded in part by the Rockefeller Foundation— was built in 1935, operational in 1942, and proved the most powerful computer available before the arrival of digital computers about 1945. In 1936 he published a paper 'Instrumental Analysis,' and immediately after he delivered his 1936 paper, Vannevar Bush started to work on the design of an electronic digital computer. He conceptualized an information storing and data retrieving machine —his later Memex (ie memory extender)— and published in 1945 the essay 'As we May Think.' This essay influenced generations of computer scientists, who drew inspiration from his vision of the future.

In 1942, the physicist John Mauchly (1907-1980) and the electrical engineer J.Preper Eckert (1919-1995), were both working at the Moore School of Electrical Engineering (University of Pennsylvania), a centre of wartime computing. The U.S. Army, meanwhile, needed to calculate complex wartime ballistics tables. In order to solve these calculations, they proposed in 1942 an all-electronic calculating machine. In April 1943, the Army contracted with the Moore School to build the *Electronic Numerical*

*Integrator and Calculator* (ENIAC) that was completed in 1945 (Figure 163, bottom). The machine was the first general purpose electronic digital computer. Like Charles Babbage's Analytical Engine (from the 19th Century) and the British World War II-computer Colossus, it had conditional branching; that is, it could execute different instructions or alter the order of execution of instructions based on the value of some data.

The ENIAC, considered the grandfather of digital computers, was enormous as it filled a 20- foot by 40-foot room and had 18,000 vacuum tubes, 70,000 resistors, 10,000 capacitors, 6,000 switches, and 1,500 relays. Completed by February 1946, ENIAC had cost the government $400,000, and the war it was designed to help win was over.

**Figure 163: Early US Computing Machines.**
Bush Differential Analyzer (top, 1931) and the ENIAC (bottom, 1945).

Source: Wikimedia Commons. Computerhistory.org; https://time.graphics/

**Figure 164: IBM Mainframes 700-series (1950s).**
Computer system: console, CPU- and tape storage cabinets.

Source: Wikimedia Commons. IBM.

US-Patent № 3,120,606 (filed on June 26th, 1947) was granted on February 4th, 1964 to Mauchly and Eckart for the ENIAC. But in the court cases of Sperry Rand Corporation versus Honeywell Corporation it was voided in 1973 as the invention of the electronic digital computer was already in the public domain.

After the Second World War had ended, the development trajectory continued. Eckert and Mauchly started a second computer project, the *Electronic Discrete Variable Automatic*

*Computer* (EDVAC) by 1945. It was ready in 1949, but became only operational in 1951. In Princeton, New Yersey, at the Institute for Advanced Study (IAS), had started the *Electronic Computer Project*.

The IAS Machine would become the origin for a range of 'IAS Machines' made in Europe (eg Zweden, Germany, Britain) and the US. In the US, IBM developed the IBM 701 in 1952: their first commercial scientific computer and its first series production mainframe computer (Figure 162). This was the start of IBM 700-series that evolved over more than a decade; the IBM 702 (1954) was followed by the IBM Model 704 and IBM 705 (1954), all based on vacuum tubes.

**Figure 165: Univac I (1951).**
Overview of Univac I (top), the Univac I internal view (middle), and recirculation board with vacuum tubes (bottom)
Source: Deutsches Museum (top), Wikimedia Commons.

Other companies followed suit, responding to the government needs. The emergence of the Cold War accelerated the government's growing awareness of the significance of digital computing and drove —funded by the Department of Defence (DOD)— major computer development projects in the 1950s.

> *As the 1950s unfolded, 14 US companies were developing electronic computers with help from the government. The Cold War prompted US President Harry S. Truman's administration into pumping money into technology in hopes of maintaining an advantage. Some of that money went to university labs. Some went to start-up companies like Engineering Research Associates. And some went to ENIAC developers J. Presper Eckert and John Mauchly, who found a backer for their business in Remington Rand. The duo built a commercial successor to their ENIAC, dubbed UNIVAC (Figure 165)—the first electronic computer to win the hearts of the new generation of scientists who adored the high-speed calculations these computers made possible. Under pressure from UNIVAC, IBM engineers felt that their pride—and possibly an important part of IBM's future—was at stake. The culture responded in a surge of esprit de c orps and created a system geared for speed: the IBM 701 Defense Calculator.*[349]

---

[349] Source: The IBM 700 Series: Computing Comes to Business. https://www.ibm.com/ibm/history/ibm100/us/en/icons/ibm700series/ (Accessed August 2021)

## Computing Machines 2.0 (1950-1980)

The preceding analysis showed the events that led up to dedicated computing machines up to the 1950s. Computing machines designed on an individual basis in a research environment of knowledge users. Some of them being sold by commercial organisations for critical applications like bulk data processing. But that commercial actively was mostly on a limited scale, not in the last place due to its price. The driving force had been the new vacuum tube technology, replacing the relay-based computing machines. A technology that made it possible to develop more reliable computers after the governmental spurred technology transfer from research institutions to industry. Such as a company called International Business Machines (IBM).

### The Era of the Mainframe Computer[350]

Jumping to the afterwar period, the 1950s saw IBM going into 'computing.' Partnering the United States government's drive to computational innovation, IBM got access to the pioneering research done at the Massachusetts Institute of Technology (MIT).

The government stimulated academic partnerships, combined with pioneering computer technology research[351] and a series of commercially successful products (the first generation of IBM's 700 series, and the IBM 1400-series, of vacuum-tube based computer systems) enabled IBM to emerge from the 1950s as the world's leading technology firm (Figure 166).

Figure 166: IBM Computers Family Tree (1950-1960s).
Source: IBM Corporate Archive.

---

[350] The history of the origins of the Mainframe and Minicomputer systems will be covered in detail in the companion to this case study: The Invention of the Computing Engines.
[351] IBM created its first laboratory in 1945; the Watson Scientific Computing Laboratory as a department of the Columbia University.

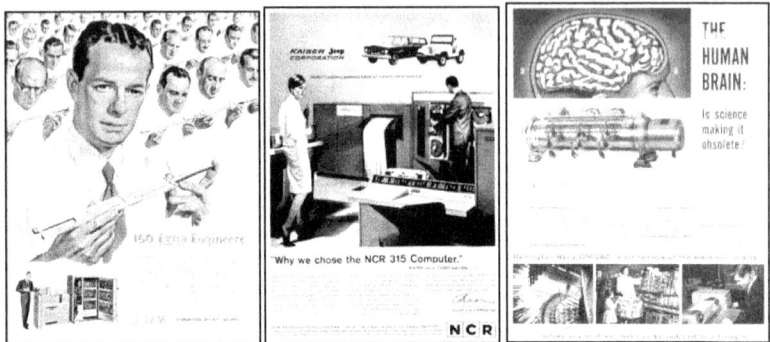

**Figure 167: Advertisement for Mainframe Computers.**
Affiche by IBM (left, 1950s), NCR (middle, 1960s) and Univac (right, 1950s).
Source: Wikimedia Commons.

IBM was not the only company that went into computer business; others like UNIVAC, Burroughs, NCR, Control Data, Honeywell, General Electric and RCA. Those commercial organisations —known as 'Snow White (ie IBM) and the Seven Dwarfs'— soon discovered that their machines (aka mainframes) could also be used by large organizations. So, they adapted their machine to a non-scientific environment. A move that reflected in their advertisement (Figure 167).

The technology of vacuum-tube based logic would become the dominant technology used for the development of a range of first-generation computers in the 1950s. They would be replaced by the second-generation computers based on the new semiconductor technology creating the transistor used in the Transistor-Transistor Logic (ie TTL) of the 1960s. It would become the next phase in the Computing Revolution that started with Era of the Minicomputers.

## The Era of the Minicomputers.

By the mid-1960s a smaller type of computer was designed on the basis of transistor technology. Although called mini-computer, it was not that small at all. Minicomputers were distinguished from the larger 'mainframes' by price,[352] function, size, use, and marketing methods. But their dimensions were still considerable (Figure 168). However, by 1968 they

---

[352] A remarkable difference was the price for an installation. While mainframes cost up to $1,000,000 ($8,630,000 in real value 2019$), minicomputers cost well under $100,000 ($863,000 in real value 2019$). Source: https://www.measuringworth.com/calculators/uscompare/relativevalue.php.

formed a new class of computers, representing the smallest general-purpose computers at that time. In addition, where mainframes required specialized conditioned rooms and computer technicians for operation, the minicomputers were operated by the —rather knowledgeable— users themselves from a desk-terminal. Although the minicomputer was created for technical computing (real-time process control, data acquisition, simulation, and interactive use), soon commercial computing (data entry, word processing, spread sheets) became significant applications. As did the rise of the early computer networks.

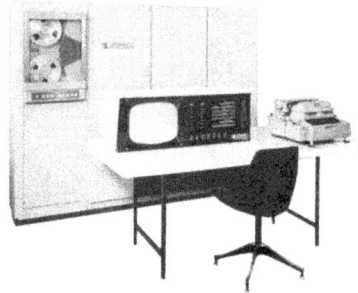

Between 1956 and 1964 groups of scientists at universities, laboratories, and entrepreneurial companies around MIT[353] in the Boston area (aka the Route 128 region) began to exploit the advantages of transistor circuitry and magnetic core memory to meet demands for computer processing by smaller organizations. They designed, assembled, programmed, and sold smaller computers that cost an order of magnitude less than traditional systems. As the success of the other industry along Route 128 led to a surplus of risk-capital,[354] while government contracts for aerospace and defence kept coming, Boston's Route 128 region became the hotbed of new companies manufacturing mini-computers.

**Figure 168: Early Minicomputers.**
DEC PDP-1 (top, 1960). Control Data Corporation Model 160A (bottom, 1960).
Source: Wikimedia Commons. Computer History Museum.

In 1960 companies like Control Data Corporation (CDC), Packard Bell, Wang Laboratories, and Digital Equipment Corporation (DEC) all introduced small-sized 'mini-computers' using

---

[353] MIT played a pioneering role in stimulating 'academic' entrepreneurship and technology transfer. The so-called *Research Row*, with the laboratories of MIT, Harvard University and industrial laboratories located to each other almost within walking distance, formed an unparalleled intellectual and technological labour pool at the end of the Second World War.
[354] Many companies were funded with capital supplied from American Research & Development (ARD)

discrete transistors for off-line processing, general computation, and real time process control (eg in power plants, chemical plants). Many were sold directly to original equipment manufacturers (OEMs) who made the systems for final end use application (eg process control in the oil-industry). [355]

Although the minicomputer was primarily created for technical computing (ie real-time process control, data acquisition, simulation, and interactive use), commercial/ business computing (data entry, book keeping, word processing, spread sheets) also became significant.

By 1965 —the beginning of the commercial minicomputer era—about a dozen start-up companies and existing companies were building computers for the control of manufacturing processes. By 1968 a total of 49 independent start-up companies and 10 merged start-up companies entered the new industry. And between 1968 and 1972 there were some hundred entrants to the market. Half of the companies were created by individuals.

As the IC-technology progressed and the chips with TTL Logic were developed, the minicomputer became even smaller. Even more when the more complex circuits like the Control Unit (CU) and Arithmetic Logic Unit were used to build single-board Central Processing Units (CPU) (Figure 209). As did DEC when it introduced its highly successful model PDP-11 (Figure 169).

However, some two decades later, as of 1983, about 53% of all entrants had failed and no longer existed. By then the oligopoly in minicomputers consisted of four firms (ie DEC, IBM, Data General and Prime).

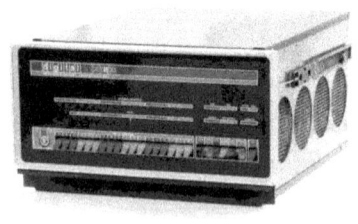

**Figure 169: The Minicomputer becomes smaller.**

DEC PDP-11 (top, 1970).

Source: Computer History Museum. Wikimedia Commons.

Only 25 percent of the entrants had success in different degrees, 75 percent did not build organizations with any longevity for a variety of reasons, including failure in engineering, failure in marketing, faulty manufacturing, or insufficient product depth or breadth. In the fast growing mini-computer market, keeping up with the technological development spurred by the IC-technology, was a crucial success factor that was hard to master.

---

[355] Source: Rise and Fall of Minicomputers.
https://ethw.org/Rise_and_Fall_of_Minicomputers. (Accessed September 2020).

By 1985 the minicomputer market collapsed due to a multiple set of reasons. One of them being the rise of the micro-computer-sets developed by the chip manufacturers. Chip sets that became incorporated in the personal computer and its box-systems. Applying the micro-computer on a chip in combination of cheap semiconductor memory, together with the concept of 'open standards'[356] and the rise of standard application software, put an axe at the roots of the minicomputer industry.

Both the Era of the Mainframes and the Era of the Minicomputers were paved with massive technological novelty. Not only the new technology of vacuum tube logic replacing the relay logic, also new architectures changing the stored program computer into the *general purpose computer*.[357] And new circuitry realizing increasing complex functionality. And all that novelty was accompanied by the inventors (and their companies) wanting to protect their intellectual property. So, they patented their contributions to the development of computing.

## Patents for Computing Machines

The patents for 'computing machines' —in analogy with the 'calculating machines'— applied for and granted, were numerous during the 1900-1980 period covering a broad range of inventions. Looking at patents titled with

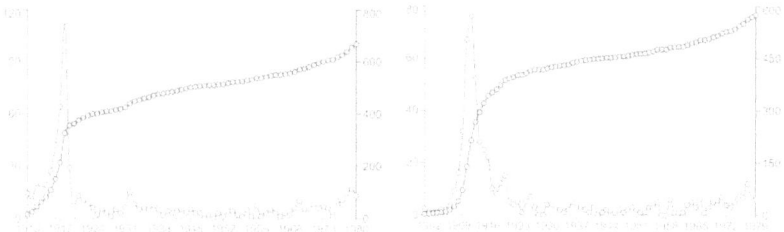

**Figure 170: Overview Patents for the Computing machines by publication date (left) and by priority date (right).**
Tables show yearly and accumulated number of patents. Left axis is the number of patents in a given year, Right axis the accumulated volume of patents.

Query (top): (ti all "machine" AND ta all "Computing") AND pd = "1910-1980".

Source: Espacenet

---

[356] An *open standard* is a standard that is freely available to the general public for adoption, implementation and updates of hardware and software. In the case of the personal computer the availability of detailed hardware specifications it gave rise to third party development of add-on hardware modules.

[357] A stored program computer is a hardwired computer designed to perform one (set of) task(s). A general purpose computer is a computer that is designed to be able to carry out many different tasks.

the word 'Computing Machine,' in the early period the granting saw a peak in 1918, just after the First World War (Figure 170, left). The filing of many of those patents however had already taken place in the period before the war (Figure 170, right), indicating that the activitities were already prewar.

This may be indicating a disruption in the patent adminstration. Between 1914 and 1918, the annual average number of patent applications fell in certain countries (France, Germany, Austria) by 40% in comparison to the years between 1910-1913. Surely, the conflict must have affected the work of national offices for intellectual property, but the belligerents also suspended the granting

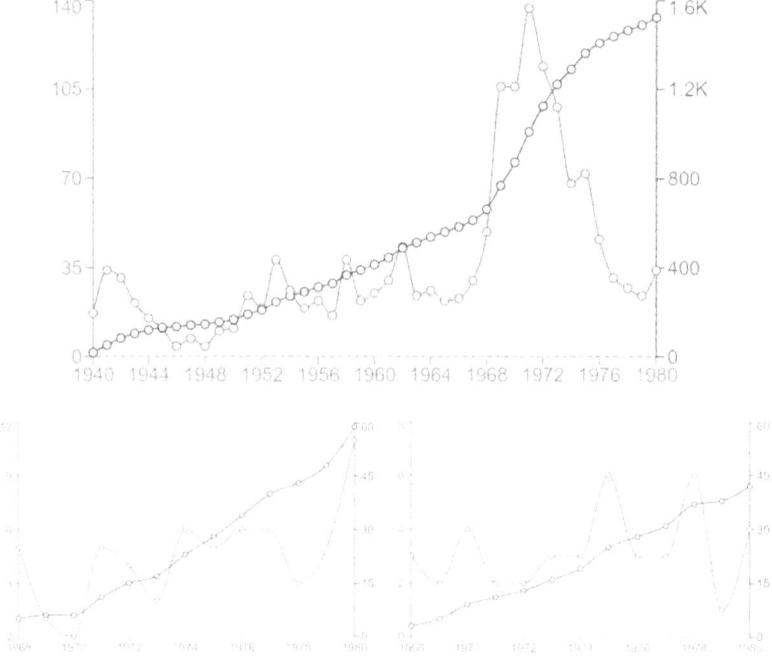

**Figure 171: Overview Patents for the Computing machines by publication date (1940-1980) and Patents issued to DEC and CDC.**

Tables show yearly and accumulated number of patents. Left axis is the number of patents in a given year, Right axis the accumulated volume of patents.

Query (top): pd = "1940-1980" AND ipc any "G06C". Query (bottom left): pn any "us" AND pd = "1950-1980" AND pa = "digital equipment corp" AND ipc any "g06F". Query (bottom right): pn any "us" AND pd = "1950-1980" AND pa = "control data corp" AND ipc any "g06F". Note: This IPC classification became effective in 1968, hence the 1968 date as marking point.

Source: Espacenet

of patents relating to national defence, or patents that were requested by nationals from enemy countries.

Nevertheless, it is clear that with the inventive activity related to computing machines —the precursor of the computer— the subject of more advanced calculating machines was already on the agenda before the First World War. This is in line with the patent explosion for calculating machines notable during the 1900-1920 period (Figure 149).

> When we consider the later patenting of the machines labelled as 'computers' (ie around and after the Second World War the mainframes and the mini-computers) and patents granted worldwide under the classification G06F15/00 (ie digital computer in general), the 1968-1974 period showed a new peak (Figure 171, top).

This was the Era of the Mainframe and the Era of the Minicomputer. In the first, and the mainframe companies contributed heavily: IBM (96 patents), Honeywell (35), Burroughs (26) General Electric (22), and General Electric (22). Among the second DEC contributed —for a broader classification area (G06F)— some 55 patents (Figure 171, bottom left) and the Control Data Corporation some 42 patents (Figure 171, bottom right).

*Pioneering Patents and Basic Innovations*

In the world of patenting inventions, some patents were more important than others, due to its nature and the impact their invention had. Its pioneering nature could be in a new field of technology (eg electricity) or application (eg telephony). Combining the patent's nature with the impact of the invention (Scheme 1) —both the technical impact and the economic impact— is an indication of the importance of the invention.

The First Computing Revolution was characterized by several milestones, some of a single computing engine, some for a specific type of computing engines (ie those realized in a specific technology). When one looks at the patents issued —or voided— in the twentieth century for computing machines the following developments are notable for their impact (Figure 175).

*The mainframe ENIAC:* The development of the vacuum-tube based ENIAC in 1945 was a landmark in the development of early computers. The cluster of innovations around the ENIAC (Figure 173) constitutes with its contributing and derived innovation a basic innovation. Its computing power was a massive leap in comparison to its peers at the time, and its general purpose nature led the way for computers being re-programmable for any number of potential use cases.

As its design was in the public domain, the patent that was filed for its invention and granted as US-Patent № 3,120,606 on February 4th, 1964, was voided in 1973.

*The IBM 700 Mainframe series:* IBM entered the computer era with the IBM Model 701 in 1952 (Figure 166); their first general-purpose computer, built with vacuum tubes. It became the first of the 700-series when in 1956, a significant upgrade to the 701 appeared: the IBM 704 and IBM 705. IBM dominated the early mainframe market for the decade with its 700 series. As a result from its massive R&D effort, IBM obtained over time more patents than any other tech-company. During the period 1930-1950 it obtained some 2,281 US patents for its improvement to mechanical and electronic systems.

This 700/7000 series were replaced in 1964 with the introduction of the IBM 360 series; a family of computers designed to cover the complete range of applications, from small to large, both commercial and scientific. It replaced all five preceding computer lines with one strictly compatible family, using a new architecture that pioneered the eight-bit 'byte' still in use on every computer today.

*The DEC PDP Minicomputer series:* With the rise of the transistor technology the Digital Equipment Corporation (DEC) created a new concept of computer use and called it the Programmed Data Processor (PDP). From the first PDP-1 model (1960) through the PDP-8 (1965) and PDP-11 series (1972) these minicomputers dominated the market. In total, more than 600,000 (1970-1990) PDP-11's were sold and it became the workhorse of the OEM market.

DEC did obtain over time a range of patents for parts of the system (eg the bus architecture, the instruction set), but a basic patent for the whole concept of the minicomputer was never applied for. However, design patents for the VT-52 video terminal, the PDP-11 system console and the PDP 11 cabinets were obtained.

These events marked a specific moment important in the development of the computer. Although they were different in nature, they had in common the impact on the further technical, application and business development. So, one can conclude that there is no single inventor of *the* computer, as so many contributed to the functional parts of the computer.

## *The First Computing Revolution (1930-1980)*

In the preceding analysis we explored the multitude of events that were part of the development trajectory of the 'computing' machines. We observed the rise of the different machine-concepts that took over a part of the mental 'computing' activity in the human brain. Not only the arithmetic activity —such as addition, subtraction, multiplication and division— and the more complex mathematical functions —such as the logarithms, rooting, trigonometric functions—, but also the logic functions of conjunction (AND), disjunction (OR), and negation (NOT). Logic functions that could make logic decisions in the process of calculation.

The result was the *stored program* computing machine. Based on the mathematical logic concepts that had developed over time, after the calculating machines grew into prominence, the 'Computing Machines' emerged. The machine that merged basic arithmetic functions of the Art of Calculus with the logic functions of the Art of Logic.

*The Invention of the Dedicated Computing Machine:* The Era of Modern Computing began with a flurry of developments before and during the Second World War. In the 1930s early electro-mechanical computing machines (later known as 'induction computers'), were using relays as logic elements. Machines, such as those designed in Germany (eg the Z-computer by Konrad Zuse), and the US (eg the Complex Number Calculator by George Stibitz). Their purpose was to realize complex and repetitive calculations (eg the trajectories of shells).[358] After the new technology of the electronic vacuum tube had matured in the 1920s, the vacuum tube also became used as an active element in logic circuits. Soon they were applied to construct computing machines. Such as the Atanasoff–Berry Computer (ABC) in 1939, and machines that resulted from Howard Aiken's proposal titled 'Proposed Automatic Calculating Machine.'[359] A proposal that became the basis for the ASCC (Automatic Sequence Controlled Calculator) resulting from agreement between Harvard University and IBM in 1939.

*The Universal Computing Machine:*[360] The early enormous computing machines were basically all 'hardwired' machines designed to perform a specific function. They performed fixed calculations, but lacked —what would later become known as— general purpose computing capabilities (ie being able to execute general purpose tasks). The concept changed when the English mathematician Alan M. Turing (1912-1954) published a

---

[358] The calculations themselves are straightforward enough: a complex multiplication requires about six simple arithmetic operations, while complex division requires about a dozen operations, and each requires temporary storage of a few intermediate results.
[359] The machine that resulted from this proposal —originally called the ASCC (Automatic Sequence Controlled Calculator)— would become known as IBM's Harvard Mark I, introduced in January 1943 and operational in March 1944.
[360] A Turing Machine was a specific mechanical device that could carry out some specific tasks in a systematic way. Each Turing Machine would work in a similar manner, using mechanisms related to the computer concepts of input, output and a program.

theoretical paper 'On Computable Numbers' in 1936. He proved that a computing machine would be capable of performing any conceivable mathematical computation if those computations were representable as an algorithm. With his concept of the *Universal Turing Machine* he introduced the notion of a 'universal machine.'

## The Machines that changed R&D and Business.

After the new technology of the electronic vacuum tube had matured in the 1920s, the vacuum tube also became used as an active element in logic circuits. Soon they were applied to construct computing machines. Such as the Atanasoff–Berry Computer (ABC) in 1939, and machines that resulted from Howard Aiken's proposal titled 'Proposed Automatic Calculating Machine.' [361] A proposal that became the basis for the ASCC (Automatic Sequence Controlled Calculator) resulting from agreement between Harvard University and IBM in 1939. [362]

Next to the workers of the Invention Rooms —the later R&D departments like the *Inventions Department* at National Cash Register

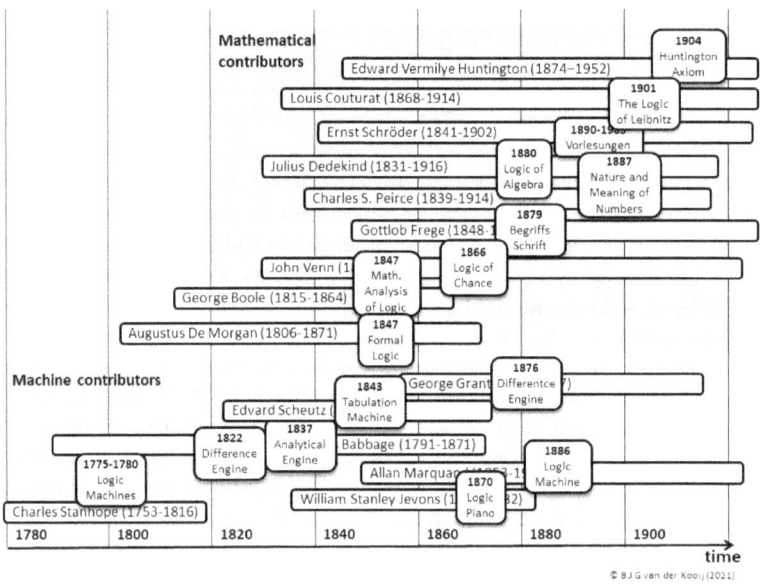

Figure 172: Overview of Contributors to the Early Development of Binary Logic and Logic Machines.

---

[361] The machine that resulted from this proposal —originally called the ASCC (Automatic Sequence Controlled Calculator)— would become known as IBM's Harvard Mark I, introduced in January 1943 and operational in March 1944.
[362] An *algorithm* is a finite sequence of well-defined, computer-implementable instructions, typically to solve a class of specific problems or to perform a computation.

Company — over time also the corporate worker started using the new tools called 'mainframes' that were maintained by the (male and female) technicians of the corporate *Computer Departments*.

## Manufacturing Early Computing Machines

The increasing use of mainframes gave rise to Mainframe industry. An industry that became known as 'Snow White and the Seven Dwarfs' (Figure 174, bottom). It sold their massive computers to the large companies that could afford their (investment- and maintenance-) costs. They became used for massive batch processing in banking and insurance, and in the military institutions involved in the Cold War and Space Race (eg by the NASA). The corresponding promotional activities of the industry emphasized the additional functionality of the new phenomenon of computing (Figure 167).

That use changed with the rise of the Minicomputer and the explosive rise of the Minicomputer Industry (Figure 174, top). The minicomputer found its application in the industry —the new application field of Industrial Automation—, so also the factory worker experienced the new computing machines. The lower priced minicomputer systems and the smart terminals —often part of a local network— were also bought by business for administrative purposes, so also the office worker experienced the new computing machines. In other words; computing left the ivory towers of the specialists, spread its wings into society and became part of the Affairs of Man. That is, the affairs related to industrial manufacturing and the office's organizational work.

## Periodizing the First Computing Revolution

Similar to the development trajectory of calculating machines, the development of (electro-)mechanical computing machine —labelled as the First Computer Revolution— can be periodized (Scheme 5).

*Computing 0.0* (1780s-1930s): A period that covers the early conceptual contributions of Charles Stanhope (1753-1816) and Charles Babbage (1791-1871) up to the Difference Engine of George Grant (1849-1917), as well as the contribution of the Mathematicians of Logic (Figure 172). In the mechanical technologies available in those days the scholars tried to convert their ideas into a mechanical reality. But their concepts were too complicated for the technology (in Babbage's case with the Difference Engine and Analytical Engine), or they had to limit their constructs to the creation of Differential Machines that the technology was able to make (such as in case of Jevons' Piano and Marquand's Logic Machine).

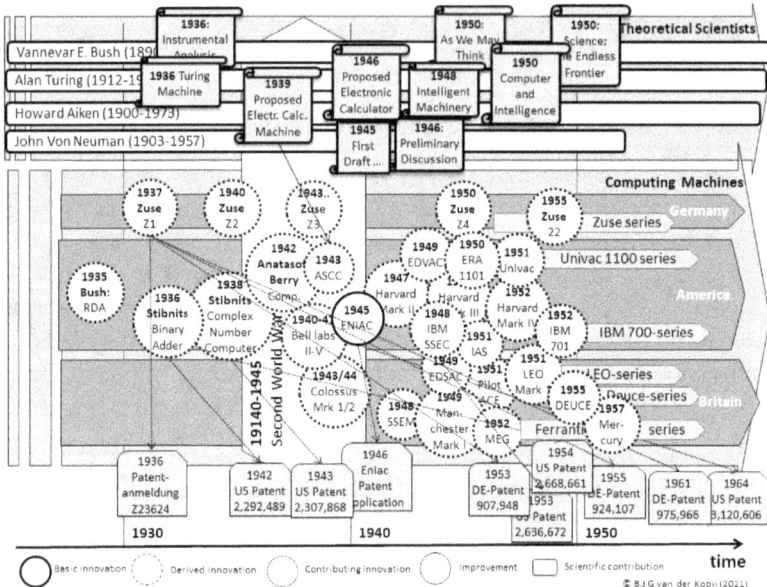

**Figure 173: Overview of Contributors to the Dedicated Computing Machines (1930-1960).**

The period of 1900-1930 was the time in which the calculating machines rose to prominence (Figure 156) and the mass market for calculating machines opened up. It was also the period around the First World War (1914-1918) and the Wall Street Crash (1929) followed by the Great Depression (1930s). The subsequent economic malaise had its effect on scientific budgets, curtailing research projects. In addition, as the 1920s were a period of drastic changes —aka the Roaring Twenties—, societies reacted with a range of popular and fundamentalists movements. Among which the Anti-Science movement 'Stop Science' in Britain that heralded a considerable public antipathy towards science.

*Computing 1.0* (1930s-1950s): A period that saw the rapid rise of computing machines for complex scientific calculation. Scientific computation that was applied at universities and research institutes in support of the war effort. After the limitations of the mechanical technologies had been surpassed by the electromechanical (ie relays) and electronic devices (ie vacuum tube), the concept of the stored program machines was replaced by the programmable computing machines. In both the US and Europe (Germany and Britain) a range of dedicated computing machines were developed at scientific institutions (Figure 173). It was the dawn of the commercial manufacturers of mainframe general purpose computers.

*Computing 2.0* (1950-1980): After the Second World War the experience build up in designing and applying the mastodonts of computing (eg the ENIAC), heralded the times of the General Purpose computers that became known as the Mainframe computers. A period after 1950 dominated by IBM and 'the seven dwarfs' (Figure 174, bottom) serving the large corporations that could effort the large investments required. Mainframes that were operated by specialized operators in the inhouse computer department. But also a period that saw in the 1960s the rise of the Mini-computers (Figure 174, top) opening up —due to their lower price level and less specialized operation— new markets. Industrial markets where the minicomputers were applied to manufacturing processes (ie the process control applications of the chemical industry), and business markets where minicomputers were used for transaction processing, file handling and database management, and bookkeeping applications. However, the rise of the Personal Computing System (eg the introduction of the IBM PC in 1981), ignited the downturn of mainframe and mini-computing industry.

*Computing 3.0* (1980-2000): After the semiconductor technology matured and increasingly complex computing systems became integrated on a chip(set), the Personal Computer emerged. This computing system heralded a new era of computing where, next to business use, also the personal use opened massive markets.

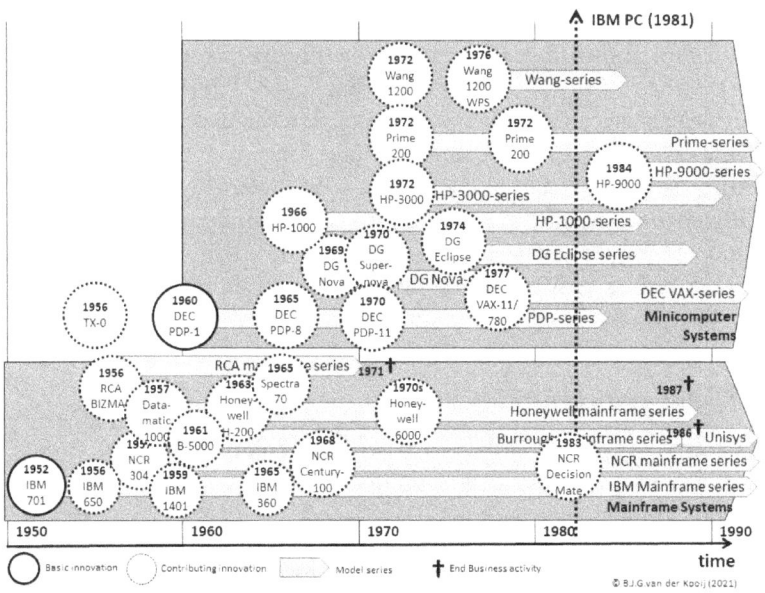

Figure 174: **Overview Computer models manufactured by the US Mainframe and Mini-computer Industry (1950-1980).**

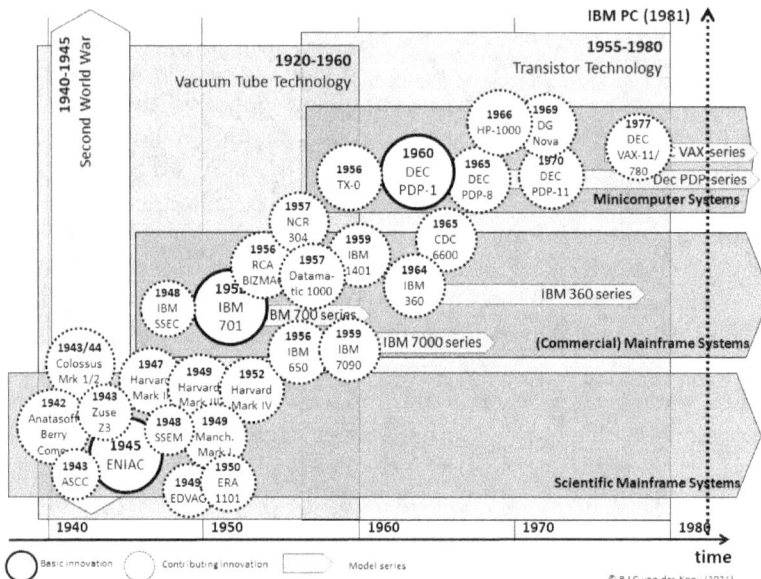

Figure 175: Overview Contributors to the US Mainframe and Mini-computers by Technology.

To illustrate the importance of the technological development in computing machines, the events in the 1930-1980 period that created the mainframe and mini-computers can also be interpreted from a different perspective; the perspective of the electronic technologies that emerged after the electro-mechanical technologies. Then, the development of the mainframe computers and the minicomputer was strongly influence by the progressing electronic technology (Figure 175). After the vacuum tube technology had been used to create logic systems, the transistor technology revolutionized logic circuits and computer design. First by using the active component of the transistor as discrete component in the electronic circuitry, later by the integrated modules of transistors aka Transistor-Transistor Logic (TTL).

*The 'I' of Information in ICT*

This computer periodization illustrates the evolutionary development of the major General Purpose Engines (GPEs) of the *Information Technologies* in the application trajectories of the GPT-Electricity (Figure 176). After the Calculation Technology that realised the Calculating Machines with their increasing functionality, it was the Computer Technology that created the Computing Machines. From the hardwired computer with their fixed (ie hardwired) functionality, to the 'general purpose' computer with its flexible

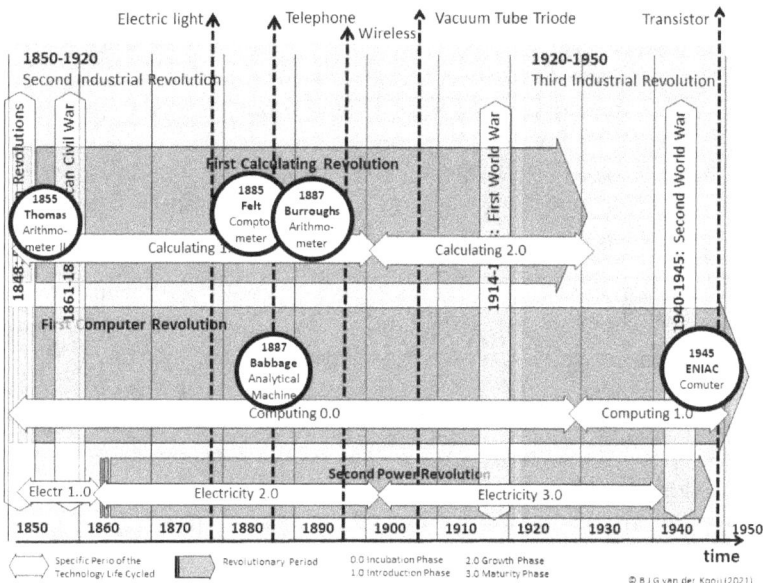

Figure 176: Overview of the Contributions to the Calculating and Computing Technologies (ie Information Technologies).

functionality contained in their programs (aka 'software'). Their combination created the Information Technologies that would become so important in later times. Together with the Communication Technologies we met before, it would become part of the General Purpose Technology of the Information and Communication Technology (GPT-ICT).

## *The IC & IT-Revolution (1850-1950)*

In the preceding exploration we identified the events during both the *Era of Communication Machines* (Figure 107), and the *Era of the Calculating and Computing Machines* (Figure 176). The first periodization focussed on the technical extension of human communication —both as the coded word and the spoken word— over distances. The second periodization was about information processing that originally belonged as an exclusive capability to the human brain. Historiography[363] labelled these developments as the processing-based Information Technologies (IT) and the transmission-based Communication Technologies (IC).

---

[363] Historiography is the study of the various approaches to historical method, the actual writing of history, and, primarily, the various interpretations of historical events.

*Era of Communication Machines:* The first periodization (Figure 177, lower half) had started with the invention of telegraphy as means to communicate over distance; it became the transmission of the 'coded word' (ie Morse code). Some decades later followed by the 'spoken word' when Bell invented the telephone. Both 'cabled' communication systems, that were complemented by the point-to-point 'wireless' telegraph and telephone systems, as well as the broadcasting systems of radio and television transmitting sound and video.

*Era of the Calculating and Computing Machines:* What both contributors to the second periodization (Figure 177, upper half) had in common, was that they were about taking over part of human intelligent activity with mathematical machines. The calculating machines performing arithmetic functions like addition, subtraction, multiplication and division; the computing machines performing logic functions of conjunction (AND), disjunction (OR), and negation (NOT). They both performed those operations on numerical data (ie numbers). And both types of information processing machines were realised in mechanical technologies.

In this exploration, we observed the events predominantly from a technological point of view, incorporating their impact. For example, in the form of the related business events. Defining technology as 'knowing how to make things,' we identified the major events (labelled as basic

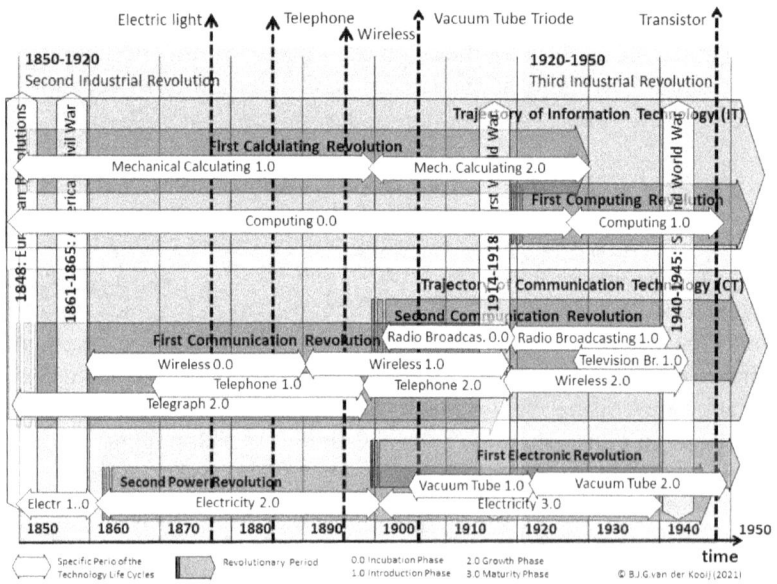

Figure 177: Overview of the CT & IT-Revolution (1850-1950).

innovations) of the Information Technologies (IT) and Communication Technologies (CT), their contributing events and the derived events that created the Clusters of Innovation (Scheme 1, Scheme 2). Among those many events we tried to identify both the evolutionary as well as the revolutionary events (Scheme 3, Scheme 4). And, to structure the events in the technological trajectories, we applied the periodization of the Technology Life Cycle Concept (TLC) (Scheme 5). After all these explorations and their explanations, we were able to place the periodized events in relation to each other in their specific time frames labelled as Industrial Revolutions, shaping up the Eras they covered (Figure 177).

Using the construct of the General Purpose Technology (GPT) enabled us to define the technologies that constitute the GPT-ICT (Scheme 6). Each of these technologies had their own technological development trajectory in the 1850-1950 period. This period encompassed two Industrial Revolutions; the Second Industrial Revolution (1850-1920) and the Third Industrial Revolution (1920-1950) both dominated by the industrialization of the manufacturing process, and the mechanization of society (eg automobile, airplane).

So, to summarize, these technological events in the 1850-1950 period can be divided into the *contributing* GPT-IT and the *contributing* GPT-CT (Scheme 13).

*General Purpose Technology of the Information Technologies* (GPT-IT): Considering the events during the Calculating Revolution and the Computing Revolution, their common factor of the technologies is 'the processing of information' by electric means. Information being the data represented by numerical systems, and the processing being the arithmetic and logic operations executed on those data.

*General Purpose Technology of the Communication Technologies* (GPT-CT): Considering the events during the Communication Revolution with the Telegraph, Telephone, and Wireless, the dominant common factor of the technologies is 'the transmission of information' by electric means. Information being the coded word as well as the spoken word.

## *The dawn of the Fourth Industrial Revolution*

Both the Second Industrial Revolution and Third Industrial Revolution were the times of mechanical technologies applied on a large scale. As a result of the Mechanical Worldview, both originated from the Scientific Revolution whereafter Science en Engineering focussed on mechanical machines (see Chapter: Context for the discoveries). Both the resulting GPT-IC and GPT-IT were based on (electro-)mechanical technologies.

Technologies that were enabling as well as limiting their development. After the pure mechanical technologies had reached the limit of their usability, gradually the electro-mechanical technologies took over. But they also had their limitations.

> The common factor in these later developments is the *second property* of 'electricity.' After its first application in power-systems as a carrier of energy, electricity became used as a carrier of data in information systems. Systems that at first were realized in electro-mechanical technologies of the electromagnet (ie the relay) in the telegraph and in the computer. As a result of this property, emerged the *IC and IT-Revolution (1850-1950)* based on the (electro)-mechanical technologies (Figure 177).

Had the natural philosophers explored the *Nature of Matter*, now their successors, the physicists, started to unravel the *Properties of Matter*. From the Classic Physics to the Solid-state Physics, concepts, theories and models developed by a range of theoretical physicists created insight in the properties of matter. From the emission of electrons (eg thermionic emission in the vacuum tube) to the way solid-state materials were conducting electrons. The Experimental Engineers took it from there, and contributed to the subsequent development of the electronic technologies. There the vacuum tube technology, and later the semiconductor technology, contributed in realizing their respective General Purposes Engines of the 'radio tube,' 'transistor' and 'integrated circuit.' And, after the technologists had created the foundations, the circuit builders took over, creating the new versions of the communication and computing machines. Establishing a period that became known as the ICT-Revolution (1950-2000).

In earlier explorations we limited the Third Industrial Revolution to the first half of the twentieth century (1900-1950).[364] Seen from the perspective of the use of power, after the First Industrial Revolution powered by steam, the Second Industrial Revolution powered by the first property of Electricity (ie the transmission of energy), the Third Industrial Revolution became powered by the Internal Combustion Engine. Continuing that perspective, the Fourth Industrial Revolution would be based on nuclear power. The construct of these Industrial Revolutions focussed on the effects of industrialization.

However, the second half of the twentieth century (1950-2000) was characterized by a feature originating from another property of electricity; the transmission of information. Already sprouting in the nineteenth

---

[364] See: Van der Kooij, B.J.G.: *The Invention of the Internal Combustion Engine.* (2021)

century with the development of the communication technologies enabling telegraphy and telephony, the calculating and computing technologies added the dimension of information processing to the period. Information processing in a broad field of application trajectories; from broadcasting information to arithmetic processing (ie calculating) and data processing. Seen in that perspective the construct of the Information Revolution would seem appropriate to label the effects of the Silicon Engine in the second half of the century.

Historians and scholars struggle with this distinction and apply the notion of Industrial Revolution as well as the Information Revolution to the twentieth century. Zooming in on industrialization, some label the whole twentieth century as an industrial revolution.[365] Others focus on the second half of the century, labelling it as the third or fourth Industrial Revolution.[366] Sure, manufacturing changed dramatically with the introduction of robotics, creating the automated factory at the last quart of the twentieth century. But the driving force was not the mechanical technologies creating the machines, but the information technologies creating the intelligent systems of production.

To summarize, in the preceding analysis we explored, zooming in on the Communication Revolution, the Calculating Revolution and the Computing Revolution, how the mechanical and electro-mechanical technologies developed up to the mid-twentieth century. Technologies that had created the telegraph and telephone, and that had created the early computing engines known as mainframe computers. All in all, it was technology that created communication machines and computing machines that already showed their profound influence on the Affairs of Man.

This exploration also illustrates that those early contributors contemplated how human thinking could be artificially mechanized and manipulated by intelligent non-human machines. From the calculating machine and logic machines to the hardwired logic of the

---

[365] Mokyr: "A third Industrial Revolution has been increasingly associated with the ICT advances of the 1980s: above all, personal computers, the internet, cellular phones, and satellite communications." Source: Brief Reflections on the 4th Industrial Revolution and the Future of Technological Progress.

[366] Davis: "The Third Industrial Revolution began in the 1950s with the development of digital systems, communication and rapid advances in computing power, which have enabled new ways of generating, processing and sharing information." Source: Davis, N.: What is the fourth industrial revolution?

Schwab: "The third industrial revolution began in the 1960s. It is usually called the computer or digital revolution because it was catalysed by the development of semiconductors, mainframe computing (1960s), personal computing (1970s and 80s) and the internet (1990s). Source: Schwab, K.: The Fourth Industrial Revolution (2016).

relay computers performing intelligent tasks. A path that would —a century later on— ultimo lead to Robotics replacing Human Labour, and Artificial Intelligence that would start to compete with Human Intelligence.

## Changing Perspective

In the preceding exploration we focussed on the major events in the technological trajectory of electricity, and more specifically, the events related to the electric transmission of information by electro-mechanical means.

Changing our perspective, zooming in on the Era of Electronics, we will now explore the technological developments that would be labelled as 'electronics.' Developments that would create a watershed between the old electro-mechanical technologies and the new technologies based on the electronic devices radio tube and transistor.

# The Electronic (R)evolutions

Volta's invention of electro-chemical electricity (ie the Voltaic Pile, Figure 178) —the start of the new GPT-Electricity as the 'electric' meta-technology— had spawned in a range of development trajectories; those of the power applications (eg the electric dynamo, the electric motor and the electric lamp), as well as the tele-communication applications (eg the telephone, the telegraph, the wireless). And all these devices were the result of the mechanical technologies combined with the new electric artifacts; among them the electromagnet with its coil creating movement as the result of 'electro-magnetic forces' (Figure 81).[367]

**Figure 178: The Voltaic Pile.**
Source: Wikimedia Commons.

After the first devices that facilitated early tele-communication (eg the Morse telegraph), over the decennia more complex systems emerged. Systems based on the electro-mechanical technology. In the course of that process, the basic component (ie the engine) and the incorporating system became intertwined as an electro-mechanical system with a specific function. Such as the *communication machines* based on the engines such as the relay, and part of a communication system for sending and receiving telegraphic coded information. And with that technical development came the industrial activity around the communication technology. The

---

[367] See: Van der Kooij, B.J.G.: *The Invention of the Electro-motive Engine.* (2015)

manufacturing industries of telegraphic equipment, as well as the service-providers —aka the Telegraph Companies— building the ever-expanding networks capable of executing that transmission. And, at the end of the line converting it for further processing; the telegram delivered by the telegram messenger boys speeding on their bikes.

This communication technology would become part of the *Era of the Information and Communication Technologies* (ICT) where the emerging new electronic engines (ie the vacuum tube and the transistor) would fuel a massive development changing the Affairs of Man fundamentally. At is foundation lay a development that had started with a rather simple device; the electro-magnet. As that electro-magnet could be used to create movement at a distance, it became the engine that drove the early relays used in a range of additional applications, such as the automatic telephone switching systems routing the call in the telephone network.[368]

The common characteristic for the electro-magnet powering the relay is —not surprisingly— the fact that it had two states. The magnet was activated (in logic terms it was ON), or it was not activated (in logic terms it was OFF). In other words, the relay as binary engine could be used to create Binary Logic Systems.[369] And that was a new insight.

## *The First Electronic Revolution (1900-1950)*

In the second half of the nineteenth century, the invention of the electric light in the form of the incandescent lamp by Thomas Edison, created a revolution on its own by lighting up private, professional, social and cultural life. With its own trajectory of development in the *Era of Electricity*, its impact on the Affairs of Man was enormous.

Next, by the early twentieth century, came the *Era of Electronics*. As a spin-off from the incandescent lamp developed by Edison, 'electronics' would start with the new development trajectory of the vacuum-bulb-based 'electronic device.'[370] And those devices would in turn create a whole new range of systems covering a broad range of functions, having a massive impact on the Affairs of Man. An impact that had its roots in earlier times when natural philosophers wrestled science from the dogmas of religion.

---

[368] In early automatic telephone exchange systems with their system of rotary dialling (1890s), a 'pulse train' of numbers created by the dial switch directs the switching system to make a connection between two parties. Each decimal number of the subscriber identifier was translated into a range of binary pulses, that drove the electro-mechanical system.
[369] Binary logic as a two-valued logic is based on two states of the truth value: true or false.
[370] The device which controls the flow of electrons is called an active electronic device. These devices are the main building blocks of electronic circuits.

## *The Era of Classic Physics*

The Scientific Revolution —the period from Copernicus' publication on the Solar System (1543) to Isaac Newton's publication on the scientific method and his Laws of Gravity (1687)— had brought scientific thinking into the Nature of Matter by natural philosophers in the learned societies and academies. The Mechanical Worldview and Newtonian mechanics dominating scholarly thinking were based on the so-called 'Classic Mechanics.'

> Classic Mechanics (aka Newtonian mechanics) is concerned with motion of objects (from projectiles to astronomical objects) and the forces that cause the motion. By means of the concept of force, Newton was able to synthesize two important components of the Scientific Revolution, the mechanical philosophy and the mathematization of nature.

Then, by the nineteenth century, the concept of 'thermo-dynamics' had sprouted from many scholarly contributions into the construct of 'work' (ie force multiplied by displacement) done by machines delivering 'motor power' (eg the steam engine). It proved essential in the understanding of the relation between heat and energy, as well as the interchangeability of mechanical, chemical, thermal, and electrical forms of work.

After the early explorations into the phenomena of electricity had resulted in the electro-chemical battery, scholarly interest focussed on the concept of electro-magnetic forces and molecular velocities; the classical *electro-thermodynamics*. With ultimo James Maxwell developing the dynamic theory of the magnetic field and formulating the distribution law of molecular velocities. The encapsulation of heat in particulate motion, and the addition of electro-magnetic forces to Newtonian dynamics established an enormously robust theoretical underpinning to physical observations.

In addition, there were the *chemical thermodynamics* investigating the relation between heat and work related to the process of chemical reactions. The development of thermodynamics both drove and was driven by the understanding of the properties and behaviour of small particles. This understanding led to the atomic theories about the structure of matter and their kinetic theory of gasses[371] relating to the forces between molecules.

---

[371] The *kinetic theory of gases* explains the macroscopic properties of gases, such as volume, pressure, and temperature, as well as transport properties such as viscosity, thermal conductivity and mass diffusivity.

## The Molecular Theory and Properties of Materials

In chemistry, the molecular theory is concerned with the existence of strong chemical bonds between two or more atoms. The late nineteenth century was the time when 'physics' was about molecules and atoms and their bonding.[372] Such as chemical bonding; the attraction between atoms, ions or molecules that enables the formation of chemical compounds. Among those chemical bonds[373] we find those that resulted from the electrostatic force of attraction between oppositely charged ions[374] as in *ionic bonds* or through the sharing of electrons as in *covalent bonds*.

> Although the properties of gasses and their molecules had been explored since Robert Boyle's 1661-hypothesis that matter is composed of clusters of particles, the real quest into this scientific field of molecular science started when, in the year 1873, a seminal point in the history of the development of the concept of the 'molecule' was reached. In this year, the renowned Scottish physicist James Clerk Maxwell (1831-1879) published his famous thirteen-page article 'Molecules' in the September issue of Nature. Maxwell showed mathematically that molecules in a rarefied gas bouncing from wall to wall also act like a solid. In other words, as pressure increases, a gas begins to behave like a fluid.

So, materials were made out of molecules. Molecules that can be combined/associated or separated/dissociated. And, as molecular chemistry deals with the laws governing the interaction between molecules that results in the formation and breakage of chemical bonds, the question was how that mechanism worked.

The Austrian physicist Ludwig Boltzmann (1844-1906) created the Theory of Valence to explain the phenomenon of gas-phase molecular dissociation. In 1898 he gave his 'Lectures on Gas Theory,' and he stated that chemical attraction, owing to certain facts of chemical valence, must be associated with a relatively small region on the surface of the atom called the sensitive region.

Several scientists, such as William Prout (1785-1850) and Norman Lockyer (1836-1920), had suggested that atoms were built up from a more fundamental unit, but they envisioned this unit to be the size of the smallest

---

[372] *Bonding* is what separates chemistry from physics. If the understanding of atoms and their component particles belongs primarily to the realm of physics, then chemistry is concerned with the aggregation of atoms into chemical entities held together by bonds.

[373] In general, strong chemical bonding is associated with the sharing or transfer of electrons between the participating atoms.

[374] An *ion* is defined as an atom or molecule that has gained or lost one or more of its valence electrons, giving it a net positive or negative electrical charge.

atom, hydrogen. In 1897 the British physicist Joseph J. Thompson (1856-1940) discovered that the atoms of a molecule are build up from specific subatomic particles; the electron. By 1908, the electron was sufficiently established as a reality in chemistry. The winner of the Noble Price for Chemistry in 1904, William Ramsay (1852-1916) gave a lecture entitled 'The Electron as an Element' as his presidential address to the Royal Society of Chemistry. Thompson's discovery gave rise to the solid-state theories.

In 1916, the American scientist Gilbert N. Lewis (1875-1946) published a now famous paper on bonding entitled 'The atom and the molecule' (1916). In that paper he outlined a number of important concepts regarding bonding that are still used today as working models of electron arrangement at the atomic level. Most significantly, Lewis developed a theory about bonding based on the number of outer shell electrons (aka valence electrons) in an atom. He suggested that a chemical bond was formed when two atoms shared a pair of electrons (later renamed a 'covalent bond' by Irving Langmuir).

The American chemist Irving Langmuir (1881-1957) —working on light bulbs during his PhD work— continued to study filaments in vacuum and different gas environments. Subsequently, he began to study the emission of charged particles from hot filaments (aka thermionic emission). Langmuir's most famous publication is the 1919 article 'The Arrangement of Electrons in Atoms and Molecules.'

So, it took a while and many scientific explorations to create insights in the molecular structure of materials and what held the particles in materials together. But there were also the explorations into the relation between molecules in metals and gasses, and their relation to heat. And what made materials emit particles. The latter encompassing the different forms of emission, such as 'thermionic emission' where particles (eg electrons) could dissociate when influenced by temperature.

Research on thermionic emission dates back to 1853, when the French physicist Antoine Henri Becquerel (1852-1908) first detected electrical current passing between two platinum wires, one hot and one cold, kept in a variety of gases. Becquerel was studying the electrical conductivity of gases but did not fully recognize the impact of thermal energy at the electrode for electrical current generation. Many others contributed to the understanding of thermionic emission in the late nineteenth century, not the least of which was the British physicist Frederick Guthrie (1833-1886) who in 1873 published his work on the relationship between heat and static electricity. Guthrie measured the discharge of a positively charged, red-hot sphere into air, helping establish understanding of the thermionic emission phenomenon.

It was Willoughby Smith (1828-1891) who discovered in 1873 the light-sensitive properties —ie the photoconductivity— of the material selenium. He presented his ideas in a paper presented at the February 12th, 1873 meeting of the Society of Telegraph Engineers as "Electrical Properties of Selenium and the Effect of Light Thereon". That knowledge led to a flood off inventions, and within two years a system for the transmission of pictures by wire was developed. What came to be known as the photoelectric cell, a device connecting light and electricity, had its beginnings in the 1880's.

From 1882 to 1889, the German physicists Julius Elster (1854-1920) and Hans Geitel (1855-1923) developed a sealed device containing two electrodes, one of which could be heated while the other one was cooled. They discovered that, at fairly low temperatures, electric current flows with little resistance between the electrodes if the hot electrode is positively charged. Experimenting with materials the found that—next to selenium— alkali metals such as sodium, potassium, rubidium and caesium were sensitive to light.

And they noted that negatively charged magnesium filaments, freshly ground with emery, are discharged not only by ultraviolet light but even by "dispersed evening daylight." The new cell, then, emitted electrons when stimulated by light, and at a rate proportional to the intensity of the light. In 1893 they modified the Cathode Ray tube into a light sensitive photo-electric cell (Figure 179). The vacuum tube light-sensitive photocathode —aka photo tube—was born.

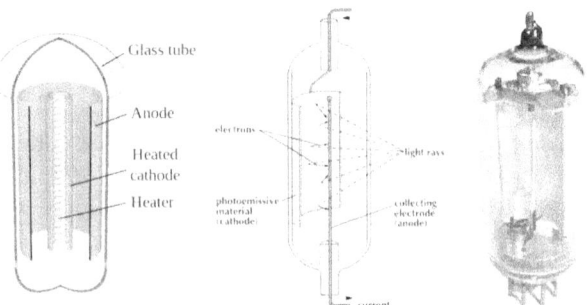

**Figure 179: Principle of Vacuum Photo Tubes and its realization.**
The Thermionic Plate tube (left) and the light sensitive Photo Tube (middle). Sylvania 350 Photo Cell (right)

Source: https://www.electricstuff.co.uk/Photocell.html

The Elster-Geitel photo tube became the traveling companion of two other types of vacuum tubes in existence at the time: (1) the *cathode-ray tube*, in which the cathode emitted electrons under the influence of a high potential, and (2) the *thermionic valve*, a valve in which electrons were emitted under the influence of heat and that permits the passage of current in one direction only.

Many of these early studies provided the foundations for Cathode Ray tubes and the X-ray tubes that were developed by Crookes and Hittorf, who we will meet later on in more detail. Their experimental research on thermionic emission that enhanced the understanding of the *Cathode Rays*, continued through the ensuing decades and became the foundation of discoveries which eventually changed the whole of chemistry and physics.

> It was the British physicist —and winner of 1928 Noble Price for Physics— Owen William Richardson (1879-1959) who, researching the emission of electricity from hot bodies, noted in 1900 that, when a metal wire becomes hot and starts to glow, the air around the wire becomes electrically charged. The electrons in the metal become so agitated that they are liberated from their atoms and cast into the air where they act as free charged particles. He formulated a law that describes the relationship between emissions of electrons and temperature. The Theory of Thermionic Emission was born.

This all was dominantly the work of the theoretical scientists. They explained the mechanisms of chemical bonding, and especially covalent bonding. A phenomenon that had already been applied by the Experimentalists when they started to work with vacuum tubes (eg the Crookes cold cathode tubes) exploring the Edison Effect. Their work resulted in the practical applications (ie the inventions) that were about to change the world. But there was more to come.

## Gas Discharge: Cold Cathode Rays

The experimenting and theorizing contributed to different strands of development. Such as the development trajectory of the heated vacuum tube in all its variations (like the Cathode Ray Tubes that evolved in the 'television' picture tubes we will meet later). And the development trajectory of the 'cold cathode ray' tubes. This special type of vacuum tube did not have a heated filament to emit a stream of electrons. Their working principle was not the 'thermionic emission,' but the 'gas discharge', in which a gas ionizes when the voltage is high enough (ie the modern neon lamps).

> This phenomenon occurs when electrically charged particles, typically electrons, move through a gas at a high velocity—roughly 2 percent of the speed of light. The speeding particles collide with the

gas atoms or molecules, ionizing them and creating an energetic plasma of charged ions and electrons. The ions, excited to higher energy states, shed their excess energy as photons of light.

The earliest experimental application of gas discharge was the Geissler Tube, invented by the German physicist—and glassblower—Heinrich Geissler (1814-1879) in 1857. Other experimental scientists like William Ramsey, Henry Cavendish and William Strutt discovered that specific gasses (aka neon, krypton and xenon) glowed with a distinctive brilliant colour when sealed in a glass tube and energized with high voltage. By the 1910s, these gas-discharge tubes —named for the electrical discharge that made them light up—would become the basis for neon lamps.

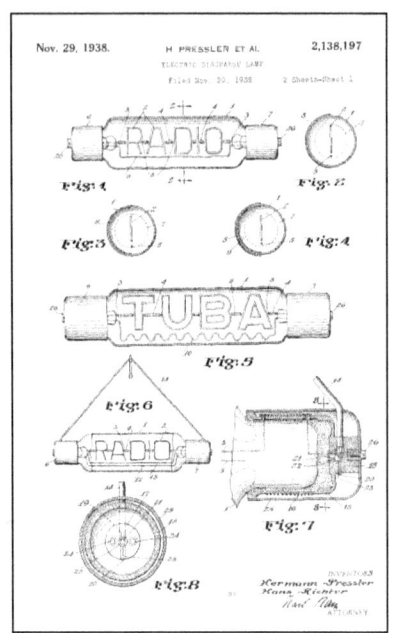

**Figure 180: Nixie Tube Patent.**

US-Patent № 2,138,197 granted November 29th, 1938.

Source: USPTO.

**Figure 181: Gas Discharge Tube.**

Source: Wikimedia Commons.

In the late 1920s and early 1930s, inventors realized that the Geissler tube's discharge glow could be extended to the cathode and that you could shape the cathode to confine the glow. By using an appropriately bent wire as the cathode, for example, you could display numbers or text (Figure 181).

One of the earliest patents for such a design went to the German inventors Hermann Pressler and Hans Richter. They filed a patent application on November 20th, 1935, and it was granted US-Patent № 2,138,197 in November 29th, 1938, for what they called an "electric discharge lamp" that functioned as a "self-luminous sign" (Figure 180).

Based on this principle the Nixie-tubes (short fort 'Numeric Indicator eXperimental No. 1') —originally developed in the mid-1950s by the American Haydu Brothers Laboratories— was further developed by the Burroughs Corporation. This small neon bulb had wires shaped like numerals, letters, or other symbols, one in front of the other, that lit up when the current was turned on.

Born in the basement of a German-American tinkerer in the 1930s and later commercialized by the business equipment maker Burroughs Corp., in the 1960s, 'Nixies' displayed not only the vacuum tube-based ANITA Mk VII calculators, but also data vital to NASA's landing on the moon, lit up critical metrics for controlling nuclear power plants, and indicated the rise and fall of share prices on Wall Street stock exchanges, among thousands of other uses.

## *The Invention of the Radio Valve*

By the end of the nineteenth century the phenomena of thermionic emission had, next to the work of the 'thinkers', attracted many 'tinkerers' who —sometimes hardly understanding the basic of the phenomenon— tried to apply its properties into mechanical artifacts.

Take as an example Thomas Edison and his invention factory in Menlo Park. After realizing that an electric current can make a metallic wire glow (ie emit light rays), he started experiments transferring the glowing wire into the incandescent lamp. He added a protective glass vessel (aka bulb) to the wire-supporting frame, he experimented with different materials for the wire, etc. He noticed the effects like the blackening of the tubes, but the theory of thermionic emission was not on his agenda.

Where the scientific 'thinkers' were concerned with their Acts of Discovery and their Acts of Invention, the 'tinkerers' were focusing with their Acts of Innovation in creating usable product that fulfilled a (latent) need in the market place. Sometime the result of deliberate experimenting, sometimes the unexpected side effect of their curiosity. Using the collected 'knowledge,' with their work they created the 'knowhow' resulting in the different technologies.

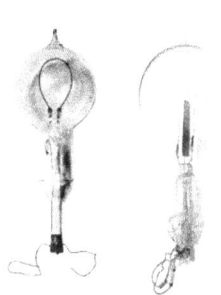

**Figure 182: The Edison Lamps.**
Incandescent Lamp (left, 1879) and Edison Effect Lamp (right, 1885).
Source: Wikimedia Commons.

## The Edison Effect

In an incandescent lamp, the thermionic emission had a negative side-effect for the incandescent lamp as it blackened the inside of the glass bulb (Figure 182). But that same thermionic emission proved essential for the development of new devices like the 'radio' vacuum tubes. To eliminate blackening of his lamps, Edison built several experimental lamp bulbs with an extra wire, metal plate, or foil inside the bulb that was separate from the filament and thus could serve as an electrode. And, depending of the applied polarity of the metal plate, he noted the diode-effect. However, after patenting the device (US-patent № 307,031 of November 15th, 1883), he did not follow up on this effect.[375] Edison continued on his trajectory to improve his incandescent lamp.

So, the phenomena of thermionic emission and the control of electric current in vacuum came into perspective. This was the beginning of a development trajectory that would lead to the vacuum tube (aka radio tube). A winding path that saw many contributions from other technologies (eg vacuum technology, glass blowing technology), and the contributions from other application fields of electricity (eg wireless telegraphy).

## Crystal detectors

With the progress of the development trajectory of systems for wireless telegraphy (Figure 105), engineers tackled the emerging problems. And one of the problems of the sparked wireless telegraphy was the detection of the weak signal. First the

**Figure 183: Early Wave Detection Systems.**

Fessenden's electrolytic detector (1902, top), Marconi Reliable Detector (middle, 1902) and the Cat's Whisker Crystal Detector (bottom, 1906).

Source: Wikimedia Commons, http://www.demajo.net/museum/page4.htm

---

[375] The diode-effect occurs in a device when the electric current is blocked/passing depending on the polarity applied on the device. In its aquatic analogy it is a valve that blocks/passes a waterflow. This explains why a diode-tube is also called a vacuum-valve.

not too sensitive coherer[376] was used for detection the electric waves, but soon many efforts were undertaken to created better performing detecting devices.

After the experimenting with crystals (eg the semi-conductor Carborundum; SiC), Karl Ferdinand Braun had conceptualized in the 1870s a new type of rectifier: the crystal detector (aka Cat's Whisker detector) (Figure 183, bottom). It started a development wave of crystal detectors. Other scientists explored different avenues to improve signal detection. Such as Fessenden's electrolytic detector (his so-called 'Barretter') (Figure 183, top) and thermal detector (the hot-wire barrater), and Marconi's Reliable Detector 'Maggie' in 1902 (Figure 183, middle).

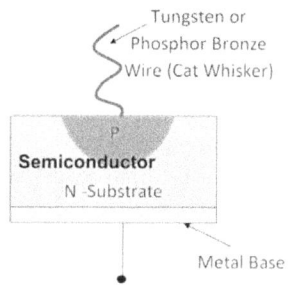

**Figure 184: Principle of the Point Contact Diode.**

Among those experimenters was the American radio pioneer G.W.Pickard (1877-1956) who began research on detector materials in 1902 and found hundreds of substances that could be used in forming rectifying junctions. He developed the silicon 'crystal detector' and was granted US-Patent № 836,531 on November 20th, 1906. The detector applied the diode effect; current would flow only in one direction from a metallic point to the base material.

Also other experimenters created similar rectifiers. Such as US military officer Henry Dunwoody (1842-1933) who was not much later granted US-Patent № 837,616 on December 4th, 1906 for a point contact detector of carborundum. Or the Indian scholar Jagadis Chandra Bose (1858-1937) who developed a diode using galena (ie lead sulphide).

This device was in fact the (accidental) invention of the (semiconductor-based) point-contact diode (Figure 184). The tungsten or phosphor bronze wire (the Cat Whisker) acting as cathode made contact with a thin substrate of semiconducting material (eg Galena) on a metal base (anode). The electric current that would only flow in one direction, heating the wire and that heat created in the semiconductor substrate a pointed P-N junction.

---

[376] Before the invention of the triode, detection was accomplished by using a 'coherer.' (eg the Branly coherer, 1890) This was a glass tube filled with nickel and silver granules. The coherer was wired in series with a battery and headphones. When the coherer received an incoming electro-magnetic wave representing a dot or dash of Morse code, the granules were drawn together due to the current flow, became a better electrical conductor, and the current passing through it powered the clicks heard in the headphones of the receiving telegrapher.

Also Henry J.Round (1881-1966), an English engineer consultant employed by Marconi in 1902, performed a number of experiments on the crystal detector using a range of different materials. He also applied a direct current to them and noticed that some actually emitted light. H.J. Round reported this in the February 9th, 1907 edition of Electrical World. This is the first known report of the effect of the light emitting diode (LED). But it would take until the 1960s before that development took off.

However, the receivers built with the Cat's Whisker were not that successful as it was very sensitive to the exact geometry and pressure of contact between wire and crystal. They needed constant adjustment. Soon the vacuum tube diode would realize the same effect in a much more reliable way, and the crystal decoder would fade away[377] as other technologies came available that created more sensitive receivers converting the received AC-signal to a DC-signal. But, this early use of crystalline materials, aka semi-conductors, heralded the sciences of the semiconductor-electronics to come. But before that was to be, quite some development in the field of vacuum physics would take place.

## Vacuum Physics: Scientists discover Electrons

Back to the Edison effect. The flow of charged particles that was emitted from the incandescent wire and that blackened the incandescent vacuum bulb, fascinated scientists. At the time, atoms were the smallest particles known and were believed to be indivisible, the electron was unknown, and what carried electric currents was a mystery. There were two theories: The 'Crookes' (eg Crooke himself, Cromwell Varley) who believed they were 'radiant matter'; that is, electrically charged atoms. While German scientists Heinrich Hertz and Eugen Goldstein believed they were 'ether vibrations'; some new form of electro-magnetic waves.

The debate was resolved in 1897 when the British physicist Joseph J. Thomson (1856-1940) —busy experimenting with the 'cathode ray' tubes— measured the mass of cathode rays, showing they were made of particles that were around 1800 times lighter than the lightest atom, hydrogen. Therefore, they were not atoms, but a new particle, the first sub-atomic particle to be discovered, which was later named the 'electron.' And the stream of electrons —the electric current— were called 'cathode rays.'

This was the start of the theories that shaped the Atomic Model of materials (Figure 222, Figure 223), as it made many engineering scientists start experimenting to discover the properties of the rays of

---

[377] Except in the 1920s and early 1930s when the cheap point contact detector came to the fore when public broadcasting started and cheap radio sets were needed for domestic use.

particles. Such as their deflection by magnetic fields, and also the light-emitting effect when the rays collided with specific materials.

The English physicist William Crookes (1832-1919) created vacuum tubes —aka the Crookes Tube (Figure 185, top)— with metal electrodes at the end; the cold —ie not-heated—cathode and anode. When a voltage was applied between those electrodes, a flow of particles (ie the 'cathode ray') originating from the cathode would be the result.

The English physicist and telegraphy pioneer Cromwell Varley (1928-1883), already in 1871 had authored a scientific paper suggesting that cathode rays were streams of particles of electricity.

The German physicist Julius Plücker (1801-1868) built a tube with an anode shaped like a Maltese Cross facing the cathode. When the tube was turned on, the cathode rays cast a sharp cross-shaped shadow on the fluorescence on the back face of the tube, showing that the rays moved in straight lines.

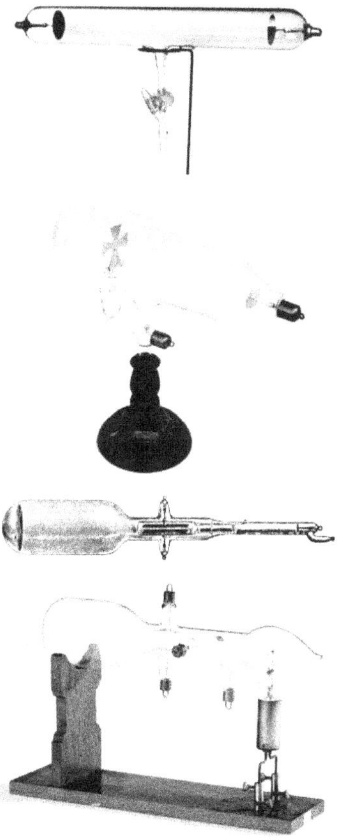

**Figure 185: Cathode Ray Tube.**
Crookes Tube (top, 1870s), Maltese Cross Tube (upper middle, 1880s), Karl Ferdinand Braun's CRT: prototype (lower middle, 1904) and tube (bottom, 1904).

Source: Science Museum Group Collection. Museum of Radiation and Radioactivity.

His student, the German physicist Johann Hittorf (1824-1914) —the German counterpart of William Crookes— did extensive research on gas discharges, discovered the electricity carrying capacity of charged atoms, and experimented with tubes that had 'absolute vacuum' as early as 1865. He also explored the shadows casted by objects in the tube (eg the Maltese Cross).

The German physicist Heinrich Herz (1857-1894) —known for his electro-magnetic wave experiments— built a tube with a second pair of metal plates to either side of the cathode ray beam. His experiment failed but later this method of deflection proved more successful. He claimed that the cathode rays were electro-magnetic waves.

The British physicist John J. Thomson (1856-1940), experimenting with the Crookes tube (Figure 186), tried in 1894 to estimate the velocity of the cathode rays, which could say much about their structure. In April 1897, Thomson had only early indications that the cathode rays could be deflected electrically (previous investigators such as Heinrich Hertz had thought they could not be). A month after Thomson's announcement of the corpuscle, he found that he could reliably deflect the rays by an electric field.

The Hungarian physicist Philipp Lenard (1862-1947) started studying cathode rays in 1888 and discovered —by creating metallic windows in the tube (Figure 185, upper middle)— that the energy (speed) of the electrons ejected from a cathode depends only on the wavelength, and not the intensity of, the incident light.

The German scientists Karl Ferdinand Braun (1850-1918) also experimented with these 'cathode rays' and created by 1897 the (cold cathode) Braun Cathode Ray tube (Figure 185, middle, bottom). The particles were attracted to the anode, that would become a 'screen' as it became coated with phosphor that lightened up when the particles collided with the coating. By 1904 the CRT started to get its function as a display tool.

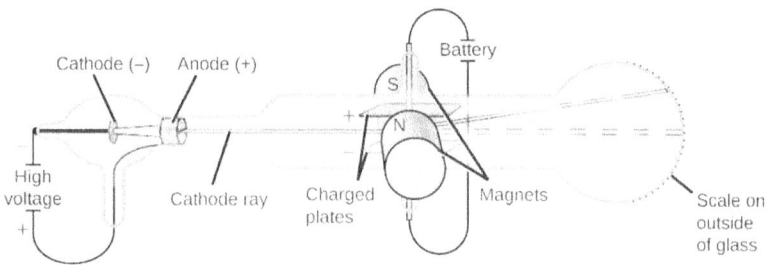

**Figure 186: Diagram of Thompson's Cathode Ray Tube.**

The ray originates at the cathode and passes through a slit in the anode. The cathode ray is deflected away from the negatively-charged electric plate, and towards the positively-charged electric plate. The amount by which the ray was deflected by a magnetic field helped Thomson determine the mass-to-charge ratio of the particles.

Source: Openstax CNX

The German physicist Wilhelm Röntgen (1845-1923) explored a different aspect of the cathode rays. In 1895 he discovered X-rays emanating from Crookes tubes and published his findings in 'Über eine neue Art von Strahlen' ('About a new kind of rays'). The many uses for X-rays were immediately apparent, the first practical application for Crookes tubes. Medical manufacturers began to produce specialized Crookes tubes to generate X-rays; the first X-ray tubes (Figure 187).

These contributions to the knowledge about that new phenomenon of flowing particles creating 'cathode rays' in the 1870s-1890s (Figure 217), paralleled the development of the incandescent lamp. The device with its heated filament emitting rays in the visible spectrum (ie light), the blackening of the bulb, and the rays of particles flowing between cathode and anode (ie electric current). It was an explosion of scientific understanding of the 'cathode rays.'

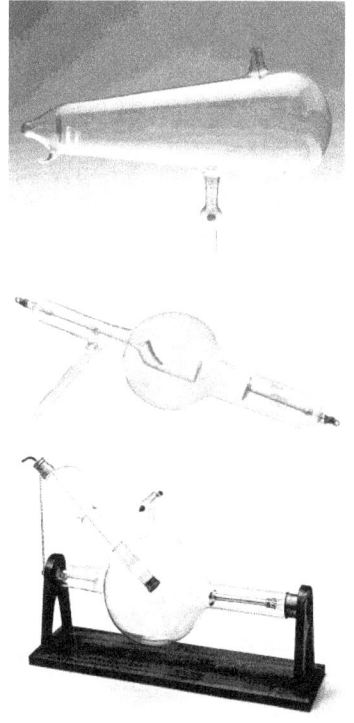

**Figure 187: Röntgen Tubes.**
German Röntgen Tube (1896, top), English Röntgen tube (1896-1900, middle), Philips Röntgen tube (1918, bottom).

Source: Science Museum, Philips.

Until the twentieth century, physicists had studied subjects, such as mechanics, heat, and electro-magnetism, that they could understand by applying common sense or by extrapolating from everyday experiences. The discoveries of the electron and radioactivity, however, showed that classical Newtonian mechanics could not explain phenomena at atomic and subatomic levels. As the primacy of classical mechanics crumbled during the early twentieth century, the scientific field of quantum mechanics came to replace it.

Among the many results from the explorations of the scientists and experimentalists —and the insight it created— would emerge the vacuum tube diode created by John Fleming and Lee DeForest. In that device a heated incandescent wire would heat a cathode that emitted particles (aka electrons) that were attracted to the anode (Figure 188). This was a fundamental occurrence, as it was the merging of (1) the knowledge and insights about the cold cathode creating the cathode rays, and (2) the experience built up with the thermionic effect in vacuum tubes.

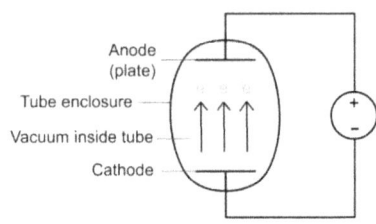

Figure 188: Principle of the Vacuum Diode.

The arrows indicate the Cathode Rays (electrons).

## The Fleming Valve

It was the Brit John Ambrose Fleming (1849-1945), consultant to the Edison Electric Light Co. and former scientific advisor for the Marconi Wireless Telegraph Co., who got involved in the problems with the transmission of long-distance signals across the Atlantic Ocean. He realised that the transmitters as well as receivers were the sources of the problem.

While working on the enhancement of the receiver, he remembered the Edison effect. Experimenting with this effect he discovered that it could be used for the control of the *flow of electrons* —ie the cathode ray— in a vacuum tube. A flow that would only go in one direction, and be blocked in the other direction; the concept of the 'diode valve' (Figure 188).

Fleming used this concept to improve the sensibility of the weak spark-induced radio-signals send by the Marconi transmitters over large distances. By 1896 he started to experiment with the diode-effect to rectify the spark-induced (multi-frequency) radio signal into a usable DC-signal. This principle of conversion of a high-frequency AC-voltage into a DC voltage for the detection of electric waves was used in the Wheatstone Bridge Rectifier (Figure 190). His subsequent research activities led to the development of the vacuum tube known as the Fleming Valve (Figure 189).

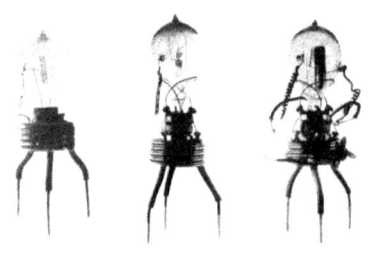

Figure 189: Fleming's Valves (1889-1904).

Source: Wikimedia Commons.

*Fleming went to work on developing a different type of detector. Since he was the victim of progressive deafness, he sought a device which could operate a mechanism by which signals would be recorded and translated by the eye rather than by the ear. The most sensible current-indicating instrument in use at that time was the d'Arson mirror galvanometer, which operated only on unidirectional current [DC]. Fleming wanted to use this sensitivity but to do so he had to rectify the high frequency oscillation of the incoming signal. [...] He then recalled the work he had done many years before on the Edison effect and decided to find out by experiment whether the know rectifying action at low frequencies would also exist at high frequencies of the oscillations used in wireless telegraphy.* (Tyne, 1977, pp: 40-41)

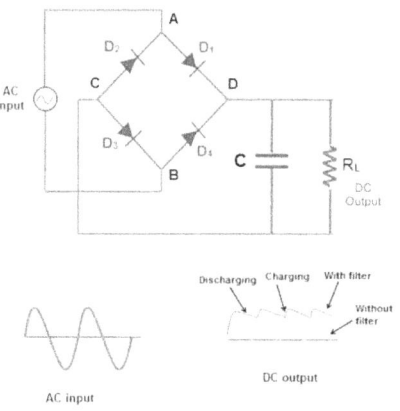

**Figure 190: A Wheatstone Bridge Rectifier converting Alternating Current (AC) to Direct Current (DC).**
The four diodes D1-4 act as electric valves.

Fleming combined the need for a more sensible detector with the technology of the vacuum tube that created the cathode ray tube (Figure 185) in the Wheatstone circuit (Figure 190).

> *To my delight I saw the needle of the galvanometer indicate a steady current passing through, and found that we had in this particular kind of electric lamp a solution to the problem of rectifying high-frequency wireless currents. The missing link in wireless was found-and it was the electric lamp.*[ibidem]

*The Fleming Thermionic Valve:* Fleming's thermionic diode (Figure 189) was an 'instrument for converting alternating electric currents into continuous currents'; a device that could be used to rectify the weak AC signal of the wireless receiver into a DC signal. He had created the vacuum rectifier (aka 'diode') and was granted GB-patent № 24,850 for "Improvements in instruments for detecting and measuring alternating electric currents", issued September 21st, 1905. Next, he applied for protection of his invention in the US and was granted US-patent № 803,684 on November 7th, 1905. The Fleming Valve essentially consisted of an incandescent light bulb with an extra electrode inside (Figure 189). It proved to be the start of a technological revolution based on vacuum tube technology.

The ramifications of the thermionic valve were myriad and far-reaching. It became a key component of radios for nearly three decades until it was replaced by the solid-state diode we will meet later on, and was integral to the development of television, telephones, and even early computers. Fleming's discovery would lead to a range of infringement cases, among which was the one based on the invention Lee DeForest patented in 1908.

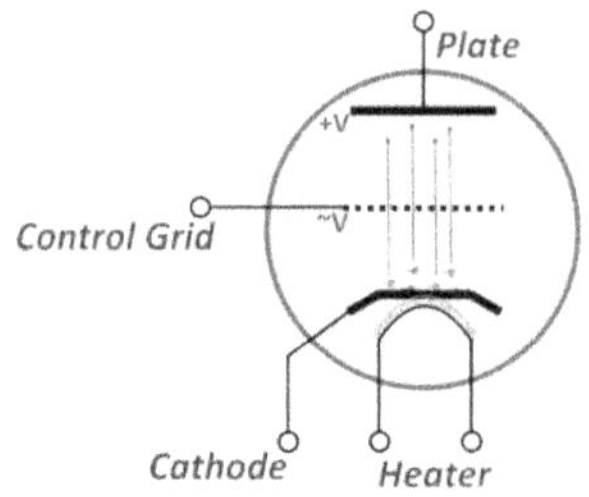

**Figure 191: Principle of the Vacuum Triode.**

## Lee DeForest's Audion

The former PhD student *Lee DeForest* (1873-1961) writing a dissertation on Hertzian waves —titled 'Reflection of Hertzian Waves from the Ends of Parallel Wires' focusing on wireless propagation of electromagnetic waves— created, after a tumultuous career[378] as inventor/entrepreneur, a device that would turn the tables in wireless communication It was called the Audion, a three-element vacuum tube aka triode (Figure 191).

DeForest was extremely creative and energetic, but often was unable to see the potential of his inventions or grasp their theoretical implications. Trying to find a solution for the insensible wireless receivers of the wireless systems available at that time, DeForest experimented —like so many others— with flame detectors (his later US-Patent № 979,275 filed on December 9th, 1905, granted in December 20th, 1910). It was the start of a range of patents (Table 6) for his Audion triode vacuum tube.

Table 6: Patents granted to Lee DeForest (Audion related).

| Patent № | Granted | Description |
| --- | --- | --- |
| US 748,597A | January 5, 1904 | Wireless Signalling Device (filed December 24, 1902). |
| US 824,637 | June 26, 1906 | Oscillation-responsive device. (filed January 18, 1906). |
| US 827,523 | July 31, 1906 | Wireless-telegraph system. (filed December 6, 1906 |
| US 836,070A | November 13, 1906 | Oscillation-responsive device (filed May 19, 1906) |
| US 841,386 | January 15, 1907 | Wireless Telegraphy (filed August 27, 1906) |
| US 841,387 | January 15, 1907 | Device for amplifying feeble electrical currents (filed October 25, 1906) |
| US 879,532 | February 18, 1908 | Space Telegraphy (filed January 29, 1907) |

Source: USPTO

---

[378] During those years, DeForest was regularly involved in patent lawsuits (spending his fortune on legal bills). He went through four marriages, had a number of failed companies, was defrauded by his business partners, and was indicted (but later acquitted) for mail fraud.

Like Fleming, he started to develop glass-tube devices using two electrodes. After some further experimenting, he applied for a patent and was granted on November 13th, 1906 US-Patent № 836,070. He called his device the "two-electrode Audion," and it consisted of partially evacuated glass bulb containing an incandescent lamp filament which was flanked by two platinum wings. In addition, on January 15th, 1907, he was granted US-Patent № 841,387 for a 'Device for Amplifying Feeble Electrical Currents'. This device shown in Figure 1 of the patent, acted as a diode (Figure 192).

Further experimenting with a range of different configurations of a third electrode (see Figure 2 in patent № 841,387) —placing it between the plate and the filament— resulted in a breakthrough. It was this third electrode (aka grid) that created a triode, and that would later show to have some unexpected properties. On January 29th, 1907 he filed for a patent that was granted on February 18th, 1908 as US-Patent № 879,532 for the device that became known as the 'three-electrode Audion.' DeForest claimed to have found a better detector:

> *"The objects of my invention are to increase the sensitiveness of oscillation detectors comprising in their construction a gaseous medium by means of the structural features and circuit arrangements which are hereinafter more fully described."*

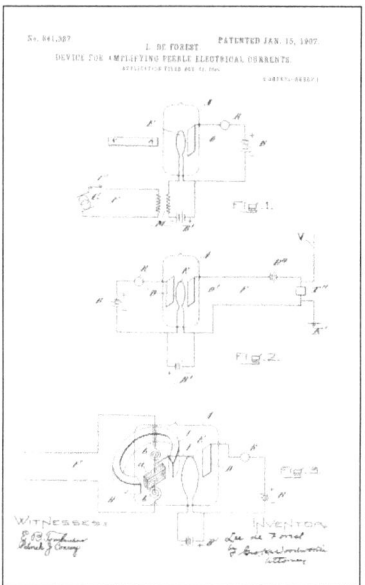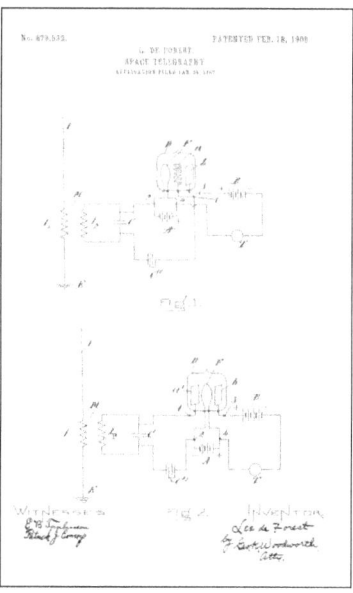

**Figure 192: Patents issued to Lee DeForest.**
US Patent № 841,387 (left, 1907) and US Patent № 879,532 (right, 1908)
Source: USPTO

It seemed that he had just added a third electrode to a device that had already been created, and that was already been patented by Ambrose Fleming: his thermionic valve. However, his new device proved to have amplifying capabilities, and that caused it to become an *active* electronic component that would also have a major impact on electronic technologies to come. The triode became the template for electronic amplification systems until the advent of the transistor in the second half of the twentieth century. But others had to discover and implement that capability of DeForest's invention.

## Improvement of the Audion

The Audion embarked on a trajectory of technical improvement. Following DeForest's instructions, the actual tubes were made by the H.W. McCandless & Company Inc., a manufacturer of customized incandescent lamps. The shape changed from cylindrical to spherical (Figure 193), and different filament configurations were applied. The Audion tube was for many years used only as a detector, and early effort to sell is as an amplifier failed. But that was about to change as he focussed on the application of his invention in broadcasting systems. Between 1907 and 1912, DeForest undertook a range of broadcasting activities using 'radio-telephony.'[379]

> *His focus was now on voice communication, and by early 1907 he was able to communicate using radiotelephony across his laboratory in New York City. Between 1907 and 1911 de Forest launched and built his radiotelephone companies; on January 13th, 1910, he broadcast a performance of several opera stars from the New York Metropolitan Opera House. His radiotelephone company went bankrupt in 1911 while de Forest was on the West Coast*

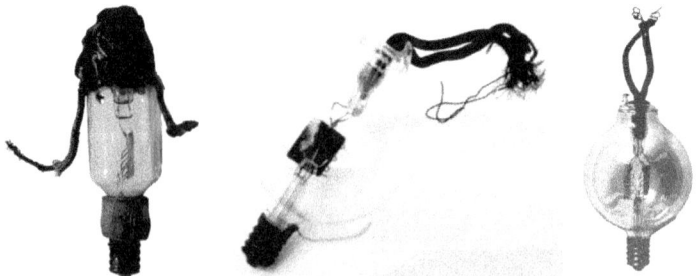

**Figure 193: Lee DeForest's Experimental Audion Tubes (1906-1908).**
Cylindrical prototype bulb (left, 1907), spherical bulb (middle, 1908), and double filament, grid and plates (right, 1908).

Source: Wikimedia Commons. National Museum of American History. Lee de Forest.org

---

[379] Radiotelephony is the wireless transmission of multi-frequency sound from one point (ie the broadcasting station) to many receiving points (the radio's).

> *supervising Signal Corps installations in Seattle and San Francisco; he remained in Palo Alto, California, to work at the Federal Telegraph Company, hired by Chief Engineer Cyril Elwell to head a team to concentrate on the development of a circuit that would cause an audio tube to amplify. It was at Federal Telegraph, through de Forest's work with C. V. Logwood and Herbert Van Etten, that the 1906 three-element vacuum tube (triode) was recognized as a detector, amplifier and oscillator of radio waves, and de Forest's career was reinvigorated.*[380]

Both the amplifier effect and the oscillator effect of the three-element vacuum tube would become the main functions for the further development of wireless communication. However, by that time DeForest had ran into problems as his second company, the *DeForest Radio Telephone Company* started in 1907, with Abraham White as business partner, began to collapse in 1909.

> *In early 1911, Lee de Forest, who supervised communications installations in Seattle and San Francisco, ended up on the west coast of the United States, California. Here he received the news that his New York partners in the De Forest Telephone Company were arrested for mail and stock fraud, as well as for placing unsecured shares in the amount of $ 1,507,505. A large jury was convened and the company declared bankrupt. Forest was in a deplorable state, to financial troubles were added to the family troubles, because of which he ended up in California.* (Pestrikov, 2019, p. 5)

As DeForest's Audion tube was used in realizing broadcasting systems, both at the sending end and the receiving end of the 'radio', the new type of vacuum tube was generalized as the 'radio-tube.'

> As a side track to his broadcasting efforts, Lee DeForest created an electronic musical instrument called the Audion Piano. The simple keyboard instrument used his electron tubes in circuits called "oscillators," which produced electro-magnetic waves at a particular frequency. The waves were fed to a loudspeaker to make them audible. As diner music —live orchestras playing in the high-end restaurants had become en vogue— DeForest demonstrated the Audion Piano to dining audiences in 1915. Unfortunately, around the same time, he became embroiled in a stock-fraud scandal and later faced many financial and legal hardships, which affected his work.

## *The Vacuum Tube in Communication Applications*

Over the following years DeForest improved his Audion triode (Figure 194, top), and others started interest in the triode for specific applications

---

[380] Source: https://oac.cdlib.org/findaid/ark:/13030/c8js9r0x/entire_text/ (Accessed June 2021)

(eg the radiotelephone repeater, the radio receiver). Among them the rising giants of the telephone industry. Both *General Electric* and the *Western Electric* subsidiary of the *American Telephone & Telegraph Company* (AT&T) that bought a license from DeForest and started creating their own radio tubes.

In their R&D laboratories researchers like Irving Langmuir and Harold Arnold engineered on the vacuum tubes improving the concept technically. Also the British and American Marconi-companies (eg the Marconi Wireless Telegraph Company) started development activities, creating their own line of radio tubes.

As by 1915 the Audion type triode was only available through the financially troubled *DeForest's Radio Telephone Company*, it was an invitation for others—already engaged in the Wireless Mania one way or another—to enter the business of tube manufacturing. So, not-licensed 'independent' businesses emerged and started to create their variants of the triode vacuum tube. And they did their business in a variation of operations, some short lived, others growing in size, with every step improving the quality of their tubes. However, those latter entrepreneurs would become confronted by legal activity protecting patent-rights.

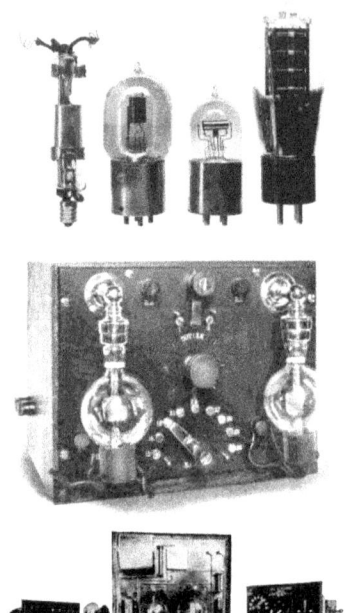

**Figure 194: Audion Improvements.**
Different versions of the Audion (1913, top), DeForest Audion Radio Receiver (middle, 1914), and Audion DeForest Audion AM radio transmitters (bottom, 1916). The Audion tubes are visible mounted on the front of the units (in the centre unit they are inside) so that the operator can see if the filament is glowing and adjust the filament current by brightness.

Source: Wikimedia Commons.

DeForest himself used his tubes to amplify electric signals in the *Audion Radio Receiver* (Figure 194, middle). By 1916 DeForest's Audion vacuum tubes had evolved into the large Oscillion vacuum tubes that he used as generators of RF signals (radio frequency). Creating DeForest's earliest transmitter circuits made with Oscillion Power triode vacuum tubes (Figure 194, bottom).

## DeForest's Business Activities

DeForest business activities started when in 1902 he and his financial backers founded his first company: the *DeForest Wireless Telegraph Company*. One of those financial backers, Abraham White, envisioned bold and expansive plans that enticed the inventor. However, White had a different agenda, was also dishonest and much of the new enterprise would be built on wild exaggeration and stock fraud.

> This was the time of the 'Wireless Telegraph Bubble,'[381] with the stock market booming. The mania for getting rich quickly through speculation in stocks affected investors in the whole country. Such as the stocks of the Marconi Wireless Telegraph Company of America, the "best investment in the world;" or the American DeForest Wireless Telegraph Company, in which "a few hundred dollars invested now and given to your children should make them independent." The DeForest promoters, as well as the brokers engaged to sell the Marconi securities, painted an alluring picture for their companies, and every wireless advertisement that appeared in the newspapers told how $100 invested in Bell Telephone stock had rolled into $200,000; and wireless stock was going to do the same.[382]

Abraham White incorporated in 1902 the *D Forest Wireless Telegraph Company of America*, with himself as the company's president, and DeForest as the Scientific Director. The wireless business would be their objective, connecting ships and shore by a string of land stations from coast to coast. Companies would be created on the Continent, in Africa, the Orient, and all these subsidiaries would pay tribute to the parent company.

> So, to create stock that could be sold, a multitude of sub-companies of the parent America company, many with DeForest's name, had been established by Abraham White, the chief promoter of the DeForest companies. Such as the *American DeForest Wireless Telegraph Company*, the *Atlantic DeForest Wireless Telegraph Company*, the *DeForest Wireless of Maine*, *DeForest Wireless of New Jersey*, and the *DeForest's Radio Telephone Company*.

---

[381] For a detailed analysis, see: Van der Kooij, B.J.G.: *The Invention of the Wireless Engine*. (2017)
[382] Source: Fayant, F.: The Wireless Telegraph Bubble. (Volume X, Number 157, June 1907) https://babel.hathitrust.org/cgi/pt?id=njp.32101079674618&view=1up&seq=397. Accessed June 2021)

White dazzled the young DeForest as they would make fortunes out of wireless telegraphy, and the name of DeForest would go down in history among the greatest names of science. However, that was not to be as DeForest eventually came into conflict with his company's management. On November 28th, 1906, in exchange for $1,000 (half of which was claimed by an attorney) and the rights to some early Audion detector patents, DeForest turned in his stock and resigned from the company. But that company with his name became the vehicle for a rather specific business development.

*The United Wireless Telegraph Company*

Abraham White organised at the end of 1906 the *United Wireless Telegraph Company* by reorganizing the American DeForest Wireless Telegraph Co. It was promoted as being a consolidation of the most prominent US and British radio firms, combining American DeForest with the worldwide holding of London-based Marconi's Wireless Telegraph Company, Ltd. American DeForest stockholders were offered the chance to exchange their now essentially worthless holdings for United stock, in a series of complicated and confusing financial transactions that were generally to the advantage of company insiders at the expense of regular shareholders.

Not long after the founding of United Wireless, Abraham White was ousted by 'reformers' headed by stock promotor C.Wilson who proved to no more honest than White. Because the primary objective of United Wireless was the sale of nearly worthless stock at inflated prices, providing profitable wireless services was not a real business objective.

> Shipboard installations and operators were provided by United Wireless at nominal rental cost or even for free. While somewhat crude by industry standards, the United Wireless equipment was efficient enough to provide adequate service for its main clientele of coastwise shipping along the United States. Subsequently United Wireless, starting with its base along the Gulf and Atlantic Coasts, expanded to dominate the Pacific coast and Great Lakes, absorbing a series of smaller firms in the process.

It took a while, but when many investors realized they were being cheated, calls for action started to appear—like the stockholders of the American DeForest Wireless companies, who applied for receivership. After years of complaints, in June 1910, inspectors from the United States Postal Department finally moved to shut down what was described as 'one of the most gigantic schemes to defraud investors that has ever been unearthed in this country'. Lee DeForest's name would be tainted in the process. In March 1912, DeForest —living in California by that time—, plus four other company officials, were arrested and charged with "use of the mails to defraud". Their trials took place in late 1913, and while three of the defendants were found guilty, DeForest was acquitted.

Crippled by the prosecution of its upper management, United Wireless declared bankruptcy and went into receivership in July 1911. In addition, the British Marconi Company sued in 1912 successfully for patent infringement. United Wireless being dissolved, the stock holders were left with $2 per share for their holdings. United's physical assets became absorbed by American Marconi.

*DeForest Radio Telephone & Telegraph Company*

DeForest, after quitting in the beginning of 1907 directly after the disclosure of his three-electrode Audion, had started a new company called *DeForest Radio Telephone & Telegraph Company* to sell the grid Audion to radio amateurs (Figure 195). Manufacturing was done by the McCandless Company that produced between 200 to 600 vacuum tubes a year for him. The devices sold for $5 – $8. But business wise, his early commercial activities in the field of radio-tubes were not too successful.

Throughout DeForest's lifetime, the originality of his more important inventions was hotly contested by both scientists and patent attorneys. In time, realizing that he could not succeed in business, he reluctantly sold his vacuum tube patents to major communications firms.

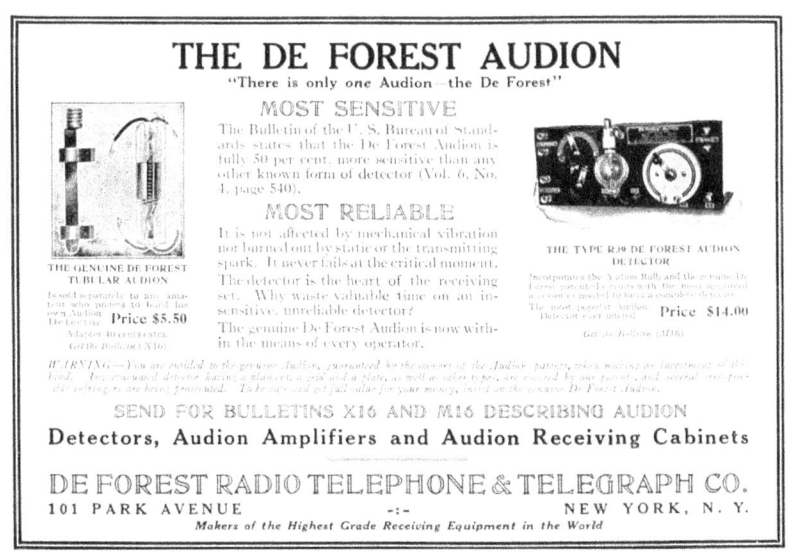

Figure 195: Advertisement for DeForest's Business Activities (1916).
Source: Wikimedia Commons.

DeForest was aware of AT&T's ambitious plan to create a transcontinental telephone line from New York to San Francisco. In 1912, he began work on converting the Audion into a telephone amplifier. Although it initially created a howling sound, AT&T was nevertheless impressed with DeForest's design, and bought the patent of the Audion for use in telephone repeaters and wireless communications in 1913 and 1914. It was Harold D. Arnold and his team of researchers at AT&T's Western Electric, who started working on improving the vacuum tube for telephony purposes. It started when in 1911 AT&T turned their laboratory into an Industrial Research Laboratory, bought DeForest's patent on the Audion and used it for amplification of telephone lines, but also thought they might get into what they called wireless telephony as a feeder for their telephone system. Their vison was already looming at the 'mobile telephony.'

And Lee DeForest? After selling the right of the Audion patents to the giant telephone company AT&T in 1914 for $90,000, he reorganized his company as *DeForest Radio Telephone Company*. But his business practises hindered its development, and in April 1917, the company's remaining commercial radio patent rights were sold to AT&T's Western Electric subsidiary for $250,000.

*Radio Patent War: DeForest's Patent Problems*

Over the course of his lifetime, Lee DeForest received more than 300 patents, the most important of which was for his Audion. And with many of his patents covering the fundamental aspects of wireless communication, came the patent infringements. His work was undertaken in the context of the much larger development of wireless communication, in which Marconi and his US-companies, other inventors like Fleming (the Fleming Valve), Fessenden (the heterodyne circuit),[383] and Edward Armstrong (the regenerative/superheterodyne circuit and Amplitude Modulation) as well as the US War Department and the US Navy were large players. And in that playing field there were the individual inventors like Lee DeForest. For DeForest the patent litigation started already early in his career when his early patent for the *responder* ( ie a detector aka receiving device) was challenged by Reginald Fessenden, who claimed priority.

> *Lee de Forest was accused on more than one occasion of dishonest business practices. One of the earliest happened in 1903 during a casual visit by de Forest*

---

[383] The heterodyne principle converted a low-frequency signal to a high-frequency wireless signal. It was the forerunner of the principle of superheterodyne reception, which made easy tuning of radio signals possible and was a critical factor for the growth of broadcasting.

*to fellow inventor Reginald Fessenden's workshop. Apparently de Forest, impressed by Fessenden's invention of the Liquid Barretter detector, stole the design and claimed it as his own invention. After three court appearances, Fessenden finally received an injunction against de Forest for patent infringement.*[384]

Whether the story is true or not, it resulted in the infringement case *National Electric Signaling Co. v. DeForest Wireless Telegraph Co.* (1905): The case was about reissues of Fessenden's US-Patent № 727,331 of May 5th, 1903 concerning Fessenden's receiver (ie the barretter) and the court ruled for Fessenden, and DeForest had to stop production. Sometime later, Marconi —who held the Fleming patent on the Fleming Valve— sued DeForest that his triode infringed on Flemings patents. But DeForest took the position that the two devices were completely different.

*Marconi Wireless Telegraph Co. v. DeForest Radio Telephone & Telegraph Co.* (1914): On November 1914 Marconi filed suit against DeForest Radio Telephone & Telegraph Co for infringement on Flemings two-element valve (US-Patent № 803,684). DeForest counteracted by claiming infringement on his Audion-patent. After demonstration of the similarity between Fleming's valve and DeForest's Audion, the court ruled in favor of Marconi, dismissing DeForest counter claim.

After his problems with Fessenden and Marconi, next Edwin Armstrong and DeForest engaged in what can be considered the most significant patent war of the early twentieth century halting the communication evolution as associated with wireless transmission. Although both inventors made clear contributions resulting in modern day radio, the courts cited in 1934 with DeForest. However, this did not help DeForest's finances, however, and in 1936, he declared bankruptcy.

> DeForest began a decades long court battle with Edwin Armstrong over who first discovered the regenerative characteristics of the Audion. Regeneration is feedback—a small signal from the output of a vacuum tube is fed back into the input— thus making weak signals very strong. Both DeForest and Armstrong claimed its discovery, and while the litigation lasted from 1914 to 1934. And while the courts would finally side with DeForest, the technical community did not.

Over time a range of small independent manufacturers infringed on DeForest's patents in effort to circumvent his rights. Among them Wallace & Company, the AudioTron Sales Company, the Electron Manufacturing Company, the Radio Apparatus Company, and many others. In February

---

[384] Source: www. LeeDeForest.org (old site) (Accessed June 2021).

1916 the DeForest Company filed suit against them and others, but the cases were settled out of court.

> At the start of the First World War, patent litigation surrounding the vacuum tube, stymied US radio technology growth. This was an obstacle for the US Navy, which sought better radio technology and equipment during the War. On April 7th, 1917, a presidential proclamation broke the patent impasse by handing over all commercial radio to the Navy, motivating research and development by offering indemnity for patent infringements. And it started to work together with three companies; General Electric, Western Electric and Westinghouse.

After the war, bootleggers flooded the market with vacuum tubes. The *Radio Corporation of America* —formed out of the American Marconi Companies and holder of Marconi patents— spent a fortune pursuing these infringers for prosecution and personal injunctions. Between 1920 and 1921, RCA reached cross-licensing agreements with the large electric industries like General Electric, AT&T, AT&T's subsidiary Western Electric, and Westinghouse concerning some 1200 patents. These cross-licensing agreements gave RCA the rights to the Marconi patents, the Alexanderson alternator, the DeForest patents on the triode valve, the Westinghouse Fessenden and Armstrong patents relating to heterodyne regeneration and feedback, and the patents regarding crystals and other detectors. It seemed to end the Patent War, although some cases took more time to finish.

*Armstrong et al. versus DeForest et al.* (1926): When Armstrong attempted to patent his new 'regenerative circuit,' DeForest filed suit claiming that he had actually devised the idea first. The resulting legal battle—the most significant patent war of the twentieth century halting the communication evolution associated with wireless transmission—between DeForest and Armstrong lasted from 1914 to 1934. Finally, at the end of two decades the Supreme Court ruled that, on the basis of a 1912 lab notebook entry, DeForest had indeed created a regenerative circuit earlier than Armstrong, even if he did not recognize its significance or file a patent application at the time. Although both inventors made clear contributions resulting in modern day radio, the courts cited with DeForest.

This exploration of the patent litigation resulting in the Radio Patent War that dominated the early days of the wireless transmission of 'radio signals' (ie sound, speech and music) before and after the First World War, illustrates the dynamics of technological development in the context of society at that time. A time where wireless communication had emerged and 'radio' technology was in its infancy, inventors-entrepreneur struggled to commercialize their inventions, and where large companies active in telephony made their first moves into the new technology of vacuum tubes. And on top of it were the governmental powers as executed by War Departments, and its institutions like their Navies.

In addition, the application field for the new phenomenon was massive. From the 'marine wireless systems' applied in commercial and military applications, to 'land based wireless systems' communications in 'point-to-point' communication and 'point-to-many' communications. The latter being a new phenomenon as it was driven by a device called 'radio' that could receive broadcasted communication.

*DeForest exploring other application areas*

And wat happened next with DeForest? After his unsuccessful activities in the tube business, in 1920 he began to work on a practical system for recording and reproducing sound in motion pictures. He developed a sound-on-film optical recording system called Phonofilm, applied for US-patents up to 1919,[385] and demonstrated it in theatres between 1923 and 1927. However, this venture became a failure as the in 1922 started company *DeForest Phonofilm Company* went already bankrupt by 1926.

From 1921-1924, his former classmate Theodore Willard Case— born in a wealthy family, fascinated by the silent movies, and having established in 1916 the Case Research Laboratory— provided DeForest with the technology of sound recording; from the patented Thalofide light sensitive vacuum tube, and the patented AEO light used to record the soundtrack, to the camera system which could record the sound onto the film in the camera itself.[386] However, that initial cooperation with Theodore Case collapsed. The dispute between Case and DeForest was due to Case not being properly credited for his lab's contributions to Phonofilms. In September 1925, Case stopped providing DeForest with his lab's inventions, effectively putting DeForest out of the sound film business, but not out of the "claiming to have invented sound film" business. The following July, Case joined with Fox Film, Hollywood's third largest studio, to found the Fox-Case Corporation. The system developed

---

[385] Such as US-Patent № 1,543,990(filed on April 24th, 1915, granted June 30th, 1925), US-Patent № 1,486,866 (filed on August 20th, 1918, granted March 18th, 1924), US-Patent № 1,554,456 (filed on August 2nd, 1919, granted March 18th, 1924).
[386] Such as US-Patent № 1,517,103 (filed on April 26th, 1921, granted November 25th, 1924), US-Patent № 1,628,822 (filed on December 19th, 1921, granted May 17th, 1927), US-Patent № 1,816,825 (filed on May 28th, 1927, granted August 4th, 1931),

by Case and his assistant, Earl Sponable, given the name Movietone, thus became the first viable sound-on-film technology controlled by a Hollywood movie studio.

Although the Phonofilm system was basically correct in principle, its operating quality was poor, and he found himself unable to interest film producers in its possibilities. Paradoxically, within a few years' time, the motion-picture industry converted to talking pictures by using a sound-on-film process similar to DeForest's. During the 1930s DeForest developed Audion-diathermy machines for medical applications, and during the Second World War he conducted military research for Bell Laboratories. Lee DeForest died in Hollywood, California, on June 30$^{th}$, 1961. Despite holding almost 180 patents and having a major impact on the radio industry, his many legal battles and poor business practices kept him from amassing an estate, and he had just $1,250 in his bank account at the time of his death.

This could be the moment to reflect, like so many contemporaries of both Fleming and DeForest have done, on an important question…

## Who invented the Radio Tube?

In the preceding part we explored how many people contributed to the development of the vacuum tube for non-lightning purposes (aka the Cathode Ray Tube) during nineteenth century and the subsequent twentieth century (Figure 217). Each contribution —from the one with a small impact to those with a big impact— had its own value. Contributions that resulted in the vacuum tube with internal grids to control the flow of electrons that became known as Radio Valves (aka Vacuum Tubes). But who then, one would be inclined to ask, can be considered as the inventor (ie the classic quest for one specific person), of the radio tube?

> This priority question has been a topic of discussion among historians of science. We will not attempt to mingle in this discussion, as it is not relevant to our explorations, but we will just try to present the different points of view within the context of that period of time.

*Facing the Big Question*

When trying to address the classic question of 'who is the inventor,' one is faced with some problems of interpretation.

Also in this case, one is facing the problem *semantics*: what exactly is an invention? What is the difference with a discovery? In what way is invention different from innovation? This is a problem that seems to be solvable by defining the notions (see Preface). However, one soon realizes that there is quite some semantic confusion as one looks up the different definitions scholars have been using.

The next problem is defining the *subject*; what is the subject of the invention? Is it the *creation* of the flow of electrons—aka the 'cathode ray'—in a vacuum tube (aka Cathode Ray Tube, CRT)? Or is it the *control* of that flow of electrons in a vacuum (Fleming's Thermionic Valve), or is it the *regulated control* of that flow of electrons (as in DeForest's Audion)?

Another problem is in the *moment* of establishing priority: the question of 'who is the first' when he did what he did: conceptualizing his idea, making his artefact by experimenting his working prototype(s), applying for a patent application or obtaining a patent for it, or selling the first commercialized products. Sometimes this time-aspect is a question that is related to the moment when someone—observing the phenomenon—made a *discovery* and the moment he/she published a paper or presented a demonstration about it. Sometimes this question is related to legal priority (related to the patent situation) of the person who created the *invention*. Was it the application date, or the date when he was granted a patent for it (that could be years later)?

In addition, the *consequences* of the priority play a role. Sometimes it is about honouring somebody who made a range of important contributions. Then it is about the honour to be 'the father of...'. But the consequences can also be more serious, as commercial interests —always translating into financial interests— play a role. Then the priority discussion can result in litigation about infringements of the patent rights.

A final problem is related to the *point of view* one uses to address the question. Is it a contemporary point of view heavily influenced by personal interests, ambitions, and prestige? Or is it a later point of view claiming a national interest (eg the national pride) or ideological standpoint (eg the Nazi-regime rewriting history), as can be observed by nearly all countries that want to honour their own national heroes? Or is it a present-day historiographic point of view that reflects on history, trying to clarify it from a historical point of view?

Clearly, there are many different —sometimes opposing and contradicting— interpretations about who is the inventor of the radio tube. They may differ, but nevertheless, all those interpretations have the process of 'Change and Novelty' in common. But again, even in that process, there are differences. Take 'the invention', the result of the process of invention as carried out by the specific person —the inventor— or a group of persons —the R&D team— that creates novelty (Figure 217). Next, there is the 'innovation', the result of the process of Change and Novelty creating a working artefact that —as the result of entrepreneurial activities— can be brought to the market by subsequent organizational activities. Both those processes, invention and innovation, make up the *Act of Invention* and the *Act of Innovation*. However, then again, confusion arises. Is that about the original 'idea' (as in the mental picture), is it the first model or prototype (as in the archetype), or is it the final product/system that is successfully commercialized?

Using our definition of the basic invention (Scheme 3) as the dominant innovation in a cluster of innovations, we let others decided on what was the 'invention.' Such as the contemporaries in the Patent Offices who decided in the invention was patentable on criteria of their own, valid in their time. In order to explore the inventions, we used the patent as an event-identifier (Figure 196).[387] And we let the historians of science and technology, who explored the Fleming and DeForest case extensively, decide who was *the* inventor.

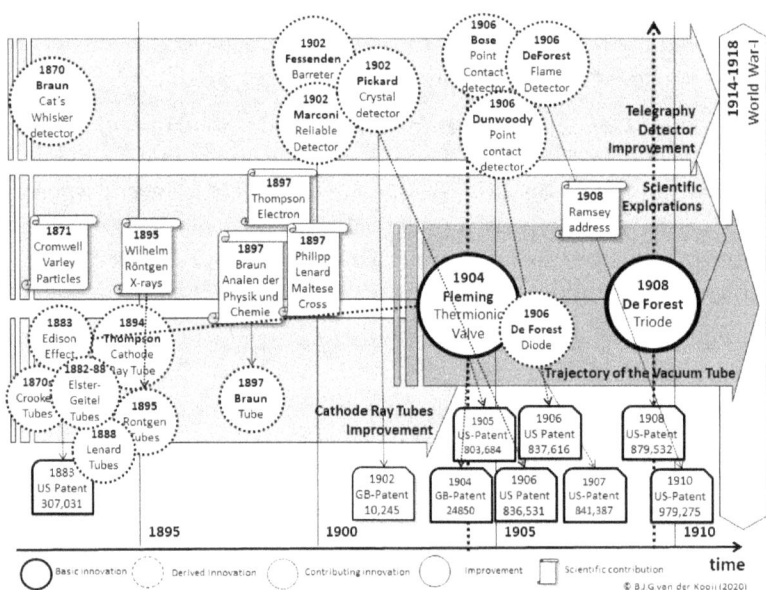

Figure 196: Contributions to the Cluster of Innovation with the Basic Inventions of the Diode and Triode Vacuum Tubes.

---

[387] During prosecution of a patent, a Patent Office examiner reviews an application to determine what is patentable. To be patentable an invention must meet all the statutory requirements for patentability: novelty, utility and non-obviousness. The first part is a specification of the invention. The second part of the patent application is a set of claims, which usually encompass more than the material set out in the specification. Claims define what the inventor considers to be the scope of her invention, the technological territory she claims is hers to control by suing for infringement.

## Early Development of Vacuum Tubes

The development of vacuum tubes by Fleming (1889) and DeForest (1906) would create the devices called diode and triode based on a 'vacuum tube technology'. Both devices made it possible to control (small) electric currents; the blocking function of the diode, and the amplification function of the triode. Thus, these vacuum tube devices became *active components* of electronic circuits. After its introduction, the vacuum tube would go and provide the necessary vehicle for applications like wireless signal transmission and detection, the sound amplifier, the wireless telephony, radio and television.

Figure 197: Evolution of Audion Vacuum tubes (1910s).
Source: Wikimedia Commons.

## Evolution of the Vacuum Tube into Radio Communication

Early Audion tubes were crude experimental devices with a limited lifespan, and electrical properties ill understood (Figure 193). From those experimental tubes DeForest developed other, improved versions. Due to his patent rights being sold to AT&T, DeForest could only sell them to amateurs. Eventually this restrictive policy was relaxed somewhat, after the Elmer T. Cunningham's AudioTron Sales Company in Oakland, California began selling a cheaper bootleg vacuum tube, and the DeForest company responded with the Type 'T' Tubular Audion Tube.

Other manufacturers were more or less taking over the further development of the concept of the glass vacuum triode (Figure 197). In their R&D departments they worked on the continues improvement of the vacuum tube. To mention some of the more prominent contributions:

Irving Langnuir (1881—1957) from Gottingen University in Germany worked on the research of the thermionic emission in vacuum bulbs. In 1909, he moved to General Electric USA to research the capabilities of Tungsten filaments in Audions. In 1913, he published the results (partly based on the research of Owen Williams Richardson (1879-1959) of the UK regarding thermionic emission). His conclusion was that no gas is required in Audions, but a high degree of vacuum.

The research of Harold D. Arnold (1883-1933) of AT&T's division Western Electric on the efficiency and life of filaments in Audion bulbs, included adding a layer of oxide over the filament coil. He also removed the cases and created a vacuum in the bulb. This improved both the electrons emission and the life of the Audion bulb.

Although DeForest had not realised it, the amplification capability of the triode would become its main feature after Fritz Lowenstein (1874-1922) showed use of the Audion as an amplifying device in 1911. In Britain, H. J. Round (1881-1966) patented amplifying and oscillating circuits in 1913. From then, the three-element vacuum tube with the signal amplification capability (later on it was called 'Triode'), became the building blocks of radio receivers in the 'Era of Broadcast Radio' that started in the1920s. Next to AT&T, the British Marconi companies joined those experimenting with the new vacuum-tube transmitters, and from there on developed their own line of Audions (Figure 198).

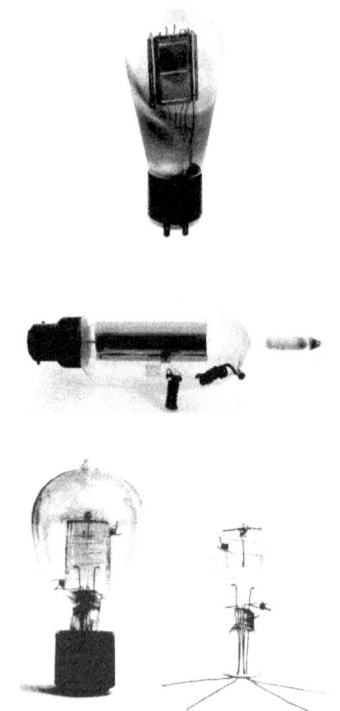

**Figure 198: Early Vacuum Tubes.**

Audion Tubular Tube: 410-series (left). Marconi's Round Valve, Type N (1914-1915) and the GE Philotron type CA with and without bulb (bottom, 1917)

Source: http://www.r-type.org/exhib/aac0042.htm, New York Heritage.

The technical evolution of the glass vacuum tube technology continued towards miniaturization (Figure 199), multifunction tubes (Figure 201), improvement of service life and reliability, increase in power and frequency to meet the requirements of the radio-transmission and electronic circuit designers that developed their end-products.

And to fulfil the increasing need for all those different forms of communication equipment made by the equipment builders, the components had to be manufactured as a mass product. The artisan time of 'glass blowing a single tube' was over. It resulted in a booming business development, both in US and Europe (Great Britain, France, Germany, Netherlands). A development that fitted in the context of its time.

## Early Vacuum Tube Manufacturing Industry

Next to the described activities of early independent manufacturers of vacuum tubes in the US (with the Wallace & Company, the AudioTron Sales Company, the Electron Manufacturing Company, the Radio Apparatus Company, and many others), also in Europe the new active component of the Audion triode had caught the attention of the scientists engaged in wireless communication. The pressure for development of wireless communication was engendered by the political pre-war tensions in Europe. Already convinced that wireless telegraphy communication had a challenging future, governments supported early experimenting with the new forms of wireless audio communication. And with that experimenting came the early participants that undertook their manufacturing activities.

In France the development of vacuum tubes was a military affair, after Lee DeForest demonstrated an arc-type transmitter from the Eifel tower. However, it took to 1914 before serious interest picked up. The military then arranged for an incandescent lamp maker, E.C. & A, Grammont in Lyon, to manufacture radio valves. With a skyrocketing demand, also other manufacturers, such as the Compagnie General des Lampes, were pressed into military service.

In Germany the Count von Arco, founder of the Telefunken Company, did send people to the US to acquire knowledge and samples of the vacuum tube. With the knowledge acquired added to their own knowledge base, their engineers started exploring the field of vacuum tubes. Telefunken rapidly became a major player in the radio and electronics fields, both civilian and military. During the First World War, they supplied radio sets and telegraphy equipment for the German military. Telefunken set up the first vacuum-tube based broadcasting transmitter in Berlin in 1923. As demand for radio spread, its parent companies Siemens and AEG built their first receivers with the Telefunken trademark. For the next forty years, Telefunken radio receivers and tubes became renowned across Europe for their quality.

In England, the British Marconi Company —already active in wireless communication after a tumultuous interaction with the British Post Office, British scientists and the British Imperial aspirations— engaged Fleming as consultant. His work was the dawn of the British radio tube manufacturing from which Marconi-Osram Valve Company (1919) emerged as a merger the valve making interests of British GEC (Osram) and the Marconi Company. It became Britain's most important tube

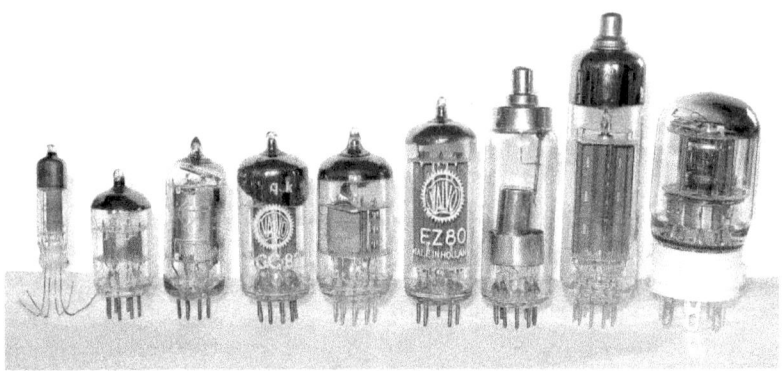

Figure 199: Interbellum Evolution of Philips Vacuum Tubes.
Source: Wikimedia Commons.

maker. Capt. Stanley R. Mullard, a research engineer for the British Admiralty, founded the Mullard Radio Valve Company Ltd. in 1920 and it became a premier manufacturer of electronic components.

In the Netherlands is was light bulb manufacturer Philips that picked up on the new device of the radio valve, and put its R&D laboratory 'Natlab' on a development project into radio technology. Next to the electric incandescent lamps, the radio-tube based receivers would become a major business activity. Later to be followed by the television.

These rather rough brushstrokes of manufacturing activities create a picture of the emerging wireless industry in all its variations from component and circuit makers to equipment and system suppliers. A widespread industry that became known as the communications industry. The development of vacuum tube in other application fields would result in the computer industry, the radio industry and the television industry.

## The Radio Tube Manufacturing Industries

As the tube technology and its manufacturing evolved, increasing the number of wires/collectors/controllers to supply triodes, tetrodes, pentodes, each new configuration was overcoming weaknesses in the previous offerings and providing improvements for signal handling and amplification. Improving its functionality and its usability, reliability and durability, that continuous evolution of the technology was dominantly realized in the R&D-facilities of industrial manufacturers.

A whole industry developed around the mass production of vacuum tubes. Often companies active in the production of incandescent lamps that were using their existing technology-base (glass blowing techniques, vacuum technologies) and the fine-mechanical wire assembly techniques for these new devices. Vacuum tube manufacturers would emerge all over the world; from the American companies RCA, Sylvania Electric Products and Westinghouse, the Marconi companies in Canada, England and Italy, to the Philips Gloeilampenfabriek in the Netherlands.

> The radio vacuum tube in all its variations became a mass-product. As an indication of this growth, figures from the RCA company show that they sold about 1.25 million receiving valves or tubes in 1922, but by 1924 this has risen to 11.35 million.

The vacuum tubes became the basic element in new electronic systems that realized the same function as their mechanical predecessors. The function was the same, but the technology to realize that function changed. Like, after the mechanical calculator and clock, the electronic calculator and clock would appear on stage. But they also beame the basic component for new applications (eg the radio and television) that did not exist before.

*Big Business Development*

While the superpowers exercised geopolitical colonial ambitions, the giant companies of wireless equipment such as Marconi, Siemens, AEG and General Electric traded, pooled, manipulated and disputed key patents with a view to creating technological monopolies that would assure them massive profits in this new line of business.

In Germany Siemens and AEG competed for German military contracts and international business against Marconi, the dominant power in wireless at the time. At the same time Siemens and AEG were locked in patent battles both with each other and ultimately with Marconi. Realising that Germany needed one powerful wireless company, the Kaiser, Wilhelm II, ordered that the two companies pool their resources and patents through the creation of a joint venture company that would focus on the development of wireless and a dominant position internationally. This company became known in 1903 as the *Telefunken Gesellschaft für drahtlose Telegraphie m.b.H.* The company rapidly became a major player in the radio and electronics fields, both civilian and military. During the First World War, they supplied radio sets and telegraphy equipment for the military, as well as building one of the first radio navigation systems for the Zeppelin fleet.[388] And it were the mother companies of Telefunken (ie AEG and Siemens & Halske) that manufactured the vacuum tubes. It was the start of massive activity in

---

[388] Telefunken did build a continuous-wave transmission system across the Atlantic Ocean in 1911 in order to evade the British monopoly on international cabled communication.

Radio Communication; both the manufacture of components (ie the vacuum tubes) and the systems (aka radios, broadcasting equipment).

In the Netherlands the *Philips Gloeilampen Fabriek*, manufacturers of the incandescent lamp (ie 'gloeilamp' in Dutch), stimulated by the scientists of the Natlab,[389] started with the development and production of vacuum tubes in 1918. They manufactured the vacuum tubes designed by Staringa Idzerda (Figure 200, left). Next to the receiver lamps, also transmission lamps were developed. Subsequent development resulted in a broad range of vacuum tubes. The period 1922-1923 saw the beginning of a great expanse in the valve manufacture, with 9 receiving, 4 transmitting and 3 rectifying types being announced and released to the public "for amateur and experimental use only". As in 1923 the idea of public radio broadcasts really sets of, the demand for radio tubes exploded. Tube manufacturers were hardly able to supply the demand, and Philips quickly changed gears. After the fabrication of light-bulbs was moved to a location just outside the centre of Eindhoven, the huge plant on "de Emmasingel" became available for the production of radio tubes. Whereas in 1921 only 320 radio tubes were fabricated in total, in 1923 the production had increased to 1000 tubes a day. Philips successfully introduced in 1924 a range of smaller vacuum tubes under the brand name 'Miniwatt.' (Figure 200, right).

**Figure 200: Philips Vacuum Tubes.**

The IDZ lamp designed by Idzerda (left, 1918) and the Miniwatt Tube (right, 1924).

Source: Invaluable.com, http://www.vintageradio.nl/

At that moment already 9 out of the 16 researchers at Philips were working on radio tubes. The popularity of radio increased explosively, and it had not remained unnoticed by Anton Philips that,

---

[389] The Philips Physics Laboratory (NatLab) was founded in 1914, partly as a response to the new Dutch patent law of 1910 and following the example of General Electric. Like the rest of the firm, it was located in the city of Eindhoven. After the war it would become the main industrial employer for young Dutch physicists. Originally the company, founded in 1891 by the Philips brothers, produced only light bulbs. But in the early 1920s it started a policy of diversification, first by adding radio tubes and X-ray tubes to its repertoire and then by producing consumer goods like radio sets. The laboratory was instrumental in the creation of these new products.

whereas the price of a radio tube on average was 10 guilders, the price of a complete radio was about 200 guilders. So obviously making radio sets was a very profitable business. This, combined with the fact that Philips now had their hands on a superior radio tube, the pentode, made them feel confident enough to start thinking about producing radios themselves. For Philips in 1926 this thinking was something of a revolution. So far Philips had been a component manufacturer. A radio was a system, and manufacturing a system would require a completely different approach. They studied the manufacturing process extensively. And the result was the Radio Set 2501 (Figure 362), in Holland nicknamed "the loaf of bread." Because of the high gain of the pentode, only three radio tubes were needed compared to four in radios from competitors. It was an enormous success. In 1927, 6000 receivers were built; in 1930 the production had already increased to half a million radios a year.

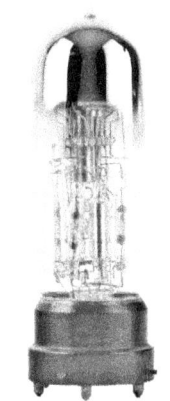

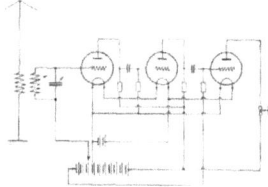

**Figure 201: The Integrated Vacuum Tube.**

The Loewe Multi-system 3NF vacuum tube (top, 1926) and its schematics (bottom).

Source: http://lampes-et-tubes.info/, Loewe.

That all happened in Europe, and in America a similar development took place. In 1919 a group of American companies formed a joint-venture company analogous to what had been created in 1904 as the German Company Telefunken. Those American companies were General Electric (GE), American Telephone and Telegraph (AT&T) and Westinghouse, and their joint-venture was the *Radio Company of America* (RCA).[390] It would become a dominant player in the Era of Broadcasting.

---

[390] RCA was actually created by GE but AT&T and Westinghouse participated by turning over to RCA the rights to patents they held in radio technology.

## Small Business Development

It were not only the big business of that period of time —with their R&D departments and large financial resources— which became involved in the development of the vacuum tube. Also many inventive individuals contributed to its technical development, creating their own invention factories. And some of them, starting to commercialize their inventions, became entrepreneurs. Such as happened in Germany during the Interbellum, during the Golden Age of German radio.

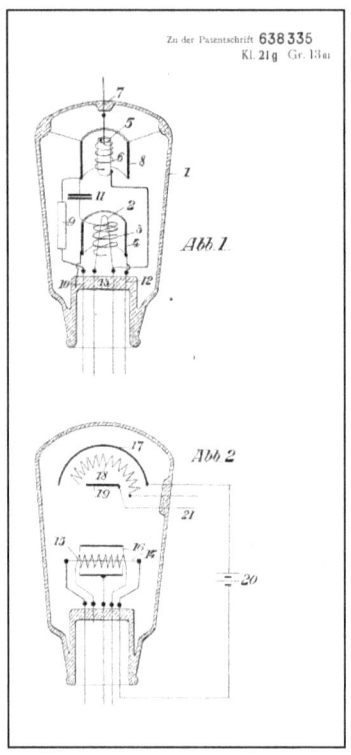

**Figure 202: Loewe's DE-Patent № 638,335.**
Source: Espacenet

The German Manfred von Ardenne (1907-1997), born in Hamburg to a wealthy aristocratic family, was home-educated by private teachers. In 1923, at the age of 15, he received his first patent for an electronic tube with multiple (three) systems in a single tube for applications in wireless telegraphy (Figure 201). At this time, von Ardenne prematurely left the Gymnasium to pursue the development of radio engineering with the entrepreneur Siegmund Loewe, who became his mentor. In 1928, at the age of 21, he came into his inheritance with full control as to how it could be spent, and he established his private research laboratory *Forschungslaboratorium für Elektronenphysik*. At the Berlin Radio Show in August 1931, Ardenne gave the world's first public demonstration of a television system using a cathode ray tube for both transmission and reception (Figure 358).

The German Siegfried Löwe (1885-1962), born into a Jewish-Protestant family, had a fascination with electrical engineering and, after having received in 1912 his doctorate magna cum laude, focussed on high frequency technology. After working for Telefunken, he founded a laboratory that dealt with radio and vacuum technology (and also with medical problems). During his stay in America In 1922, he improved electron tubes together with the American Lee DeForest. Also early as 1922, Siegmund secured the collaboration of Manfred von Ardenne. The close collaboration, which lasted until 1932, decisively fertilized the research in his laboratory.

*Circuit integration:* One result was the Loewe triple tube 3NF (Figure 201); four resistors and two capacitors were housed in one glass vessel in addition to three triode systems. This was an extension of the first vacuum based integrated circuit for which he applied for a patent in September 1924 (Figure 202).[391] In addition, he received patents for the cone loudspeaker, in which the sound-signal was converted directly through the use of membrane.

His mathematical and physical talent was fortunately complemented by his experimental and constructive skills. As a result, the Loewe factories became the nucleus of German radio and television development. In January 1923, together with his brother David, he founded Radio-Frequency GmbH for the construction of radio equipment, a pioneering company in the radio industry in Germany.

The idea of system intergration in one glass vessel was born and would result in the following years in specialized tubes; the multi-section tubes ranging from twin-triode designs and the double-diode-triode designs to the 'Multivalve' triple triode. By 1940 multi-section tubes had become commonplace in radio receivers.

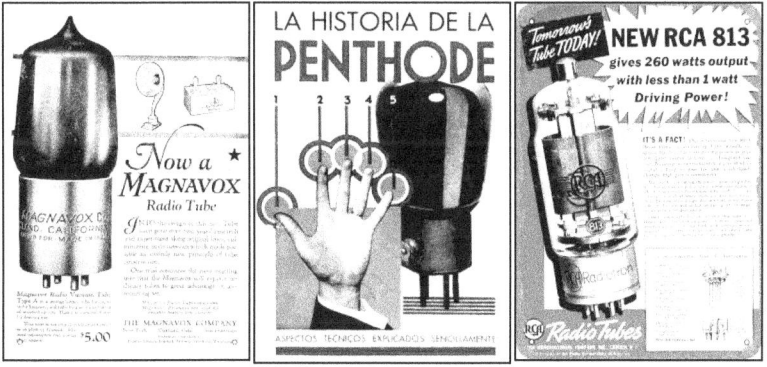

**Figure 203: Promotion for the Vacuum Tube.**
Magnavox metal sign (left, 1924), Philips Penthode Valve Brochure (middle, 1927), RCA-Tubes (right, 1938).

---

[391] He was granted DE-Patent № 638,335 (filed November 13th, 1925) on November 13th, 1936. Similar patents were US Patent №1,686,018 (filed on August 29th, 1921) granted on October 2nd, 1928

These are just a few examples of how inventive and entrepreneurial individuals contributed to the development of the vacuum tube in a broad range of variations developed for a broad range of applications. Together, the small manufacturers and the tube divisions of the large electric corporations created the vacuum-tube industry. Dominated in the US by the RCA, in Europe would arise similar manufacturers (eg Philips, Telefunken, Marconi Osram Co.) And all those radio tubes, manufactured by a range of companies, were promoted (Figure 203). From technical documentation to sales promotion, other developers were stimulated to use the vacuum tube. Their products created the foundations of the radio industry; ie the manufacturing of radio receivers that increasingly dotted the rooms of private homes.

## Patent Scope: Blocking Patent Positions

As we have explored earlier, the many contributions of the early inventor-entrepreurs as well as the scientists engineers working for the large electric companies, were resulting in a considerable patent activity afteer the war (Figure 204).

This was also the case with the inventions within the wireless technologies and the subsequent technologies for radio eguipment (eg the radio-transmission and the radio-receiving). They were patented by a range of inventors —assigning them to their companies— that applied for patents with a broad scope.[392] And the effects of those broad-scope patents was hindering the early development pace of the fruits of those technologies.

> *The earliest radio patent was a broad patent granted to the British inventor Marconi in the field of radio transmission. Marconi also invented and acquired rights to the basic technology for tuning, which he controlled until 1914, and the basic Fleming patent on the two element vacuum tube, or diode. These patents helped the Marconi Wireless and Telegraph Company establish an imposing presence in the early radio industry, which was dedicated primarily to large-scale commercial uses such as ship-to-shore communications.*
>
> *AT&T, as part of its radio operations, acquired rights to two very fundamental patents on the triode vacuum tube, an early radio wave amplification device patented by Lee De Forest. While technically only an improvement on Marconi's diode, the triode was in fact a very significant advance; it was called "the heart and soul of radio."*

---

[392] The second part of the patent application is a set of claims. Claims define what the inventor considers to be the scope of her invention, the technological territory she claims is hers to control by suing for infringement. The broader the scope, the larger the number of competing products and processes that will infringe the patent.

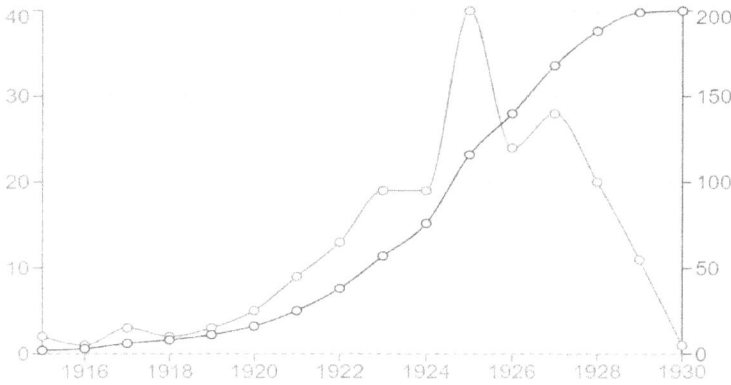

**Figure 204: Overview Patents for the Thermionic Tube.**

Tables show yearly and accumulated number of patents. Left axis is the number of patents in a given year, Right axis the accumulated volume of patents.

Query (top): (ti = "thermionic tube" OR ti = "thermionic valve") AND pd = "1900-1930"

Source: Espacenet

*Several other firms had important patent positions. General Electric entered radio as a natural extension of its expertise in electricity generating systems. It controlled the important Alexanderson patents on the electric alternator, the signal generation invention that made long-range transmission possible. Westinghouse also joined the industry, mostly on the strength of patents on receiving technology, which served as the basis of the firm's successful entrance into the inexpensive home receiver market. Other companies also held American rights of varying breadth over other important radio technologies. (ibidem*

*[...] The upshot was that no one could produce state-of-the-art radio technology without being threatened by litigation. Radio is thus a canonical instance where the presence of a number of broad patents, which were held by different parties and were difficult to invent around, interfered with the development of the technology. The various pioneers formed RCA to break the deadlock; the new company promptly acquired the American rights to the Marconi patents. The companies that owned most major radio patents became RCA shareholders. With all the constituent radio technologies under one roof, RCA established itself as the technical leader in radio and dominated its advance for many years.* (Merges & Nelson, 1990, p. 20)

And in this technological playfield, the various players fought each other in court and held their developments hostage. It took a war, and the intervention of an external party (ie the Navy) to start to find a solution.

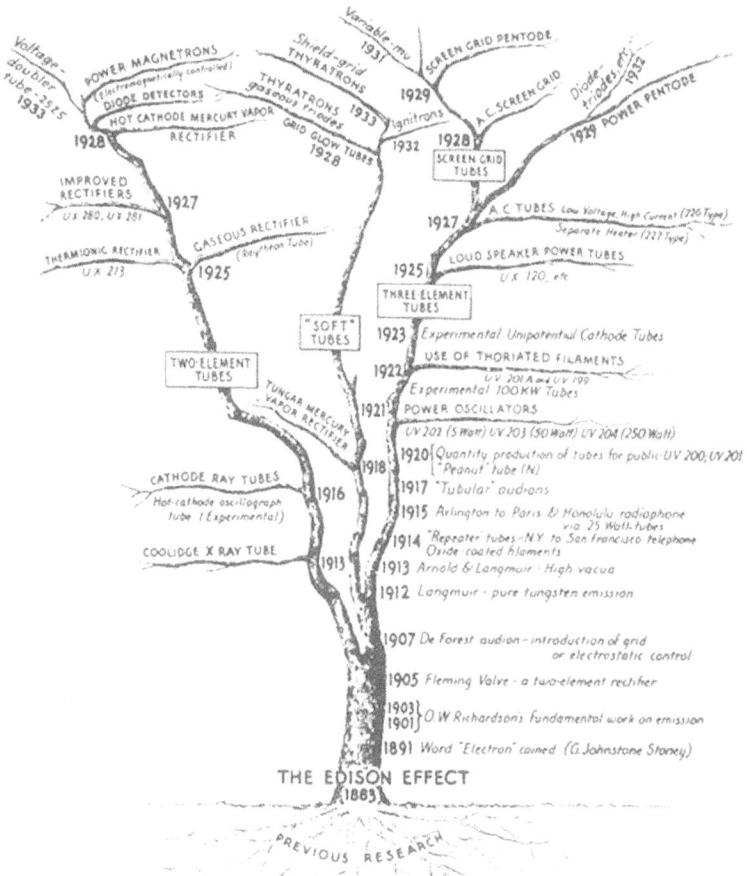

Figure 205: Family Tree of the Vacuum Tube Evolution (1900-1930).
Source: ieeeghn.org

The consequence of this development was that the Radio Corporation of America (RCA) set the scene for the developments in the following years. But that dominance did not block any more the further development of the vacuum tube device into different application fields with specialized tubes.

## *The Evolution of the Vacuum Tube*

Over a period of some three decades, from the work on the Fleming valve in the early 1900s, the development trajectory of the vacuum tube unfolded. The supporting technologies (eg the glass technology and the vacuum technology) led to improvements in performance.

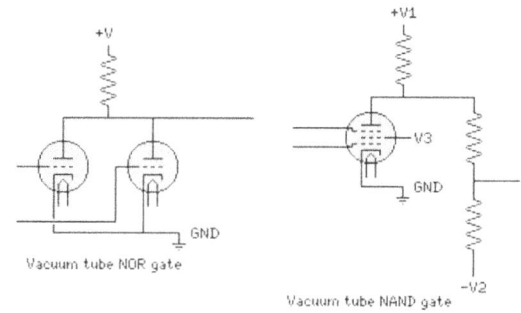

**Figure 206: Principle of the Logic Gates realised with Vacuum Tubes.**

Multiple grids placed within the tube added advanced functionality —eg the pentode— were opening new application trajectories. It was a continuous wave of improvements of the thermionic tube/valve also reflected by the number of patents granted, peaking in the mid-1920s (Figure 204).

## Evolution of the Vacuum Tube into Information Processing

The broadening range of applications required specific vacuum tubes (Figure 205) with different structural designs. A development that resulted in the multi-grid tubes such as the two-element tubes (aka diodes), the three element tubes (aka triodes) up to the five-element tubes (aka pentode). But also tubes that could handle greater power and higher frequencies.

The vacuum tube had found its breeding ground in the communication applications (aka radio). Next to that use, there was the application of information processing (aka computing). A field that had evolved from the mechanical calculating machines, and that had seen the application of relay logic to create relay-based computing machines.

> The function of the elekto-mechanical relay —a binary device with two states ON (ie activated) or OFF (ie not activated)— could be used to design logic circuits with a specific function such as a computing machine. Combinations of relays in a circuit would create more complex logic functions. Such and the AND, NAND, OR, XOR, and NOR logic function of Boolean Algebra.[393]

---

[393] *Boolean algebra* is the branch of algebra in which the values of the variables are the truth values true and false, usually denoted 1 and 0, respectively.

The same basic logic functions could be created with the vacuum tube (Figure 206). One of the early logic circuits was the 'flip-flop' circuit with a memory function, a crucial building block of digital circuitry. As the vacuum tube could be used instead of the relay, it was creating the same computing function (Figure 207).

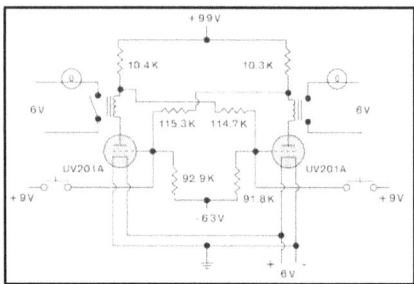

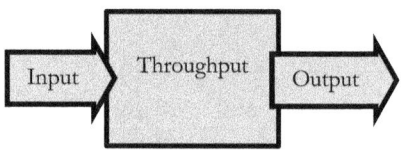

**Figure 207: From Electronic Circuit to Black Box.**

Pressing shortly the left button triggers the left-side tube to stay activated. Idem, pressing the right button, triggers the right-side tube. This is the memory function of the flip-flop.

The flip-flop circuit was used in sequential logic.[394] It was designed by the British physicists Frank W. Jordan and William H. Eccles. Jordan was granted GB Patent № 26631 (filed: 1912; published: November 20th 1913). It was the beginning of a range of logic circuits to be used in increasing complex logic machines.

A first development of an electronic calculating machine was the Atanasoff–Berry computer (ABC) that was conceived in 1937 and build by 1941. The family of logic gates ranged from inverters to circuits with two and three input gates. This electrical computing machine, with more than 300 vacuum tubes, could compute complicated algebraic equations. The system weighed more than seven hundred pounds (320 kg). It contained approximately 1-mile (1.6 km) of wire.

This was the start of the 'digital' electronic computer that arose from the early concepts of the mechanical calculating machines.

---

[394] *Sequential logic* is a type of logic circuit whose output depends not only on the present value of its input signals but on the sequence of past inputs, the 'input history' as well. This is in contrast to *combinational logic*, whose output is a function of only the present input. That is, sequential logic has 'state' (memory) while combinational logic does not.

*The Electronic Circuit as Building Block*

Constructing the Atanasoff–Berry computer was still a massive undertaking. But the foundations for logic circuits had been laid.

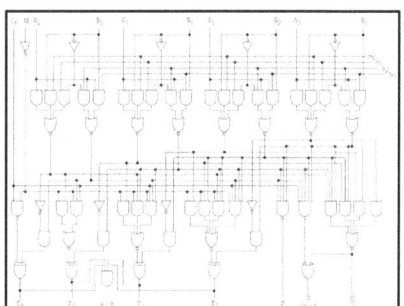

In logic circuitry, the simple basic functions AND, OR, NOT realised in the AND-, OR, and NOT gates became combined creating *building blocks* like flip-flops, counters, adders, registers,[395] and multiplexers. Sometimes the circuits were constructed out of some 100 gates. The systems designer used these circuits as a black box which needed an input signal to create an output signal

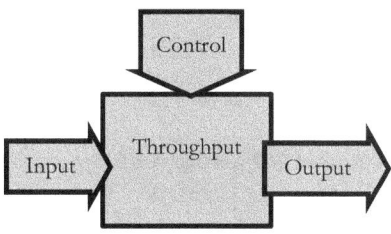

**Figure 208: Arithmetic Logic Unit (ALU) as Black Box.**

(Figure 207, bottom). These became the fixed Input-Throughput-Output *building blocks*.

From the design of simple I/O logic circuits arose the more complex electronic logic systems with the rise of the computing concepts. Over time the circuit designer using the building blocks created complex *electronic circuits* like 'control units' (CU) and 'arithmetic logic units' (ALU). Circuits that became 'controllable' (ie the function of Control in Figure 208)..Where controllable' would become 'programmable' over time.

Next, the building blocks were used to create the *electronic systems* for computing applications (aka digital circuitry). Systems with a Central Processing Unit (CPU) complemented with the Input/Output units (I/O) and a Memory unit (MEM) (Figure 209).

---

[395] A Register is a device which is used to store such information. It is a group of flip flops connected in series used to store multiple bits of data.

Originally these building blocks were used to create the early computing machines with the vacuum tube technology, like the Atanasoff–Berry computer (Figure 210). But soon the advance of the semiconductor technology trajectories, brought the circuit design to the next level of complexity. First with the transistor, next with the integrated circuits. But was to come in the near future.

## From Active Component to Electronic Circuit

The vacuum tube — having its apogee as an amplifier in communication systems—, developed over time in a range of models with specific properties. They were used as the active element in a multitude of electronic logic circuits, that became the building blocks for larger electronic systems. Among those we find the first-generation 'electronic' calculators and computers.

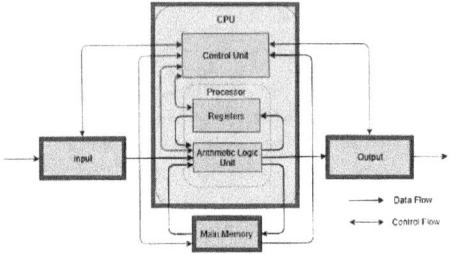

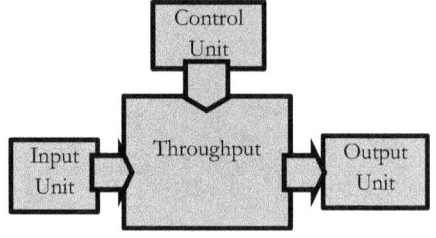

**Figure 209: Design of a Central Processing Unit from building blocks.**

*The Logic Circuit:* Originally there were the primary logic circuits —ie AND, OR, NOT circuits— realised in the electromechanical relay technology. In 1919, William Eccles and F. W. Jordan published a short paper in The Electrician describing a circuit for which they had filed a patent the previous year. The paper was titled "A Trigger Relay Utilising Three-Electrode Thermionic Vacuum Tubes." In it, they showed a configuration of cross-connected triodes that had two stable states and could be triggered to switch from one state to the other. This type of circuit, originally called the Eccles-Jordan trigger circuit, based on two triodes with a feedback mechanism, became known as a bistable multivibrator, or flip-flop.

Subsequently, in the Era of the vacuum-based computers before and after the Second World War, many more different standard vacuum-tube based modules (eg logic gates, adders, memory registers) were designed.

Placed in racks, the became the building blocks for the larger computing systems (Figure 210).

*The Operational Amplifier Circuit:* But it was not only the digital logic circuitry that rose to prominences. The operational amplifier, or 'op-amp'[396] became a building block for many non-digital systems (eg the electronic analogue computers). After the early versions of negative feedback circuits developed in the 1928s by Harold S. Black working at the Bell Laboratories, and the Brit Alan Blumlein in the 1930s (UK Patent № 425,553 granted in 1935), its further development in the early 1940s allowed unwieldy mechanical contraptions to be replaced by silent and speedier electronics. Many analogue computers relied on vacuum-tube op-amps, available commercially from George A. Philbrick's company in 1952 (Figure 211).

## Modular Circuitry

These are just a few of the electronic circuits in which the radio tube —and all its derivatives— were used by the designers of electronic circuitry. Soon the circuitry would be used in larger systems; from telephone system to computer systems, and from radio broadcasting systems to television broadcasting systems.

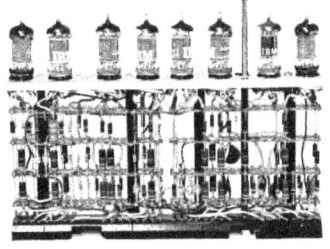

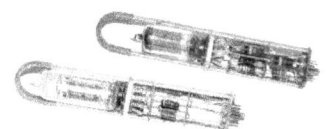

**Figure 210: Tube based Electronic Logic Circuits.**

Atanasoff–Berry computer, add-subtract module (top, 1942), IBM 700-series, pluggable Unit with 8 vacuum tubes (upper middle, 1954), Flip-Flop based memory module (lower middle, 1955), and tube module from IBM 604 (bottom, 1955).

Source: Wikimedia Commons, Computer history Museum.

---

[396] An Operational Amplifier is a specific electronic circuit acts as a differential amplifier using a negative feedback circuit.

The more complex the circuitry, the combination of components were that were needed, required increasing space. To reduce the size of the circuitry, efforts were initiated to miniaturize the building blocks.

The US government funded in 1950-1953 a study code-named Tinkertoy, in an effort to develop more efficient modular systems. It resulted in the tube based micro-modules (Figure 212). Each module consisting out of a range of stacked ceramic wafers that each supported the components. The interconnection between the wafers was realised by vertical wires.

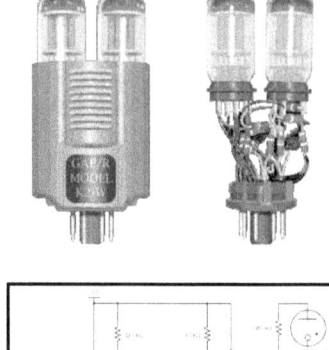

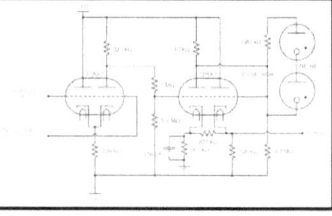

**Figure 211: Electronic Circuit Operational Amplifier.**

The operational amplifier K2-W with and without cover (top), and its circuit diagram (bottom).

Source: Microchipdeveloper.com

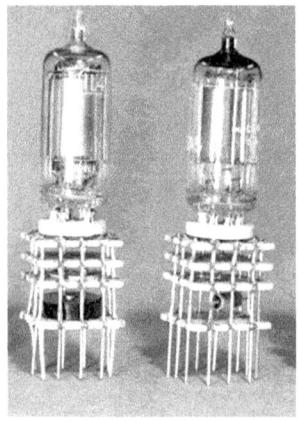

**Figure 212 : Micro-Modules.**

Vacuum based modules resulting from the Project Tinkertoy.

Source: Wikimedia Commons.
https://hackaday.com/2017/03/24/retrotechtacular-tinkertoy-and-cordwood-in-the-pre-ic-era/

## Evolution of the Vacuum Tube into Telephony

As the long copper lines of communication stretched over the continents, one of the problems became the weaking of the signal with incriminating distance. Turning a weakened signal into a strong signal was an important issue for the engineers of those days, and one of the main characteristics of the vacuum tube —its amplification capability— brought it into telephone repeaters. Although originally the Audion was used only as a radio signal detector for nearly five years after its invention, investigations of the possibilities of using Audion as an amplifier for electrical signals were soon carried out by engineers in various laboratories around the world. In Europe several engineering efforts took place to create a 'telephone relay' that would have an amplifying capability. Among those the German physicists.(Tyne, 1977, pp. 73-83)

Robert von Lieben (1878-1913) was a self-taught Austrian physicist was born to a wealthy family in Vienna in 1878. Although he did not obtain an engineering degree, his parents' fortune enabled him to pursue his scientific curiosity outside of a formal university setting in a private laboratory in the Lieben Palace[397]. There he developed the Von Lieben Valve, that was intended to amplify weak telephone signals.

**Figure 213: Lieben Valve/Love Tube (1910).**
Source: Technisches Museum Wien. Tubecollection.de

In 1906, von Lieben obtained a patent for a cathode-beam relay; DRP Patent № 179,807 granted March 4th, 1906), which purported to amplify electrical signals of any frequency by feeding the tube with electro-magnetic or electrostatic signals. By 1910 he obtained a patent for an improved design; DRP Patent № 236,716 filed September 4th 1910 and granted July 11th, 1911, making the valve more reliable. Next improvements resulted in another patent; DRP patent № 249,142 filed December 20th, 1910 and granted July 12th, 1912 for the Lieben Valve (Figure 213).

In August 1911 Robert von Lieben demonstrated his Lieben tube, developed together with Reisz and Strauss, to a group of representatives from the largest German electrical companies at the Institute for Physical Chemistry at the University of Berlin.

*This was the moment when the private laboratory of von Lieben and his collaborators had to refer to the experiences made in industry – for instance in evacuating glass tubes, in handling different materials, and in producing series of tubes of similar quality. But they could present comprehensive patents and also a perfect circuit of how to operate the valve. So von Lieben negotiated with some companies in Germany, and eventually the so called Lieben consortium was founded in early 1912. […] The Lieben consortium consisted in the Allgemeine*

---

[397] During the Gründerzeit, in Vienna the Ringstraßenpalais were created during 1860s-1890s. Among the buildings constructed along it was the Lieben Palace.

*Elektricitäts Gesellschaft (AEG), the Siemens & Halske AG, the Felten & Guilleaume Carlswerk AG, and the Gesellschaft für drahtlose Telegraphie GmbH, Telefunken.*[398][p4]

All these companies had their own business focus: eg Telefunken in wireless telegraphy, AEG in communications engineering. Stimulated by the geo-political aspirations and policies of government of the German Empire, the different companies picked up the further development of the 'Liebenröhre' (the Love Tube). The first tube laboratory was initially at AEG. A few years later, Telefunken set up its own laboratory. However, the invention and development of the high vacuum tube in the USA (1913) and in Germany (1915), soon outdated the Love Tube.

By that time, in America the DeForest triode had raised interest of a range of large telegraph/wireless companies.

In 1911, *DeForest Radio Telephone & Telegraph Co.* (New York) went bankrupt. By this time, Lee DeForest had managed to create a small group of talented engineers who were well versed in the device of the Audion and in the schemes for its inclusion. The suspension of salary payment led to the departure of this group of specialists from the company, who took the Audion technology with them and the assumption that Audion could be used not only for detection, but also for other purposes. This technology spin-off would create many new development trajectories and new business activity. Among the experts departed from Lee DeForest was an engineer named Fritz Lowenstein (1874-1922). In 1910, he organized his own company in Brooklyn (New York City), the Lowenstein Radio Company provided consulting services in the field of radio engineering. Lowenstein, experimenting with 'telephone relay circuits,' discovered that the Audion could be used as an amplifier when used in a specific electric amplifier circuit.

Lowenstein decided to sell his invention after successful testing. On January 27th, 1912, he demonstrated to the engineers of the Bell Telephone System, F. B. Jewett, and O. B. Blackwell, a sound amplifier that was placed in a sealed box, apparently to protect his invention that was not yet protected by a patent. However, his approach did not get the expected result. On April 24th, 1912 Lowenstein filed an application for a patent for an improved design of an Audion as an amplifier for telephone lines. He was granted US-Patent № 1,231,764 on July 3rd, 1917 for a 'telephone relay.' In

---

[398] Blumtritt, O.: The Lieben Valve: a German "universal amplifier" Deutches Museum, 2004. https://ethw.org/w/images/e/ef/Blumtritt.pdf. (Accessed June 2021)

subsequent years, another patent for a circuit with a negative bias acquired great importance and became the subject of litigations.

Inn 1912, Lee DeForest himself approached Western Electric, supplier of the American Telephone and Telegraph Company (AT&T).

> Western Electric's researcher Dr. Harold Arnold —a physicist who possessed the expertise in electron physics DeForest lacked— quickly grasped scientifically how the Audion worked. AT&T, facing problems with the weakening of signals on its long-distance telephone lines, quickly purchased the Audion patent from DeForest. The result was development of a high-vacuum tube for amplifying sound in telephone cables in April 1913, allowing Western Electric to span the continent in 1914.

The vacuum tube triode repeaters, invented at Bell Telephone allowed telephone calls to travel beyond the unamplified limit of about 800 miles. The opening by Bell of the first transcontinental telephone line was celebrated 3 years later, on January 25th, 1915.

## Evolution of the Radio Tube into the Television Tube

Once the electronic vacuum tube technology had started to emerge and became applied in wireless telegraphy and telephony, soon another range of applications fields opened up, each with its basic machines all based on that vacuum-tube technology. Next to the 'radio tubes' described before, in the 1920s the technical development trajectory of the vacuum tube laid the foundations for a range of different vacuum devices. Such as the 'cathode ray tubes' (CRT) used for display functions of electric signals (eg the Osciloscope, Figure 214). And it became a crucial buildings block for the development of electronic broadcasting systems known as 'television.'

The development-trajectory of the basic component called Cathode Ray Tube paralleled the display systems using the CRT; among them electronic instruments called Osciloscope, and the Television receivers set that could display—with built-in receiver—a wireless transmitted signal on the fluorescent part of the CRT. Next to the visual presentation of the signals received, the devices that created the electric signals from the objects/images they 'saw': the CRT-cameras. And these were just a few of the application on the Cathode Ray Tube technology; others would be used in medical applications (the X-ray tubes), and military applications (the Radar-display tubes).

> Basically, the CRT converts an incoming electrical signal into a light pattern. The basic principles of the CRT (Figure 214) are that: (1) Electrons can be released into a vacuum from very hot metals;

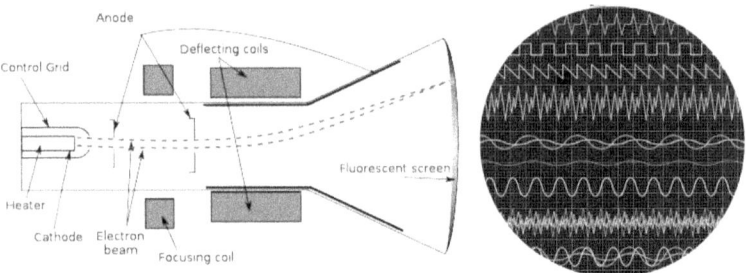

**Figure 214: Principle of the Cathode Ray Tube with Fluorescent Screen.**
The diagram (left) and the electric signals displayed on its Fluorescent Screen (right).
Source: Mark Bowers.

the heated cathode. (2) These electrons can be accelerated and their direction of movement controlled by using either a voltage between metal plates or a magnetic field from a coil that is carrying an electric current. (3) A beam of electrons striking some materials such as zinc sulphide will cause the luminescent material (called a phosphor) to glow, giving a spot of light as wide as the beam. And (4) when that electric ray could be deflected by applying an electro-magnetic field, the positioning of the ray could be controlled.

## Evolution of the Radio Tube into the Radar Tube

The development of electronic vacuum tubes greatly accelerated in the years just before the Second World War and during the war itself. The most obvious progress regarded military radio-localization systems: the system for Radio Detection And Ranging (aka Radar). The electro-magnetic waves that formed the basis for the wireless technology, proved to have—next to the before described properties—also another property; they would be reflected by a metallic object. How to use that property became a matter of technology development and engineering. It gave rise to the 'radar' systems.

A major component of the radar systems was the radar vacuum tube; the power electron tubes capable of generating strong RF pulses at the highest frequencies. Western Electric designed the 316A triode (Figure 215, top) in the doorknob shape. Philips developed in 1934-1935 the vacuum tube Model EF50 (Figure 215, bottom), that would become the workhorse of wartime radar systems as it was used on the Chain Home radars.

Most of the EF50 tubes were manufactured by the Dutch Philips company. With the approaching German invasion of the Netherlands, it was decided that the stock, production machinery and specialized tools were to be shipped by truck to Vlissingen on May 9th. The next morning the Germans crossed the border to invade Holland, and the cargo was transported to Britain by one of the ships of the 'Zeeland' steamer company (ZSM).

The idea of radar was introduced in the United States by Nikola Tesla in 1914. Eight years later, Guglielmo Marconi delivered a lecture on the principle of radar at the 1922 Institute of Radio Engineers Conference, and, in 1933, he presented his idea to the Italian Warfare Ministry to obtain financial support for an experimental development. In 1934, the French Emile Girardeau started building a radar system and installed first devices on board of the cargo ship Oregon, and the ocean liner Normandie the year after. As war clouds gathered over Britain, the likelihood of air raids and the threat of invasion by air and sea drove a major effort in applying science and technology to defence. It resulted in the Chain Home-system: the first early warning radar network in the world, and the first military radar system to reach operational status.

**Figure 215: Radar Tubes.**
Western Electric doorknob model 316A UHF power triode (top1936), and Philips EF50 (bottom, 1938).

Source: http://www.ase-museoedelpro.org/.
https://avs.scitation.org/

By the outbreak of the Second World War in 1939, a chain of early warning radar stations, called Chain Home (CH) stations, had been built along the south and east coasts of Britain (Figure 216). The CH stations were huge, static installations with steel transmitter masts over 100 metres high. Radar could pick up incoming enemy aircraft at a range of 80 miles and played a crucial role in the Battle of Britain by giving air defences early warning of German air attacks.

In the 1934–1939 period, eight nations developed independently, and in great secrecy, systems of this type: the United Kingdom, Germany, the United States, the USSR, Japan, the Netherlands, France, and Italy. In addition, Britain shared their information with the United States and four Commonwealth countries: Australia, Canada, New Zealand, and South Africa, as these countries also developed their own radar systems.

The technical development of the vacuum tube technology continued during and after the Second World War at different places.

Such as the Klystron vacuum tube —used for transmitters of electromagnetic waves at microwave frequencies— that was invented by the brothers Russell and Sigurd Varian at Stanford University. Their prototype was completed and demonstrated successfully on August 30th, 1937. Upon publication in 1939, news of the klystron immediately influenced the work of US and UK researchers working on radar equipment. The company Varian Associates, created after the war in 1948 to commercialize the klystron, would become one of the first high-tech companies in Silicon Valley.

**Figure 216: The Chain Home system.**
Coverage 1939-1940 (top), Aerial Towers at Woody Bay on the Isle of Wright, 1945 (middle) and Radar Operator plotting incoming signals (bottom).

Source: Imperial War Museum.

## Overview of the First Electronic Revolution (1900-1950)

The preceding exploration showed how the electronic vacuum tube evolved into a versatile component that could be used in a broad range of application trajectories. Trajectories that resulted in the Transmission systems, but also the Communication systems and the Radar systems with specialized tubes (Figure 218). The new electronic vacuum tube technology not only opened up totally new ways to communicate and detect over long distances, they also pervaded into the mechanical application areas. Where the Mechanical Revolution had created fine-mechanical artifacts like the mechanical calculators, now electronic circuitry replaced the mechanical parts. Realising in the calculator the basic arithmetic functionality, and extending their computational functions. The mechanical calculator became the electronic calculator that would evolve in the tube-based computing engine.[399] And at the origin of all those developments, was a rather simple device; the vacuum tube.

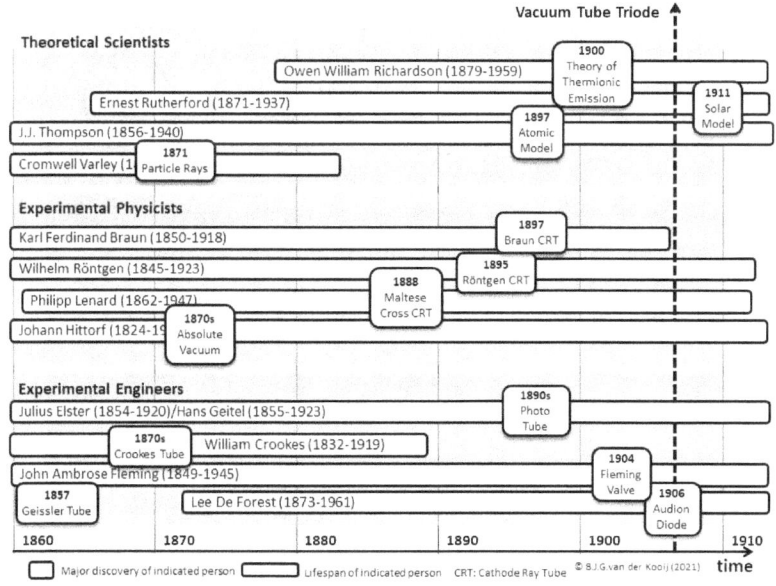

Figure 217: Overview of Contributors to the 'Cathode Ray' Vacuum Tubes.

---

[399] A similar development took place in the mechanical watch where the escapement was replaced by the quart-crystal and the hands of the (analogue) display by the (digital) Liquid Crystal display.

## Contributors to the Invention of the Vacuum Tube

The invention of the Triode Vacuum tube, attributed to Lee DeForest in 1908, was the result of many contributions in the period preceding it (Figure 217). Such as Fleming's contribution of the Thermionic Valve (aka diode vacuum tube) in 1904. His work found its base in the Edison effect, already noted in 1883. The developments leading up Fleming's contribution came from two contributing trajectories; the experimenting with the 'Cathode Ray Tube', and the contributions from the scientists that increasingly understood the physics of the new devices. The result was the mastering of the flow of electrons in vacuum (aka 'cathode rays'). So, before Fleming's valve came into existence, the experimental developments of the Cathode Ray Tube that emerged after Crooke's tubes, contributed to the creation of directional vacuum tubes known as 'diodes.' This development was paralleled by the events of the Wireless Telegraphy. There, at the same time, the improvements in the detectors for telegraphic signals, had contributed to the understanding of the diode effect.

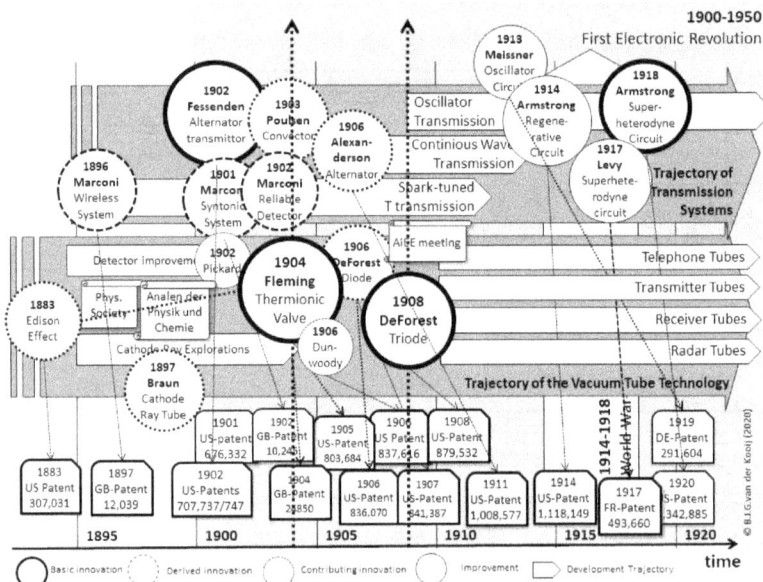

Figure 218: Overview of the First Electronic Revolution (1900-1950).
Contributions to the Cluster of Invention with the Basic Inventions of the Vacuum tube (both Diode and Triode Radio Tube).

It was the combination of all these contributions (Figure 196) that sparked both Fleming and Lee DeForest into conceptualizing a vacuum tube in which a cathode ray would transmit electric currents that could be controlled (ie by the diode), and amplified (ie by the triode). In a period of some two decades, before, during and shortly after the First World War, the radio tube technology would lay the foundations for developments during the next interbellum hat would influence the Affairs of Man dramatically. Next to all the novelty in other fields (ie the internal combustion engine and its Automobile Revolution and Aeronautical Revolution that gave society mobility), the new electronic technology based on vacuum tubes created during the First Electronic Revolution, would enable the Radio Revolution and Television Revolution that gave society news and entertainment.

## Periodization of the Vacuum Tube Technology

Looking at the development trajectory (Scheme 2) of the vacuum tube technology and seen in the perspective of the life cycle (Scheme 5), the vacuum tube technology is marked by specific periods.

*Vacuum Tube 0.0* (1870s-1900s): Early experimenters (Figure 217) with the newly invented incandescent lamp —invented by Thomas Edison in the 1870s— discovered the properties of 'cathode rays:' streams of electrons in the vacuumized space of a glass vessel. The experiments of the vacuum physicists were the basis for the views on emission from the cold cathode, the thermionic emission from a heated cathode, the luminance of the phosphor layer, and the photo-conductivity effect. The build-up scientific knowledge during its incubation phase would become the basis for the next developments.

*Vacuum Tube 1.0* (1900s-1920s): It were the contributions of Fleming and DeForest who created with the new vacuum tube technology the practical devices of the vacuum diode tube and vacuum triode tube (Figure 218). The newly introduced devices created a subsequent technical development trajectory of continuous improvements of the vacuum tube itself, as well as adaptations for specific applications in parallel trajectories.

*Vacuum Tube 2.0* (1920s-1950s): In this growth phase, the vacuum tube technology matured and its fruits —from 'radio tube' to 'cathode ray tube' (CRT)— became available to be implemented in a broad range of applications. Applications like the radio and television became revolutions of their own (ie the Radio Revolution and the Television Revolution). The industries of vacuum tube manufactures, as well as equipment builders using the new components, boomed with the unavoidable first shake-out and consolidations were notable.

*Vacuum Tube 3.0* (1950s-1970s): The rise of the transistor technology heralded the end of the meteoric rise of the application of the vacuum tube technologies. Only some specific niches survived. And with that came the adaptations and demises in its manufacturing industries. But it would take decade before the replacement had taken place, leaving niches for the tube-technology.

This periodization of the technological life cycle of vacuum tube technology during the latter part of the Second Industrial Revolution and the Third Industrial Revolution (Figure 219) became related to the additional Information Technology (IT) and Communication Technology (CT) that were derived from it along their own development trajectories (eg Computers, Wireless Broadcasting,). A period also labelled as the First Electronic Revolution.

The overview illustrates how some technologies matured, while others were just in the early phase of their life cycle. So was the period of Mechanical Calculating 2.0 quite alive and kicking when the period Vacuum Tube 1.0 started. This illustrates how the Industrial Revolution had a transition period where the 'old' GPT became complements /replaced by the 'new' GPT.

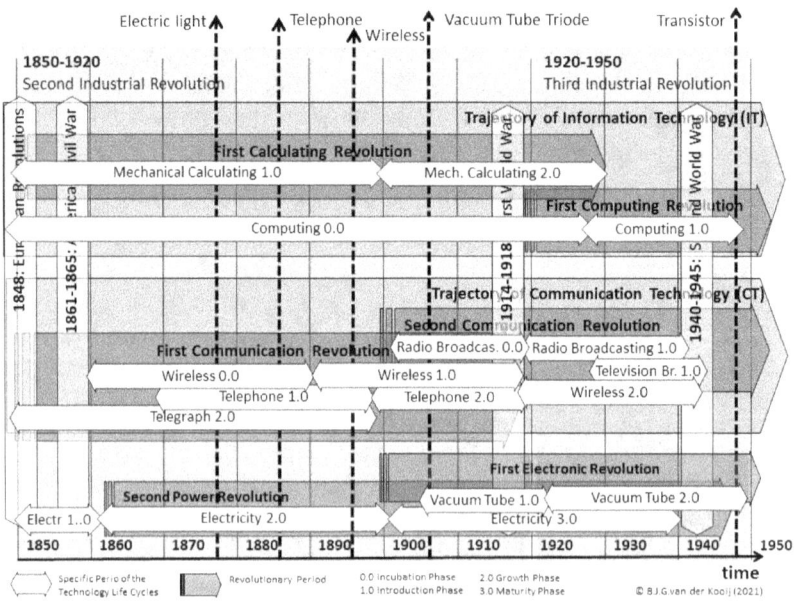

Figure 219: Overview of the First Electronic Revolution and the related Communication Technologies (CT) and Informarion Technologies (IT) (1900-1950).

## *The Second Electronic Revolution (1950-1980)*

After the rise of the electronic technologies based on the vacuum tube-technology in the first half of the twentieth century (ie the First Electronic Revolution), a new fundamental technological development took place around the 1940s-1950s that became known as the electronic 'semi-conductor' technology.[400] The vacuum tube technology in its essence was the control of electrons (ie the particles creating an electric current) *within a vacuum* (eg a vacuumized tube, or a tube filled with an inert gas like helium). The new solid-state technologies were about the control of electrons *within a solid material* (ie semi-conducting materials). It would start the Second Electronic Revolution.

### *The Era of Solid-State Physics*

During the *First Chemical Revolution* (1780s-1790s) the chemical composition of matter, materials in a gaseous (eg oxygen), a metallic (eg nickel) and a liquid (eg mercury) state were often the focus of scientific attention. Antione Lavoisier's (1743-1794) explorations had found out that water (ie $H_2O$) was a compound of 'oxygen' (ie the element O) and 'hydrogen' (ie the element H). But there were also the materials in a solid state. Like the natural mineral salt (ie NaCl), a compound of Natrium (Na) and Chlore (CL). The core of Lavoisier's 'new chemistry' work was the oxygen theory, and the work became a most effective vehicle for the transmission of new doctrines [401] that presented a unified view of theories of chemistry. And it was English chemist John Dalton (1766-1844) who introduced the *atomic theory* into chemistry.

### The Revolution in Classical Physics

For each element found, it was discovered that its basic element was the atom, and the term was used in connection with the growing number of irreducible chemical elements.

> It started in the early nineteenth century when the schoolteacher John Dalton (1766-1844) wrote in his logbook the entry "Observations on the Ultimate Particles of Bodies and their

---

[400] We use the notion of 'semiconductor technologies' to indicate both the transistor technologies (ie knowing how to make transistors) as well as the IC-technologies (knowing how to make Integrated Circuits).

[401] Naming has long been considered an important component of professional identity and discipline formation. Historians have considered the significance of affixing new names to scientific disciplines in a range of eras, specialties, and national contexts. The decline of "natural history" in favour of "biology," the rise of "physics" at the expense of "natural philosophy," and the eclipse of "alchemy" by "chemistry" all coincided with new community structures, institutions, and methodological standards.

Combinations." In September of 1803, John Dalton wrote his first table of atomic weights in his daily logbook. And in 1808 he made a periodic table of known elements (Figure 220), and he proposed that each chemical element is composed of 'atoms' of a single, unique type, and though they cannot be altered or destroyed by chemical means, they can combine to form more complex structures (ie chemical compounds).

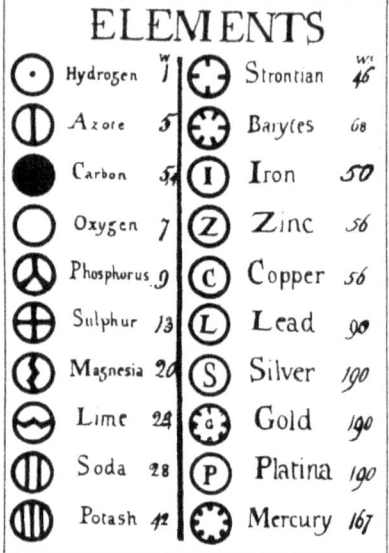

Figure 220: Dalton's Periodic Table (1808)
Source: Wikimedia Commons.

This was merely an assumption, derived from faith in the simplicity of nature. At that time, no evidence was available to scientists to deduce how many atoms of each element combine to form molecules. The essential novelty of Dalton's atomic theory was that he provided a method of calculating relative atomic weights for the chemical elements.

It took nearly a century when, by the late nineteenth century, it was discovered by other scholars that the atom was built up out of subatomic particles (ie neutron, proton, electron) that defined their properties. In the early twentieth century other physicists expanded on the developing theoretical physics.[402] They were exploring the Nature of Matter —again as they did during the Scientific Revolution— but now at deeper levels of the materials looking for elementary particles and their behaviour of atoms. Between 1890 and about 1910, physics experienced a complete change of paradigms. The fascination of experiments with 'Röntgen's rays', and of course Rutherford's radiation phenomena (the birth of nuclear physics) had set the stage for a new age in science.

---

[402] Theoretical physics can be defined as "the invention and manipulation of concepts, using mathematics where necessary, to simplify the understanding of known physical phenomena, and to predict new phenomena."

## *The Era of Modern Physics*

At the end of the nineteenth century, physics had evolved to the point at which classical mechanics —concerned about motion of objects (eg Newtonian mechanics)— could cope with highly complex problems involving macroscopic situations; thermodynamics and kinetic theory were well established; geometrical and physical optics could be understood in terms of electro-magnetic waves; and the conservation laws for energy and momentum (and mass) were widely accepted.

Then, at the beginning of the twentieth century a major revolution shook the world of physics, which led to a new paradigm, generally referred to as modern physics in which the focus was on the speed of light (special relativity), small distances comparable to the atomic radius (quantum mechanics), and very high energies (relativity). It resulted in the Atomic Theory.

## Towards the Atomic Theory of Particle Physics

By the early twentieth century, scientists had developed fairly detailed and precise models for the structure of matter, which led to more rigorously-defined classifications for the tiny invisible particles that make up ordinary matter. An atom became defined as the basic particle that composes a chemical element. Then physicists discovered that the particles that chemists called 'atoms' are in fact agglomerations of even smaller particles (aka subatomic particles). They theorized that (negatively charged) *electrons* are constituent part of the atoms, and the atom's nucleus was composed of the (positively charged) *protons* and *neutrons* (Figure 221). The concept of a subatomic particle was refined when experiments showed that light could behave like a stream of particles (called photons) as well as exhibiting wave-like properties.

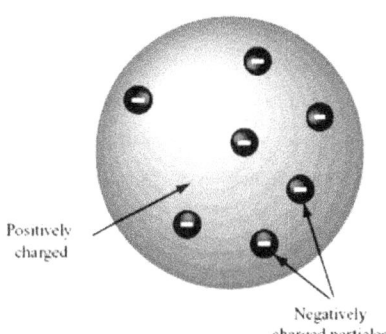

**Figure 221: Thompson's Plum Pudding Model of the Atom.**
Source: Wikimedia Commons.

Joseph J. Thompson (1856-1940) explored the 'cathode rays' in a Crooke's tube and that brought him to the conclusion of the existence of particles (aka electrons) that were negatively charged and could be deflected. In 1904 he formulated his 'Plum Pudding Model': electrons surrounding a volume of positive charge, like negatively charged 'plumps' embedded in a positively charged 'pudding' (Figure 221).

It had become known, that —in addition to the negative charged electrons— positively charged particles also emanate from the anode in an energized Crookes tube. It was the German scientist
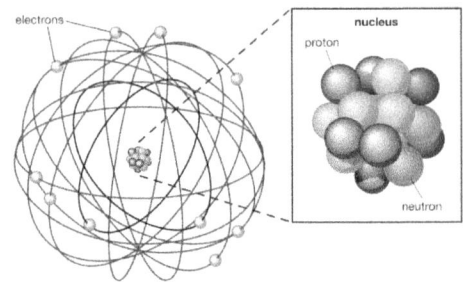

**Figure 222: Rutherford's Planetary Model of Atomic Structure.**
Source Encyclopaedia Britannica, Inc.

*Eugen Goldstein* (1850-1930) who discovered the 'Anode Rays' (aka canal rays) build up from ions —ie electrically charged atoms— colliding with atoms of the gas in the tube. His discovery would lead to the neon-tube (ie gas-discharge tube).

Following this discovery of the electron, electro-magnetic theory became an integral part of the theories of the atomic, subatomic, and subnuclear structure of matter. Thompson's student Ernest Rutherford (1871-1937), experimenting with a gold foil, investigated the atomic structure further and came up with the extended 'planetary' model of the atom (Figure 222).

> *The basic structure of the atom became apparent in 1911, when Rutherford showed that most of the mass of an atom lies concentrated at its centre, in a tiny nucleus. Rutherford postulated that the atom resembled a miniature solar system, with light, negatively charged electrons orbiting the dense, positively charged nucleus, just as the planets orbit the Sun.*

Rutherford's discovery of the nucleus meant the Thompson's atomic model needed a rethink. It was Niels Bohr who proposed a model where the electrons orbit the positively charged nucleus on stationary levels (ie orbits) (Figure 223).

> *The Danish theorist Niels Bohr refined this* [Rutherford] *model in 1913 by incorporating the new ideas of quantization that had been developed by the German physicist Max Planck at the turn of the century. Planck had theorized that electro-magnetic radiation, such as light, occurs in discrete bundles, or "quanta" of energy now known as photons. Bohr postulated that electrons circled*

*the nucleus in orbits of fixed size and energy and that an electron could jump from one orbit to another only by emitting or absorbing specific quanta of energy.*[403]

In addition to these contributions, there were other additional contributions to the understanding of atomic structures and their behaviour; the scientific field of electrodynamics.

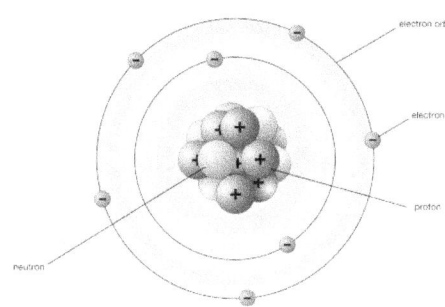

**Figure 223: Bohr's Model of Stationary States in the Shell Model.**
Source Encyclopaedia Britannica, Inc.

Already at the beginning of the twentieth century, German physicist Maxwell Planck (1858–1947) theorized that atoms absorb or emit electro-magnetic radiation in bundles of energy termed 'quanta.' On December 14th, 1900 he presented his results to the German Physical Society in a paper entitled 'On the Theory of the Energy Distribution Law in the Normal Spectrum.' It was the dawn of the quantum theory.

In parallel, the free-electron model was first proposed by the Dutch physicist Hendrik A. Lorentz (1853-1928) —in his publication 'The Theory of Electrons' (1906)— and was later refined in 1928 by the German theoretical physicist Arnold Sommerfeld (1868-1951). And it was the contribution of Albert Einstein (1879-1955) in his paper entitled 'On the Electrodynamics of Moving Bodies' (1905) that placed electrodynamics in the larger context of the special theory of relativity (originally the Lorentz-Einstein theory).

*The Shell Model: Band Structure Theory*

After the Danish physicist Niels Bohr (1885-1962) had investigated the atomic structure and the arrangement of electrons within the atom, he published his results in 1913 in the Trilogy 'On the constitution of atoms and molecules.' He found that inside the atom, the electrons spin around the nucleus on different trajectories/orbits with specific energy levels called shells or bands; his Shell Model (Figure 223). He assumed that certain numbers of electrons (eg 2, 8, 18) could occupy a closed shell.

---

[403] Source: Sutton, Chr.: Subatomic particle. https://www.britannica.com/science/subatomic-particle#ref254781 (Accessed June 2021).

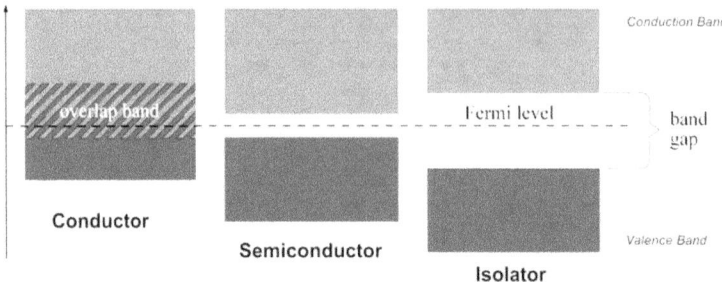

**Figure 224: Energy Band Diagram: Band Gaps determining the electrical properties of materials.**

Electrons could move between these discrete energy levels (referred to by Bohr as 'stationary states', aka 'bands' or 'orbitals'), but had to do so by either absorbing or emitting energy. In addition, he theorized that the chemical properties of each element are largely determined by the number of electrons in the outer orbits of its atoms. It gave rise to the *Band structure theory*, that was based on the way the bands of a specific material were filled with electrons. The existence of bands is mostly a feature of the outermost electrons (valence electrons) in the atom, which are the ones involved in chemical bonding and electrical conductivity (Figure 224).

*Valence Band:* The valence band is the outer side band of electron orbitals that electrons can jump out of, moving into the conduction band when excited. The valence band is simply the outermost electron orbital of an atom of any specific material that electrons actually occupy (Figure 225).

*Conduction Band:* The conduction band is the band of electron orbitals that electrons can jump up into from the valence band when excited. When the electrons are in these orbitals, they have enough energy to move freely in the material. And this movement of electrons creates an electric current (Figure 225).

The electrical properties of element were determined by the way the orbits —the bands— were filled with electrons. Some bands contained electrons, others were empty. Between these orbits were the so-called 'band gaps' (Figure 224). The material with no bandgap —or an overlapping bandgap— conducted electricity and were called 'conductors', those with a large bandgap blocked electricity and became known as 'insulators.' And in between the material with a small band gap became known as 'semiconductors.'

It became clear that, similar to the *gravitational forces* that exists in the Solar System (eg Newton's Laws of Gravity), in the atom the relative movement of particles is subject to *electro-magnetic forces*.

*The Nucleus of an Atom*

By 1920 chemical isotopes — the same chemical element with a different number of neutrons in the nucleus— had been discovered, the atomic masses had been determined to be (approximately) integer multiples of the mass of the hydrogen atom, and the atomic number had been identified as the charge on the nucleus. At the time it was believed that the nucleus consisted of protons and electrons.

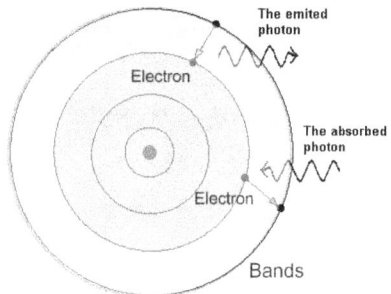

**Figure 225: Electrons changing Orbits.**
Source: Wikimedia Commons

Exploring the nucleus of the atom, the British physicist James Chadwick (1891-1974) focussed his work on the nucleus of the atom. In 1932, he postulated the existence of a 'neutron' (ie a new elementary uncharged particle)[404] (Figure 226).

This theorizing all was conceptualized on the atomic level, so mathematics started to play a dominant role as the observation of the phenomena was impossible. It was the German theoretical physicist Werner Heisenberg (1901-1976) who developed thereafter the mathematical foundation of quantum mechanics. In 1925 he published *Über quantentheoretische Umdeutung kinematischer und mechanischer Beziehungen* ('Quantum theoretical re-interpretation of kinematic and mechanical relations'). Next, Heisenberg formulated in 1927 his *uncertainty principle* that stated that the more precisely the position of some particle is determined, the less precisely its momentum can be predicted from initial conditions.

The views on the composition of the atom —ie th electrons, protons and neutrons— and the classic picture of atomic behaviour due to emission, absorption and spontaneous action were now rather complete. But the theories still reflected the classic discussion between the dual nature of waves and particles.

---

[404] Both the proton and the neutron were presumed to be elementary particles until the 1960s, when they were determined to be composite particles built from quarks.

*In the quantum picture, electromagnetic radiation has a dual nature, existing both as Maxwell's waves and as streams of particles called photons. The quantum nature of electromagnetic radiation is encapsulated in quantum electrodynamics, the quantum field theory of the electromagnetic force. […]* ibidem

## Quantum Electro-Dynamics

From the classical theories described before, a new range of theories emerged labelled as *Quantum Electrodynamics* (QED); a quantum field theory of the electro-magnetic force that is part of Quantum Mechanics.[405] In essence, it describes how light and matter interact, and it is the first theory where full agreement between quantum mechanics and special relativity is achieved. The QED rests on the idea that charged particles (eg electrons) interact by emitting and absorbing photons, the particles that transmit electro-magnetic forces in the materials.

Many physicists contributed to the development of the QED theories.[406] Such as the Austrian Erwin Schroedinger (1887-1961) who realized by 1926, based on quantum theory, that electrons were both particles and waves. He postulated that the specific location of an electron could not be known. But with his Schroedinger Equation —the counterpart of Newton's second Law of Gravity[407]— the evolution over time of that wave function could be calculated. He also determined that electrons were not located in fixed orbits, as Rutherford and Bohr had theorized, but moved about in three-dimensional space. Only the probability of their positions could be calculated. From this postulation his 'Electron Cloud' model was created (Figure 227).

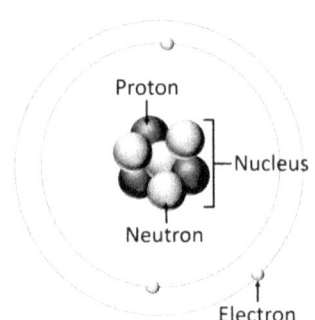

**Figure 226: The Neutron as part of the Nucleus of an Atom.**
Source: Wikimedia Commons.

---

[405] *Quantum mechanics* is a mathematical application of the quantum theory that maintains that energy is both matter and a wave, depending on certain variables.
[406] The new field of *quantum electrodynamics*, describing physical properties of electrically charged particles on an atomic scale, applied to an enormous domain to atoms, molecules, solids, and nuclei. Each specialty required different techniques and expertise, created new independent interests, and forced further divisions of scholarly efforts. The number of theoretical physicists increased as a consequence of the variety of physical phenomena under fruitful study and of the difficulty of the physics itself.
[407] Newton's second law says that when a constant force acts on a massive body, it causes it to accelerate, ie, to change its velocity, at a constant rate.

In 1928 the English physicist Paul A.M. Dirac (1902-1984) laid —as published in 'The quantum theory of the electron' (1928)— the foundations for QED-theories with his discovery of a wave equation that described the motion and spin of electrons and incorporated both quantum mechanics and the theory of special relativity. In his next publication 'A theory of electrons and protons' (1930) he postulated the existence of the positron; the equivalent of the electron with a positive charge. It was the American physics Carl David Anderson (1905-1991) who discovered the positron on August 2nd, 1932, for which he won the Nobel Prize for Physics in 1936.

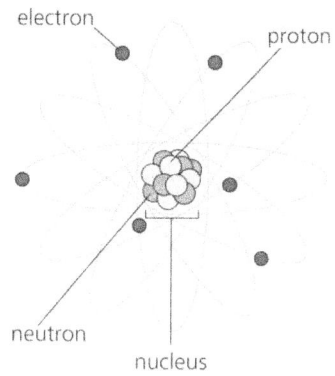

**Figure 227: Schroedinger Electron Cloud Model.**
Source Wikimedia Commons.

The preceding scientists —from Thompson to Anderson— were some of the many scientists that contributed to the insights in the structure of matter up to the atomic level and its electro-dynamics. Quantum Electrodynamics (QED), also known as the Quantum Theory of Light, eventually became one of the most precise, accurate, and well tested theories in science. It gave birth to another field of scientific inquiry; the solid-state physics.

## Solid-State Physics

As we have seen before, already in 1874, the German scientist Karl Ferdinand Braun —as part of his research in wireless communication— discovered the 'unilateral conduction' across a contact between a metal and a mineral. It led to the crystal detector used for detecting radio signals (aka the 'cat's whisker diode' or point contact diode) (Figure 183). Braun was granted a range of patents for his work.

The detector research continued, due to the fact that cheaper and more reliable devices were needed for the new 'radio' receivers that were conquering the consumer market. However, the device had its basic limitations and was ultimo replaced by the vacuum tube diode, but during the Interbellum it took to the 1930s till the crystal detector evolved in the point-contact diode detector (Figure 228).

By the mid-1930s the behaviour of semiconductors was widely recognized to be due to impurities in crystals, although this was more a qualitative than quantitative understanding by (industrial) scientists seeking solid-state alternatives to vacuum-tube amplifiers and electro-mechanical relays. Later it became to understood how impurities influence conductivity.

Over the years, the experimental scientists started to understand the mechanisms of controlling electric currents *in* solid-state materials.

Such as the Austro-Hungarian physicists Julius Lilienfield (1882-1963) who was granted the Canadian Patent № 272,437 in 1925 —and the similar US-Patent № 1,745,175 filed on October 6th, 1926 and granted on January 8th, 1930— for a 'Method and apparatus for controlling electric currents' (Figure 229).

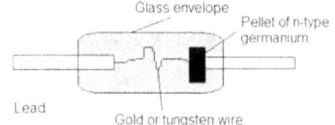

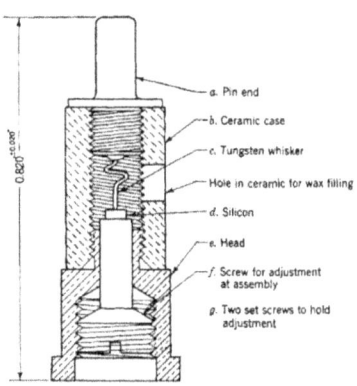

**Figure 228: The Point Contact Diode.**

The principle of the point contact diode (top), and a diagram of the Sylvania diode (1943, bottom).

Source: Wikimedia Commons.

His device became known as the Field-effect transistor (FET). And his patent was covering an alternative method to replace thermionic tubes: *"[…] the present invention has for its object to dispense entirely with devices relying upon the transmission of electrons thru an evacuated space and especially to devices of this character wherein the electrons are given off from an incandescent filament."*

A similar type of semi-conducting device was developed by the German electrical engineer Oskar Heil (1908-1994) who was granted the British Patent № 439,457 in 1934 for 'Improvements in or relating to electrical amplifiers and other control arrangements and devices.' He described the possibility of controlling the resistance in a semiconducting material with an electric field. Their work was of a scientific nature but did not result in functional devices. Later the field effect transistor (FET) would result from their ideas.

At the Bell Labs, the search for better crystal detector continued, as silicon crystal detectors became a key component of radar receivers during the Second World War. They served the crucial function of converting high-frequency, short-wave radar signals into lower-frequency oscillations that could be more readily amplified in electronic circuits. However, the basic problems of non-uniformity of the resulting devices were not solved. So, the scientists focussed their attention on understanding the basic semiconducting material of silicon. And the technologists put their attention on producing the silicon materials. At the Bell Labs, it were the metallurgist Jack H Scaff and Henry Theuerer who melted the silicon in quartz tubes. After the silicon had cooled and solidified, they cracked away the quartz to obtain black polycrystalline ingots.

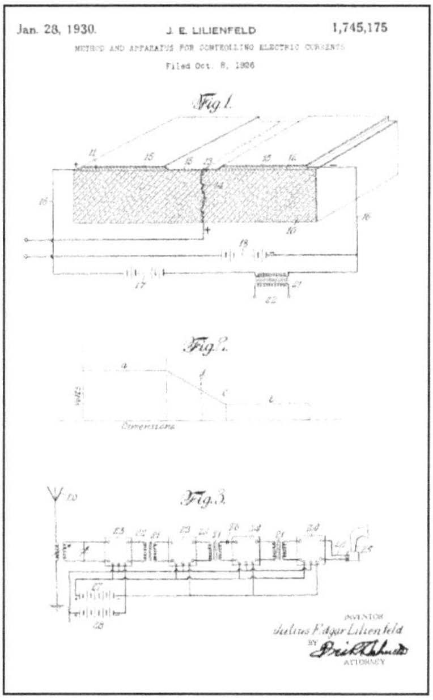

Scaff applied a patent application on April 4th, 1941, and was granted US Patent № 2,402,582 on June 25th, 1946 for 'Preparation of Silicon Materials.' (Figure 230, left)

**Figure 229: Lilienfield's US-Patent № 1,745,175 for the Field-effect Transistor (1926).**
Source: USPTO

And then came an important discovery; it was about how impurities could influence the behaviour of electrons in semiconducting materials. [408]

Russell S. Ohl (1898-1987), working at Bell Labs, in his investigation of the behaviour of certain types of crystals used for crystal detectors,[409] discovered in 1939 the mechanism behind the P-N junction. In its essence, it was the role that impurities played when introduced in semiconducting materials.

---

[408] When a foreign atom of phosphorus enters a silicon crystal lattice, for example, it occupies a site that would otherwise contain a silicon atom. But, being a fifth column element, phosphorus would have five electrons to share in forming four covalent bonds with its silicon neighbours, each of which needs one electron. Thus there is an excess of one electron that cannot find a natural home in the lattice. The doping of silicon with impurities like phosphorus creating N-type silicon enhance its capabilities to conduct electrical currents.
[409] Silicon crystal rectifiers played an important role in World War II as the key components of radar receivers.

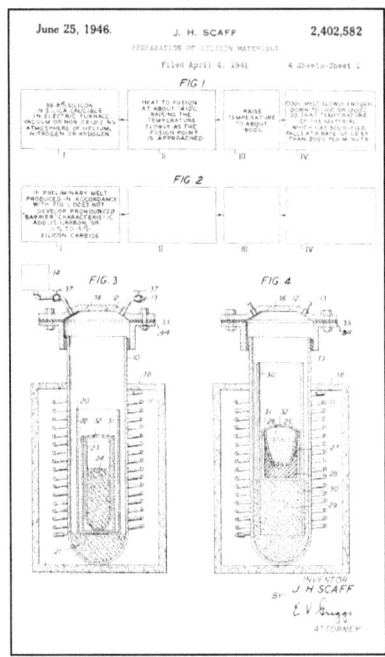 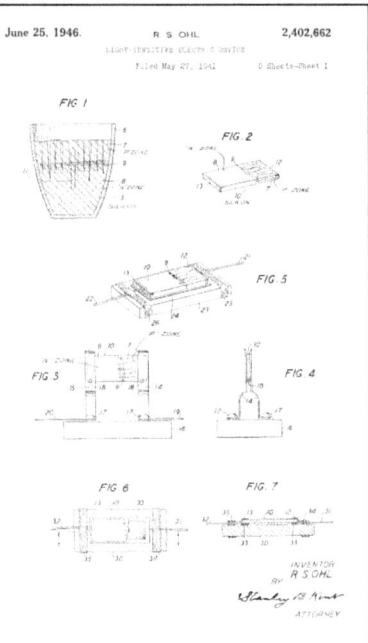

**Figure 230: Early Silicon Patents.**
Scaff's US-Patent № 2,402,582 (left, 1941), and Ohl's US-Patent № 2,402,662 (right, 1941).
Source: USPTO

As the industrial produced silicon materials available in those days were not pure, but contained impurities like aluminium that distorted the Cat Whisker's functioning, he decided to purify silicon. Designing a special furnace that could handle silicon's melting temperature of 1410 °C, Scaff and Theuerer created the rods of silicon. And indeed, the purer silicon improved the functioning of the contact-diode. However, still two different types of silicon seemed to exist in the silicon rod; the (later called) P-type and N-type material. And between those materials in a rod, there existed a barrier (later known as the PN-junction). [410]

Further experimenting of the barrier between the two different materials revealed its capability as a light-sensitive device. But that was not interesting at the time of its discovery. The was fact that introducing impurities in the intrinsic silicon, changed its electric properties.

---

[410] His work with diodes led him later to develop the first silicon solar cells.

Although Bell attorneys applied for patents in 1941, due to war conditions, the information remained classified until 1946, when on June 25th a series of patents were awarded in Ohl's name on silicon detectors and PN junctions. From US Patent № 2,402,661 (Alternating Current Rectifier), US Patent № 2,402,662 (Light Sensitive electric device, Figure 230, right), US Patent № 2,402,663 (Thermoelectric device), US Patent № 2,407,678 (Thermoelectric system), and US Patent № 2,402,839.

The concept of introducing impurities into semiconducting materials, to influence its characteristics and hence the behaviour of the electric current in the material was fundamental. But it took a lot of 'technologists' to developed the technologies that would create the actual devices.

*The Scientist Exodus*

Solid-state physics in the US had a rather slow start. That changed during the 1933-1941 period when more than 100 physicists —the refugee scientists fleeing from the Nazi-regime[411]— came to the United States from Germany, Austria, and Czechoslovakia in a scientific exodus. And even more came after Germany's defeat in 1945. This brain drain created a nucleus of knowledge on solid-state physics.

> *The American theoretical physics community came into its own after the second world war. They arrived in a new world, which they helped shaped by their contribution to the allied victory. They were instrumental in transforming the institutions of higher learning to which they returned. The physicists, and particularly the theoretical physicists, played an important part in breaking down the discriminatory barriers that had existed at many universities before the war. They also helped to establish and to exploit the new partnership between science and the government, the military, and industry, born of the war, which was to alter what it meant to do science.* (Schweber, 1986, p. 98)

The end of the Second World War heralded a wave of specialization in the chemical sciences; among those the field of solid-state physics with a range of contributors that had moved to the US. Such as the Dane Niels Bohr (1885-1962), and the Italian Enrico Fermi (1901-1951).

Solid-state physics studies how the large-scale properties of solid materials result from their atomic-scale properties. In the United States and Europe, solid state became a prominent field through its

---

[411] In April of 1933, Hitler's first anti-Jewish law was promulgated, stripping all 'non-Aryan' academics of their teaching posts. 25% of German physicists, including eleven past or future Nobel Prize winners, lost their jobs. By 1944, more than 133,000 German Jewish émigrés had moved to America; many of them highly skilled and educated. Some were even Nobel Prize winners and renowned intellectuals in physics like Albert Einstein.

investigations into semiconductors, superconductivity, nuclear magnetic resonance, and diverse other basic phenomena. A large part focussed on semiconductor materials and their properties.

Next to the many US physicists, large communities of solid-state physicists also emerged in Europe, in particular in England, Germany, and the Soviet Union. Their inventive work would disclose —next to the secrets of atomic structures— the workings of the semiconductors, and that would start the Era of the Transistor. These contributors to the solid-state physics (Figure 231), both the more theoretical as well as the experimental physicists, laid the foundations for the contributions of Bardeen, Brattain and Shockley.

## Overview contributors to the Invention of the Transistor

In the preceding exploration we identified some important events in the development of solid-state physics that led to the conception of the transistor. The essence of the new device being the control of the flow of electronic within (semi-conducting) material.

The control of the flow of electrons in vacuum —the cathode rays that were observed during the early experimenting with vacuum tubes— had shifted to solid state physics where it was about the control of the flow of

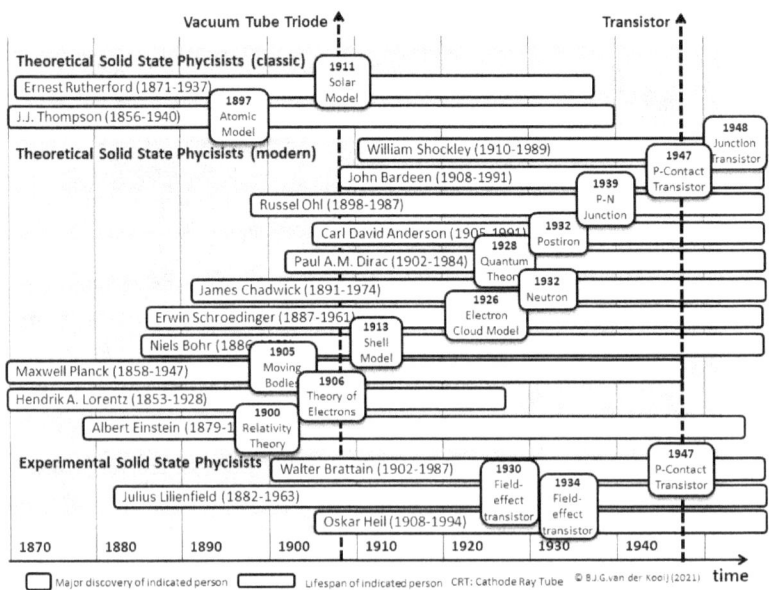

Figure 231: Overview of the Contributors in Solid State Physics leading to the Transistor.

electrons in semi-conducting material. And many theoretical solid-state physicists had developed their theories and models (Figure 231, top) by the turn of the century. Their views on the composition of the atom — including their ideas about electrons, protons and neutrons— and the classic picture of atomic behaviour due to emission, absorption and spontaneous action were rather complete. But the theories still encompassed the classic discussion between waves and particles.

That changed after the German theoretical physicist Albert Einstein revolutionized the view on the Macro-Cosmos (aka the Universe) with his general *Theory of Relativity*. And the German physicist Max Planck revolutionized the view at the Micro Cosmos of the quantum mechanics of electrons, protons, neutrons, photons and positrons. Niels Bohr, Erwin Schrodinger and Paul M. Dirac, advanced Planck's theory and made possible the development of quantum mechanics. Their insights gave rice to the Quantum Electro-Dynamic theories (QED-theories). It took the discovery of the components of the atom—from neutron to positron—operating in Schroedinger's Electron Cloud Model of the atomic structure, to create the foundation for the Point-Contact transistor as invented by the theorist Bardeen and the experimenter Brattain. The breakthrough in the

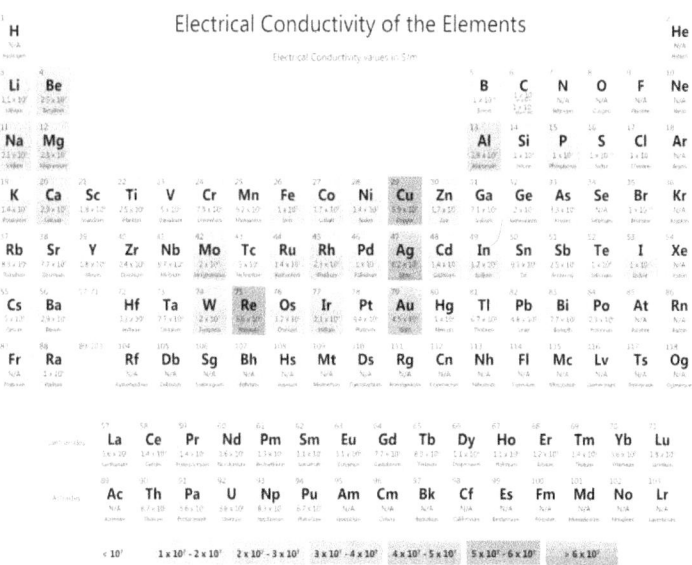

**Figure 232: Electrical Conductivity of the Elements.**
The darker the shading, the better the conductivity (eg Cu, Ag, and Au).
Source: Wikimedia Commons.

concept of the transistor, however, was Shockley's use of silicon in creating the bipolar/junction transistor.

### The Basic Workings of Semiconductors[412]

Solid materials have a crystal structure (aka lattices) in which the atoms, ions or molecules are arranged in a three-dimensional way.[413] In the material the electric current —ie the charged particles known as electrons— travels as a result of the behaviour of the electrons in the atoms. Some materials are conducting the electrons easily (eg metals), other materials resist their movement (Figure 232). That conducting or isolating property is the result of what happens in the atoms that constitute the materials.

The band structure theory placed the electrons of a single, isolated atom in bands around the atom's nucleus (Figure 233). The electrons occupying the atomic orbitals each have a discrete energy level. They create the electronic band structure of the atom; a range of energy levels that electrons can occupy. Conduction takes place when an electron jumps from valence band—and becomes the valence electron— to the conduction band creating an 'electron hole' —ie the absence of an electron from a full valence band— in the valence band. When they change their orbit, this results in an energy emission of a quantum of discrete energy called 'photon.'[414] The process is called stimulated emission. When they absorb energy of a photon, electrons can change between orbits (Figure 225).

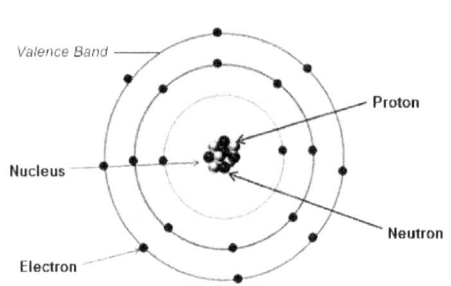

**Figure 233: Band Structure of an Atom.**
Source: Wikimedia Commons.

*Photons are exchanged whenever electrically charged subatomic particles interact. The photon has no electric charge, so it does not experience the electromagnetic force itself; in other words, photons cannot interact directly with one another. Photons do carry energy and momentum, however, and, in transmitting these properties between particles, they produce the effects known as electromagnetism.*(ibidem)

---

[412] In this Chapter, technical in nature, we focus on the fundamental aspects of the semiconductor. It explains how the mechanisms in semiconductor material work. It is intended to illustrate the essence of the technology, but can be skipped by the casual reader.
[413] A system of one or more electrons bound to a nucleus is called an atom. If the number of electrons is different from the nucleus's electrical charge, such an atom is called an ion.
[414] These theories are now considered to be the old quantum mechanics theories. We will use them for illustration purposes.

In addition to the stimulated emission, there was spontaneous emission of an atom when it transits from an excited energy state to a lower energy state (e.g., its ground state) and emits a quantized amount of energy in the form of a photon.[415]

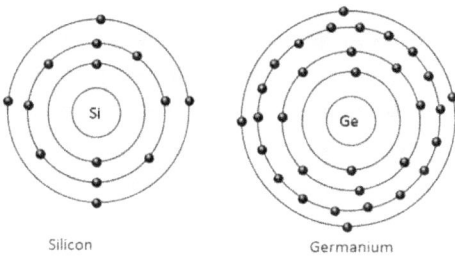

**Figure 234: Electrons in Intrinsic Semiconductor Materials.**

Notice the four electrons in the outer band.
Source: Physics and Radio Electronics.com

The outer most band of an atom —the valence band— defines the electrical properties of the atom; from insulation to conduction, depending of the width of the band gap (Figure 224).

So, in the atom the electrons circle around the nucleus. When a small amount of external energy is applied to the valence electrons, they gain enough energy and break the bonding with the parent atom. The electron, which breaks the bonding with the parent atom, moves freely from one place to another place. These electrons are called *free electrons* (Figure 236).

Depending on the material, that valence band can contain three, four or five atoms. The *pentavalent materials* like phosphorus (P), arsenic (AS), antimony Sb), have five electrons in the valence band (Figure 235, left).

*Semiconductor materials* like Silicon (Si) and Germanium (Ge) have four electrons in their valence band (Figure 234). And the *trivalent materials* like Boron (B), Gallium (G), Indium (In), and Aluminium (Al) have three valence electrons. (Figure 235, right). Both the trivalent and pentavalent materials can act as a

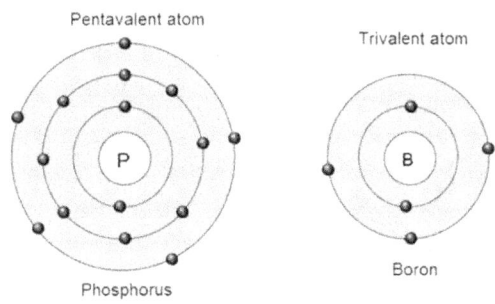

**Figure 235: The Occupation of the Valence Band in different materials.**

Source: Physics and Radio Electronics.com

---

[415] If atoms (or molecules) are excited by some means other than heating, the spontaneous emission is called luminescence.

doping to influence the behaviour of the (intrinsic) semiconductor material Silicon and Germanium.

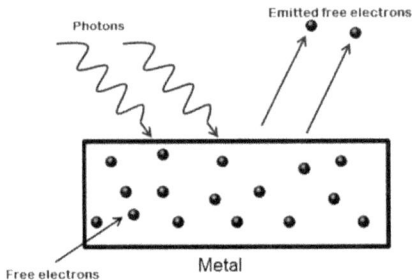

In materials, the atoms are arranged in a crystal structure know as lattice. The free electron cannot escape from the crystal lattice; the arrangement of the (groups of) atoms in a crystal. To escape the atom, the energy of the electron must be increased above its binding

**Figure 236: Photo-electric Emission.**
Source: Physics and Radio Electronics.com

energy to the atom. When the external energy in the form of heat, light, or electric field is applied to the metal, the free electrons gain enough energy and break the bonding with the metal. This *thermal emission* creates the 'cathode ray' in a vacuum tube. Also an (external) electric field and light—ie photons—can make a free electron escape; the *electric field emission* and *photo-electric emission* (Figure 236) of a semi-conductor device.

The intrinsic semiconductor materials (eg Silicon and Germanium) have an crystalline atomic structure in which the free electron can travel. By introducing impurities in the (intrinsic) semiconductor materials (aka dopants), the conductivity properties of the silicon lattice change as it influences the band gap. If the dopant is a pentavalent impurity (five electrons in the valence band, Figure 235) the surplus of electrons makes its conductive. Since the fifth electron has nothing to bond to, it is free to move around, allowing an electric current to flow through the silicon; it becomes the majority carrier (Figure 239, top). If the dopant in trivalent impurity (three electrons in the valence band, Figure 235) there is an abundance of free holes. This means the electrons in the valence band become mobile, and the holes move in the opposite direction to the movement of the electrons (Figure 237). Because the dopant is fixed in the crystal lattice,

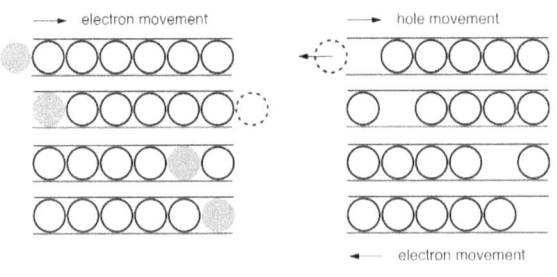

**Figure 237: Electron movement and Hole movement in semiconductor material.**
Source: Wikimedia Commons.

only the positive charges can move. They become the majority carrier (Figure 239, bottom).

*Electron movement and hole movement:* In a pure (ie intrinsic) semi-conductor, each atom is surrounded by four atoms (Figure 234) and is engaged in chemical bond with all of them. This is because a pure semiconductor has four valence electrons. A trivalent impurity has 3 valence electrons (Figure 235). When it is added as a dopant, one atom of it forms bonds with three atoms of the semiconductor. Only three atoms are engaged in bonding. There is an empty space between the atom of impurity and the fourth atom of pure semiconductor. Such an empty space is called a hole which is available for conduction.

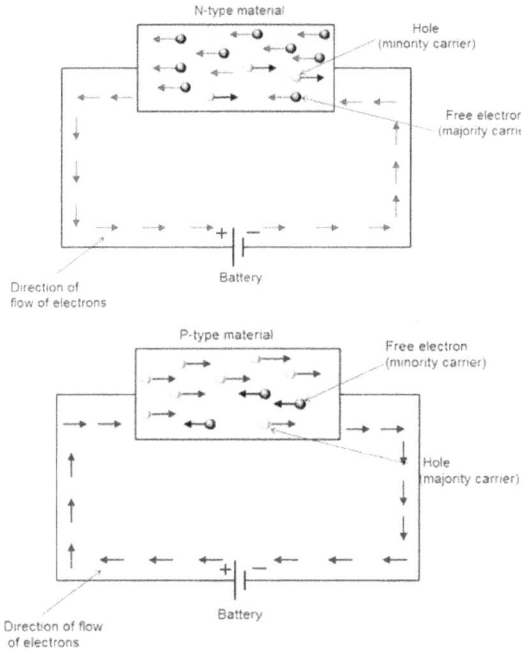

**Figure 239: Doped Semiconductor Material creating the N-type or P-type semiconductor.**
Source: Physics and Radio Electronics.com

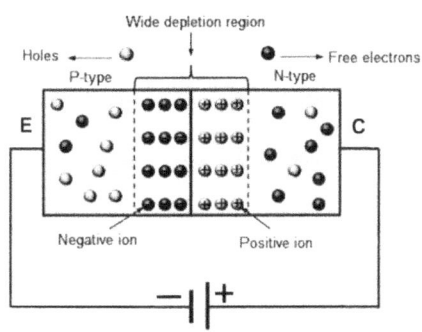

**Figure 238: The P-N Junction.**
Source: Physics and Radio Electronics.com

So, the doping with impurities makes from an *intrinsic semi-conductor material* an *extrinsic semi-conductor material*. When a semi-conductor material like Silicon or Germanium is 'doped' with impurities of another material, it creates an (extrinsic) semiconductor; either the N-type semiconductor (pentavalent doping) or the P-type semiconductor (trivalent doping) (Figure 239). The difference in the two materials is the way electricity is transported through the material. In N-type material electrons are majority carrier, in P-type material holes are the majority carrier.

*The Junctions of P- and N-semiconductor material*

Given these two types of materials, their combination is the next step. When the P-type semiconductor material and the N-type semiconductor material are joined, the P-N junction is created (Figure 238). A depletion region forms instantaneously across a P-N junction. In the junction, free electrons in the N-side conduction band migrate (diffuse) into the P-side conduction band, and holes in the P-side valence band migrate into the N-side valence band. And that junction does have specific properties as the flow of electron can pass the junction only in one direction (aka the diode effect) from emitter (Connector E in Figure 238) to collector (connector C).

The combination of P-material between the N-P junction and P-N junctions creates an additional effect when a voltage is applied on the P-material (connector B, Figure 240). The electron behaviour can now be controlled by applying a voltage on the B-connector of the 'base layer.'

The free electrons at the left side n-region (emitter) and right-side n-region (collector) experience a repulsive force from each other. As a result, the free electrons at the left side and right-side n-regions (emitter and collector) will move into the p-region (base). During this process, the free electrons meet the holes in the p-region (base) near the junction and fill them. As a result, the depletion region (positive

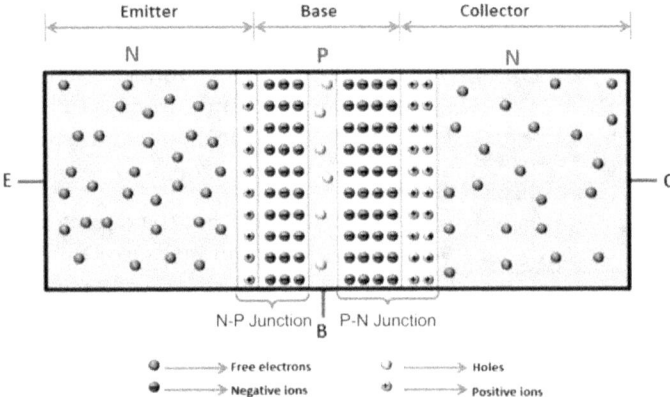

**Figure 240: The P-N-P Junctions (transistor).**
Source: Physics and Radio Electronics.com

and negative ions) is formed at the emitter to base junction and base to collector junction.

The combination of junctions became the PNP-transistor and the NPN-transistor. Each with its own characteristics that can be used in a range of electronic circuit configurations. Such as the amplifying configuration that became the main feature of the transistor (Figure 241). It were these P-N junctions (and their counterpart the N-P junctions) that created the devices like the PN-diode and PNP/NPN-transistor that would become the building blocks of electronic circuits during the Era of the Transistor.

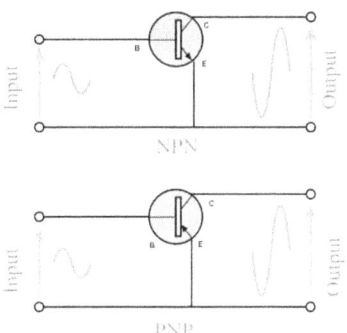

**Figure 241: Symbolic Representation of the Transistor.**
Source: https://www.electronics-notes.com/

## *The Invention of the Transistor*

Had the invention of the vacuum tube been largely the work of the individual inventors, since the rise of organised industrial R&D laboratories, inventive work had become a collective effort. Such as organized by the laboratories of the General Electric Company (GE) [416] and the American Telephone & Telegraph company (AT&T) —but also the Philips Nat Lab in the Netherlands and the Siemens' Labs in Germany— that brought together many scientific and engineering capabilities, and that were funded by the financial resources from their parent companies.

> One of those industrial laboratories was the Bell Telephone Laboratories (aka Bell Labs). As subsidiary of AT&T, already in the 1930s it was one of the largest industrial research laboratories of that time. And by the mid-1950s it had become a renowned research facility in a range of different scientific fields and their application. One of those fields was solid-state physics.[417]

The Bell Labs' work on the transistor emerged from war-time efforts to produce extremely pure germanium 'crystal' mixer diodes, used in radar

---

[416] GE was the commercial continuation of Thomas Edison's activities resulting in the electric lamp. AT&T was the outgrowth of Alexander Bell's invention of the telephone. Both had earlier entered the radio field and took up the research of vacuum tubes.
[417] Solid-state physics is the study of rigid matter, or solids. It studies how the large-scale properties of solid materials result from their atomic-scale properties.

units as a frequency mixer element in microwave radar receivers. It was the physicist *William Shockley* (1910-1989), —with a PhD from MIT in 1936— who was given in 1945 the task to find an alternative to the shortcomings of the vacuum tube technology (eg its power consumption, heat dissipation, vulnerability to shocks).[418]

## The Point-Contact Transistor

The key to the development of the transistor was the further understanding of the process of electron mobility in a semiconductor. The work of the theoretical physicist —the thinker— John Bardeen (1908-1991) and the experimental physicist —the tinkerer— Walter Brattain (1902-1987), under the management of the theoretical physicist William Shockley (1910-1989), resulted in the 'point-contact transistor' (Figure 242).

Shockley had started working in 1936 on the solid-state physics theory that was the basis for the transistor. Inspired by the cat's whisker principle (Figure 183) with a germanium crystal, and the diode/triode concepts of the vacuum tubes, he worked on the concept of a solid-state detector. The war disrupted the explorations but after 1945 the development was picked up again by a newly formed research group of scientists headed by Shockley. By 1947 a working prototype had been created by Bardeen and Brattain, and was demonstrated at December 23rd to Bell's management.

> The demonstrated point-contact transistor consisted of a sliver of germanium with two closely spaced gold point contacts held in place by a plastic wedge (Figure 242, right). They selected germanium material that had been treated to contain an excess of electrons, called N-type. When they caused an electric current to flow through one contact (called the emitter) it induced a scarcity of electrons in a thin layer (changing it locally to P-type) near the germanium surface. This changed the amount of current that could flow through to the collector contact. A small change in the current through the emitter caused a larger change in the collector current. They had created a current amplifier that became known under the name 'transistor.'

---

[418] This chapter is based on multiple resources. Among them: Riordan, M. et all: The Invention of the Transistor. https://www.pks.mpg.de/~yast/Articles/MyArt/pS336_1; The story of the transistor (Bell Telephone Laboratories, 1958) http://wedophones.com/TheBell System/pdf/tsott.pdf; Riordan M.: From Bell Labs to Silicon Valley. https://www.electrochem.org/dl/interface/fal/fal07/fall07_p36-41.pdf Riordan, M. The Silicon Dioxide Solution. https://www.d.umn.edu/~htang/ece4311_doc_F11/04390023.pdf. (Accessed August 2021).

The Invention of the Silicon Engine

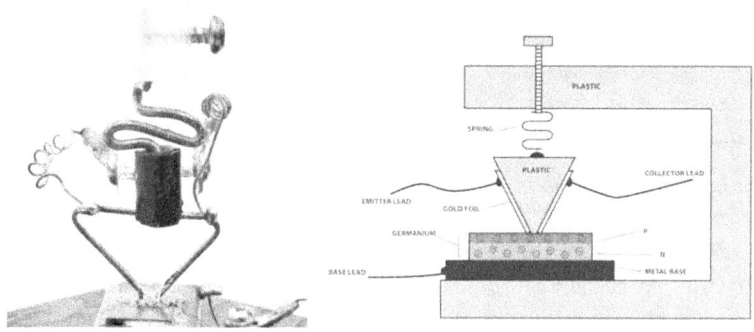

**Figure 242: The Point-to-Point contact Transistor (1947).**
Replica (left) and diagram (right) of the point-to-Point contact transistor.
Source: https://www.computerhistory.org/revolution/digital-logic/12/273

*Patenting the Invention*

At the Bell Laboratories put a high value on patenting, it was obvious the invention of the transistor would have to be patented. So, Bell Lab's patent attorneys started working, and explored several different patent applications. They were facing a problem of an earlier patent issued to Lilienfields; US-Patent № 1,745,175 filed on October 6th, 1926 and granted on January 8th, 1930. However, his concept was never built and had no practical value. Ultimo, the patenting work of the attorneys would result in a range of patents; US-Patents № 2,524,033; 2,524,035; 2,524,034; 2,560,792; 2,569,347.

For the point contact transistor, a patent was granted to the 'Three electrode Circuit Element utilizing semiconductor materials' as US-Patent № 2,524,034 (filed on February 26th, 1948, and granted on October 3rd, 1950) mentioning Bardeen and Gibs as inventors. The second patent was filed on April 17th, 1948 and granted as US-Patent № 2,524,035, issued to Bardeen and Brattain on October 3rd, 1950, under the title 'Three-Electrode Circuit Element Utilizing Semiconductor Materials' (Figure 243).

*Their device featured two closely spaced metal points jabbed delicately into a germanium surface—hence its name, the "point-contact" transistor. They called one point the "emitter" and the other point the "collector," while a third contact, known as the "base," was applied to the back side of the germanium sliver. A positive electrical bias on the emitter enhanced the conductivity of the germanium just beneath the collector point, amplifying the output current that flowed to it from the base. [...] Their brilliant pioneering work has overshadowed much of the subsequent development years of the transistor, including the crucial change from germanium to silicon in the mid-1950s. That shift in semiconductor material proved essential to the device's glorious future as the fundamental building block of virtually all of today's integrated circuits. For germanium, to put it simply, was just not up to the task.*[419]

Bell Labs announced the invention of the point-contact transistor at a press conference on June 30th, 1948 to create publicity for the invention (Figure 244). The attendees were given headphones to hear the device amplify and oscillate and to listen to a broadcast received on a radio set that used transistors instead of valves. After the unveiling, the public showed only lukewarm interest, but Bell Labs was besieged by requests for sample devices from engineers in the electronics industry and the armed forces. By the summer of 1949 the lab had fabricated 4000 working germanium transistors.[420]

The point-to-point contact transistor was —according to the patents issued— invented by Bardeen and Brattain. Shockley was not even mentioned in the patents. By December 23rd, 1947, the two had improved the device and demonstrated that version to Bell Labs management, including Shockley. Realising the importance of the discovery, Shockley started manoeuvring himself as part on the invention and, as time went on, Shockley became more aggressive in claiming credit for the invention.

*Shockley had launched a media offensive. As Bardeen and Brattain's supervisor, he designated himself Bell Labs' spokesman on transistors. He approved a nationwide publicity tour that featured himself as well*

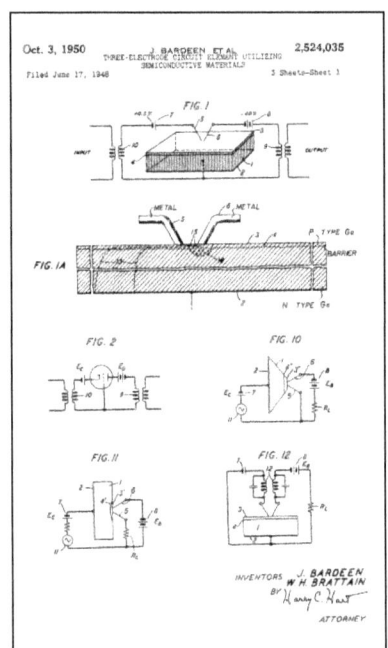

**Figure 243: The Patent for the Point-contact Transistor.**

US Patent № 2,524,035 granted to Bardeen and Brattain.

Source: USPTO

---

[419] Riordan, M.: The Lost History of the Transistor (IEEE Spectrum, 2004). https://spectrum.ieee.org/tech-history/silicon-revolution/the-lost-history-of-the-transistor (Accessed September 2020).
[420] The three scientists were jointly awarded the 1956 Nobel Prize in Physics for "their researches on semiconductors and their discovery of the transistor effect."

*as the other two. For the stock Bell Labs photos of the inventors, Shockley arranged to place himself in the center of an experiment with Bardeen and Brattain looking on, and the caption said that all three had invented the transistor. [...] Articles based on interviews with Shockley, and Shockley's own writing as well, tended to minimize Bardeen and Brattain and emphasize Shockley's role [...]*[421]

**Figure 244: Shockley (centre) surrounded by Bardeen (left) and Brattain (right).**
Official press photo of Bell Labs.

## Parallel Invention of the Transistor

Less known is the fact that in Europe at the same period of time a similar parallel development of the transistor had occurred. Only four years after the Second World War had ended in Europe, a shining technological phoenix had miraculously risen from the still-smouldering ashes of the devastation.

*In late 1948, shortly after Bell Telephone Laboratories had announced the invention of the transistor, surprising reports began coming in from Europe. Two physicists from the German radar program, Herbert Mataré and Heinrich Welker, claimed to have invented a strikingly similar semiconductor device, which they called the transistron, while working at a Westinghouse subsidiary in Paris. [...] After the Bell Labs revelations, Mataré and Welker had little difficulty getting the PTT minister to visit their lab. Thomas urged them to apply for a French patent on their semiconductor triode; he also suggested they call it by a slightly different name: transistron. So the two physicists hastily wrote up a patent disclosure and passed it on to the Westinghouse lawyers. [...] But the French government and Westinghouse failed to capitalize on the technical advantages in semiconductors that they then appeared to have. [...] What is arguably the most important invention of the 20th century remarkably occurred twice—and independently.*[422]

In France they were granted FR-Patent № 1,010,427 (filed on August 13th, 1948) on June 11th, 1952. They also filed for a patent in the US on August 11th, 1949 and were granted US-Patent №

---

[421] Kessler, R.: Absent at the Creation (Washington Post, April 6th, 1997) https://www.washingtonpost.com/archive/lifestyle/magazine/1997/04/06/absent-at-the-creation/2a432ee5-b1e3-49b9-93f2-ad821d1832dd/ (Accessed June 2021)
[422] Source: Riordan, M.: *How Europe Missed the Transistor*. https://spectrum.ieee.org/tech-history/silicon-revolution/how-europe-missed-the-transistor

2,673,948 on March 30th, 1954 for a "Crystal Device for Controlling Electric Currents by Means of a Solid Semiconductor" (Figure 245).

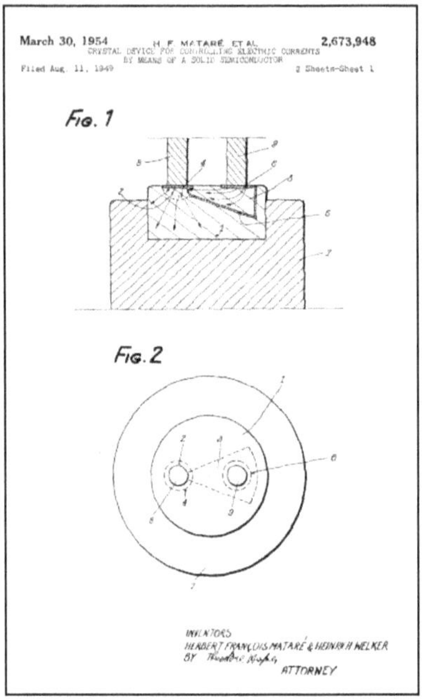

## The Bipolar Junction Transistor [423]

The point-contact transistor was a problematic design. And, to Shockley's displeasure, he was not considered as a part of inventors on the patents issued.[424] So, he went on to work on his own design; it would become known as the bipolar junction transistor, a superior device that took over from the point-contact type.

After the point-contact transistor had proved the idea that an active device could be made with specific solid-state materials, Shockley began a month of intense theoretical activity from December 24th, 1947 to January 23rd, 1948 that came to be referred to as Shockley's 'magic month.'

**Figure 245: Patent for the Transistron.**
US-Patent № 2,673,948
Source: USPTO/Espacenet.

> He [Shockley] *knew that one way to get his name on the Nobel Prize would be to make a significant contribution of his own to transistor development. He also knew, from discussions with his subordinates Bardeen and Brattain, that the next improvement lay in what would become known as the junction transistor, a device that could be more reliably manufactured than its predecessor. Shockley removed Bardeen and Brattain from the project and kept it for himself, consulting them only when he ran into obstacles.* (ibidim Kessler)

Professional jealousy and bruised ego spurred Shockley into a frenzy of independent work, which he initially kept secret from the rest of the group, thus starting to alienate them. Shockley locked himself in a hotel room in Chicago's Bristol Hotel with pen and paper in hand. During this period he

---

[423] Source: Riordan, M. et all: The Invention of the Transistor.
https://www.pks.mpg.de/~yast/Articles/MyArt/pS336_1 (Accessed August 2102).
[424] The three scientists received the 1956 Nobel Prize in Physics for their invention.

would develop the concept of what is now known as the junction transistor. By the end of January 1948, he had come up with a theoretical transistor design that worked quite differently from the design of Bardeen and Brattain, being composed of a sandwich of p-type germanium between two n-type regions. He publishes his theory in articles and a book.[425]

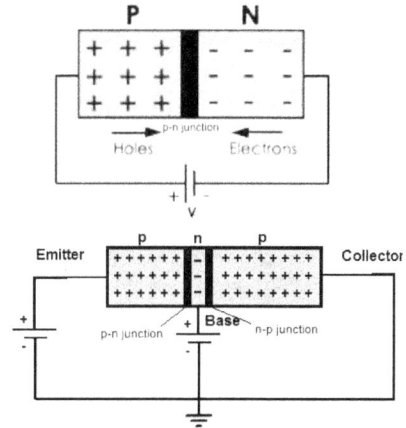

**Figure 246: Principle of P-N Junction in Solid State Material.**

The P-N junction creating the diode effect (top), and the P-N-P junctions creating the triode effect (bottom).

Shockley's invention was profound, providing a much more durable and practical design than Bardeen and Brattain's point-contact transistor, and making it easy to manufacture.

Convinced as he was of the fact that his principle of his 'junction' transistor was better than that of the 'point contact' transistor, he constantly confronted his colleagues with it. As a result, the relationship with Bardeen and Brattain didn't get any better. On the contrary, eventually boiling with restrained anger, Bardeen left the Bell laboratory in 1951. Brattain was put to work elsewhere in the laboratory —far away from Shockley— and eventually became a professor who received a second Nobel Prize for his later work. The original fruitful cooperation of the 'thinker' (Bardeen), the 'doer' (Brattain) and the 'visionary' (Shockley) came to a controversial end.

During the work on the point-contact transistor, there was a discussion going on how such a device did work? The question was if the electricity flowed over the surface of the material or through the semiconducting material. Inspired by that discussion, on January 23rd, 1948 —now working at home— Shockley conceived a distinctly different transistor that was

---

[425] Shockley, W. (1949). "The Theory of p-n Junctions in Semiconductors and p-n Junction Transistors". *Bell System Technical Journal*. Shockley, W. (1950): Electrons and holes in semiconductors, with applications to transistor electronics. Van Nostrand Company, Inc. (1950).

based on the p-n junction discovered by Russell Ohl in 1940.[426] It became the 'junction' three-layer sandwich of germanium or silicon (Figure 246, bottom).

On June 26th, 1948 Shockley applied for a patent for his bipolar transistor. His "Circuit element utilizing semiconductor material" was granted as US-Patent № 2,569,347 on September 25th, 1951. (Figure 247). He also was granted US-Patent № 2,502,488 for a "Semiconductor amplifier" which was filed on September 24th, 1948 and issued April 4th, 1950. And for a "Semiconductor translating device having controlled gain" he obtained US-Patent № 2,623,105 which was filed on September 21st, 1951 and issued on December 23rd, 1952.

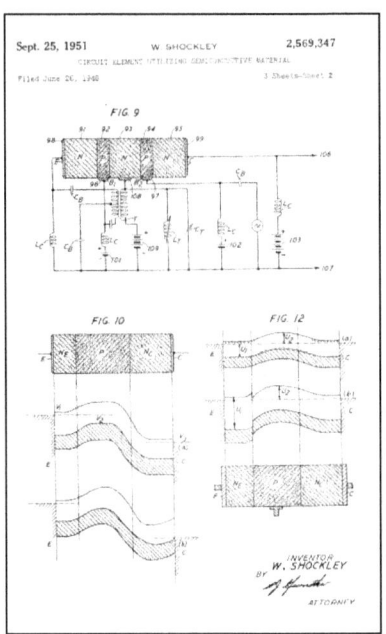

**Figure 247: The Patent for the Junction Transistor.**
Shockley Patent № 2,569,347.
Source: USPTO

*Transistor Technology Development: the Quest for Silicon*

While germanium was the go-to material for the early time in creating transistors, it had both its pros and cons. When compared with the quality of silicon, it was much less reactive and had a lower melting temperature. This allowed electrons to flow through the germanium way quicker than silicon and provided a higher frequency response. However, outside of those two advantages, the need to find a suitable replacement for germanium was clear. Current just leaked way too high, and as the temperature increased, so did the very delicate balance between the junctions in a transistor. This made it hard to control the free electrons.

---

[426] The *p-n junction* is the interface/boundary between two different semiconductor material types; p-type and n-type. The difference is in the impurities in the material The p-side or the positive side of the semiconductor has an excess of holes and the n-side or the negative side has an excess of electrons.

The new device may have been functioning in the laboratory, bringing it to a design that could be manufactured took some time. The main problem was lack of sufficiently pure, uniform semiconductor materials. When that was solved, the next issue was how to introduce impurities in the material to create the p-n junctions.

> An issue at the time was the production of pure silicon. Russel S.Ohl (1898-1987) came up with methods for producing high purity silicon. He was granted US-Patent № 2,402,661 (filed on March 1st, 1941) on June 25th, 1946, as well as US-Patent № 2,402,839 (filed on March 27th, 1941) on June 25th, 1946. Both patent applications were filed before the United States entered the Second World War, and they were issued after it was over.

> Another method for purifying Silicon was developed by *Gordon Teal* and *Keith Storks*. They were granted US-Patent № 2,441,603 filed on July 28th, 1943 and issued May 18th, 1948. It was because of people like Ohl, Teal, and Storks that, in 1947, high purity germanium and silicon were available.

Step by step the solid-state technology of creating the transistor was developed. There were major breakthroughs in 1954 and 1955 with the development of diffused-base transistors; this technology allowed for creation of a very thin base region through the controlled diffusion of impurities into the grown germanium or silicon crystal during the manufacturing process. It would evolve in the meta-transistor and the planar transistor we will meet later on.

## Who invented the Transistor?

The invention of the transistor was the second fundamental invention in the Era of Electronics. It would become basic for a broad range of electronic systems that were implemented in a myriad of applications. We explored how many people contributed to the development of Solid-State physics and the device that could control the flow of electrons in semiconducting material (aka the Transistor) during twentieth century (Figure 231). Each contribution —from the one with a small impact to those with a big impact— with its own value. Contributions that resulted in the semiconductor device that could control the flow of electrons within semiconducting material. But who then, can be considered as the inventor (ie the classic quest for one specific person), of the transistor?

*Facing the Big Question*

When trying to address the classic question of 'who is the inventor,' one is faced with some problems of interpretation, as we already noted before in the case of the Radio Tube; from semantics to the subject, from moment in time to the consequences to the point of view (see chapter: 'Who invented the radio tube?' for details).

All those interpretations have the process of 'Change and Novelty' in common. But again, even in that process, there are differences. Take 'the invention', the result of the process of invention as carried out by the specific person —the inventor—, or a group of persons —the R&D team— that creates novelty (Figure 231). Next, there is the 'innovation', the result of the process of Change and Novelty creating a working artefact that —as the result of entrepreneurial activities— can be brought to the market by subsequent organizational activities. Both those processes, invention and innovation, make up the *Act of Invention* and the *Act of Innovation*. However, then again, confusion arises. Is that about the original 'idea' (as in the mental picture), is it the first model or prototype (as in the archetype), or is it the product/system that is successfully commercialized?

Using our definition of the basic event (Scheme 3) as the dominant innovation in a cluster of events, we let others decided on what was the 'invention.' Like the contemporaries in the Patent Offices who decided if an invention was patentable on criteria of their own, valid in their time. So, to explore the inventions, we used the patent as an event-identifier (Figure 248). And we let the historians of science and technology, who explored the

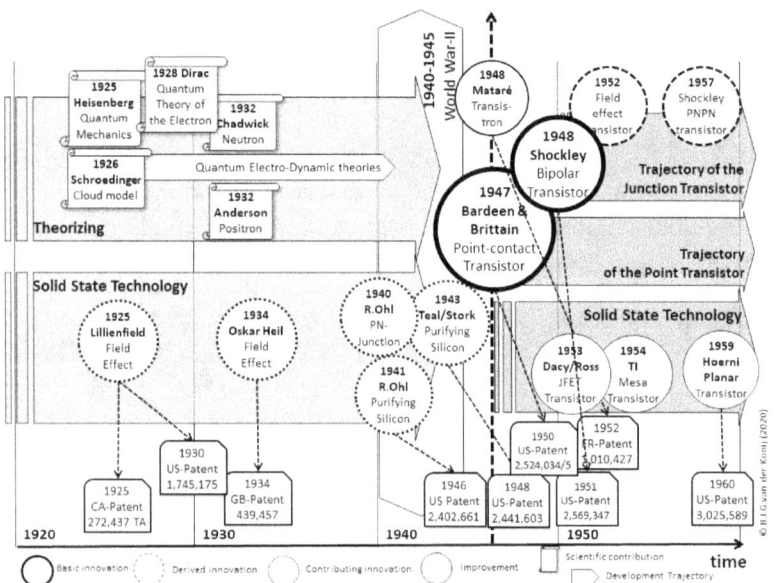

Figure 248: Contributions to the Cluster of Invention with the Basic Inventions of the Transistor (1920-1960).

Shockley, Bardeen and Brattain case extensively, decide who was *the* inventor.

## *The Transistor and its Manufacturers*

Luckily, Bell Labs had an important realization: development of the transistor was going to move a lot more quickly if they opened up the field to other companies.[427] And licensing their patent rights to others may create an interesting source of revenues.

> Led by electrical engineer Jack Morton, this program fostered technologies of zone-refining [...] and growing large single crystals of germanium and silicon [...]. It also developed techniques for forming p-n junctions, preparing semiconductor surfaces and attaching metal leads - plus logic circuits and systems involving transistors. Morton advocated sharing this transistor technology with other researchers and companies because Bell Labs and its parent AT&T could benefit from advances made elsewhere. So during the 1950s they sponsored three gatherings at which other scientists and engineers visited Bell Labs to learn the new semiconductor technology first hand. Held in September 1951, the first meeting specifically addressed military users and applications. In April 1952, over 100 representatives from 40 companies that had paid a $25,000 patent-licensing fee came for a nine-day Transistor Technology Symposium, including a visit to Western Electric's ultramodern transistor manufacturing plant in Allentown, PA. There were participants from such electronics titans as GE and RCA, as well as from then-small firms like Texas Instruments and Sony.[428]

So, in September 1951, Bell Labs hosted a symposium to spread the gospel about what the transistor could do. Attending the conference were some 300 scientists and engineers. The attendees all went home to their respective companies with a great sense of what the transistor could do, but little idea of how to build one. As some hundred signed up for a patent license, they got the book 'Transistor Technology,' and after the symposium the two volumes of the 'Ma Bell's Cookbook.'

Each chapter in parts 1 through 4 of the Cookbook described in detail a stage of transistor manufacturing, from crystal preparation to

---

[427] AT&T's generous patent licensing policy was a consequence of an anti-trust lawsuit in 1949 by the Department of Justice demanding the separation of WE from AT&T. In an effort to avoid the split, the company renounced the idea of monopolizing the transistor market. The company restricted itself to the manufacture of transistors for military use, and the patent was licensed through WE. In 1956, AT&T reached a settlement with the Department of Justice that prohibited AT&T from entering any markets other than the telecom industry and required AT&T to provide other companies with its patents freely or for reasonable royalties.

[428] Source: 1952: Bell Labs Licenses Transistor Technology. Computer History Museum. https://www.computerhistory.org/siliconengine/bell-labs-licenses-transistor-technology/

the measurements of the fabricated transistor's characteristics; and each chapter in parts 5 and 6 explained the design of transistors for manufacture and the manufacturing procedure, respectively.

The attendees represented companies that were both big (such as IBM and General Electric), and small (such as then-unknowns like Texas Instruments). And they all started making transistors for their markets; some even started focussing, next to the military market, on the consumer markets. From the component manufacturing companies, they would become the vertical integrated industries that used transistors to manufacture systems as an end-product.

For the military, the urgent needs of national security overrode the somewhat high-cost considerations in the early transistors. Almost immediately upon disclosure of the invention, Bell Laboratories and the military initiated a joint program to develop transistors and circuits for military uses. And many of the early manufacturers (eg Fairchild, Sylvania) stepped in their shoes.

Western Electric, part of the Bell Telephone System, as a manufacturer of telephone switching systems and long-distance carrier systems, implemented the new devices in new tone-frequency dialling and automatic selection of routes for long distance telephone calls.

And there was the industry that was either active in the vacuum tube manufacturing, originated from other business fields, or were new start-ups. They, becoming aware of the enormous potential of the latent markets, stated to develop the new magic transistor in a range of product families. Their commercial transistors began to roll off production lines during the 1950s. Transistors that found applications in lightweight devices such as hearing aids and portable radios.

**Figure 249: Early Transistor Radios.**
Regency Model TR-1 with open chassis (left, 1954) and Sony Model TR-55 with open chassis (right, 1955).

Source: Wikimedia Commons.

Texas Instruments Inc. —working with the *Regency Division* of Industrial Development Engineering Associates that was granted US-Patent № 2,995,652 for the transistor radio— manufactured in late 1954 the first transistor radio. Selling for $49.95, the Regency TR1 employed four germanium junction transistors in a multistage amplifier of radio signals (Figure 249, bottom left). The very next year a new Japanese company, Sony,[429] introduced its own transistor radio and began to corner the market for this and other transistorized consumer electronics (Figure 249, right). With these portable radios, both companies would become leading in consumer electronics in the following decades.

And there were a range of companies that served the niche-markets where the application of the transistor made new products possible.

The hearing aid —from its mechanical version of the 'ear trumpet,' to its tube based versions (1920s)— had a limited performance. That changed when the hearing aids became transistorized in the early 1950s (Figure 250) and a range of companies manufactured them; such as Acousticon , Beltone, Radioear Corp, Siemens, Sonotone and Zenith. The first transistorized consumer product in the US was a $229.50 hearing aid from Sonotone in 1952. The Model 1010 used two vacuum tubes

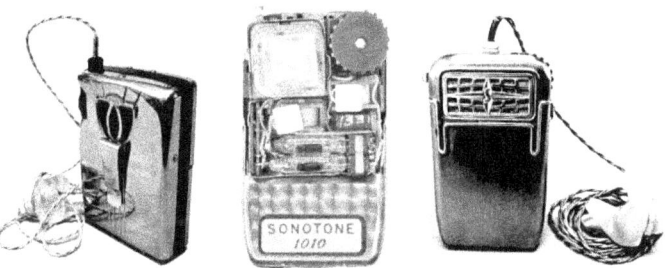

**Figure 250: Transistor Hearing Aids.**
Beltone Model TR (left, 1953), Sonotone Model 1010 (middle, 1953), and Sonotone Model 1111 (right, 1953).
Source: https://hearingaidmuseum.com/, https://www.computerhistory.org/

---

[429] About the same time, Masaru Ibuka and Akio Morita had formed the company Tokyo Tsushin Kogyo. Totsuko was a technology intensive company, employing many highly-educated scientists and engineers who made up a third of Totsuko's employees due to its intensive research and development (R&D) activities developing the tape recorder. Totsuko was the first Japanese company to obtain a transistor manufacturing license from Western Electric, and in 1955 they renamed their company Sony.

and one transistor made by Germanium Products Corporation. In 1953 the Sonotone Model 1111 was an all-transistor model patented by US-Patent № 2,789,160 (granted April 16th, 1957). By 1954, 97 percent of all hearing aids used only transistors.

## Patent Scope: No Blocking Patent Positions

Had the development of the wireless technology and the radio transmission being held up by the many broad-scope patents owned by different companies, in the case of the semiconducting devices was faced by a different situation. Although the inventors also put their broad-scope claims forward, the effect was differetn from the earlier broad-scoped radio-patents (ie regarding both radio transmisson and radio receiving).

> *There are two instances in the history of this technology where a broad-gauged patent was issued which could have given its holder control over a large "prospect," but in fact did not. One involved the initial transistor patents held by AT&T.[430] Because of an antitrust consent decree, AT&T was foreclosed from the commercial transistor business. Some have argued that it is not clear whether AT&T would have gone into the merchant transistor business even in the absence of a consent decree. In any case, given that it was not going to do so, AT&T had every incentive to encourage other companies to advance transistor technology because of the value of better transistors to the phone system. AT&T entered into a large number of license agreements at low royalty rates. Many companies ultimately contributed to the advance of transistor technology because the pioneer patents were freely licensed instead of being used to block access.*

This explains why the Bell Laboratories were following their generous policy on patenting we described before.

> *The second instance involved the parallel inventions of the integrated circuit (by Texas Instruments) and the Planar process for producing them cheaply (by Fairchild Instruments). Both of these companies obtained patents on their own inventions, which meant that each had to license the other to produce integrated circuits effectively. Cross licensing was favored by the government; the Department of Defense, which for some time had provided the lion's share of the market for semiconductors, had a strong interest in seeing these important technologies become broadly available throughout the industry. Again, the absence of a single, broad patent assisted the rapid development of an industry. The second recent cumulative technology developed without strong, broad patents is electronic computers. Although original computer inventors Eckert and Mauchley did file for and receive a patent on their basic ENIAC design, the patent was ruled invalid because of a judgment that the prior art included much of what they claimed.*

---

[430] The Bell Telephone Laboratories were part of the American Telephone and Telegraph (AT&T) company.

*Since this ruling, patents have played only a very minor role in the computer industry, and where patents are concerned, cross licensing is common. As a result, the pace of technical change has been rapid.* (Merges & Nelson, 1990, p. 21)

## Technology Transfer and the Legal Framework

That rapid pace of development had also another cause, as the people who were involved in the development of the semiconductor technology (both technologists and circuit designers), were the carriers of the know-how. And when they moved from company to company, they took that know-how with them. A technology transfer that could be limited by the legal framework of state-legislation and employment policies of companies.

Many companies had in their employment contracts a clause, that restricted the employee to be active in a similar activity when he left a company (eg a non-competition clause or a non-disclosure clause). But it depended on the regional business culture how this was implemented. And in California the state legislation influenced the business culture of Silicon Valley fundamentally. That Californian State legislation had a specific background that originated from the time of the Gold Rush (1848-1855) and had resulted in the 1872-Law.

> The Californian Gold Rush started as a boiling pot with violent eruptions. But after a while the need for laws arose, and the creation of the law in 1872 was clearly motivated by a desire to minimize the number of shootings in and around the gold mines of El Dorado and other California areas. The objective was to create a law that would dramatically reduce the likelihood of disputants resorting to violence to resolve misunderstandings between gold mine owners and hired hands. The new law in question stated that a contract between an employer and employee should in no way be limited in his freedom to change employer, even if that meant competition.

> *16600. Except as provided in this chapter, every contract by which anyone is restrained from engaging in a lawful profession, trade, or business of any kind is to that extent void.*[431]

> In other words, an agreement in which a non-competition clause limited the freedom of the employee, was null and void.

The effects of this law are still noticeable today where many employment agreements have their specific agreements. Like the *Non-Competing Agreement* (NCA) and *Non-Disclosure Agreement* (NDA)[432] in case

---

[431] California Code - Section 16600, also known as CAL. BPC. CODE § 16600.
[432] NCA: Agreement that limits an employee's ability to work with competing firms in a specific geographic area and for a specific period of time. NDA: A legal contract between at

the employee leaves the job to work for another employer. The signing upon employment of a standard pair of NCA and NDA documents is binding in America. Except in California, where the 1872 law restricts the NCA/NDA rules so that they essentially in no way limit the ability of the employee of any company at any time to go work for another company. Not even when that company is a direct competitor of the organization he just left. In other words, when California's legislators adopted a law in 1872 they still guaranteed in later time residents of California state full freedom to choose and change jobs.[433]

## The Transistor Manufacturing Process

Those companies that signed up for a license to manufacture the transistor were facing an enormous task. Many had build up experience in the vacuum tube technology, but the semiconductor technology required a whole new set of techniques as well as new basic knowledge. And many obtained both the basic knowledge from the knowledge the Bell Labs had buildup over the years.

The transistor was realized with different techniques; from the crystal-growing technique creating the ingots (Figure 251, left), the IC fabrication technology (Figure 251, middle) with its the metal deposition technigue up to the alloy-junction growing technigue. And finally (Figure 251, right), the packaging of the chip. And all were based on the semiconducting material of germanium and silicon obtained with a crystal-growing technique creating the substrates. From around 1953 this all resulted in a transistor manufacturing process that can described as follows.

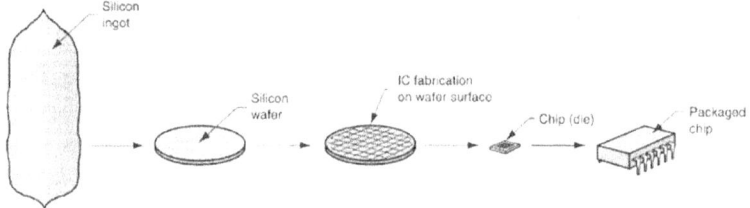

Figure 251: Overview Semiconductor Manufacturing Technology from basic material to packaged chip.
Source: https://sites.google.com/site/ icmanufacturing2204/products-services

---

least two parties that outlines confidential material, knowledge, or information that the parties wish to share with one another for certain purposes, but wish to restrict access to by third parties.
[433] Gromov, G: Silicon Valley History. Chapter: Main Difference of the Legal Framework of Silicon Valley. Source: http://www.netvalley.com/silicon_valley_history.html

The photo-lithographic techniques of creating the patterns on the substrate using masks (one for each step), had a cyclic nature (Figure 252). The photolithographic process made it possible to create patterns on the substrate according to the patterns on the masks. Each phase used a mask to create a different pattern by exposing UV-light on

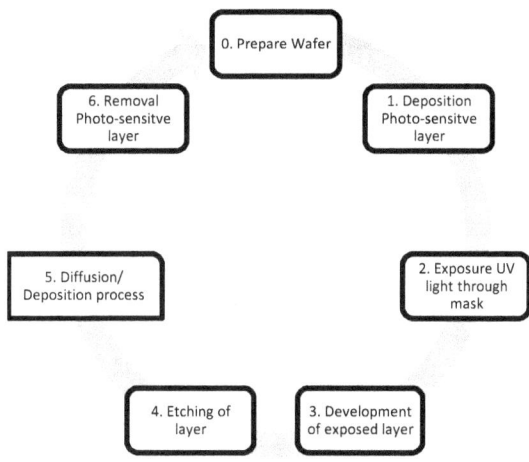

Figure 252: The Photolithographic Technology.

the substrate (Figure 253). And when the pattern was exposed on the photosensitive layer, it was etched to remove the (not)-exposed areas creating an equivalent pattern of protected and not protected areas on the substrate. And then those process the areas are treated (ie etched, diffused, grown) in the next phase of the total process.

Without going into the details of the transistor technology, basically the grown-junction techniques resulted in the 'Mesa transistor' (developed at Texas Instruments in 1954) and the 'Planar transistor' (developed by J.Hoerni in 1959). What they have in common were the operational techniques (eg the photo-lithography operations), what distinguishes them is the way the junctions were created.

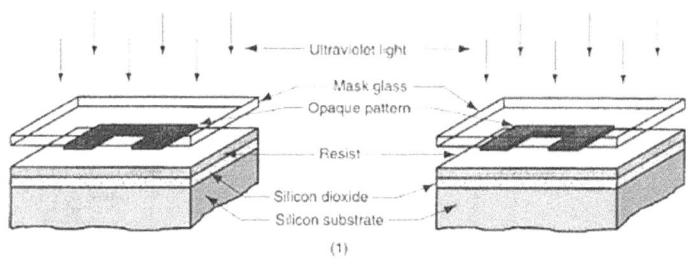

Figure 253: Photo-lithographic Techniques.

Source: https://sites.google.com/site/icmanufacturing2204/products-services

The *Mesa Process* (resembling plateaus, aka mesa) creates transistors that basically consisted of three impurity layers piled up vertically, each rich in either electrons (n-type) or electron deficiencies, better known as holes (p-type). The main drawback of the mesa structure is that its p-n junctions, the interfaces between layers where the transistor's electrical activity occurs, are exposed at the edges (Figure 254, left).

The *Planar Process* creates by etching and diffusion junctions within the diffused material, in which all of the p-n junctions terminate at approximately the same geometric plane on the surface of the semiconductor. And there it can be protected by a Silicon Oxide layer (Figure 254, right). The process involved forming transistor junctions through a series of photographic exposure, chemical etching, and doping techniques on a slab of polished silicon. The planar process also made it easy to interconnect neighbouring transistors on a wafer, paving the way to another Fairchild achievement: the first commercial IC's.

The Planar process was a breakthrough and became the backbone of transistor manufacturing and virtually made the germanium transistor and all other semiconductor manufacturing techniques obsolete within months.

> *Initially, the planar process yielded only a few working transistors in every 100—much worse than the mesa process. But as various problems, such as pinholes in the oxide layer, were resolved, yields rose and doubts evaporated. In April 1960, Fairchild sold its first planar transistor, the 2N1613—a metal cylinder about half a centimeter in diameter and almost as high, with three little metal legs sticking out beneath it.* (Michael Riordan, 2007, p. 54)

The Planar process made better and more reliable transistors, but most importantly, it transformed the production of semiconductors from a handcrafting operation into a high-volume, electronic lithographic "printing" process that lowered their production costs.

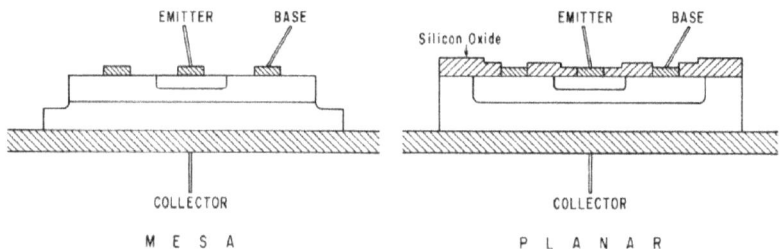

**Figure 254: Mesa technology versus Planar technology.**
Side views of a mesa (left) and a planar transistor (right), from a 1960-report by Hoerni.
Source: Riordan, M. The Silicon Dioxide Solution.
https://www.d.umn.edu/~htang/ece4311_doc_F11/04390023.pdf

Based on Hoerni's work, Fairchild developed the 2N1613 transistor in 1960 and sold it commercially. The teardrop shaped chip (Figure 255, top) packaged in a TO-5 housing (Figure 255, bottom) became highly successful. Fairchild licensed rights to the process across the industry.

With the Planar process and its photolithographic techniques, Fairchild could batch-process hundreds of identical transistors on a single wafer. And the Planar process opened the way to the Integrated Circuit. It took some creative thinking, however.

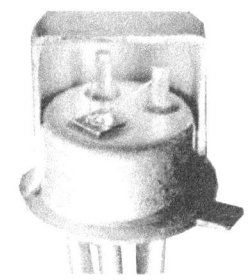

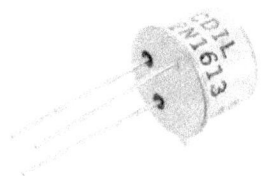

*Out of this creative stew emerged another crucial concept, which historians have so far overlooked. With the planar transistor, it was now easy to put all three electrical contacts—to the emitter, base, and collector—on one side of the silicon wafer. At first glance, it might seem just a marginal improvement, but this feature, plus the fact that a single metal such as aluminium could be used to form the connections, meant that Fairchild could now, in effect, print electrical circuits—transistors and all— on silicon. Like the typographic patterns of ink impressed onto paper by a printing press, the patterns of the individual semiconductor devices and metal interconnections could now be imposed photolithographically on a single side of a wafer.* (Michael Riordan, 2007, p. 55)

**Figure 255: The 2N1613 transistor.**

The teardrop chip (top), open TO-package showing die (middle), and packaged chip (bottom).

Source: https://sudonull.com/ Wikimedia Commons.

### Packaging the Die in its Housing

The preceding exploration concerns the process of creating individual transistors. In its essence the manufacturing of integrated circuits is similar, with the difference that the number of transistors in a circuit is increasing. Both the individual transistor and the circuit end up as a 'die' being part of the wafer (Figure 251). After the processed wafer (eg 2,5" in diameter in

1969) with its multitude of transistors (aka Chip or Die) was finished, it was sliced into individual dies (aka die-singulation), and these were packaged in a protective housing (Figure 255).[434]

## *The Evolution of the Transistor*

With the invention of the transistor started a development trajectory of continuous technological improvement resulting in families of transistor (Figure 256). Depending on where they were going to be used (eg in telephone systems, military applications, consumer products), they followed a trajectory of continuous improvement. Next to the improvement of the technical characteristics of the devices themselves, the technology developed as more advanced photo-lithographic techniques were introduced to create the devices.

### The Breakthrough of the Silicon Transistor Technology

For the first six years of their existence, transistors had all been made with germanium. But germanium has serious limitations and in the mid-1950s the semiconducting material 'silicon' came in the picture. It was just a matter of time before the first silicon transistor would be made.

Morris Tanenbaum (born 1928) presented on January 26th, 1954, a silicon transistor while working as a member of Shockley's research group at Bell Telephone Laboratories. But the world's dominant semiconductor company kept this achievement under wraps.

A similar development had taken place at Texas Instruments, but the Texas upstart rushed to announce it.

Gordon Kidd Teal (1907-2003), a chemical physicist, had worked at Bell Labs for 22 years, and worked there on the growing of silicon crystals. In 1953 he moved to Texas Instruments, to become director of research of TI's Central Research Laboratory, modelled after Bell's Labs. By the time Teal arrived, the firm had almost 1800 employees and was generating about US $25 million in annual sales. The TI-culture being more publicity minded, in 1954 he presented a paper "Some Recent Developments in Silicon and Germanium Materials and Devices," at the Institute of Radio Engineers (IRE) National Conference. It described how they had managed to create a NPN-structure using a new technology; the grown-junction single-crystal technique.

---

[434] That packaging of the transistor underwent over the years a development of its own. It started with the Transistor Outline Package (TO), followed by the Small Outline Transistor (SOT) and thin small outline transistor (TSOT) package.

So, on two different places, a similar development had taken place. But it was the company culture that shaped the next developments.

> But the [Bell] Labs did not pursue the process further, thinking it unattractive for commercial production, which allowed Texas Instruments (TI) to claim credit for this breakthrough several months later. Having left Bell Labs to organize a research lab at TI, Teal hired a team of scientists and engineers led by chemist Willis Adcock to work on silicon transistors. Employing high-purity Dupont silicon, they made their first successful silicon transistor — an n-p-n structure using the grown-junction technique — on April 14, 1954. Unaware of Tanenbaum's work, Teal presented this achievement on May 10 at an Institute of Radio Engineers conference in Dayton, Ohio, creating a sensation by announcing that silicon transistors were in production and available for sale. With little competition, TI dominated the silicon-transistor market for the next few years and made significant inroads into Raytheon's position as the largest merchant market supplier of transistors. By the end of the 1950s, silicon had become the industry's preferred semiconductor material.[435]

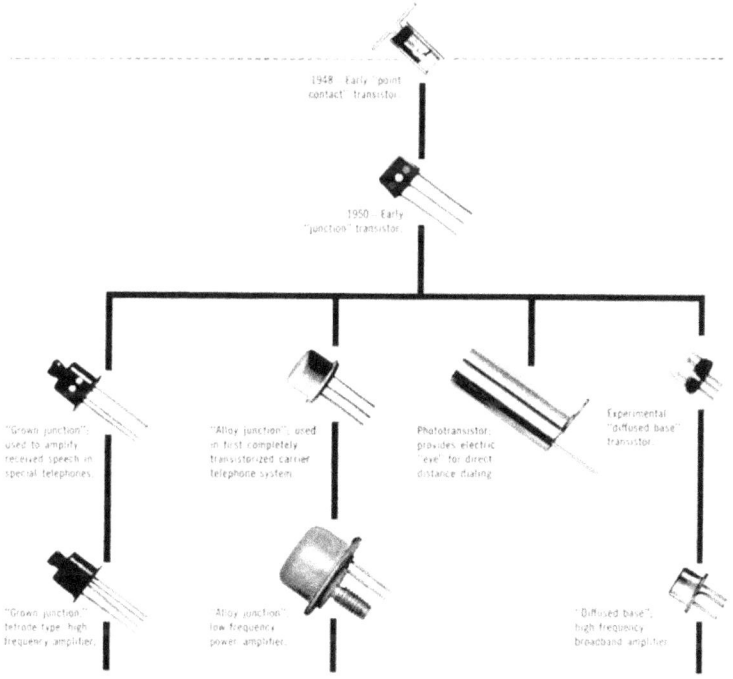

**Figure 256: Overview First Families of Transistors.**
Source: https://www.chipsetc.com/the-transistor.html (Accessed September 2021).

[435] Source: *1954: Silicon Transistors Offer Superior Operating Characteristics*. Computer History Museum. https://www.computerhistory.org/siliconengine/silicon-transistors-offer-superior-operating-characteristics/ (Accessed June 2021)

Remarkably enough, neither development was patented. Bell Labs patent lawyers pointed out that the Bell-team had used well known techniques and the combination of those with the new material, silicon, was not patentable. The silicon transistor technology —and especially the planar process with the addition of the epitaxial process[436]— proved not only a breakthrough in the manufacturing of single transistor. Adding a deposition of a metal pattern (ie Aluminium) for the connector pads on the substrate —ie the base of the bonding wire connecting the chip and its housing— proved another breakthrough (Figure 257, middle).

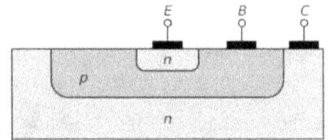

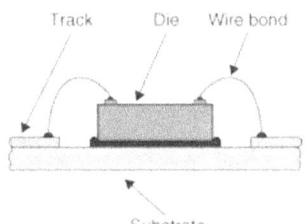

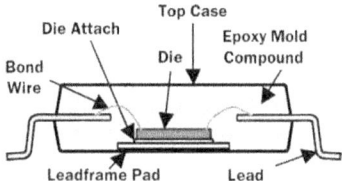

*All the key development and engineering problems were either solved or on course for an elegant solution. There was a sound foundation for the long-term manufacture of semiconductor devices. Silicon, the semiconductor of choice, could be produced with crystalline perfection and purity more than adequate to the task. Critical dimensions in all three*

**Figure 257: Connecting the Die and Packaging it.**

Top: The side view of the die of a NPN Bipolar junction Transistor. Middle: The Wire bond between the die and substrate with the connector pads (ie track). Bottom: The packing in a protective housing.

*directions could, if necessary, be controlled to a fraction of a micrometer. Electrical contacts could be made with a single metal and without the need for microscopic precision. The resulting devices would eventually be solidly reliable. And all this could be done with batch processing with the promise of high yield and low unit cost. Some 13 years after its invention, the transistor now had a sound engineering foundation. This provided the base for the next giant step. The integrated circuit (IC) [...]* (Ross, 1998, p. 22)

---

[436] Epitaxy refers to a type of crystal growth or material deposition in which thin, crystalline layers are formed with one or more well-defined orientations with respect to the crystalline seed layer. It improved the switching speed.

After the bipolar technology, other technologies such as the Metal Oxide Semiconductor (MOS) technology, developed. In 1959 M. M. (John) Atalla and Dawon Kahng at Bell Labs achieved the first successful insulated-gate field-effect transistor (FET) in this technology (ie MOSFET) (Figure 258). A technology that had been long anticipated by Lilienfeld, Heil, Shockley and others.

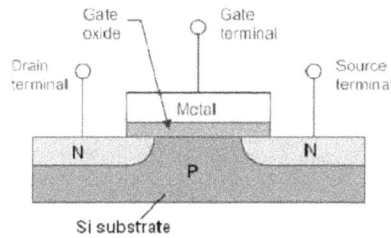

**Figure 258: Diagram of MOSFET.**
Source: Wikimedia Commons

Atalla and Dawson filed a patent application on March 8th, 1960 and were granted US-Patent № 3,102,230 on August 27th, 1963 for an 'Electric field-controlled semiconductor device.' Their fabrication process became known as the Metal-Oxide -Semiconductor (MOS) technology. It would become a dominant technology for the further developments.

## From Active Component to Application System

This was the beginning of a broad range of technological development trajectories. Silicon proved to be a fertile material and the solid-state technology unleashed many of its properties; it was light-sensitive (the photo-sensory effect), it could produce light (the light emitting effect) used by the Light Emitting Diode (LED), it could uni-direct currents (the diode-effect) and it could amplify currents (the triode-effect).

The transistor itself was improved upon in subsequent technologies creating the Junction gate field-effect transistor (JFET) in 1953, the Bipolar Junction Transistor (BJT), a range of Germanium transistors, and the silicon Planar Transistor in 1959. The MOSFET, also developed in 1959, became the 'workhorse' of the electronics industry. The new material of silicon would become the backbone of further transistor development.

But that was not all as also the system integration improved as the result of different technologies.

## The Hunt for Micro-miniaturization of Electric Circuits.

The transistor —together with the passive components like resistors and capacitors— became a dominant part of electronic circuitry being mounted on the so-called Printed Circuit Boards This mounting method replaced the metal chassis (Figure 259, top) with the rather messy point-to-point interconnection in use up till then for early electronic assemblies (eg radio assemblies) (Figure 259, upper middle).

The boards, originally made from an isolation material like bakelite had a thin copper layer on which the interconnection was created (ie in an etching process) (Figure 259, lower middle). In addition, it had holes drilled through the board supporting the components placed on one side mechanically and soldering them to the interconnection on the other side (Figure 259, bottom).

By the early 1950s the use of photolithographic and silk-screening techniques enabled mass production of PCBs. And a new batch-oriented dip-soldering techniques (aka dip-soldering) enabled the mass production of the soldering process fixating the components to the interconnection

This was the micro-miniaturization trajectory in which not only the components packaging became smaller (eg the mini-tube, the transistor and the mini passive components), but also the improved interconnection techniques made

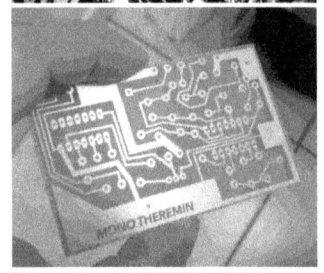
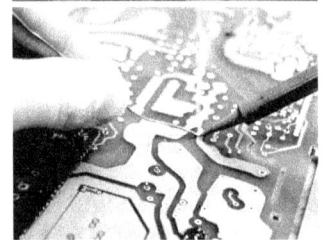

**Figure 259: Supports of electronic circuits.**

Radio Assembly Chassis (top and upper middle, 1942), PCB after etching (lower middle), and hand-soldering components on board (bottom).

Source: Wikimedia Commons. Science Museum.

higher packaging densities possible. And the boards became more complicated with the rise of the single layer PCBs into the multi-layer PCBs of increasing complexity.

It were Moe Abraham and Stanislaus Danko from the US Army Signal Corps who came up with a circuit board automatic assembling process and transformed the production of circuit boards. They used a copper interlinking design and dip soldering technique to mount leads into boards. Besides, they sketched the wiring design and snapped it into a zinc plate to make boards. In 1950 they filed for a patent application and on July 31st, 1956, Moe and his colleague were granted US Patent № 2,756,485 for their concept.

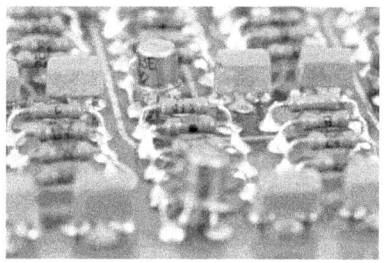

**Figure 260: Electronic Components mounted on a Printed Circuit Board.**

Printed Circuit Board with components (top, middle), and radio-receiver circuit board (bottom).

Source: Istockphoto.com (top, fair use claimed). Unknown (bottom).

All these techniques were focusing on the micro-miniaturization of the electric circuitry realised with discrete components on Printed Circuit Boards. A process that would be influenced by another miniaturization process that created the Integrated Circuits.

*The Application of PCB-based circuitry*

With the rising complexity of the circuits mounted on the PCBs, the PCB-based circuitry found its use in a range of applications.

*Military Applications:* The military were among the first customers as they began using junction transistors almost immediately in electronic systems for airplanes and missiles, where engineers were trying to squeeze in complicated communication and guidance systems. Transistors were perfect for these military systems, because they were much smaller and used much less electrical current than vacuum tubes.

*Consumer applications:* In addition, there was the mass market of consumer applications. Soon the new devices were used in hearing aids, portable radios, and all sorts of other electronic devices directed at the consumer needs.

*Logic Applications:* And there were the calculating and computing devices developed in other technologies where the transistor replaced the relay or vacuum tube.

Every electronic circuit that had used a relay of vacuum tube to realize a specific function, saw the active device being replaced by a transistor. So the vacuum-tube based building blocks now became the transistorized building blocks (Figure 168). The development of the transistor technology itself steadily moved on. But that was not anymore in the East (with exceptions like Texas Instrument located in Texas), but in the West of America where a region below San Francisco became the centre of solid-state manufacturing companies.

## Overview of the Invention of the Transistor

We explored in the preceding analysis the events that created the new electronic engine called transistor as the result of both the theorizing of the 'solid state scientists' and the technology development of the 'technologists.' We noted the invention of the Point Contact Transistor and the invention of the Bipolar Transistor (Figure 261). The Point contact transistor proved to be a dead-end development as it was quickly superseded to the Bipolar Junction Transistor.

The preceding exploration illustrated how the transistor evolved into a versatile component that could be used in a broad range of application trajectories. Trajectories that encompassed the Calculating and Computing systems, but also the Communication and Broadcasting systems. The new electronic technology not only improved communication over long distances, it also pervaded into the mechanical application areas. The

Mechanical Revolution had created fine-mechanical machines like the mechanical watches and the calculators. It soon turned out that the time-function could also be realized electronically, as well as the basic arithmetic functionality and scientific functions.

In a period of about a decade, before, during and shortly after the Second World War, the transistor technology would lay the foundations for developments that would influence the Affairs of Man dramatically. Next to all the novelty in another fields, the new electronic technology based on transistor created during the first part of the Second Electronic Revolution, would enable the Personal Computer Revolution and Mobile Phone Revolution to come.

### Periodization of the Transistor Technology

Looking at the development trajectory (Scheme 2) of the transistor technology and seen in the perspective of the life cycle (Scheme 5), the transistor technology is marked by specific events creating the basic innovation that became the start of the Semiconductor Revolution.

*Transistor 0.0:* (1920-1940): With the rise of insight into the molecular and atomic properties of matter (Figure 231), also the electro-dynamic theories emerged as part of the Modern Physics. The behaviour of free

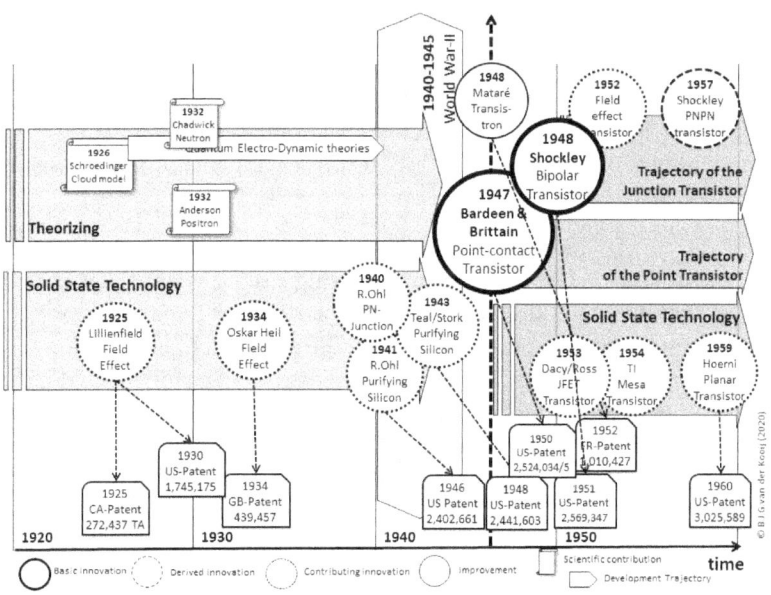

**Figure 261:** Contributions to the Cluster of Invention with the Basic Inventions of the Point Contact Transistor and the Bipolar Transistor.

electrons created by thermionic emission (as in the tube), was complemented by the understanding of what happened in a specific group of materials; the semiconducting materials. The control of the flow of electrons in vacuum —the cathode rays that were observed during the early experimenting with vacuum tubes— had shifted to solid state physics where it was about the control of the flow of electrons in semi-conducting material.

*Transistor 1.0* (1940-1960): The invention of the transistor, building forth on the insights created by the many contributors to the solid-state physics, started the development-trajectory of the solid-state based technology for creating these new active devices: the transistor technology. The point-to point transistor did not make it, but the bipolar junction transistor became a breakthrough. The more when the semiconducting material of Germanium was replaced by Silicon, it led to a technology that would replace and complement the vacuum tube technology in a rapid pace. Had the vacuum tube needed some three-four decades to mature from its moment of birth (Figure 196), the transistor would become in two decades during and after the Second World War the dominant technology.

*Transistor 2.0* (1960-1980): In a rapid pace the silicon based bipolar junction transistor became the dominant component used in a multitude of discrete applications.[437] The MOS planar technology creating the MOSFET transistor became the working horse of electronic circuits. Different versions, oriented at different applications, were developed. The newly introduced devices created a subsequent technical development trajectory of continuous improvements of the transistor itself, as well as adaptations for specific applications in parallel trajectories (Figure 256).

The transistor-technology creating the single devices of the transistor used in discrete circuits (Figure 260), was the dawn of the subsequent development into the semiconductor based Integrated Circuits. From creating a single active component, the transistor technology would expand into the semiconductor technology[438] creating circuits with a multitude of components during the next phase of the Second Electronic Revolution.

---

[437] A discrete circuit in this context is an electronic circuit build up from individual components; resistors, capacitors, inductors and transistors. An integrated circuit is a comparable electronic circuit that is integrated in a single piece of material creating a 'chip.'
[438] So, in this view, the transistor technology is the early part of the semiconductor technology. The IC-technology forms the second part.

## Industrial Dynamics in the Valley: from Start-up to Spin-off

Back to William Shockley who, after his 'magic month' in 1948 resulted in the bipolar junction transistor, in the following years became alienated from his co-workers. In the midst of a mid-life crisis, he divorced from his wife in the early 1950s. At the same time, he was nominated for the Nobel Prize (which he received together with Bardeen and Brattain in 1956).

> *By 1954 William Shockley was becoming increasingly frustrated by his lack of professional advancement at Bell Labs. Although head of transistor research, he had been passed over for higher positions and seemed mired at this middle-management level despite his many publications and patents, including the all-important ones on the junction transistor. Thus he began casting around for other possible positions, taking a leave of absence to serve as a visiting professor at Caltech and as a military advisor in the Pentagon's Weapons Systems Evaluation Group. But neither option proved satisfying. So by early 1955 Shockley was seeking to get back into industrial research—this time as the head of his own company. [...]* (Michael Riordan, 2007, p. 36)

So, by 1956 Shockley started in Silicon Valley (California) his own laboratory *Shockley Semiconductor Laboratory* as a division of Beckman Instruments in Mountain View, California. By that time his demeanour started to change, as evidenced in his increasingly autocratic, erratic and hard-to-please management style. Then, a core group of Shockley employees, later known as the 'Traitorous Eight,' became unhappy with his management of the company. Late 1957 they left en bloc and created *Fairchild Semiconductor*, a division of Fairchild Camera and Instrument Co. to make silicon transistors. They were quite successful as by 1963 they had a turnover of $130 million.

### The Birth of Silicon Valley

The invention of the point-contact transistor as well as the bipolar junction transistor was epoch making. It was the birth of a complete new electronic technology: the solid-state (aka semiconductor) technology that was able to manipulate electric currents with solid state material. But there was more then only the technological development, also the industrial consequences were epoch making as it created a whole new industry on the Western Coast. As there, at the Stanford University things were happening that coincided with the budding industrial activities that paralleled the research efforts.

As a result of the aspirations of Fred Terman —Dean of the Stanford University— to create an industrial park around university, his Stanford Research Park (SRP) started in 1951 (Figure 262). Terman began to convince companies to come to Palo Alto and set up shop in the first university-owned industrial park in the world.[439]

First aboard was Varian Associates, a growing company with ties to Terman and Stanford, which obtained a park lease in 1951. Terman then convinced Hewlett-Packard to head out to Page Mill Road. Soon a flood of other corporations —eg General Electric, Eastman Kodak and the Lockheed Missiles System Division— would make Stanford Industrial Park one of the most respected addresses on the west coast, and eventually give the Peninsula a collection of big name companies to rival any conglomeration back east. Also the photocopier company Xerox established its *Xerox Palo Alto Research Centre* (PARC) in 1970.

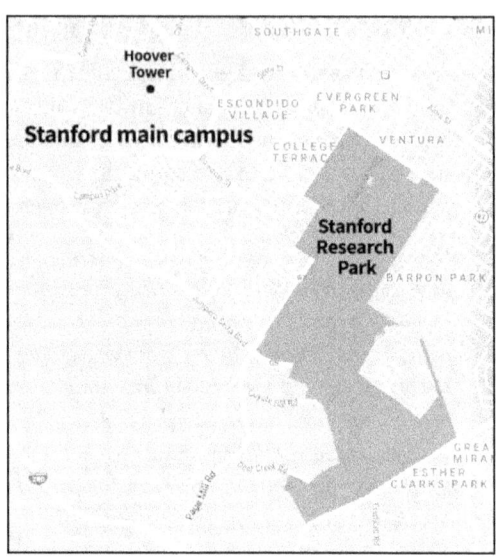

**Figure 262: Stanford Research Park (SRP).**
Source: Stanfordresearchpark.com

> It was a double-edged sword, as a mutually beneficial relationship developed: professors consulted with the rent-paying tenants, industrial researchers taught courses on campus, and companies recruited the best students. And entrepreneurial students like William Hewlett, David Packard and the Varian brothers Rusell and Sigurd would start up business activities from the university. The Park was Silicon Valley in miniature.

As a result of these university-business dynamics in the Park, the phenomenon of the high tech start-up company had become part of the culture of the Valley.

---

[439] Terman was faced with dominance of East Coast Universities in military research at the universities. He had been running the 800-man Electronic Warfare Lab at Harvard university. The governmental policies concerning funding military research was stimulated by the emerging Cold War. When he became Dean of Stanford University, he wanted a piece of the cake for Stanford, and created the Stanford Electronics Research Lab (ERL.). It became the start of Stanford's involvement in military research on a large scale.

## The Creation of Shockley Semiconductor Laboratory

As mentioned before, through the early 1950s a series of events led to William Shockley[440] becoming increasingly upset with Bell's management, and especially what he saw as a slighting when Bell promoted Bardeen and Brattain's names ahead of his own on the point-contact transistor's patent. Also his abrasive management style became more and more problematic and limited his promotion prospects with Bell Labs. These issues came to a head in 1953, and he took a sabbatical and returned to the California Institute of Technology as a visiting professor on the University where he had started his academic education.

> *And yet despite this resume of accomplishment and his indisputable intelligence, William Shockley always seemed to find a way to allow his personality to spoil his own success. Throughout his 79 years, Shockley threw tantrums, fought petty disputes and turned his friends into enemies. He divorced his first wife, severed relations with his children and forced his brightest employees to leave his company and outshine him elsewhere. Finally, as an angry and bitter man in his retirement years, Shockley became the champion of an angry and bitter theory. His association with it would bring him scorn, ridicule and deep contempt from the scientific community and the broader public. In the end, his reputation in tatters and his accomplishments largely overshadowed, William Shockley would die friendless and ignored, his own worst enemy.*[441]

Still pursuing his dream of exploiting his invention of the junction transistor, exploring alternative opportunities, Shockley took a leave of absence as a visiting professor at Caltech, followed by a year in Washington, D.C., as a deputy director at the Pentagon. In August of 1955, Shockley flew to LA to spend a week with his new friend Arnold Beckman, a California chemist and businessman. He wanted him on the board of his new company.

> *Over the next few months, Shockley began talking with executives at RCA, Raytheon, Texas Instruments and elsewhere about the possibility of starting a firm expressly devoted to producing diffused-silicon devices. For most of that summer, he met with only passing interest until he spoke with his friend and fellow Caltech alumnus Arnold Beckman, founder and president of Beckman Industries, Inc. In early September they met in Newport Beach, California, and agreed to found a new division led by Shockley whose purposes included "the*

---

[440] Shockley was born (1910) and raised in Palo Alto, California in a family where his father was a mining engineer and his mother a mining surveyor.
[441] Source: William Shockley: Paranoia Runs Deep. Palo Alto History .Org.; http://www.paloaltohistory.org/william-shockley.php

*development of automatic means for production of diffused-base transistors."* (Michael Riordan, 2007, p. 36)

Arnold Beckman (1900-2004), born in a farming community where father was blacksmith, had

**Figure 263: The original building of Shockley Semiconductor Laboratory in Mountain View.**
Source: Computer History Museum, Chemical Heritage Foundation Collections.

been both a doctorate student and a professor at the California Institute of Technology (CALTECH). As the founder and president of Beckman Instruments —a pre-war supplier of medical and research instruments and a producer of electronic analog computers for industrial automation— Beckman believed that improved silicon transistors would benefit his company, and so he agreed to fund Shockley's work. They discussed how they would manufacture and develop new transistors and other types of semiconductors and revolutionize the electronics field, which was then dominated by the vacuum tube.

As a result of their shared vision, in February 1956 Shockley and Beckman announced the formation of the *Shockley Semiconductor Laboratory*,[442] established with the financial support of Arnold O. Beckman as a Division of Beckman Instruments on a location in Mountain View (Figure 263). Beckman agreed to fund a laboratory under the condition that its discoveries should be brought to mass production within two years.

> Shockley was lured to the Palo Alto area in Santa Clara Valley by Beckmann's friend and Stanford's provost, Fred Terman who thought that a solid research institution in the area would benefit Stanford University. And Shockley's plans fitted in that view.

With a location picked out, Shockley just had to find the people. After failing to attract people from Bell Lab's moving to the West, he started scouring universities for the brightest graduates, as well as an advertisement campaign, and travelled the country to attract extremely capable engineers

---

[442] As a company, Shockley Semiconductors company was insignificant except for the people he hired and the fact that it is considered to be the origin of Silicon Valley.

and scientists. By September 1956, the lab had 32 employees, including Shockley. And the people he hired would change the course of the semiconductor technology.

> Among those early employees were Jean Hoerni, Gordon Moore and Robert Noyce, who learned about and developed technologies and processes related to silicon and diffusion while working there. The group began the process of preparing the facility and designing, fabricating and debugging its equipment, including hand-crafted crystal growers and diffusion furnaces. While teaching them how to grow pure crystals and diffuse basic P-N junctions, Shockley trained a cadre of engineers and technicians in the art and science of working with silicon. In Moore's words, he "put the silicon in Silicon Valley."

Shockley initially planned to work on the mass production of diffusion bipolar transistors, but then set up a 'secret project' on the development of Shockley's PNPN diodes, and in 1957 stopped all works on bipolar transistors. Noyce, Moore, and others thought that this diode was much too difficult a project for the young company to attempt at such an early stage in its evolution. But Shockley was not listening and began devoting more and more of the company's resources to this pet project. He eventually set up a separate R&D group working mostly in secret on the four-layer diode, which further angered technical staff members left out of the loop. This would also spark the trouble in his relation with Beckman. But before that was to happen, he got in trouble with his employees.

> *On a hot summer morning in San Francisco in 1957, eight of the most talented young scientists in America convened for a clandestine meeting at the Clift Hotel. They gathered over breakfast in the famed Redwood Room, a bastion of the city's old guard. A nervous energy consumed the table, fueled by uncertainty, possibility, and fresh-brewed coffee. [...] After considering numerous options, the men decided they must defect. They planned to establish their own company under the leadership of MIT graduate Robert Noyce, a charming, personable twenty-nine-year-old electrical engineer from smalltown Iowa.* (ibidem)

It was the start of a range of meetings with Arnold Beckman, and the focus of the discussion was William Shockley.

> *Despite his scientific gift, William Shockley was no director, no manager. They could work with him, and valued his advice, but they no longer could directly for him. Among themselves, the senior staff decided that Gordon Moore should act as their spokesperson and that he would contact Arnold Beckman—the ultimate boss— directly. Moore phoned Beckman, who immediately grasped the reason for the call. Instead of skirting around the matter, he tackled reality head o. "Things aren't going all that well up there, are they?" That simple question reflected not*

> *only Beckman's straightforward dedication to solving problems but also his innate grace, as he saved Moore the embarrassment of being the first to bring up the reason for his call. No, not really was Moore's replay.* (Thackray & Myers, 2000, p. 246)

Moore and his colleagues met with Beckman over several dinners, struggling to find a solution that would keep the Laboratory together. But their discussion did only come up with the idea replacing Shockley with another director. That proved a bridge to far for Beckman, and he ultimately said he was planning to stick with the senior researcher. He met with Shockley over dinner, and brought him the news of the resignation threat in person. Shockley was stunned by the news and —refusing the proposal of a new director— he undertook action.

> *Shockley and several staff members simply left the dissidents behind, moving into a space in the new Spinco building now ready at Stanford Industrial Park. Shockley and his group continued to pursue the four-layered diode, while the dissidents stayed in the old location in Mountain View, working directly on semiconductor production. The gender solution had only served to split the organization apart.* (ibidem)

As Beckman, in view of Shockley's Noble Prize nomination, did not want to change leadership but instead installed a new manager, the dissenting senior staff felt that they could not stay since the events of the preceding month had produced an unbridgeable rift. Shockley's managing style drove eight of the employees away within a year.

> *Eventually, the paranoia would lead to the 1957 defection of the so-called "Treacherous Eight," who fled Shockley Superconductor to form rival Fairchild Semiconductor --- eventually producing the first integrated circuits and beating their former boss to the punch. Many of the "Treacherous Eight" would later start spinoff companies, including Intel and National Semiconductor, ushering in the next generation of Silicon Valley. They would also manage to get rich while Shockley's company tanked.* (ibidem)

The Shockley Semiconductor Laboratories, renamed in 1958 as Shockley Transistor Corporation, never made a profit and Beckman sold the operation at a substantial loss in 1960. Shockley himself returned to the safe haven of academia on Stanford University, relegated to the side-line as Silicon Valley enjoyed its meteoric rise in the years that followed.[443]

---

[443] Shockley, after winning the Nobel Prize, became interested in questions of race, human intelligence, and eugenics. As he became a scandalous figure on the American scene —and he revelled in his infamy— his extreme views damaged his reputation. He died in 1989.

## The Creation of Fairchild Semiconductor

So, in late 1957, eight of Shockley's researchers, had resigned and decided to create their own company. Several of the eight had —after being introduced by the thirty-year-old New York financier Arthur Rock— met with Sherman Fairchild and had described the situation at the laboratory.

The fabulous wealthy[444] Sherman Fairchild (1896-1971) was a colourful, prominent entrepreneur and investor known for unconventional thinking. He wasthe owner of Fairchild Camera and Instrument Corporation (Fairchild C&I Corp.). Emerging from Fairchild Aviation in 1944 (a builder of airplanes), and Fairchild Camera (manufacturing aerial photography equipment for military applications), during the 1950s Fairchild C&I invested heavily in R&D. And then the discontent group of Shockley employees came by with the promise of a challenging new technology.

*Semiconductors were a logical choice for the expansion-minded company. The devices were increasingly making their way into the very missiles and satellites that used Fairchild Camera and Instrument's products and into the commercial and industrial markets [...] Camera and Instrument had closely studied the possibility of entering the semiconductor field [...] and had decided the easiest way around years of basic research would be to expand through acquisition.* (Berlin, 2001, p. 74)

**Figure 264: Office Building of Fairchild Semiconductor.**
Office at 844 E.Charleston Road (photo teken in 2010).
Source: http://www.computerhistory.org/semiconductor/timeline/1960-FirstIC.html

With the financial backing of Fairchild, the eight men —called the 'Traitorous Eight' by Shockley— founded *Fairchild Semiconductor* in 1957 and leased a location just twelve blocks from Shockley's facility (Figure 264). From the cramped housing they came into a spacious environment to set up their manufacturing laboratories.

---

[444] Sherman, the largest stockowner of IBM, was an inventor and founder of a range of companies based on his inventions. In 1920 Sherman had invented and developed a novel camera shutter/lens system and associated timing mechanism that enabled accurate aerial photography for the first time.

They each got 100 shares in the venture from the total of 1,325 (225 shares went to the corporate investors Hayden, Stone & Co., 300 shares were kept in reserve). The parent company got an option to buy of the venture's stock for $3 million at any moment before it had three successive years of net earning greater than $300,000 per year. And Fairchild provided a loan of $1,38 million. [445]

By August, 1959, Noyce had organized a team at Fairchild Semiconductor and by May, 1960, Fairchild Semiconductor was ready to produce transistors. Often building their own equipment from adapting other technologies,[446] the founding group developed an improved silicon transistor that found immediate application in aerospace and military defence systems. In addition, Noyce, aware of the US Space Program's problems with wired interconnection of the discrete components, found different kind of solutions for the circuit-problem.

The new company started out producing transistors (ie the 2N679) in 1958, selling their first 100 units to IBM at $150 apiece. Soon, their transistors were sold to the United States military and used in missiles and military airplanes.[447] In September 1959 they sold $6,500,000 worth[448] of high-speed silicon devices.

> *Fairchild Semiconductor grew astoundingly quickly in its first six years [...] By 1961, Fairchild Semiconductor had more doubled both its share of the world semiconductor market and the size of its product line. It operations had expanded to seven locations that together occupied more than 200,000 square feet. [...] One year later [...] more than 40 percent of Camera and Instruments worked for the semiconductor division.* (Berlin, 2001, p. 81)

---

[445] Fairchild invested in the new electronics company on condition that the parent company could take over the shares of the daughter company's founders for 3 million dollars in case the new company was to make a profit greater than $300,000 within three years. Since Semiconductor made profit within three years, the founders-entrepreneurs had lost their claim to shares in 'their' company. This time it was the traitorous eight who felt betrayed.
[446] The IC-production process is based on photolithography. The process creates pattern parts on a thin film or the bulk of a substrate (also called a wafer). It uses light to transfer a geometric pattern from a photomask (also called an optical mask) to a photosensitive (that is, light-sensitive) chemical photoresist on the substrate. A series of chemical treatments then either etches the exposure pattern into the material or enables deposition of a new material in the desired pattern upon the material underneath the photoresist.
[447] In January 1958, the parent company Fairchild C&I had its contract with the US Air Force —for photographic camera reconnaissance systems for their B-58 bomber— cancelled. Quite a blow as the contract created half of the firm's sales. However, Fairchild C&I's loss was Fairchild's Semiconductor gain, as it started delivering transistors to IBM's Federal Systems Division, a major supplier for the computers on the B-70 bomber.
[448] $1 in 1959 had a real price worth of $8.88 in 2020 (based on Consumer Price Index) (source: Measuringworth.com).

Fairchild Semiconductor swung from wild success to dramatic failure. In the period October 1957-August 1960 it grew to 1400 employees. Nine years after its founding, the company employed some 11,000 people and generated more than $12 million in profits. However, one year later, the firm was struggling for survival.

*The Meteoric Rise and Abrupt Collapse*

Fairchild Semiconductor pioneered new products and technologies together with an entrepreneurial style and manufacturing and marketing techniques that reshaped the semiconductor industry. The planar process —conceived just three months later and developed in early 1959— revolutionized semiconductor device production. At the peak of its influence in the mid-1960s, the division was one of the world's largest producers of silicon transistors and controlled more than 30% of the market for ICs. Fairchild Semiconductor grew to be about a $150 million business and some 30,000 employees by the late sixties (Figure 265).

> But by then, history had repeated itself, as in 1960 Fairchild Camera & Instrument decided to exercise its contractual option to buy the stock of the venture. The result of a tax-free stock swap was that each of the founders owned $300,000 worth of stock in the mother company. The financial interest of the Semiconductor founders was now linked to the welfare of the mother company.

The enormous growth caused problems and tensions of its own between the mother company and the start-up Semiconductor company.

**Figure 265: Early Semiconductor Manufacturing.**
The Fairchild Semiconductor diffusion area in 1960.
Source: The Computer History Museum Collection, courtesy of National Semiconductor.

One factor was the matter of business culture between the informal research-oriented group and the more formal management style of the mother company. And, as usual with rapid growth, problems arose. The transfer of new projects from R&D in Semiconductor, to manufacturing in Fairchild C&I plants being one them. Early 1965 brought the first signs of management problems.

> By 1961, Fairchild Semiconductor's R&D and manufacturing operations were not only in different buildings, but in different towns: R&D was in Palo Alto, and the manufacturing fab was in Mountain View. The tensions between the two sites ran deep [...] This gap between development and manufacturing contributed to a rapid proliferation of small companies spinning off from Semiconductor. By 1961 half of the founding team had departed to create their own firms. By the end of the decade some two dozen companies had been started by former employees who left the company. (Berlin, 2001, p. 84)

And this all happened in the times when semiconductors were a booming business that attracted many to start on their own. Semiconductor's focus —and the accompanying management style— was on innovation in the fertile semiconductor technology attractive to electronic companies. And in that business climate the management of the mother company Fairchild C&I undertook a frenzy of acquisitions.

> The company [Fairchild C&I)] expanded into graphic arts, space research, oscilloscopes, office equipment, and even home movie cameras. It bought a cathode tube company, and another that manufactured offset printing presses and other printing supplies. None of these acquisitions proved successful, and Semiconductor employees watched with disgust as the profits their division generated went to shore up foundering operations elsewhere in the newly sprawling company. (Berlin, 2001, pp. 85-86)

## Stock Options for Employees

In Silicon Valley, in order to attract and bind employees to companies, next to the salary the system of stock options was implemented as part of the employee's remuneration package.[449] Not surprisingly, an important

---

[449] Next to the (often more limited salaries) the employee would be allowed to participate in a stock option plan. This gave him the right to buy stocks in the company at a predetermined price at a later moment (hopefully when the stock price would be higher, but that depended also on the employee's efforts). The big trick is that, the moment the option expires and the actual stock price is higher (eg $25/stock) than the issue-price (e.g. $10/stock), the employee cashes the difference of $15/stock. And it is not the company that pays for this 'remuneration', but the 'stock market'. The stock options could have specific conditions. Like the condition that the employee would have to be employed by the company for a specified term of years before "vesting", i.e. selling or transferring his stock or

element of the rapid growth of Fairchild's Semiconductor Division was the employee stock option plan (ESO). To renumerate the many technologists — including those others than the founders—, received stock options in the company. As competition for well-schooled employees increased, the stock option became was important instrument to attract or bind them to a company. But the Eastern New England business culture[450] of that time — not used to the business dynamics that accompanied the new technologies— proved to limit Fairchild.

> *At the same time, Camera and Instrument began a new stock option plan that was less generous than the old. Specifically, the company reduced the number of shares available for options by 25 percent, even while adding employees at a rate of 2,000 per year [...]* (Berlin, 2001, p. 91)

This policy proved also problematic to keep the technologists on board. In the period 1959-1962 the first ones started to leave and created their own start-up companies. Followed by a second wave in 1963-1967 (Figure 268). Then came 1968 as a year of troubling times for their company and country. Fairchild Camera and Instrument had been caught up in a recession that had started in 1967. It affected the whole industry, and heralded problems for Fairchild's Semiconductor.

## The Early Semiconductor Manufacturers

From the early days of the semiconductor devices like the transistor, they were produced by the established electric companies like Sylvania and General Electric. Many had a history in manufacturing incandescent lamps, followed by the radio-tubes. Or they had been active in radio, television and radar sets. After the Second World War had ended their specific wartime production efforts, and business returned to normal, they looked for new activities and their 'electronics' divisions were created to implement the new developments. For many that was in Germanium transistors (Figure 266).

> Take General Electric (GE), an old company from the Electric Era, that was the result of a massive merger of many companies, being one of the pre-war television manufactures and after-war mainframe computing companies. It was also was active in semiconductor R&D during the Second World War as a contractor to the Radiation Laboratory at MIT. Germanium based diodes were developed in its General Electric Research Laboratories in

---

stock options. Thus, stock options became both a tool to stimulate employee's motivation to stay and a tool to reward him, almost without cost for the company.

[450] The Boston area known as Route 128 was dominated by large, vertically integrated, and secretive minicomputer producers like DEC, Wang, Prime, and Data General. Technology, skill, and know-how were trapped within the boundaries of the large corporations.

Schenectady (New York State). So, when the Bell Labs published their invention of the junction transistor, it was picked up by GE rapidly (Figure 266, top).

Another example would be Sylvania Products Company (1924), that merged with many incandescent lamp manufacturers. Wartime production brought an explosive growth. After the war was over, the manufacturing of television and radio sets —and many other activities— was expanded. Sylvania was a major producer of germanium transistors and diodes in the 1950s and 1960s, and competed successfully with a broad range of device types, including transistors designed for high-speed switching and computer use. In 1957 the Electronics Division of Sylvania was divided into two new organizations, the Semiconductor Division and Special Tube Operations. This was done primarily due to intensified efforts to establish a meaningful position in the growing semiconductor industry (Figure 266, bottom).

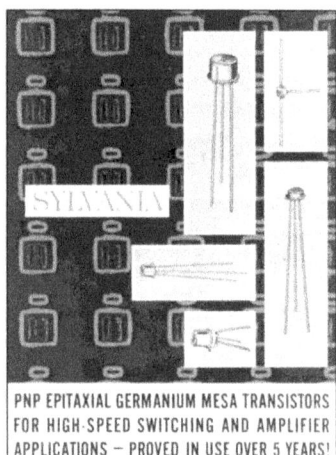

**Figure 266: Germanium Transistor Manufacturers.**

Advertisement from General Electric (top) and Sylvania (bottom).

Source: Transistor History, Semiconductormuseum.com

By the early 1950s, many U.S. companies were actively engaged in the manufacture of germanium diodes and transistors, including CBS, Hughes, National Union, Radio Receptor, Raytheon, RCA and Transitron. Many of them located the North Eastern 'Route 128 Belt.' But some had risen in other regions from totally different activities (eg Texas Instruments originating from seismic exploration services to the petroleum industry).

The availability of a broad range of transistors for circuit-design on a Printed Circuit Board, became challenged by the new phenomenon of Integrated Circuit realising a similar circuit on a piece of silicon (aka die).

> *The same electronic circuit could be designed in dozens of different ways, using different combinations of transistors, resistors, and diodes. Silicon and germanium products competed for market share, as did various kinds of integrated circuits. The early 1960s also witnessed competition between integrated circuits and discrete components, such as transistors and diodes. One result of the great flexibility of circuit design was a highly differentiated marketplace, with thousands of different types of transistors, other components, and circuits for sale-and seemingly endless opportunities for new firms in search of niche markets.* (Berlin, 2001, p. 84)

So many of the start-up's found their customers either with the military (with their Cold War and Space Race projects), or the electronic industry that wanted the ICs for their end-products. Especially the military-industrial complex —not short on state-funding R&D— was an important customer.

> *[...], the federal government was an important customer for Semiconductor, even when the firm did not directly contract with the military. The vast majority of Fairchild Semiconductor's early customers were aerospace firms buying products to use in their own government contract work, and these customers developed specifications that reflected military requirements. Fully 100 percent of the early integrated circuits (which Semiconductor pioneered) went to military uses. While Semiconductor assiduously courted the industrial and commercial markets, its products nonetheless could also be found in surveillance radar and transmitters for space vehicles; in Polaris, Minuteman, and Advent missiles; and in the MAGIC airborne inertial guidance*

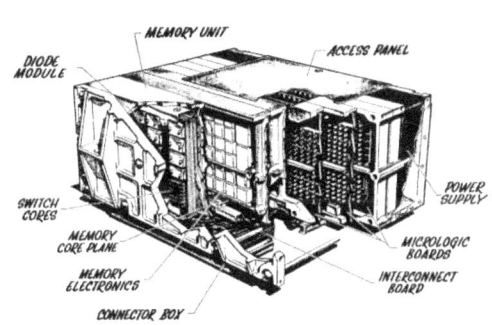

**Figure 267: Drawing of the Magic I, the Missile Guidance Computer.**

The MAGIC I (1961-1963) was designed for ballistic missile guidance and was the "first complete airborne computer to have its logic functions mechanized exclusively with integrated circuits." It used 2,098 Fairchild Micrologic integrated circuits.

Source: http://www.righto.com/2020/03/the-delco-magic-line-of-aerospace.html

*computer* [Figure 267], *as well as the MARTAC missile control computer.* (Berlin, 2001, p. 90)

Then came the silicon technologies. It heralded also the industrial move from the East Coast to the West Coast. Sand—ie silicon, the basic material for semiconductor manufacturing—was cheap. But, as the development of the semiconductors proved to be dominantly technology-driven, and that technology was very capital intensive, the creation of new business activity was depending on two factors: the technologists and the venture capitalist. And both were to be found in 'Silicon Valley.'

Especially in the region near San Francisco today known as Silicon Valley, the new electronic technologies spawned massive business activity that settled on the orchard lands of the fertile San Francisco Peninsula. The access to scientific knowhow available at universities like Stanford University, the easy access to venture capital, and the massive spending of US Department of Defence in the time of the Cold War and Space Race, fuelled that technology driven development. And it gave rise to the business creation by

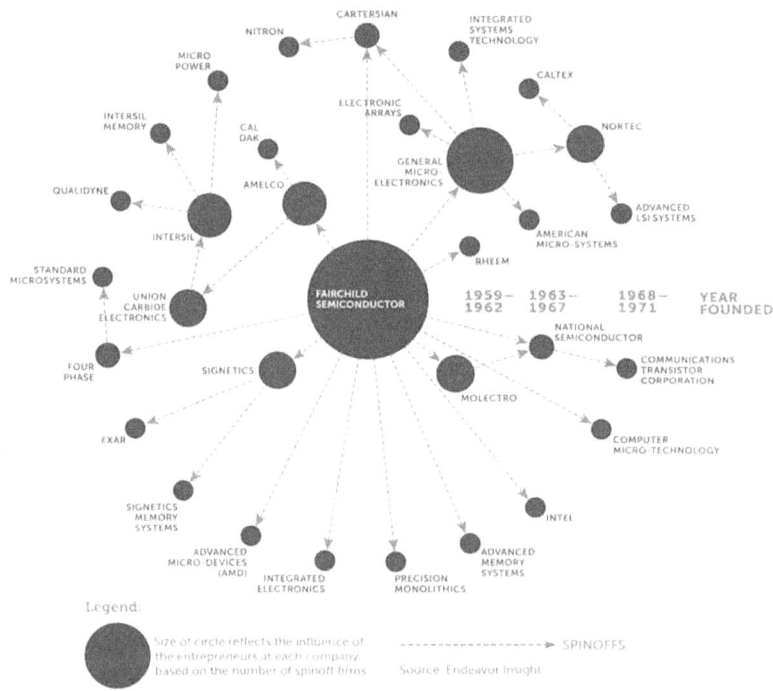

**Figure 268: The Spin-off of Fairchild Semiconductor.**

Source: https://siliconvalley.corriere.it/files/2014/01/endeavor-insight-sv-2-retina.png

corporate spin-offs that originated from the 'Traiterous Eight' leaving William Shockley's Semiconductor Laboratory to create Fairchild Semiconductor. Next, followed by the 'Fairchildren' leaving Fairchild Semiconductor (Figure 268). As did also the engineer-inventors of the Xerox Palo Alto Research Centre (aka PARC, 1970) that left for Apple Computer, Microsoft, and Adobe.

The spin-offs of Fairchild Semiconductor came in waves. The first Fairchild spin-off Rheem Semiconductor in 1959 was followed in quick succession by founder Jay Last's Amelco Semiconductor, where he was joined by Jean Hoerni and Eugene Kleiner, and by Signetics.

*Rheem Semiconductor Corp:* The brothers Richard and Donald Rheem started in 1925 with production of oil drums. Expanding their business over time—eg into water heaters, heat pumps, air conditioners—in 1959 they decided to fund Ed Baldwin. This first general manager of Fairchild quit together with four of his staff in 1959 to start Rheem Semiconductor in Mountain View, CA to produce diffused silicon mesa transistors in competition with his former employer.

*Amelco Semiconductor:* This spin-off was founded in 1961 in Mountain View, California by Fairchild founders Jean Hoerni, Eugene Kleiner, Jay Last, and Sheldon Roberts, with financing from Teledyne Corporation arranged by the venture capitalist Arthur Rock. As a division of Teledyne (that was focussing military applications), it was renamed as Teledyne Semiconductor in order to supply mono-lithic integrated circuit in support of Teledyne's military business.

*Signetics:* At the same time, one of the first companies to focus exclusively on integrated circuits, the Signetics Corporation, was founded in Mountain View, CA in 1961 by David Allison, David James, Lionel Kattner, and Mark Weissenstern, all of Fairchild. Based in Sunnyvale, CA, Signetics became a major supplier of high-volume bipolar digital logic and memory devices. Signetics was an early innovator in digital logic, with many of its logic circuits becoming industry standards that found initial application in military and space systems. After is acquisition in 1962 by Corning Glass, throughout 1964 Signetics dwarfed Fairchild and anybody else in the manufacturing of integrated circuits. The parent company, Corning Glass, sold the operations to the Dutch Philips in 1975.[451]

---

[451] On a personal note, your author, being a student mastering in micro-electronics at the University at Technology in Delft, the Netherlands, stayed in 1976 three months as a Philips sponsored trainee at Signetics.

But there were also others that profited from the brain drain of Fairchild-people leaving the company. Such as there were:

*Molectro Science Corporation:* Founded in 1962 by Fairchild's engineers J. Nall and D. Spittlehouse, it started developing high-performance bipolar ROMs, PROMs, and specialty LSI logic devices. In 1965 it was acquired by National Semiconductor.

*National Semiconductor:* Already in 1959, as a spin-off of the Sperry Rand Corporation, National Semiconductor was founded. By 1965 it had—after a lawsuit from the mother company, acquired venture capital from Peter J. Spraque's family funds. In 1967, Sprague hired five top executives away from Fairchild, among whom were Charles E. Sporck and Pierre Lamond who took some thirty-five Fairchild people to join them. Immediately after becoming CEO, Sporck started a historic price war among semiconductor companies, which then trimmed the number of competitors in the field. Among the casualties to exit the semiconductor business were General Electric and Westinghouse.

This was the beginning of the departure of many talented people who were going to start a business themselves, joined other start-up's or were hired by larger firms for their technical expertise. In the subsequent waves (Figure 268) others left Fairchild Semiconductor, created not only a brain drain of knowledge, but also a new wave of competitors for Fairchild.

## The Role of Venture Capital in Silicon Valley

Originally the funding of many of the wireless business in the 1930s and 1940s often was one way or the other, determined by the commissions of the Department of Défense. Before the Second World War (1939–1945) investment in high-risk business (aka venture capital) was primarily the domain of wealthy individuals and families. J.P. Morgan, the Wallenbergs, the Vanderbilts, the Whitneys, the Rockefellers, and the Warburgs were notable investors in private companies.[452]

After the war, New England was the first industrial region in USA (leather goods, textiles and machine tools) and its industrial base was rather diversified. Also, in the Route 128 region of New England, capital to finance business came from rich individuals having accumulated their wealth, creating informal venture investing.[453] Banking was another way to

---

[452] They created investment companies like J.H. Whitney & Company (Whitney being one of the wealthiest Americans at that time), Bessemer Securities (grandpa originating from the steel industry) and Rockefeller Brothers Co. (who earned their capital in oil business).

[453] The walking companions of Capitalism are monopoly creation and extreme individual wealth. From the robber barons of the Second Industrial Revolution (the steel and railroad era), up to high tech moguls and soviet oligarchs of modern times.

finance those companies. And many large corporations financed their R&D through retained earnings. Then the American Research & Development Council (incorporated 1946, funded largely by individuals and institutions) started financing the region's start-up's (eg the minicomputer company DEC). The concept of venture capital had emerged.

*American Research and Development Corporation:* The first modern venture capital firm was formed in 1946. Then MIT president Karl Compton, Massachusetts Investors Trust chairman Merrill Griswold, Federal Reserve Bank of Boston president Ralph Flanders, and Harvard Business School professor General Georges F. Doriot started American Research and Development (ARD). Dorit was originally a banker at Kuhn, Loeb & Company, the premier investment bank of its time (along with JP Morgan & Co.). He became later a professor and dean at Harvard Business School. The goal of the company was to finance commercial applications of technologies that were developed during World War II and to encourage private sector investments in businesses run by soldiers who were returning from World War II. Doriot was the heart and soul of ARD and is justifiably called the "father of venture capital." Doriot's focus was on adding value to companies, not just supplying money. Companies funded by ARD were considered to be 'members of the family.' ARD's staff under Doriot's direction began providing industry expertise and management experience to the companies they backed in order to increase their chances. ARD raised $3.5 million through a public equity offering, of which $1.8 million came from nine institutional investors like MIT, Penn University, and the Rice Institute. ARD's significance was that it was the first institutional venture capital firm that raised capital from sources other than wealthy families although it had several notable investment successes as well.

The rise of the fast-growing, capital-intensive semiconductor manufacturers, created a need for another type of venture investment. Soon, venture capital firms focused their investment activity primarily on starting and expanding companies. Companies that were aided by the Small Business Administration (SBA).

> In 1953 the Small Business Administration (SBA) was created by an act of Congress. It was a response to a study during the 1950s by the U.S. Federal Reserve that revealed a large gap in the ability of small businesses to gain needed credit. They could not compete with the big companies for the lucrative after-war military market. In 1958 one of its programs was the *Small Business Investment Company* (SBIC) program that authorized the creation of private venture capital firms.

Silicon Valley's investment banking community was firmly established in the late 1960s and early 1970s. In the 1960s the common form of *private-equity fund*, still in use today, emerged. Enabled by the Small Business Investment Company (SBIC) Act of 1958, private-equity firms organized *limited partnerships* to hold investments in which the investment professionals served as general partner and the investors, who were passive limited partners, put up the capital. During the late 1960s the SBIC-program helped fuel the creation of today's formal venture capital industry by creating hundreds of *venture capitalist firms*[454] overnight.

> *The first venture capital firm in California, Draper, Gaither and Anderson, was founded in 1959 as a limited partnership. The following year saw the establishment of two federally leveraged SBICs, Continental Capital Corporation and Small Business Enterprises. Another SBIC, Draper and Johnson, was set up in 1962. The important firm, Sutter Hill, was founded as the venture capital arm of the real estate development firm. Bank of America and a number of other commercial banks also provided venture financing for expanding businesses during this early period. In 1961, New York investment banker Arthur Rock formed a model limited partnership with Tommy Davis of Kern County Land Company [...] The Bay Area venture capital industry thus emerged from a period of active experimentation with different types of organizations for providing venture capital. The seminal period, 1956-1963, witnessed the establishment of more than a dozen important venture capital firms in San Francisco and Silicon Valley. Faced with acute difficulties mobilizing funds and the need to share information and expertise, these early venture capitalists gradually evolved into an interactive community trading information and participating together in rudimentary co-investments. Venture capital in the Silicon Valley area evolved gradually alongside the high technology enterprises that spring up there. Venture capital thus became an integral part of [...] a social structure of innovation: an interactive system comprised of technology intensive enterprises, highly skilled human capital, high caliber universities, substantial public/private research and development expenditures, specialized networks of suppliers, support services such as law firms and consultants, strong entrepreneurial networks, and informal mechanisms for information exchange and technology transfer.* (Florida & Kenney, 1988, p. 310)

At Davis & Rock, the partner Arthur Rock helped finance the Shockley Semiconductor Laboratory, and later Fairchild Semiconductor. His biggest deal was funding Intel in 1968 when he raised $2.5 million for Gordon Moore and Robert Noyce's team.

---

[454] A venture capital firm (VC firm or venture firm) is a collection of legal entities formed for the purpose of generating substantial returns for its investors by investing in high-risk companies that have yet to prove that their business models works.

*Fairchild's role as Innovation Machine*

Fairchild Semiconductor may have been the mother of many companies, it had an impact of its own. For example, on the development of the IC-technology. Fairchild as a company in the 1960s was the birthplace of planar process by Jack Hoerni and the Integrated Circuit by Robert Noyce. It was also the pioneer of its use in a range of digital, linear and memory applications. From a simple start-up company in 1957 making transistors, it grew into an industrial mastodont of the nascent IC-Industry. It developed the MOS-semiconductor technology, licensing it to other companies.

As we will explore further on, its technological leadership gave it a dominance in the marketplace. By 1965, Fairchild's process improvements had brought low-cost manufacturing to the semiconductor industry, making Fairchild nearly the only profitable semiconductor manufacturer in the United States. Fairchild dominated the market in diode-transistor logic (DTL), op-amps and custom mainframe computer custom. Although Fairchild established a Memory Products business unit it was never able to leverage pioneering efforts in MOS semiconductor memory into the significant market position achieved by its most important spin-off Intel.

> Fairchild expanded its operations rapidly, and opened its first plant outside the Bay Area in South Portland, Maine, in 1962. As one of the first US technology companies to expand into Asia, C. E. (Ed) Pausa led the construction of a factory in Hong Kong in 1964. Two years later Hong Kong employed 5,000 workers versus 3,000 in California. Sales rose from $21 million sales (1960), to $90 million (1965), and $150 million by the late 1960s. From the eight founders of 1957, it had expanded to 30,000 employees by that time.

But in the late 1960s came the downturn, and Fairchild Semiconductor for the first-time lost money, and the founders left the company.

## The Creation of INTEL Semiconductor

Fairchild Semiconductor had proven to be a success and grew rapidly. Sherman, however, pulled money from Fairchild Semiconductor to solve financial problems at the mother company Fairchild Camera & Instrument. This happened after he had invoked his purchase right stipulated at the start of the company in 1957 and had become the full owner of Fairchild Semiconductor. Then came 1968 as a year of troubling times for their company and country.

> Fairchild Camera & Instrument had been caught up in a recession that had started in 1967. It was bleeding badly. The country was undergoing protests against the war, with assassinations of both

Martin Luther King Jr and Robert Kennedy before the summer was over. In that context Fairchild Camera & Instruments implemented a cost-saving and restructuring program to face the problems.

*The changes came too late. By the end of 1966, Fairchild began to miss its promised deliveries, at times meeting only about one-third of its customer commitments.' At the same time, the company failed to market several new products developed in R&D because the process of transferring from development to manufacturing was so inefficient that the devices were never manufactured in volume. [...] By the end of 1966, Semiconductor's festering troubles had become apparent even to outsiders. In the fourth quarter, Camera and Instrument's profits dropped below those of the third quarter, and the parent company, according to one report, "placed the blame at the door- step [of Semiconductor], saying the division was having serious problems in production of integrated circuits and its product mix was top- heavy with low-profit items. ...] A command to reduce costs infuriated the Semiconductor employees. For years, the Semiconductor division had been more profitable than the company as a whole, with other divisions losing money and serving as net drains from the Semiconductor bottom line.* (Berlin, 2001, pp. 93-94)

Already grumbling about the 'Eastern large company management style,' the employees at Fairchild Semiconductor started to leave through every possible exit door. Among them Robert Noyce who resigned on June 25th, 1968. Together with Gordon Moore, and with help of the venture capitalist Arthur Rock, he created N(oyce)-M(oore) Electronics, a semiconductor memory company that would be renamed as Intel.

> They called the company INTEL (short for INTegrated ELectronics). To raise the start-up capital, Arthur Rock was approached who already knew the two men because he had also helped finance Fairchild Semiconductor. He provided $3 million in venture capital. Over the next two years, another $2 million was obtained. In 1971, the company went public and raised $6.8 million ($23.50 per share). An investor who had bought 100 shares for $2,350 at that time received more than $2 million for it in 1996.

Soon after Intel's founding, a third visionary joined the team: Andrew S. Grove, a Hungarian émigré who had played a critical role in the development of metal oxide semiconductor (MOS) large scale integrated (LSI) technology. Not long after that, top engineers Ted Hoff, Federico Faggin, and Stan Mazor joined the group. They would become essential contributors to the rapid development of INTEL.

*A Dynamic Company in a Dynamic Industry in Dynamic Times*

And that development was meteoric, as already in 1975 the company boasted a profit of $137 million and employed some 4,600 people. By 1979 —a decade after its start—, Intel's net revenues[455] were $663 million and it employed more than 14,000 people.

> The first target of INTEL was the memory market, by then dominated by magnetic core memories. At the time, sales of mainframe computers were expanding. While these machines used integrated circuits to perform logic calculations, programs and data were stored on magnetic devices. Intel targeted that market with their memory chips. These memory chips were known as dynamic random-access memories (SRAM and DRAMs). In April 1969, Intel introduced its first product: the 3101 static random-access memory (SRAM). By October 1970 Intel introduced the 1103 DRAM chip that could store 1Kb. The 1103 put Intel firmly on the map. The chip soon became the memory technology of choice for computer makers, and by the end of 1971, 14 out of the world's 18 leading mainframe computer makers were using the 1103.

Intel started in 1969 with an initial business plan —described on one page— that focused 'on the needs of the electronic systems manufacturers.' Intel created in 1972 the Memory Systems Division selling complete memory systems. After the first product lines created an entry in the memory market, Intel subsequently was drawn into another market; the micro-computers systems they helped to create. But there was more, as was proven by other manufacturers like Texas Instrument that started making pocket calculators for the consumer market (in direct competition with their clients the Japanese calculator manufacturers.)

> Not long after the initiation of the microprocessor and its related products,[456] Intel started to diversify as it bought in 1972 a company called Microma, a digital watch company making solid state quartz watches with liquid crystal displays. Intel thought it could make the watch's chip and the display as complementary items, but it had production issues. More importantly, Intel failed at the consumer marketing game and didn't like to pay for ads (one advertisement by

---

[455] Net revenue (aka net sales) is a company's gross sales (aka gross revenue) minus discounts and returns. Net revenue minus the costs of goods produced (aka operating expenses) gives the gross profit (ie gross sales minus operating expenses). When also taxes and interests are deducted is becomes the net profit.
[456] When you are able to make the complex IC-set of a micro-computer system, it is not that hard to put the ICs on a board and sell at board-level; the single-board computer (aka SBC). And when you are able to do that, putting the board(s) in a box is not that difficult at all.

itself could cost as much as $600,000). By 1978 Intel shut down Microma at a loss of $15 million.

In 1972 Intel opened its first international manufacturing facility in Malaysia, which would host multiple Intel operations. Among them the back-end operations of chip assembly and packaging. The company was joining the wave of 'off shore operations'[457], and the later division between the designers (ie the factory-less manufacturers aka fab-less) and the producers (ie the companies aka fabs or foundries).

> INTEL followed a trail set out when the earliest offshore investment in semiconductor assembly was made in 1961 by Fairchild Semiconductor in Hong Kong for the assembly of discrete transistors. Over the next fifteen years, this pioneering investment was followed by assembly investments by other companies in seven other economies of the region. By the mid-1970s, there were dozens of US-owned assembly plants throughout the region employing about 1,000 workers each. By 1977, US companies employed close to 100,000 workers in offshore assembly plants, compared to 114,000 domestic employees, of whom 64,000 were directly involved in production. By 1978 around 80% of US semiconductor production was assembled abroad.

That was happening at the back-end of the fabrication process. But also at the front end (ie the wafer processing) a similar process was starting as design-operations were separated from wafer processing in the clean rooms. At the 1980s a handful of US chip companies, such as Xilinx and Chips & Technologies, were already outsourcing all of their manufacturing, primarily to integrated Japanese manufacturers. In Taiwan[458] the foundry-model would start in the mid-80s.

But before that all was to happen, a crucial invention had to take place. An invention that changed not only the semiconductor industry, but would have a profound effect on the Affairs of Man in the coming times.

---

[457] Assembly was the easiest stage of production to be moved offshore. It was functionally separate from the other stages of production even when performed in close proximity to fabrication. Furthermore, assembly began with a relatively high use of less-skilled direct labour. Also, the value-to-weight ratio of an IC is very high, so transporting light-weighted products over distances is not an issue.

[458] The first foundry, Taiwan Semiconductor Manufacturing Corporation (TSMC) was founded by Morris Chang, a Chinese-born, MIT-educated executive with 25 years' experience at Texas Instruments who moved to Taiwan in 1985.

## *The Invention of the Integrated Circuit.*

The dynamics in the Valley ran high as the importance of the invention of the transistor became more obvious to an increasing group of people. Such as the group of technical insiders; the researcher themselves and their lab-managers who could envision where the new components could be used for. And their technical-active clients like the military, who became enticed by the transistor in realising their own projects (eg the Space Race projects and the Cold War projects). Also the non-technical people —such as the business owners, marketing managers and investors who were involved— understood that this was 'something big.'

> *The problem the integrated circuit was designed to solve had been vexing the semiconductor community for a number of years. Given the transistor's inherent small size, low power dissipation, and potential for high reliability, it had long been appreciated that the transistor should make it possible to build systems with thousands of active devices working together and operating at high speed. There were pressing applications for such systems in computers, telephone switches, and in several military projects.* (Ross, 1998, p. 22)

But the transistor was still a component that was used in the discrete electronic circuits mounted on Printed Circuit-boards (Figure 260) that contained the interconnection between the components. So, the search to integrate electronic circuits into something smaller (aka micro-miniaturization), was evident. It had already been tried in the vacuum tube technology (Figure 358) without much success (Figure 212) outside military applications where money was not a consideration.

To improve the component density of electronic circuitry, the interconnection between the components assembled on the 'board' was compacted. In the 1950s the PCB evolved into double-sided boards, followed by the multi-layer boards. Also the micro-modules —which we will meet further on— had their heyday for a while. Nevertheless, the discrete circuitry still was quite bulky. But with the new semiconductor technologies becoming available, that was about to change. It would start the Era of the Integrated Circuit within a decade after the invention of the transistor itself.

## The Era of the Integrated Circuit (1960-1975)

The new solid-state components (the 'active' components of the diode and transistor) were used by the 'circuit designers' to build electronic circuits: from the analogue amplifier- and radio-circuits, to the digital logic for hardwired logic circuits like electronic calculators. The *analogue circuits* (aka linear circuits) would become the building blocks of analogue electronic systems, amplifying, filtering and limiting analogue signals. The *logic circuits* —from Resistor-Transistor Logic (RTL) to Transistor-Transistor Logic (TTL)— would become the building blocks of the mini-computer.

However, the electronic circuits with their discrete active components (ie diode, transistor), passive components (ie resistors, capacitors and inductors), and their interconnecting wiring printed on the board, were still quite bulky. So on many places the search was on to look for a smaller alternative; the hunt for micro-miniaturization. Then, the idea to integrate electronic active and passive components *in* a single substrate emerged and would lead to the Integrated Circuit (IC).

Early concepts of an integrated circuit go back to 1949, when German engineer Werner Jacobi (1904-1985) —a physicist working at Siemens AG— filed a patent for an integrated-circuit-like semiconductor amplifying device; his 'Halbleiterverstärker.' He was granted German Patent № 833,366 on May 15th, 1952.

Between 1953 and 1957, Sidney Darlington and Yasuro Tarui (working at the Electrotechnical Laboratory in Tokyo) proposed similar chip designs where several transistors could share a common active area, but there was no electrical isolation to separate them from each other.

The English engineer Geoffrey Drummer (1909-2002), working for the Royal Air Force's Telecommunications Research Establishment, conceptualized the idea of an 'integrated circuit' realized in the semi-conducting technology. In September 1957 he presented a design exercise; a model to illustrate the possibilities of solid-circuit techniques. He began a campaign to encourage substantial UK investment in IC development, but was met largely with apathy. The UK military failed to perceive any operational requirements for ICs, and UK companies were unwilling to invest their own money. This in contrast with US companies that, next to the military markets created by governmental policies (eg the Space Program and the Cold War projects), envisioned the latent commercial mass markets from calculating machines, time devices to computers and radio.[459]

*The Hunt for Integration (1)*

In the times before the semiconductor Integrated Circuit, the electronic circuits were built up from discrete components (eg vacuum tube,

---

[459] Sharing information about research progress was part of Bell Labs policy. This practice was followed in the case of the invention and development of the transistor, as evidenced by the offering, in 1951 and 1952, of two symposia and one summer school. The spirit of communication probably was further sustained by the institutional climate of Silicon Valley. Movement of key people among companies was so easy and occurred so often that open communication was inevitable and was difficult to avoid.

transistor, resistors, capacitor, etc) mounted on a chassis or board (Figure 259, Figure 260). Notwithstanding the micro-miniaturization of the electronic circuits, they took still a lot of space and were labour intensive to make. So the hunt was on for smaller, more reliable —and, if possible, cheaper— electronic systems. The challenge was to find cost-effective, reliable ways of producing 'integrated' modules.

**Figure 269: Micro-Modules of the early 1960s.**
Open Micromodule for the Burroughs B5000 computer (top, 1961)

Source: https://www.chipsetc.com/the-rca-micromodule.html

One of the most interesting developments was an RCA proposal for integrating the components into a single, standardized micro-module. The Army loved the concept, and in April 1958, it awarded RCA a $5 million contract in what became known as the Micromodule Program. While RCA was the prime contractor, more than 60 producers participated in developing individual microelement components and wafers.

Known as the hybrid micro-module system, it called for piling tiny ceramic wafers or miniature PCB's stacked with discrete components on top of each other like dishes. Connecting wires ran up the sides of the stacks through holes in the wafers. And on top, the active element (ic the transistor) was placed. Micromodules were not only somewhat easier to make than conventional electronic systems, they were also a good deal smaller.

It seemed that the micromodule was the circuit package for the future. It was more expensive than conventional printed-circuit packages, but follow-on costs, like maintenance and logistics (including storage, handling and training) were lower. Nearly any type of circuitry could be made, and the production of a wafer element could be automated. And, the reliability of circuits, compared to traditional circuits, skyrocketed.

In the following years, millions of additional funding were poured in the project. In 1963, micromodule production at RCA alone reached a peak of 10,000 units a month. Military use of the module soured (eg helmet radio system). Other companies, like Burroughs, applied the micromodules in their computers (Figure 269). But the years of glory for the micro-modules were limited, as another development to solve the integration problem was already underway. A project that integrated the components in one 'wafer,' but still needed wiring to connect them. A project that would trigger the demise of the micro-module program. By late 1964, the initial enthusiasm for discrete micromodules was gone and there were no further major commitments.

*The Hunt for Integration (2)*

By the late 1950s organizations as Bell Labs, the RCA Labs, the Diamond Ordnance Fuse Laboratory, Sprague Electric, Texas Instruments, and Fairchild Semiconductor began to take seriously the possibility of making real a long-held ideal: creating an entire circuit in a single piece of semiconductor material.

> The transistor function made in silicon had been the start. Now it was the question if other (passive) components could also be made in the silicon structure during the diffusion process; the integrated passive devices. Such as the conductivity function (ie the resistor element), the storage function (ie the capacitor element) and the inductive function (ie the inductor element). The answer, given by the circuit designers and technologists, was positive. As we will see further on.

But that development trajectory had obstacles of its own. Next to the institutional conservatism encapsuled in certain cultures —such as the Bell Labs that did not believe in integration of complete circuit in silicon—, the creation of the integrated circuit was hindered by three fundamental technical problems: (1) the integration problem, (2) the isolation problem and the (3) inter-connection problem. And each problem was solved by another inventor.

*The integration problem:* It was Jack Kilby (1923-2005), an electrical engineer working at Texas Instruments, who integrated a prototype of an electronic circuit (a single-transistor oscillator with a distributed RC feedback) on September 12th, 1958, followed by a second prototype (a two-transistor trigger circuit) on September 19th, 1958. The circuits were introduced on a March 1959 conference. Although expensive ($450) compared to their predecessor ($50), within a year TI had developed a line of six solid circuits including gates and flip-flops.

Although it showed that the electronic (active and passive) components could be integrated in the semiconductor material, none of these patented prototypes solved the problem of isolation and interconnection as the components were separated by cutting grooves on the chip and connected by gold wires. The result was a hybrid IC.

Between February and May 1959 TI advocates filed a series of patent applications for Kilby: US-Patent № 3,072,832, US-Patent № 3,138,743 (Figure 270), US-Patent № 3,138,744, US-Patent № 3,115,581 and US-Patent № 3,261,081.

*The isolation problem:* Then there was the issue of isolate the components from each other. Next to applying design structures, this problem was solved with the new manufacturing process developed by Jean Hoerni (1924-1997), a physicist working at Fairchild Semiconductor. He came up with the planar process.

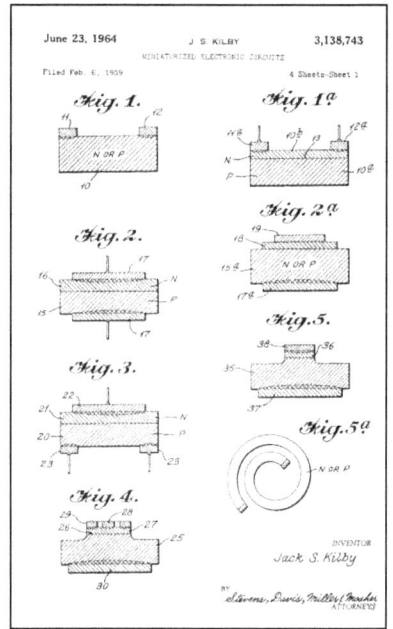

**Figure 270: Kilby's Patent for an Integrated Circuit.**
Jack Kilby's US Patent № 3.138,743 (1964).
Source: USPTO

Hoerni's 'Planar' process became widely used as it used isolation layers on the surface of the chips to isolate the component *in* the material from the interconnection *on* the material. His solution was filed on May 1st, 1959, and implemented in US-Patent № 3,025,589 (the planar process) granted March 20th, 1962, and by US- Patent № 3,064,167 (the planar transistor) granted November 13th, 1962.

The process was extended to independent transistors on a single piece of silicon by Kurt Lehovec (1918-2012), a Czech physicist that came to the US with the operation Paperclip and worked at Sprague Electric Co. in 1959. He created the P-N junction isolation. Lehovec applied for a patent on April 22nd, 1959 and was granted US-Patent № 3,029,366 on May 10th, 1962 (Figure 271) for a 'multiple semiconductor assembly.'

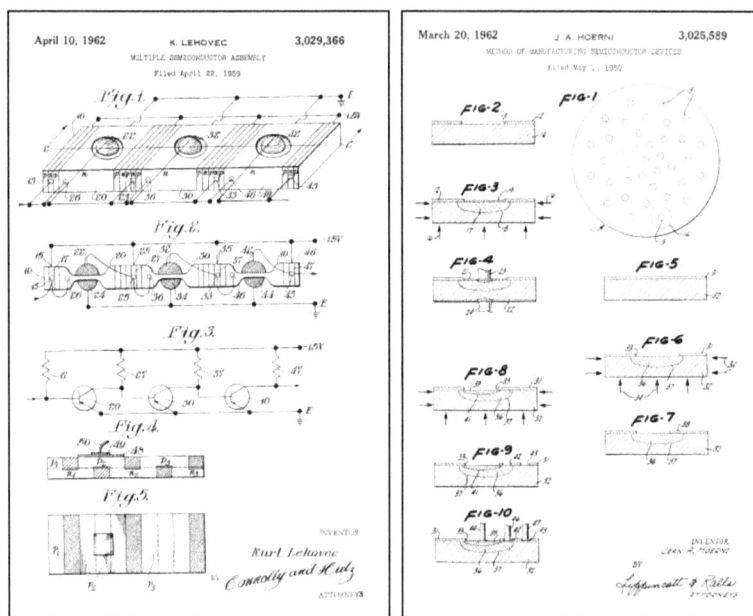

**Figure 271: Patents contributing to the Integrated Circuits.**
Lehovec's Patent № 3,029,366 (left, 1962), and Jack Hoerni's Patent № 3,025,589 (right, 1962).
Source: USPTO

*The interconnection problem:* And then there was the problem of interconnecting the components with each other. The interconnection problem itself was solved by Robert Noyce at Fairchild later the same year. On top of the semiconductor substrate, he applied an extra layer of metal in which the interconnection could be etched.[460]

These contributions proved to be fundamental to the implementation of electronic designs in a single monolithic integrated circuit. In addition to this technical breakthrough, also the IC-technology itself had been developed into a technology geared for mass production.

---

[460] He deposited very thin paths of metal (usually aluminium or copper) directly on the same piece of material as their devices. After etching away the non-used copper, these resulting small paths acted as wires. With this technique an entire circuit could be 'integrated' on a single piece of solid material and an integrated circuit (IC) thus created.

## The Invention of the Monolithic Integrated Circuit

The *monolithic integrated circuit* (IC) using Silicon was developed by Robert Noyce (1927-1990), a physicist with a PhD from MIT, when he was working at Fairchild Semiconductor. He developed a way to interconnect the IC components (ie by aluminium metallization) and proposed an improved version of insulation based on the planar process technology developed by Jean Hoerni (Figure 273, bottom).

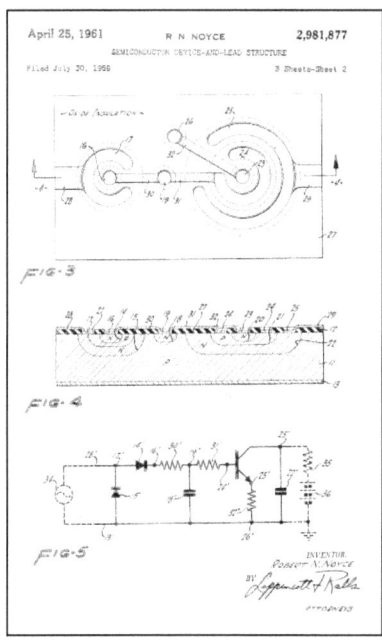

**Figure 272: Noyce's Patent for an Integrated Circuit.**
Robert Noyce's US Patent № 2,981,877 (1961).
Source: USPTO

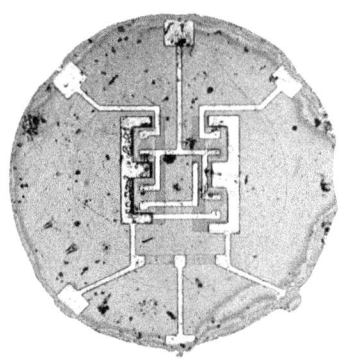

**Figure 273: Kilby's and Noyce's IC's (1959).**

The Germanium slice holds a transistor and other components (top, Kilby). The silicon substrate holds the interconnection, active and passive elements of a flip-flop circuit in one construction (bottom, Noyce).

Source: Texas Instruments, Fairchild.

Noyce applied for a patent on July 30th, 1959 and was granted US-Patent № 2,981,877 on April 25th, 1961 for his "Semiconductor device-and-lead structure," which would come to be known as the monolithic integrated circuit (Figure 272). After months of work a team at Fairchild Semiconductor led by Jay Last succeeded in making the first planar integrated circuits.

At the same time Jack Kilby (1923-2005), working on the integration of components, created a hybrid integrated circuit (hybrid IC) using the material Germanium, and 'flying wires' as interconnection (Figure 273, top).

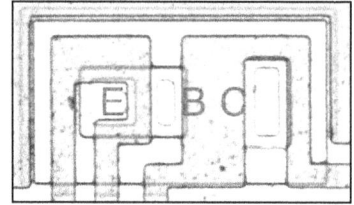

> *It was Kilby, in 1958, who first demonstrated that it was possible to produce transistors, diodes, capacitors, and resistors in one piece of semiconductor and interconnect them to create functioning circuits. His early circuits had about ten components. Kilby used wire bonding to interconnect the components within the chip.* (Ross, 1998, p. 23)

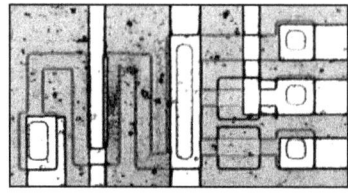

Kilby filed for a patent on February 6th, 1959 and was granted US-Patent № 3.138,743A on June 23rd, 1964 for "miniaturized electronic circuits" (Figure 272). As he applied in February 1959, this was a few months before Noyce applied for his patent.[461]

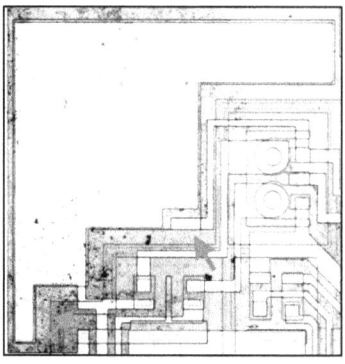

Although it took a while and much discussion among historians, it became clear that this had been a parallel invention of a monolithic Integrated Circuit (Noyce) and a hybrid Integrated Circuit (Kilby). Both contributed to the invention of the Integrated Circuit that was the end of a long development trajectory (Figure 306) and became a basic innovation with an impact that would echo in solid-state technology in the coming decades.

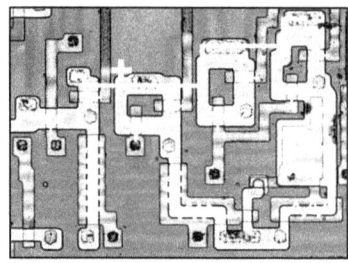

**Figure 274: Die shots of Elements in the Chip Layout.**

The transistor area with emitter E, collector C and base B (top), the resistor area (upper middle), the capacitor area (lower middle), and the Die layout with elements indicated (bottom).

Source: http://www.righto.com/

---

[461] Kilby was awarded the Nobel Prize in Physics on December 10th, 2000.

*Further development of early Integrated Circuit Design*

The development of semiconductor technology progressed with new IC-technologies such as the Metal-Oxide-Semiconductor technology (MOS-technology). This technology enabled the MOS-FET transistor that came in two technological versions: P-channel MOS (PMOS) and N-channel MOS (NMOS).[462] And with these technologies, the circuit designers started to implement their circuit designs into monolithic semiconductor integrated circuits. Creating in the semiconductor material both the active and passive elements.

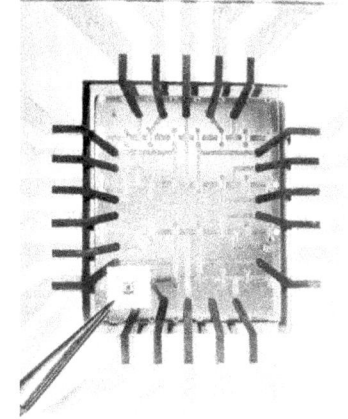

The circuit designer of an integrated circuit would designate specific areas on the die for the passive elements. And, by applying the doping and insulation techniques in P- and N-type material, would create the desired functionality (Figure 274).[463]

The earliest experimental MOS IC to be fabricated was a 16-transistor chip built by Fred Heiman and Steven Hofstein at RCA in 1962 (Figure 275, top). General Microelectronics later introduced the first commercial MOS integrated circuit in 1964, a 120-transistor shift register developed by Robert Norman. These ICs from the early 1960s were the start of the development of different ranges of integrated circuits.

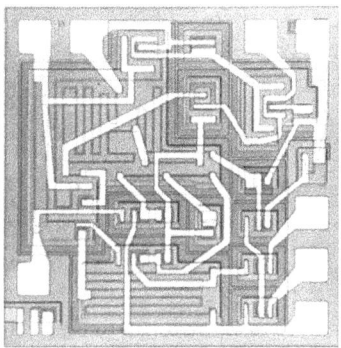

**Figure 275: Die-shots of Early Integrated Circuits.**

Logic: 16-transistor logic IC by Heiman and Hofstein (top, 1962), Linear: 15 transistor linear IC by Talbert and Widlar (bottom, 1965)

Source: Computer History Museum.

---

[462] P-channel MOS (PMOS) logic uses p-channel MOSFETs to implement logic gates and other digital circuits. N-channel MOS (NMOS) logic uses n-channel MOSFETs to implement logic gates and other digital circuits.

[463] A 'die shot' is a photograph of (a part of) the 'chip' itself without its packaging. It shows the surface of the chip.

*Logic Integrated Circuits:* By 1963 the logic-ICs were developed in the form of Resistor-Transistor Logic, the Diode-Transistor Logic and Transistor-Transistor Logic ICs. Patented by James Buie of Pacific Semiconductor in 1961, TTL (Transistor-Transistor-Logic) emerged as the most popular logic configuration of the next two decades.

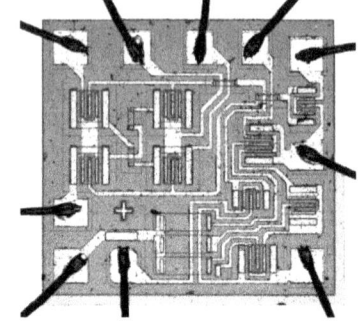

Buie applied on September 8th, 1961 a patent application for 'Coupling Transistor Logic and other circuits.' He was granted US-Patent № 3,283,170 on November 1st, 1966.

Fairchild came with the uLogic 923 RTL Integrated Circuit, a Flip-Flop housed in a TO-package (Figure 255). By 1965, three engineers from Fairchild: Don Forbes, Rex Rice, and Bryant Rogers invented a 14-lead ceramic Dual-in-Line Package (DIP) with two rows of pins. This semiconductor package came into volume production in early 70's.

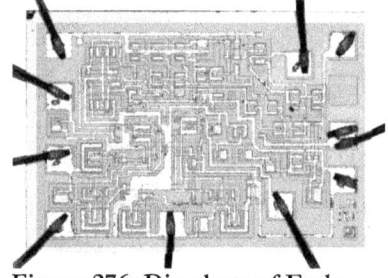

**Figure 276: Die-shots of Early Linear Integrated Circuits.**
Motorola MC1596 chip (1970, top), and the Signetics LM361H chip (bottom, date unknown),
Source: Mikhail Svarichevsky, ZeptoBars.com

As a result of their commercial success, a range of IC-manufacturers jumped on the bandwagon and started manufacturing TTL-series; Sylvania, Texas Instrument, Fairchild, and Signetics.

*Linear Integrated Circuits:* Next to the digital IC's, analogue IC's (aka linear IC's) were developed by Amelco, Fairchild, RCA, TI, and Westinghouse. Such as the first widely-used commercial product —the Fairchild μA702 operational amplifier— created in 1964 by process engineer Dave Talbert and designer Robert Widlar (Figure 275). Their 1965 successor, the fourteen transistor and fifteen resistor μA709, established a mass market for linear ICs. For a few years, Fairchild was the leader in the field of linear ICs. Demand for its products exceeded its production capacity by a factor of ten.

Soon other manufacturers followed and created their specific circuitry. Motorola offered the MC1596 Balanced Modulator/ Demodulator Chip containing eight transistors, three resistors and one diode (1970) (Figure 276, top). And Signetics offered the LM-series with the LM361H analog Differential Comparator IC (Figure 276, bottom).

*Memory Integrated Circuits:* By 1965 the first Read-Only Memory chips (ROM) appeared. These non-volatile memory devices were used to permanently store information, such as micro-program code, look-up tables, character generation

In 1965 Sylvania produced a 256-bit bipolar TTL ROM for Honeywell that was programmed one bit at a time by a skilled technician at the factory who physically scratched metal link connections to selected diodes. Production orders were satisfied with custom-mask programmed devices. In 1965 General Microelectronics developed slower but four-times larger 1024-bit ROMs using MOS technology.

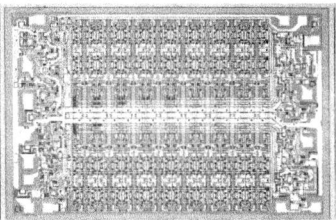
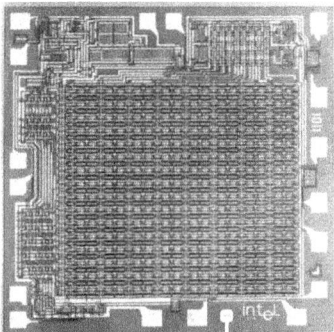
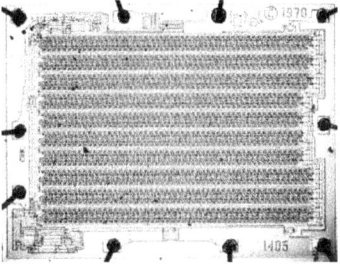
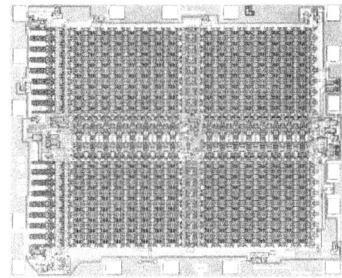

**Figure 277: Die-shots of Early Memory Integrated Circuits.**

Intel 3101 SRAM (top, 1969), Intel 1101 SRAM (upper middle, 1969), Intel 1405 (lower middle, 1970) and Intel 1103 DRAM (bottom, 1971).

Source: Intel.com, http://www.righto.com/, cpu-galaxy

By the early 1970s Fairchild, Intel, Motorola, Signetics, and TI offered 1024-bit TTL ROMs, while AMD, AMI, Electronic Arrays, General Instrument, National, Rockwell and others produced 4096-bit (4K) MOS devices. The ROM-ICs found their mass application in desktop calculators and game consoles. Next to the Read Only Memory, in addition, the genus of the volatile memory device called Semiconductor Active memory (SAM) was developed. Such as the Random Access Memory (RAM) and its derivates.

Intel, that started as a manufacturer of logic circuits by becoming a memory-chip manufacturer, had introduced in 1969 the bipolar 3101 SRAM chip with 64 bits of storage. In the same year, Intel also produced the 1101 SRAM chip use a metal-oxide semi-conductor process and rely on silicon gates rather than metal.

Next it introduced the 3301 Schottky bipolar 1024-bit read-only memory (ROM) and the first commercial metal–oxide–semiconductor field-effect transistor (MOSFET) silicon gate SRAM chip, the 256-bit 1101. It was followed by 1405 shift-register memory, and the 1103 DRAM memory chip that became the bestselling semiconductor memory chip in the world by 1972 (Figure 277).

> Robert Norman filed for a patent application for a semiconductor static RAM design at Fairchild on March 5th, 1963, and was granted US-Patent № 3,562,721 on February 9th, 1971 for a 'solid state switching and memory apparatus' (Figure 278).

The Random Access Memory circuitry[464] came in several versions: from the Dynamic RAM (DRAM) to the Static RAM (SRAM). Each with its own properties that made it suitable for specific applications. In computer applications it would replace the 'Core Memory' devices from the late 1940s. With the explosion of the Digital Age (the period of time when the IC-based computers emerged), the volume of the manufactured memory chips exploded too.

*Packaging the Chip*

Creating the chips was done at the 'front end' of the production process. Packaging the chip in a protective housing was done at the back 'end' (Figure 251). Early integrated circuits were packaged in ceramic flat packs, which the military used for many years for their reliability and small size. The other type of packaging used in the 1970s, was called the ICP

---

[464] A Random Access Memory (RAM) is a type of memory in which data can be stored (by writing) and collected (by reading) in a random matter. For these read/write functions is had additional circuitry.

(Integrated Circuit Package). Over time a range of different housing concepts developed (Figure 251); among them the popular Dual in Line Package (DIP, Figure 257) in a lot of variations; the quad plat package (QFP), and many others (Figure 279).

*IC Technology Development: The MOS-Revolution*

As the MOS-FET transistor was to become the most used semiconductor device and a vehicle of IC-development, the MOS-technology was fuelling the technological and economic growth of the early semiconductor industry. The MOSFET as the most widely manufactured device in IC-history, became the basis of modern electronics, and it became the enabler of the subsequent Computer Revolution in the Digital Age to come.

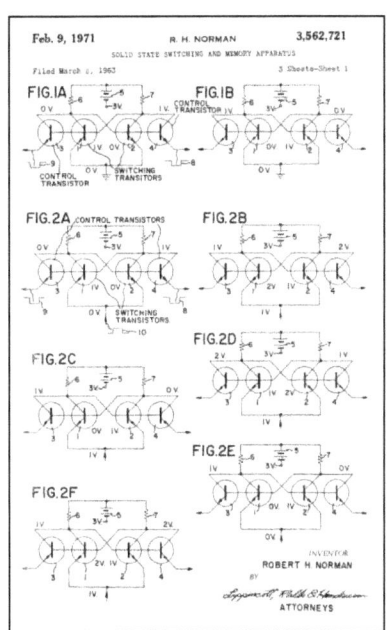

**Figure 278: Patent for Memory Integrated Circuits.**
Robert Norman's US Patent № 3,562,721.
Source: USPTO

**Figure 279: Different Chip Packages.**
Source: Wikimedia Commons.

There were some major development forces notable in the development of IC-technology and its specific equipment. One of those forces concerned the basic structure of the transistors in the semiconductor material as a result of a specific production process.

*During its first decade the semiconductor industry went through five basic transistor structures: point contact, grown junction, alloyed junction, surface* *barrier, and diffused-base. Manufacturers built their own equipment to support each generation. Jack Kilby noted that "probably the most expensive piece of equipment that we used cost less than $10,000." As production moved to high volumes with the planar process, techniques were standardized across the industry and independent equipment producers emerged.*[465]

Next, the MOS-technology (1960s) —in its two versions P-MOS and N-MOS— developed into the C-MOS-technology in the 1965s at Fairchild Semiconductor.

Frank Wanlass, working for Fairchild, combined P-MOS and N-MOS. He filed for a patent application on June 18th, 1963 and was granted US-Patent № 3,356,858 on December 5th, 1967 for 'Low stand-by power complementary field effect circuitry' (Figure 280, left). The first CMOS integrated circuit was later built in 1968. On

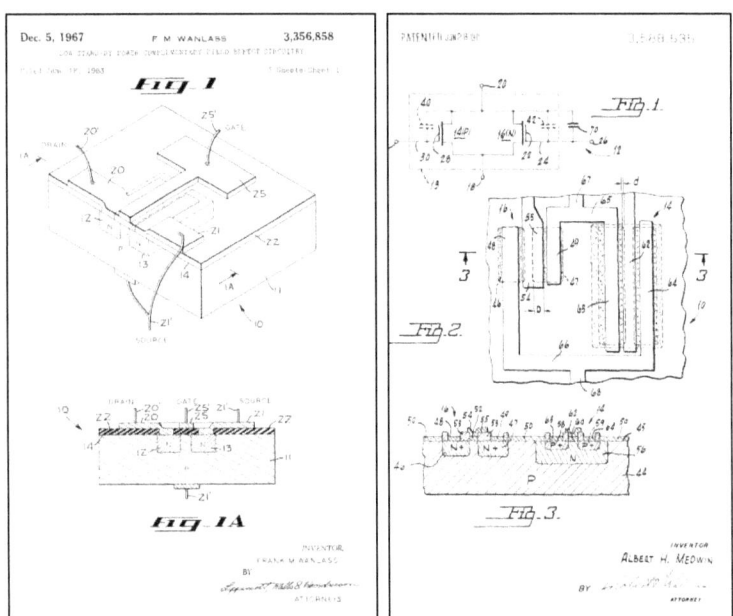

**Figure 280: Patents for C-MOS Integrated Circuit.**
Robert Wanlass's US Patent № 3,586,858 (left) and Albert Medwin's US Patent № 3,588,635 (right).
Source: USPTO

---

[465] Source: 1967: Turnkey Equipment Suppliers Change Industry Dynamics. https://www.computerhistory.org/siliconengine/turnkey-equipment-suppliers-change-industry-dynamics/ (Accessed January 2022)

April 2nd, 1969 Albert Medwin working for RCA a patent application for an 'Integrated Circuit' —which employed the insulated gate Field Effect Transistors and in which both P-channel and N-channel devices were included— that was granted as US-Patent № 3,588,635 on June 28th, 1971 (Figure 280, right).

The CMOS IC technology was commercialised by RCA in the late 1960s. RCA adopted CMOS for the design of integrated circuits (ICs), developing CMOS circuits for an Air Force computer in 1965 and then a 288-bit CMOS SRAM memory chip in 1968.

## Overview of the Invention of the Integrated Circuit

We explored in the preceding analysis the events that created —after the invention of the bipolar transistor had paved the way— the new electronic device called Integrated Circuit. The result of both the developing solid-state technology as the result of the contributions of the 'technologists,' as well as the design efforts of the 'circuit designers' (Figure 281).

We noted that the Integrated Circuit, with its origins in the discrete Logic, Linear and Memory circuit, was a breakthrough in the development path of small, compact, and reliable systems. The photo-lithographic based manufacturing process (Figure 252) enabled the step from the single circuit

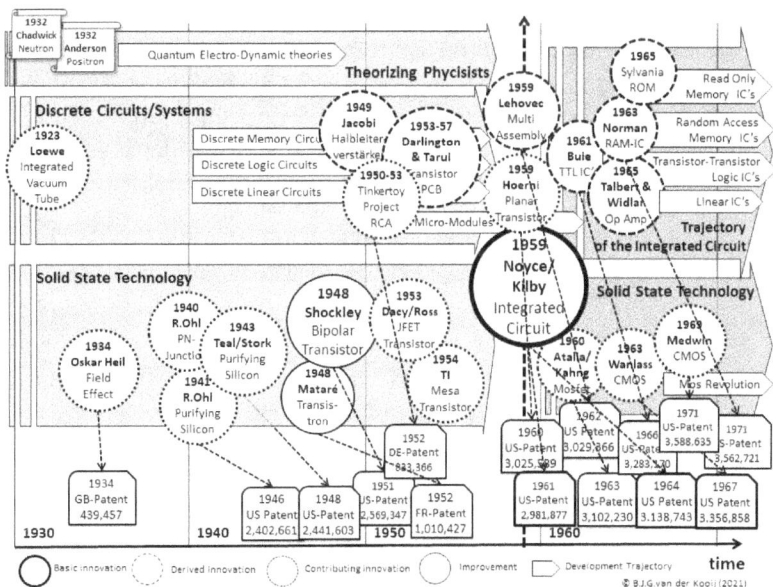

Figure 281: Contributions to the Cluster of Invention with the Basic Invention of the Integrated Circuit.

realized on a small wafer (Figure 273) to the multi-die process on increasingly larger wafers (Figure 282, Figure 283).

Again, the *scientist*s that contributed to the transistor, were contributing to the understanding of Solid-State physics. The *engineers* developed from their work in the transistor-based circuits in the development trajectories of the Circuits Systems that created the memory ICs, the Logic ICs and the Linear ICs that emerged from the MOS Revolution. And the *technologist* developed the technologies. It were Robert Noyce and Jack Kilby who integrated the circuits on a single die; the Integrated Circuit. From that moment the technological development trajectory of the Solid-State Technology and the complementary trajectory of the circuit went hand in hand, creating a wealth of ICs: from the linear ICs to the memory ICs (Figure 281).

## Who invented the Integrated Circuit?

We explored how many people contributed to Solid-State physics and the device that could control the flow of electrons in semiconducting material (aka the Transistor) during twentieth century (Figure 231). We also explored the many contributors to the integration of circuits into semiconductor structures (aka Integrated Circuits) a decade later (Figure 304).

Each contribution —from the ones with a small impact to those with a big impact— had its own value. Contributions that resulted in the semiconductor devices with internal structures able to control the flow of electrons. First in small circuits, but later in increasingly complex circuitry. Just like we reflected on earlier in the case of the vacuum tube and the transistor, also now the question arised who can be considered as the inventor (ie the classic quest for one specific person), of the Integrated Circuit? Let's try and explore that topic in a similar way.

*Facing the Big Question*

When trying to address the classic question of 'who is the inventor,' one is faced with some problems of interpretation, as we already noted before in the case of the Radio Tube; from semantics to the subject, from moment in time to the consequences and to the point of view.

Clearly, there are many different —sometimes opposing and contradicting— interpretations about who is the inventor of the internal combustion engine. They may differ, but nevertheless, all those interpretations have the process of 'Change and Novelty' in common. But again, even in that process, there are differences. Take 'the invention', the result of the process of invention as carried out by the specific person —the inventor—, or a group of persons —the R&D team— that creates novelty (Figure 217). Next, there is the 'innovation', the result of the process of Change and Novelty creating a working artefact that —as the result of entrepreneurial activities— can be brought to the market by subsequent organizational activities. Both those processes, invention and innovation, make up the *Act of Invention* and the *Act of Innovation*. However, then again, confusion arises. Is that about the original 'idea' (as in the mental picture),[466] is it the first model or prototype (as in the archetype), or is it the product/system that is successfully commercialized?

Using our definition of the basic invention (Scheme 3) as the dominant invention in a cluster of innovations, we let others decided on what was the 'invention.' Such as the contemporaries in the Patent Offices who decided if the invention was patentable on criteria of their own, valid in their time. So, to explore the inventions, we used the patent as an event-identifier (Figure 196). And we let the historians of science and technology, who explored the Noyce-Kilby case extensively, decide who was *the* inventor.

## From Small Scale (SSI) to Very Large Scale Integration (VLSI)

Going back to the evolution of the IC-technology, it was clear that the overall development of the IC-technology was ruled by some specific mechanisms: (1) the technology push to finer resolutions (ie transistor density), (2) the ever increasing wafer[467] size and die-size, and (3) the production yield. Each mechanism contributed in its own way to the meteoric rise of the IC. A development that started from the first, rather simple circuits realised in a silicon substrate (ie those with two transistors). From there the number of components on a chip progressed with the increasingly sophisticated litho-graphic technologies.

**Figure 282: Wafer with individual dies (ie IC's).**

The squares on the wafer above are the die's that contain an integrated circuit.

These technologies made finer inter-connection possible, creating integrated circuits that held as many as ten diodes, transistors, resistors and capacitors, making it possible to fabricate one or more logic gates (eg

---

[466] See our working papers: Deep origins of Innovation (Part I): creation of Novelty, and Deep Origins of Innovation (Part II) Memes of Novelty.
[467] A *wafer* is a thin slice of semiconductor, such as a crystalline silicon (c-Si), used for the fabrication of integrated circuits. The wafer serves as the substrate for microelectronic devices built in and upon the wafer. From the 2-inch wafer (1960), wafer size increased to 12 inch (2002).

a Flip-Flop) on a single die. In addition, as wafer size increased (Figure 283), more 'chips' (aka die) were produced collectively in a single lithographic process on one silicon wafer (Figure 282). And with the implementation of the 'clean room' manufacturing facilities, the number of good dies on a wafer (ie the yield) improved dramatically. And that had its effect on the price.[468]

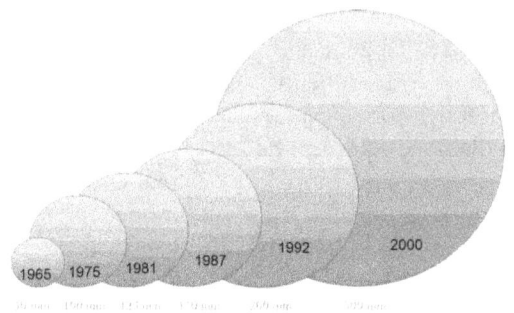

Figure 283: Incresing wafer size (1965-2000).
Source: Wikimedia Commons.

*Small Scale Integration (SSI):* Driven by the demands and massive budgets of the early aerospace projects (eg the Minute Man Missile project and the Apollo Program), the Small Scale Integration (SSI) technologies developed rapidly. The financing by the US Government's military supported the nascent integrated circuit market until costs fell enough to allow IC firms to penetrate the industrial market and eventually the consumer market.

As the litho-graphic technologies for chip manufacturing improved constantly, the circuitry on a die became more compact thanks to the smaller dimensions of the components on the chips (aka structure size, or device geometry) caused by finer resolutions creating smaller lines (aka feature size) (Figure 284, left axis), while in addition the chip size increased (Figure 284, right axis). The result was that increasingly complex systems could be designed on a single die. In addition, the increasing wafer size made it possible to implement more dies on a wafer (Figure 282).

Each next step in the integration process made it possible to create larger memory chips and more complex circuitry (such as the later central processing units) in a single Integrated Circuit. It resulted in generations of circuits, such as the subsequent generations of memory chips that grew from 256 bit to 1 kb, 4 kb and 16 kb. The first generation introduced chips which contained hundreds of transistors on each chip, a phenomenon called 'Medium-Scale integration' (MSI; 100-1,000 transistors). Further development, driven by the same scaling technology and economic factors, led to 'Large-Scale integration' (LSI; 1,000-10,000 transistors) by the mid-

---

[468] The average price per integrated circuit dropped from $50.00 in 1962 to $2.33 in 1968.

1970s, followed by 'Very Large Scale integration' (VLSI, 10,000-1 million transistors) in the 1980s.

With the increasing component count with a factor of 1,5/year (as predicted by Gordon Moore in 1965),[469] the chip-technologies made it possible to manufacture increasing complex systems. From the memory chips with increasing data storage, to the increasing functionality of TTL Logic chips and the micro-computer CPU-chips, it became a trajectory of continuous improvement was followed (Figure 285).

*Medium Scale Integration (MSI):* Improvements in technology at the early 1970s soon led to manufacturing logical circuits with hundreds of components: the 'medium-scale integration' (MSI). Such as the logical circuits like the SN7400 series. With TTL logic, the first minicomputers were realized. Such as the DEC PDP-series minicomputers, the Data

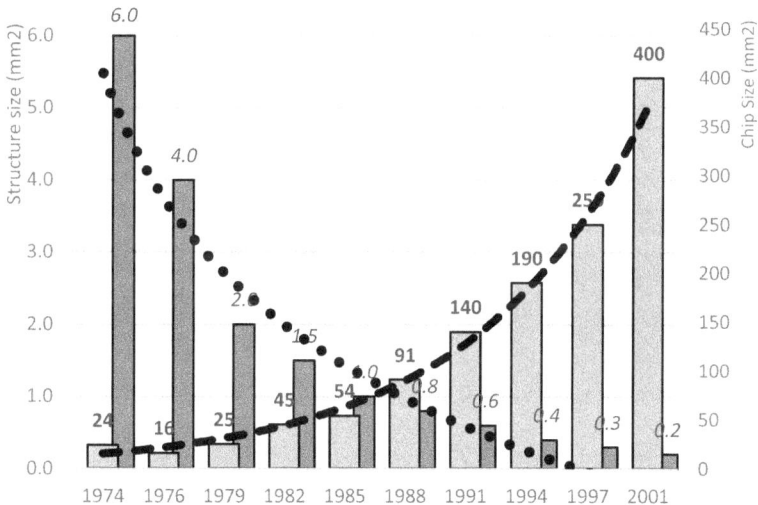

**Figure 284: The Technological Driver: smaller structures/features and larger die/chip (1974-2001).**

The combination of the development of the structure sizes on a chip (left) decreasing, and the size of the chips (ie die size) increasing (right).

Source: https://www.tf.uni-kiel.de/matwis/amat/elmat_en/kap_5/backbone/r5_4_1.html

---

[469] This phenomenon became known as Moore's Law. The original observation made in 1965 by Gordon Moore, was that the number of transistors per square inch on integrated circuits had doubled every year since the integrated circuit was invented. Moore predicted that this trend would continue for the foreseeable future. In subsequent years, the pace slowed down a bit, but data density has doubled approximately every 18 months, and this is the current definition of Moore's Law, which Moore himself has blessed.

General Nova-series and the Hewlett Packard 21MX, 1000 and 3000-series. Companies that made these MSI-ICs were Sylvania, Motorola, RCA, National Semiconductor and Signetics. Around 1965, the first memory ICs appeared that replaced the previously used magnetic toroidal memories. It was Intel, which started in 1968 and with its 1103 memory IC, a 256-bit RAM memory, that opened up the memory market in 1969.

*Large Scale Integration (LSI):* Again improvements in technology at the mid-1970s led to manufacturing logical circuits with hundreds of components: the 'large-scale integration' (LSI). There were two niches for these kinds of ICs; the memory chips and the processes chips. The size of the memory ICs increased from 4 Kb (price $18/piece, 1974), 16kb (price $33/piece, 1976) to 64kb (price $47/piece, 1979). In addition to the relatively simple memory circuits, there were the much more complex microprocessor circuits. With, in addition to the 4004 microprocessor (1971) with chip count of 2,250 transistors, soon the 8080 CPU (1974) with chip count of 6,000 transistors, the 8086 CPU

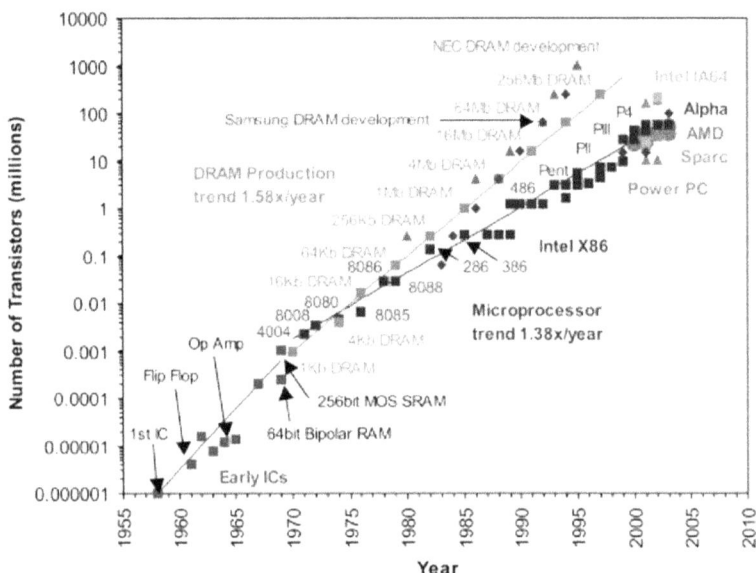

**Figure 285: Increasing number of transistors of Memory-chips and CPU-chips (1950-2005) illustrating Moore's Law.**

The increasing size of computer memory chips (RAM, SRAM, DRAM) and microprocessors (from the 4004 through X86 series to A64).

Source: https://www.researchgate.net/figure/Evolution-of-transistor-count-of-CPU-microprocessor-and-memory-ICs_fig2_228913859

(1978) with a chip count of 29,000 transistors, and the 80836 CPU (1985) with a chip count of 275,000 transistors were manufactured.

*Very Large Scale Integrations (VLSI):* The increasingly sophisticated techniques made it possible to integrate more than a million logical circuits on a 'chip' in the eighties/nineties. The size of the memory ICs rose from 64kb (price $47/piece, 1979), 1 Mb (price $100/piece, 1986), 4 Mb (price $124/piece, 1988), 16 Mb (price $275/piece, 1991) to 256 Mb (price $575/piece, 1998). In addition to the relatively simple memory circuits, there were the much more complex micro-processor circuits. From the chip count of 134,000 transistors (80286 CPU, 1982), to a chip count of 275,000 transistor (80386 CPU, 1986) until 1989 the 80486 CPU with a chip count of 1.2 million transistors. Followed by the Pentium microprocessor family in the nineties which grew from a chip count of 3.1 million transistors (Pentium 1, 1993), to a chip count of 5.5 million transistors (Pentium Pro, 1995) and a chip count of 7.5 million transistors (Pentium II, 1997), up to a chip count of 28 million transistors (Pentium III, 1998).

## *The Invention of the Computer on a Chip: the Micro-computer*

We described in the preceding exploration the development trajectories of the technology. Mentioning from time to time the conceptualization by the *circuit designers* of electronic logic circuits: from the many logic TTL-circuits up to the more complex Control Unit and Arithmetic Logic Unit. A development that resulted in many logic circuits realised by the *technologists* (aka chip designers) in the at that moment in time available IC-technology. And those circuits were used by the *system designers* to build the electronic systems; for the calculating systems (aka calculator) to the computing systems (aka computer). Remarkable enough, it was a rather simple system —the calculator— that played a large role in the further development of the electronics circuits: [470]

> Companies like desktop calculator manufacturers —eg Canon, Olympia, Olivetti and Busicom— came to Texas Instruments and Intel with a set of specific requirements defining the functionality of their calculators. The chip designers converted those specifications to a chip set, normally four, five or six chips to execute or implement the specifications. For the calculator companies, it was really amazing that the chip designers provided this ultimate service of compressing

---

[470] On a personal note, much of the following exploration is based on an earlier research project conducted by the author for his Master thesis (1976) at the University of Technology in Delft, The Netherlands. The thesis (in Dutch) was published as 'Micro-computers, Innovatie in de Elektronica' (Microcomputer, Innovation in Electronics) (Kluwer, 1978).

so many units on to just four, five or six chips because the previously used technology called the TTL would use hundred to two hundred chips. The cost savings were enormous, enabling lower sales prices and opening up the mass consumer market. No wonder for the IC manufacturers the business was doing great, notwithstanding the bottleneck of the limited number of chip designers that spent up to six months working on one project.

So, to solve the design bottleneck, the search was for more efficient use of the designer's time by finding a different architecture of the logic design. And to integrate that design on a single chip. It resulted in two different development trajectories that would have a great impact.

## The Single Chip Computing Circuits

Inspiration for the designers to create a 'single chip computing circuit' came from the early concepts and architectures of computing systems developed for the mainframe and mini-computers. So, circuit design efforts focussed on the circuitry's logic architecture in a way that it could fit on a single chip.

A contributing factor was the development of the 'System Bus'; the pathway that connected the major components of the system

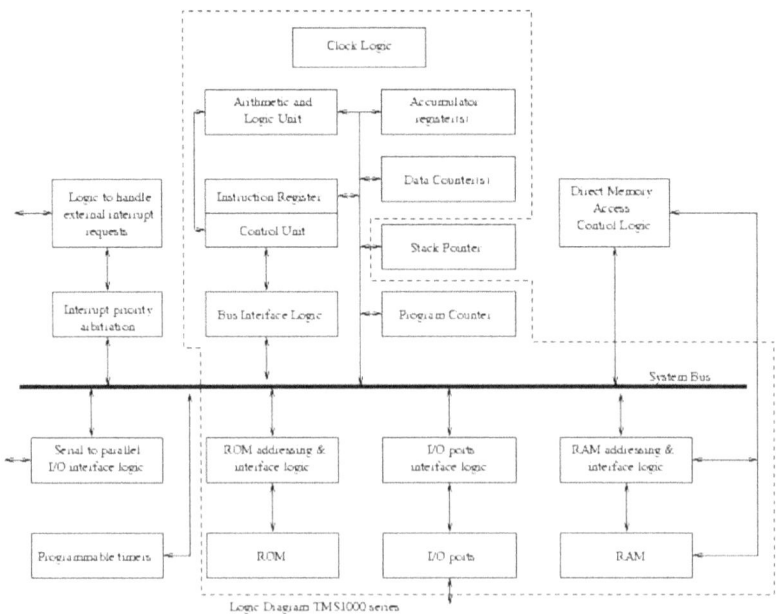

**Figure 286: Logic Architecture of the Single Chip Processing Unit.**

(Figure 286). It combined the function of a data bus to carry information, an address bus to determine where it should be sent or read from, and a control bus to determine its operation.

It was the combination of the work from the circuit designers creating the functional and architectural design, and the IC-technologist realising the actual design of the chips, that would create the microprocessor on a chip.

*The Invention of the Microprocessor.*

With the new solid-state technology capable of creating increasingly complex integrated circuits, the focus came on creating a programmable logic circuit consisting out of a CPU (ie the Central Processing Unit) and its Memory on a single silicon substrate. Intel, together with MOSTEK,[471] was in 1968 approached by the Japanese calculator maker Busicom, who had developed ideas for implementing ICs in its calculators. Although reluctantly —Intel was focussing on Memory-ICs at that time— Intel took

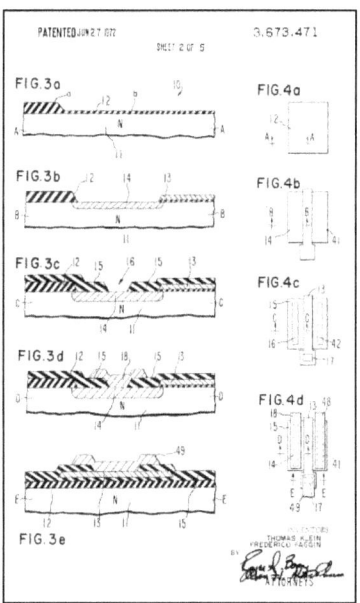 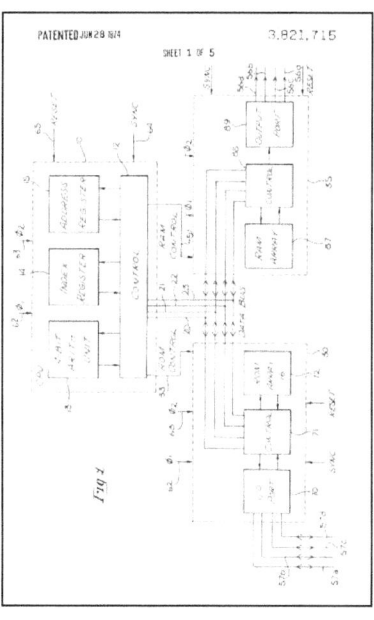

**Figure 287: Patent for Silicon Gate Technology MOS Integrated Circuit and for the MCS-ICs**
Federico Faggin's US Patent № 3,673,471 (left), and Intel's US Patent № 3,821,715 (right
Source: USPTO

---

[471] MOSTEK, located in Worcester, Massachusetts, was set up by ex TI-employees. They became world leader in memory chips. For Busicom they developed the single chip calculator solution MK6010, which was used in Busicom's LE-120A calculator.

on the project and created a project team of four man.

On the Busicom side of the Busicom Project, it was Mastoshi Shima, an electronic engineer, who participated. In April 1968, Shima was asked to design the logic for what was intended to become a future calculator chipset to be designed and produced by a semiconductor company. Shima went to Intel in June 1969 to present the proposal. Intel assigned Marcian Edward Hoff (aka Ted Hoff, Intel's twelfth employee) to the team.[472] With his experience in computer architecture, Hoff suggested putting the more complex steps into programs in memory, reserving the basic logic circuits for a single chip. Working with team member Stanley Mazor, Hoff designed the instruction set and architectural specifications of the chip.

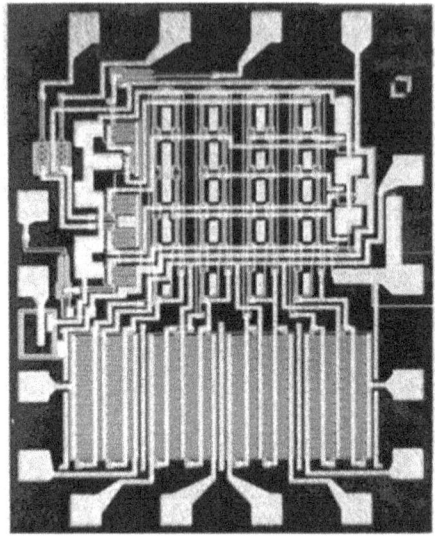

**Figure 288: Die-shot of the Fairchild 3708.**

Source: Historyofinformation.com.

Federico Faggin, who had already developed earlier at Fairchild the silicon-gate technology (SGT) by 1968 and implemented it in the Fairchild 3708 logic circuit (Figure 288). When INTEL was founded in 1968, he was hired by Intel and his SGT became the core technology for the fabrication of MOS-IC's. Faggin became the technologist for the project. He accomplished what no one had achieved before: to fit a general-purpose CPU into a small, commercial silicon chip. He invented a technique to design and layout 2300 random-logic transistors into a single chip with 5 times the speed and twice the circuit density (for half the cost) of the incumbent metal gate technology.

---

[472] On a personal note, your author had the opportunity during the research for het Master thesis, to interview Ted Hoff in 1977.

On October 8th, 1970 the patent application was filed and on June 27th, 1972 US Patent № 3,673,471 was granted for "Doped Semiconductor Electrodes for MOS type Devices' (Figure 287).

The result of the team effort became known as the 'microprocessor (μCPU) on a chip'; the Intel 4000 chipset (1971) consisting out of the 4004-CPU (Figure 289) and the accompanying memory chips 4001, 4002/1/2, and 4003 (Figure 290). It was the result of a development that was spurred by needs of (Japanese) manufacturers of desktop calculators.

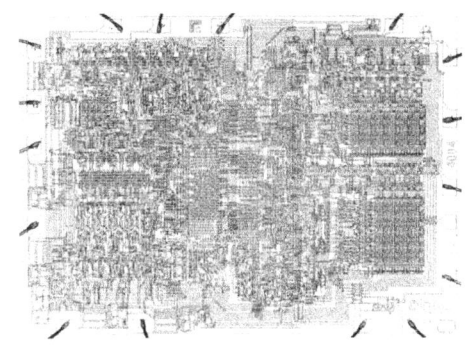

> *Although Intel began as a memory chip company, in 1969 we took on a project for Busicom of Japan to design eight custom LSI chips for a desktop calculator. Each custom chip had a specialized function-keyboard, printer, display, serial arithmetic, control, etc. With only two designers, Intel didn't have the manpower to do that many custom chips. We needed to solve their problem with fewer chip designs. Ted Hoff chose a programmed computer solution using only one complex logic chip (CPU) and two memory chips. In 1970 Intel designers implemented a 4-bit computer on three LSI chips (CPU, ROM, RAM) housed in 16-pin packages.* (Mazor, 1995, p. 1601)

**Figure 289: The Intel 4004 CPU (1971).**
From die to board: actual chip 4x3 mm (top), the packaged chip (middle), and the complete board with the 4000 chip set (bottom).

Source: Wikimedia Commons (top), Wikichip (middle), Nigel Tour (bottom).

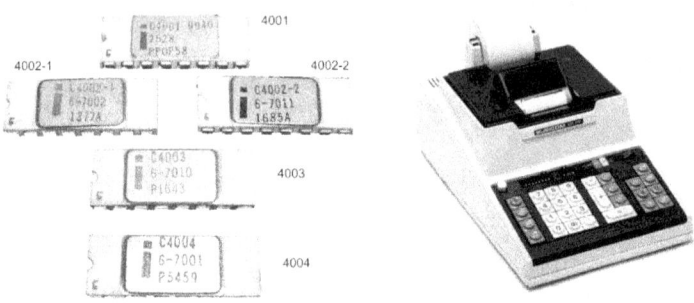

**Figure 290: The MCS-4 chipset and its calculator application.**

The MCS-4 chipset left, 1971), Busicom 141-PF Calculator using the Intel 4004 CPU (right, 1971).

Source: CPU-Galaxy, http://www.vintagecalculators.com/html/busicom_141-pf_and_intel_4004.html.

On January 22nd, 1973 they filed a patent application for the external bus organization (granted on June 28th, 1974 as US Patent № 3,821,715 to Faggin, Hoff and Mazor (Figure 287).

So, the first Intel micro-processor chipset (aka micro-computer) was custom-built for the calculator manufacturer Busicom; it became known as the Busicom Chipset as it had resulted from the 'Busicom Project.' The chipset became used in the Busicom 141-PF Calculator (Figure 290, right).

By that time it was obvious that the design had many advantages and offered considerable flexibility for the application designers. After Intel had bought back the rights for the design from Busicom,[473] the 4000-family was completed by March 1971, in production by June 1971 and introduced to the market in November 1971 under the name MCS-4 (Figure 290, left).

## Evolution of the Microprocessor

After that chip set for the first 4-bit micro-computer, soon more powerful processors[474] were introduced (Table 7). Starting with the 8008 (1972, Figure 291), this 8-bit CPU was followed by the 8080 (1974), 8085 (1976), 8086 (1978), 8088 (1979). The 8086, was Intel's first commercial 16-

---

[473] Intel acquired the rights by offering to return the $60,000 development cost and to produce the chip at a lower cost.
[474] A microprocessor can become more powerful (ie capable of processing more data in a given period of time) by increasing the clock speed (creating the Turbo-versions), by widening its data bus (from 4 bit to 8 bit and further) and by faster memory access through enlarging its address bus.

The Invention of the Silicon Engine

## Table 7: Evolution of Intel Microprocessors (1971-2000).

| Processor | Year Of Introduction | No. Of Transistors | Intial Clock Speed | Address Bus | Data Bus(in bit) | Addressable Memory |
|---|---|---|---|---|---|---|
| 4004 | 1971 | 2300 | 108 kHz | 10 bit | 4 | 640 bytes |
| 8008 | 1972 | 3500 | 200 kHz | 14 bit | 8 | 16 k |
| 8080 | 1974 | 6000 | 2 MHz | 16 bit | 8 | 64 k |
| 8085 | 1976 | 6500 | 5 MHz | 16 bit | 8 | 64 k |
| 8086 | 1978 | 29000 | 5 MHz | 20 bit | 16 | 1 M |
| 8088 | 1979 | 29000 | 5 MHz | 20 bit | 8 | 1 M |
| 80286 | 1982 | 134000 | 8 MHz | 24 bit | 16 | 16 M |
| 80386 | 1985 | 275000 | 16 MHz | 32 bit | 32 | 4 G |
| 80486 | 1989 | 1.2 M | 25 MHz | 32 bit | 32 | 4 G |
| Pentium | 1993 | 3.1 M | 60 MHz | 32 bit | 32/64 | 4 G |
| Pentium Pro | 1995 | 5.5 M | 150 MHz | 36 bit | 32/64 | 64 G |
| Pentium II | 1997 | 8.8 M | 233 MHz | 36 bit | 64 | 64 G |
| Pentium III | 1999 | 9.5 M | 650 MHz | 36 bit | 64 | 64 G |
| Pentium 4 | 2000 | 42 M | 1.4 GHz | 36 bit | 64 | 64 G |

Source: The World of Microprocessors. http://microprocessor7.blogspot.com/2013/05/intel-microprocessors-historical.html

bit CPU and is considered to be the chip that launched the era of X86 processors.

These processors were followed by 16/32-bits chips 80286 (1982, Figure 291, top), the 80386 (1985), and the 80486 (1989) (Figure 291, bottom). Each version having more capabilities due to the change from the 8-bit architecture to the 16/32-bit architecture, processing power due to higher clock rates and addressable memory. The 80486 drove Intel through its greatest phase of growth. By 1993 a new line was introduced; the Pentium-series of 32/64 bit micro-processors. As a result of constant improvements in design, architecture, component density and production technology, each new generation of microprocessors became more powerful.

**Figure 291: Early Evolution Intel's CPU on a Chip.**

From the 8008 chip (top) to the 80486 chip (Bottom)

Source: cpu-world com

The circuitry of the microprocessor became increasingly complex, resulting in increasing complex circuitry of the die. The 80286 microprocessors contained some 134,000 transistors (Figure 292), the 80486-microprocessor introduced seven years later, had some 1,2

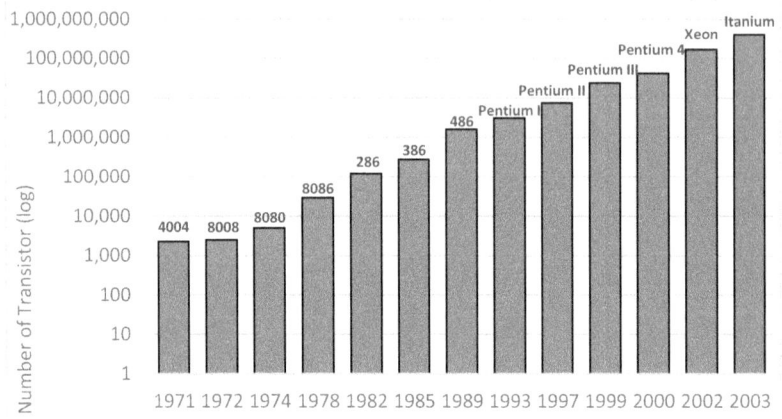

**Figure 293: Microprocessors: Number of Transistors (1971-2003)**
Source: http://www.singularity.com/charts/page75.html;
https://www.intel.com/pressroom/kits/quickrefyr.htm

million transistors. And the Pentium 4 chip contained the staggering number of 42 million transistors of a single die (Figure 293).

**Figure 292: Die shot of the 80286 processor.**
Enlargement of the actual die (aka chip). On a surface of some 50 mm² the circuit wit 134,000 transistors is realized. It was packaged in a 25x25mm housing.
Source: Wikimedia Commons.

The 80286 was employed for the IBM PC/AT, introduced in 1984, and then widely used in most PC/AT compatible computers until the early 1990s. One of the earliest complete systems to use the i486 chip was the Apricot VX FT, produced by British hardware manufacturer Apricot Computers.

But that was not the end of it, as Intel would introduce new product lines of microcomputers (ie Xenon and Itanium-processors) in the next years. Technology was the driven force behind this development of the computers on a chip. It showed in the numbers of transistors that were incorporated on a chip (Figure 293). Their performance also increased as a result of architectural and functional improvements, and they became the driving force of the Personal Computers.

So, IC-technology fuelled the development of the microcomputer chips. Complementary to that development, also the companion memory chips increased in capacity. After Intel's 3101 SRAM memory chip with 64 bit storage (1969) and the 1101 SRAM memory chip with 256 bit storage (1969), their capacity exploded. From the 256-bit single chip memory of the late 1960s to the 256 Mb single chip memory of the late 1970s/early 1980s. The later having a million times more storage capacity. And with the continuous improvements of the IC-technologies the memory size of the single memory IC exploded also by the late 1980s (Figure 294).

Over time the new 'general purpose' microcomputers build with these processors followed their own development trajectory; each generation was more powerful, prices dropped as volumes rose with each new application.

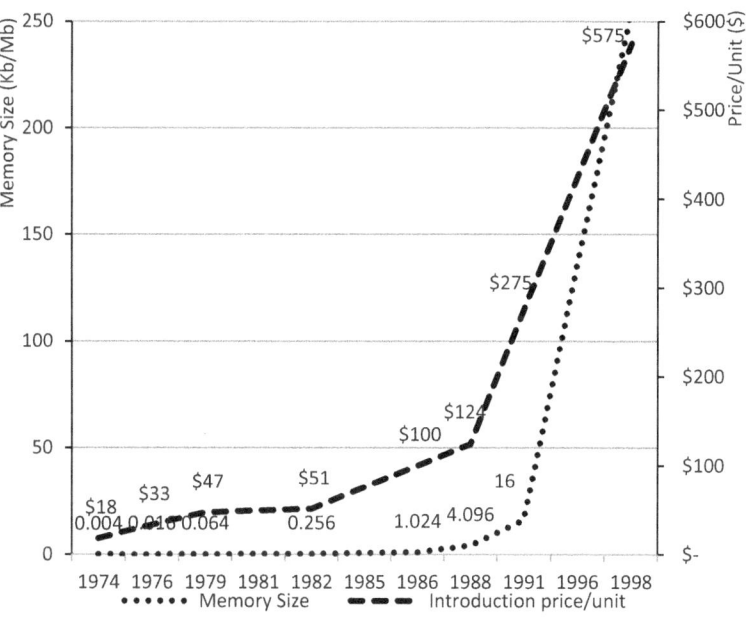

**Figure 294: Memory Size Development including Introductory Price.**
Source: Wikimedia Commons.

*The Integrated Circuit Industry*

Next to the start-ups like INTEL, from some of the classic manufacturers of transistors, the IC-manufacturers emerged. Large firms created their R&D-intensive semiconductor divisions, new spin-offs grew rapidly. A whole new 'high-tech' industry with its own chains of suppliers (of materials and machines, but also of the related services) emerged.

> IC's were manufactured in highly specialized factories by a growing range of range of semiconductor companies. Together they formed the *Integrated Circuit Industry* manufacturing microprocessors and memory-chips. Next to Intel Co. companies like Motorola, Zilog Co., American Micro Devices and MOS-Technology, designed and manufactured microprocessors; resp. the 6800-series (1974), the Z-80 series (1976) and the Intel 'second sources' made by AMD. And MOS-Technology created the 6501/6502 microprocessor (1975) that would become the founding processor of a company named Apple.

The Microprocessor industry had dynamics of its own. Individuals or groups of developers —for a range of reasons— would split off from their parent company, to pursue their own objective.

> Such as the company MOS-Technology that was founded because the 6800-development team at Motorola Inc. could not convince their management to build a low-cost microprocessor (the latter 6502 microprocessor). Frederico Faggin and Ralph Ungermann, both developers of the 4004/8080 microprocessor decided to start the company Zilog in 1974. All microprocessor that became used in the larger systems like the 'Personal Computer.'

As the IC industry supplied such essential components to the manufacturers of the end products, these required that the IC manufacturers —the 'first source'— licensed their technologies and designs to other manufacturers. It was the practice of the 'second source' supplier.

> The company Advanced Micro Devices (AMD), started in 1969 by former Fairchild employee Jerry Sanders, became a second source supplier of microchips designed by Fairchild, National Semiconductor and Intel.

*The Invention of the Micro-controller.*

The first micro-processors may have been designed for dedicated calculator applications, soon they became used for more general applications as its function could be defined by the application software that was loaded in its RAM-memory. That was different with another type of processor where the application would be stored in ROM-memory.

*The Microcontroller for Calculators:* In parallel to the microprocessor development at Intel, Texas Instruments —since 1952 active in the electronic industry, from manufacturing transistors up to consumer products like transistor radios and transistor calculators— entered into IC-manufacturing when its employee Jack Kilby invented the Integrated Circuit in 1958. Subsequently they explored how the devices could be used in specific applications.

One of those projects was the CAL-TECH project focusing on a miniature calculator. After two years of development, TI-designers Merryman and Van Tassel completed the first hand-held calculator in 1967 (Figure 295, top). The battery-powered device could accept six-digit numbers, perform the four basic arithmetic functions, and print results as large as 12 digits on a thermal printer,

**Figure 295: Early IC-based Calculators.**

TI's Prototype (top, 1967), and Canon Pocketronic (bottom, 1970).

Source: Wikimedia Commons.

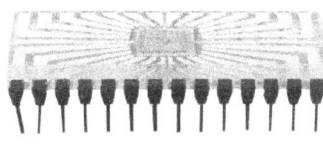

**Figure 296: Packaged Calculator Chips.**

The TMS 1802 (top) and TMS 1000.

Source: Wikimedia Commons. Computerhistory.org

and was prices at $20 in large quantities. At its hart was a single-chip microcontroller optimized for use in a calculator.

Merryman and Van Tassel filed on December 21st, 1972 a patent application and were granted US Patent № 3,819,921 on June 25th, 1974 for a Miniature Electronic Calculator (Figure 297, left).

This early result led to the development of a 'Calculator on a Chip' —originally called the TMS1802

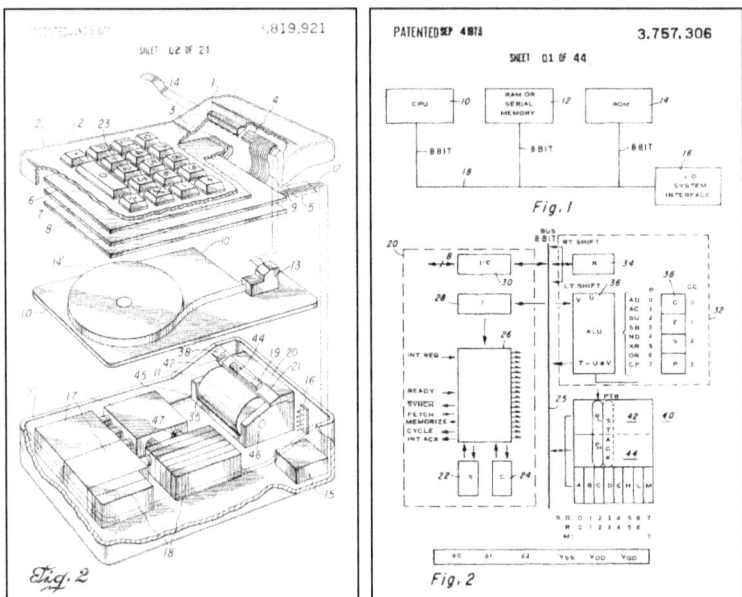

**Figure 297: Patents for Early Integrated Circuit- based Calculators.**
Patent № 3,819,921 for the 'Miniature Electronic Calculator' (left) and US Patent № 3,757,306 for a 'Computing Systems CPU' (right).
Source: USPTO

(Figure 296, top)— that contained some 5,000 transistors on a single chip and was 9.9 cm x 4 cm x 2 cm as a packaged chip. The TMS1802 chip was the first in what became the TMS 0100-series of calculator on a chip ICs that were the heart of hundreds of different calculators.

Texas Instruments contacted Canon Inc. in Japan and arranged a co-production of a pocket calculator. In April 1970, the Pocketronic (Figure 295, bottom) appeared on the Japanese market; it was a four-function, entirely electronic calculator that retailed for about $400.

In September of 1971, TI finished the design for their TMS0100 single-chip calculators. Designs were done by the Texas Instruments engineers Gary Boone and Michael Cochran. Gary applied for a patent for his single-chip processing machine on August 31st, 1971. On Sep 4th, 1973, he was awarded US Patent № 3,757,306 for a 'Computing Systems CPU' (Figure 297, right).

Thus began Texas Instruments' meteoric rise in the calculator industry. Some historians mark the development of what became the TMS0100 series "calculator on a chip", which Cochran made a major contribution to, as the first commercial microprocessor. It certainly created the foundation for the exploding electronic calculator industry.

*The Embedded Microcontroller:* Gary Boone's and Michael Cochran's 1971 design of Texas Instruments TMS1802 single-chip calculator device — later renamed to TMS0102— provided the foundation for a new family of micro-controllers. It was the TMS1000 general-purpose 4-bit MCU family announced in 1974 (Figure 296, bottom for packaged version and Figure 298 for the chip).

After, in September of 1971, they had finished the design for their TMS0100 single-chip calculators, the designers stared looking for a different solution. It resulted in the *micro-controller* (aka micro-computer) that could easily be programmed to other applications than calculators. Their design became the *TMS 1000 micro-controller* series to be used in pre-programmed embedded applications.[475]

With Gary Boone and Michel Cochran as inventor, TI filed a patent application on December 2nd, 1974 for a 'Variable function programmed calculator,' that was granted as US Patent № 4,074,351 on February 14th, 1978.

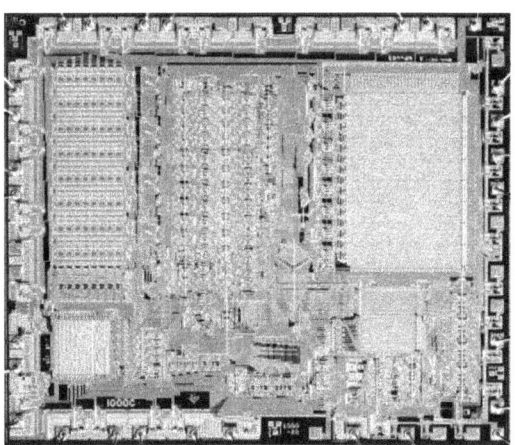

**Figure 298: The actual 'chip' of the TMS1000 processor.**
Enlargement of the actual die (aka chip).
Source: Wikimedia Commons.

The TMS 1000 processor was produced in some forty different versions. Although the microcontroller was used at the Texas Instruments internally in its calculator products between 1972 and 1974, the device was further refined over the years. It became a 'Computer on a chip' as it combined the CPU,

---

[475] The software which is programmed into the microcontroller is capable of handling only a limited range of applications: such as burglar alarms, garage door openers, toys and video games.

memory and I/O onto one chip. TI offered this microcontroller for sale to the electronics industry from 1974 onward, and by 1983, nearly 100 million TMS 1000 had been sold. Over time microcontrollers became used in automatically controlled products and devices, such as automobile engine control systems, implantable medical devices, remote controls, office machines, home appliances, power tools, toys and other *embedded systems*.[476]

This short overview of the Integrated Circuit illustrates how the early evolutions in semiconductor technology spawned of a range of (general purpose) *computing engines* (aka micro-computer), and (dedicated) *control engines* (aka micro-controller) that evolved in different application technologies. A development that started the trajectory of both the general-purpose micro-computer, as well as the dedicated micro-controller. And it was the work of Faggin, Hoff and Mazor that resulted in the programmable device that created the foundations for the development of the micro-computer that would cause the Personal Computer Revolution.

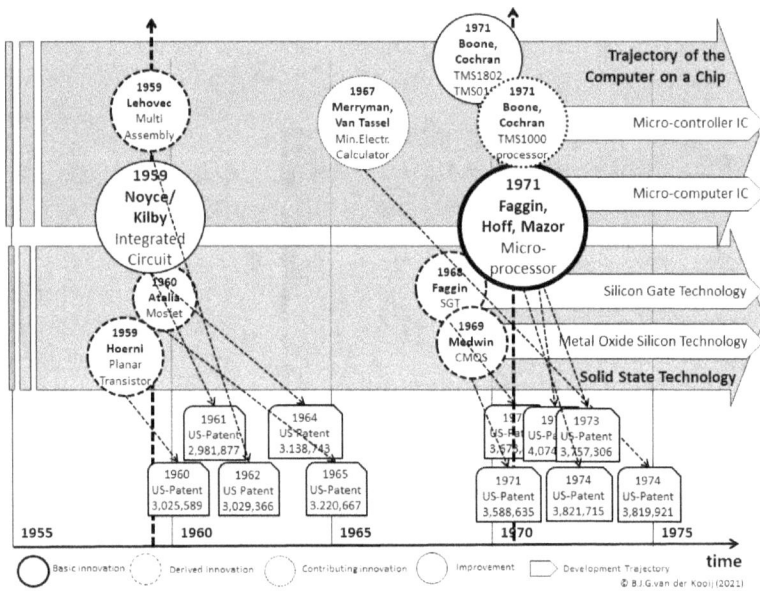

**Figure 299: Overview Invention of the Micro-processor.**

---

[476] An embedded system is a microprocessor-based computer hardware system with software that is designed to perform a dedicated function, either as an independent system or as a part of a large system. It is the next stage for many home-appliances. For example: hand-rotating drum washing machine (1850s), electric motor-powered washing machine (1900s), automatic washing machine (1980s).

## The Silicon Engine as a Technology Driving Force

With invention of the multi-element electronic circuit on a single chip (aka the Integrated Circuit) came the development of the Semiconductor technologies described before. Consisting out of many sub-technologies — eg the techniques of phosphorus diffusion, oxide masking, photolithography, aluminium evaporation— it is labelled as the 'IC-technology' (Figure 251).

> *The IC manufacturing process can be summarised briefly. It all starts with a suitable semi-conductor material, such as silicon. It is refined to a highly pure state and then sliced into thin sheets. From these sheets, a photoresist film is added to imprint the circuitry as it bakes and is then etched to create a permanent configuration. The Si wafer is then doped with specialised impurities. As mentioned, multiple layers can be etched onto a single IC chip. The chips are then tested, separated and packaged, ready for delivery.* [477]

And an important part of that technology is the photolithographic process (Figure 252).

> *The chip is built upwards, layer by layer. Each layer is made by putting masks with particular patterns over the silicon and then altering the qualities of the silicon -- or perhaps putting down metal or insulators -- in the exposed parts. [...] The chip starts out as a thin wafer of P-type silicon. This is then coated with a layer of silicon dioxide -- kind of a silicon rust, which doesn't conduct electricity. On top of this is placed a chemical called photoresist. Flashing a pattern of light (like the grid of light and dark that's formed by a window screen) on the photoresist turns any parts exposed to the light hard. The bits left in shadow stay soft. When an etching chemical is applied those soft parts, and the silicon dioxide underneath them, are removed. The hard photoresist is then dissolved, leaving a pattern of raised silicon dioxide along the surface. Since the silicon dioxide doesn't conduct electricity, it keeps different parts of the final circuit separated from others.* [478]

---

[477] Source: https://sites.google.com/site/icmanufacturing2204/products-services (accessed August 2021).
[478] Source: The Integrated Chip. The American Institute of Physics. http://www.pbs.org/transistor/science/info/chip.html (accessed August 2021).

## The Clean Room and its Production Equipment[479]

In 1968, standard manufactured practices were lax. Facilities were kept reasonably clean, but employees wore their own clothes and the environment was not cleaned from dust and particles (Figure 265). When it became obvious that the production process for semiconductor was sensitive to teeny dust particles, in the early 1970s the manufacturing started to take place in 'Clean Rooms'; manufacturing areas in which particle concentration and environmental conditions could be controlled within specified limits. In this area, all the production equipment was placed. More rigorous clean suit protocols began at Intel in 1973. From the simple clean rooms of the early days, the production spaces became more sophisticated over time. Even more when the MSI- and LSI-integrated circuit with the more miniature structures became vulnerable to small airborne particles (aka dust) that influenced the yield of the wafer.[480]

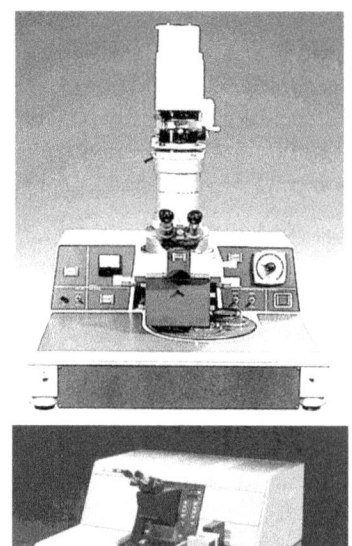

Figure 300: Clean Room Equipment (1970s).
Canon PPC-1 (top, 1970), Perker Elmin's Micralign PE-100 (botttom, 1975).
Source: Canon

For the first two decades of the industry, semiconductor manufacturers custom-built most of their own tools and equipment in-house. Because the increasing precision was problematic for manual handling, (semi)-automated machinery (eg step-and-repeat cameras, testing equipment, wafer-handling equipment) were developed by specialized equipment manufacturers.

> The early efforts to make tools for the photo-lithographic production of IC were by the IC-technologists themselves. Such as Jay Last and Robert Noyce who built —taking advantage of Fairchild's knowhow in optical and camera technology— one of the

---

[479] The subject of semiconductor manufacturing technology is massive, and too large to include in our case study. So, we just touch on the surface of this subject
[480] The IC yield is the ratio of the number of good chips versus bad chips on a wafer.

first 'step-and-repeat' cameras at Fairchild in 1958. By the early 1960, specialized companies started making commercial 'step and repeat' cameras that projected the mask-pattern on the wafer, step by step.

Among the manufactures of lithography equipment and wafer handling equipment that emerged in the 1970s were a range of American and Japanese companies entering semiconductor manufacturing. Many started to commercialize the machines they build for their own use in the photo-lithographic process. Or they cooperated with independent manufactures that supplied 'Front end' equipment (ie used in wafer processing) and 'Back end' equipment (ie used in packaging).

Such as the camera manufacturer Canon, that started building semiconductor lithography equipment. Utilizing technology originally developed for camera lenses during the mid-1960's, Canon developed high-resolution lenses for photomask manufacturing. With the aim of expanding its operations, the company began developing semiconductor lithography equipment for wafer fabrication, and in 1970, entered the business with the introduction of the 'PPC-1,' Japan's first domestically produced semiconductor lithography equipment (Figure 300, top).

Around the semiconductor manufacturers a completely new equipment industry arose. Perkin Elkin, an optical-instrument company, introduced the Micralign PE-100 projection scanning aligner for $98,000 (Figure 300, bottom). By the mid-seventies Perkin-Elmer had become the largest equipment supplier in the entire industry, and by the early 1980s some 2000 machines were be in use worldwide. Also the American company Tencor Instruments Inc. that in 1977 entered the semiconductor equipment market, grew into a major supplier of the semiconductor manufacturing industry.

Or take the Dutch Royal Philips NV that created with Advanced Semiconductor Material Int. (ASMI, est. 1964) the company ASM-Lithography in 1984. They manufactured machines like the ASML PAS 5500, Step & Scan System (1990) automating the wafer handling process. Two decades later ASML was the world leader in clean room equipment.

Other early front-end tool suppliers include Thermco (diffusion furnaces) and its Japanese licensee Tokyo Electron Ltd, DW Industries (deposition systems), and GCA/Mann (photo-lithography). Electroglas, a 1961 Fairchild spinout, built the EG 900 probe testing equipment for wafer testing. In 1965 Kulicke & Soffa introduced commercial contact aligners. Varian Associates built evaporators, vacuum pumps, and ion-implantation systems.

Over time all the steps in the processing of the wafers became highly automated, reaching precision levels manual-handling could not achieve. Also the design process of the increasingly complex circuitry on the chip became supported by advanced systems. The CAD (ie Computer Aided Design) and EDA (ie Electronic Design Automation) programs were used to speed the design task and eliminate design errors.

With the decreasing structure size (Figure 284) the 'cleanness' of the manufacturing spaces became more and more an issue. It resulted in the semiconductor Clean Room manufacturing area where all the equipment to process the wafers into chips (Figure 251) was placed and operated. The employees operating the equipment had to wear protective clothing (ie a cleanroom suite) (Figure 301).

## Patenting: Protecting Intellectual Property

With the rapid development of the Semiconductor technology came the issue to protect the intellectual proprietary of development devices, tools, techniques and processes. And all the work that went into the design of the circuitry on the IC.[481] Patenting was an issue to establish that intellectual property. Especially for some patents that became key-patents (Table 8).

**Figure 301: Clean Room.**
Operators of the machinery wearing specialized Clean Room suites.
Source: Unknown (date unknown).

---

[481] As the ruling Patent Laws of the 1970s did initially not recognize intellectual property rights related to topology (ie circuit layouts), it were mainly devices and methods that could be patented.

Prior to 1984 to copy an IC (by means of reverse engineering) to produce a competing chip with identical layout —aka the pirating of chips (Figure 302)— was not illegal under Patent Law. It was the Semiconductor Chip Protection Act of 1984 that recognized the importance of the topology (ie the three-dimensional layout of all the elements constituting an integrated circuit and their interconnections fixed on a physical medium) of Integrated Circuits.[482]

*Patent Conflicts*

In the years 1959-1961 the competition among the US semiconductor manufacturers did not have an aggressive character. True, they were competing for the same military contracts, and cooperated with the same (Japanese) companies to get their calculator-contracts. That changed however in the 1962s when Texas Instruments management started a stricter policy of patent enforcement with their largest competitor: Fairchild and its CEO Robert Noyce.

**Table 8: Overview of some Important US Patents for Semiconductor Devices (Transistor and Integrated Circuit).**

| Year[1] | Subject | Patent № | Inventor/Assignee |
|---|---|---|---|
| 1926/1930 | Method and Apparatus for Controlling Electric Currents | US 1,745,175 | Lilienfield |
| 1948/1950 | Three-Electrode Circuit Element Utilizing Semiconductor Material | US 2,524,035 | Bardeen & Brattain/Bell Labs |
| 1948/1950 | Semiconductor Amplifier | US 2,502,488 | Shockley |
| 1948/1951 | Circuit element utilizing semiconductive material | US 2,569,347 | Shockley/AT&T Corp. |
| 1959/1961 | Semiconductor device-and-lead structure | US 2,981,877 | Robert Noyce/ Intel |
| 1959/1962 | Method of manufacturing... | US 3,025,589 | Jaen Hoerni/Fairchild |
| 1959/1962 | Multi Semiconductor Assembly | US 3,029,366 | Kurt Lehovec/Spraque |
| 1959/1962 | Semiconductor Device | US 3,064,167 | Jaen Hoerni/Fairchild |
| 1959/1964 | Miniaturized electronic circuits | US 3,138,743 | Jack Kilby/TI Inc. |
| 1960/1963 | Electric Filed controlled semiconductor Device | US 3,102,230 | Dawon Kahng/ Bell Telephone Labor Inc. |
| 1963/1967 | Low stand-by power complementary field effect circuitry | US 3,356,858 | Frank Wanlass/Fairchild |
| 1969/1971 | Integrated Circuit | US 3,588,635 | Medwin/RCA |
| 1970/1990 | Single-chip integrated-circuit computer architecture. | US 4,942,516 | Gilbert Hyatt |

1) The first year indicated is the year of patent application, the second year is the year the patent was granted.

---

[482] On a personal note, your author as Member of the Dutch Parliament, was his faction's spokesman when a similar Chips Act was proposed in the Netherlands.

For the Integrated Circuit, there were two major patents, that of Kilby (US-Patent № 3,138,743 for a 'Miniaturized Electronic Circuits') and that of Noyce (US-Patent № 2,981,877 for a 'Silicon based IC'). Kilby had invented a germanium version of the circuit, while Noyce developed the silicon version of the integrated circuit. Both had their method patented, but who was the infringer? Kilby's 743-patent was filed six months earlier than Noyce's 877-patent, yet it was issued three years later. It seems that at the time the U.S. Patent Office wasn't quite sure what to make of Kilby's patent application. Texas Instruments, which held the patent for Kilby's invention, started an infringement case that would have crippled license-income of both companies. It took some years, but Texas Instruments and Fairchild settled the issue in 1966 by an agreement on cross-licensing.[483]

On April 10th, 1962, Sprague's Lehovec received a patent for isolation with the PN-junction. Texas Instruments immediately filed a court case claiming that the isolation problem was solved in their earlier patent filed by Kilby. The court decided otherwise and Lehovec's priority on the isolation patent was finally acknowledged in April 1966.

On May 20th, 1962 Hoerni's patent on the planar technology was challenged by Raytheon. Hoerni found the court on his side and Raytheon withdrew their claim and obtained a license from Fairchild.

Those were some of the internal US patent conflicts. But then the American semiconductor manufacturers, having built up tight relations with Japanese companies, came in conflict with the Japan Ministry of International Trade and Industry (MITI). Both Fairchild and Texas Instruments saw their investment plans in IC-manufacturing plants blocked, obstructed and delayed on ministry-level.

> The problems reached state-level in 1965 when Texas Instruments retaliated by threatening with embargo on the import of electronic equipment that infringed their patents. Notwithstanding a secret deal with Sony, for almost thirty years many Japanese companies were producing ICs without paying royalties to Texas Instruments. Till in 1989 the Japanese court acknowledged the patent rights to Kilby's invention. In 1993, Texas Instruments earned $520 million in license fees, mostly from Japanese companies.

---

[483] Texas Instruments' Semiconductor division developed the transistor radio in 1954. From 1954 to 1958 TI was the only firm capable of producing silicon transistors in quantity. In 1967, Texas Instruments developed the electronic desktop calculator (the calculator) using an IC. In Japan, electronic equipment manufacturers released calculators one after another, and fierce 'calculator wars' continued until the end of the 1970s.

But before that was to happen, the whole IC-industry was still faced with a problem as many 'pirates'[484] copied their designs. They were buying the chips that companies had spent years developing, reverse engineering them, manufacture the new design, and then selling the exact copy at a fraction of the price (Figure 302).

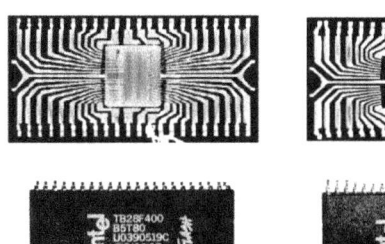

**Figure 302: Pirated Chips.**

The original INTEL IC (bottom left in DIL-package) when X-rayed (top left), compared with the pirated version (packaged, bottom right) and being x-rayed (top right).

Source: Wikipedia Commons.

Essentially, they were counterfeiting the chips and the Patent Law proved inadequate to protect their inventions.[485]

> *This infringement fell outside the realm of patent because these pirated designs were not inventions, but rather designs etched onto the chips of already patented devices. But they also fell outside the scope of copyright protection, because these designs were functional and not aesthetic, as they designed to control the conductivity of the products they powered. And trade secret protection obviously failed as well once the design was introduced to the public. After much lobbying by the semiconductor industry, Congress settled on creating a new type of copyright for a "mask work" for these semiconductor designs. In 1984, Congress passed the Semiconductor Chip Protection Act and created an entirely new IP right, the first new IP right in 100 years.*[486]

The pirates were not the only ones who troubled the IC-industry. But that surprise would emerge some two decades later.

---

[484] The development of a chip design into a chip is a labour-intensive and capital-intensive activity that can take month. As the layout of a chip is defined by its set of photo-lithographic masks, 'reverse engineering' is a method of recreating these masks (aka the circuit netlist).

[485] On a personal note: The author as a member of the Dutch Parliament contributed in 1984 to the Dutch equivalent of the American *Semiconductor Chip Protection Act*; his contribution can be found at the 'Chipswet' (Handelingen Tweede Kamer der Staten Generall, Kamerstuk 19919, ondernummer 5).

[486] Source: https://www.jdsupra.com/legalnews/the-top-ten-patent-wars-semiconductors-7-27555/

## A Submarine Patent surfaces

A startling event was a 1970 patent-application of the electronic engineer Gilbert Hyatt (1938) The American engineer Hyatt was a profound contributor to the development of calculating and computing systems, who —after working for companies like Teledyne— had created his own company Micro Computer Inc. in 1968.[487] Being a logic circuit designer, he focussed on designing a computer that could fit on a single chip and used as machine control systems in Numerical Control Machinery. Hyatt contracted Intel and Texas Instruments to build his chips, while simultaneously raising venture capital. Micro Computer Inc. also developed applications for its technology, including the operation of a machine tool control system and an integrated circuit drafting system. On December 28th, 1970, Hyatt filed his first patent application covering this microcomputer.

In early 1971, his investors attempted a coup but realized they could not take the technology with them, as Hyatt had built in protections by licensing his patents to his own company. However, the revolt crippled Micro Computer Inc., which subsequently closed its doors in September 1971. Next, Hyatt started to work in the aeronautic consulting business and educated himself in Patent Law and became registered patent agent. A capability that would come to use in the next years.

>On December 28th, 1970 Hyatt had filed for a range of patent applications. Among them for a Single-chip integrated-circuit computer. Most of those were granted by 1977/78, but one application for a 'Single-chip integrated-circuit computer' took more time. In the meantime, he continued filing patent applications for computer system architecture (filed on December 28th, 1970 and granted as US Patent № 4,371,923 on February 1st, 1983), and a data processor architecture (filed on February 22nd, 1974 and granted as US Patent № 4,523,290 on June 11th, 1985).

However, for one of his 1970-patent application it took some more time. After Hyatt had been filing continuation[488] after continuation for many additional claims[489], it took till June 17th, 1988, when he filed its final

---

[487] At Teledyne he was granted a range of patents for logic circuits; Widely recognized for his achievements in computer technology, inventor Gilbert P. Hyatt received nearly 75 patents between 1970 and 1997. But then the Patent and Trademark Office (PTO) stopped awarding him patents, instead blocking action on his pending applications.
[488] A 'continuation' application is a new patent application allowing one to pursue additional claims based upon the same description and priority date as a pending 'parent' application.
[489] Prior to June 8th, 1995, a patent's term was measured as 17 years from the date it was granted (ie date of issuance). The fact that patent term was keyed to the date of issuance, rather than the date of filing, incentivized certain patentees to delay prosecuting their patents

version. Subsequently, the US-Patent №. 4,942,516 granted in July 17th, 1990 for a 'Single-chip integrated-circuit computer architecture' was describing the architecture and logical design for a micro-controller that could be housed on a single chip (Figure 303). This so-called 'submarine patent' (ie a patent whose issuance and publication are intentionally delayed by the applicant for a long time), was all that time kept secret —as was the custom for patent applications— by the Patent Office.

The patent application's forty-four 'single chip' and 'Integrated Circuit' claims made it a broad-based patent that could affect the IC-Industry (eg invalidating the patents of Gary Boone for a computer on a chip, and Ted Hoff for the 4004-computer chip).[490] News of this granting was not made public until late August, when it hit the computer industry like a time bomb. Its implications were far-reaching, as it gave Hyatt rights to the technology used in most computer applications, especially those that fuelled the Personal Computer Revolution of the 1980s.

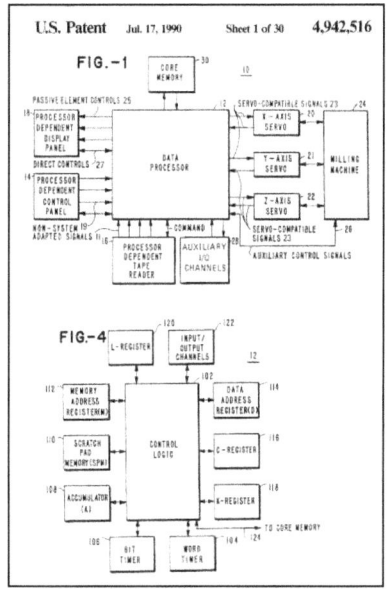

Figure 303: The Hyatt Patent № 4,942,516.
Source: USPTO

> Imagine, as the microprocessor was already widespread in thousands of applications, and millions of it were sold by a range of vendors (such as Intel, Motorola, Texas Instruments), those suppliers suddenly would have to pay (also overdue) license fees because the 'filing date' of their patent was later. They could therefore be held liable for patent damage. But there's more. Their own patents, with the associated license fees, were also at risk. So that inevitably led to action by the stakeholders. All the more so since the 'architecture'

---

by abandoning applications and filing continuing applications in their place. This delay strategy was found in many application fields, especially in the case of broad-based patents.
[490] Hyatt's applications focus on fundamental aspects of microchip and integrated circuit technology could be extremely valuable as enforceable patents. Others shared that opinion as Philips' efforts netted Hyatt over $150 million.

patent granted was a very 'broad-based' patent (like Alexander Graham Bell's patent for the telephone concept).[491]

In 1990 Hyatt licensed that patent and 22 of his 69 other patents to the multinational Royal Philips Electronics NV. Part of the deal was Philips would assist him in licensing the patents to other companies. Philips then began enforcing the patent rights on the rest of the electronics industry and collecting royalties. By 1992, Hyatt had received some $70 million from American, Japanese, and European companies. However, others protested, such as Texas Instruments and Intel who appealed Hyatt's 1970-patent claim. The result was that the Patent Office's Board of Appeals and Interferences partly overturned Hyatt's claim to the single-chip technology on a technicality. The industry breathed a sigh of relief. In the meantime, another issue popped up as Hyatt's 1991/1992 income from the licencing became an issue with the California Franchise Tax Board (FTB).

In 1993, the FTB conducted a residency audit and ultimately concluded that Mr. Hyatt owed $13.6 million in back taxes on the royalty income received in 1991 and 1992, fraud penalties, and interest. After 26 years, the accrual of interest brought the total liability to nearly $55 million by 2017. However, in that year, Hyatt scored a major win before the Board of Equalization —essentially California's tax court— which rejected the FTB's claims of fraud. It ruled that Hyatt owed no state tax for income in 1992 but did for a part-year residence in California in 1991. It reduced his potential tax bite from $55 million to $1.9 million plus interest.

But that ruling was not the end of it. Hyatt's harassment complaints about the FTB's tactics prevailed in Nevada state courts, including that state's supreme court. The jury, which concluded that the audit process was wrong and invasive. declared that Nevada owed damages to Hyatt. Although he was originally granted $389 million in damages by a Nevada jury for fraud, invasion of privacy, intentional infliction of emotional distress, and in punitive damages; several additional rulings by the Nevada Supreme Court and the U.S. Supreme Court reduced the amount to $100,000.[492] This was a big victory for Hyatt, especially since the court decided in a 4-1 vote that he did not intentionally mislead the California tax collector. Hyatt claimed to be the victim of tax collectors that had a vendetta against him.

---

[491] See: Van der Kooij, B.J.G.: *The Invention of the Communication Engine 'Telephone'* (2016).
[492] Based on: https://www.reedsmith.com/en/perspectives/2019/06/california-v-hyatt-the-story-draws-to-a-close. https://calmatters.org/commentary/2019/05/california-hollow-victory-tax-battle-hyatt/ (Accessed September 2021).

And last but not least, after getting 75 patents, Hyatt did not get another one since the 1990s, and for a long time he didn't know why. Soon he found out that the Patent Office had since 2014 a special way of flagging potentially controversial patents; the secret Sensitive Application Warning System (SAWS). Those flagged applications went all the way to the leadership of the Patent Office, and they were often never granted a patent by the office.[493] Although the SAWS was retired in March 2015, by 2018 Hyatt's patent applications are still in patent purgatory, and Hyatt was still fighting court battles.[494]

## *Overview of the Second Electronic Revolution*

In the preceding analysis of important events in the development of semiconductor engines (ie both Transistor and Integrated Circuit), we focussed on the results of a technical process used in the R&D environment and perfected for the manufacturing environment we labelled as the semiconductor technology. The foundation for these engines was laid by the many contributions to implement the Solid State physics; the new way to control the behaviour of electrons creating electric currents. Had the vacuum tube been about the control of the behaviour of electrons *in the vacuum* of the 'vacuum tube', now the transistor was about the control of the behaviour of electrons *in (semiconducting) material.*

## Contributors to the Invention of the Transistor

The *Classic Physics* with the theories of both the macro-world (ie Universe) and micro-world (ie the Nature of Matter and its molecular properties), had been replaced by the *Modern Physics*. A large part being the Solid-State Physics that developed theories about solid 'matter' itself; its atomic structure and the constituting elements like electrons, protons and neutrons. Focusing on the electric properties, the theories and models of electrons moving within the material created insight in materials conducting, resisting or 'semi-conducting' the movement of electrons within the material. That understanding created the basis for the invention of the transistor.

> The development of the insight of the electric properties of solid materials based on their atomic structure was the result of the work of the *Theoretical Solid-State Physicists.* It took some decades, from the early pre-

---

[493] This case shows that the USPTO is acting unlawfully and criminally to implement ultra vires a policy that certain organizations may successfully obtain patents while others are too disruptive to be allowed to have patents. In other words, the USPTO is making innovation solely a prerogative of big corporations and of big capital.
[494] Based on: https://venturebeat.com/2018/08/31/why-80-year-old-inventor-gil-hyatt-says-patent-office-is-waiting-for-him-to-die/ (accessed September 2021).

war contributions being interrupted by the First World War to the Interbellum contributions, to create the foundations of Solid-State Physics (Figure 304, middle).

The development of the Semiconductor Technology was largely the result of the contributions of the *Experimental Solid-State Physicists*. They created the techniques and processes to make the artefacts that would ultimo become known as the 'transistor.' (Figure 304, bottom).

The invention of the transistor may have been a fundamental technological event, it was only the beginning of a technical development trajectory of continues improvement. A trajectory that was interrupted by another event; the invention of the Integrated Circuit.

## Contributors to the Invention of the Integrated Circuit

After the invention the transistor, it had taken a decade from Shockley's the bipolar transistor to the invention of integrated circuit. From the original 10-component chip (Noyce's invention in 1959) to the 2,300-transistor chip (Hoff's invention of the micro-processor on a chip in 1971) had taken a little over a decade. It was the result of the combination of two types of expertise:

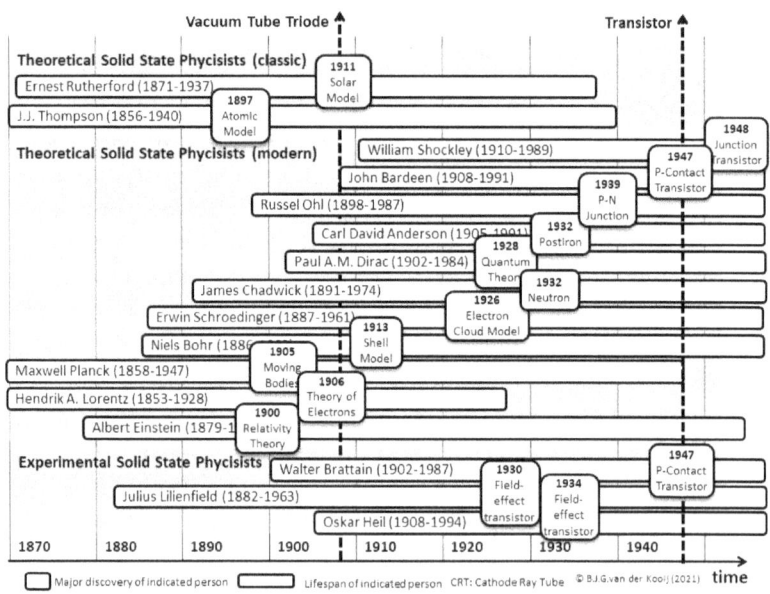

Figure 304: Overview of the Contributors in Solid State Physics leading to the Transistor.

The development of the Solid-State Technology was largely the result of the contributions of the *technologists*. The people that processed the semiconductor material of the single wafer in a range of processing steps into an Integrated Circuit, and who invented new variations to improve those processes ((Figure 305, bottom). Much of their work was based on the insights of the theoretical physicists (Figure 305, top). The further development of the *technology* to create the circuitry in semiconducting material itself was the result of the experimental work of the technologists. After the bipolar technology with its many variations, they created the highly successful Metal Oxide Silicon (MOS)-technology and the Silicon Gate Technology (SGT) (Figure 306).

The development of the functionality of the Integrated Circuit was the result of the work of the *circuit designers*. Among the many circuits they designed, was the 'computing systems on a chip'[495] that found its inspiration in computer architectures of the mainframes and minicomputers, and developed along the application trajectories of the Micro-controller IC and Micro-Computer IC.

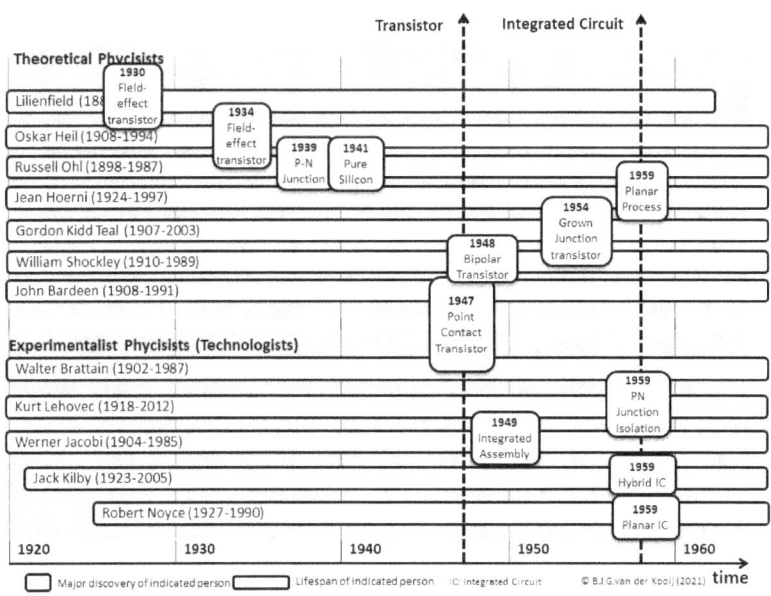

**Figure 305: Overview of Contributors to the Integrated Circuit.**

---

[495] Not to be confused with the concept of the System-on-a-chip (SOC) that refer to more complex systems.

It is not surprising that the pull in the development came from the market, in this case the manufacturers of the electronic calculators that envisioned the new IC-based calculating machines and spurred the IC-manufacturers to develop the ICs they needed. They saw the market potential of cheap, functional calculating machines that could be realized with the new IC-technology.

In some two-and half decade the Second Electronic Revolution had exploded from the single bipolar transistor element to the multi-element circuitry of the single chip micro-processor.

## Periodization of the Semiconductor Technology

Looking at the development trajectory (Scheme 2) of the Semi-conductor technology and seen in the perspective of the technology life cycle (Scheme 5), this technology is marked by specific events creating the basic innovations to which many thinkers and tinkerers contributed.

*Semiconductor 0.0* (1900-1930s): The Era of Modern Physics started with exploration of the structure of solid materials (Figure 305). After the conceptualization of the electron, neutron and proton, the atomic theories laid the foundation for understanding and theorizing the

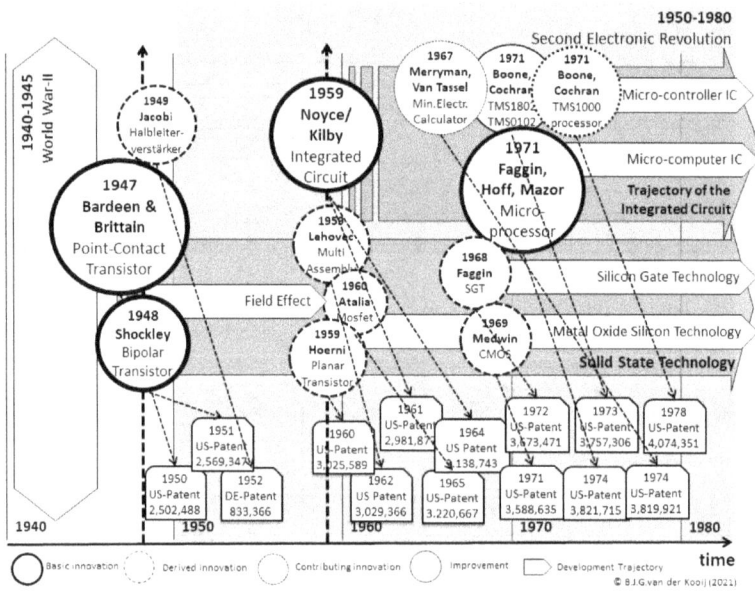

**Figure 306: Overview of the Second Electronic Revolution (1950-1980).**
Contributions to the Cluster of Invention with the Basic Inventions of the Transistor and the Integrated Circuit.

behaviour of semi-conducting materials. The result was Dirac's quantum theory of the electron; the Quantum Electro Dynamic theory that became important to the Solid-State physics. It took up to the 1940s — and a Scientific Exodus— before domain this became a separate scientific field.

*Semiconductor 1.0* (1930s-1960s): It were the contributions of both the theoretical solid-state physicists and the experimenting engineering scientists that led to the creating of the devices that would be known as transistors. Interrupted by the Second World War, the highly experimental technological development of semi-conducting materials resulted in the invention of the Point-contact transistor (1947) and the Bipolar transistor (1948). From there it took a decade to integrate the monolithic components in increasing numbers on a single chip: it became known as the Integrated Circuit (1959) (Figure 273).

*Semiconductor 2.0* (1960s-1980s): In this growth phase, the semiconductor technology matured and its fruits became available to be implemented in a broad range of applications. Even more when the electronic circuitry became increasingly integrated. First as Small-Scale Integration IC's (up to 100 transistor/chip), soon followed by Medium-Scale Interrogation

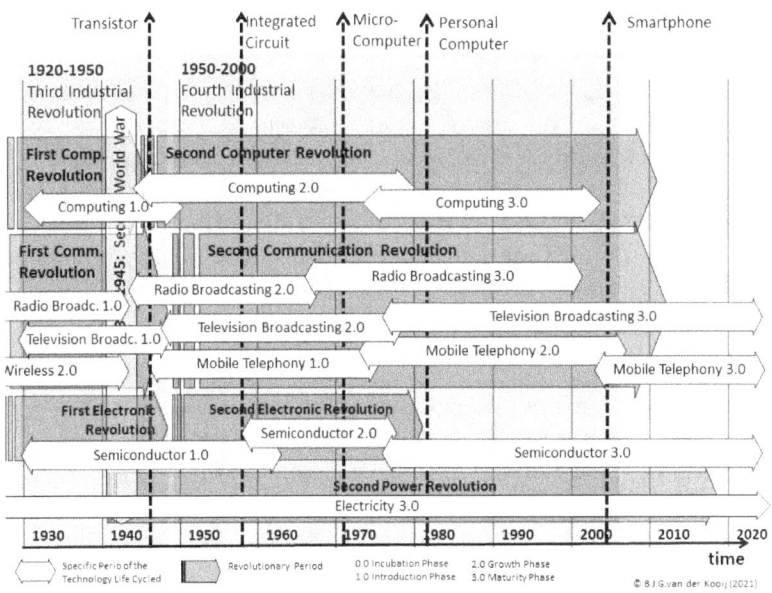

Figure 307: Overview of the Second Electronic Revolution and related Communication Technologies and Information Technologies (1950-2000).

IC's (mid 1960s, up to 1,000 transistors/chip), and the Large-Scale IC's (mid 1970s, up to 10,000 transistors/chip). The industries of semiconductor manufacturers, as well as equipment builders using the new (system) components, was booming (Figure 268) and subsequently the unavoidable first shake-out and consolidations were notable.

*Semiconductor 3. 0* (1980s-present): The rise of the semi-conductor technology followed an evolutionary path. The more when the application trajectory of computing stimulated increasing densities and growing numbers of transistor/chip. It became the time of Very Large-Scale integration with the single chip computer.

This periodization of the technological life cycle of semiconductor technology during the latter part of the Third Industrial Revolution and the Fourth Industrial Revolution (Figure 307) became related to the additional Information and Communication technologies that were derived from it (eg Mobile Telephony, Broadcasting and Computing). A period also labelled as the Second Electronic Revolution.

## *Overview of the Electronic Revolutions*

The late nineteenth century and the first half of the twentieth century were a time of explosion of novelty. Not only in the mechanical technologies (eg the automobile, airplane, the calculating machines, typewriters, etc. etc.), but also in the electric technologies. The telegraph and telephone systems matured, both in the cabled and the wireless version. And the Sciences of Physics had added the insights in the elementary properties of electricity, resulting in electronic devices like the radio tube and the transistor. Both engines that controlled the behaviour of electrons; one the behaviour in vacuum, the other the behaviour in semiconducting material. Each creating a technological revolution of its own.

### The Inventions in a Revolutionary Perspective

We explored the major events that constituted the time when the Electronic Revolutions occurred. Two different periods each marked by a specific invention; the *First Electronic Revolution* marked by the invention of the vacuum tube, the *Second Electronic Revolution* marked by the invention of the transistor and the integrated circuit (Figure 308). Both revolutions started with the invention of their electronic engine (ic the radio tube and the transistor). These components followed a range of development trajectories and were used in a multitude of application trajectories. The vacuum tube made the radio and television receiver possible, and enabled the first general purpose vacuum-tube based computers. The transistor became both a replacement for the vacuum tube as well as a device enabling new applications of its own. The more when the single transistor became incorporated in integrated circuits made with the semiconductor technologies.

The consequences of this 'technology push' were massive. Had the electric light and the electric motor influenced both private life (lightening up from the darkness and powering the household machines) as well as business life (lightening the factories and powering the industrial machines), now the human intelligence was complemented with tools (eg by the calculating and computing machines) and the Era of Information got a new dimension (by radio and television broadcasting). News and entertainment were brought into the private home, shaping lifestyle and cultures.

The mechanical computing machines were to embark on a development of their own when electronic logic circuits were applied; the vacuum tube creating the mainframe for 'scientific computing,' and the transistor the minicomputer for 'business computing.' A development that continued even more when the Integrated Circuit technology made the single-chip microprocessor possible. A device that would enable micro-computing systems like the micro-controllers and micro-computers.

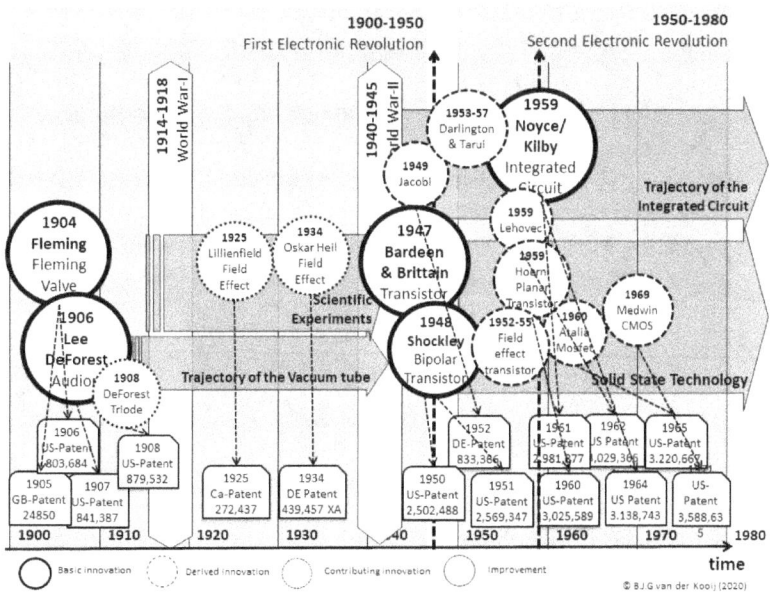

Figure 308: Overview Clusters of Innovation and their Basic Innovations in both the Electronic Revolutions.

That micro-processor laid the foundation for two development trajectories that each got an evolution of its own. It developed into the technical trajectory of the *micro-controller*. A device that became the controlling part of larger systems like tooling and machinery. And it developed into the technical trajectory of the *micro-computer* that would enable 'personal computing.' The standalone computing machines that were became the personal tools in a range of general applications (eg word processing) and dedicated applications (eg design applications). Both were the computing machines that would become the workhorse of the application trajectories they opened up; from the robotics to the personal computer.

It were the basic devices of the Transistor, Integrated Circuit and Microprocessor —shaping the metaphor of the Silicon Engine— that lay at the foundations of the revolutionary development in the second half of the twentieth century. These GPE's would become the backbone of the GPT-ICT that would herald the Information Age and the Information Machines to come.[496]

## Changing perspective

In the preceding exploration we focussed on the major events in the technological trajectory of vacuum tube and semiconductor technologies. The first trajectory with the events that culminated in the vacuum tube after the 1910s, the second with events that occurred up to the 1960s and culminated in the invention of the Microcomputer on a Chip. Each of the technological development trajectories had its own application trajectories. And in order to understand their impact on the Affairs of Man, changing our perspective, we now will focus on other domains of interest. We will go and explore the events that created other major application trajectories of the electronic technologies; the machines that would arise in the application field of radio, television and mobile telephony.

---

[496] This development will be explored in the companion case study.

# The Radio Revolution

The *Era of Telegraphic Communication* (ie the electric transportation of the coded word over distance), followed by the *Era of Telephonic Communication* (ie the transportation of the spoken word over distance) had caused a revolution in communication over distance by means of electric currents in a copper cable. Next the *Era of Wireless Communication* had seen the cabled infrastructure dwindle away and to be replaced by the electro-magnetic waves travelling the ether (Figure 106). Electro-magnetic waves that transported information in a revolutionary way that baffled both young and old (Figure 309).

**Figure 309:** *Grandpa Listening in on Wireless* **by Norman Rockwell (1920).**

Source: https://www.norman-rockwell-france.com/ Courtesy Norman Rockwell Estate, www.img.com (Mike Mueller, July, 2018).

Parallel to the development of the wireless communication system, the new electronic technologies of the 'Cathode Ray' tube emerged. The subsequent development trajectory of the 'radio tubes' not only resulted in the active components (ie the diode and triode vacuum tubes), but also in artifacts based on those components. Such as the device that became known as 'Computer' and that originated from the development trajectory of calculating machines into computing machines. And the device known as 'Radio receiver';[497] used to receive wireless broadcasted information (ie speech, sound). Followed not much later, as part of the development trajectory of the cathode ray tubes, by the devise called 'Television receiver.' And last but not least, the development trajectory of the 'Mobile Phone' that would emerge from wireless telephony. These artifacts 'radio,' 'television' and 'mobile phone' would become the driving forces of the *Era of Radio Communication*, the Era of *Television Communication* and of the *Era of Mobile Communication*.

## *The Second Communication Revolution (1900-1950)*

Together the Era of Radio Communication, the Era of Television Communication and the Era of Mobile Communication had a massive influence on the Affairs of Man. They had in common the transmission of the 'spoken word' by the auditive channel; more in general labelled as 'sound'. The sound that could be 'live' (ie the spoken word) or be 'recorded (ie on a gramophone record). And television added the transmission of the 'visual image' by the video channel. And that visual image could be a 'life picture' created by a camera, or a 'moving picture' (aka movie) that was recorded with other means (ie film- and video-technology).[498]

### *Communication Systems*

The Era of Communication in its different forms, had at its base the 'communication system.' In a similar way as the written word was recorded by the sender (eg on a paper enclosed in a letter) and transmitted (by mail

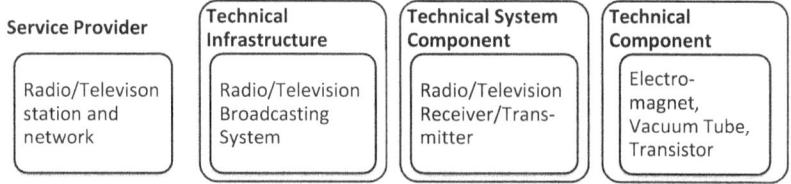

Figure 310: Elements of the System of Broadcasting Communication.

---

[497] The word 'radio' originates from the Latin 'radius', meaning 'ray' or 'beam'. However, it is used with different meanings. It can used to indicate the radio-spectrum (ie the frequencies used to transmit radio-signals), or the broadcasting of radio-programs by service-providers (aka radio stations). And it can be used to identify the artifacts (eg the radio receiver). Here the word 'radio' is mostly used for devices known as the vacuum tube-based radio receiver as it became available for the consumer markets.
[498] Film-technology involves recording on film in a chemical process (ie a strip of celluloid) and playing back of moving pictures. Video-technology involves the recording and playing back of moving pictures and sound with electronic means.

courier) to be read by the receiver, also other means of communication[499] always had a triple configuration: the sending part, the transmitting part and the receiving part. So, we will start our exploring by describing the elements of our perspective and by defining some basic elements of communication systems.

Technical based communication — in the sense of the transfer of information— is supported by a communication system.[500] The information to be shared has to be generated. In biological human systems it originates in the human mind and emerges as speech created by the language capabilities of the brain and the mechanics of the human vocal system. In technical systems it can be the information captured as sound (ie speech, music), or as a picture created by technical means (ie a camera). Information contained in a single picture (ie a photograph) or stored as a moving picture (ie a cinematograph).

So, 'Radio' and 'Television' are technical systems. Similar to the systems 'Telegraph' and 'Telephone' with a) their communication infrastructure, b) the telegraph and telephone machines, and c) their components like the electromagnet, microphone and loudspeaker. A technical system that enabled the service providers of radio and television programs (Figure 310).

Both the *radio receiver* —a device that received radio-signals and converted them in an auditable form—, and the *television receiver* —a device that received radio- and video signals and converted them in an auditable/visible form— were part of a technical system (Figure 310). The

**Point-to-Point Communication**

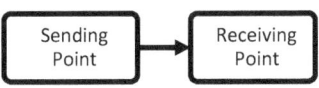

**Point-to-Multipoint Communication**

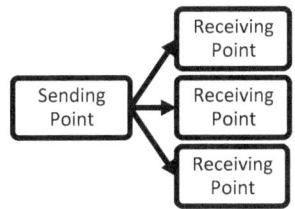

**Multi Point-to-Point Communication**

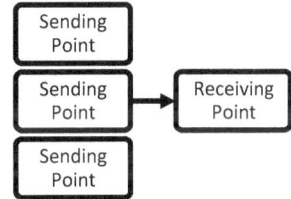

**Figure 311: Basic structure of Communication Systems.**

---

[499] The word Communication comes from the Latin 'communicare' meaning 'to share.'
[500] In biologic systems like humans, it is facilitated by the vocal cords speaking (sending) and ears (receiving). That system can facilitate one-way communication between two parties, or two-way communication in which both parties can send and receive information.

*radio broadcasting system* where the technical radio infrastructure (ie recording studio, radio transmitter equipment, the radio spectrum of frequency bands, and the radio receiver) was complemented by the service-providers (the radio stations) that broadcasted their content (the radio programs, news programs). And the comparable *television broadcasting system* where the television infrastructure (ie recording studio, television transmitter and receiver) was complemented by the service-providers (the television stations) that broadcasted their content (the television programs, news programs). In their totality they created the Broadcasting System.

Radio (ie wireless) communication basically is communication between a sending party and a receiving party by means of electro-magnetic waves. In telegraphy and telephony in the form of *point-to point* communication (Figure 311, top). Another part of the total radio communication structure is the mono-directional *point-to-multipoint* communications called 'broadcasting' (Figure 311, middle).[501] And at the receiving end the receiving part selects one of the sending parties (Figure 311, bottom).

## Wireless Communication

The wireless transmission takes place along electro-magnetic waves in the *radio spectrum* of frequencies (aka the ether) from 30 Hz to 300 Ghz. To prevent interference between different sending parties, the generation and transmission of radio is arranged in frequency bands (eg HF = High frequency band, UHF = Ultra High Frequency Band) (Figure 312).

In the early twentieth century the radio spectrum was divided in shortwave bands (SF), medium wave (MF) bands and a long wave (LFW) bands. With the technological progress, and the increasing use of radio-signals, over time the higher frequencies (eg VHF, UHF) became used.

For information (voice, music, television) to be transmitted, it must be attached to a radio-frequency carrier wave, which is then transmitted in a given frequency channel. So, each radio station transmits at a specific frequency (aka radio channel) in such a band, and the receiving party selects the frequency (ie the radio station) he wants to listen to by 'tuning'[502] the circuit (Figure 313).

---

[501] The notion Broadcasting denotes (a) as a verb, to transmit programs or signals intended to be received by the public through radio, television, or similar means, and (b) as a noun, the radio, television, or other program received by the public through the transmission.
[502] Tuning may be accomplished by varying the capacitance, which consists of interleaved metal plates separated by air spaces with one set of plates movable. Another method of tuning involves varying the inductance by insertion or withdrawal of an iron dust or ferrite core in a cylindrical coil of copper wire.

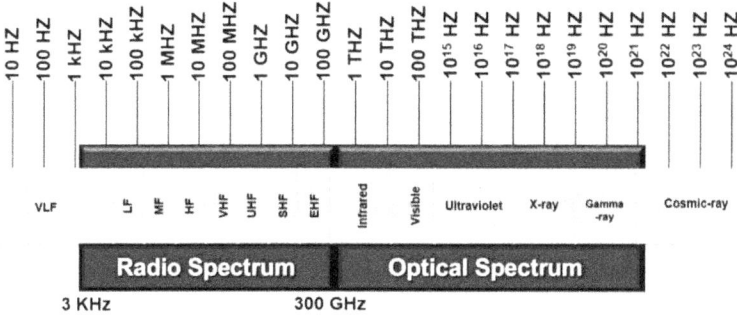

Figure 312: The Radio Spectrum and Optical Spectrum of Electro-magnetic waves.

In the case of television broadcasting, there is next to the sound (aka audio signal registered by a microphone), the picture (aka video signal registered by the camera). Both signals are converted into the transmittable frequency by the 'Transmitter' (Figure 314, top). At the receiving end the 'Television Receiver' tunes to a specific frequency (aka tv-channel) and converts the combined signal back to the picture (ie by the screen) and the sound (ie by the loudspeaker) (Figure 314, bottom).

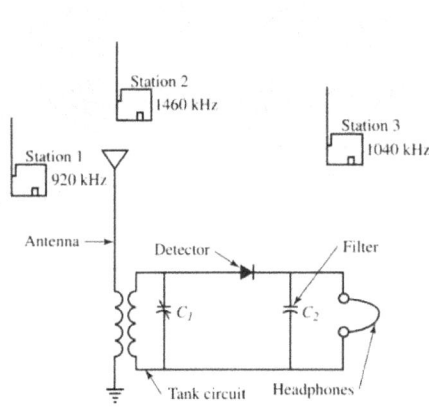

Figure 313: Tuning in Broadcasting

The Receiving Party selects the frequency of the Sending Party by tuning the tank circuit (ie LC circuit).

These are some of the basic configurations of Communication Systems. They realize their specific function; from the one-way and two-way direct communication systems (ie wireless telephony), to the broadly distributed multi-receiver system (ie radio and television broadcasting). Systems that are build-up of system components (eg sending equipment and receiving equipment), that in their turn are consisting out of components (eg the radio tube). Communication systems that were the result of many curious, inventive and creative minds that contributed to its developments.[503]

The nuclei of the technical system-components are our General Purpose Engines (GPE) like the vacuum tube and transistor. Each with its own technical development trajectory, and the application trajectories of the machines that use them. We will go and explore those application trajectories (Scheme 2) where the General Purpose Engines of Electronics became implemented into the radio and television each causing a revolution of their own.

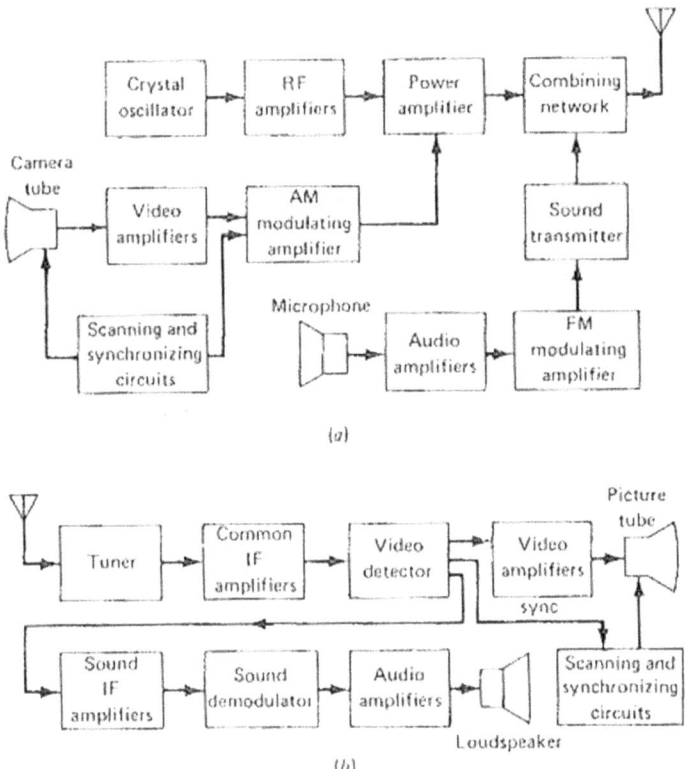

**Figure 314: Block diagram of the Transmitter and Receiver of Television Broadcasting.**
The system of transmitting (a, top) and receiving (b, bottom) television broadcasts.

---

[503] When one keeps the structure of Communication Systems in mind, the work of the contributors to radio becomes more transparent. Some experimented with the transmission signal, others worked on sending equipment, or worked on the receiving equipment. And some focussed on all these parts creating the communication systems.

## The Spectrum Problem: Chaos in the Air

Radio Waves were part of the radio spectrum that became divided in frequency bands (Figure 312). These different parts of the spectrum have different characteristics. With the evolution of wireless radio —that followed the evolution of wireless telegraphy— that spectrum became crowded by a multitude of experimental transmitting stations each using a frequency in the radio spectrum. It resulted in a fierce competition for air space. The result was 'chaos in the air' with signal interference[504] and the distortion of the radio signals. Soon interference was recognised as one of the biggest problems of broadcasting. The diverse and complex problems resulting from the fast-growing number of broadcasting stations needed to be solved. So, initiatives were undertaken to regulate the radio spectrum.

In Britain with its traditional dominance of the General Post Office (GPO) as regulator of wireless telegraphy, the Wireless Telegraphy Act of 1904, gave regulatory power to the GPO. They used that to create a monopoly for the British Broadcasting Corporation (1922/1927).

In the chaos of afterwar Germany, the government took control of the sprawl of amateur broadcasting. The Deutsche Reichspost created, next to the ban of reception of radio broadcasts for private individuals, a licensing system for radio broadcasting by 1923. It regulated the regional broadcasting initiatives.

In the US, the Radio Act of 1927 —successor of the Radio Act of 1912 regulating wireless telegraphy— increased the government's regulatory powers over radio communication, and adopted a standard that radio stations had to be shown to be "in the public interest, convenience, or necessity".

These are just a few of the regulatory efforts of governments that took place to solve the spectrum problem of early radio broadcasting. Later in time —with the advent of television broadcasting and mobile telephony— the problem would arise again.

## *Radio Broadcasting 0.0. (1900-1920)*

The notion 'wireless radio' —aka 'Radio'— was originally referring to military and marine applications that emerged after Marconi's experimenting early twentieth century. The technology —aka 'Radio at Sea'— was widely adopted in the maritime sector. By 1912, some 327 Coastal Radio Stations (CRS) and 1,924 Ship Radio Stations (SRS) for wireless telegraphy had been established. The distress signal SOS send by

---

[504] Interference is the undesired interaction of emissions from different sources.

sinking steamer Titanic —that collided on her maiden trip on an iceberg in the North Atlantic on the morning of April 15th, 1912— emphasized the importance of wireless communication.[505]

Next to wireless telegraphy transmitting the coded word, wireless telephony transmitting the spoken word appeared on scene and became a sensation. Wireless Radio captured the imagination of thousands of ordinary persons who wanted to experiment with this amazing new technology. The number of amateur radio enthusiasts in US started to expand, especially in the industrial northeast. Next to the verbal communication, they started to experiment with broadcasting music. Soon the pre-war Radio Craze ignited the after-war Radio Boom all over the western world. A development fuelled by commercial radio broadcasting.[506]

Already in the 1910s the first receivers of radio-signals were built. These *crystal radios* were the first widely used type of radio-signal receivers. Sold and homemade by the millions, the inexpensive crystal radio was a major driving force in the introduction of radio to the public, contributing to the development of radio as an entertainment medium with the beginning of radio broadcasting around 1920. At that time, the rather limited embryonic crystal receiving sets were superseded by the first amplifying receivers which used vacuum tubes.

## The Birth of Amateur Radio

In its essence, wireless transmitters and receivers were not that complicated to build (Figure 315, top). True, the wireless technology as a whole was in its infancy at the start of the twentieth century, but soon the curiosity and ingenuity of many creative people resulted in the construction of amateur-made devices: the 'ham radio' was born (Figure 315, bottom). Often these pioneering amateurs —in modern parlance called 'nerds'— that already had been experimenting with private telegraph lines, were now starting to experiment with the newly discovered electric waves. Educated by the hobby-magazines, influenced by the publications in the press, some would become professionally involved in the world of wireless

---

[505] This chapter is based on information from many sources in the public domain. Much data (without citing them, sorry) are obtained from websites by dedicated individuals like (radio): antiqueradio.org; philcoradio,com; xroads.virginia.edu; radiomuseum.org; and (television): tvhistory.tv; earlytelevison.org; historymuseum.ca, etc.

[506] One of the great attractions to the radio listener was that once the cost of the original equipment was covered, radio was free. Stations depended on contributions free of charge, or made money by selling air time to advertisers. The possibility of reaching millions of listeners at once had advertising executives scrambling to take advantage. By the end of the decade advertisers paid over $10,000 for an hour of premium time.

communication, not only as wireless operators but also in the early manufacturing of the early systems.

It was not too long before thousands of radio amateurs were active, especially in the industrial northeast part of America, though they were soon followed by amateurs in California. They organized in amateur radio clubs. In 1908, students at Columbia University formed the Wireless Telegraph Club of Columbia University. The Junior Wireless Club of New York City formed in 1909 and the Wireless Association of America with some 3,000 members.

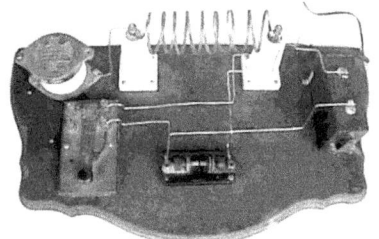

**Figure 315: Homemade amateur Radio Devices.**

Radio Spark Transmitter (top, Date unknown) and Radio Receiver (bottom, 1917).

Source: http://w1tp.com/imperad.htm, https://www.invaluable.com/

The ham radio-amateurs were found in countries with a cable-telegraph tradition and the early marine wireless tradition such as the US and Great Britain, but also on the continent Australia.

> *The large body of radio amateurs in the USA, augmented by those in Europe, resulted in a co-operative movement that benefited by a cross-fertilisation of ideas and thoughts motivated to exploit the wavelengths below 200 m. From the apparently random experiences of the first experimenters, a more definite picture was emerging. Amateurs soon learnt that certain wavelengths, at certain times of the day and night, could give a predictable quality of service. Generally, it seemed that the lower the wavelength, the greater the distance over which communications became possible.* (Wood, 1992, p. 22)

In the United States in 1913 there were 322 licensed amateur radio operators (aka 'hams') who would ultimately be relegated to the seemingly barren wasteland of the radio spectrum called Short Wave. By 1917 there were 13,581 amateur radio operators, illustrating how building a radio receiver had become a rage. The typical builder was a boy or young man. Many older people thought that all radio would ever be was a whim and certainly so long as the public had to build its own radios, put up with poor reception, and listen to dots and dashes. But after a few experimental broadcasts of music and speech over earphones, relatively more people were going to be interested in having a 'radio.'

Sharing knowhow in 'radio'-clubs and through hobby magazines, the hams did build their own radio sets, some even starting business activities and selling complete radio-systems. The rapid expansion of amateur radio during the 'Ham Radio Mania' (Figure 316), with many thousands of transmitter/receiver units set up by 1910 for *point-to-point communication*, led to a wide spread problem of inadvertent and even malicious radio interference with commercial and military radio systems. Notwithstanding the ban on ham radio during the First World War, the amateur radio picked up rapidly again after the war. And, not surprisingly in the entrepreneurial climate of the US, many of the amateurs went into the radio business.

**Figure 316: Radio Amateur and his Ham Radio (1923).**

Source: Ballantine, S.: Radio Telephony for Amateurs

Such as the company *Adams-Morgan Co.* (AMCO), the country's foremost maker of ham receivers, established by Alfred Powell Morgan partnering with Adams as silent partner in 1909. They build complete radio receiver sets under the brand names Paragon and Paradyne as well as components and sold by mail-order (Figure 317).

As radio communication had experienced great expansion, efforts to form a technical organization of wireless practitioners in 1908–1912, resulted in the *Institute of Radio Engineers* that was founded in 1912. To serve the needs of growing professional community, they started professional journals,

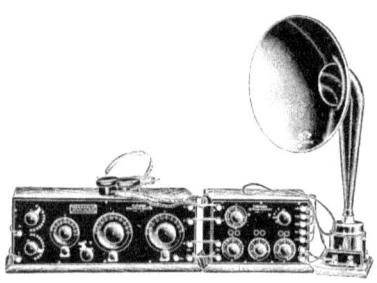

**Figure 317: Early Ham Radio Set.**

Drawing of Adams-Morgan Paragon receiver and, amplifier (left) and Magnavox loudspeaker (right, 1922),

Source: Wikimedia Commons.
https://antiqueradios.com/superhet/

organized regional and professional groups for cooperation and exchange between members. This was also the beginning of a range of radio institutions like the Federal Radio Commission (FRC), and the Radio Manufacturers Association (RMA) created in 1924 by fifty radio manufacturers in Chicago.

Also in Europe, the sparks of the US Radio Craze had ignited interest, and soon lead to early ham radio and broadcasting activity. Next to the amateurs making their own ham radios, some early manufacturing activity popped up (eg in Britain and Germany). Although —after the First World War had silenced much amateur radio activity in Europe— the governmental attitude of restriction remained active after the war, some early pioneering work was done.

In the Netherlands, the *Nederlandse Radio Industrie* (Dutch Radio Industry) was founded in 1914 by the engineer Hanso Idzerda (1885-1944) in the neutral Netherlands. Idzerda experimented with wireless, designed and made his transmitting and receiving devices (Figure 318) while the Philips company manufactured for him the vacuum tube (aka IDZ-vacuum triode). On November 6$^{th}$, 1919 he held the first public airing of a radio programme he called 'Soireé Musicale' that could be received over a distance of 500 km up to England. His station PCGG continued to broadcast music and radio plays till 1924 when he, unable to raise revenues from selling equipment, bankrupted. Next, the *Nederlandsche Seintoestellen Fabriek* (NSF) took over the role as public broadcaster for the Netherlands.

In Germany, where losing the First World War had created specific economic circumstances, also ham radio amateurs became active building their radio sets and organizing in clubs. To operate a radio for private use however, they needed a permit from the 'Deutsche Reichspost' (Imperial Mail) that had extended its postal regulating control from the cabled telegraph systems to wireless systems. The employees of the German Reichspost themselves were also exploring new radio technologies in their laboratory. And on December 22$^{th}$, 1920, the

**Figure 318: Dutch Radio Broadcasting (1919-1924).**

The transmitter (1919) used for the radio station PCGG.

Source: https://www.beeldengeluid.nl, https://www.nvhr.nl/

first radio broadcast in Germany hit the airwaves. "Attention, attention — this is Königs Wusterhausen on radio wave 2700" announcing a Christmas concert. It took a while, but the first official radio entertainment program in Germany, organized by the record company Vox, was broadcasted on October 29th, 1923 to 467 listeners. One year later, there were already one million listeners within the Reich's entire territory. In 1925, a central *Reichs-Rundfunk-Gesellschaft* (Reich Broadcasting Corporation), came into being, merging nine regional broadcasters. And in 1932, there were more than four million paying radio subscribers, and at least as many non-paying listeners. By then, radio had developed into state broadcasting.

As the Allies had lifted the after-war ban on listening to radio waves, 'entertainment broadcasting' became permitted. However, ham operators needed approval for the construction and operation of a radio reception system for private use (eg the Audion Permit). This restrictive policy from the Reichs Post Ministerium stimulated amateurs to organize themselves. By June 1925 the German Radio Technology Association (DFTV) was founded to represent the interest of the radio fans. On July 18th, 1925, radio amateurs from all over the world came together in Paris to set up the *International Amateur Radio Union* (IARU). But the attitude of the Reich Ministry of Post stayed restrictive. Radio amateurs were forbidden from connecting the transmitters to antennas, for example, by decrees of the Reich Postal Minister of 1930 and 1931, which made amateur broadcasting almost impossible.

## The Dawn of Military Radio Communication

Notwithstanding Marconi's success in long distance wireless transmission and marine wireless transmission, at the beginning of the First World War, wireless military communication was in its embryonic state. As the war progressed there was a growing appreciation for improved electrical communications of much greater capacity for the larger military units and within regiments for electrical communications. A need which had heretofore been regarded as unessential and impractical as carrier-pigeons and message dogs did a good job. It had resulted in cumbersome and vulnerable cabled telegraph and telephone infrastructures during the Trench War of 1914-1918. But it gave the impetus to all the belligerents to develop wireless as an alternate means of communication on the battlefields.

During the First World War, early wireless communication emerged in the Army, Navy and Airforce. However, their use was limited. Wireless telegraph communication was only employed extensively by the navies of the world and had a major influence on the character of naval warfare. High-powered shore and ship stations made wireless communication over long distances possible. A business that was controlled by the British, and neither the Americans nor the Germans liked that.

When the United States entered the first World War in April 1917, the government took control of most civilian radio stations in order to use them for the war effort. The US Navy, aiming at retaining a monopoly after the war, contrary to instructions it had received, began purchasing large numbers of radio stations. Among which Marconi shore-stations. When the war ended, Congress rejected the Navy's efforts to have peacetime control of the radio industry and instructed that the Navy return the stations it had taken control of to the original owners. However, as the British dominance in wireless communication raised national security issues, the Navy looked for an alternative that would result in an "all-American" company taking over the American Marconi assets. They found an alley in General Electric and on November 20th, 1919 transformed the American Marconi activities into the *Radio Corporation of America*. And with the activities came all the patents rights that were the base of Marconi's monopoly. The decision to form the new company was promoted as a patriotic gesture. The corporate officers were required to be US citizens and a majority of the company stock needed to be held by US citizens.[507]

At the onset of First World War, wireless radio communication was still in its infancy. The early mobile radio equipment was primitive, had a very short range, and often negotiated atmospheric interference. Military radio equipment also used early vacuum radio tubes, which were heavy and bulky. As a result, the equipment was difficult to tote around on the battlefield, even on mules and horses, which were still the military's primary mode of equipment transportation. But radio development progressed, soon man-held portable sets and airplane sets were developed. By 1916 all three British armed services became depending upon wireless. In the last two years of the war, the British Army invested more heavily in wireless, and by mid-1917, they viewed wireless as a valid means of communication. Although the British War Office initially rejected investment in airborne wireless, by 1917, airborne valve-based wireless played a vital role in

---

[507] RCA's dominance as holder of some 4,000 patents lasted a decade, as in 1929 it was decided that the conglomeration of these companies was a monopoly, and they were forced to disband. The U.S. Department of Justice brought antitrust charges against the three companies. An agreement was reached by 1933 between RCA and on one side and Westinghouse and General Electric on the other side. RCA continued as a separate organisation possessing manufacturing facilities, Westinghouse and General Electric were to divest of their RCA stock, and licensing agreements were modified. In addition, the agreement covered also financial issues, the transfer of assets, and broadcasting agreements.

spotting artillery and enabling ground-to-air communications. It influenced the warfarin considerably.

When the First World War had ended, and the soldiers went back to civilian life, many of the wireless telegraph operators became radio-amateurs. The organised in radio societies and hobby clubs, some already started before the war. The interest in wireless transmission spawned a large number of magazines and books giving practical details of how wire-less equipment could be built (Figure 319).

These are just a few impressions of the early days of radio within the dynamic social context around the First World War. That changed after the war with technical developments that saw the amateur ham radio in the 1920s expanding into professional activity, simple crystal-based receivers into the more elaborate valve-based *Tuned Radio Frequency* (TRF) systems, and experimental broadcasts into state/public and private organized broadcasting services. Times in which the radio amateur contributed to further development of radio broadcasting that would change the Affairs of Man drastically.

Figure 319: Amateur Radio Magazines of the 1920s.
Source: Wikimedia Commons.

And at the foundation of that elaborate system of organizations —from equipment manufacturers to service providers called radio stations— were a range of technical innovations that were related to 'radio circuitry'; combinations of interconnected active and passive electronic components that realize a specific function.

## The Conception of the Radio Receiver Circuit

Like Fessenden contributed to the signal transmission systems for telephony signals with the heterodyne principle, Lee DeForest contributed to the active electronic component with the vacuum tube (aka radio tubes). Next to Fessenden's and DeForest's contribution, it had taken the contribution of another creative mind, Edwin H. Armstrong, to discover that that the vacuum tube had much more potential: *'The action of the Audion as a detector of radio frequency oscillations is very different from its action as a simple amplifier.* [508]

> This extended property of the Audion, when configured in a proper circuit with positive feedback (ie the regenerative property), made it possible to use it both as a transmitter and as a receiver of a continuous wave. It was the result of the evolution in the reception of radio signals with help of vacuum tubes that had started with Fleming's diode (Figure 320).

Armstrong's discovery and his subsequent experimenting resulted in four principal inventions.

(I) His work started with the receiving station of a wireless circuit as he developed the *Regenerative detector circuit:* the combination of an amplifier circuit with a positive feedback circuit that improved the gain of an amplifying circuit.

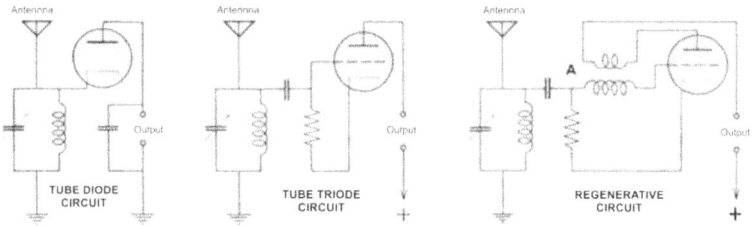

**Figure 320: Wiring diagrams of Receiver Circuits illustrating the evolution into the Regenerative Circuit.**

The Tube Diode circuit (left) is similar to the crystal diode and has no amplification. The Tube Triode circuit (middle) amplifies some 100 times, and the Regenerative Circuit (right) with the feedback coil (A), amplifies up to 1,000 times.

Source: Lesing, L.P.: The Late Edwin H. Armstrong. Scientific American. Vol. 190, No. 4 (April 1954), pp. 64-69

---

[508] Armstrong, E.A.: 'Some Recent Developments in the Audion Receiver: The Audion as Detector and Amplifier. Proceedings of the Institute of Radio Engineers', September, 1915, pp. 215–238: Source: http://earlyradiohistory.us/1915reg.htm

On October 6th, 1914, US-Patent № 1,113,149 was issued for his invention of the Regenerative Circuit (Figure 321). But two years later, Lee DeForest challenged Armstrong's patent rights in court. It would become the beginning of Armstrong's tragic fight for his patent rights, which continued the rest of his life.

(II) Next he developed the *Superheterodyne Circuit* (1918): the use of frequency mixing to convert and amplify a received short wave signal. In December of 1919, he presented his circuit

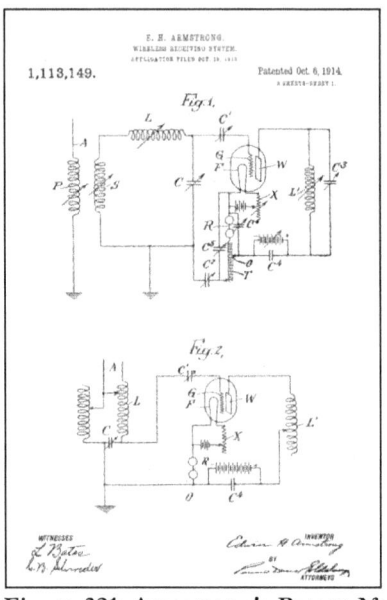

**Figure 321: Armstrong's Patent № 1,113,149 (1914).**

Source: USPTO.

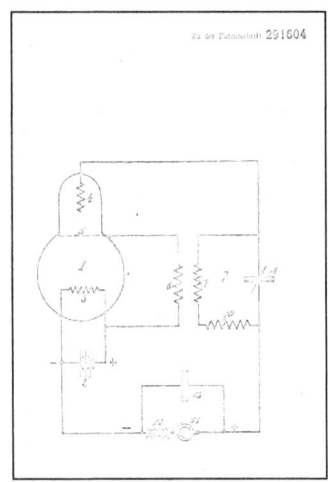

**Figure 322: Meissner DE Patent № 291,604 (1919).**

Source: Espacenet

in his paper, "A New System of Short Wave Amplification," to the Institute of Radio Engineers. His invention brought him fame, but he also became involved in Patent litigation and the U.S. Patent Office eventually assigned primary credit for the idea to a French engineer, Lucian Lévy.[509] For his invention he was granted several patents after the war (Figure 328). Among them US-Patent № 1,342,885 (filed February 8th, 1919) for a "Method of receiving high frequency Oscillations" that was granted on June 8th, 1920.[510]

---

[509] Lucian Lévy had filed on March 29th, 1920 patent GB 160,799 for an improved electric oscillation generator.

[510] Another application of the regenerative circuit is the creation of a single frequency when the circuit starts oscillating. This would become the base for the electronic oscillator which became the basis of radio transmission by 1920.

(III) In 1922, he developed the *Super-Regenerative Circuit* that used —next to the regenerative circuit— an extra oscillator. However, the circuitry was rather complex and this limited a broad adoption. He was granted US-Patent № 1,424,065 (filed June 27th, 1921) on July 25th, 1922, and which he presented on June 7th, 1922 before the Institute of Radio Engineers.

(IV) And in 1933 he realized his contribution to the transmission system of *Frequency Modulation* (FM) that was to complement Amplitude Modulation (AM). He was granted four patents among which US-Patent № 1,941,447 on December 26th, 1933 for this invention. His contribution would become the start of FM-radio.

*A Fury of Heterodyne Circuits*

The invention of the triode vacuum tube made it possible to apply 'amplification' in circuits (Figure 320). The principle of feedback in the heterodyne circuits made its amplification increase. And an important contribution of the heterodyne circuits was the creation of single frequency electro-magnetic wave by oscillation.

It was the Austrian engineer Alexander Meissner (1883-1958), a physicist from the Technical University of Vienna busy improving the design of antennas for transmitting at long wavelengths, who devised new vacuum-tube circuits and amplification systems, and developed—independently from Armstrong—the heterodyne principle for radio transmission. In 1907 he worked for Telefunken in Berlin where he invented the 'Meissner oscillator'; a high frequency circuit that became used in vacuum tube transmitters, to create a single frequency wave that could be used as a signal carrier (Figure 326). This oscillator transmission paved the way for the development of the short-wave radio band. Meissner applied on April 10th, 1913 for a patent, that was granted as DE-Patent № 291,604 on June 23th, 1919 (Figure 322).[511]

In England the engineer Henry. J. Round (1881-1966), joined the British Marconi Company in 1902, and worked on improving radio tubes. In 1913 and the following year, Round patented a number of ideas for valve improvements including that of an indirectly heated

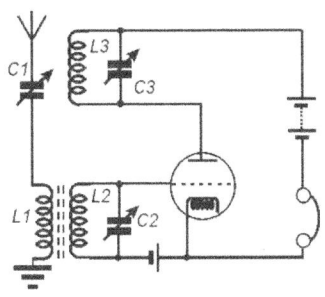

**Figure 323: Round's Autodyne Receiver circuit.**
Source: Wikimedia Commons.

---

[511] The patent does not mention Meissner but was issued to the Gesellschaft für Drahtlose Telegraphie GmbH (aka Telefunken) where he worked.

cathode. This was a major step forward and it paved the way for enabling valves to be used far more widely. Also during this time, he patented his auto-heterodyne (autodyne) receiver: the receiver in which the same valve was used as a mixer and an oscillator (Figure 323).

In total Round was granted some 117 patents. Among which several patents for his improvements in vacuum tubes: GB-Patent 1914 13,247 (filed May 29th, 1913) granted on May 27th, 1915; US-Patent № 1,649,489 (filed July 9th, 1921) granted on November 15th, 1927. And he also received patents for an improved receiver for wireless telegraphy. Such as the GB-Patent № 1913 27,480 granted November 28th, 1913 and GB-Patent № 1913 28,413 (filed December 9th, 1913) granted December 6th, 1914. And the US-Patent № 1,472,092 (filed July 9th, 1921) that was granted October 30th, 1923 (equivalent to GB № 1913 27,480) (Figure 324).

In France, the radio engineer *Lucien Lévy* (1892-1965) was working during the First World War in the Eiffel Tower Military Radio Telegraphy laboratory on telegraph emissions from the Eiffel Tower. As a result of his experimenting Lévy filed a patent application for the superheterodyne principle in August 1917 and was granted French

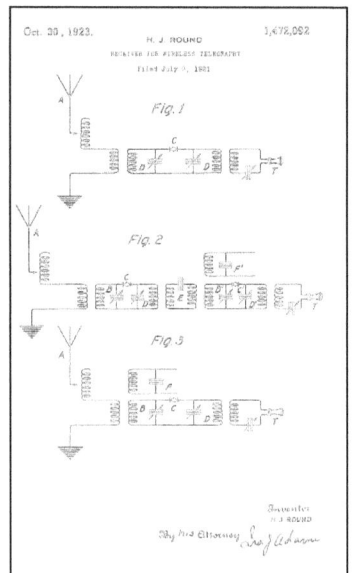
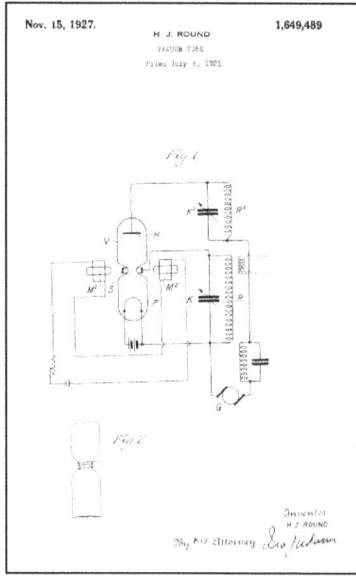

**Figure 324: Round's Patents for Tube and Circuitry.**
US-Patent № 1,472,092 (1923) and US-Patent № 1,649,489 (1927).
Source: USPTO.

Brevet № 493,660 on August 4th, 1917, and in 1918 an upgraded version of the principle followed, described in a second patent. Later, after the conflicting US-Patent issued to Armstrong, Levy was awarded US-Patent № 1,734,938 on November 5th, 1929, with a priority date of August 4th, 1917 (so he had filed 6 months prior to Armstrong).

After the war, in 1920, Levy founded the Etablisssements Radio LL, specializing in construction of radio receivers. Radio LL made the first tube receivers, and in 1922 it produced a receiver with high-frequency amplification with circuits tuned by adjustable iron cores. In 1923 he built his first portable transmitter. In 1924, Radio LL produced the first mass produced superheterodyne receiver in Europe. And in 1926 he launched the broadcasting station Radio LL.

Between 1912 and 1930 some 300 patent were granted in Great Britain, Germany, France, Switzerland and the US to improvements in circuits producing continuous electrical oscillations; creating a fury of the heterodyne circuits (Figure 327).

The contributions of the experimenter with heterodyne circuits would create the foundations of wireless radio; both its transmission and its reception. Together with Armstrong's technique of mixing information signals and carrier signals, it became possible to modulate the wireless

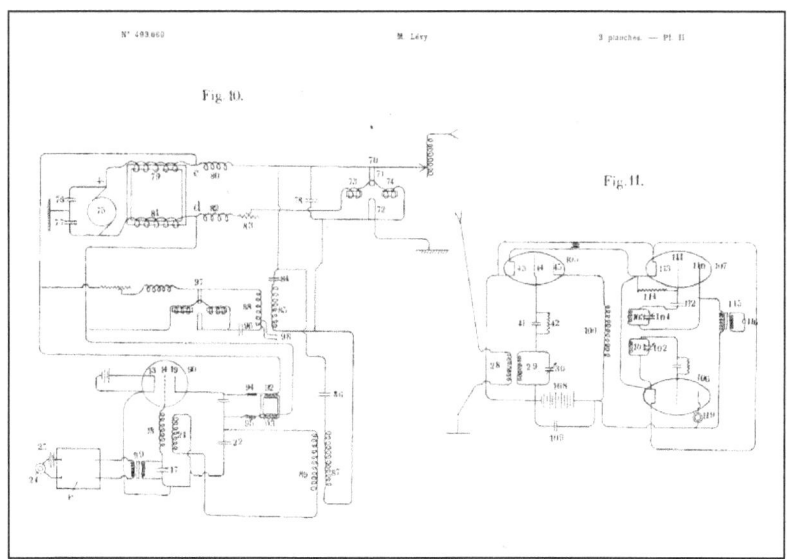

**Figure 325: Levy's French Patent 493,660 (1917).**
Source: Espacenet

*transmission frequency* with another frequency containing information like music and speech. Amplitude modulation (AM) was born (Figure 326).

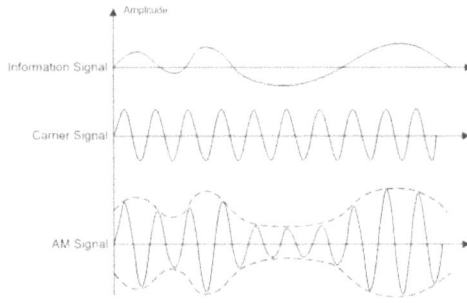

> *The first acknowledged radio transmission of human voice was indeed made in 1915, but was done by the Western Electric Company, a subsidiary of AT&T, from the US Navy Station at Arlington, Virginia to the Eiffel Tower in Paris.* (Raboy, 2016, p. 383)

**Figure 326: The Principle of Amplitude Modulation (AM).**

The transmission frequency of the carrier signal is combined with the information signal (eg the audio signal) to create the modulated AM Signal.

That was the birth of the new form of wireless telephony that made the transmission of speech and sound possible. Except for some

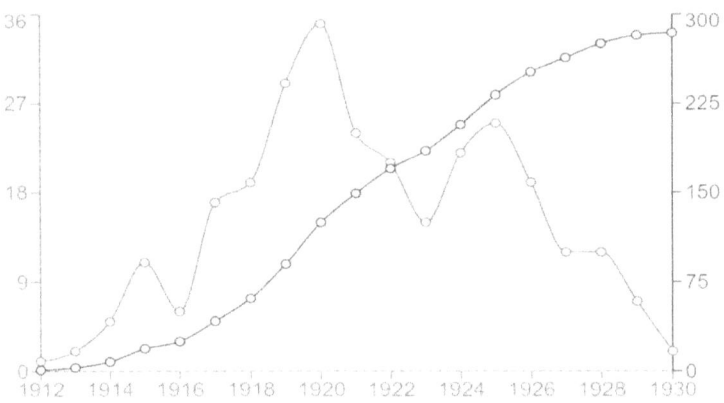

**Figure 327: Overview Patents granted for Heterodyne Circuits.**

Tables show yearly and accumulated number of patents. Left axis is the number of patents in a given year, Right axis the accumulated volume of patents.

Query: pd = "1914-1930" AND cl any "H03B5/10" Priority date 1912-01-01 to 1930-12-31.
Class: H03b5/10: Generation of oscillations using amplifier with regenerative feedback from output to input with frequency-determining element comprising lumped inductance and capacitance active element in amplifier being vacuum tube.

Source: Espacenet

experimenting, the transmission of music would follow later. Such as a broadcast in 1910 when Lee DeForest, aired experimental programs from New York's Metropolitan Opera House. But it was not until 1916, when a Westinghouse engineer named Frank Conrad played records for his friends over the air, that the idea of radio as a public medium took shape.

On the receiving end of the transmission, another development took place. After the simple detectors of early spark-gap telegraphy (ie the coherer), more advanced detectors had been developed (eg Fessenden, DeForest et al). These detectors became part of the receiver circuits — detecting and amplifying the often rather weak signal— that improved over time the separation of sound frequencies from carrier frequencies (eg the 'heterodyne circuit'). It was the birth of the 'radio receiver.' A development that got a boost with the 'superheterodyne circuit' using the vacuum triode/pentode as amplifying element.

*The Invention of the Superheterodyne Circuit*

During the war, Edwin Armstrong developed his superheterodyne circuit that would become fundamental to the development of radio (and television later in time). It would also become part of the later patent controversy around the first invention of the principle. Its development trajectory, however, was rather special.

When the United States entered the First World War, Armstrong joined the Army Signal Corps in charge of the Radio Group of the Research Section,[512] and was posted to France. On his way to France, stranded for three days when fog closed the Channel, Armstrong had taken the opportunity to visit London. Stopping at the Marconi Co. offices, he met Captain H.J. Round, for the war's duration in charge of a chain of wireless direction-finding stations for the Admiralty. Round, working on the vacuum triode tube with indirectly heated cathode, had also experimented with regeneration and developed his autodyne circuit (Figure 323).

Once in Paris by March 1918 he conceptualized the superheterodyne circuit, first as a means to locate the position of bombing aircraft, that were sending weak signals by the sparks of their motor ignition. By the time the prototype of the receiver was completed and tested, and it was ready for

---

[512] Armstrong's activities were part of the dawn of the military use of wireless communication. For example, the communication from recognisance aircrafts, communication from scouts behind enemy lines, or communications between headquarters and the troops would benefit enormously when cabled infrastructures could be eliminated. But as the communication machines realized with the technology of the vacuum tubes were still in the infancy, the involvement of the military in its development was an important stimulant; both for the technology itself (improving the radio tubes) as well as for its application (eg two-way radio communication systems).

trial at the front, it was the time of the signing of the armistice. It would take the thread of the next world war to revive the interest of the military again.

Armstrong filed for a range of patents for wireless receiving systems before and after the war. Such as US-Patent № 1,113,149 (filed on October 29th, 1913, granted October 6th, 1914). Based on his pre-war patents he was granted US-Patent № 1,342.885 on June 8th, 1920 (filed September 2nd, 1919) for his method of receiving high-frequency oscillations by means of a super-heterodyne receiver (Figure 328).

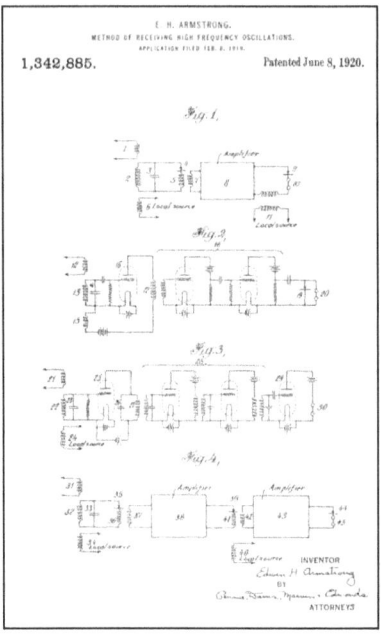

**Figure 328: Armstrong's US Patent № 1,342,885 (1920).**
Source: USPTO

With the war over, the development push of the military slacked. However, the 'radio' principle was established and realised in a prototype radio set (Figure 329). It was the start of radio receiving devices when the cash-poor Armstrong found his ally in the *Westinghouse Electric & Manufacturing Company* (WE), part of the RCA-conglomerate. Having become involved in radio during the war, and wishing to set up a world-wide communications business like the British-controlled American Marconi company, the Westinghouse company invested heavily in Fessenden's old company and its valuable patents. And in the same spirit,

**Figure 329: Experimental Radio Set.**
Prototype of Armstrong Receiver (1920).
Source: RadiolaGuy.com

on October 5th, 1920, Westinghouse took out an option of $335,000 for the commercial rights for both the regenerative and superheterodyne patents, with an additional $200,000 to be paid if Armstrong prevailed in the regenerative patent dispute with DeForest. He also received sixty thousand shares of RCA stock, which he sold just before the Wal Street Crash in 1929 for $114/share. It made him a millionaire.

**Figure 330: Early RCA Radio Sets.**

Westinghouse RC-radio set of the RA tuner and DA detector-amplifier (left, 1922), and RCA version of Radiola Grand Radio (right, 1923).

Source: Wikimedia Commons. https://Antiqueradios.com/superhet/

Westinghouse's entry in the radio business was checkmated by its rival *General Electric* (GE). GE had obtained, with the governments blessing, the Marconi patents from the American Marconi companies to set up the Radio Corporation of America (RCA). Westinghouse cross-licensed with GE/RCA in 1920, offering the Superheterodyne patent and the Regenerative Detector patent as their end of the agreement. In that business context, the newly formed Westinghouse Broadcasting Co. was to set up broadcasting stations and sell radio sets. And the mother company organized to manufacture the 'radios' (Figure 330, left). First by adapting the versions for the ham radio market, next for the consumer market of non-technical users. And the headset for individual use became replaced by the 'acoustic horn' that enabled more people to listing to the broadcast.

*The Invention of the Frequency Modulation (FM)*

The transmission of the radio signal was realized originally with the technique of Amplitude Modulation (AM) where the amplitude of the single frequency carrier signal was modulated with the audio signal (Figure 326). However, AM broadcasting is susceptible to distortion and harbours background noise making the sound quality poorer overall.

Among others, Edwin Armstrong tackled in the early 1930s the problem of AM-broadcasting and came up with a new solution that would become known as Frequency Modulation. With this technique the carrier frequency itself (not its amplitude) would be modulated with the audio signal, creating a FM-signal that moves up and down in frequency; the modulation is carried only as variations in frequency and the amplitude remains unchanged (Figure 331). The advent of FM was considered a very significant technical improvement because of immunity to electrical noise interference, as well as an improvement in the quality of the reproduced sound. The drawback was that the bandwidth that an FM station occupies is twenty times bigger than an AM station. And, thousands of AM transmitters and millions of AM radios could become obsolete.

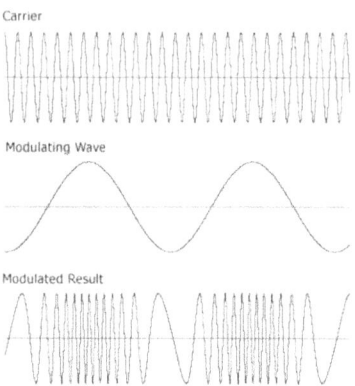

**Figure 331: The Principle of Frequency Modulation (FM).**

The transmission frequency of the carrier signal is combined with the information signal (eg the audio signal) to create the modulated FM Signal.

Armstrong filed on May 28th, 1927 his patent application for *"a new method of transmission in which the frequency of the transmitted wave (not its amplitude) is varied in accordance with the voice frequency to be transmitted and by a new method of reception in which only the frequency variations of the received wave arere-translated into voice frequencies, the effects of fading being thus prevented from appearing in the translated signals."*

He was granted US-Patent № 1,941,447 on December 26th, 1933 (Figure 332). Together with US-Patent № 1,941,068 and US-Patent № 1,941,069.

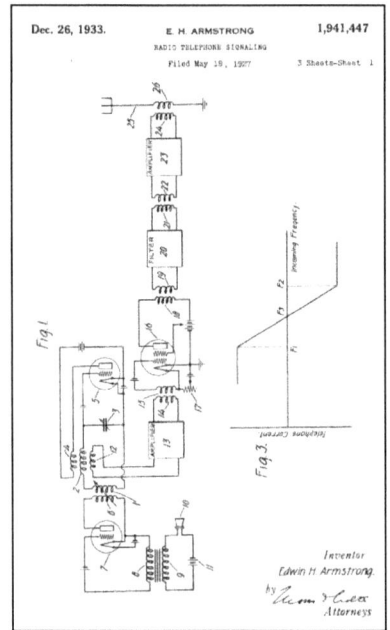

**Figure 332: Armstrong's US Patent № 1,941,447 (1933).**
Source: USPTO

On June 17th, 1936 Armstrong formally demonstrated his FM system to the FCC, and set up his FM station W2XMN in Alpine, New Jersey. Soon interest in FM soared and by the fall of 1939 the FCC had received more than 150 applications to build FM stations.

But not everybody was excited about the FM-development. RCA had at that point made vast investments in AM, both in the transmitting equipment and radio receivers. All of their transmitters and all of the millions of radios that they had sold used it, and the investment was not yet depreciated. Radios were now commodities sold entirely based on price, not quality. RCA told itself that consumers didn't care what the music sounded like from their radios, they just wanted to get it as cheaply as possible. So, they took some years evaluating the new technique of FM and then declined to license it. Subsequently Armstrong licensed his patents to smaller companies like General Electric and Zenith (Figure 333). He also got the FCC to allocate a band for this new kind of radio with 40 channels in the 42-50 MHz range. RCA tried to block further development by petitioning the FCC to reserve those radio bands to television broadcasting (the new business they were focussing on). But that move failed.

**Figure 333: Early US FM Radio Receivers.**
Zenith Model 5S319 (top, 1939), General Electric model HM-80 (bottom, 1940).
Source: Wikimedia Commons.

For the new FM-band, Armstrong designed the complete system of transmitter, antennas, and receivers, and set up pilot broadcasting services in New York and New England in 1939. Then came the Second World War, and everybody used FM during the war. Armstrong allowed the military to use his patents royalty-free for the duration, a gesture that no company could make and that even he, with his lab expenses, could barely afford. Mobile FM communications were of tremendous value in the thrusts and parries across Europe and the Pacific.

After the war had ended, seeing that it was losing the technological race, RCA tried to license Armstrong's patents offering him a cash payment of million dollar, but no subsequent royalties, for his patents. Armstrong refused. Every other licensee paid 2% royalties, and he felt that giving such a deal to RCA would be unfair to the companies that had actually worked with him instead of against him. Then RCA circumvented his patents with (claimed) non-infringing patents. RCA moved again to cripple FM broadcasting, now succeeding to move the FM band from 44-50 MHz to 88-108 MHz, immediately obsoleting all the transmitters and FM-radio receivers that had been built for the old band. Nearly 200 FM stations, which, despite the handicaps, had opened in post-war enthusiasm, found it necessary to close up shop. Before the war the invisible hand of the market had pointed to a rosy future, but the iron fist of the state controlled by the big companies nearly choked FM at birth, then pushed it into the broadcasting wilderness for decades.

As RCA had been building FM receivers using his patents for the previous eight years without paying him a dime. His patents only had two years left to run. It was time for them to pay. He brought a patent infringement suit against RCA and NBC in 1949. Armstrong himself was called on to be the first witness. RCA's lawyers kept him on the witness stand for an entire year with niggling and irrelevant questions. Another two years elapsed when RCA was called upon to reveal the mountain of research it had done on FM in the 1930s. The

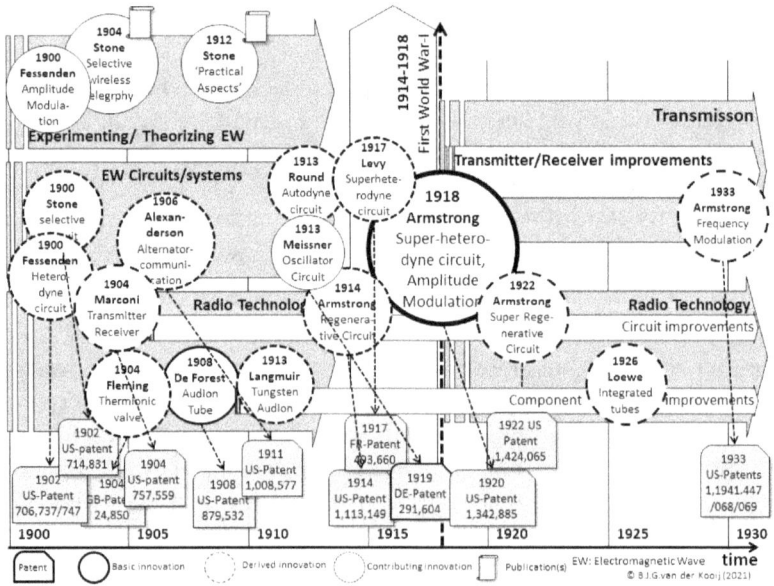

Figure 334: Cluster of Innovations around Armstrong's Super-Heterodyne Circuit (1918).

capstone of this was when RCA's director David Sarnoff himself claimed that RCA had invented FM all by itself without any help from Armstrong. After a claim like that, Armstrong would hear no talk of settlement. By 1953, Armstrong's licenses and patents had all expired. His crushing legal bills and research expenses brought him to near bankruptcy. On January 31st, 1954 he committed suicide.[513]

Armstrong left with his basic innovation (Figure 334) a legacy to radio broadcasting and television broadcasting. And RCA succeeded to delay the development of FM-radio for at least a decade.

## The Radio Broadcasting Boom

Westinghouse and General Electric were the two 'big boys' moving in the radio business. In addition, the new phenomenon of radio gave rise to the former ham-radio suppliers that started selling complete systems; next to the receiver unit, they offered the amplifying unit and the loudspeaker. The latter device made it possible to hear the broadcasted signal by a larger group of people (Figure 337). The character of the radio market changed from a nerd/technical user, to a consumer market of the non-technical user who was only interested in the radio broadcasting as realised by the early radio broadcasters.

After the First World War radio ban was lifted with the close of the conflict in 1919, a number of small stations began operating using technologies that had developed during the war. Many of these stations developed regular programming that included religious sermons,[514] sports, and news. During the 1922's 'Radio Broadcasting Boom', most programming was commercial-free, and entertainers, caught up in the excitement of this revolutionary new invention, performed for free. Meanwhile, a few people wondered how to pay for all this. In early 1922, the American Telephone & Telegraph Company began promoting the controversial idea of using advertising to finance programming. Initially AT&T claimed that its patent rights gave it a monopoly over US radio advertising, but a 1923 industry settlement paved the way for other stations to begin to sell air-time. And eventually advertising-supported private stations became the standard for U.S. broadcasting stations.

---

[513] After his death, his ex-wife Marion settled with RCA for a million dollars. She also pursued other court cases, defending his patents and receiving infringement awards from other manufacturers.
[514] The use of the new media radio and television to communicate Christianity, is called Televangelism. Their ministers, official or self-proclaimed used the media to preach gospel, some reaching millions of people. The funding of Christian Radio and Television came from donations of the listeners.

In early 1922 the *Radio Broadcasting Boom* hit America. Radio broadcasting, which earlier had interested mainly amateurs, now began to appeal to a far wider and more affluent audience. Armstrong himself termed his new circuit "the Rolls Royce method of reception" and, as with the automobile, the superheterodyne attracted many patrons precisely because of its expense and complexity. It was the start of the 'radio craze' phase of the Radio Revolution.

> *That winter, however--the winter of 1921-22--it came with a rush. Soon everybody was talking, not about wireless telephony, but about radio. A San Francisco paper described the discovery that millions were making: "There is radio music in the air, every night, everywhere. Anybody can hear it at home on a receiving set, which any boy can put up in an hour." In February President Harding had an outfit installed in his study, and the Dixmoor Golf Club announced that it would install a "telephone" to enable golfers to hear church services. In April, passengers on a Lackawanna train heard a radio concert, and Lieutenant Maynard broke all records for modernizing Christianity by broadcasting an Easter sermon from an airplane.*[515]

## *Radio Broadcasting 1.0 (1920-1945)*

In the 1920s, following the end of the First World War, across the US and Europe, from the ham-radio amateur circles and the nascent radio-tube industry, the first 'broadcasting stations' arose. Such as KDKA in Pittsburgh, Pennsylvania, England's British Broadcasting Company (BBC), and in Germany the Reichs-Rundfunk-Gesellschaft (RRG). Starting off as small-scale operations mostly, they sparked public interest and civilians began to purchase 'radios' for private use. Almost overnight it seemed that everybody wanted to jump on the bandwagon of radio and everyone went into broadcasting: from newspapers, banks, department stores, churches, universities and colleges, to cities and towns. And…, political ideologies and movements that became dominant in government policies.

### Early Radio Broadcasting in America

In the US, by 1919, after the end of the First World War, radio broadcasting pioneers across the country resumed transmissions. The early stations gained new call signs. The year 1920 was the year of the first 'commercial' AM broadcasting by XWA in Montreal, Quebec, Canada. Correspondingly, the first commercial broadcasting in the US took place in Pennsylvania by Frank Conrad, who was also responsible for the founding of the first licensed broadcast station in the world, KDKA. Conrad was an

---

[515] Source: Allen, F.L.: Only Yesterday, an Informal History of the 1920s.(1931) https://gutenberg.net.au/ebooks05/0500831h.html. (Accessed November 2021)

Assistant Chief Engineer in the Westinghouse Electric Corporation, and KDKA broadcasted for the first time on November 2nd, 1920. KDKA was a huge hit, inspiring other companies to take up broadcasting. From that transmission, entertainment broadcasting snowballed in the USA.

> *Americans were spellbound. Before radio, the only means of mass communication had been via the print media, and long-distance communication was only achieved by mail, telegraph or telephone. The receipt of news, weather or crop reports took hours, days, or even weeks in some places. If you wanted music, you could only hear it by attending live concerts, listening to scratchy phonograph records, or performing it yourself. But then suddenly, with just a modest investment in some radio parts, wires and batteries, any ordinary citizen could hear a live orchestra playing in a city a thousand miles away; farmers could get instant weather alerts or the day's market prices; and politicians could talk to voters across an entire region at once instead of making dozens of whistle-stop speeches.*[516]

The issuing of broadcasting license began as a trickle, but soon exploded. In a period of four years, there were 600 commercial stations around the country. To keep up with the cost of improving equipment and paying for performers, stations turned to advertisers. The flagship of AT&T, the broadcasting station WEAF, became the first to sell airtime, setting a precedent that would reverberate throughout the broadcasting world. In August 1922, the first radio ad, for a real estate developer, was aired in New York City.

> *The public, however, was overcome by a radio craze after these initial broadcasts, and radio became a product of the mass market. Manufacturers were overwhelmed by the demand for receivers, as customers stood in line to complete order forms for radios after dealers had sold out. Between 1923 and 1930, 60 percent of American families purchased radios. [...] The rapid spread of radio listeners and programs lead to inevitable confusion and disruption. Radio waves were up for grabs, as stations competed with one another for time and listeners. Many programs overlapped. Listeners of one program were frequently interrupted by overlapping programs. In addition, the public, the government, and emerging radio corporations viewed radio as a means of public service, rarely as a vehicle for personal profit. Radio manufacturers alone experienced financial gain from the radio boom.*[517]

---

[516] Source: Schneider J.F.: Radio Broadcasting's First Years - What Was It Like?. http://www.theradiohistorian.org/first_radio/first_radio.html (Accessed November 2021)
[517] Source: Emergence of Radio in the 1920s and its Cultural Significance www.xroads.virginia.edu/~ug00/3on1/radioshow/1920radio.htm. (Accessed November 2021)

## The Rise of Broadcasting Networks

A fundamental shift in American broadcasting came with the realization by the late 1920s that individual stations could easily share the cost of providing programs as a part of a broader network service with national appeal. So, broadcasting networks of local stations developed to share programming and became big businesses.

First the regional broadcasting networks expanded; the *National Broadcasting Company* (NBC) —founded by the RCA— began regular broadcasting in 1926, with telephone links between New York and other Eastern cities. NBC became the dominant radio network. Groups of stations that carried syndicated network programs along with a variety of local shows, soon formed RCA's Red and Blue networks (Figure 335). Two years after the creation of NBC, in 1927, the *United Independent Broadcasters* became the *Columbia Broadcasting System* (CBS) and began competing with the existing Red and Blue networks.[518] On September 29th, 1934, four AM radio stations —WXYZ in Detroit, WGN in Chicago, WOR in New York, and WLW in Cincinnati— agreed to form a cooperative, program-sharing radio network under the name of the Mutual Broadcasting System (MBS). It opened the way to national chain broadcasting and by 1924, it was possible to broadcast from coast to coast over a chain of 26 radio stations.

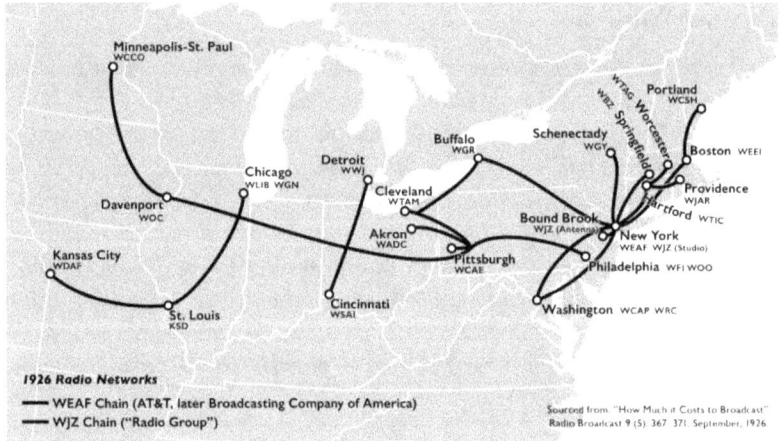

Figure 335: Radio Network Connections (1926).
Source: Wikimedia Commons.

---

[518] After the Federal Communications Commission (FCC) declared in 1941 that no company could own more than one radio network, NBC in 1943 sold the less-lucrative Blue Network to Edward J. Noble, the millionaire maker of Life Savers candy, who initially renamed it the American Broadcasting System before settling on the name the *American Broadcasting Company*, Inc. (ABC).

## The Explosion of Broadcasting Stations

After the First World War, when many people were looking forward to good times, the new phenomenon of radio introduced a whole new practice of entertainment to people's everyday lives. The number of listeners justified the establishment of radio-stations especially for the purpose of broadcasting entertainment and information programs (Figure 336).

Between 1921 and 1922 the sale of radio receiving sets and of component parts for home construction of such sets, began a boom that was followed immediately by a large increase in the number of transmitting stations. Starting with five stations in 1921, by 1925 some 571 broadcasting stations had been licensed in the US. By 1940 there were some 765 broadcasting stations.

The radio broadcasting stations, identified by their 'call signs,' popped up in a short period of time. Many of the early stations were highly experimental or speculative in nature, and some quickly disappeared, while others proved to be more permanent. Radio stations merged, or ceased operating and new parties entered the scene. Such as Chicago's KYW, a

**Figure 336: US Broadcasting Stations.**
Detroit based WWJ Station (top, 1920), New Jersy Station WDY (middle 1922), and San Francisco based KPO Station (bottom, 1922).

Source: http://www.theradiohistorian.org/

Westinghouse station that experimented with an opera format. The Detroit-based WWJ Station broadcasted a news format. The warehouse Sears & Roebuck started the WLS station —ie World Largest Store— in 1924 and set the standard in the Midwest for farm broadcasting.

> *For the listener, radio programming was like a continuous vaudeville show, with changing acts being presented by different stations every hour or less. For one hour, the listener might enjoy live entertainment and a clear signal from one of the more powerful Class "B" stations. But then that station would then sign off and be replaced with the scratchy signal of one of the weaker Class "A" stations, frequently suffering from poor audio quality, AC hum, and a wandering frequency. Music from these smaller stations would be provided by a phonograph, player piano, or the occasional live volunteer amateur vocalist. Some stations came on the air several times daily, while others transmitted only a few hours a week. A station might come on the air for just a few minutes at noon to give a local time check and weather forecast, and then shut down in favor of another station at 12:05. [...] In those early years, there were no station "formats" – every station broadcast a potpourri of programs, and listeners scanned the newspaper schedules to select the ones that interested them. Program fare on the higher-quality stations consisted mostly of live music concerts interspersed with news bulletins, sports scores, and weather forecasts. Added to this was the occasional lecture or poetry recital, and church services on Sunday mornings.*[519]

To cater the public interest in radio, in several cities a range of activities were developed to promote the new phenomenon. Chicago emerged as a broadcasting centre because its first radio stations, thanks to geography, were heard from the eastern seaboard to the Rockies and beyond.

During November 18-24th, 1924 the *Chicago Radio Show* was organized. As the primary goal of the show was to exhibit new radio designs and products, it featured over 250 exhibits of radio receivers, parts, and receiver accessories, including exhibits from Italy, Japan, Germany, and France. But the show wasn't just an exhibit of radio products; it was a true fair filled with entertainment and special attractions. Four broadcast studios aired special programs, that could be observed by the public. An amateur receiver builder's contest awarded 25 cash prizes. The public interest was overwhelming as 25,000 people/day visited the show. The second day, 10,000 people were denied admission by the Fire Marshal to 'preserve safety.' During that week the total attendance reached 139,902 people.

---

[519] Source: Scheider, J.F.: *Radio Broadcasting's First Years - What Was It Like?* http://www.theradiohistorian.org/first_radio/first_radio.html (Accessed June 2021).

The early US radio broadcasting industry was a mixture of the local experiments. As we have noted earlier, after the First World War had ended, in 1919 came the creation of the state-sanctioned radio monopoly under the name of Radio Corporation of America (RCA). RCA was owned by a GE-dominated partnership that included Westinghouse, American Telegraph and Telephone Company (AT&T), Western Electric, United Fruit Company, and others. RCA would go and dominate the US manufacturing industry as well as the broadcasting industry. Next to marketing the radio-sets manufactured by the owners of RCA, RCA also created the first nationwide American radio network, the National Broadcasting Company (NBC).

As the number of radio stations outgrew the available frequencies, interference between stations became problematic, and the government stepped into the fray. Regulation of radio broadcasting started with the Radio Act of 1927, and created the *Federal Radio Commission* (FRC) to regulate the multitude of broadcasting stations (732 by 1927) cluttering the airwaves and competing for the scarcity of frequencies. The Radio Act of 1927 allowed major networks such as CBS and NBC to gain a 70 percent share of US broadcasting by the early 1930s.

Next, the Communications Act of 1934 created the *Federal Communications Commission* (FCC) and ushered in a new era of government regulation. The organization quickly began enacting influential radio decisions. Among these was the 1938 decision to limit stations to 50,000 watts of broadcasting power.

*Life Style: Listening to Radio*

As radio broadcasting was the cheapest form of entertainment, it provided the public with far better entertainment than most people were accustomed to.

**Figure 337: Family Life around Radio Broadcasting (1920s).**
Source: Wikimedia Commons.

As a result, its popularity grew rapidly in the late 1920s and early 1930s, and by 1934, sixty percent of the nation's households had radios. During the Great Depression, stimulated by its low-cost accessibility, it was a cheap form of home-entertainment for the masses. It became part of the Interbellum lifestyle (Figure 337).

> *During the Great Depression, radio became so successful that another network, the Mutual Broadcasting Network, began in 1934 to compete with NBC's Red and Blue networks and the CBS network, creating a total of four national networks. As the networks became more adept at generating profits, their broadcast selections began to take on a format that later evolved into modern television programming. Serial dramas and programs that focused on domestic work aired during the day when many women were at home. Advertisers targeted this demographic with commercials for domestic needs such as soap. Because they were often sponsored by soap companies, daytime serial dramas soon became known as soap operas. … Popular evening comedy variety shows such as George Burns and Gracie Allen's Burns and Allen, the Jack Benny Show, and the Bob Hope Show all began during the 1930s. … By the late 1930s, the popularity of radio news broadcasts had surpassed that of newspapers. Radio's ability to emotionally draw its audiences in close to events made for news that evoked stronger responses and, thus, greater interest than print news could.[520]*

With the new phenomenon of radio broadcasting, Americans from coast to coast could listen to exactly the same programming. This had the effect of smoothing out regional differences in dialect, language, music, and

**Figure 338: Radio Advertisement 1920s.**
RCA promoting the Aeriola Senior (left, 1922) and Radiola models (middle, 1923), and Atwater Kent (right, 1926).

Source: Wikimedia Commons, https://www.saturdayeveningpost.com/2017/06/rcas-home-entertainment-center/

---

[520] Source: Evolution of Radio Broadcasting. https://opentext.wsu.edu/com101/chapter/7-2-evolution-of-radio-broadcasting/

even consumer taste. Radio also transformed how Americans enjoyed sports. The introduction of play-by-play descriptions of sporting events broadcast over the radio brought sports entertainment right into the homes of millions. Although most radio listening was music and entertainment in the form of comedies and dramas, the various radio networks developed news departments. These began to rival print media, if not in depth of coverage, certainly in immediacy.

And with radio broadcasting market exploding, the manufacturers started en masse making—and advertising (Figure 338)— their radio receivers. But the early radio did not become cheap, and to justify its investment early radio advertisement focussed on its advantages. In its theme-advertisement RCA introduced a whole new practice of entertainment to people's everyday lives focussing of sports, news and cultural events.

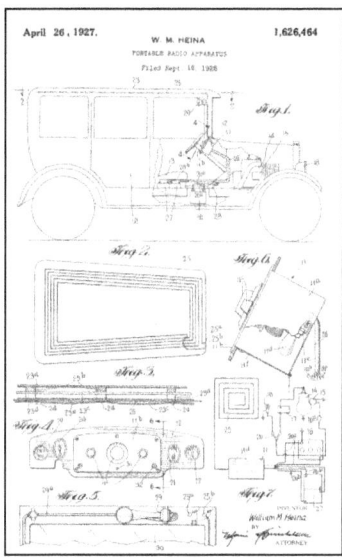

**Figure 339: Car Radio.**
Heina's US Patent № 1,626,464 issued on April 26th, 1927.
Source: USPTO.

*The Dawn of the Car Radio*

In the short period of time after the end of the First World War, a period that became known as the Roaring Twenties, the phenomenon of radio broadcasting exploded along a range of development trajectories: eg the technical trajectory of the radio receiver design, the application trajectory of the broadcasting stations (ie service providers and their formats). And the explosion of its manufacturing. Overnight radio had become big business.

Basically, the radio receiver was a device for home-entertainment. But soon other applications popped up. Take for example the car radio. Its use in automobiles started with professional used as a means for wireless communication, such as police broadcasts. But with the progress of technology, it also reached the mass market of car owners. Initially, the battery-powered radio was individually adapted for installation into a car. But soon special models came on the market, often included as an (expensive) option of the automobile (eg Buick, Chevrolet).

In mid-September 1926, a patent was filed by William M. Heina for a portable radio that could be adapted to an automobile (Figure 339). Granted as US-Patent № 1,626,464 on April 26th, 1927, his claims dealt primarily with the location of a radio chassis in combination with the dashboard. The patent drawings pictured an installation which served as a guide for most all future auto radios.

In 1930, the American *Galvin Manufacturing Corporation* marketed a Motorola branded car radio receiver for $130 the Model 5T71 (Figure 340, top). As it was expensive —the contemporary Ford Model A costed $540— they began selling Motorola car-radio receivers to professional users; police departments, fire stations, taxi companies and municipalities in November 1930. And in 1933 Ford began to offering Motorola's as an option pre-installed in the factory. In 1934 the price of the radio, installation included, had dropped to $55.

The next step was 'Vehicle Audio' where the radio receiver could receive broadcasted radio signals. Many manufacturers adapted their models for the after-car market, often separating the control part (aka as Head Unit) from the rather bulky receiving unit. The German company Blaupunkt introduced the model AS 5 medium-wave and long-wave radio in 1932. The Dutch company Philips introduced in 1937 its Model 247B car radio (Figure 340, middle). By the late 1930s, push button AM radios were considered a standard feature. In

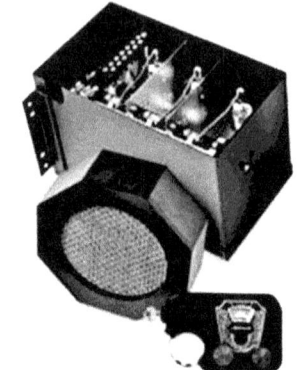

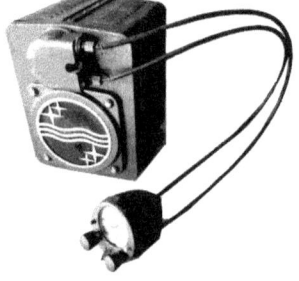

**Figure 340: Early Car Radio.**
Motorola Car Radio Model 5T71 (top, 1930), Philips Model 247B car radio (middle, 1937), and Blaupunkt Model A610B FM car radio (bottom, 1952).

Source: https://www.doctsf.com/ philips-247-b/f4566, https://infinigeek.com/

1946, there were an estimated 9 million AM car radios in use.

It took to the early 1950s before the FM band became available on car radios. In 1949 Blaupunkt advertised the first FM-capable car radio. And Blaupunkt offered model 7A650 in 1951, and model A610B in 1952 (Figure 340, bottom) that created the foundations for their dominant position in car radios.

## Radio Broadcasting in Europe

The development of radio broadcasting, followed a similar path as in America. But in the European context of the different national cultures of that period of time, it was shaped in a different way.

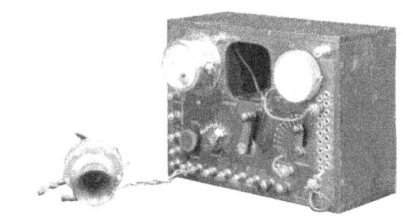

**Figure 341: Early British Broadcasting**
Homemade three-valve receiver (top, 1922).
Marconi Chelmsford Studio (bottom, 1922).

Source: Science Museum. https://www.mhs.ox.ac.uk/

*Radio Broadcasting in Britain*

Also in Britain, in the years following the First World War, many former military radio operators became amateur radio enthusiasts, tinkering with their home-made sets to pick up transmissions, and transmitting their own talks or music. They used radio to share their discoveries, forming a community of fellow experimenters.

The popularity of the new phenomenon was notable on the *First All-British Wireless Exhibition* held in September 30th to October 7th, 1922 in London. Some sixty stands were taken by companies anxious to display their wares and a record number of paying visitors attended. It attracted some 25,000 people.

Then the British Marconi Company started experimental broadcasts at its Chelmsford factory. On June 15th, 1920, the Australian soprano Dame Nellie Melba sang in a makeshift studio at that factory, using a microphone created with a telephone mouthpiece and wood from a cigar box. Sponsored by the Daily Mail, she opened her recital at 19:10 by singing 'Home Sweet Home' and, after other popular favourites and several encores, closed with the national anthem. Her voice, carried from an aerial with towering masts, was heard as far away as Iran and Newfoundland. However, the devices we today call radios —the wireless receivers that could receive the AM signals carrying human speech and music— with their vacuum tubes, would not until sometime later become available to the general public as the Chelmsford broadcast were banned in 1920 by the General Post Office (GPO).[521] It stopped abruptly radio broadcasting in Britain, raising massive protest from the amateur experimenters. The government partly yielded, and applied restrictions that resulted in limited broadcasting. The Marconi company was allowed to resume experimenting under call sign 2MT, and soon other manufacturers followed.

But the government restriction went far beyond technical issues Although the state-owned General Post Office (GPO) had received by 1922 nearly 100 broadcast licence requests, anxious to avoid the same chaotic expansion experienced in the United States, the GPO proposed that it would issue a single broadcasting licence to a company jointly owned by a consortium of leading wireless receiver manufacturers. That became known as the *British Broadcasting Company Ltd.* So, and radio broadcasts began in 1922. The company was to be financed by a royalty on the sale of BBC wireless receiving sets from approved domestic manufacturers.

> On November 14th, 1922, the London BBC Transmitting Station 2LO transmitted its first broadcast; a news bulletin. On the April 23rd, 1924, a speech by King George V was broadcasted for the first time. Millions heard his voice as he opened the Wembley Empire Exhibition and traffic was even stopped on London's Oxford Street as crowds listened on loudspeakers.

However, the number of electronics enthusiasts who began to create their own sets or bought rival unlicensed sets, meant that the sales revenues did not produce adequate funds. The BBC-monopoly also restricted presenting news bulletins before 19:00—to avoid competitions with the Paper Barons— and was required to source all news from external wire services. And, as the licence revenue was split between the BBC and the GPO, and the ban on advertising limited the revenues, wireless manufacturers became anxious to exit the loss-making consortium.

---

[521] The British Government licensed the BBC through its General Post Office, which had original control of the airwaves because they had been interpreted under law as an extension of the Post Office services (ie Telegraphy and Telephony).

Although the broadcasts quickly spread across the UK, it failed to usurp newspapers until 1926 when the newspapers went on strike as part of the May 1926 General Strike.[522] At this point the radio and the BBC became the leading source of information for the public. [523]

At the end of 1926 the BBC was nationalised and became the *British Broadcasting Corporation* with a state-monopoly on broadcasting. Till then British people could only listen to the early broadcasts that fit the policies of the BBC.

> *From the point of view of the working man, radio programmes became even more dismal. British broadcasting was staffed and run by men who had not come from the working class and did not understand the working man, but presumed to understand what the working man wanted. Britain occupies a unique place in broadcasting history in that it was the first nation to see this new medium as an instrument of social manipulation rather than for entertainment. It was also the first country to perceive radio broadcasting as a means of dissemination of propaganda on a national level. The seeds had been sown for the perception of radio in an even more important role: as an instrument of foreign policy whose function would be to disseminate informational propaganda to the rest of the world.* (Wood, 1992, p. 35)

The BBC soon developed the world's most emulated model of public-service radio broadcasting.

*Radio Broadcasting in Germany*

Also in Germany the radio tube soon became the driving force behind the wireless receivers called 'radio.' Remember, society at the time of the after-war Weimar Republic was in transition. It was the time of the 'Glückliche Zwanziger Jahre'[524] (ie Roaring Twenties) and —after the Allies had lifted the ban[525] on listening to radio waves— radio broadcasting became large part of it. Radio was welcomed in Germany like a liberating miracle, especially at a time of intense emotional and economic hardship. The first radio clubs and radio newspapers came in existence. The birth of

---

[522] The May 1976 General Strike was a nationwide strike involving some 1,5-1,75 discontented workers. It was called by the General Council of the Trades Union Congress (TUC) in an unsuccessful attempt to force the British government to act to prevent wage reductions and worsening conditions for 1.2 million locked-out coal miners.
[523] To provide a different service from the domestic audience the Corporation started the BBC Empire Service on short wave in 1932 (later renamed in the BBC World Service).
[524] In Germany the afterwar period was not so 'glücklich' as it was the period of revolt, separatism, hyperinflation and massive unemployment during the Weimar Republic. But it was also the time of decadence with cabaret dancing in the 500 night entertainment clubs (ie Cabaret Clubs) in Berlin.
[525] After the war, in accordance with the Treaty of Versailles, private citizens in the occupied Rhinelands of Germany were forbidden from listening to radio broadcasts.

radio in Germany is October 29th, 1923. On that day, the first entertainment program was broadcasted from the studio in the Vox-Haus[526] in Berlin.

In 1924 regional broadcasters emerged rapidly; the 'Mitteldeutsche Rundfunk AG' started in Leipzig, followed by the 'Süddeutsche Rundfunk AG' in Stuttgart, the 'Südwestdeutsche Rundfunkdienst AG' in Frankfurt. and the 'Nordische Rundfunk AG' in Hamburg. In 1925, a central *Reichs-Rundfunk-Gesellschaft* (Reich Broadcasting Corporation), came into being, merging the regional broadcasters. Its task was to regulate finances (the license fees for owning a receiver), perform joint administrative tasks and coordinate programming. Radio broadcasting developed into state-broadcasting.

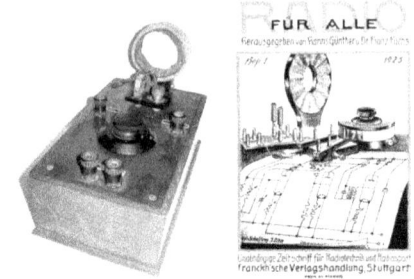

**Figure 342: Early Radio in Germany (1920s).**

Homemade detector (top left, 1925), and the Magazin 'Radio fuer Alle' (top right, 1925) .Bargfield Exhibition Stand on the Radio Exhibition Stuttgart (bottom, 1924).

Source: https://www.dokufunk.org/

The new miracle 'broadcasting' fascinated the masses. Stimulated in 1924 by the first Great Radio Exhibition in Berlin, Frankfurt and Stuttgart (Figure 342), the 'radio amateurs' joined by the thousands the radio clubs that were emerging in leaps and bounds throughout the Reich. Amateurs that were tinkering with a crystal detector or a simple tube device, exchanging experiences, and obtaining their knowhow from magazines. With the result that by 1932 there were more than four million paying radio subscribers. And at least as many non-paying listeners. The daily broadcasting time also increased steadily. In 1923, it was 60 minutes; by 1932, there were already 15 hours of radio programs every day.

---

[526] Building owned by the record company Vox.

All over Germany, the radio clubs emerge —some with an Audion tube permit— followed by early radio manufactures. The Reichs Post Ministerium (RPM) tries to get an increasing grip on radiobroadcasting in order to eliminate undesired developments:

> *A movement has arisen which is systematically proceeding against the registers of the empire; it is directed by people who are directly or indirectly interested in mass sales of radio equipment and who want to achieve that radio reception is generally released for any purpose. [...] Propaganda is made openly about it; entire radio receivers as well as individual parts are offered; it is shown how such devices can be produced by oneself, and how they can be hidden and removed from surveillance by the way they are constructed. All of this takes place in a form that, for the time being, legal action could only be taken with great difficulty and only by combating each individual case.*[527]

After the founding of the umbrella organization for all radio associations, the German Radio Technology Association (DFTV), on July 28th, 1925 in Munich, 22 regional associations joined it. The separation of the radio listeners and the KW amateurs had already begun.

After the National Socialists came to power on January 30th, 1933, the Reichs-Rundfunk-Gesellschaft was strengthened and the radio companies that had been independent until then were dissolved. The radio became the most important propaganda instrument of the Nazis, controlled and directed by the Reich Ministry for Public Enlightenment and Propaganda under the direction of Goebbels.

*Radio Broadcasting elsewhere in Continental Europe*

The preceding development took place in some of the larger countries at that time. But all over Europe radio broadcasting systems emerged.

After the first experimental broadcasting from the Eiffel Tower by Radio Tour Eiffel in 1921, in France Radio Paris began operations in 1922, followed by Radio Toulouse and Radio Lyon. The first French radio station, the post of State broadcasting on the Eiffel Tower (FL), started broadcasting on December 22nd, 1922. Soon, other private broadcasting stations appeared, on the initiative of manufacturers of receivers, such as Radiola and its eponymous station, launched in 1923. The National Broadcasting Office was created in 1929. By then France had eleven state radio stations dependent on the Radio Paris (PTT) and fourteen private radio stations.

---

[527] Source: "Funkfreunde" versus authority. https://www.dokufunk.org/ amateur_radio/ history_dl_1/index.php?CID=26266 (Accessed July 2021).

The government exerted tight control over radio broadcasting. Political debate was not encouraged. Radio was a potentially powerful new medium, but France was quite laggard in consumer ownership of radio sets with 5 million radio receivers in 1937. Compared to over 8 million in both Britain and Germany, and 26 million in the United States. As war approached, Frenchmen learned little or nothing about it from the radio. The government thought that policy wise, because it wanted no interference in its policies.

In the Netherlands, the first broadcasting station became operational from as early as November 6th, 1919, which belonged to a Dutch ham radio operator Hanzo Idzarda. Hanzo started public broadcasting with the "PCGG" station from Den Haag ('The Hague') with his homebrewed transmitter (Figure 343, top).and his programs became popular around the globe as "Hague Concerts" (Figure 343, bottom). When Idzarda started broadcasting in 1919 his audience was rather small. They were mainly technical enthusiasts who tuned into the regular transmissions.

**Figure 343: Early Dutch Broadcasting**

Deka de Luxe Receiver made by the Dutch Radio Industry (top, 1922), and musicians in PCGG studio (bottom, 1922).

Source: https://www.beeldengeluid.nl, https://www.nvhr.nl/, Wikimedia Commons.

Then, the NRI (Netherlands Radio Industrie/Dutch Radio Industry) started broadcasting from 1919 until 1924 a transmitter from The Hague and broadcasts regular music programs. It took until 1923 before others became active. The second party was the NSF (Netherlands Seintoestellen Fabriek) in Hilversum. The NSF operated a transmitter from July 21th, 1923. On April 1st, 1924 the NSF founded the HDO (Hollands Draadlooze Omroep), a special organization to make radio programs.

From 1924 many Dutch broadcasting associations —based on the different ideological sections of Dutch society called 'Verzuiling' (pillarisation)— were founded: the Protestant NCRV, the Roman Catholic KRO, the Socialist VARA and the liberal Protestant VPRO. It was an important step to bring radio to the people. When Philips introduced a simple to operate radio set in 1927— in Holland nicknamed "the loaf of bread"—, (Figure 362) radio began to boom. During the 1920s, sport club owners cautiously embraced radio broadcasting of sports games. Many owners feared radio would dissuade fans from attending the games in person. Actually, since not many people could visit stadium to watch games, radio started to entertain people by enabling them to enjoy sports by sound.

On March 11th, 1927, Philips went on the air with shortwave radio station PCJJ (later PCJ) which was joined in 1929 by a sister station (Philips Omroep Holland-Indië, later PHI). PHOHI broadcast in Dutch to the Dutch East Indies.

In Luxembourg a first transmitter was built by 1924 by the brothers Francois, Aloise and Marcel Anen. In 1925, after 2 years of preparatory work and tests, they created the *Association Radio Luxembourg*, intended to manage the future Grand Ducal radio station. With its central location in Western Europe, the Grand Duchy was an ideal site for broadcasts to many nations, including the United Kingdom, the more when they used high-powered transmitters (ie 200,000 Watt). Next, the *Compagnie Luxembourgeoise de Radiodiffusion* (CLR) — Luxembourg Broadcasting Company— with a majority Franco-Belgian capital was founded in 1931 and it obtained a monopoly to operate commercial broadcasting.

In May 1932, Radio Luxembourg began high-powered test transmissions aimed directly at Britain and Ireland. The British government accused Radio Luxembourg of "pirating" the various wavelengths it was testing. In the years from 1933 to 1939, the English language service of Radio Luxembourg gained a large audience in the UK and other European countries, with sponsored programming aired from noon until midnight on Sundays and at various times during the rest of the week.

> *From the first day Radio Luxembourg came on-air it showed that it had found the formula. It took Europe by storm with its popular style of programmes, transcending class barriers and national frontiers. Such was the huge success enjoyed by Radio Luxembourg that within two years it had captured a 50% share of the listening audience throughout Europe. Even more remarkable was the fact that out of 4 million listeners to Radio Luxembourg, 2 million were in the British Isles. This was twice as many listeners as the BBC. [...] The coming of Radio Luxembourg was like a breath of fresh air sweeping through the air*

*waves of Europe. Its entertainment programmes were like nectar to the working man; they turned a dismal day [ie Sunday] into a pleasure to be looked forward to.* (Wood, 1992, pp. 43-44)

## Radio Broadcasting during the Second World War (1940-1945)

Up to the Second World War, the radio as a means of mass-communication had matured and the medium of commercial broadcast radio grew into the fabric of daily life in the United States. The start of the war had interrupted that process in different ways. One of them was the government-influence of the broadcasted content aka censorship. The entertainment industry during this period was often controlled by the government to depict the war positively and keep civilian spirits high. The governments also censored the majority of mass media entertainment across popular platforms such as radio, film, and music. And …, they practised the dark arts of psychological warfare.[528]

In America, broadcasting was offered twenty-four hours a day in an effort to keep citizens engaged, and 90% of American families owned a radio during the Second World War. However, propaganda was not well received, reminding citizens of the propaganda tactics used in World War I. This led President Roosevelt and others to attempt to convince the people that the government was not out to censor information but inform the public. In his fireside chats he spoke about current themes. For example on his 'New Deal' policies, but during the war the content of the chats moved from bolstering Roosevelt's New Deal policies to discussing various aspects of America's involvement in the Second World War. Other radio programs were focussing on the enemy describing the evil of the Germans and giving very detailed descriptions of concentration camps and stories heard from them.

On October 30th, 1938, Orson Wells broadcasted a radio play that emulated an attack by the Martians on Earth. In a time when the world was preoccupied with war, and the radio was (mis)-used for propaganda, his play threw the entire United States into a panic by simply adapting a classic H.G. Wells science-fiction novel 'The War of the Worlds,' and presenting it in such a way as to sound like it was unfolding through a progression of on-air radio news reports. Within

---

[528] Based on numerous sources, among which: Concho, Chr.: Radio Propaganda during World War II, https://web.stanford.edu/class/; Andres, E.: 6 World War II Propagnada broadcasters, https://www.history.com/news/6-world-war-ii-propaganda-broadcasters; The Fake British Radio Show That Helped Defeat the Nazis, https://Smithsonianmag.com; as well information in the public domain.

a few weeks the radio play was written about nearly 12,000 times in newspapers around the world and became the most infamous radio broadcast in history.

Radio was not only for the general public, as it was used to entertain the troops after in 1942, when the War Department created the *Armed Forces Radio Service* (AFRS). From the beginning Hollywood A-List performers such as Bob Hope, Dinah Shore, Fred Allen, Frances Langford, Spike Jones, Frank Sinatra, Burns and Allen, Vincent Price, Ginger Rogers, Gary Cooper, Tallulah Bankhead, the Andrew Sisters, Bette Davis, Judy Garland, Bing Crosby, and Margaret Whiting all appeared free of charge, allowing AFRS to produce extremely high-quality shows for far less than expected.

**Figure 344: Pre-war US Broadcasting.**
Fireside chats by President Roosevelt (top, 1933-1944). US Panic after Radio play 'War of the Worlds' (bottom, 1938).

Source: https://www.britannica.com/event/fireside-chats, https://theradionerd.com/

Also at the home-front the commercial radio started to include war-related content. Special radio series like 'This is War'[529] was broadcasted throughout the country. Radio —next to the printed media— brought the information from the Theatres of War to the home-front.

> One of the most successful radio program methods the American government used was the "you technique". This method put the listener directly in situations like battle or being in a military camp by addressing them personally. Radio shows brought the war home to the American people in a way that had never been imagined before.

---

[529] 'This is War' was the first dramatic and didactic series on America's war effort, and it wove dry statistics on military production and conscription into moving tales of global war and national mobilization.

*Radio as a Tool of Propaganda*

Most people in America where completely unaware that the radio was designed with a master propaganda plan, by both the government, advertising agencies and members of Hollywood. Overall, the radio propaganda effort during Second World War was highly persuasive. In addition, the in 1942 newly established government sponsored 'Voice of America,' broadcasted worldwide from a range of stations in a range of languages. But the radio broadcasting as a tool of propaganda had been discovered already before the war.

In Russia the People's Commissariat for Posts and Telegraphs supervised radio broadcasting. After the October Revolution of 1917, in eventually adopting broadcasting to 'reset the mental horizons of its population', the Soviet authorities faced particular problems. One was the poverty of much of the population, for whom ownership of a domestic radio set was an unjustifiable extravagance. Radio listening became a collective activity in village squares, community halls and factories. Radio offered the opportunity to bring Soviet philosophy and ideology to the masses. The first live direct broadcast came on November 7th, 1925 during the 8th anniversary of the Revolution. The first broadcasting station opened in 1922 but the first public broadcast was not made until 1924. Over the next ten years, the number of stations continued to rise with 23 stations in 1929, 60 stations in 1932, and 90 stations by 1933. By 1928, thirty-six foreign firms were engaged in giving radio support to the Soviet Union.

In Nazi Germany, radio was an important propaganda tool. German propaganda helped shape Germany into the efficient war machine that it began during the war. Just a few short months after the outbreak of the Second World War, German propagandists were transmitting close to 11 hours of programming a day, with the majority of the transmissions in English as well. Their focus was, next to influencing their own population, on eroding the pro-British sentiment. And propagandists also focused on certain groups such as capitalists, Jews, and specific newspapers and politicians. Joseph Goebels, the German propaganda minister, started the project in which millions of cheap radio sets —the 'people's receiver' (volksempfänger)— were subsidized by the government and distributed to citizens.

In Britain, with the blackout and lockdown closing public venues, the radio was one of the few means for entertainment. However, the BBC with its broadcasting monopoly stopped broadcasting sporting events. What listeners got was censored war news, announcements, lots of gramophone records and live music. In addition, in response to the Nazi-German propaganda, he British joined the propaganda fight as soon as the war began, launching their own black propaganda campaign from fake German radio stations (eg Gustav Siegfried Eins, SG1) that broadcasted to both Germany and German occupation troops.

Everywhere, the radio was used as a propaganda tool for the government's own policies, but also as psychological weaponry to demoralize the enemy; both its populations and its troops. Famous were Tokyo Rose (English-speaking broadcasters of Japanese propaganda), Axis Sally (English speaking female broadcasters of Axis propaganda aimed at American troops), Lord Haw-Haw (the nick applied to announcers who broadcasted Nazi-propaganda) and BBC's Tom and SG1's 'Chief'

The quarter century to about 1950 was also *Radio's Golden Age* in most industrial countries, where —despite wartime setbacks— radio flowered before the advent of television. The emphasis was on high-quality culture, education, and music, often with a strongly nationalistic, political and religious tone. Radio created an auditive window to the world; news, music, and radio plays, and would become a creative era in which dramatic radio thrived and was a vital part of Western Culture (Figure 365). Till the rise of television diminished its impact during the *Golden Age of Television*.

## *Radio Broadcasting 2.0 (1945-1970)*

Building on its wartime experience, radio broadcasting expanded exponentially after 1945, with many countries adding new languages and services, and a number of fairly small nations playing a prominent role on the air. In Allied occupied territories, it started with rebuilding or replacing pre-war radio stations and network links. In 1945–1950 Allied occupation authorities in Germany and Japan required dramatic changes in both programs and management, chiefly in order to diminish centralized control and excessively nationalistic content. Reconstruction in other European countries, made the number of transmitters in Europe increased from 566 in 1950 to more than five times that number by 1962.

> Among those influential nations, Britain's BBC played a prominent role. After the *Home Service program* — devoted to both serious and lighter music, educational and children's broadcasts, most of the BBC's talk and discussion programs, and 60 percent of its news—, and the *Light Program* —focussing more on popular (light and dance) music, drama, and outside (remote) broadcasts— it inaugurated the *Third Program* in September 1946 to provide erudite talk and high-quality music programming "for the serious minded, for the educated and those who wish to be so."

*Radio Free Europe* (aka radio Liberty), a radio broadcasting organization created by the United States government in 1950 and directed at Eastern Europe, was started to provide information and political commentary to the people of communist Eastern Europe and the Soviet Union. By the late 1950s, sixty different foreign language stations were broadcasting to the Soviet Union.

## Radio Broadcasting in the US

Untouched by the Second World War, American radio stations rapidly expanded in number to more than 2,000 AM outlets by the early 1950s. Most were in smaller markets gaining local radio service for the first time. Beginning with the 1948–1949 season, however, network television in the East and Midwest (with national service by 1951) doomed American radio networks. Because American commercial television expanded faster than many expected, radio listeners of 1945 would find a dramatically different system and programs within a decade. The number of network radio affiliates declined by slightly more than half, and network drama and variety programs (which had shifted to television or left the air) were replaced by music-driven local programming. Public-service-oriented radio systems changed more gradually, their mission continuing into television; because of its high cost, however, public-service television grew slowly, thus extending the importance of educational radio.

**Figure 345: Early US FM-radio.**

Zenith 7H820 Dual FM band radio aka 'Armstrong radios' (top, 1948), RCA Victor AM/FM Radio Model 1R81 (bottom, 1951).

Source: Wikimedia Commons, Radio Attic Archives (James LeFevre).

As electronic technology had progressed, the use of radio shifted. On the one side the new electronic device called 'transistor' had made inroads in radio receiver design. Radio listening outside the home was expanded dramatically by the sale of portable transistor radios and cheaper car radios. (In 1951 half of American cars had radios; 80 percent had them by 1965.). On the other side, the new phenomenon of 'television' boomed. Being able to have a visual experience—next to the auditive experience of radio—fascinated people.

*Further Development of Radio Technology*

Next to the AM-radio technology (Figure 326) the new FM-radio technology (Figure 331) emerged in the late 1930s. After a slow start in the late 1940s —due to RCA's machinations to protect their AM-interests— during the 1960s FM-radio became the fastest-growing segment of the broadcast business in the United States. By the late 1960s, FM had been adopted for broadcast of stereo, but still AM-broadcasting dominated the radio stations. It was —again— a parallel development where the broadcasting stations and receiver equipment d each other's development. The more stations were broadcasting in FM-technology, the more radio receiver manufacturers included the FM-band on their radios (Figure 345). The more FM-radios were sold, the more stations were FM-broadcasting.

In continental Europe, where the medium wave band (known as the AM band because most stations using it employ amplitude modulation) was overcrowded, FM became adapted easier (Figure 367).

## Radio Broadcasting in Continental Europe

Traditionally, the technology of Amplitude Modulation (AM) was the carrier radio signal that transmitted the broadcasted radio programs (Figure 326). A new technology called Frequency Modulation (FM) had been developed by Edwin Armstrong in the early 1930s (Figure 331). But commercial FM service in the US faltered for a time after 1949 as broadcasters focused on developing the more popular television and AM radio services. In Europe, however, FM was soon perceived as a means of reducing horrendous medium-wave overcrowding and interference problems. It also helped serve regions largely unreached by existing stations.

As part of the rebuilding of its industry after the devastation of the war, Germany led Europe in early FM broadcasting. The first FM transmissions were on the air by 1949, and most of West Germany was covered with FM signals. By 1955, 100 FM transmitters were in operation in West Germany. Italy, facing a severe shortage of medium-range frequencies, followed suit, providing its first FM services in the early 1950s. A decade later, multiple FM transmitters were operating in Belgium, Britain, Norway, Finland, Switzerland, and Sweden.

> Probably the most extreme examples of the potential of local FM radio took place in the 1970s in both France and Italy. A number of unlicensed small FM stations went on the air in Italy in late 1974 into 1975. When an Italian court held that the state broadcasting authority did not have a monopoly on local radio, hundreds of new stations followed, and by mid-1978 some 2,200 were on the air, providing Italians with the most radio stations per person of any country.

France went through a more limited version of the trend: by the early 1980s there would be more than 100 such stations in Paris alone.

*Offshore Radio: Pirates ruling the Waves*

In reaction to monopolized public service radio establishments, and their rather fixed programming, a new form of broadcasting emerged during the Swinging Sixties of the twentieth century. They were labelled as 'Radio Pirates' as they were using transmitters built into small ships moored beyond territorial limits. Or mounted in former abandoned British Second World War Sea Forts before the coast of England in the Thames Estuary. And they fought the monopolistic programming practises of the radio establishment.

The first, *Radio Murcur*, began service of the coast off Denmark in July 1958; it was followed by *Radio Veronica* two years later. The Swedish pirate station *Radio Nord* began operating in 1961, and Radio Veronica provided transmissions into Britain the same year. *Radio Caroline* began popular music broadcasts into Britain in 1964 and *Radio Atlanta* managed only to broadcast less than two months in 1964. A similar fate was assigned to *Swinging Radio England* that operated from May 1966-November 1966. The commercial station *Wonderful Radio London* operated from a ship —a former warship— anchored in the North Sea before England from December 1964.

The main theme in their programming was music. Disc jockeys presented their program; from Rock 'n' Roll to the music listed in the Top 40 list of popular songs. And their jingles —eg 'This is Veronica'— became famous. The practise spread rapidly and shipboard radio stations were soon also stationed off Italy, France, and New Zealand.

The Dutch Radio Pirate Radio Veronica had started broadcasting in April 1960 from a ship called Borkum Riff. Notwithstanding the jamming by the Dutch Postal Office, it became quickly popular. Especially its program known as the *Veronica Top 40* became popular among the younger audience. Following the actions taken by the Scandinavian Governments, the Dutch Governments started in 1962 working on ways to close the station down, but such was the popularity of the station that nothing was done about it. In 1964 a new ship, the Nordeny, was purchased that took over in 1965. Nevertheless, because of the Dutch Marine Offences Act, the station was to close down on the 31st of August 1974.

Another form of radio pirates operated from abandoned Second World War Sea Forts in the territorial waters of the Thames Estuary (Figure 346). *Called Radio City* they started in 1964, but it did not last long as the Marine Broadcasting Offences Act forced its closure in February 8th, 1967. And on an abandoned oil rig called the REM Island, the Dutch station *Radio Nordsee International* started in 1964, and broadcasted five years when it became *Radio Caroline International*. As the British conservative government jammed the signal that action was short-lived, and also a hijacking and bombing operation in 1971 failed. From June 1971 until the end of August 1974 Radio North Sea International was a regular and reliable broadcaster from international waters, four miles from the coast of the Netherlands.

The pirates were outlawed over time, but their influence and popularity lived on and the radio establishment picked up their concept.

**Figure 346: Radio Pirates (1960s).**
Radio pirate Caroline on the ship Mi Amigo (top), and Radio forts in the Shivering Sands defence forts (bottom).

Source: Wiimedia Commons. Offshoreradio.co.uk.

In the Netherlands it created the public service channel 'Hilversum 3.' And some of the former pirates became even part of the public broadcasting establishment. Radio Veronica became the Veronica Broadcasting Organisation, Radio Nordsee became the Television Radio Broadcasting Foundation (ie TROS) both receiving a license to broadcast.

The radio pirates heralded a change in the format of radio broadcasting that was also a response to the rise of television broadcasting.

## Radio Broadcasting 3.0 (1970s-2000s)

With the new medium of television replicating and expanding the functions of radio broadcasting, the radio was losing the battle of public attention. Many radio-stations went of the ether. Others — narrowing and shifting their formats to all music, all news, all sports, and ever-more niche topics like conservative talk and the religious programming— struggled.

### Demise of the Radio Broadcasting Industry

In the 1980s and 1990s, radio stations countered threats posed by personal video recorders and digital compact discs through a greater emphasis on listener-driven programmes. By the late '90s and early 2000s, radio stations were reinventing themselves to cater to niche audiences. There were stations dedicated to specific genres of content —talk radio, punk rock stations, classic radio, and even stations that played music by a single band 24 hours a day— anticipating the emergence of Spotify and iTunes by a decade or more. Depending on regional situations —such as in India where television did not penetrate the rural households for a long time— the traditional AM/FM Community and private radio stations stayed longer in the air. In developing countries radio expanded its role as educational medium.

AM/FM radio stations did not disappear overnight, but adapted and transformed their formats in many ways. The Community Radio[530] with its solid base in local communities even flourished. But the general trend was downsizing and reorientation. And the airwaves were going to be emptied of the analogue AM/FM radio signals over time.

> Specialized radio stations such as Radio Free Europe/Radio Liberty (RFE/RL), switched of their transmitting in steps. As in the early 1990s many countries from the former Soviet sphere became democracies and some joined NATO and the European Union, Radio Free Europe (RFE/RL) considered its mission fulfilled in several places. The Hungarian service was closed in 1993, the Polish service in 1997, and Czech broadcasts (produced in cooperation with Czech Public Radio since 1995) ended in 2002.

With the rise of the Internet and the mobile communication devices, and the transformations of the Digital Era, new digital media platforms (eg Internet radio)[531] took over the classic functions of air wave radio.

---

[530] Community radio is when local people produce and broadcast their own programs and participate in operating the station.
[531] Internet radio is a digital audio service transmitted via the Internet. This form of broadcasting is called webcasting.

## *The Radio Manufacturers (1900-1945)*

Not surprisingly, the rise of the radio broadcasting was paralleled by the rise of the number of radio receivers sets. Originally made by the ham radio amateur, soon the entrepreneurial spirit resulted in the first commercialized products. In the early days, the idea of radio design did not exist; the receivers were technical artifacts and they showed it. However, with the progress of its popularity as a home-based artifact, radio receivers were more and more housed in wooden enclosures; from the cabinet design and cupboards, to the table-top tombstone and cathedral design. Glossy decorations, elaborated wooden inlets, veneer works and textile coatings made splendorous products out of radios.

### Radio Manufacturing in the US

As had been true of earlier 'high-tech' industries such as the telegraph and electric lighting in their formative years, what was accomplished in the early years of the radio industry was primarily brought about by individual people. None of the major electrical and telephone companies played a significant and innovative role in the formative years of the radio industry. So this industry's early history is a story of individuals, many of whom were both inventor and entrepreneur. Over time that changed and after the 1920s this industry's history would become largely one of organizations.

Much of that early development of the radio industry (especially the manufacturing) took place in the industrial Nord-East of the US.[532] A region that had already based earlier mechanical industries (eg the bicycle industry, the automotive industry). There, the 'skyrocketing companies of the radio boom days,' followed by its almost equally speedy demise, dramatically illustrated the perils of unchecked corporate exuberance and risk-taking. Companies that operated in the context of the American society with its capitalistic culture. Companies big and small, established or starting up, contributing in small and large ways to the Radio Boom of the Interbellum. Companies that already practised new mass production techniques; from the single station manufacturing to the assembly lines. In other words, it was a period later characterised as the Third Industrial Revolution,[533] that nourished the development of the Radio in all its dimensions.

---

[532] After the American Civil War (1861-1865), during the *Gilded Age* (1870-1900) industrialization had grown largely in the northern and north-eastern states: it became *America's Manufacturing Core*. By 1890 Philadelphia had hundreds of bicycle manufacturers, and by 1910 a massive automobile assembling industry had emerged in the Detroit Region.
[533] See: Van der Kooij, B.J.G.: *The Invention of the Internal Combustion Engine*. (2021)

**Figure 347: The Acoustic Horn.**

The Magnavox loudspeaker (top, 1920), Dictogrand Radio Horn (middle, 1924), the Amplion AR4 Dragon Loudspeaker with wooden horn (bottom, 1925)

Source: Wikimedia Commons, Science Museum. http://www.antik-radio.dk/laudspeakers/loudspeaker.htmlE

And among those early manufacturers sprouting from the ranks of the radio amateurs, we find the manufacturers of the Crystal receiver set. Simple constructions for the detection of radio waves without any amplification, that used 'headphones' to create an auditive signal. With the advent of the Radiophone (Figure 3), the need to cater a broader audience was apparent.

*The Invention of the Loud Speaker*

One of the seemingly minor developments of early radio was related to making the sound available to more than one person. Loudspeaker-design started with copying the hearing aid of the horn, and transferring it into a device that —for example in Edison's Graphophone — amplified mechanically (Figure 347). Next, using the electric coil as a means to vibrate a cone, the electro-mechanical production of sound from an electric signal, emerged. The so-called 'acoustic horn' was based on the moving coil electrodynamic loudspeaker (Figure 86). Soon it became integrated in the radio design, or was offered in

**Figure 348: Box-shaped Loudspeakers.**

GE's Rice-Kellogg Dynamic Loudspeaker (left, 1924), RCA 100B Loudspeaker (right, 1929).

Source: New York Heritage. https://www.radiolaguy.com, worthpoint.com. http://www.antik-radio.dk/laudspeakers/Wood%20-%20Metal%20speaker.html

box-shaped encasings for the speakers (Figure 348).

It were two Danish, Edwin S.Pridham and Peter L.Jensen, who developed in 1915 a sound device that became their acoustic horn; a moving coil speaker connected with a gooseneck horn. They made their first public demonstration in Golden Gate Park on December 10th, 1915, and another on December 25th playing music in front of San Francisco City Hall to a crowd of 100,000 people. Jensen and Pridham merged with the Sonora Phonograph Corp. and formed the *Magnavox Company* in San Francisco August 3rd, 1917.

Soon others European and US manufacturers (eg Amplio, Atwater Kent, Figure 347) followed and started manufacturing loudspeakers. Also the large companies (eg RCA, Philips) manufactured their own speakers based on the invention of GE-employees Chester Rice and Edward Kellogg (Figure 348).

## The Rise and Demise of the Atwater Kent Manufacturing Company

The *Atwater Kent Manufacturing Company* was one of the early leading radio manufacturers. It was founded by Arthur Atwater Kent (1873-1949), an inventive engineer by nature. He started his entrepreneurial activities making electric motors and fans at the *Kent Electric Manufacturing Company* in 1895. By 1911, Kent's company had grown to 125 employees, and moved to larger

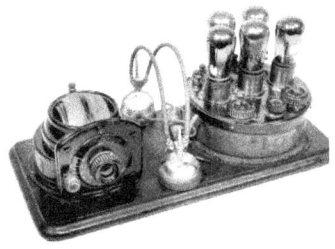

**Figure 349: Atwater Kent Radio Models.**

Model 5 breadboard (top, 1923), Model 12 Breadboard (upper middle, 1924). Coffin style Model 20C (lower middle, 1925), and Model 55 Pooley floor cabinet (bottom, 1929).

Source: Worthpoint, Atwaterkentradio.com, http://www.sparkmuseum.com/

manufacturing facilities. Automobile starters and lighting systems were added to the product line. The company continued to grow due to success in the automotive electronics field and government contracts during the first World War.

After the war had ended, and with demand dwindling, he started in 1919 the second business, the *Atwater Kent Manufacturing Works*. In 1919 the company began building headsets for the nascent radio industry. In 1921, Kent produced his first radio components, selling the do-it-yourself kits consisting of 'breadboards'[534] that could be assembled by early radio enthusiasts, and the first complete radio set was shipped from the factory in November 1922. Next, he put radio systems in wooden boxes and cabinets (Figure 349), and promoted them widely (Figure 338).

In 1924, the company moved to a new $2 million plant North Philadelphia. This plant, constructed in sections, would eventually cover 32 acres (13 ha) (Figure 350). Atwater Kent was the largest manufacturer of radios in the country from 1926 through 1929. At its peak in 1929, the company employed over 12,000 workers manufacturing nearly one million radio sets.

**Figure 350: Atwater Kent Radio Assembly.**
Atwater Kent Radio set Assembly Rooms, 1925.
Source: https://www.shorpy.com/ node/3542

The meteoric rise of the company was followed by an equally rapid fall during the years of the Depression. Along with the economic downturn came overpowering competition from other

---

[534] To create an electronic circuit like a radio receiver, the components (tubes, coils, resistors, etc) have to be interconnected by wiring. To support the components mechanically, a support is needed. Originally a wooden breadboard, later a metal chassis.

radio manufacturers such as Philco, and increasing demands from labour unions. The company adjusted to consumer demands by building smaller, tabletop radio sets, but could not keep up with the competition from the cheap 'All American Five'[535] radio sets that were manufactured in the millions by hundreds of manufacturers from the 1930s onward.

Kent dissolved his design engineering facility in 1931, and closed his radio factory in 1936. With a fortune estimated to be in the millions, Kent retired to Bel Air, California, where he pursued social and philanthropic activities until his death in 1949.[536]

## The Rise and Demise of Majestic Radio

*Majestic Radios*, created in 1927 in Chicago and trademarked as "The Mighty Monarchs of the Air" had originally started in 1921 as the *Grigsby-Grunow-Hinds Company*. Being an automobile aftermarket manufacturer, it profited from the Automobile Boom and reached $5 million sales in 1927. One of their successful products was the Majestic Battery Eliminator; a replacement for AC-chargers for battery-powered radios. But the market for chargers collapsed so they had to find new business.

After first manufacturing the battery-eliminators, but already in 1927 they began making 'Majestic radios featuring dynamic speakers with moving-coils and advanced circuitry employing screen-grid tubes for improved reception.

**Figure 351: Grigsby-Grunow Radio Receiver Manufacturing Plant (1930).**
Final Assembly in Cabinet department.
Source: https://chicagology.com/

The expansion plans needed financing. So, in March of 1928, Grigsby and W.C. Grunow (Hinds had been sold out) raised $1.1 million in a public offering IPO of 29,000 $40 shares in order to diversify into the radio business.[537] By April, 1928, Grigsby-Grunow had purchased the 500,000 square foot plant of the Yellow

---

[535] The design of the "All American Five" super-heterodyne receiver typically used five vacuum tubes to receive AM-broadcasting.
[536] Source: Atwater Kent Manufacturing Company, Philadelphia Pennsylvania. http://www.historic-structures.com/pa/philadelphia/atwater_kent.php
[537] The stock, the "wonder of Wall Street," rocketed to over $1,116 in 1929.

Truck & Coach Company in Chicago, converting it in a radio manufacturing plant (Figure 351).

Organization was the name of the game. Steady mass production, planned in advance with extreme care, checked hourly in each department, with inspectors and testers at hundreds of different stages in the making of each radio set. For the 13,000 employees, 1,300 inspectors checked on their work.

The following month, the company announced their line of 'Majestic' receivers highlighted by a $137.50 console —the Majestic Model 71 tabletop receiver— with excellent styling and a powerful electro-dynamic loudspeaker incorporated in the cabinet. Backed by an aggressive advertising campaign, Majestic shot up to the number one position in industry sales in 1929.

In April of 1929, Grigsby-Grunow entered the 'radio tube' business by acquiring the La-Salle Corporation, an RCA-licensed maker of radio tubes, commencing manufacturing under the Majestic name at 5,000 tubes a day.

The *Grigsby-Grunow Company* at first enjoyed meteoric success, buoyed by its superior loudspeaker technology. By some accounts, in 1929 it outsold all other radio manufacturers. Total revenues did rise from $5,000,000 in 1927, $40,000,000 in 1928 to $70,000,000 in 1929. By the end of May 1929, after just 12 months, profits had soared to around $5,000,000. Their stock, the 'Wonder of Wall Street' had rocketed from $40 to $ 1,116 in the first half of 1929. And by September 1930 some 7,000 radio sets were produced daily.

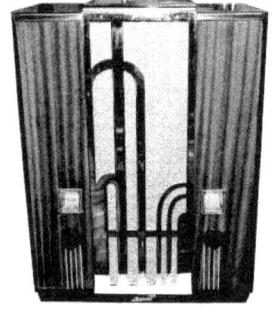

**Figure 352: Majestic Radio Receivers.**

Majestic Model 52 (top, 1931) and Majestic Model 161 (middle, 1933), and Majestic Model 174 Tombstone (bottom, 1933).

Source: Tuberadioland.com

The rapid pace of meteoric growth, however was interrupted by the 1929 Wall Street Crash. At its peak in 1930, Grigsby-Grunow employed 11,000 workers at its Chicago factory and sales reached $61 million annually. In April, 1930, as a divarication of business, the company announced the formation of *Majestic Household Utilities*, a new subsidiary to manufacture refrigerators and other appliances, such as vacuum cleaners and washing machines. Deliveries of Majestic refrigerators began in October.

In their radio business, the product range was extended rapidly with a range of models (Figure 352). The Majestic "Smart Set" line of less expensive, but stylish, table radios was introduced in mid-1933 and enjoyed strong sales. Grigsby-Grunow introduced their 7-tube Majestic Model 174 'smart-set' tombstone radio in late-1933/early-1934. It embodies an innovative 'tune-o-stat' feature that turns the radio on and off at chosen times during the day, automatically tuning the radio to the station carrying a desired broadcast at the programmed time.

In the aftermath of the October 1929 stock market crash, as the Great Depression began to bite, things were at first slow to change at Grigsby-Grunow. Next to economic issues, Grigsby's problems were compounded by litigation and royalty issues. In early 1930, Magnavox filed suit alleging infringement of their Jensen and Pridham dynamic speaker patents. The Magnavox dispute was only one of Grigsby-Grunow's ongoing litigation woes. On June 26th, Grigsby-Grunow filed a $30,000,000 suit against the RCA 'radio trust' claiming triple damages from alleged restriction of trade through the operation of an illegal pool of 4,000 patents. But that claim was dismissed as the RCA trust was dissolved in 1933.

Nonetheless, Grigsby-Grunow declared bankruptcy in November, 1933, and ended production of Majestic Radios in February, 1934 followed by its liquidation. The trademark 'Majectic' was acquired by a new investment group by forming the public company the Majestic Radio & Television Corporation. But also that company got into financial problems and filed for bankruptcy in October, 1939. It survived the war, but the post-war efforts to rekindle the radio business failed and the next bankruptcy loomed in February 1948, and the court ordered liquidation in November 1948.

## The Rise and Demise of Philco

PHILCO was another leading radio manufacturer in the same region. In 1892, Thomas Spencer, his brother Frank G. Spencer, Frank S. Marr and two business associates started a small company in Philadelphia, Pennsylvania to produce carbon arc lamps. However, after by 1899 the carbon arc lamp business had come to a standstill, they reorganized and created by 1906 the Philadelphia Storage Battery Company, manufacturing storage batteries for the rising number of automobiles. The replacement battery business was going well by 1910 and the company made a profit, and by 1911 the chemist James M. Skinner was hired. On his suggestion, the company started making an electric start motor replacing the former crank handle to start the internal combustion engines of automobiles. It worked out well and by 1913 they had reached $576,000 in sales, growing to over $1 million in 1917. By 1919 the trademark Philco was created.

As the 1920s dawned, a new curiosity was beginning to be more widely known: the radio that soon experienced the radio boom. By 1923, the number of broadcasting stations had grown to over 580, and was still growing. More than 200 manufacturers were producing radio receivers. And they were all battery powered.[538]

So the company went into radio-batteries, and their chargers as well. Philco's sales for 1924 rose to $4,700,000. Another smart move was to offer Socket-Powers (DC rectifiers), that made it possible to power the radio from the AC-home infrastructure of light sockets. Sales rose to $15 million in 1927 and over a million units had been sold. But that not last, as other radio companies designed radios powered from light sockets and having tube rectifiers on board. Radio batteries became overnight obsolete and Philco faced business collapse. Philco's management team headed by Skinner had to make some important strategic decisions.

**Figure 353: Philco Radio Receivers.**

Philco 511 "Coffin" Radio & Model 211 Speaker (top, 1928), Philco Model 70 Baby Grand Cathedral-style Radio (middle, 1931), and Philco Model 16X (bottom, 1933).

Source: www.tuberadioland.com/, https://philcoradio.com/gallery2/1934a/

---

[538] A cumbersome array of lead-acid batteries powered the early radio; the so-called A and B batteries.

In 1926, Philco decided to go and make their own radio receivers. To get a license from RCA and Hazeltine Laboratories (holders of relevant radio patents), they bought two other companies; the Murdock Company and the Timmons Radio Products Corp. So, they started their own product line of radio receivers; the 511 Series (Figure 353).

The first Philco radios were introduced in mid-1928, and 96,000 units were produced that year, making Philco radios 26th in the nation in production volume. In 1930 the company sold 600,000 radios, grossed $34 million, and was the leading radio maker in the country. By 1934 they had captured 30% of the domestic radio market. In the next years a flood of new models was introduced.

Volume was now the name of the game. As most early radios were individually hand-made around a chassis supporting the components, and a cabinet, their pricing was oriented at wealthy customers. To serve the lower-priced mass market, the company switched to mass production. It cost Philco $1 million to recondition its plant, another $1 million to install a conveyor system, and an additional $2.5 million for supplies and materials. The new assembly line was finished in April 1929 (Figure 354), but had to be shut down in May due to many technical problems, creating financial distress. After being refinanced by the Philadelphia National Bank and the technical problems solved, they introduced a new product line in June 1929 based on a new type vacuum tube.

**Figure 354: Philco Manufacturing (1933/1936).**
Source: https://philcoradio.com/, Science Museum

By the end of 1929, Philco had moved up to third place in the radio industry, behind Majestic and Atwater Kent. They sold over 400,000 radios that year and were able to repay all of their bank loans by March of 1930. The gamble had proven to be successful. In 1930, Philco sold 616,000 radios, grossing close to $34 million[539] achieving the number one position in radio sales in the U.S. The year 1930 also marked the introduction of Philco's first 'cathedral' design —the Model 20— which was an instant sales hit and the first of a long series of very popular Philco cathedrals. By 1934 they had captured 30% of the domestic radio market.

Philco's product strategy resulted in a wide line of radios beginning with five-tube sets all the way up to high-fidelity consoles with 20 tubes in 1937-38. The 'cathedral' style of different model sold over 1930-1938 some two million unts. Philco also made more simple battery-powered radios which were by then called "farm radios", most of which had cabinets identical to their AC powered versions. Novelty created was part of Philco' product strategy. Not only in styling of the cabinet design by leading industrial designer, on the technical side the new super heterodyne circuit was adapted. And the new rage of automobile radio receivers was met by acquiring a car radio manufacturer. Philco began doing research into high fidelity sound in late 1931. Its first efforts at high fidelity were made

**Figure 355: Philco Promotion and Advertisement.**
Philco News Magazine for dealers (top), and Philco Advertisement (bottom, 1933).
Source: Wikimedia Commons.

---

[539] Based on https://philcoradio.com/library/index.php/philco-history/. The real wealth of that income be $527 million in 2020. Source: https://www.measuringworth.com/

available in January 1932, incorporated into two new consoles, Models 112X and 90X. The firm also introduced smaller low-cost tabletop models adopting Alfred Sloan's philosophy "A model for every purse and purpose."

The product strategy was complemented by marketing strategy in which promotion was an important element. The dealer franchise network was supported by a comprehensive distribution and sales network, sales incentives and sales support (eg the Philco News magazine). And the market was flooded by advertisement running in popular magazines (Figure 355).

The year 1932 was the worst year of the Depression. Things were getting worse for everyone, and Philco did not escape unscathed. License fees for RCA became an issue brought to court. Philco's competitors were claiming that Philco's pricing of its sets were ruining the radio industry. New developments in broadcasting were on the horizon, and Philco started operating a TV station W3XE in 1932.

At that time the 'Radio with Picture' (ie the television) came on scene, capturing the business interest of many manufacturers. So, Philco began work on all-electronic TV in 1931, with TV pioneer Philo T. Farnsworth joining the company that year, bringing his patent rights along. But that cooperation went into personal problems and Farnsworth left.

By 1933 the US economy improved, sales went up again. Philco's gross profit for 1934 was up to $33,000,000 on sales of 1,250,000 sets. Philco sold more than two and a half times as many radio sets that year than did its nearest competitor RCA. Nevertheless, by then the tide was turning and in 1937 Philco's sales was $40 million, having sold 1,500,000 radios, with a loss of $100,000. In May 1938 the United Electrical, Radio and Machine Workers went on strike against Philco. The production was halted for month, and the Union 'won the battle but lost the war'. After settling the strike, Philco began to purchase most parts from outside sources, cutting back on its labour force. As a result, by 1941 Philco's employment had gone from a peak of 10,000 to 12,000 down to around 5,000 people. As unit sales slipped in 1938 due to the strike, dollar sales fell to $22 million, and the loss rose to $222,000. On the positive was that in 1939 that Philco won its lawsuit with RCA over patent royalties. RCA was forced to return $750,000 in excess royalties, and to draw up a new licensing agreement with Philco.

In the meantime, Philco had diversified into the 'white good' business by acquiring an air-conditioning manufacturer (1938) and refrigerator manufacturing division (1939). Dropping the old product

Cool Wave air conditioner line for a new design, with a new advertisement campaign they sold 89,000 refrigerators in 1940.

Philco had been a privately held company, its executives owning stock and receiving their dividends from the profits the company made. However, those shares would be worth a bundle when an IPO (Initial Public Offering) on the stock market was considered. It is not hard to imagine an interest in such a move, but there seemed to have been an opponent in the form of Skinner, the largest stockholder. Before 1939 was over, however, Skinner resigned from Philco, and in 1940 the company went, after a reorganization, public as the *Philco Corporation*. A total of 1,221,100 shares were sold by the company and by its officers.

Then came the Second World War, and Philco went, next to the radios and refrigerators, into radar. By early 1943, Philco had Government contracts totalling $50 million and letters of intent for $100 million more. During the war, also several important organizational changes took place at Philco. In April, 1943, the Philco International Corporation was formed to directly handle the distribution of Philco products throughout the world. With the end of the war, returning to civilian production, came also a redefinition of the business. In 1947 the 'Home and Auto Audio' segment was reduced to 24% of total sales, while the new 'Radio-Phonograph and Televisions' segment had exploded to 30%.[540] Philco entered a new era in its corporate business development. And the radio business was a diminishing part of it.

> *Thus, it is clear that Philco's success in the 1928-1937 period was due to an innovative but integrated strategy. Philco's executive stock ownership plan attracted and kept a high-quality management team and motivated then towards the common goal of profits. This team designed a clever production and control strategy which kept manufacturing costs low, thus, allowing Philco to offer the public more value for its money. Philco's product strategy and aggressive advertising campaign were tailored to augment the "selling-up" scheme which Philco used to move expensive sets when the competition could only sell midgets. The volume that was achieved, as well as Philco's tight production control program, allowed Philco to attract and keep the best radio dealers in the country. It was a well thought out, carefully honed strategy.* (Wolkonowicz, 1981, p. 47)

---

[540] Philco was a late-entrant in the television business. Its early television research in the 1930 had not resulted in production models.

## The Radio Corporation of America

An important player in the Era of Radio Broadcasting was the Radio Corporation of America, formed on October 17th, 1919 from the American Marconi Co. Over the next few months, it became a patent trust between AT&T, General Electric, Westinghouse, and United Fruit Co. so radios could be freely manufactured by those companies. RCA-branded radios during the 1920's were manufactured by GE/Westinghouse with a 60/40 split between the two companies. Both Westinghouse and General Electric would be become the dominant parties in the first two years of the broadcast era, and RCA would act as their sales agent. And as the holder of key patents of wireless communication, RCA was in an industry-dominant position, and being reluctant to licence its patent rights.

In 1922 RCA began selling receivers under the 'Radiola' name. The product line Radiola had its origin in the 'music box' of the Aeriola line (Figure 356, top), that originated from earlier RA and DA receiver sets manufactured by Westinghouse (Figure 330). In 1922 RCA introduced the Radiola product line of radio receivers. Starting with the Radiola I (Figure 356, middle), it manufactured ten models up to 1924. From 1927 on, it introduced a new line of Radiola's ranging in price from $69.50-$$895. Starting with Models 16 (Figure 356, bottom) through 86 (1930). They were manufactured in volume; Model 16 sold 85,000 units, Model 18 some 250,000 units and Model 33 some 236,000 units.

**Figure 356: RCA Radiola Radio Receivers.**

The Aeriola Jr receiver with headphones (top, 1921), RCA Radiola Model I (middle, 1922), and the Radiola Model 16 (bottom, 1927).

Source: https://www.radiolaguy.com/

Although the patent cross-licensing agreements had been intended to give the participants domination of equipment sales, the tremendous growth of the market led to fierce competition, and in 1925 RCA fell behind Atwater Kent as the leader in receiver sales. In addition, its monopolist position gave rise to governmental scrutiny, resulting in a forced break up by 1933.

In 1929 the company made its first moves into consumer electronics products when RCA purchased the Victor Talking Machine Company, then the world's largest manufacturer of phonographs (including the famous "Victrola") and phonograph records.

These are just a few of the many examples of the development of Radio in the US in the 1920s-1930s; the time known as the Roaring Twenties. A period of time in which the Goldrush of the Radio Craze resulted in massive industrial dynamics with hundreds of pioneering participants in the mid-1920s. Participants that saw their booming business flourish and perish, as the shake-out reduced their numbers. Only eighteen radio manufacturers and assemblers remained in business by 1934. So, the Roaring Twenties and the Great Depression became a period where the new means of communication broadcasting entertainment and news influenced people's life style. Creating the new institutions of 'Radio Stations' broadcasting their 'programs influencing American culture.

> *An amalgam of nearly every public institution and a trusted guest in the private homes of millions of Americans, radio had by the 1930s announced itself as a new social space unifying the nation in the face of daunting social and economic uncertainty.* [541]

The developments that took place in the US, were echoed in Europe in a quite similar, sometimes in a little different way, depending on the local context.

## Radio Manufacturing in Europe

Also in Continental Europe, the making of radio equipment was a separate development from the radio broadcasting. Like other developments in the Era of Electric Communication, this also was the work of creative and entrepreneurial individuals. Individuals that worked on their own, or individuals that worked in the R&D departments of the communications industry. It was the period of time in which the inventor-entrepreneur rose to prominence.

---

[541] Hilmes, H., Loviglio, J.: Radio reader: Essays in the cultural history of radio. Psychology Press 2002) p. IX

*Radio manufacturing in Britain*

Britain had known a society of ham radio amateurs that were gagged during the war. But as soon as the war had ended, they were back and the first complete wireless receivers appeared in the UK in 1920. Some were manufactured in the factories of the big American Electric Companies that had established their own subsidiaries in Britain. Such as the receivers made by the British Thomson Houston Company (BTH), and the Edison-Swan Electric Ltd. Also the British Marconi Company started to make receivers, and introduced the Marconiphone in 1922 (Figure 357, top). Their later receivers complied with Post Office specifications and tests, and was therefore awarded the BBC authorisation stamp. Other receivers introduced in 1921, were manufactured by local Home-kit manufacturers like Burndept Electronics (Figure 357, middle), and L. McMichael Limited. Even the Marconiphone was offered in kit-form.

Although many different valve receivers were available from about 1923, crystal sets remained the most popular receivers for several years. This was due to their low cost and freedom from the expensive high-tension batteries, and the re-charging of the low-tension accumulators in the valve receivers. Valves that had a relatively short life, and needed frequent replacement.

**Figure 357: Early British Radio Receivers.**

Marconiphone V2 Radio & A2 Amplifier (top, 1925). Burndept 'Ethodyne' superheterodyne receiver wiht aerial (middle, 1925). Cossor Empire Melody Maker Model 234 (bottom, 1931)

Source: https://www.technogallerie.com/; collection.sciencemuseumgroup.org.uk; https://cossor.com/.

Next to the big companies and early pioneers, by the mid-decade a lot of new entrants of the wireless receivers entered the radio business. Such as the battery manufacturer Ready Company that acquired radio-maker Lissen Ltd in 1928 to enter the radio-business. Or Roberts Radio (founded 1932) that introduced in 1933 its first cabinet style radios as kit and became a British icon. However, the failure rate was among small-scale manufacturers, was also high.

In the 1920's and 30's there were introduced a large numbers of radio kits. These were popular as they were a lot cheaper than the ready built receivers. Such a supplier was A.C. Cossor Ltd., a manufacturer of radio tubes that sold the Cossor 'Melody Makers' Kits also completely assembled (Figure 357, bottom). Introduced in 1927, more than 120,000 units had been built by March next year.

By 1930 there were five million radio sets in Britain, but listeners had little alternative but to tune in to the BBC. Demand however existed for more programming of popular music; especially for dance band music and hot jazz. To exploit this interest, a private company, the *International Broadcasting Company* (IBC) was set up by Leonard Plugge. IBC hired air-time from overseas stations and transmitted popular programmes aimed at the UK market from Radios Lyon and Normandy, Radios Athlone, Méditerranée and Radio Luxembourg. These programmes were perfectly legal —the BBC itself was transmitting to the continent— but the attitude of the BBC and the government towards the IBC was continuously hostile.

The scene was set for ten years in which UK airwaves became a battle zone between public service broadcasting and commercial radio interests.

*Radio Industry in Germany*

In the economic, social and political upheaval of the after-war period, also the radio industry grew fast as all that radio equipment had to be manufactured. On the one hand it were the large companies like Telefunken and Siemens & Halske that expanded their radio department into manufacturing radios. On the other hand, there were the smaller companies like Nordmende, Saba, Blaupunkt, and Radiofrequenz.

## The Rise and Rebirth of Loewe Radio

The company Radiofrequenz GmbH was started in 1923 in Berlin, when Siegmund Walter Loewe and his brother David Ludvig Loewe established a radio manufacturing company. After the development of a new type of vacuum tube 3NF—a vacuum container containing the frames of three smaller radio tubes, aka the multi-element tube (Figure 201)—he used it for the OE 333 Regional receiver (the Ortsempfänger) (Figure 358). On March

3rd, 1923 he organized a radio demonstration in his laboratory for President Friedrich Ebert[542], that opened later many doors. This radio, which was soon to be built in millions, sold for 39.50 RM, a third of the price of comparable devices. In September 1926, he introduced a similar radio for middle and long wave bands.

Despite its rapid advances, in the mid-1920s Loewe was forced to cease production of tubes. The action was initiated by a consortium of European manufacturers, including Siemens & Halske, AEG, Felten-Guillaume, and Telefunken, who controlled rights of certain essential electronics parts invented by the Austrian Robert von Lieben (Figure 213). As a result of these restrictions, for approximately one-year Radiofrequenz endured serious financial hardships. Siegmund Loewe, however, did not intend to surrender. In 1926 he filed a suit in the German patent office claiming the rights to six patents. The legal action prodded Telefunken to accept a compromise. On August 15th, 1926, Radiofrequenz was able once again to produce and sell multielement tubes in Germany.

**Figure 358: The Integrated Vacuum Tube used in the Loewe Radio Receiver.**

The Loewe Multi-system 3NF vacuum tube used in the Loewe Regional Receiver OE 333 (1927).

Source: http://lampes-et-tubes.info/, Loewe

By 1926 the German radio industry had experienced a thorough consolidation, shrinking from about 150 companies in 1924 to only 40 in 1926. However, Radiofrequenz survived and demand was so great that within a year the company had outgrown its first small workshop in Berlin. By the end of 1928, the company moved to a large, new factory in Berlin and increased its workforce to more than 1,100 people.

In 1929 the company was reorganized as a holding company, Berliner Radio-Handels-Aktiengesellschaft. At the same time, it began setting up its own specialty companies for the production

---

[542] Ebert was so impressed by the invention that Radiofrequenz was able to persuade the German Post Office—the public body that had authority for regulating broadcasts—to form an association for making commercial radio broadcasts. Radiofrequenz was one of the first companies to receive government permits to produce and sell radio receivers.

various devices and parts. For example, Loewe Audio GmbH was founded to produce radio tubes, Orthophon-Apparatebau GmbH for the production of radios, loudspeakers and resistors, and Eudarit-Pressgut GmbH for spare parts for radios. The products of all these companies were being sold under a single brand name; Loewe-Radio.

German radio was in such a strong growth phase at the end of the 1920s that the industry was essentially unaffected by the onset of the Great Depression in 1929.At its base was the explosion of the German radio audience —which expanded from three million in 1929 to over four million in 1932— and the introduction of mass-produced radio parts which lowered the cost of sets considerably. But the Interbellum was also the time that saw the rise of nationalism, and the power of the Nazi Party. And the Nazi government, from the beginning of its regime in 1933-1934, forced changes on the company.

>It pressured the partly Jewish Loewe brothers into emigrating. So, after Nazi-instigated demonstrations against David Ludwig Loewe, he moved in 1933 to England. His brother Siegmund Loewe remained in Berlin where he was forced first from his leadership of the German Broadcasters Association in 1933 and then from the chairmanship of Radio AG D.S. Loewe in 1938. By then he fled to Switzerland, having all his companies confiscated by the Nazis. [543]

As the radio production was considered an essential war industry, the by then confiscated company received practically unlimited financial support from the Bank der Deutschen Luftfahrt (German Aviation Bank). Subsequently, between 1937 and 1943, the firm's revenues rose from 8.4 million RM to 62 million RM. After the end of the war, the Loewe plants in the Russian zone (ie Berlin) was completely stripped by the Russian Army, and the other factories (ie East-Germany) were nationalized. By 1947, Siegmund returned to Germany, and Allied forces returned his ownership. Notwithstanding the seven-year ban of wireless broadcasting after the war, the Loewe-companies blossomed during the Western German Economic Miracle (aka Wirtschaftswunder). It went into commercial television in 1952, and diversified into other consumer products like the Optaphon (a tape recorder). By the mid-1960s, the company went through a complicated chain of restructuring and consolidation.

In 1963 Dr. Siegmund Loewe passed away. Following his death, the Loewe family divested itself of all of its shares in the company, selling 15% to the Dutch company Philips Electronics.

---

[543] Source: Loewe AG. https://www.encyclopedia.com/books/politics-and-business-magazines/loewe-ag; https://www.deutsche-biographie.de/sfz53906.html#top

## The 'Volksempfänger' as Vehicle for Propaganda

Notwithstanding the Great Depression, the new phenomenon of radio had transferred into part of 'home entertainment.' Also in Germany range of companies offered the steeply priced furniture-style radio models selling for 200-400 RM. In 1925 there were some sixty German companies manufacturing this kind of radio sets. But the low-priced mass market was still untapped, while the public interest was booming.

It started in 1924 with the first 'Große Deutsche Funk-ausstellung' (Great German Radio Exhibition). More than 170,000

**Figure 359: Early German Radio Receiver.**
The 'Siemens D-Zug' radio (top, 1924). Loewe Opta EB100 (bottom left, 1928). The bakelite Volksempfanger (bottom right, 1930s).
Source: Wikimedia Commons. Newsiemens.com

visitors saw the rather sober presentation without glitz and glamor. Siemens & Halske presented their first, three-level tube receiver, known as the 'Siemens D-Zug' with a horn loudspeaker (Figure 359, top).

By 1926 the German radio industry had experienced a thorough consolidation, shrinking from about 150 companies in 1924 to only 40 in 1926. But the market continued to grow. By the early 1930s, everyone wanted a radio as the new invention of radio broadcasting brought news, music, dramas, and comedy right into the home. Propaganda Minister Joseph Goebbels saw its potential to transmit Nazi messages into the daily lives of Germans. The only hurdle was producing and disseminating the devices on a mass scale. Then the National Socialists 'invited' companies to manufacture a low-cost, mass-produced radio receiver for the emerging mass market. On the one hand, a few small radio equipment manufacturers were unable to meet the capped prices imposed by political power, so that several of these manufacturers went bankrupt or were absorbed by larger competitors; but on the other hand, as the number of radio listeners increased sharply, some companies rose to prominence.

Under Goebbels's direction the 'Volksempfänger', or 'people's receiver,' was born (Figure 359, bottom right); the model "VE 301", presented in August 1933 at the 10th Great German Radio Exhibition in Berlin. More than 100,000 units sold during the first two days of the exhibition.

The purpose of the Volksempfänger program was to make radio reception technology affordable to the general public. Ads positioned the Volksempfänger as the intermediary for the greater German community that would make the country strong and prosperous again by bringing political, cultural, and economic ideas into every household (Figure 360). Regular programming included operas, classical concerts, light dance music, games, jokes, and popular arts. To be sure, programming was highly censored. Forbidden content included 'corrupted music,' such as American jazz, pop, and swing.

**Figure 360: German Radio Broadcasting.**
Poster for the Volksempfänger (top) and German family listening to it (bottom).
Source: https://spectrum.ieee.org/, Ullstein Bild.

The Model VE301 was available at a price of 76 German Reichsmarks (RM) equivalent to two weeks' average salary. However, it only could receive 'Deutschland-sender' and the local

'Reichssender.' Despite a comparatively high monthly radio license fee of RM 2, the number of radio equipment in German households increased from 25 to 65 percent between 1933 and 1941.

All radio companies in the German Reich were obliged to produce the radio equipment developed at the instigation of the Propaganda Ministry in the same way. In addition, precise instructions from the Propaganda Ministry regulated the media coverage down to the smallest detail. In order to counteract the listeners turning away from the weariness of Nazi propaganda, at Goebbels' instructions, numerous request concerts, entertainment programs, radio plays and additionally the Wehrmacht reports got a permanent place in the programs.

## The Rise and Bombing of Nordmende

The German Otto Hermann Mende (1885-1949), born in a family of forest workers, and the businessman Karl Rudolf Müller founded the company Radio H. Mende & Co in 1923 in Dresden. Two years later, the engineer Ulrich Günter and the nephew of the founder Martin Mende joined them and the company quickly developed into a flourishing business. Under Ullrich, the Mendes company rose to become one of the largest German radio manufacturers. Like the other companies (eg Telefunken, Schaub-Lorenz, Radione), also Nordmende manufactured a 'Volksempfänger (Figure 361, top).'The production numbers rose from 2,000 devices/year (1925) to 200,000 devices/year (1937). In 1938 the annual production was 250,000 radios with a workforce of 3,000. Mende offered broad range of table-top and cabinet models (Figure 361, bottom).

**Figure 361: Nordmende Radios.**
The Nordmende Volks (top, 1938) and product advertisement (bottom, 1938).

Source: https://www.radiomuseum.org/, http://www.hifi-archiv.info/Nordmende

During the Second World War the company focussed on the production of military radio equipment. After the bombing of the factory in 1945, Martin Mende (son of the founder) created a new company in Bremen, the "North German Mende Broadcast GmbH" whose name was later changed into NORDMENDE. In June 1945 the Russians dismantled the factory and transported everything to the Soviet Union.

Notwithstanding these disasters, Nordmende became one of the leading German manufacturers of radios, televisions, tape recorders and turntables in the post-war period.

*Radio Industry in Continental Europe*

In some other European countries, the invention of the vacuum tube had spirited electric manufacturers to manufacture vacuum tubes. Then, the use of those tubes into specific wireless equipment was a logic step. As was the manufacturing of the radio receivers.

In the Netherlands Philips Gloeilampen Fabriek NV (Philips Lightbulb Factories Ltd.), already prominent in the manufacturing of incandescent lamps, entered also into the manufacturing of vacuum tubes. As a logical extension of the build-up knowhow, the next strategic

**Figure 362: Early Philips Radios.**

Philips Model 2501 (top three parts, 1927), Model 930 (middle, 1931), and Model 536 (bottom, 1935).

Source: Wikimedia Commons. Royal Philips. https://www.tsf36.fr/ ephilips.htm

step was manufacturing radio receiver sets. And the result was the Radio Set 2501 (Figure 362), in Holland nick-named "the loaf of bread." Because of the high gain of the pentode, only three radio tubes were needed compared to four in radios from competitors. The receiver was an enormous success. In 1927, 6,000 receivers were built; in 1930 the production had already increased to half a million radios a year.

Also in the Scandinavian countries ham radio broadcasting emerged after the First World War. After the Swedish Broadcasting Organisation (Svenska Radioaktiebolaget, SRA) in 1925 started broadcasting, its founders of the Swedish electric industry

**Figure 363: Danish Radio Receivers: Bang & Olufsen.**
B&O Eliminatoren (top, 1927) and B&O four lamp's receiver (bottom, 1928)
Source: http://www.antik-radio.dk/

(eg L.M. Ericsson) became involved in radio manufacturing. The SRA Radiola brandname was born. And the Firma Radiofabriken Luxor, that started in 1923, introduced the radio receiver brand Motala. Peter Bang and Svend Olufsen started the company Bang & Olufsen, and developed the B&O Eliminator —not a radio receiver but a power unit replacing the radio batteries — in 1927. Soon a range of radio receivers followed (Figure 363).

Their manufacturing facilities were the workshops where the radio receivers were assembled individually (Figure 364).

In France, where wireless radio was a military affair, the early radio industry also emerged after the Era of Amateur/ Ham radio with its crystal sets of the early 1920s. It was characterized by companies like Société des Établissements Péricaud S.A., the compagnie Radiotechnigue, the Ducretet-Thompson Company, the Etablissements Radio-LL (all in Paris), and the company Radio-Techna from Châlons-sur-Marne.

In many cases the precursor of these companies had been active in the manufacturing of crystal radio receiver sets, early radio tubes, telephony

**Figure 364: Nordic Radio Manufacturing.**
Bang & Olufsen Radio manufacturing in Struer, Danmark (left, 1928) and Luxor Radio Factory in Motala, Sweden (1934)
Source: Wikipedia Commons.

and gramophone equipment and expanded their business by building radio receivers.[544] Others emerged from their broadcasting experiments; such as the Etablissements Radio-LL of Lucien Levy launching Radio LL in March 1926.

In the period of a decade, they followed a similar trajectory as the manufacturers in other countries; from the crude receivers of the 1920s, to the cabinet style designs of the 1930s. And all those manufacturers were competing for the same customer; the family that looked for home-entertainment (the music), wanted to know what happened elsewhere (the news), followed the progress of their favourite sport team and was entertained by the radio-stars of those days (Figure 337). So, they radio as furniture obtained a prominent place in family-life, which was emphasized by the advertisement for radio (Figure 365).

### Early Radio Receiver Advertisement

Among the experimenters of ham-radio, the component suppliers promoted their 'radio' products in the hobby magazines. During the Roaring Twenties, there was a huge demand for receivers by the general public. Between 1923 and 1930, 60 percent of American families purchased radios. However, as those radio receivers did no become cheap, the manufacturers had to convince their prospect buyers of the quality of their models. So they referred to lifestyle; they showed in their advertisement the

---

[544] In 1928, French manufacturers were confronted for the first time to the competition with a major foreign radio company. Previously specialized with the production of vacuum tubes, the Dutch society Philips began the production of radio receivers with a level of industrialization neither met in Europe before.

**Figure 365: Advertisement for Radio Receivers.**
Affiche for US Atwater Kent Radio (left, 1933), French Radio Monopole (middle, 1934) and Dutch Philips for the British market (right, 1938).
Source: Librarie Elbe, Paris. Wikimedia Commons.

modern, sportive and educated people (often female). Notwithstanding the 1929-Wall Street Crash and the subsequent economic depression, sales of complete receivers sets started booming in the 1930s. As the competition became fiercer, the manufacturers put some more effort in marketing them, by advertising in journals like the Saturday Evening Post Journal and Ladies Home Journal, and in the newspapers (Figure 365).

That was the advertisement *for* the radio, but the 1920s also saw the rise of advertisement *on* the radio. The time of the radio commercials that became part of the business model for the radio stations. It became a solution for the classic funding problem of public broadcasting services.

> The first recognized form of radio advertising came in early 1922 when AT&T began to sell 'toll broadcasting' opportunities, in which businesses could underwrite or finance a broadcast in return for having their brand mentioned on air. Later that year, the New York radio station WEAF was the first station to run an official paid advertisement. These forms of radio advertising become more and more popular, leading stations into the golden age of radio.

During radio's golden age, advertisers sponsored entire programs, usually with some sort of message like "We thank our sponsors for making this program possible", airing at the beginning or end of a program. While radio had the obvious limitation of being restricted to sound, as the industry developed, large stations began to experiment with different formats. Advertising became a hot commodity and there was money to be made.

## Radio Design 1930s-1950s

The first consumer radios in the 1920s were a complex mix of separate elements consisting of a receiver with multiple controls, a battery, and a loudspeaker (or headphones). They were technical artifacts and they looked like it (Figure 356, Figure 357, Figure 358).

But that changed as the radio receivers of the 1930s became a part of the home-entertainment, and thus became parts of furniture. The result was wooden radio's with Victorian styling, housing and hiding the radio components intended for the lounge room in cabinets and tombstone-radios (Figure 352, Figure 353).

With the progress of technology enabling compacter designs and increasing functionality —next to shortwave and longwave AM, FM had become available— also the design of radios changed into a whole new type of radio. Under the influence of the new Art Moderne (aka Art Deco movement), as well as the availability of new materials (eg bakelite) the radio receivers got their table-top designs, some quite futuristic (Figure 366). As the trend for modernism evolved, the distinctive Art Deco style of the late 1920s was overtaken by Streamline Modern design (Figure 366, bottom). Radio became an

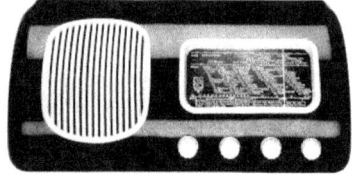

**Figure 366: Art Deco Radio Design.**

The Spartan Table Radio Model 557 (top, 1936), B&O Beolit 39 (upper middle, 1939), Firestone Air Chief Radio (lower middle, 1939) and Stewart Warner 03-5B1 Art Deco Bakelite Tube Radio (bottom, 1939).

Source: Brooklyn Museum, Bang and Olufsen, Deutsches Technikmuseum. Smithsonian. https://artdecocollection.com/

object which consumers chose for their home not just based on what they would hear but how it would look. Radio became a visual as well as an aural experience.

The preceding —rather fragmented—exploration, shows how the manufacturing of the early radio-receivers in the Interbellum-period was massive, widespread and fragmented in the western world. The manufacturing of radio equipment (both sending and receiving equipment) started when —in a rather low-entry business[545]— many inventor-entrepreneurs with a background in ham radio jumped on the bandwagon, and formed their own radio-companies. Soon followed by other entrepreneurs that —after being manufacturing other electric products such as the AC/DC rectifier units— moved into manufacturing radio receivers to serve the exploding radio receiver market. This latter was business that developed in parallel with an explosion of radio broadcasting stations.

## Radio Manufacturing (1950-1970)

Before the Second World War, a radio receiver was part of the furniture (Figure 352, Figure 353) and the wooden cabinet design was dominant. After the war, with the rise of new materials and

**Figure 367: European Table-top Radio Receivers (1950s).**

Radio Telefunken Model Operette 50 (top, 1950), Grundig Model 2043W 3D (upper middle, 1954), Saba Swatzwald 6-3D (lower middle, 1955), and Philips Serenata Model B5A73A (bottom, 1957).

Source: Deutsches Technikmuseum. https://www.anticaradio.com/

---

[545] A business that does not have technical, financial, monopolistic, or regulatory barriers. In technological terms, radio was not complicated, in financial terms setting up a production shop did not require massive funding.

extended functionality, the radio receiver found its 'dominant design.' As it became a house-hold item, the functionality gained from the appearance. The experimental futuristic designs of the late 1930s (Figure 366), heralded the change in radio design to come after the Second World War. A change into smaller models in the US, and to the specific European table-top models (Figure 367)

The radio manufacturing industry had matured, together with the progress of the vacuum tube technology and the functionality of the radio receivers. But then the advent of the television broadcasting, dramatically changed the home-entertainment landscape. Radio, which had been America's favourite form of at-home amusement, declined in importance in the 1950s. Variety, comedy, and dramatic shows left the airwaves to make room for TV-broadcasting. Radio increasingly focused on news actuality, talk shows, and sports broadcasting. And audio survived in specific niches.

## Music Entertainment

One such a niche was entertainment offered by the gramophone industry with recorded music (ie concert, opera, musical) reproduced from wax-rolls or flat shaped discs played on the 'gramophone' machines (Figure 368). It was a combined development of the mechanical carrier of music (the record) and the mechanical reproducer of the music (the record player).

**Figure 368: Music Players.**
Edison's Wax Cylinder Phonograph (left, 1899), Berliner's flat disc gramophone (right, 1890).
Source: Wikimedia Commons. Science Museum.

*The Record Player*: Originally, the record player (aka phonograph, later turntable) was a rather clumsy mechanical device. First the wax-based players came on scene, soon followed by the disc-based players (Figure 368). Following an evolutionary path of development, the record player, equipped with a crystal stylus, was mechanically decoding the content of rotating vinyl discs, and reproducing it mechanically in the loud-speaker horn. The record players, originally housed in box-style housing, became housed in cabinets. Sometimes together with the radio (ie the Radiogram) (Figure 369).

*The Music Disc:* The carrier of the recorded sound had soon evolved from the wax-cylinder to the flat record disc (Figure 370). Starting with the 78-rpm disc, that came in different sizes (7, 10, and 12 inch), emerged the later 'extended playing disc' (ie EP) that rotated at 45 rpm. To be followed by the 33 rpm 'long play' discs (LP).

A whole industry emerged following the business practises of that time. Early start-ups (eg Bell's Volta Graphophone Company, the American Graphophone Company,) that either faded away, or merged with others to become the pillars of a new industry. Such as the *Columbia Phonograph* company that by 1912 was split up in two companies, one to make records (ie Columbia Records) and one to make players (Columbia Phonograph). Or the *United States Gramophone Company* (1894) created by the German-born American Emile Berliner, with dependences in England, Canada, and Germany (ie *Deutsche Gramophon*). And the spin-off companies like the

**Figure 369: Early Gramophones.**
Pathe Reflex Coq table-top gramophone (left, 1915), HisMastersVoice Cabinet model 161, (middle, 1926) and EMI Marconiphone Model 575 (right, 1938).

Source: www.grammophon.ch, dogramofonu.pl

*Viktor Talking Machine Company* (1901). The latter using the brand named His Master's Voice (Figure 370).[546]

From this early development the record industry would follow a not too spectacular development over the next decades. After the Second World War, the record player became quite popular. The stylus converted the movements in an electric signal that could be amplified. And next to the 'turntable,' combined with the audio amplifier and the loudspeaker boxes (Figure 348) became part of the 'Modern Home.'

## Home Audio Systems

Radio broadcasting had brought the world into the private living environment. By the role of audio became even broader as both the 'hardware' (ie the audio systems) and the 'software' (ie the sound) found their own development trajectory in the Home Audio Systems. Basically, a record player combined with a radio-receiver.

**Figure 370: Recorded Music.**
Logo of His Master's Voice (top) and Red Label vinyl record disk (bottom, 1928).
Source: Wikimedia Commons. Science Museum.

*The Vinyl Audio Disk*

Notwithstanding the radio itself lost terrain to the television, the 1950-1970 period was the time of the play-back machines; the record players and tape recorders. And the music they reproduced was created during the Golden Age of the LP (1970s) by the music industry. The vinyl record business, mass producing their 'albums' under a range of different labels during the Album Era (1960s-2000s), bloomed.

> In 1948 Columbia Records introduced the long-playing (LP) record, which, with a rotational speed of $33^{1}/_{3}$ RPM and the use of very fine grooves, could yield up to 30 minutes of playing time per side. Shortly afterward RCA Corporation introduced the 45-RPM disc, which could play for up to 8 minutes per side (Figure 371). These LP's and 'singles' supplanted 78 rpm records in the 1950s, and

---

[546] After that company became acquired by RCA in 1929 it produced gramophone records under of the label of his Master's Voice.

**Figure 371: Early Records and Audio Cassette Player with tape cassette.**
33 RPM LP from Columbia Records (top row left, 1948) and 45 RPM Single from RCA Victor (top row right, 1949). Cassette player (lower row left) and cassette (lower row right).
Source: Wikimedia Commons, Mainatework.com, Dutchcharts.nl. Royal Philips.

stereophonic (or 'stereo') systems, with two separate channels of information in a single groove, became a commercial reality in 1958.

The record disc created a whole new competitive industry —the music industry— with large manufacturing plants. And on the horizon loomed the next generation of the audio-media; the Tape Cassette (1963) and the Compact Disk (1979) each with its own player.

*The Tape Cassette*

The magnetic tape as a carrier for information was already known for a long time, and used in the 1950s. By the 1960s it resurfaced in a different application area; the audio tape encapsulated in a cassette (Figure 371). The two-spool cartridge to be used in a cassette player was introduced in 1963. Within five years, 85 manufacturers sold over 2.4 million players. During the 1970s and 1980, the cassette's popularity grew as a result of being a more effective, convenient and portable way of listening to music.

## HI-Fi in the 1950s-1960s

The radio-receiver (aka tuner) was one element of the early home entertainment sets, the record player as part of the music reproduction system (aka the Audio set) another. To compete with the new medium of television broadcasting, audio had to find is new place at the home. And that niche was characterized by responding with better sound quality. A niche that became known as High-Fidelity (ie Hi-Fi) sound reproduction. And the sound systems developed accordingly by adding modular amplifier sets (eg pre-amplifiers and power amplifiers) (Figure 372, top), tape decks (Figure 372, bottom), and multi speaker sound consoles.

Especially the British and German manufacturers concentrated on the complete high-end quality audio systems (aka HiFi) for domestic sound reproduction.

The British *Acoustical Manufacturing Co. Ltd.* introduced their "Quality Unit Amplifier Domestic" (QUAD) line of audio sets. By 1959 the QC II and 22 control unit was complemented with AM/FM tuners (Figure 372, top). The Quad 33, introduced in 1967 and manufactured up to 1982, sold 120,000 units (Figure

**Figure 372: Hifi Audio sets.**

Quad II Amplifier (top, 1959) and Quad 33 audio set of pre-amplifier and tuner (right) and amplifier (left) (upper middle, 1967). Revox T26 (1951, lower middle) and Revox A77 tape recorder (bottom, 1967).

Source: Wikimedia Commons. Studer.

372, middle). Becoming known for their QUAD range of products, the company changed its name to QUAD Electroacoustics Ltd.

Next to the tuner and the amplifiers, there was the 'tape' recorder where the sound information was stored on a magnetized tape. Leading the tape past a magnetic head, the tape recorded (or replayed) the sound.

> In 1952, the Swiss/ German company *Studer/Ela AG* introduced for the consumer market the Revox T26 tape recorder in different versions for a hefty 1,240-1,650 CHF. By 1967 they introduced the A77 Tape Recorder series — in 186 different special versions for hospitals, air control tower and parliaments— of which more than 460,000 units were build up till 1977. The basic unit costed 2,000 CHF. (Figure 372, bottom).

## *Overview of the Early Days of Wireless Radio*

The evolutionary development of Wireless Radio around the turn of the twentieth century took place on several levels; among them the component level, the system-component level and the infrastructure level (Figure 310). Next to the component and system-component level developments, there were —in modern parlance— the service providers of the radio stations and radio network infrastructures.

- On the *component level* it was the development of the Vacuum Tube Technology that sprouted into a range of application trajectories; from the Receiver tubes and Transmitter tubes, to the Telephone tubes. Figure 373, middle)

- On the *system-component* level it was the development of the radio transmitter and radio receiver as wireless communication apparatus/equipment. The result was the continuous development of electronic circuits; heterodyne circuits that were designed both as oscillator circuit and mixing circuit. Circuit that became used for creating the transmission signal, and for the modulation and demodulation of the radio signal. Systems of radio transmission that became known as the system of AM and FM radio transmission (Figure 334, Figure 373, top).

- On the *infrastructure level* the use of 'Electro-magnetic waves' in wireless radio communication over an expanding radio spectrum (Figure 312), and the progress in the radio technology improving the system circuitry (Figure 334), paved the 'airways' (ie the frequency bands) for the broadcasting industry to use.

The period 1890-1920 was a dynamic period for wireless radio. Such as the early 1900s with the conflicts about 'radio'-patent rights (Figure 373, bottom). After the sparked-induced transmission, the single wave transmission and the modulated single-wave transmission emerged. And that opened new trajectories of transmission by designing circuits to handle the 'electric waves.' Trajectories in which the ham radio amateurs, some of them with entrepreneurial aspirations, started applying the novel ways of wireless transmission for other purposes than just experimenting. They started their companies with varying success, creating the early pioneering days of radio transmission.

This was also a period of time in which major economic interest became involved. Like the great telegraph and telephone companies that soon recognised the value of the new way to realize the radio communication. Also the military had recognised the importance of wireless 'radio' communication; like the US Navy that needed communication with their ships at sea. The more when the First World War disturbed the preceding peacetime developments and when during the war the Wilson-led US government took over the nation's radio, cable, telephone, and railroad companies. It the early aftermath of the First World War, the cards became shuffled as big businesses in telegraphy and telephony became interested (eg Westinghouse, AT&T). The next few years would see a battle for dominance by some of the largest companies in the United States, with the main battle 'players' being AT&T vs. RCA.

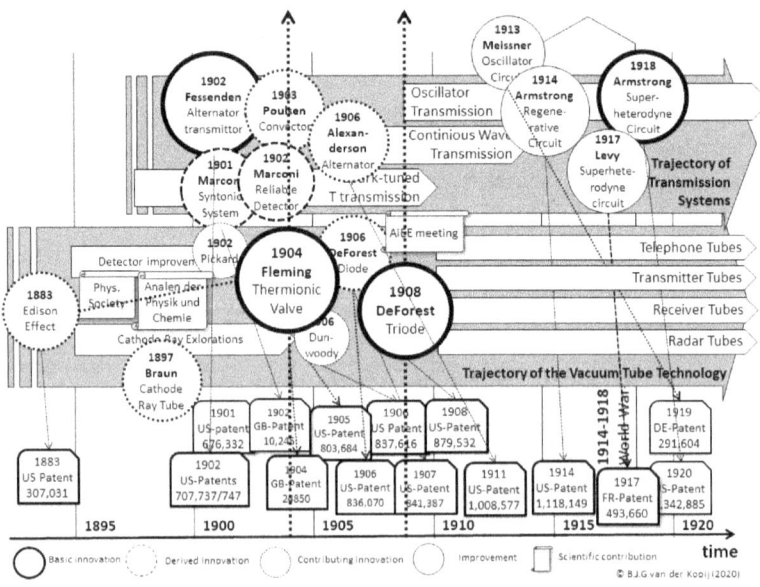

Figure 373: Overview of Vacuum Tube development in relation to Wireless Transmission (1890-1920).

## Periodization of the Radio Broadcasting Era

Looking at the development trajectory (Scheme 2) of the Wireless communication technology and seen in the perspective of the life technology cycle (Scheme 5), the radio technology is marked by specific events creating the basic innovations to which many thinkers and tinkerers contributed.

*Radio Broadcasting 0.0* (1900-1920): By the early 1900s, the developments of the wireless marine communication, as well of the rise of the ham radio and their hobby clubs, sparked of the broadcasting of sound to a general audience. The *Radio Craze* ignited the Wild West of the Air-waves, and governments stepped in with regulations of the radio spectrum. First manufacturers of hobby kits and complete sets became active. Engineering efforts introduced, after the crystal radio the heterodyne circuits. Also in Europe, the sparks of the US Radio Craze had ignited interest, resulting in a similar pattern of development.

*Radio Broadcasting 1.0* (1920-1945): In early 1922 the *Radio Broadcasting Boom* hit America. Technology was the driving factor. First the electro-mechanical technologies from the spark creation to the single wave generators. Followed by the early electronic technologies. And both Fleming's diode (1904) and DeForest Triode (1907) as active electronic component (Figure 188, Figure 192) were essential contributors to this development in electronic radio circuits (Figure 104). A development trajectory that saw an explosion after the First World War and created the radio's Golden Age in the second quart of the twentieth century.

> *When, after years of losing money, radio finally started to become profitable in the late teens, then grew explosively with the broadcasting boom in the early twenties, the "wise capitalists" at major industrial corporations like G.E. began to enter and dominate the industry, in particular by buying up most of the major patents. In contrast, after nearly two decades of pioneering work and struggling companies, in 1921 Lee De Forest abruptly sold most of his radio interests and moved on to other fields. De Forest later explained that he felt the time had come when "the building up of this technique and institution might better be left in the hands of those with greater capital, influence and personnel to carry on" and further noted that broadcasting "grew amazingly, once the large organizations with ample capital took hold of it".*[547]

A striking period in the development of wireless radio was the development of the heterodyne circuit. Of the many contributors to this type of circuit, it was Armstrong's superheterodyne circuit that proved

---

[547] Source: White, Th. H.: Big Business and Radio (1915-1922). https://earlyradiohistory.us/sec017.htm

to be essential. It spurred both the development of the system-components of the radio transmitters and radio receivers (Figure 334).

*Radio Broadcasting 2.0* (1945-1970): After the Second World War —with broadcasting technology maturing and adding FM-broadcasting next to AM-broadcasting— broadcasting stations grew in number, and radio broadcasting expanded exponentially after 1945, with many countries adding new languages and services. As it was the time of the Cold War and the radio became a tool for spreading ideologies among society (eg Radio Liberty). It was also the time of opposition to the state-dominated broadcasting as the free radios and the radio pirates emerged. From the land-based broadcasting stations (eg Radio Luxembourg) to the sea-based broadcasting stations (Radio Caroline and Radio Veronica). From a tool of propaganda, radio broadcasting became a tool of entertainment. Like the Community radio, where local people produced and broadcasted their own programs and participate in operating the station, serving local communities. Sometimes fitting in the regulations/legislation, but also as an illegal form of broadcasting.

*Radio Broadcasting 3.0* (1970-2000): In the last quarter of the twentieth century, the rise of television broadcasting heralded the demise of radio broadcasting. Starting in the 1930s in the US, the new medium of television soon conquered also the continent. The role of radio broadcasting in entertainment melted away, as news source it survived a while and as community radio it stayed on.

## Changing Perspective

In the preceding exploration we focussed on the evolution of broadcasting technologies, characterized by events that were initially of an electro-mechanical nature, but soon dominated by the new electronic device of the vacuum tube (aka 'radio' tube). In large brushstrokes we painted a picture of the technology-induced radio broadcasting developments in the first half of the twentieth century.

The subsequent technological development trajectories in the second half of the twentieth century, saw events that influenced radio broadcasting even more. Events that brought broadcasted 'audio' with news, culture and entertainment in homes. And 'Radio' influencing both individual life-styles as well as cultures, opposing state-dominated efforts to stay in control over the new medium. Broadcasting became part of society, reflected cultures, and created novelty on a large scale.

Now, changing our perspective, we will zoom in on the specific events that resulted in the development of the Television Technologies that brought 'video' in homes.

## The Television Revolution

After the marine-radio telegraphy that conquered the ether in the early 1900s, the audio spectrum became in the 1920s cluttered with radio broadcastings. A technical development that started with the simple ham amateurs, fuelling a cultural development called 'radio' that was to influence the Affairs of Man during the first half of the twentieth century.

With the development of the radio-engines —both the transmitters and the receivers— came the development of the radio broadcasting system and its infrastructure of regional and national radio stations. A development that took place in a couple of decades in a rather dynamic period of time (aka the Third Industrial Revolution). An Interbellum period that was flanked by two World Wars, and a period in time that saw both the cultural dynamics of the Roaring Twenties as well as the economic dynamics of the Great Depression. A period of time that brought a shock of massive novelty, both cultural and technical, and changed traditional lifestyles (Figure 374).

**Figure 374: The Culture Shock; Flintstones watching TV.**
Source: Wikimedia Commons.

In the decades before, technological progress had brought western societies new ways to communicate; the *Era of Telegraphic Communication* (ie the electric transportation of the coded word over distance), followed by the *Era of Telephonic Communication* (ie the transportation of the spoken word over distance), and the *Era of Wireless Communication* where electro-magnetic waves travelling the ether (Figure 106) replaced the copper cables. In addition, the rather simple artifact of the internal combustion engine unleashing the Power of Explosion, had ignited the *Automobile Revolution* and the *Aviation Revolution* that had revolutionized human mobility. This all paralleled by the artifacts that shaped the *Era of Entertainment*; the gramophone playing recorded music, and the cinema showing recorded film movies.

## The Era of Mass Entertainment

By the midst of the Interbellum, a new technical development emerged that would —like the Radio Revolution— influence the Affairs of Man in a fundamental way: it became known as the 'Television Revolution.' But before that was to happened, there was a development related to it, that had created the foundations for the future broadcasting developments.

### Mass Entertainment

This was the *Era of Mass Entertainment* in which large urban populations experienced rising real wages that enabled them to spend money during their increasing leisure time. Leisure time where the new automobile would bring them to the drive-in cinema, the vaudeville theatres and the concert halls. And an era where earlier technological novelty, many originating from Edison's and Bell's inventive activities (eg the Gramophone, the wax cylinder Phonograph, the Kinemascope camera)[548] started to reach the masses at their homes. Those fruits of their creativity reached the masses in two ways.

**The Film Entertainment**

Next to music entertainment reaching the home of the masses, another medium emerged to entertain the masses; it was called the 'film movie.'

*The Film Movie equipment:* The entertainment offered by the 'film movie' that came into existence when —from the still picture camera aka photography— emerged the film camera capable of shooting multiple pictures in a row. When played back at a certain speed before one's eyes, this action was creating the illusion of motion in the human brain.

The earlier invention of the recording of sound on a wax-cylinder in Edison's Inventory Factory at Menlo Park, was followed by efforts to record images on a wax cylinder. The contributors experimented also

---

[548] The Kinemascope allowed a singular watcher to look through a peep-hole to view a motion picture loop that could be presented for about half a minute before starting over.

with other media —eg with a disc containing pictures—, but ultimo the wax cylinder was replaced by a 35 mm wide celluloid strip containing that sequence of subsequent picture shots (aka a scene). And moving that celluloid strip through a projector —a device equipped with a lightbulb and a lens— with a specific speed (eg 16 frames/minute)— created the early single scene movie (Figure 375).

*The Film Movies:* This animated photography, creating the 'movies' (ie multiple scenes of moving pictures on a longer filmstrip), became the start of the film movie and the film-movie industry. A development that began with the showing of the short filmstrips —containing scenes with circus performances, dancing women, cockfights, boxing matches— to individual persons viewing it in a Kinetoscope Parlour (Figure 376, bottom).

This new form of entertainment was an instant success, and a number of mechanics and inventors, seeing an opportunity, began toying with methods of projecting the moving images onto a larger screen. First on a screen for small audiences, soon screened to larger audiences in movie-theatres. In doing so, they developed the movie projector and the film reel to be used in the 'cinema' (Figure 375). It became a new form of public entertainment offered by the movie-industry.

The French brothers Auguste and Louis Lumière invented a lightweight photographic camera that could record, develop, and project film (Figure 375, right). With their Cinématographe camera they filmed the arrival of a train in one shot. The 50-second silent film showed the entry of a train pulled by a steam locomotive into the Gare de La Ciotat, France. Projected to the public on a screen (Figure 377), as the story goes, the audience was so overwhelmed by the moving image of a life-sized train coming directly at them that

**Figure 375: Early Equipment for Moving Pictures.**
Home Projecting Kinetoscope (left, 1900s). Principle of film strip contained on a film reel (middle). The Cinematograph for shooting the film and screening the film (bottom, 1895), Source: Wikimedia Commons.

people screamed and ran to the back of the room. Nevertheless, the early Lumiere presentations in Paris delighted people, drawing huge crowds.

Soon, the early films[549] were projected on a screen by a film projector (aka the Bioscop) invented by the German brothers Skladanowsky. In the US, Edison acquired the right for an improved projector; the Vitascope. The Lumiere brother soon developed with their Cinématographe the concept of the Cinema (ie the movie theatre) (Figure 377).

The movies had a profound impact on its earliest viewers. The moving image was a dramatic improvement upon the still photograph, a medium with which viewers were already familiar. While cinema initially competed with other popular forms of entertainment —circuses, vaudeville acts, theatre troupes, magic shows, and many others— eventually it would supplant these various entertainments as the main commercial attraction. It would result in the *Nickelodeon Craze* (1904-1908) when investors, recognizing the motion picture's great moneymaking

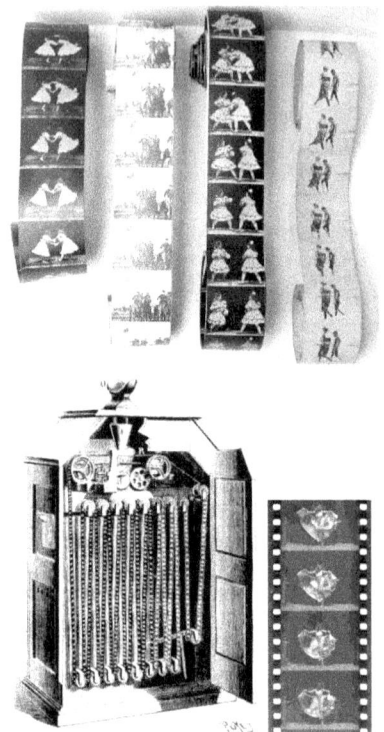

**Figure 376: Moving Pictures.**

Sequence of still photographs creating a scene (top). The Kinetoscope (middle left, 1894) and the 35mm filmstrip 'Madame Butterfly (middle right, 1894-95), Kinetoscope Parlour (bottom, 1895s)

Source: Wikimedia Commons.

---

[549] A (movie) film is a sequence of subsequent pictures (aka frames) that create the impression of motion when projected a certain speed before the human eye. It was also known as 'animated photography.'

potential, began opening the first permanent film theatres around the country, where the admission charge of 5ct (ie a nickel) brought movies to the masses. Between 1904-1908, around 9,000 nickelodeons opened.

*The Dawn of a new Entertainment Industry*

This was the dawn of a new industry that mirrored the development of the radio industry. Not only with early technical contributions from the manufacturers of the recording and projection equipment, but also with the studios that created the moving pictures and owned the movie-theatres where they were presented. Labelled as the movie industry, it would undergo extraordinary transformations, some driven by the artistic visions of individual participants, some by commercial necessity, and others by mere accident.

In the US, the early films were produced by a handful of small companies just outside of New York City. Companies like Biograph, Essenay, Lubin Manufacturing, American Pathé Brothers, Selig, Polyscope, Vitagraph, Edison, Kodak and Melies. In order to establish more control over the business, they created the *Motion Picture Patents Company* (MPPC) in December 1908. The MPPC eliminated the outright sale of films to distributors and exhibitors, replacing it with rentals. It also established a monopoly on all aspects of filmmaking. And it owned the patents on the hardware: the motion picture cameras.

Clearly, newcomers were hindered in their expansion by these industry owners that monopolized the industry. Consequently, these independent filmmakers left the East Coast completely and moved about as far away as they could get; to Los Angeles on the West Coast in the State of California. Soon they clustered together in the

**Figure 377: Early Moving Pictures as Screenplay.**
Advertisement of the Lumière Brothers Cinema (top, 1895). Promotion of Edison's Vitascope projection (bottom, 1896)
Source: Wikimedia Commons.

1910s in the region called Hollywood, Los Angeles. By 1915, Hollywood had become the centre of the American film industry as more independent filmmakers relocated there from the East Coast.

By the mid-1920s, a handful of Hollywood production companies had evolved into the wealthy motion picture industry; conglomerates that had their own studios, distribution divisions, and movie theatres, and contracted with performers and other filmmaking personnel. Movies were big business, by the end of the decade, there were 20 Hollywood studios, and the demand for films was greater than ever, giving rise to new movie format of the 'feature' film.[550]

A similar development took place in Europe, where cinema was dominated by France and Italy. In France the 'films d'Art' emerged; films that were long on intellectual pedigree and short on narrative sophistication. Pathé Frères released the French silent movie 'The Hunchback of the Notre Dame' (1911) (Figure 378, left). The Italian cinema's lavishly produced costume spectacles brought it international prominence in the years before the war with super-spectacle films like 'The last Days of Pompei' (1908) and 'Qua Vadis' (1912).

**Figure 378: Early Silent Film Movies.**
Poster for The Hunchback of the Notre Dame (left, 1911), The Big Parade (middle ,1923) and The Ten Commandments (right, 1925).
Source: Wikimedia Commons.

---

[550] A feature film tells a fictional or fictionalized story, event or narrative. The earliest narrative films, around the turn of the 20th century, were essentially filmed stage plays.

## Golden Age of Hollywood

Hollywood became a dream machine, and with the movies came the film stars: from Rudolph Valentino, Douglas Fairbanks and Mary Pickford, Clark Gable and Greta Garbo, to Bing Crosby and Bob Hope. With comics like Charly Chaplin, Buster Keaton and Laurel & Hardy. Actors staring in the silent movies like the blockbuster of the biblical story of the Exodus 'The Ten Commandments' (1923) and the war agony of the trenches themed in 'The Big Parade' (1925) (Figure 378, middle and right).

In 1920, the filmmaking landscape was revolutionized when sound was introduced. From the silent movies emerged the 'talkies' with their soundtracks,[551] both creating the hits of the late 1920s: The 'Jazz Singer' (1927),[552] the 'Singing Fool' (1928), the 'Lights of New York' (1928), —the first full-length all-talking film—, and 'On with the Show' (1929), the first all-talking colour movie (Figure 379). By 1929, most movies used sound.

Talking pictures proved to be disastrous for the Vaudeville performances that could not compete with the technology of the talkies and

**Figure 379: Early Talking Movies during the Roaring Twenties.**
Poster for The Jazz Singer (left, 1927), The Lights of New York (middle, 1928), and On with the Show (bottom, 1929).
Source: Wikimedia Commons. IMDb

---

[551] Lee DeForest was awarded several patents related to the optical sound-on-film system technology called Phonofilm. After his stay in Germany (1921-1922), he demonstrated his system in 1923, after establishing the DeForest Phonofilm Company in 1922. Lack of interest from the big movie companies combined with Lee's awkward business practises lead to its bankruptcy in 1926.
[552] The Jazz Singer" triggered the talking-picture revolution and its profit of $3.5 million caused Warner Bros. to begin its rule as one of Hollywood's top studios. By the 1930s Warner Brothers was producing about 100 motion pictures a year and controlled 360 theatres in the United States and more than 400 abroad.

many of its actors were unable to adapt to the format of sound motion pictures. Talking films also hurt the careers of the many orchestra musicians who provided the live score to many of the original silent movies. But some actors rose to fame thanks to the film movies.

*Charlie Chaplin* (1889-1977), who started his career as a vaudeville actor, became a film actor by 1914. The Englishman *Stan Laurel* (1890-1965), born in a family active in theatre business, began his career in music halls, and became a film actor by 1917. The American *Oliver Hardy* (1892-1957), a comic actor that became obsessed with the motion picture industry, worked by 1913 as a cabaret and vaudeville singer and was in more than 250 productions. By 1915, Hardy had made 50 short one-reel films.

These three actors, became icons of the slapstick genre of movies. Chaplin started in 1926 producing a silent film 'The Circus' featuring him as the Tramp. It premiered on January 8th, 1928 and became the seventh-highest grossing silent film in cinema history taking in more than $3.8 million in 1928. Stan Laurel and Oliver Hardy became a comic duo, and their first feature film was 'Pardon Us' (1931). They appeared as a team in 107 films, starring in 32 short silent films, 40 short sound films, and 23 full-length 'feature' films.

**Figure 380: Slapstick as Movie Genre.**
Poster for The Circus (top, 1928), and poster for Pardon Us (bottom, 1931).
Source: Wikimedia Commons, IMDb.

The short one- and two-reel films were shown in small movie theatres known as Nickelodeons. By 1910 they numbered 10,000, and by 1916 there were more than 21,000 operating in the US. Following the *Nickelodeon Boom* era, the 1920 became also the decade of the Picture Palaces (aka Dream Palaces); large urban theatres that could seat 2,000-3,000 guests at a time, with full orchestral accompaniment. The largest theatre in the world, the Roxy Theatre (dubbed "The Cathedral of the Motion Picture"), opened in

New York City in 1927 with a 6,200-seat capacity.

The spike in theatre attendance that followed the introduction of talking films changed the economic structure of the motion picture industry, bringing about some of the largest mergers in industry history. By 1930, eight Hollywood studios produced 95 percent of all American films, and they continued to experience growth even during the Depression.

**Figure 381: The Auditorium of the Roxy Theatre (1927).**

Source: rivingfordeco.com/tag/hildreth-meiere/#jp-carousel-3726

*The Big Five:* The greatest output of feature films in the US-studios occurred in the 1920s and 1930s (averaging about 800 film releases in a year). And among those studios we find the 'Big Five' of vertical integrated companies produced more than 90 percent of the fiction films in America and distributed their films both nationally and internationally: Warner Bros. Pictures (1923), Paramount studios (1927), RKO (Radio-Keith-Orpheum) Pictures (1928). Metro-Goldwyn Pictures (1920), and Fox Film Corporation/Foundation (1915).

This all was part of the Era of the Film Movie where the novelty of the 'moving picture' was converted into a mass entertainment industry. Our rather limited exploration painted in large brushstrokes a picture of an artist/cultural and technical/economic development that influenced the Affairs of Man. From its early days of the Kinemascope and its Parlour, the film industry had grown over half a century into a gigantic industry supplying society's entertainment needs. After the silent movie, the sound movies of the 1920s created the Golden Age of Hollywood that saw movie attendance peak in 1946. But that dominant position of the film industries in the entertainment industries was not to last. As it would be challenged by a new technological event called Television Broadcasting.

## *Television Broadcasting 0.0 (1900-1930)*

More or less in conjunction with the technological development of the radio as a listening device for auditive information (aka radio broadcasts), another device for distributing auditive and visual information (aka television broadcasts) came on scene during the Interbellum. The device was called 'television'[553] and it would —after a slow start— contribute and shape the 'home-entertainment industry.'[554]

## The Home Entertainment Industry

People were used to be entertained, but they had to go out of the home to enjoy 'life entertainment.' From home to the vaudeville theatre and the music hall, the wealthier members of society travelled in the new mobility device of the automobile. For those materially less favoured that stayed at home, the inventions of the gramophone brought Music to the Masses and ignited a whole new branch of the 'entertainment industry.'

> This branch was originally an industry making the hardware of the gramophone and disk, combining it by providing its content. This resulted in the many gramophone companies recording music performances and reproducing them on disc,[555] to the manufacturers of the spring-powered—later electromotor driven— 'gramophones'; the machines that reproduced the music from the recording discs.

Next to this new industry creating music that could be played at home, another industry of companies arose creating the moving pictures (aka film movies), as well as the manufactures of the equipment to make and present the movies either at home or in a theatre. These technological developments, especially in the area of mass media, meant that entertainment could be produced independently of the audience, packaged and sold on a commercial basis by an emerging entertainment industry, and consumed in the private home; 'home-entertainment' was born.

And while these players were building the entertainment scene, a new spin-off trajectory of the Era of Electricity emerged. A development trajectory that created a device that broadcasted recorded sound (ie audio) as well as recorded movies (ie video) electrically to a special device. A device influencing the social lifestyle of the masses of common man, who could now enjoy being entertained at home. It was the television receiver

---

[553] The word television is both used in (1) the domain of technical devices (ie the television receiver set), the total technical system and infrastructure (eg the camera, wireless transmission equipment, receiver and the television studio), as well as (2) the content domain of broadcasted information (as in films, programs, news). We will focus on developments in the technical domain.

[554] This is the totality of businesses devoted to mass entertainment; from musical and theatrical performances to cinema showing film. It covers the production side (the radio/television/movie studios) as well as distribution side (eg the movie-theatres).

[555] Originally, the early acoustic recording was done by a cutting stylus connected to the diaphragm of the acoustic horn and scratching the wax of disk with a music track in a circular way. From that master disk, copies were made, first of shellac, later (ie 1950s) of vinyl, in a mould pressing process.

set becoming the home-cinema. A development that became one of the catalysts for the downfall of the once so powerful film industry.[556] And the rise of a new member of the home-entertainment industry; the television becoming to influence societies and its members with a new culture.

## Seeing over Distance: Tele-Vision

The receiving device for the television-broadcasts known as the 'Television' (ie television receiver set) was to produce the sound by means of the loudspeaker, as well as the moving image by means of the picture display unit. Both signals were received (ie transmitted wirelessly) from the video recording device (aka picture camera) and a microphone for the sound (Figure 314). During the development of television, engineers and scientists applied the technologies available to them at that period of time, to realize those functions. And they started with the electro-mechanical technologies.

*The Electro-mechanical Solutions*

At first, scientists experimented with a mechanical scan system (eg the Nipkov disk) for dissecting the picture.[557] Soon followed by others theorizing and experimenting in development trajectories mixing the mechanical technologies with the electric technologies.

*The Mirror Wheel/Drum Television:* The Frenchman Lazare Weiller (1858-1928)[558] theorized a mirrored wheel for the dissection of pictures. The rotating wheel scanned the images projected on a selenium cell that created the electric signal. This idea was used by the German Otto von Bronk (1872-1951) who obtained German Patent № 155,528 (filed on June 12th, 1902) that was granted October 22nd, 1904. In addition, the Polish/Austrian inventor Jan Szczepanik (1872-1926) (aka the Polish Edison) used small mirrors in his device named Telephoto.

The second television device using an oscillating mirror drum is the Telephote of the French Georges Rignoux, for which he filed

---

[556] Another factor were the governmental anti-trust actions against the monopoly of film studios showing only in film theatres their companies produced films: eg United States v. Paramount Pictures (1948).

[557] In 1883 developed the idea of dissecting pictures into spots of light by means of a rotating disk with holes in it. These spots could then be captured by a selenium cell and transformed into electrical impulses. He applied for a patent in 1884. However, only a year later the patent had lapsed, because Nipkow failed to pay the annual fee.

[558] Weiller visited the United States in 1901 and was very impressed by the booming economy and the metallurgical, electrical and mechanical industries. He wrote a book on the subject, *Les grandes idées d'un grand people* (The Great Ideas of a Great People). In 1889 he published the article *Sur la vision à distance par l'électricité* (On vision at a distance by electricity) where he proposed a way to scan, transmit and project images.

several patent applications. Such as the application filed on February 10th, 1906, that was granted as FR-Patent № 364,189 on May 25th, 1906. Another was filed on May 20th, 1908 and was granted French Patent № 390,435 on October 5th, 1908.

The next contribution was from the American Gilbert Sellers who filed on July 18th, 1908 and November 25th, 1908 a patent application for electrical transmission of (graphic) messages. He was granted US-Patent № 939,338 (Figure 382) and US-Patent № 939,339 on November 9th, 1909.

These were some early contributors to the electro-mechanical mirror-drum-based dissectors in television systems. The actual conversion from the photons (ie the light beam) to the electrons (ie the cathode ray) was realized with a photo tube (Figure 179).

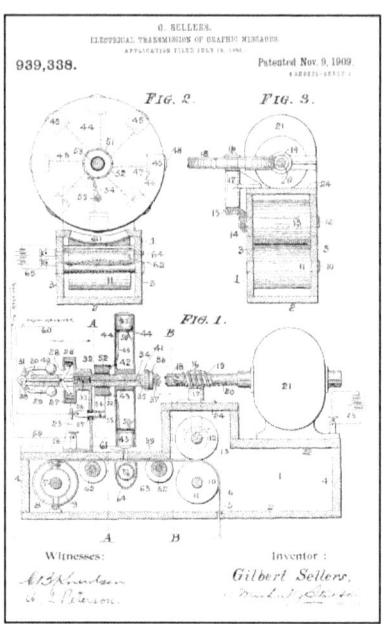

**Figure 382: Sellers US-Patent № 939,338 for a Mechanical Television System.**
Source: USPTO

> Others tried different development trajectories, such as the Englishman Henry Sutton (1855-1912) who worked with electrostatically charged liquids in his 'Telephane.' And the German Ernst Ruhmer (1878-1913) applied the light-sensitive properties of selenium in line-of sight-telephony; his photophone).

Thus, the development trajectory of television systems started with a system created with a mix of technologies: from the mechanical technologies to the electro-mechanical technologies and early electronic technologies. From about 1900, engineers and scientists increasingly tried not only to produce ideas and concepts, but they also tried to develop functioning technical devices and to experimentally explore the problems of image dissection, transmission and synthesis.

*The number of inventions and technical devices phenomenally increased in European countries in the period leading up to the First World War, but two approaches clashed. On the image construction side, the dissection of pictures and their transformation into electric signals was carried out in an electro-mechanical way by means of a Nipkow disk or a variation on it. On the image reproduction side, the synthesis of the impulses back into pictures, on the contrary, could be carried out electronically by means of the Braun tube. This discrepancy between mechanical and electronic technologies persisted until the 1930s and hindered the development of television for a long time.* (Bignell & Fickers, 2008, p. 106)

The development of these hybrid systems by experimenters —the heritage of mechanical thinking— impeded for a period the development of television technology. But after the First World War had ended, the experimenting continued, and among those experimentalists we find the following contributions.

The Scottish electrical engineer and television pioneer John Logie Baird (1888-1946), experimented with the (mechanical) Nipkov disk in the after-war period. In February 1924, he demonstrated to the Radio Times that a semi-mechanical analogue television system was possible by transmitting moving silhouette images. Baird's system used the Nipkov disk for both scanning the

**Figure 383: Early Mechanical Television.**

Baird's Electro-mechanical Falkirk Transmitter (top, 1924). Jenkins Radiovisor system (upper middle, 1925). and motor driven Nipkov-disk signal converter (lower middle, 1929). Baird's Televisor (bottom, 1929).

Source: TVhistory.tv, https://americanhistory.si.edu/, Science & Society Picture Library.

image and displaying it (Figure 383, top). By 1932 Baird had developed a scanner with a mirror drum, that became the first commercially viable television system —the Televisor— and sold 10,000 sets. However, by 1932, the Nipkov disk and mirror drum had both pretty much had their day. This was also the end of the 'low definition' era of television.

Baird applied for several patents: US-Patent № 1,707,935 (filed on August 6th, 1925) granted on April 2nd, 1929 for a 'Television Apparatus' (Figure 384), and US-Patent № 1,735,946 (filed on October 7th, 1927) granted on November 19th, 1929 for improvement in 'Television and like system.' In total he obtained some 24 US-patents, that were complemented by foreign patents.

In the same decade others had also experimented, and patented, electro-mechanical systems.

The English consulting engineer Archibald M. Low (1888-1956) —active from combustion engines to bedpans, writing numerous books with the character of 'technology forecasting' (eg Wireless Possibilities, 1923)— had developed an early tele-vision system he called TeleVista. Low's invention was crude and under-developed, but the idea was there. On October 4th, 1917 he applied for a patent that was granted as GB-Patent

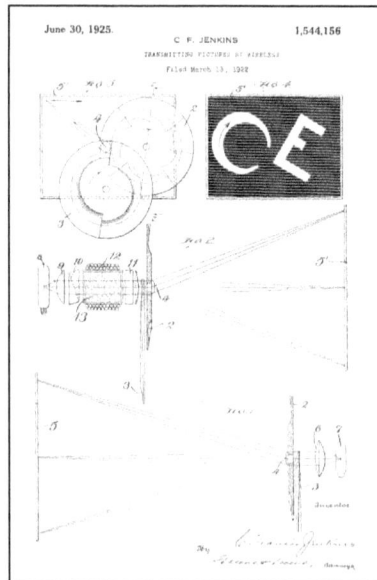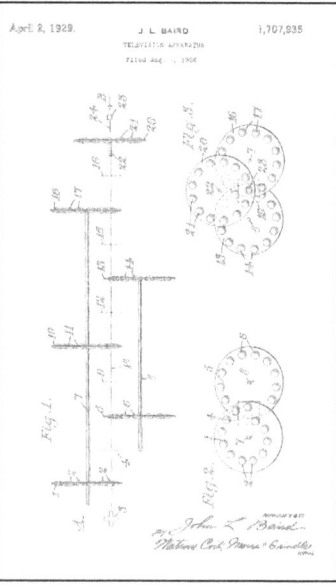

**Figure 384: Early Patents for Electro-mechanical Television Dissectors.**
Jenkins US-Patent № 1,7544,156 (left,1925), Baird's US-Patent № 1,707,935 (right, 1929).
Source: USPTO

№ 191,405 on February 25th, 1923 for an 'Improved apparatus for the electrical transmission of optical images.' Without much of a scientific background, but thanks to his affinity with publicity, he was able to show his system in the famous Selfridge Store Exhibitions in 1925. His system employed an electro-mechanical scanning mechanism, with its matrix detector (camera) and mosaic screen (receiver). The war brought him in military service and he never found time to develop it further.

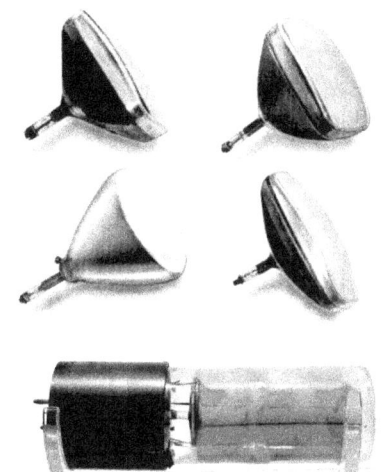

**Figure 385: CRT as Picture Tube and Camera Tube.**

Picture Tube CRT (four tubes on top). The Farnsworth experimental image dissector CRT (bottom).

Source: Wikimedia Commons, National Museum of American History.

The American engineer Charles F. Jenkins (1867-1934) also experimented with mechanical television. Jenkins made the first public presentation of his 'Radiovisor' (Figure 383, bottom) on June 23rd, 1925, and was granted the US-Patent № 1,544,156 for 'Transmitting Pictures by Wireless' (filed on March 13th, 1922) on June 30th, 1925 (Figure 384). In December 1928, Jenkins formed the Jenkins Television Corporation in New Jersey to manufacture his Radiovisors. Over the course of his life, Jenkins received more than four hundred patents. Many of these patents were in the fields of motion pictures and television.

After the failing efforts of applying the mechanical devices, the Cathode Ray Tube —originally used to display electric signals on two 'axes' in electronic instrumentation like Oscilloscopes[559] —became dominant as a picture tube in a television display unit (Figure 385, bottom). Growing in size it would become the device to display more complex 'video' signals; from then it became known as the television picture tube. But before that was to happen, many early experimenters contributed to the concept of 'seeing over distance.'

---

[559] An Oscilloscope—aka oscillograph—is a test/measuring instrument of electric signals. They are displayed with the CRT using the time-variable on the x-axis, and the amplitude of the signal on the y-axis, making 'electricity' more or less visible.

*The Electric Solutions*

The mechanical systems were limited in their performance and did not lend themselves well to the dissection of pictures. The application of electronics at the camera side of the system provided a first breakthrough. Scanning the pictures —that arrived on the screen of a reception tube— with an electron beam, led the way to the development of an electronic *camera tube*. In addition, applying electronics at the presentation side of the system, led to the development of the *picture tube*. Both would be the fruits of the development trajectory of the specialized vacuum tubes know as Cathode Ray Tubes.

> Starting with the basics, one has to realize that a television system involves (1) equipment located at the source of production (among which the television-camera),[560] (2) equipment located in the home of the viewer (the display unit aka television set), and (3) equipment used to transmit the television signal from the producer to the viewer. The total system realised the concept of 'seeing over distance' (aka 'tele-vision') that extended the human vision over long distance.

*The Invention of the Camera Tube*

In addition, research started with the question how to convert an image (aka picture that emitted photon of light) into an electric signal. The solution was the image dissector; a camera tube that creates an 'electron image' of a scene from photocathode emissions (electrons) which pass through a scanning aperture to an anode, which serves as an electron detector.

The German professor Max Diekmann (1882-1960) and his student Rudolf Hell (1901-2002) developed such a device. They filed for a patent application on April 5th, 1925, and were granted DE-Patent № 450,187 for 'Photoelectric image splitting tubes for Televisions' on October 3rd, 1927. The device had a limited performance, and it was Philo Farnworth who patented in 1927 the first practical version of a fully electronic imaging device for television: his Iconoscope (Figure 387).

*The Invention of the Picture Tube*

Originating from the Braun Tube of the 1897s (Figure 185), many scientists had started working on the Cathode Ray Tube in the early 1900s. From the early concepts described before, the cathode ray tubes with a phosphor coating on the inside of (part of) the tube became the basis for 'display' units (aka picture tube). That way the technological trajectory of

---

[560] The developments in the area of the television studio and camera-technology are outside the scope of our exploration as we will focus on the developments on the receiver-end.

the specialized tubes and systems started a range of new application trajectories.

The Scottish engineer Alan Campbell-Swinton (1863-1930) began experimenting around 1903 with the use of cathode ray tubes for the electronic transmission and reception of images. In a letter published in the June 18th, 1908-issue of Nature, called 'Distant Electric Vision', he proposed modification of the cathode ray tube that allowed its use as both a transmitter and receiver of light, and developed a theoretical basis for electronic 'tele-vision'.

These were the important devices — invented at different moments in time— that contributed both to the invention of the systems of the Black and White Television.

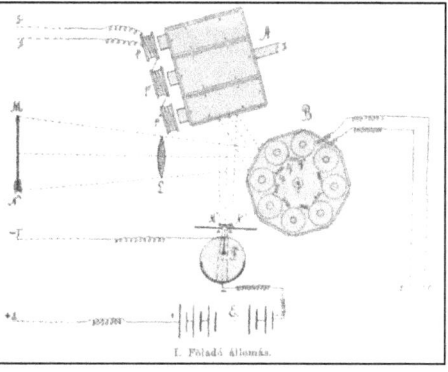

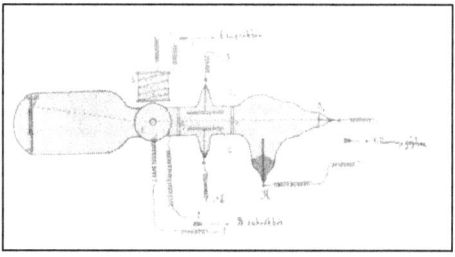

**Figure 386: Drawings of the Television System of Boris Rosing (1911).**

The mechanical detector unit (top), and the receiver unit with a CRT (bottom).

Source: Wikimedia Commons.
http://epa.oszk.hu/00000/00030/03006/pdf/VU_EPA00030_1911_29.pdf. (Hungarian (?) newspaper article edited by author.

## The Invention of the Black & White Television

All over Europe, next to the writing over distance (tele-graph), and speaking over distance (tele-phone), now seeing over distance (tele-vision) became the focus of scientific interest.

The Russian physicist Boris Rosing (1869-1933), descending of the Dutch immigrant Peter Rozing, became interested in the 'electric telescopy' in 1897 while working at the St. Petersburg Institute of Technology. Using a cathode ray tube, he succeeded in displaying an image that was created electro-mechanically by a mirror-drum scanner projecting light on a photocell (Figure 386, top). In 1907, Rosing used both the CRT and the mechanical scanner system in an experimental television system. In the receiver, he displayed crude geo-metrical patterns onto the phosphor screen of a CRT. He demonstrated his system in 1911, and newspaper articles (eg in The Scientific American) in that same year published about his 'Telegraphic eye.'

Rosing patented his system in Russia, Germany, France and Britain. The Russian patent № 18,076 (filed: July 25th, 1907) granted on October 30th, 1910) was for a "Способ электрической передачи изображений на разстояние" (Method for the electrical transmission of images over a distance). In addition, he obtained Deutsches Reich Patent № 209,320 (filed: November 26th, 1907) granted April 24th, 1909 for "Verfahren zur elektrischen Fernübertragung von Bildern" (Method for the electrical transmission of images). The British patent № 27,570 (1907) (filed: December 13th, 1907) was granted on June 25th 1908 for "New or improved method of electrically transmitting to a distance real optical images and apparatus therefor." And in France the French patent № 430,453 (filed: March 11th, 1911) was granted on October 17th, 1911 for "A method of electrically transmitting light images at a distance").

By that time, the idea of being able to transmit images wireless over distance —aka electric telescopy (aka distant vision)— fascinated many engineers. Both in Russia, Britain and America, and many experimented one way or the other with the phenomenon of the Cathode Ray tube, both its development and its application.

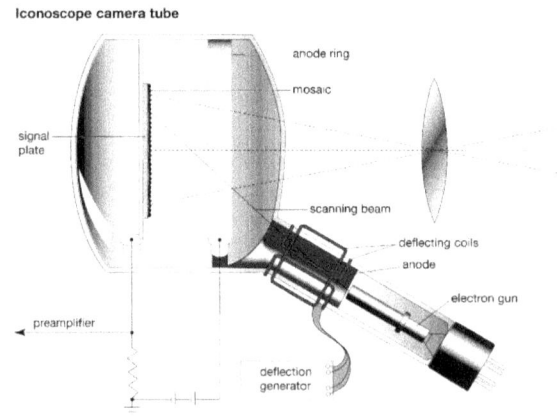

**Figure 387: Diagram of the Iconoscope.**
Source: Encyclopedia Britanica.

The Russian-American inventor Vladimir Zworykin (1888-1982), who studied at the St. Petersburg Institute of Technology under Boris Rosing, used the Cathode Ray tubes to build a camera tube called 'Iconoscope' used to scan an image (Figure 387).[561] Later followed by the picture tube

---

[561] This photomultiplier tube converts light (ie photons), into an electrical signal.

known as the 'Kinemascope.' Rosing and Zworykin exhibited a hybrid television system in 1910, using a mechanical scanner in the transmitter and the electronic Braun tube in the receiver.

As this was the time of massive social upheaval (ie the Russian Revolution of 1917) their development became interrupted. Fleeing from Russa, Zworykin emigrated to the United States in 1919, started working for Westinghouse Electric Corp. in Pittsburgh in 1920, and became a naturalized US citizen in 1924.

In 1923 he filed a patent for an all-electronic 'television system,' which had cathode-ray tubes for both transmitting and receiving images. It took a while but on December 20nd, 1938, after many patent revisions, he was granted US-Patent № 2,141,059 (Figure 388). By 1924 he began building a television system based on his patent, and in 1925 he demonstrated an almost entirely electronic system for several Westinghouse executives, who were not impressed. He was told by management to "devote his time to more practical endeavours," yet he continued his efforts to perfect his system.

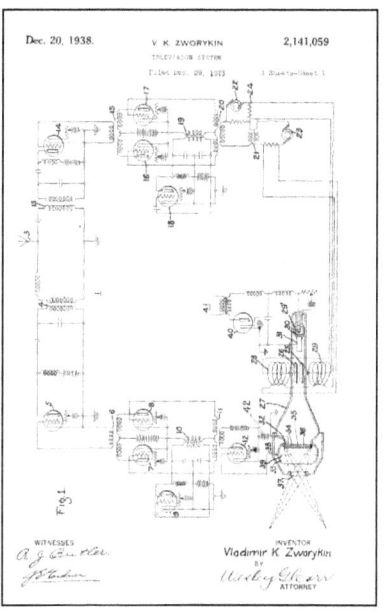

**Figure 388: Zworykin's US Patent № 2,141,059 for a Television System.**
Source: USPTO

*Goliath versus David*

The management of the large electric large companies of that period in time had more considerations than just adopting a new phenomenon with uncertain prospects. And they were part of the larger group of the Radio Corporation of America, that developed more and more into a multi-business mastodont based on a range of new technologies.[562] But the new

---

[562] David Sarnoff (1891-1971), a Jew from Russia who emigrated in 1900 to New York, America. Working as telegraphist he picked up the distress signal from the SS Titanic after it collided with the iceberg. By 1914 he worked for the American Marconi company, where he became a protégé of radio inventor Guglielmo Marconi. Rewarded by the Marconi company

concept of television was too promising to ignore and the large corporations became further engaged.

In 1930 Westinghouse's television research activity was transferred to RCA, and Zworykin became head of the television division at RCA's Camden, New Jersey, laboratory. In the later part of his life, Zworykin played an important role in RCA's further TV development, and he eventual became director of RCA, in those days the leading CRT producing company in the US. RCA made large progress in developing different and better tubes for TV, radar and oscilloscopes.

General Electric in the meantime, also had started a television project with developing both experimental television sets and using it for broadcasting experiments. An early prototype was the Octagon (1928) using a mechanical, rotating disc technology to display images on its 3-inch screen. But it took till after the Second World War before the firsts TV-models were introduced in 1948.

With the emergence of radio as a new mass medium in the late 1920s, radio became the dominant point of reference for the development of television. The success of radio as a mass medium for information and entertainment now shaped the conception of television as a medium for television programmes. But the real impetus for the further development of television however, had to come from the television pioneers with an entrepreneurial inkling.

The American Philo T. Farnsworth (1906-1971), fascinated already at a young age by electricity, was educated at the Brigham Young High School in Salt Lake City. After graduating from high school, Philo had no trouble finding work. He worked on logging crews, repaired and delivered radios, sold electrical products door to door, and worked on the railroad as an electrician. In Salt Lake City —being a Mormon like his father— met two prominent San Francisco philanthropists, Leslie Gorrell and George Everson, and convinced them to fund his early television research. With an initial $6,000 in financial backing, Farnsworth was ready to turn his dreams of an all-electronic television into reality. It started when he realized that an electronic beam could scan a picture in horizontal lines, reproducing the image instantly.

---

with rapid promotion, he became chief inspector, and in 1915 or 1916 he wrote the famous 'radio music box' memo, in which he proposed the development of a commercially marketed radio receiver for use in the home. After the Radio Corporation of America had absorbed the American Marconi company in 1919, Sarnoff became RCA's general manager, and brought RCA to dominate the radio-business.

*According to surviving relatives, Farnsworth dreamed up his own idea for electronic-rather than mechanical-television while driving a horse-drawn harrow at the family's new farm in Idaho. As he plowed a potato field in straight, parallel lines, he saw television in the furrows. He envisioned a system that would break an image into horizontal lines and reassemble those lines into a picture at the other end. Only electrons could capture, transmit and reproduce a clear moving figure. This eureka experience happened at the age of 14.[563]*

Already intrigued by the properties of cathode ray tubes, he developed by 1927 a video camera tube (aka the *Image Dissector*) (Figure 389), followed by the display unit (aka the Image *Oscillite*), both all-electronic devices. Using both components he developed a television system without electro-mechanical component, that could transmit live images over distance (Figure 392).

By January 7th, 1927 he filed for a patent that was granted as US-Patent № 1,773,980 on August 26th, 1930 for his camera tube (Figure 390). A second patent was needed to begin the whole television story; US-Patent № 1,773,981 which he obtained for the cathode ray tube (CRT) providing the display tube aka the receiver; the television screen. They were the first of a range of patents he obtained during his lifetime. By the age of 64, Farnsworth held more than 300 United States and foreign patents, most of which formed the foundation of the television industry as it swept the world and changed the nature of modern civilization.

Farnsworth developed a system —the analogue video signal— in which the information of the object was scanned along horizontal lines and converted to a video-signal. That signal could, after transmission,

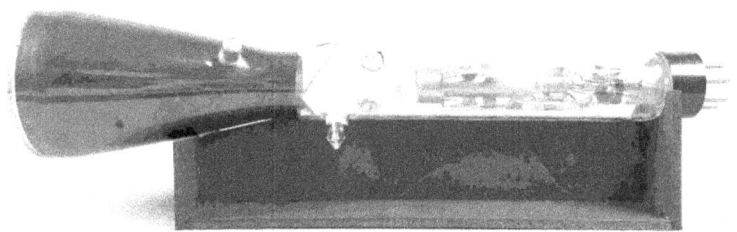

**Figure 389: Farnsworth's CRT Camera 'Image Dissector.'**
Source: HistroyofInformation.com

---

[563] Source: Schwarz, E.I.: Who Really Invented Television? (September 1, 2000). https://www.technologyreview.com/2000/09/01/236187/who-really-invented-television/

reproduce the information contained by projecting the horizontal scanning lines on the receiver tube in the format of frame (Figure 391). It would become the basis for the three transmission standards to come.[564]

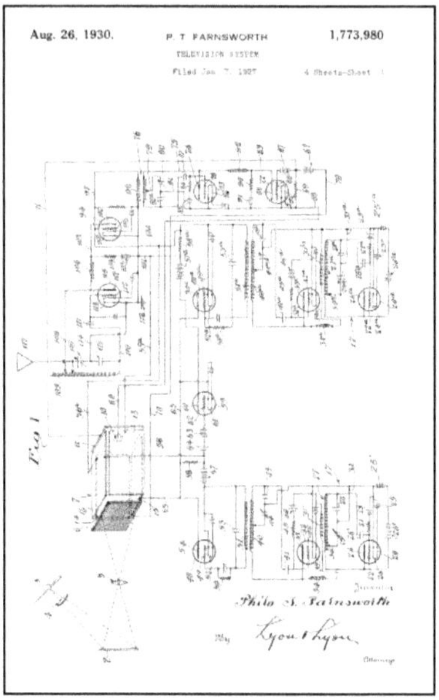

By 1928, Farnsworth —then 22 years old, and his priority rights established by the patent application— had developed the system sufficiently to hold a demonstration for the press. That gave him publicity, and in 1929, Farnsworth further improved his design by eliminating a motorized power generator, thus resulting in a television system using no mechanical parts (Figure 392).

By then, Farnsworth's financial backers were putting pressure on him to sell the whole company, rather than just a license, to Westinghouse. After all, the stock market had recently crashed, and stocks had taken a dive. By 1930 he was visited by Vladimir Zworykin at the behest of Farnsworth's financial backers.

**Figure 390: Farnsworth's US Patent № 1,773,980 for a Television System.**
Source: USPTO

> Zworykin, found Farnsworth's image dissector camera tube superior to his own. As a result of Zworykin's advice, David Sarnoff, director of RCA offered in 1931 to buy Farnsworth's patents for US$ 100,000, with the stipulation that he become an employee of RCA. But Farnsworth refused.

---

[564] In the NTSC-system (ie National Television System Committee, 1941) the video signal contains frames of 525 interlaced lines, and the frames are projected at 30 frame/second. In the PAL-system (ie Phase Alternating Line, 1962) a picture is made up out of 625 interlaced lines and displayed 25 frames per second.
And in the French SECAM-system (ie Séquentiel couleur à mémoire color/sequential with memory 1961) a colour picture is made up out of 625 interlaced lines and displayed 25 frames per second. However, the way SECAM processes the colour information, it is not compatible with the PAL video format standard.

*The rejection brought the full wrath of the mogul down on the inventor. "Sarnoff decided to break him in patent court," says Pem Farnsworth. In other words, Sarnoff would do to Farnsworth what he did to those who had developed key radio inventions but had refused to cooperate fully with RCA. Sarnoff and his team of lawyers would launch a legal assault aimed at overturning the patents on appeal, which would tie up the inventors emotionally and financially for years.* [565]

The result was a patent battle over priority rights, with Zworykin's patents that RCA owned, at the centre.

*RCA brought interference proceedings against Farnsworth at the Patent Office, claiming that inventor Vladimir Zworykin (now an RCA employee) had predated Farnsworth patents with his own filed in 1923. The main issue was whether Claim 15 of the '980 patent for an "electrical image" was disclosed previously in Zworykin's application, which later became US Patent No. 2,141,059, entitled "Television System." The PTO found, in Interference #64,027, that "priority of invention awarded to Farnsworth."* (ibidem)

So, the U.S. Patent Office rendered a decision in 1936 awarding priority of the invention of the image dissector to Farnsworth. RCA lost a subsequent appeal, but litigation over a variety of issues continued for several years till 1939 it lost the final appeal. Then Sarnoff finally agreed to pay Farnsworth royalties.

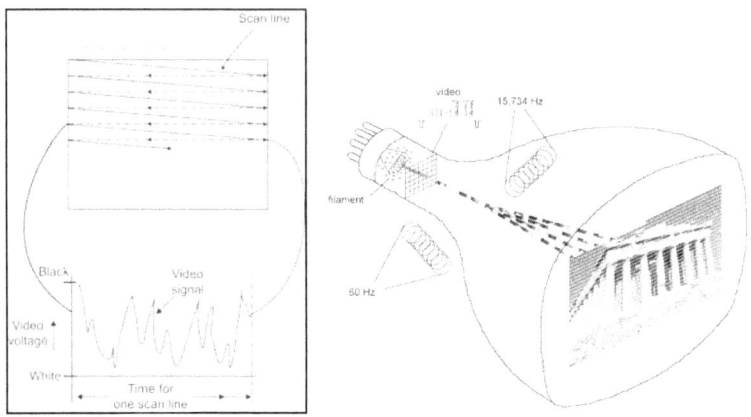

**Figure 391: Working of Analogue Television.**
Principle of scanning the image (left) and displaying the video-signal (right).
Source: https://www.danalee.ca/ttt/analog_video.htm

---

[565] Source: https://www.jdsupra.com/legalnews/the-top-ten-patent-wars-television-10-73080

In the meantime, after an initial contact to sell his licenses and working for Philco[566] in 1931 —by that time the largest manufacturer of radio sets in America, selling more units than RCA— Philco backed out by 1933 after been pressured by the RCA. So, in 1933, Farnsworth left Philco and set up his own company, *Farnsworth Television* in Philadelphia.

**Figure 392: Farnsworth's TV-set and Camera (1927).**
Replica of the camera (top left) and receiver unt tube (top middle).
Source: https://www.thehistoryoftv.com/

In 1932, while in England to raise money for his legal battles with RCA, Farnsworth met with John Logie Baird, a Scottish inventor who had given the world's first public demonstration of a working mechanical television system in London in 1926, using an electromechanical imaging system, and who was seeking to develop electronic television system. As a result, the picture camera became available to Baird's company via a patent-sharing agreement.

In the summer of 1934, Farnsworth staged at Philadelphia's Franklin Institute the first public display of his television system. Following his success at the Franklin Institute, Farnsworth set up his own television broadcasting station. From 1937 to 1939, a stream of cartoons, sports, and live entertainment poured from the station transmitter to about fifty television sets throughout Philadelphia.

In 1938 his investors scoured the nation for a manufacturing plant that would allow them to profit from Farnsworth's invention. The company *Farnsworth Radio and Television Corporation,* went public in 1939. By that time RCA mounted a publicity campaign, claiming they had invented the television. To support that claim, in 1939, RCA televised the opening of the New York World's Fair, including a speech by President Franklin Delano Roosevelt, who was the first president to appear on television (Figure 393).

---

[566] Farnsworth and his investors struck the deal with Philco that brought the young inventor to Philadelphia. Philco would foot the bill for Farnsworth's experiments. In return, Farnsworth would grant Philco a license to manufacture television receivers and help the company establish a station for television broadcast.

Then started the Second World War and many companies, including RCA and General Electric, turned their attention to military production. During the war years —television was officially put on hold as the US government banned all nonmilitary electronics manufacturing—, Farnsworth won government contracts to develop new technologies for military use, including radar tracking devices and electronic surveillance equipment. After the war, the inventor moved with his family to Fort Wayne, Indiana, where the Farnsworth Corp. began to produce television sets. But after his patent protection ended in 1947, he could not compete with RCA. His tv-business took a downturn and he was forced to sell it to the International Telephone and Telegraph Corporation (ITT) in 1949.

**Figure 393: Roosevelt TV broadcast from NY World Fair (1939).**
Source: worldhistoryproject.org/

## *Television Broadcasting 1.0 (1930-1950)*

By the 1930s, many early contributors had shaped the dawn of the Era of Television. It started with a multitude of television systems used for broadcasting television signals; from the described electro-mechanical systems to the full electronic CRT based systems (Figure 391). And with those systems came a range of non-standardized techniques for building the video pictures on the screen.

> Many of those early television sets were still in the prototype stage, and the broadcasting systems were still highly experimental. Also a number of experimental broadcast stations began producing some special television programming. The radio-broadcasting powers NBC and CBS built in New York their first television-broadcasting stations. However, the Second World War impeded the development of the medium, slowing it as people and materials were directed to this major world conflict.

At the beginning of the1930s, television was not only imagined as a programme medium, but the institutionalisation of television began with regular broadcasts of experimental programmes.

## Early Experimenting/Demonstration of Television Broadcasting

The 1930s were the infancy of television broadcasting. In 1927 the American Telephone and Telegraph Company (AT&T) gave a public demonstration of the new technology, and by 1928 the General Electric Company (GE) had begun regular television broadcasts. In 1929 Baird convinced the British Broadcasting Corporation (BBC) to allow him to produce half-hour shows at midnight three times a week. At the Berlin Radio Show in August 1931, Manfred von Ardenne gave a public demonstration of a television system using a CRT for both transmission and reception. Philo Farnsworth gave the world's first public demonstration of an all-electronic television system, using a live camera, at the Franklin Institute of Philadelphia on August 25th, 1934. The British company Electric and Musical Industries, Ltd. (EMI) broadcasted in November 1936 a television signal.

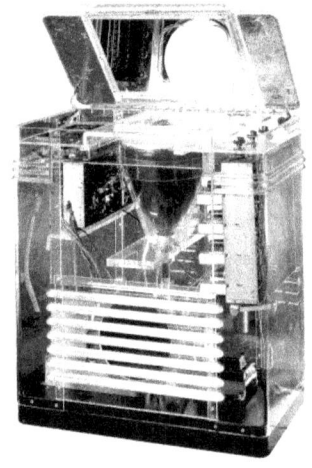

**Figure 394: RCA Hall of Television World Fair New York (1939).**
RCA' Hall of Television with TRK-12 sets, Model TRK-12 Television Demonstrated in transparent cabinet.
Source: MZTV Museum of Television.G665

In the year 1937, television was already prominently staged at the Exposition Univerelles des Arts et Techniques in Paris, both in the French 'Palais de la Radio' and in the German Telefunken Pavilion. Financed by the Reichspost, Telefunken offered two television attractions: a system combining telephony and television, called 'Video telephony', and a television program consisting of live transmissions, from a camera placed on the roof terrace of the German pavilion, and films of fifteen minutes.

In April 1939, RCA televised the opening of the New York World's Fair,[567] including a speech by President Franklin Delano Roosevelt (Figure 393). RCA was prominently present at the New York World Fair. It had its own pavilion dedicated to radio and television. The Hall of Television was equipped with a television studio and the television sets (the TRK-12 receivers) to give television broadcasting demonstrations (Figure 394). Daily from 11 a.m. to 9 p.m., there were filmed news, variety pre-recorded at Radio City studios and live broadcasts made in NBC's mobile studio car, compilations of visitor interviews, and reports on special events of the exhibition and other exhibitors.

The World's Fairs in Paris (1937) and New York (1939) —where the regular broadcast of television programmes in the United States was inaugurated— mark the phase of staging television as a politically and culturally important medium and industrial reality rather than a technical or scientific attraction. The chosen mottos for the Fairs – 'Arts et techniques dans la vie moderne' and 'The world of tomorrow' respectively, illustrate that the perspective was on the Modern World to come.

> *Immediately after its 1939 World's Fair premiere, RCA tried to sell television receivers to the public. But there were few programs to watch and the tiny sets were expensive, so this time the market did not take off. [...] With their experience in radio, Philco and Zenith were quick to follow RCA into the television business. These competitors, along with the second major broadcast network Columbia Broadcasting System (CBS), skirmished with RCA over the technical standards*

Figure 395: Promoting Television for the Home.
Newspaper (left, 1927), Magazine (middle, 1928) and RCA promotion (right, 1939).
Source: http://www.tvhistory.tv/pre-1935.htm

---

[567] The 1939 New York World's Fair was dominated by an optimistic vision of a technological future.

*for television, but RCA's system won out when the National Television System Committee (NTSC) adopted the RCA approach.* [568]

This was the decade in which the new phenomenon of television became promoted in popular magazines and manufacturers (Figure 395). In Paris and New York, however, it was no longer the 'technical miracle' that was presented but the promise of a new mass media for the modern world. Then, that 'Modern World' was disrupted, as the US was drawn into the Second World War after the attack on Pearl Harbour, and the manufacturing of televise sets was banned by government and the large companies turned over to wartime production.

## The Invention of Colour Television

Basically, television sets are particle —ie electrons— accelerators. Electro-magnetic fields propel beams of charged electrons across vacuum tubes and toward their intended targets, the light emitting phosphors on glass screens of the picture tube. Before the war, the picture tube had followed a technical development trajectory of continuous improvement. Improvements in tube size (from 5, 7, 12 to 19 inches in 1938 to 21 inches in 1955), tube form (from round to square), improving the electron guns, the magnetic deflection of the cathode rays, and the phosphors on the screen where the electrons created the picture. And…, a large part of that development effort was directed at creating a colour picture tube as the market was asking for colours (just like in the Technicolor film movies that baffled its audiences).[569]

So many inventors focussed their attention on the subject. John Logie Baird created in 1938 his *Telechrome*, the first all-electronic single-tube colour television system. (Figure 396). Other developments were General Electrics' *Penetron tube*, and Paramount Pictures' *Chromatron tube*. However, the

**Figure 396: Baird's Telechrome.**
Two-gun Telechrome tube: Long-neck model (1944).
Source: http://www.hawestv.com/, Science Museum

---

[568] Source: Hoover, G. Tech Wars: RCA and the Television Industry (February 5th, 2021); https://americanbusinesshistory.org/tech-wars-rca-and-the-television-industry/
[569] The Technicolor three-colour process, developed by Kodak in the early 1930s, was a technology for colour film.

principle used in these designs proved to be unpractical and it became a dead-end technology. But one design survived and became successful; the Japanese Sony's *Trinitron tube*.

- The *Telechrome Tube* used two electron guns, aimed at either side of a thin, semi-transparent mica sheet (Figure 396). One of the sides was covered in cyan phosphor and the other red-orange, producing a limited colour gamut. Baird applied for a British patent on July 25th, 1942, and was granted GB Patent № 562,168 on June 21st, 1944.

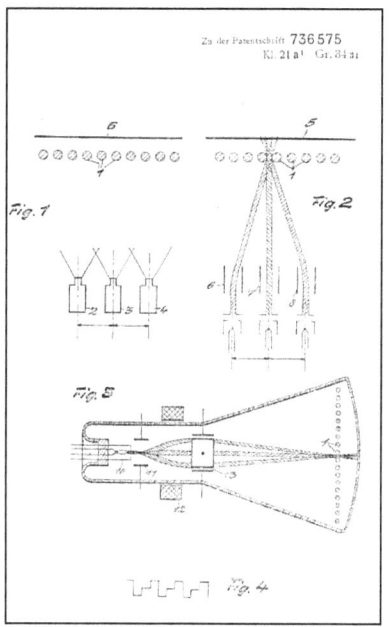

**Figure 397: Flechsig DE Patent № 736,575.**
Source: Espacenet

- The *Penetron Tube* (aka Presentation Tube) is a two-colour system (and their combinations). Similar to the B&W-tube, it has a single electron gun, but the phosphor layer is applied in three layers of different colour-emission. Colours were selected by increasing the power of the electron beam, which allowed the electrons to flow through any lower layers to reach the proper colour. The Penetron Tube was original designed by Koller and Williams while working at General Electric (GE). They were issued US-Patent № 2,590,018 on October 24th, 1950 for 'Production of colored Images.'

- The *Chromatron Tube* used an electronic focusing system for the electron beam that consisted of a series of thin metal wires or plates —aka aperture grill—placed about 1/2-inch before the phosphor screen. The three phosphor layers with their specific colour were arranged in vertical stripes. And the deflection of the beam would hit a specific stripe. This tube—aka Lawrence Tube—was developed by *Ernest Orlando Lawrence* (1901-1958) in 1951. But instead of RCA's three electron guns, he used one gun in combination of a web of charged wires located before the phosphor screen. He filed a patent application on January 8th, 1953, and was granted US-Patent №

2,854,504 on September 30th, 1958 for a 'System of color television transmission.' The tube, however, was hard to mass-produce.

The *Trinitron Tube*, developed by the Japanese B&W TV manufacturer Sony, was based on an aperture grille of the Chromatron system. But instead of using each of the beams of three electron guns, it applied a single gun producing three electron beams from three cathodes. The combination with the aperture grill resulted in a new patentable system Sony called Trinitron.

Already by 1938, the German inventor Werner Flechsig (1900-1981) developed the seemingly simple concept of placing a sheet of metal just behind the front of the tube, and punching small holes in it. Using three electron guns for the red, green and blue colour, creating the colours on screen.[570] Flechsig had invented the principle of colour picture generation with shadow-mask picture tubes. A prototype TV with Flechsig's visionary colour tube appeared at the 1939 Berlin Radio Show. However, the war put an end to technical implementation of Flechsig's shadow mask picture tube.

Flechsig filed on July 12th, 1938 for a patent and was granted DE-patent № 736,575 on June 22nd, 1943 for a "Cathode Ray Tube for the Development of Multicolored Pictures on a Fluorescent Screen" (Figure 397).

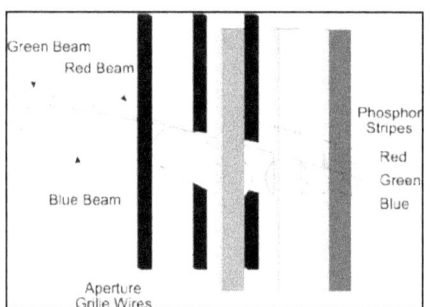 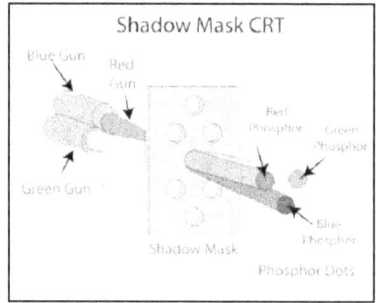

**Figure 398: Principles of Colour Television**
The Aperture Grille (left) and the Shadow Mask (right) creating colour by mixing RGB-signals.
Source: Wikimedia Commons.

---

[570] In a Black and White television (B&W) the electrons emitted from the electron gun a drawn to the other end of the tub by a high voltage. When they hit the phosphor layer, they cause a white light on that spot. In a colour display, the uniform coating of white phosphor is replaced by dots or lines of three coloured phosphors, producing red, green or blue light (RGB) when excited.

The Shadow Mask Tube (SMT) was born (Figure 398). Flechsig's contribution however was hardly noted in the turmoil of the German army invading Sudetenland in the summer of 1938 bringing Europe on the brink of the Second World War.

*Colour TV War*

Among the early experiments with colour tv in the US, we find the large companies such as RCA, General Electric and broadcasting networks like CBS. In their R&D laboratories the researchers all worked feverish in developing colour tv sets, and a *Colour TV-War* developed between RCA and CBS. Ultimo RCA did win after suing CBS in court. But there was more that cemented that victory, not in the least because RCA's design was compatible with the 8 million black-and-white sets already in use; a feature that CBS missed.

> Also in 1938, the French telecom engineer M. Georges Valensi (1889–1980) worked in RCA on colour television. He filed on January 17th, 1938 a patent application and was granted the French patent № 841,335 on February 6th, 1939 for his "Procédé de télévision en couleurs", and was granted the equivalent US-Patent № 2,375,966 on May 25th, 1945. Then Alfred Christian Schroeder (1915-2006) contributed to the further invention of the shadow mask tube and several other major inventions on colour tubes. He filed for a patent on February 24th, 1947 and was granted US-Patent № 2,595,548 on May 6th, 1952 for a 'Picture reproducing Apparatus' (Figure 399).

The General Electric (GE) Research Laboratory had also been working on a variety of systems that would allow them to introduce colour sets that did not rely on the patented shadow mask principle. Through the 1950s they had put considerable effort into the Penetron concept, but were never able to make it work as a basic colour television. So, they started looking for alternate arrangements. GE eventually improved on the basic shadow mask system with a simple change to layout that circumvented RCA's patents. General Electric scientist Lewis R. Koller (1895-1993) filed on October 24th, 1950 for a patent and was granted US-Patent № 2,590,018 on March 18th, 1952 for the 'Production of Coloured Images."

Also the broadcasting company CBS was developing in its laboratories a colour television set. It was Dr. Peter Goldmark (1906-1977) —inspired by the film movie 'Gone with the Wind' in Kodak's Technicolor— who developed by 1940 a field-sequential colour technology based on the Nipkov Wheel, a mechanical device. By June 1940 he was able to show still pictures from a colour slide on a 5-inch colour monitor. This led to the first disclosure of the CBS Colour Television System to the public on August 28th, 1940, its first demonstration to the press on September 4th, 1940, and on February 20nd, 1941 the first colour television pictures were broadcasted.

Working for CBS, Goldmark received several patents for the colour tv: from US-Patent 2,304,081 (filed on September 7th, 1940) granted on December 8th, 1942; US-Patent № 2,480,571 (filed on September 17th, 1940) granted on August 30th, 1947; US-Patent № 2,406,760 (filed on September 19th, 1940) granted on September 3th, 1946; US-Patent № 2,435,962 (filed on November 20th, 1940) granted on February 17th, 1948; and US-Patent № 2,435,963 (filed on December 13th, 1940 granted on February 17th, 1948. They were all for a 'Color Television' based on a mechanically rotating disk.

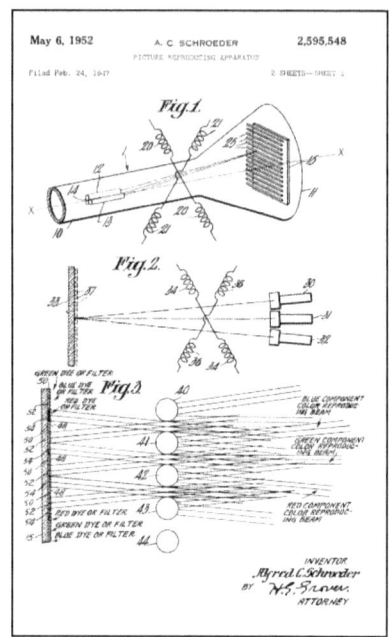

**Figure 399: Schroeder's Patent for Shadow masked Picture Tube.**
US Patent № 2,595,548 of May 6th, 1952)
Source: USPTO

While the CBS colour broadcasting schedule gradually expanded to twelve hours per week (but never into prime time), and the colour network expanded to eleven affiliates as far west as Chicago, its commercial success was doomed by the lack of colour receivers necessary to watch the programs, the refusal of television manufacturers to create adaptor mechanisms for their existing black and white sets, and the unwillingness of advertisers to sponsor broadcasts seen by almost no one.

In desperation, CBS bought a television manufacturer, and on September 20nd, 1951, production began on the first and only CBS colour television model. But it was too little, too late. Only 200 sets had been shipped, and only 100 sold, when CBS pulled the plug on its colour television system on October 20th, 1951, and bought back all the CBS colour sets it could to prevent law suits by disappointed customers. RCA was the obvious winner of the Colour TV War. The success of RCA's shadow mask tube (SMT) for colour TV was overwhelming; many companies followed and started production of SMTs as well, not only in the USA, but also in Japan and Europe.

Despite these early successes with colour programming, the adoption of colour television was a slow one. It wasn't until the 1960s that the public began buying colour TVs in earnest and in the 1970s, the American public finally started purchasing more colour TV sets than black-and-white ones.

*Manufacturing Television sets in the US*

Next to the television broadcasting, this was also the time of commercial electronic television-set manufacturing by the early pioneers of television. Television sets that appeared in a range of designs; desk, console and cabinet model. And sets that used different display techniques as there were no uniform technical standards applicable yet. Televisions that were realized by the inventor-entrepreneur as well as by the large (radio) companies.

Allen B. DuMont (1901-1965), a graduate of the Rensselaer Polytechnic Institute in Troy, New York in 1924, started his career working for the Westinghouse Lamp Company, building up experiences in vacuum-tube production. Later he became chief-engineer for DeForest Radio Corp. By 1935, after founding his own company Allen B. DuMont Laboratories, Inc. he was working on improvements in long-lasting CRTs for use in television receivers. For his 'electron turbine' he filed a patent application on January 1$^{st}$, 1930, and was granted US-Patent № 1,999,407 on April 30$^{th}$, 1935.

In the late 1930s his Dumont Laboratories manufactured the all-electronic Model 180 desk-model television receiver (Figure 400). To finance his business, he sold a 26% interest in his fledgling company to a giant in the film industry Paramount Pictures in 1939 for $212.000.

> DuMont's involvement with Paramount ultimately proved to be a big mistake. Eager to hinder the development of television, which it perceived as a serious threat to the motion picture industry, Paramount thwarted DuMont's plans on many occasions.

Next to manufacturing, he started broadcasting. However, the DuMont Television Network (1946) was not an unqualified success, being faced with the major problem of how to make a profit without the benefit of an already established radio network as a base. DuMont had more success with his television receivers. By the year of 1951 his manufacturing company was doing gross business of about $75 million a year. DuMont sold his manufacturing operations in 1960 to the television manufacturing division of Emerson Radio. He died in 1965.

The dominant company Radio Corporation of America (RCA) pioneered —next to radio broadcasting, radio and gramophone manufacturing— also with the creation of gramophone discs and film movies through its subsidiaries like RCS Photophone (1928). Soon followed by broadcasting activities by the NBC in the emerging field of television. Under directorship of David Sarnoff (1891-1971), RCA began development of television in early 1929, and demonstrated an all-electronic B/W television system —Model TRK-12— at the 1939 New York World's Fair (Figure 394).[571] RCA began regular experimental television broadcasting from the NBC studios to the New York metropolitan area on April 30th, 1939 via station W2XBS, channel 1.

From the contribution of the inventor entrepreneurs like Dumont, to the large industrial conglomerates like RCA, emerged the television industry; manufacturers of broadcasting equipment, service providers broadcasting programs and news, and the makers of television sets. The American airwaves became

**Figure 400: Early Tabletop, Console and Cabinet design Televisions.**
Dumont Model 180, (top, 1939), and RCA TT-5 model (middle, 1939), and Baird Model T-14 (bottom, 1939).
Source: National Museum of American History, MZTV Museum of Television.

---

[571] Also other manufacturers, such as Westinghouse Electric, General Electric, DuMont Laboratories, etc. were exhibited their television models.

cluttered with the transmission of the television stations. But that was after the eruption of the Second World War.

The year 1940 looked promising at first to the television industry. But, unfortunately, television sets were expensive, with little programming, and with the prospect of world war and uncertainty over jobs, few sets were sold. The year 1941 was even more dismal than 1940, for makers of television sets. Although some of the trade articles were positive and upbeat, the reality of the situation was that no one was buying the sets. And in 1942, after the attack on Pearl Harbour, and the United States becoming engaged in Second World War, all commercial production of television equipment was banned for the remainder of the war.

*The US Television Boom of the late 1940s*

At the end of the Second World War, in the US of 1946, television sets went on sale again, and network television began to provide programming, although there were only ten licensed television stations in the country. At that time, radio was the dominant broadcast medium, already in almost thirty-four million US homes, but it would soon experience a mass exodus of its audience. By 1948, only two years later, almost one million homes had televisions, and by 1950 there were 107 licensed television stations. By 1952 over twenty-four million sets were in use served by 225 stations, their sales stimulated by dropping prices.

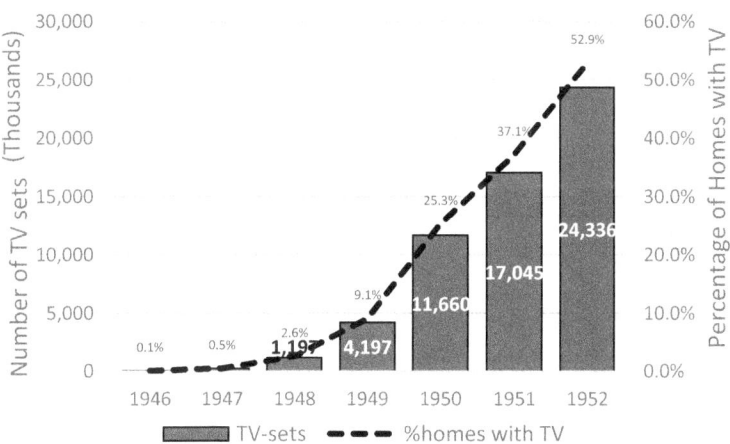

**Figure 401: US Television Boom (1946-1952).**
Production of Television Receiver set in the US (1946-1952).
Source: http://www.earlytelevision.org/us_tv_sets.html

The after-war years saw the first 'Television Boom' (Figure 401), with thousands of viewers buying or constructing primitive black & white television sets to watch primitive programs. It was the Wild-West of different technical systems without any basic standard, manufactured by a multitude of manufacturers.

> Before the Second World War, in Europe there were several companies active with the early CRT techniques like, Leybold & Von Ardenne, Emil Gundelach, Philips, Lorenz AG, Telefunken AG, Otto Pressler, Gladitz GmbH, Fernseh AG, Radio AG, D.S.Loewe, Baird Television, Marconi-EMI and Cossor. Some of them were closely connected to RCA or the Farnsworth company. Most of these firms had also long reputations in early vacuum tube technology. In the US General Electric, the Bell laboratories, National Union, Westinghouse, RCA Victor, Farnsworth Television, and Dumont developed their pre-war CRT's.

In the period 1947-1960 some fifty manufacturers of TV-sets were active in the US (Table 9). From the small companies that did not survive the decline of the market due to the broadcasting crash of the late sixties, to the larger companies that survived. And they all made their separate television sets in a crowded market place. Some went for the elaborate cabinet design (equivalent to the early radio), others manufactured table-top models, or went for the (trans)-portable mini-TV (Figure 402). And with their advertisement campaigns they tried to position their product differences: from sound and picture quality, and entertainment life style to overall reliability (Figure 403).

**Figure 402: Television Receiver sets (1947-1960).**
Capehart-Farnsworth 661-P Television (left, 1948), Zenith H2445R The Tennyson television (middle left, 1951), Hoffman M143U Easy Vision television (middle right, 1954) and the Emerson Model 1232 Television-Radio (right, 1956).

Source: https://www.sfomuseum.org/, https://antiqueradio.org/

## Table 9: Manufacturers of TV sets (starting 1947-1970)

| | Manufacturer | Years of TV-set Production |
|---|---|---|
| 1 | Admiral | 1947- 1979 |
| 2 | Andrea | 1947-1978 |
| 3 | Base | 1947-1965 |
| 4 | Calbest | 1947-1968 |
| 5 | Capehart | 1947-1965 |
| 6 | Certified Radio Labs | 1947-1966 |
| 7 | Clairetone | 1967 |
| 8 | Colonial Radio (Sylvania) | 1949-1983 |
| 9 | Color Electronics | 1964 |
| 10 | Conar | 1962 |
| 11 | Cortron Ind (Hoffman) | 1948-1970 |
| 12 | Curtis-Mathes | 1960 to date |
| 13 | Delmonico | 1967 |
| 14 | DuMont | 1938-early 1970s |
| 15 | Emerson | 1947-1973 |
| 16 | Farnsworth | 1947-1965 |
| 17 | **General Electric** | 1947-1986 |
| 18 | Heath - Heathkit | 1964-1989 |
| 19 | Hoffman (Cortron Ind.) | 1948-1970 |
| 20 | Kane Electronics | 1967 |
| 21 | Kent | 1951-1964 |
| 22 | Magnavox | 1948-1976 |
| 23 | **Matsushita (National, Quasar, Technics, Ramsa)** | 1959 to date |
| 24 | Mattison | 1949-1968 |
| 25 | Mercury-Pacific | 1950-1964 |
| 26 | Motorola | 1947-1974 |
| 27 | Muntz (Howard Radio) | 1948-1973 |
| 28 | Nivico (Nippon Victor Corporation) | 1960s |
| 29 | North American Audio | 1962 |
| 30 | Olympic | 1948-1971 |
| 31 | Orion | 1981 |
| 32 | Packard Bell | 1948-1974 |
| 33 | **Panasonic (Matsushita)** | 1975 to date |
| 34 | Philco (Philco-Ford) | 1947-1976 |
| 35 | Philmore | 1948-1965 |
| 36 | Pilot (Jerrold) | 1964 |
| 37 | Ravenswood | 1962 |
| 38 | Radio Corporation of America | 1946-1986 |
| 39 | Sanabria | 1960 |
| 40 | Setchell Carlson | 1950-1970 |
| 41 | Sonora | 1962 |
| 42 | **Sony** | 1961 to date |
| 43 | Sylvania | 1949-1983 |
| 44 | Symphonic | 1960 |
| 45 | Tech-Master | 1948-1968 |
| 46 | Telequip | 1949-1968 |
| 47 | Teletronics | 1970 |
| 48 | TMA | 1948-1973 |
| 49 | Transvision | 1947-1967 |
| 50 | Trav-Ler | 1948-1965 |
| 51 | Warwick | 1951-1977 |
| 52 | Wells-Gardner | 1948-1990 |
| 53 | Westinghouse | 1947-1969 |
| 54 | Zenith | 1948 to date |

Source: http://www.tvhistory.tv/1960-2000-TVManufacturers.htm
Names printed in bold are Japanese companies.

Throughout this decades-long evolution of television from discrete pieces of scientific invention to integrated commercial applications, technological innovation was at the heart of the chain of events. Television was going to have wide ranging and profound influence on the Affairs of Man. Offering a 'window on the world,' it would influence societies in multiple ways. With its fast-moving, visually interesting, highly entertaining style, once the TV matured, it was to command many people's attention for several hours each day. In a way it competed with other sources of human interaction such as family, friends, church, and school. The broadcasted content (eg the talk shows and sitcoms) influenced the way that people thought about social issues as race, gender, and class. And, it played an important role in the political process, particularly in shaping national election campaigns.

## Television Broadcasting in Europe (1930-1950)

The first half of the 1930s was marked by a period of economic depression in the United States that spread to Europe. With market for television-sets holding back, and the limited performance of the television-sets, the development of TV-broadcasting halted. Notwithstanding experimental broadcasts in Germany by the Reichspost, in Great Britain by the BBC, in France by the ministry of Postes, Télégraphes et Téléphones, and in the US by the networks, European tv-broadcasting was fragmented. Not in least because the was no real standardization in broadcasting formats.

The first official channel of French television appeared on February 13th, 1935, the date of the official inauguration of television in France. In

Figure 403: Television Advertisement (1950).

Capehart (left), Zenith Black Magic (middle) and Raytheon (right).

Source: http://www.hifi-archiv.info/Radio-Werbung/1950, https://clickamericana.com/

1939, there were about only 200 to 300 individual television sets, some of which were also available in a few public places. During the German occupation, the Fernsehsender Paris was broadcasted from the Eiffel Tower in Paris under Nazi-control. On October 1st, 1944, television service resumed after the liberation of Paris.

In Britany the BBC began transmitting the world's first public regular high-definition service from the Victorian Alexandra Palace in north London on November 2nd, 1936. Already during the 'Radio-Olympia-Weeks' of August 1936, the BBC had presented a series of television programmes to the public. The BBC transferred successful variety and entertainment formats from radio to television. During the war the broadcastings were suspended and by June 7th, 1946 resumed. At the end of 1947 there were 54,000 licensed television receivers.

In Germany, the new medium of television broadcasting was used by the Nazis for propaganda purposes. It had started with the 1936 Olympic games when Nazi TV fed live, real-time images to 28 new viewing parlours throughout Berlin. After the Olympics, the station increased its programming and focused, as before, on health, order, and obedience. In November 1943, Allied bombing destroyed Rundfunk's transmitter and by late 1944, Nazi TV ceased operations. And after the War, the occupying powers established their own television stations, reflecting their own particular biases.

However, the 1930s in continental Europe were the period of early Americanization resulting from the American cultural imperialism, that brought not only technological progress to Europe, but also created there a mass consumption economy. Next to the production of automobiles in Europe, the American motion pictures conquered the European cinema. But television broadcasting was lagging behind and had to wait after the Second World War to pick up.

## *Television Broadcasting 2.0 (1950-1970)*

The Second World War, with its freeze on commercial television, delayed the rise of the new medium television. But after 1945 begins the period of the definite implementation of television, leading to its introduction over nearly all of Europe within a period of twenty years. [572]

---

[572] In the post-war period, American dollars (through the Marshall plans for reconstruction), American cinema, cartoons, and icons of modernity from jeans to milk shakes and Coca-Cola arrived on a massive scale all over Europe. American films, film stars, music and consumer goods represented modernity and a new age after war, poverty and depression. But in many cases the establishment saw the liberators and economic helpers of Europe as a

The post-war era marked the start of television as a mass medium. Although television first struggled to become a national mass media in the 1950s, it became a cultural force — for better or worse— in the 1960s. And much of that development was originating from the other side of the Atlantic Ocean. In the US, before these two decades were over, the three national networks were offering programs that were alternately earth shaking, sublime and ridiculous.

In 1950, there were some eleven million TV sets in the US, and in 1952 there were 24 million. By 1960, the cost of TV sets had come down considerably being present in 87 percent of the homes, taken the place of radio as the family's evening focal point. Between 1959 and 1970, the percentage of households in the U.S. with at least one TV went from 88 percent to 96 percent. By 1970, there were around 700 UHF and VHF television stations. This was the *Golden Age* of US television broadcasting.

**Figure 404: Television Broadcasting as a Lifestyle and Cultural Event.**

Family Life around the Television (top/middle, 1950s), and Leonard Bernstein in the program Omnubus (bottom, 1954).

Source: Wikimedia Commons.

---

threat to national culture, the more when the young in the Swinging Sixties saw it as a kind of liberation from a paternalistic national culture.

TV signals that could reach into the most remote corners of the US broke down the last vestiges of isolation in rural America. Common national carriage of popular TV shows, news and sports events meant that there was a shared national experience. The day after major televised events, researchers found that almost everyone was talking about the event. They weren't saying the same things, but there was a sense of national dialog.

> Families unified around their new TVs, and it helped parents and children socialize and connect in new ways and helped the family spend more time together. Some described their new TV sets as members of the family (Figure 404, top and middle).

As the households with TVs multiplied and spread to other segments of society, more varied programming came in. Many of the genres that today's audiences are familiar with were developed; westerns, kids' shows, situation comedies, sketch comedies, game shows, dramas, news and sports programming. Those early productions changed after the adoption of videotape in 1957. Many live dramas were shot 'live to tape,' still retaining a 'live' television look and feel but able to both preserve the program for later broadcast and allowing the possibility of retakes.

High-culture dominated commercial network television programming in the 1950s with the first television appearances of orchestra conductor Leonard Bernstein on November 14th, 1954 in the culture-oriented CBS-program *Omnibus* (Figure 404, bottom). It was soon followed by the first telecasts from Carnegie Hall, the first live American telecasts of plays by

**Figure 405: Sitcoms, Newscast shows and Animated sitcoms.**
Sitcom I love Lucy (left, 1951-1957), the Tonight Show (middle, 1954), and The Flinstones (right, 1960-1966).

Source: https://movieposters2.com/, https://www.senscritique.com/, https://pics.alphacoders.com/

Shakespeare, the first telecasts of Tchaikovsky's ballets *The Sleeping Beauty* and *The Nutcracker*. Many lightweight television programs of this era evolved from successful radio shows, such as the *I Love Lucy* shows that drew heavily from both film and radio (Figure 405, left). By the late 1950s, as television began reaching larger portions of rural America, rural sitcoms and Westerns boomed.

## The Rise of the Golden Age of Television (1948-1959)

In the 1950s, most US television entertainment programs ignored current events and political issues. Instead, the three major networks (ABC, NBC, and CBS) developed prime-time shows that would appeal to a general family audience. Between 1953 and 1955, television programming began to take some steps away from radio formats. It was to become the Golden Age of Television.

*Entertainment:* NBC television president Sylvester Weaver devised the 'spectacular,' a notable example of which was *Peter Pan* (1955), starring Mary Martin, which attracted 60 million viewers. Weaver also developed the magazine-format programs *Today,* which made its debut in 1952 with Dave Garroway as host (until 1961), and *The Tonight Show* (Figure 405, middle), which began in 1953 hosted by Steve Allen (until 1957). The third network, ABC, turned its first profit with youth-oriented shows such as *Disneyland,* which debuted in 1954 (and has since been broadcast under different names), and *The Mickey Mouse Club*.

*News:* Television news first covered the presidential nominating conventions of the two major parties, events then still at the heart of America politics, in 1952 (Figure 406, bottom). The term 'anchorman' was used, probably for the first time, to describe Walter Cronkite's central role in CBS's convention coverage that year. In succeeding decades these conventions would become so concerned with looking good on television that they would lose their spontaneity and eventually their news value. The power of television news increased even more with the arrival of the popular newscast, *The Huntley-Brinkley Report*, on NBC in 1956.

By the mid-1950s, television programming was in a transitional state. In the early part of the decade, most television programming was broadcast live from New York City and tended to be based in the theatrical traditions of that city. That changed when the centre of the television production industry was moving to the Los Angeles area, and programming was transforming accordingly: the live theatrical style was giving way to shows recorded on film in the traditions of Hollywood. The Hollywood studios like Walt Disney film studio and Warner Bros film studio grasped the new opportunities by supplying programming to networks.

In the 1950s and 60s, television news produced perhaps some of its finest performances. Edward R. Murrow exposed the tactics of innuendo and unsubstantiated charges that Sen. Joseph McCarthy used to exploit the country's fear of Communism. The televised debates between Kennedy and Nixon (Figure 406, top) were credited with giving JFK a slim election victory. And there was the new phenomenon of animated television series (aka Cartoons) in the late 1950s. Starting as short animations oriented at children, it became half-hour sitcoms for the whole family (eg the *Flinstones*) (Figure 405, right).

During the 1960s, television news broadcasts brought the realities of real-world events into people's living rooms in vivid detail. The CBS *Evening News* with Walter Cronkite (Figure 406, middle), which debuted in 1962, quickly became the country's most popular newscast. Soon, journalist Walter Cronkite was known as the most trusted man in America.

**Figure 406: TV-News Events (1960s).**

Kennedy-Nixon debate on TV (top, 1960), anchor-man Walter Cronkite for the CBS Evening News (middle, 1962), and Democratic National Convention (bottom, 1960).

Source: lefigaro.fr, www.biography.com/media-figure/walter-cronkite, Smithsonian Magazine.

The assassination of President Kennedy on November 22th, 1963 glued the American population to the TV-screen. The networks devoted days and days of airtime to coverage of the tragedy, the funeral and the aftermath. Around the same time as Kennedy's assassination, horrific images from Vietnam were streaming into people's living rooms during the nation's first televised war.

In addition to the devastation caused by the president's death and the Vietnam War, Americans were also feeling the pressure of the Cold War; the clash between the United States and the Soviet Union in the years following the Second World War. This pressure was especially great during periods of tension throughout the 1950s and 1960s, such as the 1962 Cuban Missile Crisis, a confrontation that caused many people to fear nuclear war. In reaction, TV broadcasters and viewers turned to escapist, rural comedy-themed programs focussing on the American heartland. Such as the sitcom of the *Beverly Hillbillies* (CBS, 1962–71), a show about a poor backwoods family who move to Beverly Hills, California, after finding oil on their land. None of the 1960s sitcoms mentioned any of the political unease that was taking place in the outside world, providing audiences with a welcome diversion from real life.

## The Demise of Radio and the Golden Age of TV

In spite of changing attitudes toward the medium, by 1960 there was no question that television was the dominant mass medium in the United States. That year, average daily household radio usage had dropped to less than two hours; TV viewing, on the other hand, had climbed to more than five hours per day and would continue to increase annually. Between 1960 and 1965, the average number of daily viewing hours went up 23 minutes per TV household, the biggest jump in any five-year period since 1950. At the movie theatres, weekly attendance plunged from 44 million in 1965 to 17.5 million by the end of the decade.

> *In 1959 two key events underlined the demise of television's Golden Age. The first was the quiz show scandal, which reached its apex that year. The quiz show, which awarded large cash prizes to contestants who answered questions posed to them by a host, had become a dominant program type on prime-time TV by 1955. In the fall of 1956, the networks aired 16 evening quiz shows, 6 of which were among the 30 highest-rated shows of the season. By 1958, however, widespread allegations were circulating that many of these shows, in order to maintain dramatic tension, had been fixed—that contestants were told the answers before appearing on the air. […] The second event of 1959 was the appearance of* The Untouchables *(ABC, 1959–63), a series about organized crime activity in Prohibition-era Chicago. Although the series had only a casual relationship to actual events, this film noir-influenced historical drama is now*

*considered a minor classic. However, the frequent machine-gun fire and pre-Miranda warning speakeasy raids that characterized the show contributed to the protests of the depiction of violent acts that were becoming increasingly common among parents' groups, educators, and other cultural watchdogs.*[573]

The 1960s saw a rural-comedy boom with the rural sitcom. For many viewers these programs brought hours of escapist pleasure; to others they came to identify American TV as a cultural wasteland catering to the lowest common denominator of public taste. However, over the course of the late 1960s/early 1970s, concurrent with the development of colour television, the evolution of television led to an event colloquially known as the 'rural purge' of television networks. Genres such as the panel game show, the western, variety show, barn dance and rural-oriented sitcom all met their demise in favour of newer, more modern series targeted at wealthier suburban and urban viewers (ie the young baby-boomers who were more of a target group for the advertisers on TV).[574] And in contrast to the single camera setup without a studio audience, the newer shows that came to television in the early 1970s were multiple-camera setups with live studio audiences, a trend that would become the norm throughout the 1970s.

The backlash from the purge prompted CBS to commission a rural family drama, 'The Waltons,' for its fall 1972 schedule. It went on to run for nine seasons, starting a trend for family dramas throughout the 1970s. Among the new shows was the war comedy-drama television series M*A*S*H (ie the Mobile Army Surgical Hospital in the Korean War) from 1972-1983 that had one of the highest-rated shows in US television history.

## Television Broadcasting in Europe (1950-1970)

The European television broadcasting —flooded with experiments in the mid-1930s but interrupted/limited during the war— became only in prominence after the Second World War. In 1950 the *European Broadcasting Union* (EBU) was created to coordinate national interest and conflicts, standardization issues and to help with the implications of the unprecedented technological changes.

---

[573] Thompson, R. Television in the United States. https://www.britannica.com/art/television-in-the-United-States/The-year-of-transition-1959. (Accessed December 2021)
[574] In a move uncharacteristically bold for an American television network, CBS scrapped an assortment of its hit series and launched what turned out to be an unprecedented updating of prime-time television programming. Within four years, entertainment TV would look nothing like it did in 1969. The "real world" of social, familial, and national dysfunction, which had been ignored by TV for so long, was about to break into prime time. With the spectacular success of three strikingly new programs —'All in the Family,' 'The Mary Tyler Moore Show,' and 'M*A*S*H'—CBS redefined the medium.

Notwithstanding the national practises and approaches, the influence of the US-television broadcasting —aka Americanization—was notable in the national programming. American networks sold their recorded programs to Europe's public and state TV-stations. The introduction of television now became a symbol for a new, modern lifestyle in a Western world during the turmoil of the Swinging Sixties. At the same time the Cold War also affected in Europe its development as a political medium.

In Germany, the first regular television service began in Berlin on March 22$^{nd}$, 1935, as Deutscher Fernseh Rundfunk. As just a few people owned television receiver sets, they went to Fernsehstuben (ie television Parlours) to watch television. During the war, radio broadcasting was the dominant —although state controlled— medium, and after the war the Allied Forces controlled television broadcasting. In 1948 the British occupation forces authorized the Nordwestdeutscher Rundfunk (NWDR) to make plans to broadcast television programs for the British zone reflecting the BBC-practises. Other regional networks also started to launch television in their own areas. And in each occupation zone, the occupying forces stamped their own broadcasting practises mark on the broadcasting. The companies in the American occupation zone — confronted with the soviet ideology over the border— were more determined to promote TV as a "window to the world", rather than mere "pictured radio." In 1950 the 'Arbeitsgemeinschaft der öffentlich-rechtlichen Rundfunkanstalten der Bundesrepublik Deutschland'(ARD) was created by joining regional public-service broadcasters. The new entity was financed by an obligatory fee which every German household with at least one radio receiver paid. After starting with a schedule of a mere two hours per-night, television became more widespread in Germany in the 1960s. Colour broadcasts were introduced in 1967. Without competition from private broadcasters, the ARD stations made considerable progress in becoming modern and respected broadcasters. In order to create competition, a second nationwide network was established in 1961; the Zweites Deutsches Fernsehen (ZDF) tha started broadcasting in 1963.

France resumed its television service very early, on March 29$^{th}$, 1945, and used the facilities of the 'Fernsehsender Paris' left by the Germans. Copying the British in radio-broadcasting, by 1948 French television broadcasting became a state monopoly with the creation of the French Broadcasting Television ('Radio Television de France,' RTF). From 1964 to 1975, French radio and television was monopolized through an organization known as the Office de Radiodiffusion Télévision Française (ORTF). Over time that monopoly came under attack —also from social-liberal political sides where the 'New Society' was advocated— and limited reforms were initiated. But it took the Law of 1974 before major reform occurred with the abolition of the ORTF and seven independent companies resulted from this dismantling. Among them the three national television broadcasting companies TF1, Antenna 2, and France Regions (FR3).

In Britain, where paternalistic sentiments shaped the radio broadcasting practises[575] of the BBC, television —launched in 1936 before the war as BBC Television Service (BBC-TV)— returned in 1946 in the air. Competition to the BBC was introduced in 1955, with the commercial and independently operated television network of Independent Television (ITV). The BBC television broadcasting —renamed in 1960 as BBC-TV— showed popular programming, including comedies, drama, documentaries, game shows, and soap operas, covering a wide range of genres and regularly competed with ITV to become the channel with the highest ratings for that week. In 1964 the BBC-TV was restructured in 1964 into BBC-One and BBC-Two. And the latter was "to broadcast programmes of depth and substance." BBC Two's historical scope was arts, culture, some comedy and drama, and appealing to audiences not already served by BBC One or ITV. Both ITV and the BBC used American imports (within quota limits) to provide relatively inexpensive programmes with wide audience appeal that allowed them to focus funding on more prestigious factual and cultural programming.

The rest of Europe followed a similar pattern; the experiments followed by the establishment of regular TV-broadcasting in the second half of the 1950s. After a long incubation period, television finally became the leading mass medium of social self-reflection in Europe at the end of the 1960s and beginning of the 1970s.

## The TV-Manufacturers (1950-1980)

The rapid integration of television into American society coincided with the explosive rise of this post-war consumer culture for several reasons. As a household appliance that was finally affordable to a large part of the population, TV provided hours of entertainment for what suddenly seemed like a reasonable price. The CRT -based TV-sets were not so much more a part of the furniture anymore, but had found their dominant box-shaped and console design (Figure 407), realised in the new Bakelite technology. And design became more important within a decade.

---

[575] Television was not to be dedicated to popular entertainment, but aimed to elevate citizens to a higher level of culture and information, in short television should produce national citizens.

## Television Manufacturing in the US

After the Second World War had ended, in 1948, about 800,000 television receivers were sold in the US. Two years later, 1950 sales were 7.5 million and remained at 5-6 million per year for the next decade. Another goldrush was on. Radio receiver brands like General Electric, Westinghouse, Philco, Zenith, and Motorola all jumped on the bandwagon and entered the TV set business alongside RCA. And there were also the small companies that entered the television business (Figure 407).

In the US, the *Admiral Corporation*, a late entrant into the radio industry, was the brainchild of aggressive Chicago entrepreneur Ross Siragusa. In 1939 revenues in radio topped $9 million. Entering after the war ended the television business, in 1949 Admiral produced already 400,000 television sets. The company used plastic cabinets instead of the wood used by other makers, bringing out the 10-inch Consolette TV in 1949 at a price of only $249.95, a hundred dollars less than competitors (but even then, the Consolette cost almost $3,000 in 2021 dollars). The Admiral Corporation's revenues tripled between 1948 and 1950. It made more than one million television sets in 1950 and in the following year it manufactured five million. Along with companies such as Dumont and Philco it became one of the leading makers of televisions.

The 1940s and early 1950s were among the best years for Admiral. The growth of the company mirrored

**Figure 407: US Television models in late 1940s.**

Admiral Consolette TV (top, 1948), General Electric Model 807 (middle, 1949). Bush TV 22 (bottom, 1950).

Source: www.radiomuseum.org, collection sciencemuseumgroup.

the prosperity of post-Second World War America. Sales grew from $67 million in 1948 to $251 million in 1953. But with the rising Japanese competition in the 1960s — notwithstanding that Admiral's sales grew from $347 million in 1967 to more than $600 million in 1973— profits were falling sharply. In 1971 the picture tube manufacturing company was acquired by RCA, in 1973 the rest of the company was acquired by Rockwell International.

With maturing television technology and the progressing radio casing technology (ie Bakelite), the aesthetic design of the television set became an issue. Already from the late 1930s, prominent architects and designers were enlisted in the design of early television receivers and their display at national exhibitions. As the rate of introduction of new models was high, by 1939, an average-sized manufacturer was producing annually 'about five new table radios and three television sets, there was a focus on 'good design.' As the television set got a place in the living room, where domestic life took place, it became a 'device of modernity.'

Some manufacturers created design departments. Such as the models designed by Philco. Its design director, Herbert V. Gosweiler presented America with a forward-thinking design that would be applicable decades into the future. (Figure 408).

In the 1960s the sales of colour television sets took off. By 1962, a million colour sets were in use, generally

**Figure 408: Design Models of Television Receivers.**

Philco Console Model Predicta (top, 1958), Predicta Model Chalet (middle, 1959), and transportable tabletop Model Predicta Princess (bottom, 1959).

Source: www.cooperhewitt.org/

selling for $6-700 ($5-6,000 in 2021 dollars), over three times the price of a black-and-white television. In 1964, US makers produced 94% of the colour TVs sold in the US.

RCA was the market leader for years, but that was to change as the political climate was turning against RCA, as well as the management decision to diversify into a broader danger of corporate activities and become an international business conglomerate. In 1960 it went into the mainframe business challenging the market leader IBM (to leave it again in 1971) In 1965 it went in the publishing business by acquiring Random House Publishers (to leave it in 1980). In 1967 it went into the rental car business acquiring Herz (to leave it again in 1985). Just to mention a few of its unsuccessful diversifications. An effort to return to its original wireless business failed. And in 1985 General Electric bought RCA remaining activities and dismantled in the following years its assets. And thus, RCA's glorious history of innovation, changing the lives of millions, came to an ignominious end.

Notwithstanding the efforts of individual manufacturers to add 'unique selling points' to their television sets (eg the square CRT), they operated in an overcrowded market. Although by the 1970s, some of the television sets were part of the furniture (eg the console style model), but it did not take long before the television receiver had found its dominant design of the (table-top) box that, when supported by (four) legs, became an floor-based console tv set (Figure 409). The 1970s TV-broadcasting's rise in popularity went hand in hand with the sales of television sets. Each year millions of TV sets were made (Figure 410), the industry matured and followed the classical pattern of mergers and acquisitions with a shake-out of the smaller manufacturers.

**Figure 409: Television Design in the 1960s.**
Motorola Model 23SF3 Console TV (left, 1960), Gratez Burggraf (middle, 1965) and Murphy Radio Ltd television set (right, 1960s)
Source: tvhistory.tv, 1stDibs.

*Television Manufacturing outside the US*

In Europe's Germany, the economic revival in the early afterwar period had other priorities then radio and television broadcasting. Nevertheless, when the occupying forces relaxed the rules, early experimental television broadcasting —in many cases (ie Germany, France) under state control— took off slowly. Much of the industrial infrastructure being destroyed, the availability of television sets for the consumer market was limited. But that changed when during the German Economic Miracle, former radio-manufacturers (eg Loewe, Saba, Nordmende) entered the television business with table-top designs (Figure 411). The large US TV-manufacturers, in order to evade the Europe's Tariff Wall, invested heavily in technology transfer to European production facilities. RCA formed alliance with national manufacturers; from tube- and radio-manufacturers to makers of television sets.

But also the large corporation like Royal Philips (Netherlands), recuperated after the Second World War to enter the mass-market of television. Expanding rapidly, it established manufacturing facilities all over Europe (eg Philips UK). In Britain, with its rich history in wireless broadcasting, the large companies were moving into the television business when the wartime production dissolved instantly. Stimulated by the broadcasting of the Coronation of Queen Elizabeth II on June 2nd, 1953 some 536,000 sets were sold due to the coronation fever.

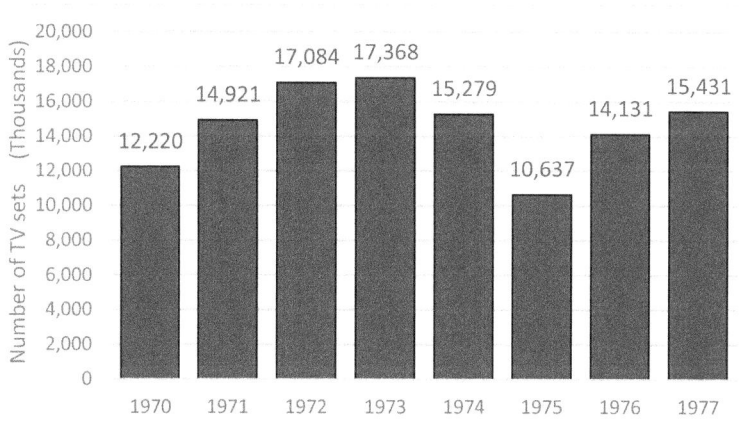

**Figure 410: Number of TV set sold (1970-1977).**
Source: http://www.tvhistory.tv/Annual_TV_Sales_70-77.JPG

The development of television manufacturing is quite diverse in those years, but the big surprise for the industry was the unexpected massive competition of Japanese manufacturers.

*The Rise of the Eastern Manufacturers*

In the 1960s, the American TV-manufacturing industry was turned upside down by the Japanese acquisition of its technology through licensing arrangements. The Japanese acquired transistor and B&W TV production rights from the Radio Corporation of America (RCA) and General Electric (GE). This enabled Japanese firms to rapidly achieve economics of scale in B&W TV production. By the late 1960s, their major factories were on average twice the size of those of West Germany and no less than six times those of the biggest UK production unit. In 1968, there were 28 American-owned TV manufacturers in business. By the end of 1976, only six remained.

*The Big Squeeze*:[576] Japan's raid on the American market dates back to 1956, when the largest Japanese manufacturers formed the Home Electronic Appliance Market Stabilization Council, a production cartel. The intent

**Figure 411: European Television Sets (1950s).**

Philips TX1410 (top, 1952), Bush TV 31 (middle, 1953), Saba Schauinsland W II (bottom, 1953/1954).

Source: https://www.maximus-randd.com/, Radiomuseum.org, www.thevalvepage.com.

---

[576] Source: Choate, P.: Japan and the Big Squeeze, https://www.washingtonpost.com/archive/opinions/1990/09/30/japan-and-the-big-squeeze/0fb1617e-8756-4390-a776-f1619d59869a/ (accessed December 2021).

of the cartel was to monopolize the domestic market for television receivers, radios and other home electric products and to exclude foreign imports. Once their home market was secure, they would launch a drive against the far richer American market.

As part of the broader foreign policy,[577] the US government forced US companies to license their technology to members of the newly formed cartel. RCA, GE and Westinghouse licensed and then transferred monochrome technology to members of the cartel. In 1962, RCA went one step further and licensed its colour technology.

> Japanese government officials (ie the MITI) helped by ensuring that US exporters were harassed by import safety inspectors; the Electronics Industries Association of Japan persuaded Japanese distributors not to handle certain American TV products. US television exports to Japan soon fell precipitously. Japanese manufacturers were thus able to sell a set for more than twice as much in Japan as they could abroad.

In the US they sold their sets through importers. With dumping practises, they were able to under-price the American manufacturers. The result was that jobs in the U.S. television manufacturing industry fell 50 percent between 1966-1970. They dropped an additional 30 percent between 1971-1975, and 25 percent more between 1977-1981.

Next to Japan, other countries in the Far-East (eg Taiwan, Korea) would also start manufacturing TV-sets. It heralded the end of television manufacturing in the US.

## Television Broadcasting 3.0 (1970-2000)

By the 1970s the television industry, both the manufacturers of the hardware as well as the broadcasting industry) were on full steam (Figure 410). But technological progress was in the air. The long and continuous quest for improved video quality —called Quality High-Definition Television (HDTV)— had resulted in the advanced television explorations initiated by the Federal Communications Commission (FCC) in 1987 at the request of American broadcasters. Stimulated by the Japanese efforts of analogue HDTV development in the mid-1980s, the FCC started in 1987 a project to explore the *Advanced Television Service*. A project that in its essence at the end of the 1980s was a death knell for most analogue high-definition technologies that had developed up to that time. And by early 1993, after a

---

[577] In the broader context of the Cold War in the Pacific, and the US opposing the rising communist influence, the US was allowed to maintain airbases on Japan's territory (eg on the island Okinawa).

rigorous technical review of four digital HDTV standards and one analogue proposal, the FCC affirmed the superiority of digital over analogue. It resulted in the Telecommunications Act of 1996, assigning new DTV licenses to broadcasters. Subsequently, digital HDTV technology was introduced in the United States in the 1990s by the 'Digital HDTV Grand Alliance,' a consortium of television companies[578] and MIT. The alliance came to an agreement on the transmission standards, and the partners started developing their systems.

Notwithstanding the development of the Digital Television Adaptor (DTA) prolonging the technologic lifecycle of the massive installed base of vacuum-based television receivers, the new transistor-technology moved on in the design of television receivers. It heralded the fall of the CRT-based analogue television and the rise of Digital Television.

## The Portables: Fall of Radio-tubes, and Rise of Transistor

In the mid-1950s the new transistor as a component had already replaced the vacuum tube in the radio receiver. It had been the start of the development of a development trajectory of whole new range of portable radio receivers. From the first Regency TR1 transistor radio receiver in 1954 (Figure 249), soon followed an explosion of 'transistor radios.' With the transistor radio, music and information suddenly became portable. From a collective experience (Figure 337, Figure 360) the radio became an individual experience. It was the start of the Portable Electronic Age where the word 'Transistor' became the label for the small radios (Figure 412).

Transistor radios became extremely successful because of three social factors; a large number of young people due to the post–

Figure 412: The Portable Transistor Radio.
Sony TR-63 (left, 1957), Realtone (middle,1962), Orion (right, 1965).
Source: Wikimedia Commons.

---

[578] The Grand Alliance included AT&T, General Instrument, MIT, Philips, Sarnoff, Thomson, and Zenith.

Second World War baby boom, a public with disposable income amidst a period of prosperity, and the growing popularity of rock 'n' roll music.[579]

Transistorizing the radio receiver might have been a first challenge, doing the same for the television receiver was the next. Already in 1959 Toshiba developed a (rather bulky console style) CRT television receiver that incorporated transistors. But in the mid-to-late 1960s and through the 1980s it seemed as if there was sort of an informal battle between manufacturers to see who could make the smallest television set.

The advent of the transistor had not pass unnoticed by the Japanese television set designers. Sony introduced already in 1959, next to the TR-63 transistor radio, its all-transistor 8-inch portable television Model TX8-301 (1959) still equipped with a CRT-display (Figure 414). Too early for the US market with a frail performance record, this hybrid television set became a commercial failure after its introduction in 1960. It was succeeded by the 'Micro TV'; Sony Model TV5-303 in 1962 (Figure 413, left).

Other Japanese companies (eg Matsushita Denki with the Panasonic brand) followed suit with a long series of product launches in the United States (Figure 413). The television receiver became smaller and smaller.

The rising Japanese dominance in television manufacturing paralleled the dawn of the Digital Era as the semiconductor technologies created, next to the analogue circuits, the digital circuits with increasing complexity

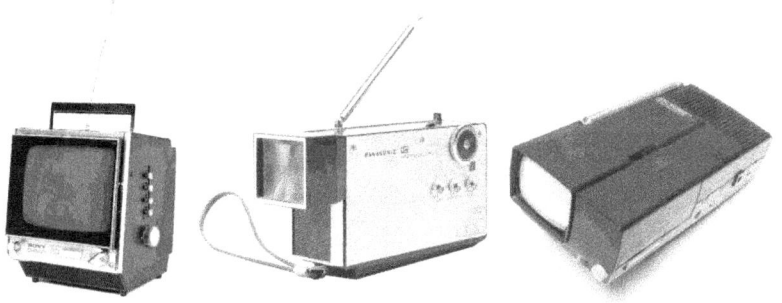

**Figure 413: Portable Television Receivers.**
Sony Model TV5-303 (left, 1962) Panasonic (middle, 1972) and Sinclair Microvision MTV1B (right, 1967).

Source: Wikimedia Commons. www.centrepompidou.fr, mztv.oncell.com, nightfallCrew.com.

---

[579] Another device of the portable rage would be Sony's Walkman, a portable cassette player introduced in 1977.

realized in the semi-conductor technology. And one of the technological developments made the television display larger and larger.

## The Digital Television Revolution

Originally the television system used analogue technologies in which the combined video and audio signal was transmitted. However, with the rise of the digital technologies in the domain of computing, came also their use in television broadcasting; it became known as Digital Television (DTV). It represented the first significant evolution in television technology since colour in the 1950s. A transition from analogue to digital broadcasting that began in the decade before the year 2000.

The digitalisation started at the receiving end. In the mid-1980s, Toshiba released a television set with digital capabilities, using integrated circuit chips such as a microprocessor to convert analogue television broadcast signals to digital video signals, enabling features such as freezing pictures and showing two channels at once (aka Picture in Picture).

By the 1990s the Telecom institutions started to develop the standards for the new methods of transmission. Such as the US Digital Satellite System (DSS) standard, followed by several standards for digital terrestrial television broadcasting. In June 1990 the General Instrument Corporation (GI) surprised the industry by announcing the world's first all-digital

**Figure 414: Portable Television.**
Advertisement for the Sony TX8-301 (1959).
Source: Wikimedia Commons.
www.antiqueradio.com,

television system. By 1993 a consortium of European broadcasters, manufacturers, and regulatory bodies agreed on the Digital Video Broadcasting (DVB) standard, and efforts were begun to apply this standard to satellite, cable, and then terrestrial broadcasting. In the late 1996 the FCC approved the new ATSC standards in the US, but the acceptance was slow. DVB standards were adopted worldwide and became the benchmark for digital television worldwide.

## The Digital Television Display

Digital Television, with the digital recording, transmission and decoding techniques, took off in combination with a new type of television display; the Flat Panel Display that saw an evolution from the Plasma-display, the LCD-display to the LED-display (Figure 415). These flatscreen TVs were based on the Liquid Crystal Display technology (LCD), the Plasma Display Panel technology (PDP) and the Light-Emitting Diode technology (LED). Each of these technologies having its own technological development trajectory that fuelled a range of application trajectories.

*The Rise of the Plasma-TV:* The plasma video technology to create an image, started in the 1970s realizing small displays for adding machines, cash registers, pinball machines, etc. replacing the Nixie-tubes. By the 1980s larger version became used in portable computer screens and as stand-alone monitors. By the 1990s the TV-receiver got its plasma-display. In 1992, the Japanese company Fujitsu introduced the 21-inch (53 cm) full-colour plasma panels. In 1997, Philips introduced the first large commercially available flat-panel TV, using the Fujitsu plasma panels. It was the dawn of increasing panel seizes, and further market penetration, notwithstanding the high cost of the TVs. In 2010, the shipments of plasma TVs reached 18.2 million units globally, but the reign of plasma would be short, and from then on it lost terrain to the LCD-TV.

*The Rise of LCD-TV:* The Japanese companies that headed the development of LCD, starting with small LCD displays. Inherently to the technology,

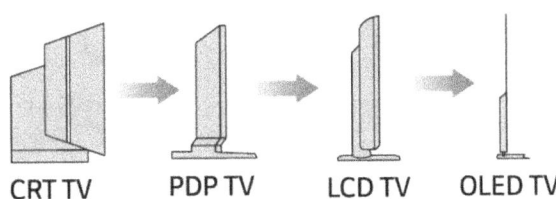

**Figure 415: The Evolution of the TV-Display (side views).**
PDP and LCD TVs need a backlight.

those LCD-displays needed a fluorescent light as backlight to create the image —and colours— though polarized filters (Figure 416). In the 1980s the first prototype of a flat panel television TV screen was developed as forerunner to the LCD screen with a LED powered backlight.[580]

In 1982, Seiko Epson released the first LCD television, the Epson TV Watch, a small wrist-worn active-matrix LCD television. Sharp Corporation introduced the dot matrix TN-LCD in 1983, and Casio introduced its TV-10 portable TV. In 1984, Epson released the ET-10, the first full-colour pocket LCD television. That same year Citizen Watch introduced the Citizen Pocket TV, a 2.7-inch colour LCD TV, with the first commercial TFT-LCD display.

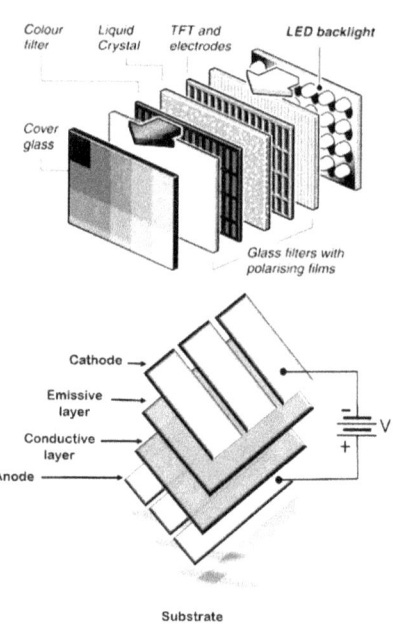

That was the beginning, but it took till the early 2000s before the larger flatscreen LCD-TV entered the market. In 2006, LCD prices started to fall rapidly and their screen sizes increased, and several vendors were offering 42" LCD TVs.

*The Rise of the OLED-TV*: Then the organic light-emitting diode (OLED) technology, that emitted light directly, was used to create the array of individual points that displayed the pixels of the video signal. Basically (Figure 416), it uses two layers of polymers; the emitting layer and the conducting layer. When a voltage is applied between the anode and cathode, an electric current of electrons flow from

**Figure 416: Diagram of LCD Display with LED backlight (top) and OLED Display (bottom).**

LCD: The backlight emits white light that is filtered to direct the liquids crystals creating the image. OLED: The light is created in the polymer material as the result of an electric current.

---

[580] Confusingly it was labelled by Samsung as QLED (ie Quatum dot LED TV).

cathode to anode. The result is the emission of a bright light. And using the anode and cathode layout, on the substrate (ie glass) small islands (aka pixels) are created of individual OLEDS.

As the OLED-technology leads itself to creating large display with many pixels, the display sizes and pixel density increased continuously. In October 2007, Sony announced the XEL-1 TV, the world's first 11" OLED TV for $2500.

**Figure 417: Smart TV.**
HP SLC3760N MediaSmart 37" LCD (top, 2007), Pavv Bordeaux TV 750 (bottom, 2008).
Source: Wikipedia Common.

The Digital Television with its flatscreen display would herald the Second Television Revolution that took off in the twenty-first century. A revolution that was not anymore about the broadcasting and reproduction of HDTV digital video signals, but would herald the *Resolution Race* as well as the *Display Size Race* where the displays were getting sharper, brighter and bigger. And it would create the Smart TV that contributed to the Smart Revolution to come.

*The Smart Television*

Basically, a TV-receiver displays the TV-signal from the traditional broadcasting media that is selected and decoded by its receiver circuit. But with the rise of the Internet, the traditional television display was also used as a display for the emerging Internet services. Even more, in a technological convergence of computers, televisions, and digital media players, a new type of television developed; the Smart TV.

> In the first place, a Smart TV is a digital television receiver build with digital circuitry to decode and display the digital video signal. Next, a Smart TV is also a computer, has a processor/memory and an operating system that can display 'application programs.'

The development trajectory of the Smart TV may have started with the Home Entertainment Centres and the Media Centres; integrated in the sets or as external electronic devices (ie boxes) with capabilities to add additional facilities to the traditional TV-functionality.

*Set-top box:* Starting as a converter for analogue tv-signals into digital video, the functionality of these devices expanded into internet access. Combined with the functionality of a digital media player, they added smart functionality to the standard television set.

*Media Centre:* As an external device —aka home server— to enable the user access to terrestrial, satellite or cable broadcasting or streamed from the internet, these devices became a network media centre. The received data could be stored on a hard disk, and played back at a later moment. And users could stream television programs and films through selected services such as Netflix.

*Smart TV:* The facilities of the set-top box/media centres became integrated into the television set itself. And with all their functionality the became the 'intelligent TV' (Figure 417).

> Such as the HP's Mediasmart TV (SL3760) introduced in 2007 where the MediaSmart element —essentially a small A/V box— was attached to the back of the TV. The internet connection was either Ethernet or WIFI. Or the Samsung Bordeaux TV 750 introduced in 2008 that connected to the Internet.

The mass acceptance of digital television in the mid-late 2000s and early 2010s greatly improved Smart TVs. Smart TVs became the dominant form of television during the late 2010s.

## *Overview of the Early Days of Television*

The transmission of the coded-word, followed by the spoken-word and analogue sound, had created a wave of novelty labelled as 'radio broadcasting.' Technological innovation created radio equipment for broadcasting, and the radio-receivers for the consumer-reception. But that was not the end of the new forms of broadcasting as the next step would be the transmission of visual information; video graphics broadcasted to television receivers for consumer-reception that became known as 'television broadcasting.'

The development of television took place in the context of its time, as we explored in earlier chapters. Times that, after the Scientific Revolution (Figure 8), had seen the rise of mechanical thinking and the rise of (fine) mechanical technologies. In the same period of time the Renaissance Revolution had changed the social structures and infrastructures. The combination had created the Mechanical Worldview. By the second half of nineteenth century, it had fuelled massive economic progress (Figure 12). By the early twentieth century the world was flooded with mechanical novelty; from automobile and airplanes, telegraph and telephone, clocks and calculators. A development that had its impact outside the domain of technological development.

> *The cultural developments of the nineteenth century were also crucial for the emergence of television. The new techno-scientific thinking was deeply rooted in cultural traditions influencing the perception, experience and conception of the modern world. The most important phenomenon in changing cultural experiences was acceleration – at first the acceleration of vision. Thanks to the invention of the railway, people could travel faster through the landscape than by horse or carriage, they saw 'faster' and the landscape passed by smoothly. Because people could travel faster it seemed that distances decreased, places moved closer together and the world became smaller. [...] This acceleration of perception was most prominently experienced in Western industrial countries like Great Britain, Germany and France. As 'laboratories of modernity', big cities like London and Paris were the most prominent loci of accelerated rhythms of life.* (Bignell & Fickers, 2008, pp. 101-102)

One could say that the 'pace of life had fastened'; physical distances were bridged faster, communication over distance became an instant affair. But that was only part of it, as the conversion from an agricultural-based to an industrialized society during the Industrial Revolutions had also brought changing working circumstances, and the increasingly growing middle class emerged next to the working class of the former peasants. And with that came the rise of leisure time: time that could be spent on physical and cultural entertainment (from sports to theatre).

> *Accelerated communication over long distances did not only concern language and written texts, but increasingly the transmission of pictures too, because of a growing demand for visual representations linked to the expansion of mass media consumption. [...] In the middle of the century, the demand for visual entertainment was answered further as the illustrated press developed and picture stories were reproduced, comic strips were invented and early experiments with motion picture film took place.* [ibidem]

The phenomenon of leisure time gave rise to many technical and cultural developments that fulfilled latent needs. The automobile of the urban class responded to its mobility needs, the moving pictures to its entertainment needs. And then the radio broadcasting brought entertainment to the home. With the introduction of television, the Modern Age seemed to have arrived in a new way.

So, television as a medium of visual communication was obvious, but it was unclear where the technical trajectory of the new communication medium would lead to. However, its technological development trajectory did not emerge in a development vacuum.

> *With the emergence of radio as a new mass medium in the late 1920s, radio became the dominant point of reference for the development of television. The success of radio as a mass medium for information and entertainment now shaped the conception of television as a medium for programmes. This new conception had an impact on the research agenda of television engineers too. [...] The medial character of television therefore corresponded rather to radio than to the cinema. Television simply added the moving picture to the wireless transmission of sound. In the same way that film turned into an audio-visual medium in 1928 when sound was added to pictures, the sonic medium of radio seemed to turn into television thanks to the addition of moving images.* (Bignell & Fickers, 2008, pp. 109, 116)

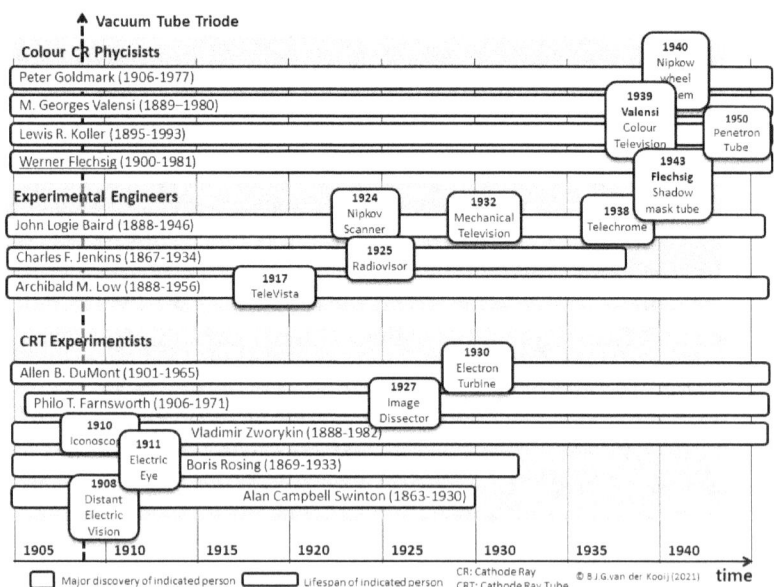

Figure 418: Contributors to the Development of the Television-system.

## Contributors to the Development of the Television-system

The development of television-system (ie recording, transmission, and receiving video-content), was not the work of one single person as we have seen in the invention of telegraphy, telephony and wireless. Television was the result of a common team effort (eg in the R&D laboratories) and a combination of a number of technical innovations. Similarly, although nations made various claims for their pioneers afterwards, television-broadcasting was not the invention of one single person or nation, but the result of international activity.

The same was the case with the technology that created the television artifacts (ie the picture camera, transmission system, and receiving units). These inventions did not come from a single parson, place or country, but they did emerge in the nineteenth century in the European and North American context of industrialization. It was a development to which both the 'CRT-experimentist' and the 'physicists' contributed (Figure 418). Over a long period of time, engineers, scientists and radio amateurs —but also the leaders of the large radio corporations that grasped the new opportunities— worked on the accomplishment of television as it later became realised in the form of a broadcasting medium.

However, there were contributions in specific areas that laid the cornerstones for further developments. Contributions that can be attributed to individuals. Such as in the development of the television receiver set; a cluster of innovations (Figure 419) in which the contribution of Philo T. Farnsworth with his all-electronic television system stands out due to its impact. His contribution was based on those of numerous others. Such as the thinkers experimenting with the electromechanical technologies available to them, and theorizing about the use of electric waves. Next there were tinkerers constructing the components and circuitry of the (hybrid) television systems. In addition, there we the contributions of the component-builders and the circuitry designers.

## Periodization of the Television Era

After the radio-broadcasting started transmitting sound, the next step was the transmission of graphics in motion; such as realized by the motion of the pictures in a movie-film. It needed equipment to record the motion (aka the video-camera in modern times) and equipment to receive and present the transmitted motion (aka the television set). And both were realized in the vacuum tube technology; the *photo tube* to record, and the *cathode ray vacuum tube* (CRT) to present the motion. Both were the result of their own technological trajectories of development, much of which was executed in Europe.

During the interbellum this initiate the experimental broadcastings with the system of analogue television.[581]

*Television Broadcasting 0.0* (1920s-1940s): Early experimenting before the First World War with mechanical technologies were doomed to fail facing the complexity of the transmission task (Figure 419). However, after the war, with the advent of the Cathode Ray Tube as medium to register and display electronic signals, the many contributors —from the CRT experimentalist to the experimental Engineers— developed their early transmission systems for video (Figure 418). First on Black and White (BW) displays, but by the end of the Interbellum on colour displays.

*Television Broadcasting 1.0* (1930s-1950s): This was the time of electronic television-set manufacturing by the early pioneers of television. Television sets that appeared in a range of designs; desk, console and cabinet models that used different display techniques as there were no uniform technical standards applicable yet. Next to the technology part realising the all-electronic television, the Interbellum saw the rise of the television 'providers;' the company owned television studios

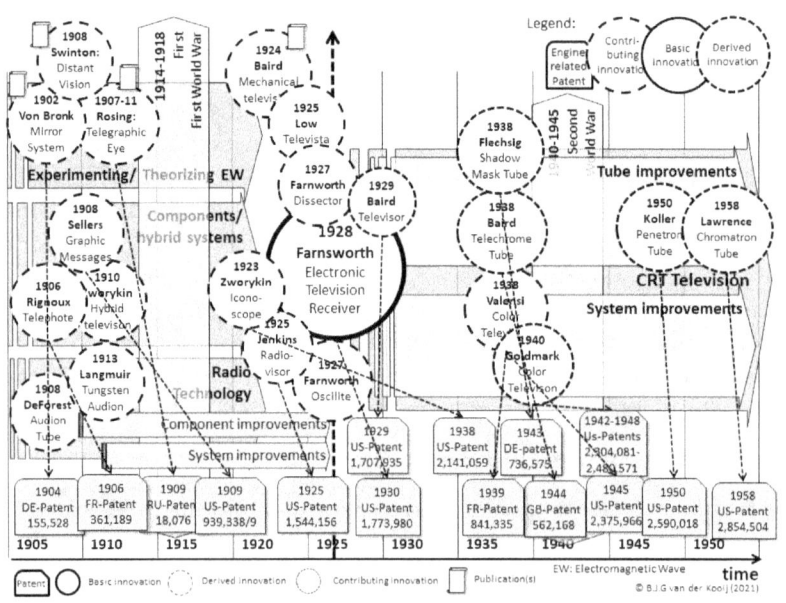

Figure 419: Overview of Cluster of Innovations for the Television Receiver.

---

[581] Analog television is the original television technology that uses analogue signals to transmit video and audio. In an analogue television broadcast, the brightness, colours and sound are represented by amplitude, phase and frequency of an analogue signal.

broadcasting their programs. World's Fairs in Paris (1937) and New York (1939) marked the phase of staging television as a politically and culturally important medium and industrial reality rather than a technical or scientific attraction. The 1930s were the infancy of television broadcasting, mainly by the manufactures setting up experimental broadcasting station to introduce the public to the new medium.

*Television Broadcasting 2.0* (1950s-1980s): The *Golden Age of Television* (late 1940s-late 1950) with its pent-up demand, saw the television set penetrate the homes. With early television broadcasts being limited to live or filmed productions. Although all-electronic colour was introduced in the US in 1953, high prices and the scarcity of colour programming greatly slowed its acceptance in the marketplace. During the following ten years most network broadcasts, and nearly all local programming, continued to be in black-and-white. It wasn't until the 1960s that the public began buying colour TVs in earnest and in the 1970s, the American public finally started purchasing more colour TV sets than B/W ones. The number of television providers exploded. By 1970, there were around 700 UHF and VHF television stations.

*Television Broadcasting 3.0* (1980s-2000s): The CRT-based television was part of an analogue television system. But with the advent of digital circuits, these became more adapted to, and incorporated into television transmission systems. And soon, after a transition period in which broadcasters could send their signal on both an analogue and a digital channel, the advent of digital television broadcasting DTV[582] meant the end for analogue TV broadcasting systems. The rise of digital television had to wait for the advent of new technologies for the recording, transmission and presentation of the digitalized video signal. And those technologies were based on the IC-technologies with their microcomputer concepts. A process was paralleled by the development of new digital coding and transmission standards to enable digital TV broadcasting that resulted in the concept of High-Definition TV (HDTV). By the early 1990s the first experimental transmissions started the process of digital migration where the analogue television broadcasting technology became converted to, and replaced, by digital television.

## Changing Perspective

To illustrate the impact of the Silicon Engines —the metaphor for the Transistor, Integrated Circuit and Microprocessor — on society we explored the application areas of the General Purpose Engines (GPEs)

---

[582] Digital television (DTV) is the transmission of television signals using digital encoding.

during the First Electronic Revolution (1900-1950) and the Second Electronic Revolution (1950-1980). To illustrate that development in an historic context, we started with the events that contributed to the audio broadcasting technologies during the Radio Revolution.

The subsequent exploration of the Television Revolution focussed on the evolution of the early video broadcasting technologies, characterized by events that were both of an electro-mechanical as well as electronic nature. Events that occurred up into the second part of the twentieth century and that brought 'video' with news, culture and entertainment in the Affairs of Man.

Now, changing perspective again, we will zoom in on the specific events that resulted in the development of the Mobile Telephony that brought 'connectivity' to the Affairs of Man. From the early mobile telephones of the 1950s, to the cell phones of the 1970s. The latter developing over time from a mere cell phone to a feature phone and the smartphone.

## The Mobile Telephone Revolution

In the short period of time after the end of the First World War, a period that became known as the Roaring Twenties, the phenomenon of radio broadcasting exploded along a range of development trajectories: eg the technical trajectory of the radio receiver design, the application trajectory of the broadcasting stations (ie service providers and their radio formats). And the explosion of manufacturing activities. Overnight, radio had become big business.

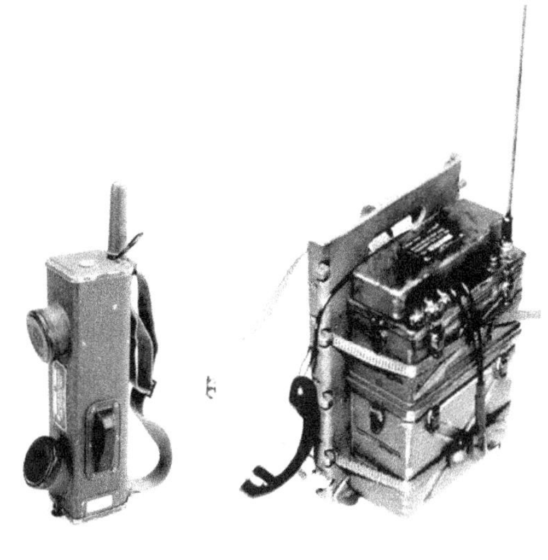

It had started with the rise of the wireless technology into maturity during the 1920s, and the progress of the vacuum tube technology, when towards the Second World War a new communication system emerged the two-way wireless systems for bidirectional person-to-person voice communication.

**Figure 420: Second World War Military two-way Mobile Communication.**
The SCR-536 Handie Talkie (left) and SCR-300 backpack Walkie Talkie Radio (right).
Source: Victorian Collections, Worthpoint.com.

## Personal Communication

The early development of bidirectional communication took place within different domains. One of them being the US military that had become interested in short distance communication before the second World War erupted. Next, the industrial pioneers were experimenting with the concept of mobile radio. And third, similar to the telegraph craze, the ham radio amateurs were fascinated by radio communications.

> The military had already installed radio receiver sets in military reconnaissance planes to help pilots communicate. Expanding their use to tanks and field artillery was next on the military agenda, as it was giving soldiers the ability to coordinate whilst separated.

> The nascent radio industry of the 1930s had their focus of the two-way car radio (eg in police cars for emergency coordination). Now they were expanding this concept into new designs for two-way communication.

The result were the early devices for mobile bidirectional communication. Such as the Motorola SCR-536; a five-tube battery powered receiver/transmitter that originated from the military backpack unit SCR-300 (Figure 420). Developed by the Galvin Manufacturing Co. and commercialized under the name Motorola, by July 1941 it was in mass production. When the United States entered the war in December 1941, the company stepped up production to ship thousands of radio units to the front lines.

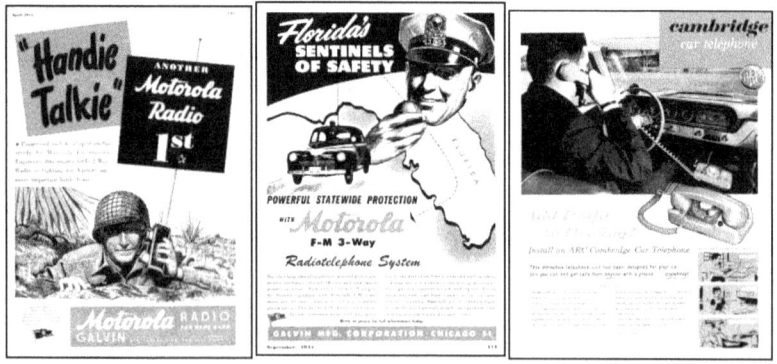

Figure 421: Mobile Telephony: Early Advertisement.

Motorola advertisement for military two-radio (left), and Galvin advertisement for community services (middle). Cambridge promoting the car telephone (right).

Source: Wikimedia Commons, Reddit.com

Handy-Talky radio-telephones became standard equipment for infantrymen as well as for paratroopers. By the time Second World War ended, and with 130,000 units produced, Motorola's handheld SCR-536 Handy-Talky two-way radio was an icon. And surplus Motorola talkies found their way into the hands of ham radio operators and former military personnel immediately following the Second World War. The first commercial operations started, and soon commercial advertisement began to appear (Figure 421).

Then there were the ham radio amateurs. As they were excluded from the right to use the airwaves during the Second World War, the focus of their interest shifted from long distance communication to short distance communication. Even more when, after the war, the military surplus came into their hands. With their contributions the concept of the hand-held portable two-way transceivers/receivers (aka 'handie talkie' or 'walkie talkie') was developed further into the mobile phone.[583]

## *Mobile Telephony 1.0 (1950-1975)*

Mobile telephony started as an expansion of the network of fixed telephony companies using a widespread network of landlines. They decided to offer a new (expansive) communication service; mobile telephony. It became a system of radiophones; mobile two-way radio systems connecting ships, planes, and mobile land units. Either as a standalone system (ie private networks), but often connected to the existing telephone infrastructure of the fixed land lines. The system connected the 'mobile' units with a fixed base-station that was in turn connected to the existing telephone network of landlines (Figure 428).

The mobile two-way radio sets —the Handy Talkie[584] and the combat backpack radio sets (Figure 420)— had proven their use in the Second World War, and subsequently were improved to be used after the war by (non-military) community's civil emergency services (eg the police, fire brigade, ambulance). From there they penetrated the market of the business professional (eg doctors), and later the market of the wealthy private consumer (Figure 421).

---

[583] The most common method of working radiotelephones is half-duplex operation, which allows one person to talk and the other to listen alternately. If a single frequency is used, both parties take turns to transmit on it, known as simplex. Dual-frequency working or duplex splits the communication into two separate frequencies, but only one is used to transmit at a time with the other frequency dedicated to receiving. The user presses a special switch on the transmitter when they wish to talk; the "press-to-talk" switch.
[584] The nickname of a two-way backpack AM radio set SCR-536 using vacuum-tube technology. Its successor, the SCR-300 was called the 'Walkie Talkie.'

This was the beginning of a range of generations of mobile telephony.[585] Generations that saw the development of both the networks of wireless transmission stations, as well as the telephone devices that used the networks.

*Zero Generation of analogue cellular telephony (0G):* In the US mobile telephony had started by 1946 with networks using the radio waves (AM and FM) as carrier, and that were an addition to the fixed land lines: the Highway System —intended for trucks and barges on inland waterways rather than private vehicles— and the Urban System for mobile private subscribers. By 1948, the commercial service was available in 60 cities in the United States and Canada, with 4000 mobile subscribers, handling 117,000 calls per month. It was a rather bulky system —placed in the trunk of a car— of the fixed car phones (Figure 422, top). Bell Labs and Motorola were the main competitors in the US; Bell Labs did most of the work developing the cell technology, but Motorola was ahead in phone-engineering development.

In the UK, there was also a vehicle-based system called 'Post Office Radiophone Service,' which was launched around the city of Manchester in 1959. It was

**Figure 422: Mobile Radio Telephony.**

Trunk part of the Motorola telephone system (top, 1950s), PYE PTC 116 Reporter Mobile Phone (middle, 1951-1953), Motorola TLD-1000 MTS (bottom, 1963).

Source: Wikimedia Commons. Science Museum.

---

[585] In the field of mobile communications, a "generation" generally refers to a change in the fundamental nature of the service, non-backwards-compatible transmission technology, higher peak bit rates, new frequency bands, wider channel frequency bandwidth in Hertz, and higher capacity for many simultaneous data transfers.

not until 1965 that service was made available in London. The launch coincided with the opening of the Post Office Tower, which was the location for the base station transmitters covering the central London area. They were sold as Reporter Mobile Phones (Figure 422). By the early 1970's the state of the art was the Nine Channel Radiophone. There were only three Post-Office authorised suppliers of equipment, Storno, Pye and Marconi. By the early 1970's the state of the art were the multichannel mobile phones: such as the Marine Seaphones and the Nine Channel Radiophones.

*First Generation of Analogue Cellular Telephony (1G):* Those were the early developments of a new phenomenon headed by American and Russian[586] companies. But both the US and Russia lost out to developments in Japan and Northern Europe. With technology progressing, so did the mobile phones.

The first bulky 'attaché case' and 'suit case' type mobile telephones were offered by a range of companies (eg the Finish company Nokia, the Swedish company Ericsson, Figure 423). They used a wireless radio-network. In 1956, the MTA system (Mobiltelefonisystem A, Mobile Telephone system A) was launched in Sweden, but that soon faced capacity and connection problems. In Finland, car phone service was first available in 1971 on the first-generation ARP (Autoradiopuhelin, or

**Figure 423: Early Mobile Radio Telephony.**
Nokia Talkman (left, 1970s), Ericsson Hotline (middle, 1980s), Vodafone Transportable VT1 (1985).
Source: Wikimedia Commons. Science Museum.

---

[586] In 1957 in Russia, the Soviet engineer Leonid Kupriyanovich developed the LK-1 Radiophone. A mobile device/station that could communicate with a base station connected with the system of landlines.

Car Radiophone) service. This was succeeded in 1982 by the *Nordic Mobile Telephone* (NMT) system, used across Scandinavia and Finland, and in other often remote areas.

In North America, car phones typically used the Mobile Telephone Service (MTS), or the *Improved Mobile Telephone Service* (IMTS) before giving way to analogue cellular service of the *Advanced Mobile Phone System* (AMPS) in 1984.

Network services by the Nippon Telegraph & Telephone company (NTT) began in Tokyo in 1979. Within 5 years the fully automated cellular network covered the whole of Japan and became the first nationwide 1st Generation (1G) analogue network.

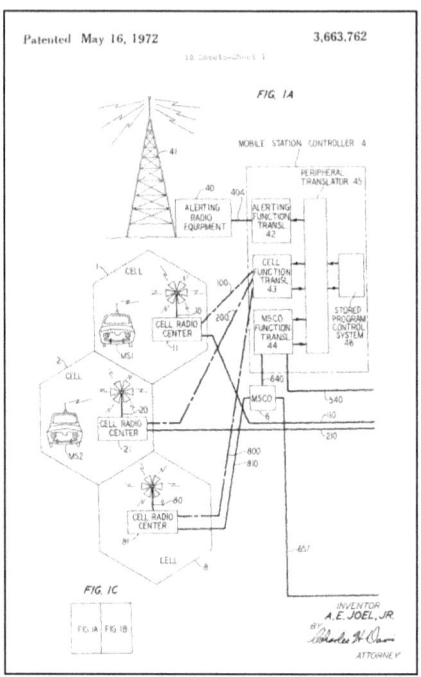

**Figure 424: Mobile Telephony Patent № 3,663,762.**
Source: USPTO

It was Amos E. Joel (1918-2008), an MIT-educated electric engineer working at Bell Labs, who headed the team that developed the cell structure and was granted US-Patent № 3,663,762 on May 16th, 1972 (Figure 424).

The early battery power guzzling mobile phones realized in vacuum-tube technology, opened the market for mobile telephony (Figure 422, top). But their acceptance in the market was only to happen after the new transistor technology emerged, And after the new concept of digital telephony for the mobile network had replaced analogue telephony.

The analogue 1G network was not perfect, but it remained until 1991 when it was replaced with the digital 2G network that combined analogue and digital frequencies. In the subsequent generations of mobile phones, not only the capacity (voice and data) of the network became expanded, also the facilities and performance of the mobile phone were enhanced.

## Mobile Telephony 2.0 (1975-2000)

With the rise of digital electronic circuitry during the Digital Era, also in mobile telephony the digitalization took place. The device underwent a convergence of different technologies; the Liquid Crystal Display (LCD) technology for the display, the Integrated Circuit (IC)-technology for the circuitry, the Microprocessor technology for the embedded systems, the NiCd battery-technologies for longer standby and speaking times, and the network technologies (aka network protocols) enabling the transmission of increasing voice and data volumes. The introduction of 2G systems saw telephones move from historic 1G telephones to the 2G hand-held items, which proved to be much more portable.

### Second Generation of Digital Cellular Mobile Telephony (2G)

Those first 2G telephone systems were, seen from a present-day point of view, rather primitive. Moving around with the phones caused problems when the mobile station came 'out of reach' of the base station. Already in 1947, Bell Labs was the first to propose a cellular radio telephone network with two-way radio telephones.[587] The search was on for a solution by the early 1990s when the Silicon Engine reached mobile telephony.

> In Europe, the European-developed 'Global System for Mobile Communications' (GSM) standard —using the 900 Mhz band and a digital standard— was adapted when Finland created the first 2G-network in 1991. In the United States, American carriers favoured the 'Code Division Multiple Access' (CDMA) system. The 2G networks differed from the previous generation by using digital transmission techniques instead of analogue transmissions and also by fast out-of-band signalling.

The hand-in-hand development of the networks as well as the handheld phones caused an explosion in cell phone usage in the 1990s with the onset of 2G connectivity.[588] The advent of prepaid cell phones also emerged during this time.

The first cell phones, created with the transistor technology, used an infrastructure of wireless cells; hence the name cell-phone. That network infrastructure —a terrestrial cellular network of base stations (cell sites)— made it possible to move around, and the network took care of the connection (as well as the financial aspects). From this origin, the cellular

---

[587] The primary innovation was the development of a network of small overlapping cell sites supported by a call switching infrastructure that tracks users as they move through a network and passes their calls from one location to another without dropping the connection.
[588] In 1990 there were some 12,4 million subscribers to mobile telephony worldwide.

network protocols developed over time in different 'intermediate generations' (eg 2.5 GPRS).[589] Each generation being characterized by new frequency bands (eg GSM 900, 1800, 1900 Mhz), higher data rates, and additional facilities.

With the technology progressing, from those bulky mobile phones of the first-generation (Figure 422), soon the smaller handheld cell phones were developed (Figure 425). The first —publicly promoted— call on a handheld mobile phone was already made as early as April 3rd, 1973 by Martin Cooper, then of Motorola, to his opposite number in Bell Labs who were also racing to be first. That prototype evolved —after ten years of development and massive investment— in the DynaTAC series.[590]

In 1983 the world got the first ever portable cell phone in the shape of the Motorola DynaTAC 8000X that weighed 1.75 pounds, was 13 inches long, and only provided about 30 minutes of talk time. It cost an eye-watering $4,000 and was a huge status symbol at the time. Other manufacturers soon followed with their own designs. In

**Figure 425: The Hand-held Mobile Phone.**
Motorola DynaTAC 8000 (left, 1973/1983), Nokia Mobira Cityman 450 (middle left, 1987) and Ericsson Hotline 900 Pocket (middle right, 1987) and Sony CM-H33 (right, 1992).

Source: Wikimedia Commons, Reditt, Science Museum Group Collection.

---

[589] GPRS (General Packet Radio Service) technology is a cellular wireless technology developed in between its predecessor, 2G, and its successor, 3G. GPRS could provide data rates from 56 kbit/s up to 115 kbit/s. It could be used for services such as Wireless Application Protocol (WAP) access, Multimedia Messaging Service (MMS), and for Internet communication services such as email and World Wide Web access.
[590] TAC stands for Total Area Coverage, DynaTAC for 'dynamic total area coverage.'

1987 Nokia, which had been developing bulky phones for the Nordic Mobile Telephone (NMT) service that began operating in 1981, unveiled the Mobira Cityman in 1987. It cost the equivalent of €4,560 and weighed 800g. And Ericsson entered the market also in 1987 with its Model Hotline 900 Pocket. In the east, the Sony CM-H333 introduced in 1992 became one of the most desirable handsets (Figure 425).

*The Digital Revolution in Mobile Telephony*

The switch from the old analogue cellular networks to new digital networks in the early 1990s allowed European handset manufacturers to flourish, as the fact that the European Union had accepted the GSM standard across Europe meant they suddenly had one vast market of eager professional and private consumers of mobile tele-communication on their doorstep. It resulted not only a boom in network providers, but also in booming handset and network manufacturing companies. Companies that introduced cell phones in a multitude of physical designs; from the brick model and flip model to the clamshell model (Figure 426).[591]

In 1992 Nokia launched the Nokia 1011 (Figure 426, left), the first so-called 'candy bar' phone for GSM networks. Compared to Motorola's preference for 'clam shell' phones, the 1011 was to define the standard shape of Nokia phones of those days. In 1996 Motorola hit back at the new European handset upstarts with the StarTAC (Figure 426). At the time it was the world's smallest and lightest

**Figure 426: Handheld GSM Mobile Phones.**
Candy bar design Nokia 1011 (left, 1992), clamshell design Motorola StarTAC3000 (closed and open ,1996), and slide design Nokia 7110 (right, closed and open, 1999).
Source: Wikimedia Commons.

---

[591] Many GSM phones support three bands (900/1,800/1,900 MHz or 850/1,800/ 1,900 MHz) or four bands (850/900/1,800/1,900 MHz), and are usually referred to as tri-band and quad-band phones, or world phones; with such a phone one can travel internationally and use the same handset.

phone —at 88 gr.— and had four hours of talk time, 47 hours of standby battery power and a vibrating ringer. A must-have for the business man at that time, and worldwide some 60 million were sold over its lifetime. Subsequently, Nokia introduced a range of improved mobile phones with extended functionality. Such as the Nokia 7110 introduced in 1999 that offered a mobile browser; the WAP Browser (Wireless Application Protocol) aka Micro-browser for Internet access (Figure 426).

With the descending prices and increasing performance, smaller and lighter designs with a longer battery life, at the end of the twentieth century the cell phone was ready to take off in the marketplace. But there one major roadblock; the capacity of the networks.

## Third Generation of Digital Cellular Telephony (3G)

The boom in mobile telephony caused capacity problems for 2G-technologies. So, the next generation (3G) was introduced to tackle that problem, and the first commercial 3G networks were introduced in mid-2001 when in Japan the Nippon Telegraph & Telephone Company (NTT) offered it on its network. They were soon followed by variations; the W-CDMA and the CDMA2000.

### The Packet Switching Revolution

The main advantage of moving from 2G to 3G was that 3G used the technique of packet switching instead of circuit switching to transmit data along transmission channels (Figure 427). This allowed for faster data transmission speeds (eg 2 Mbit/s data rates). And it opened the door for Internet access and media streaming over mobile networks.

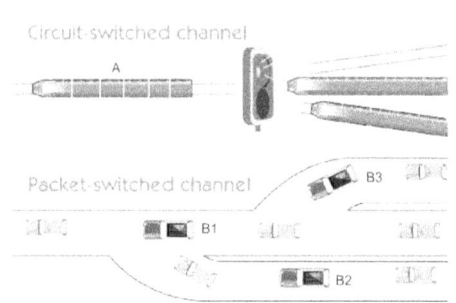

**Figure 427: Circuit Switching versus Packet Switching.**

Instead of sending complete data-trains over reserved point-to-point lines (A), individual data-cars (B1-3) find their way to their destination.

Source: https://www.computerworld.com/article/2593382/networking-packet-switched-vs-circuit-switched-networks.html

Soon new facilities were added to the network services next to the wireless telephony; mobile Internet access, fixed wireless Internet access, video calls and mobile video technologies. And all this development was based on massive R&D efforts, resulting in a myriad of patents. It has been estimated[592] that there are almost 8,000 patents declared essential related to the 483 technical specifications which form the 3G standards. Twelve companies accounted in 2004 for 90% of the patents; Qualcomm, Ericsson, Nokia, Motorola, Philips, NTT DoCoMo, Siemens, Mitsubishi, Fujitsu, Hitachi, InterDigital, and Matsushita.

## *The Mobile Networks and its Operators*

The technological progress of the 1G, 2G and 3G-generations of mobile telephony (both in terms of mobile telephone networks as well as in terms of the mobile handsets) was paralleled by the business development of the so-called mobile *Public Telephone Operator* (PTO)[593] that installed and exploited the mobile networks. They were complemented by the non-network-owning companies —aka *Virtual Network Operator* (VNO)— that leased bandwidth on those networks to offer similar services (aka service providers). The rise of these mobile network operators started with the early offering of mobile services by the large *Public Switched Telephone Network* (PSTN)-companies.[594] These were the companies that were the owners of the local and regional telephone exchanges (ie Private Branch Exchange, PBX), as well as the international 'trunk' connections. A system with a copper cable-based infrastructure of

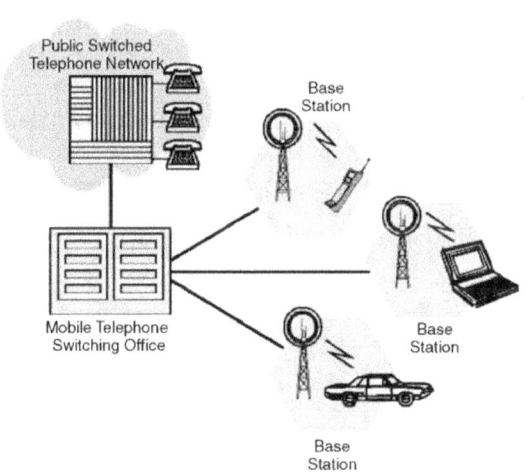

**Figure 428: Diagram of Mobile Cellular Network.**

Source: http://what-when-how.com/ networking/ cellular-telephone-networks-networking/

---

[592] Goodman, D.J.; Myers, R.A.: 3G Cellular Standards and Patents. IEEE WirelessCom 2005, June 13th, 2005. (Accessed March 2022).
[593] A mobile network operator aka Network Service Provider (NSP) is a telecommunications company that offers mobile telephone services or mobile Internet access. The operator provides a SIM card to the customer who inserts it into his mobile phone or touch pad to gain access to the operator's cellular network. An independent service provider (ISP) rents space on the network
[594] See: Van der Kooij, B.J.G.: *The invention of the Communication Engine 'Telephone'* (2016)

fixed lines. Since Alexander Graham Bell had invented the telephone, a whole system of these companies was created around the monopoly of the Bell System; each with their local subscribers, their local switch boards and local exchanges.

It had started with the local systems where a (often female) switchboard operators would make the connection between two parties on a 'party line' (Figure 2). Then the automated switching systems took over and the calling party 'dialled' the receiving party, assisted by the automatic switch exchanges in the Carrier Company facilities connected by a network of 'trunk' lines. There the call would be forwarded through the PSTN network, ending up at the receiving party (Figure 429).

The companies in the PSTN infrastructure started offering mobile services such as the mobile car radio operating on the AM and FM channel in the audio-spectrum. Wireless networks that were connected through *Mobile Telephone Switching Offices* (MTSO) to the traditional fixed landline network of the PSTN-companies (Figure 428).

In the US, originating from the nationwide Bell System monopoly on cabled telephony,[595] associated companies had been licensing user rights and offering telephones made by the giant American Telegraph & Telephone Company (AT&T); a holding company for the many Bell Companies, as well as a long-distance carrier. These affiliates then leased the telephone as part of their subscriber services. Not surprisingly, following the technological developments, the giant AT&T commercialized the *Mobile Telephone Service* (MTS) in 1949. Its telephone operators were serving on three channels some 5,000 customers placing 30,000 calls a week. In 1959 it introduced its *Improved Mobile Telephone Service*

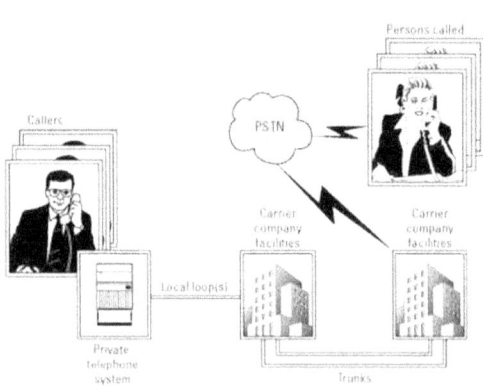

**Figure 429: Diagram of Public Switched Telephone Network (PSTN).**

Source: https://what-when-how.com/voip/ private-telephone-systems-reduce-pots-line-costs-voip/

---

[595] See: Van der Kooij: B.J.G.: *The Invention of the Communication Engine 'Telephone' (2016).*

(IMTS), still the capacity of twelve channels was a limiting factor for its 40.000 subscribers.

In Europe the introduction in 1960 of the *Mobiltelefonisystem* (MTA)-network in Sweden, was succeeded by 1971 the *Mobiltelefonisystem D* (MTD) that had some 20,000 users at its peak. In Norway, in 1966 the *OLT* network (Norwegian for Offentlig Landmobil Telefoni, Public Land Mobile Telephony) was introduced. By 1981 they had 30,000 mobile subscribers. In Finland the *Autoradiopuhelin*, "car radio phone" (ARP) was launched in 1971. By 1977 they had over 10,000 subscribers, with a peak in 1986 of 35,560 subscribers. The *Nordic Mobile Telephony* (NMT) replaced it in 1990.

In Britain, the (tele)-communication monopoly of the Post Office was exploited by the state-owned Post Office Telecommunications. Thi I institution was reformed as the state-owned *British Telecom*, and that became in 1984 the British Telecommunications PLC (aka BT). In 1982 the government announced the availability of two licenses. They went to the BT owned Cellnet (Telecom Securicor Cellular Radio, forerunner or the O2 network) and to Racal-Vodafone (forerunner of the tradename Vodafone). By 1988 they had some 500,000 subscribers using the networks based on the Total Access Communication System (TACS).

> By the end of 1986 there were 100,000 mobile phone subscribers within the UK which doubled to 200,000 by the summer of 1987 and by the end of 1995, 7% of the UK population owned a mobile phone. After ten years of service, the mobile phone had started to become accepted as a normal part of (high) society.

These were the early days of the wireless telephone networks. Next to the large corporations expanding into the mobile telephony field, small start-up companies had emerged and were serving local areas. With technology progressing and creating the handheld mobile phones, public interest exploded. As did the telecom business that was dominated by the mobile network providers that became —after a massive process of mergers and acquisitions— the giants of mobile telephony services.

In the US, driving on a wave of technology-driven business dynamics of mergers and acquisitions —after the breakup of the Bell System into the Baby Bells— the company Bell Atlantic was created. Next Bell Atlantic merged in 1998 with the GTE Corporation —America's leading provider of local telephone services—; the new entity was named *Verizon Communications*. The mergers that formed Verizon were among the largest in U.S. business history, culminating in a definitive merger agreement, dated July 27th, 1998. The new company became a major player in US telecommunications with 63 million fixed telephone lines and 25 million mobile telephone subscribers in forty states representing $60 billion in turnover.

In 1999 Bell Atlantic, together with the British Vodafone Airtouch PLC, also created *Verizon Wireless*; a new wireless business —with a national footprint, a single brand and a common digital technology— composed of Bell Atlantic's and Vodafone's US wireless assets (ie Bell Atlantic Mobile, AirTouch Cellular, PrimeCo Personal Communications and AirTouch Paging).[596]

In Britain, after a range of mergers, Orange Personal Communications Services Ltd was created on March 28th, 1994. A holding company structure was adopted in 1995 with the establishment of Orange PLC. After it went public in 1996 it grew to 1 million customers in 1997. In the Disruptive Nineties,[597] after a turbulent period of take-overs by the German Mannesman company (in 1999) that in turn became acquired by Vodafone (in 2000), some government intervention, and due to EU regulations, it ended up as a division of France Telecom that subsequently rebranded all its mobile telecommunications as Orange.

Also in Britain, the company Racal Electronics, had entered into a partnership with the American Millicom Inc, in order to enter the mobile phone market. The newly formed company Racall-Millcom Ltd. obtained a Post Office license and Vodafone was launched on January 1st, 1985 under the new name, Racal-Vodafone (Holdings) Ltd. That was the start of the *Vodafone Group* that next acquired Talkland (1996), Peoples Phone and Astec Communications. On June 29th, 1999, the Vodafone Group completed its purchase of American service provider AirTouch Communications, Inc. and changed its name to Vodafone Airtouch PLC. On September 21st, 1999, Vodafone agreed to merge its US wireless assets with those of Bell Atlantic Corp to form the Verizon Wireless Corporation.

In France, after the Age of Minitel (ie the French videotext online service) during the 1980s-1990s, the first car phone service Radiocom 2000 in 1986, and the Bi-Bop (ie a wireless pocket telephone) in 1991, in 1992 France Telecom entered the GSM market with its *Itineris* mobile phone service. In December 1998, Itineris covered 97% of the French population with 7,700 relay antennas and had 5.5 million customers, or

---

[596] For a more detailed expose, see: The history of Verizon Communications; https://www.verizon.com/about/sites/ default/files/Verizon_History_0916.pdf

[597] This was the time of the Internet Bubble followed by the DotCom Crash, and the Telecom Crisis followed by the Telecom Crash when numerous telecommunication companies bankrupted.

49.6% market share. In 2000 France Telecom bought through a set of complicated mergers the network operator Dutch Equant NV, became 100% owner of the venture Global One, and integrated its activities as Orange Business Services by 2006. In 2006, the company's activities as an Internet Service Provider (ISP), previously called Wanadoo, were also re-branded Orange.

In Japan mobile telephony was launched in Tokyo by the Mobile Communications Network (later renamed as DoCoMo) of the Nippon Telegraph and Telephone (NTT). Its network soon expanded to cover the whole population of Japan by 1979. Established in 1992, NTT DOCOMO launched its first digital cellular phone service the next year and the world's first mobile Internet-services platform in 1999. It helped to establish the W-CDMA standard for mobile communications and then the first 3G service based on this standard in 2001. NTT DoCoMo offered in 1999 a new mobile network service called i-Mode. It became an instant hit —especially among the younger hyper-fashion-conscious Japanese teenagers— and 10 million subscriptions had been sold in 1999, a number that doubled next, and trebled in the 2 following years.

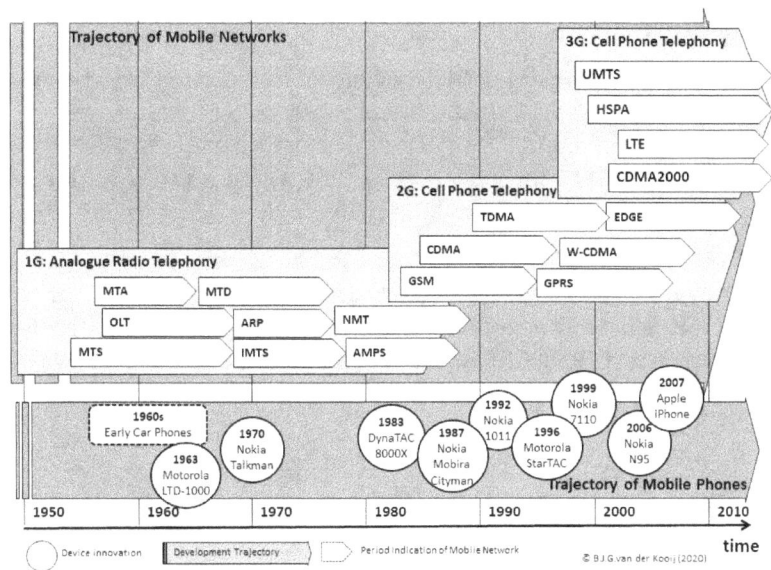

**Figure 430: Overview Three Generations Mobile Networks (1960-2000).**
The networks (upper trajectory arrow) are related to some iconic mobile phones (lower trajectory arrow). The periods of the Networks are just indications as their introduction date and lifespan varies from region to region. For details of abbreviations see text.

These examples illustrate the business dynamics of the exploding mobile telecommunication market. In a wild-west atmosphere of mergers and acquisitions, national monopolies were created and cutthroat competition ruled the business.[598] The technology-driven three generations of mobile telephone networks (Figure 430) would lay the foundations for the emerging 3G Smartphone in the early twentieth century.

## Development of the Cell Phones into Feature Phones

The before described development of network protocols and of network *infrastructures* became intertwined with the development trajectory of the cell phone *devices*. A development in which the activities of a mobile phone manufacturer and a network provider became interconnected. Connected by technology in developing the generations of cell phones; such as the 2G multiband cell phone supporting multiple radio frequency bands and dual modes of 2G networks.[599] Connected by the market as they were serving (potential) subscribers with the facilities of their connection networks accessible on the handsets. So, a hardware manufacturer and service-provider often worked together.

In order to attract more subscribers, Mobile Network providers (aka Telecommunications Service Provider) —similar to the landline providers— offered a range of subscription contracts; from the standard subscription with afterward periodic billing (ie Bill pay) to the pay-as-you-go contracts (Pre pay). They also offered subscription contracts with a specific contract period (eg 12, 24 months) in which the cell phone was included as part of a commercial bundle. And, as the network operators were protecting their interests to keep that bundle intact, they locked the mobile phone to their services.[600]

> Verizon had by the early 2000s contracts with several hardware suppliers based on Apple's iOS and the Android OS. In 2011 the Verizon iPhone (ie iPhone 4) became available to Verizon's 93 million customers. Either to be sold or to be leased as part of the subscriptions.

---

[598] A development that reflected in the stock market as from April 1997 to March 2000, the Nasdaq index of telecommunications stocks rose spectacularly, from 198 to 1230, an average annual increase of approximately 84 percent.
[599] The multiband phones facilitated international 'roaming' where physically travelling brought the user outside his native network. He needed then connection with another visiting network that could be using a different radio band. In order to be able to switch from carrier mode (eg GPS to CDMA networks), he needed a dual mode device.
[600] SIM-Lock: To ensure that the buyer of a specify phone stayed with the network provider, the SIM-lock restricted the use of the phone to specific countries and/or networks. SIM locking became common for subsidized phones sold with prepaid contracts.

## The Battles of Telecom Giants

As the mobile telephone rapidly expanded on the network infrastructure of the fixed telephone system of the PSTN (Figure 428), it became a more important part of the development of the telecom industry. And it saw the rise of large companies; from giant landline operators (eg AT&T) to the large mobile operators (eg Verizon)[601] serving millions of subscribers.

Technological advance dramatically changed the cast of players involved in the telecommunications infrastructure. It grew from the original private and publicly owned telephone monopolies to include a host of new entrants such as competitive access providers, resellers, value-added carriers, new interexchange carriers, new local exchange carriers, cable companies, wireless carriers, direct broadcast satellites, media conglomerates, and specialized brokers.

As a result, the industrial ecosystem[602] of the telecommunications industry became made up of a complex set of hardware suppliers, mobile service providers, institutional regulators, and different groups of customers creating the value chain of mobile telephony. The key stakeholders in this *telecommunication value chain* included the tele-communication handset and equipment manufacturers, and the telecommunication service providers. In each segment individual companies rose to prominence, and battled with each other for market share.

**Figure 431: Promotion of the new Smartphone.**
Advertisement for LG Prada design phone (upper left), Samsung advertisement comparing with Apple's iPhone upper right.
Source: Wikimedia Commons.

*The Battle for Market Share:* In such a fast-growing market as the cell phone market, the stakes were high. And continuous novelty creating was part of the corporate strategy. Novelty in the form of features, in the form of design, and novelty in the form of marketing practises. The result was the continues introduction

---

[601] Verizon had in 2007 some 235,000 employees, AT&T some 309,050.
[602] Industrial ecosystems encompass all players operating in a value chain: from the smallest start-ups to the largest companies, from academia to research service providers to suppliers.

of new Feature Phone models by companies like Nokia, RIM Blackberry, and Ericsson. Next to new features, the existing features improved constantly; such as the system components like the mobile camera that saw dramatic rise in resolution and techniques (eg autofocus, zooming, flashlight).

It was all about market share. Market share of the phone suppliers, market share of the network providers. One way of attracting customer was by promotion (advertisement and tv-commercials). Verizon launched its 'Can you hear me now' campaign (2002) stressing the quality of its network. Samsung put up a controversial campaign directly confronting Apple (2008). LG promoted the luxurious aspects of their Prada-line with well-known actors like Edward Norton and model Daria Werbowy (Figure 431).

*The Rise of the Mobile Alliances and Partnerships*

In order to survive in that dynamic ecosystem, corporations created strategic alliances and partnerships.[603] Alliances between the existing powerful players, as well as alliances with newcomers in the mobile business. And mixed alliances when the partners found a common goal (eg a new generation of mobile phone standards).

In 2006 the *Next Generation Mobile Networks* (NGMN) *Alliance* was established as a mobile telecommunications association of mobile operators, vendors, manufacturers and research institutes. Its objective was the development of 3G mobile telephone standards.

But the alliance-creation often had fewer common interests and the mobile telephone market was highly competitive. So, another form of the alliance creation with existing players often took place between the hardware suppliers and the network providers. Hardware manufacturers like Nokia, Sony, LG supplied their phones to the network operators who branded them under their own names.[604]

*The Battle of the Operating Systems:* Also known as the Platform Wars, the different operators and manufacturers adhered to a specific operating system for the devices. So there was, just after the announcement of the disruptive iPhone, a fierce battleground opening up. It saw the creation of *Limo Foundation* (short for Linux Mobile OS) by Motorola, NEC, NTT DoCoMo, Panasonic Mobile Communications, Samsung Electronics, and Vodafone. Soon followed by the *Open Handset Alliance*

---

[603] A strategic alliance, or partnership as it is often referred to, is a business relationship that exists between two or more firms.
[604] This practise was known as Private Label of With Label. The hardware manufacture HTC had the tradition of supplying private label phones for individual carriers like Orange.

(OHA) in 2007 by 34 companies (eg NTT DoCoMo, T-Mobile, Sprint Nextel, HTC, LG) led by Google. They promoted the use of Android OS against platforms like Apples iOS, Nokia (Symbian OS) and Blackberry OS. And not much later response came in the form of the creation of the *Symbian Foundation* (2008) to promote the use of the Symbian operating system. The Symbian Foundation was founded by Nokia, Sony Ericsson, NTT DoCoMo, Motorola, Texas Instruments, Vodafone, LG Electronics, Samsung Electronics, STMicroelectronics and AT&T. It lasted till 2010.

*The Battle with Newcomers:* The alliances of the old guard with newcomers were often the result of a defensive action in the fast-changing mobile ecosystem. Nokia, faced with drastic losing market share after Apple's Phone was introduced, teamed up with another newcomer in the *Nokia-Microsoft Alliance* in order to compete with the dominant players Apple and Samsung. Nokia being a dominant player in the Feature Phone market before 2007, and Microsoft being a dominant player in the operating systems for the personal computer market. Both Nokia and Microsoft had underestimated the impact of the iPhone. However, after the announcement of the Microsoft partnership, Nokia's market share deteriorated; this was due to demand for Symbian dropping when consumers realized Nokia's focus and attention would be elsewhere.

One of those newcomers was Google, basically a company that exploited a search engine (Google Search using web crawlers[605]) to browse the Internet. Its revenues came from advertisements associated with search keywords and 'clicks' of viewers on the presented websites. Its meteoric rise, due to new services (Google News, 2002; Google Books, 2004; Google Maps, 2005; Google Patents, 2006) and acquisitions (You Tube, 2006; Double Click, 2007) had made it an interesting player. Even more when it acquired Motorola Mobility in 2011. One reason for the acquisition was Motorola's wealth of patents, another Google's goal of creating more traffic for its existing cash creating activities (eg Google Ads).

*The Battle for Internet Access.* The access to the Internet was one of the big features for cell phones. The Wireless Application Protocol (WAP) — introduced in 1999— facilitated the access to the content of the Internet from a mobile phone. The WAP solution was great for handset manufacturers. They could write one WAP browser to ship with the handset and rely on developers to come up with the content users wanted. The WAP solution was also great for mobile operators. They

---

[605] Google's web crawling function scours online data and feeds results back to the machine so that they can be indexed and ranked for relevance on search engine results pages.

could provide a custom WAP portal directing their subscribers to the content they wanted to provide, and wallow in the high data charges associated with browsing. Most of the early WAP sites were extensions of popular branded sites, such as CNN.com (24-hour news network) and ESPN.com (sports network).[606] Users accessed the news, stock market quotes, and sports scores on their phones.

The WAP browsers were not user friendly, typing in long URLs with the numeric keypad was a tremendous pain. Most WAP sites were one version and did not account for individual phone specifications. As a result, WAP fell short of commercial expectations, except in Japan where NTT DoCoMo scored high, especially with their young subscribers. By 2013, WAP had largely disappeared.

In that context of network development from the 1980s to the early 2000s, the manufacturers of the cell phone created a massive stream of mobile devices.[607] In the 80s, the race was on to produce the smallest, lightest mobile phone. Each new model was smaller, lighter or boasted more features than its competitors. Their designers used the advantages of the technological progress to increase their functionality, and at the same time decrease their size and price. As the technology advanced, companies figured out how to pack all the features their customers wanted into a smaller, portable, more affordable model. The VLSI-Integrated Circuit technology (Figure 284, Figure 285) offered them ample opportunity.

At the hardware front, the physical design went through a rollercoaster of changes; from the candy bar design to the clamshell and flip design (Figure 463). And the device underwent a convergence of technologies; the housing, the circuitry, the display, the user interface, the antenna and the battery. On the front of functionality, the system designers realised the circuits enabling the new product 'features,' and the circuit designers translated those designs in the mobile Integrated Circuits.

This rather limited overview illustrates the dynamic business environment of the mobile phone industry where 'innovation' was the dominant theme.

---

[606] Among the first companies to launch a WAP site was Dutch mobile phone operator Telfort BV in October 1999.
[607] A cell phone handset contains a radio transmitter, for sending radio signals onward from the phone, and a radio receiver, for receiving incoming signals from other phones. All a cell phone has to do is communicate with its local mast and base station; what the base station has to do is pick up faint signals from many cell phones and route them onward to their destination, which is why the masts are huge, high-powered antennas.

## The Evolution of the Mobile Phone Circuitry

Mobile bidirectional communication had sprouted in the 1950s the Mobile Telephony. Propelled by the Silicon Engine, in the 1970s the Mobile Cell Telephony emerged that would follow a trajectory of rapid, technology driven development. And part of that development took place in the mobile communication devices (aka mobile handsets). Before exploring the more detailed application areas of the mobile phone, let's from a bird's eye perspective explore the influence of the Silicon Engine along the technological development trajectory.

*Mobile Phone Hardware*

Next to the basic components of a cell phone (ie battery, antenna, microphone and speaker, the keyboard and display screen) (Figure 432), the mobile phone needs electronic circuits for receiving and transmitting. Inside the mobile phone those electronic circuits (IC's) were placed on motherboard holding the circuitry and creating the interconnection. From the early brick phones with analogue (1G) circuity to the smartphone (3G) the Silicon Engine spurred a massive development on component level (Figure 433). This development was focussing on integration and prolongation of usability.

*Integration:* To become a real mobile telephone, the hand-held device had to be small and light weighted. To pack all the needed circuitry in the housing, the 'motherboards' of the mobile phone became increasingly cramped with component and circuitry (Figure 433, top). Next, embedded IC's (SoC) realized the basic punctionality on a single,

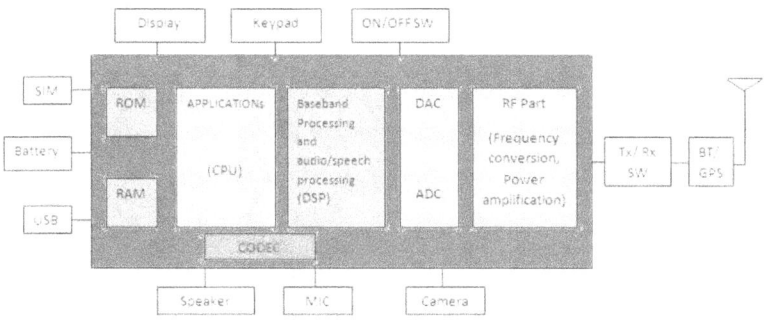

**Figure 432: Basic Components and Circuitry Diagram of a Smartphone.**

The circuitry is placed on a motherboard (dark square). Basic components are connected to that board. Tx/Rx: Transmit/Receive, BT/GPS: Bluetooth/Global Positioning System. SIM: Subscriber identification Module, CODEC: coder-decoder or compression-decompression.

Source: RF Wireless World

double-sided motherboard (Figure 433, middle). And the smartphone became packed with VLSI-IC's (Figure 433, bottom).

*Prolongation Usability:* With the increasing miniaturization, the power consumption dropped. Combined with improved batery technology, that gave longer standby time and talk time.

### The Road to the Smartphone

Starting with the brick model of the DynaTac portable phones (1973), all that receiving and transmitting circuity was 'hardwired' on a motherboard. And an important part of that board became occupied with the dedicated Systems on a Chip (SoC) microprocessors.

The IBM Simon (1992) Personal Communicator, which we will meet further on in more detail, was in fact a handheld computer with the functionality of a Personal Digital Assistant (PDA) and having a mobile phone capability. The two-board system was equipped with a Single-Chip PC platform (Figure

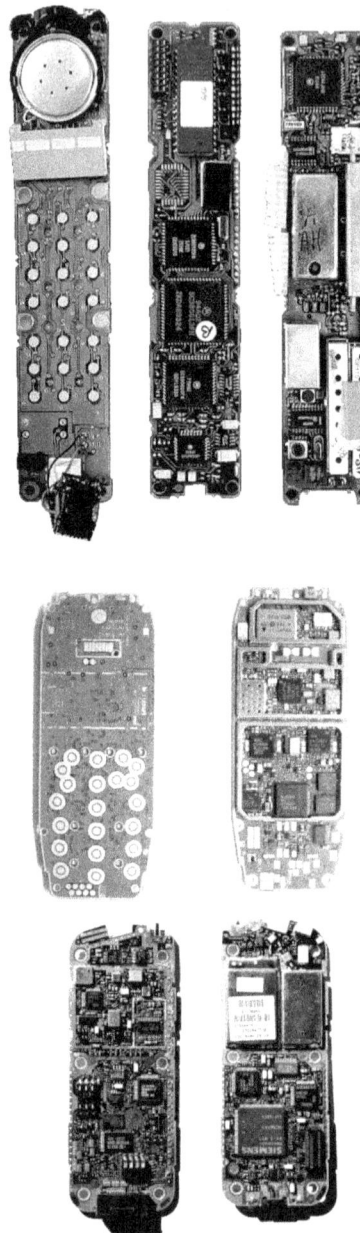

**Figure 433: Mobile Phone Motherboards.**
Boards of Motorola DynaTac (top), the NOKIA 3310 motherboard (middle, front and back), and smartphone board (type unknown, bottom front and back).

434, top)[608] as the phone came with its own microprocessor (the Vadem VG 230 chip, 8086-compatable) and an operating system (ie the Datalight ROM-DOS).[609] The phone circuitry (on a separate board) contained the chips of a modem and fax, touchscreen operation and receiver/ transmitting circuitry (Figure 434, bottom). The device was able to send, store and receive packet data in the form of fax, email and pager.

From the early 1G systems developed the more flexible systems that used more advanced dedicating processing units. Such as the Baseband Processor (BP) that manages all the wireless radio functions of a cellular device in order to connect to a wireless cellular network (Figure 435, bottom). Baseband processors typically run a real-time operating system (RTOS) as their firmware. And with digital capabilities of the 2G systems came the circuitry Digital Signal Processing (DSP)[610] and their Analogue-to-Digital (AD) and Digital-to-Analogue (DA) converters. And when the smartphone implemented micro-computer functionality a separate CPU with its RAM/ROM memory was used for the processing of Applications (Figure 432).

**Figure 434: Inside look on the motherboard of the IBM Simon.**

Microprocessor Controller board (top), and Cell Phone Board (bottom).

Source: Sudonull.com

---

[608] It included 11 built-in programs: to-do list, calendar, calculator, appointment scheduler, electronic sketch pad, world time clock, input screen keyboards, handwritten annotations, address book, and address book.
[609] ROM-DOS is an MS-DOS compatible operating system designed for embedded systems
[610] A digital signal processor (DSP) is a specialized microprocessor chip, with its architecture optimized for the operational needs of digital signal processing.

## Mobile Phone System Software

Each computer-based device is equipped with 'software' that creates its functionality. The smartphone, based on dedicated computers, is equipped with system software complemented by application software.[611]

In the modern cell phone, the mobile processor was implemented with a dedicated instruction set. Each instruction resulting in a 'command' to perform a task. In their totality they create the mobile phone Operating System (mobile OS).[612] The OS that ran its operation; from starting up the device, and making the wireless connection, to executing the Applications (in modern parlance Apps).

For 2G mobile telephones it started with the Symbian OS that emerged in the 1980s from the development of the pocket computer PSION Organizer. The Psion organizers used the EPOC operating system. Such as the PALM OS that was used by Pilot PDAs: the Pilot 1000 and 5000 (1996). From the EPOC OS, the Symbian Mobile OS developed in a range of device related system software supporting their feature functionality. The Nokia-spurred Symbian OS powered by 2009 some 250 million devices.

**Figure 435: System of a Chip Embedded Processors.**

NEC V40 equivalent of V20 processor (top). Baseband Processor (bottom).

Source: Wikimedia Commons.

---

[611] The two main types of software are system software and application software. System software controls a computer's internal functioning, chiefly through an operating system, and also controls such peripherals as monitors, printers, and storage devices. Application software, by contrast, directs the computer to execute commands given by the user and may be said to include any program that processes data for a user.

[612] Operating systems are the set of low levels programs (eg for file management, data retrieval and storage) that are required for providing the interface between the different applications and the hardware of the computer systems or mobile phones.

Later, originating from the evolutionary development of the microcomputer (Figure 437) with the operating systems like Microsoft NT, Microsoft CE, and Microsoft's Pocket PC 2000 operating system, came Windows Mobile OS. Microsoft, as part of the development of their Windows CE operating systems for desktop personal computers, introduced Window Mobile in 2003. It came in different editions, among which 'Windows Mobile 2003 for Smartphone.'

## The Rise of Android OS

Android began in 2003 as a project of the American technology company Android Inc., to develop an operating system for digital cameras. In 2004 the project changed to become an operating system for smartphones in response to those manufacturers that did not want to be hooked on the available operating systems, each with their own limitations (and their license fees). Android's management wanted to provide the world's first complete open handset platform solution, and they wanted to make it available without license fees to the phone manufacturers. But they struggled to find the venture capital to realise that vision. So, they started looking elsewhere, and found a listening ear at Google, Samsung and HTC. Google was interested in the mobile market as potential users for their search services,[613] Samsung and HTC were makers of mobile phones. The end of the negotiations being that

**Figure 436: Graphic User Interface of Android 1.0 (2008).**
Source: Wikimedia Commons

---

[613] Founded in 1998, Google exploited a search engine for browsing the Internet. Google indexes billions of web pages to allow users to search for the information they desire through the use of keywords and operators. Their revenues come from the sales of advertisement, and from 'clicks' to access to identified pages. The search and advertisement activity were a profitable business as for the year 2005, Google reported total revenues of $6.139 billion, an increase of 92.5 percent over revenues of $3.189 billion in 2004, and net income for 2005 increased to $1.465 billion, from $399 million in 2004.

Android Inc. was bought by the American search engine company Google Inc., in 2005 for some $50 million.[614]

On November 5th, 2007, Google Inc. announced the founding of the *Open Handset Alliance*, a consortium of dozens of technologies and mobile telephone companies, including Intel Corporation, Motorola, Inc., NVIDIA Corporation, Texas Instruments Incorporated, LG Electronics, Inc., Samsung Electronics, Sprint Nextel Corporation, and T-Mobile (Deutsche Telekom). The consortium was created in order to develop and promote Android as a free open-source operating system with support for third-party applications. It was to compete with the other major smartphone platforms of the time, such as Symbian, BlackBerry OS, and iPhone OS.

After its introduction in 2008 with a quite advanced Graphic User Interface[615] (Figure 436), over the years the Android OS would improve and released in several new versions; Android 2.0 (2009), Android 3.0 (2011), and Android 4.0 (2011).

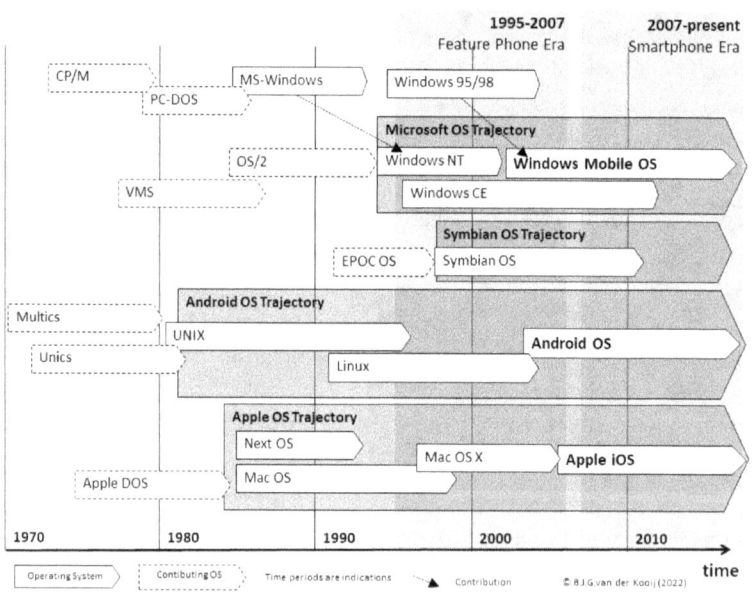

**Figure 437: Evolution of Mobile Operating Systems.**
From the Mainframe OS (VMS), Mini-computer OS (Unix) and Micro-computer OS (PC-DOS, Apple DOS, Windows) emerged Windows Mobile OS, Android OS and iOS.

---

[614] The cash rich Google Inc. acquired in 2005 eight more companies.
[615] The User Interphase (UI) gives the user of the device access to its functions and facilitates user data-input.

## The Rise of Apple's iOS

After its phenomenal success in the personal computer business with the Macintosh PC's, Apple Computer was looking into other applications for tablet-like Macintosh computers (ie handheld PCs). And the concepts of the emerging Personal Digital Assistant (PDA) and Personal Information Manager (PIM) looked promising businesses. So, its CEO John Sculley initiated the development of the Newton MessagePad with its own Newton Operating System, and revealed it in 1992 (Figure 449, middle). However, other manufactures had also rushed to develop PDAs, and notwithstanding later upgrade models, by 1998 the product line was discontinued by Steve Jobs.

But, not-withstanding its failure, the heritage of the Newton OS would create the foundation of the operating system for following Apple's mobile devices. Such as the iPhone introduced in 2007 that used Apple iOS; a closed-source code used by Apple-only products (such as the iPhone, iPod, and iPad).

The Newton offered a selection of icons to access its functions (Figure 438,

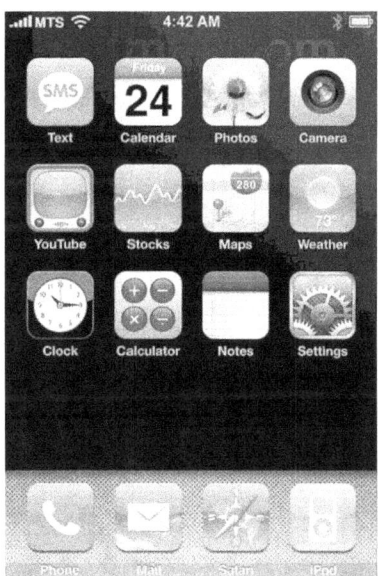

**Figure 438: Screenshots Apple Graphic User Interface (GUI).**

Text based UI of the Newton OS (left) and graphic UI Apple iOS (right). Compare the icons for 'Clock', 'Calculator', etc.

Source: Wikimedia Commons.

top). For text input it contained handwriting recognition software as well as a QWERTY virtual keyboard. With the advancement of graphic display resolutions, the Apple iOS (Figure 438, bottom) offered a more esthetical User Interface.

## The Rise and Fall of Windows Mobile Phone OS

The smartphone world was already split into two when Microsoft entered the mobile phone market in 2010; Apple's iOS for the brand-conscious user, and Google's Android for the practical user. When Microsoft came on scene, the world was about to see one of the largest technology companies make a real attempt to break into the duopoly of one of the biggest and fastest growing tech markets.

Since the introduction of the IBM personal Computer, Microsoft had become a dominant player in software for the microprocessor-based computing with its DOS operating System. Even more after it introduced the Windows-based operating system. Microsoft was continuedly developing its Windows-versions for different platforms (eg Windows 95, Windows Vista, Windows 7-11, Windows NT, Windows CE). And by the mid 1990 they started to focus on mobile telephony as a spin-off of the Windows version for handheld PC's.

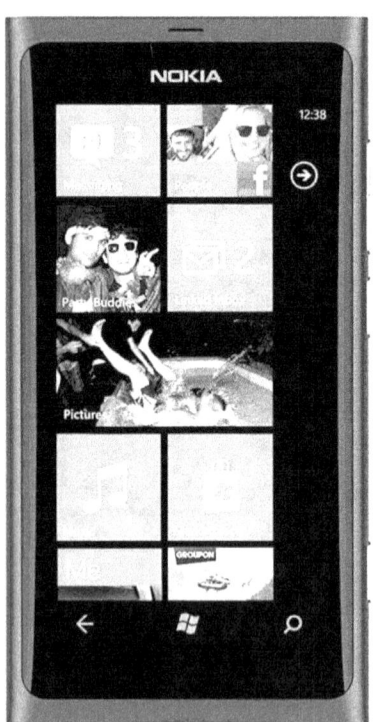

Aware of the rise of the mobile computing devices, Microsoft offered Windows Mobile OS. Its origin dated back to Windows CE in 1996 (Figure 437), though Windows Mobile OS itself first appeared in 2000 as Pocket PC 2000 which ran on Pocket PC PDAs. It was renamed 'Windows Mobile' in 2003, at which point it came in several versions (similar to the desktop versions of Windows). By 2010 it was succeeded by Windows Phone OS. Microsoft realised it was late to the game and began working on a new operating

**Figure 439: Graphic User Interface of Nokia Lumia 800 (2011).**
Source: Wikimedia Commons

system in 2008. Time was limited and the product had to be launched quickly, so the brand-new Windows Phone OS cut off compatibility with its predecessors.In 2011, Microsoft and Nokia jointly announced their partnership to a world which was finally sitting up and taking notice. It was exciting news after all; Microsoft's software was marrying Nokia's hardware. The same year, they unveiled the Lumia series which ran on Windows, starting with the Lumia 800 (Figure 439).

> After being introduced at the Nokia World Conference in London with a massive sound and light show, Nokia shipped between 800,000 and 1.3 million Windows Phones in 2011.

That was the start of a short-lived alliance. New models like the Lumnia 1020 were introduced, Nokia's Maps and Store had been merged with their Microsoft counterparts (ie Windows Phone Marketplace) and Bing was made the common search engine. In September 2013, Microsoft bought Nokia's mobile phone division outright for €5.4 million and changed the brand name to a rather drab 'Microsoft Mobile.' However, for all its innovation and novelty, Windows Phone never really made it big. There were well-publicised updates and there were advertisement campaigns, but to little avail. In 2016, the tech giant finally accepted the fact. It cut thousands of jobs in the smartphone division and ceased advertisement. Microsoft went back to its original business; the software for Personal Computers.

## The Development of Feature Phones

Back to early days of the cell phone and the mobile phone networks. It was not only the mobile networks that improved (in terms of capacity as well as capabilities), also the 'features' —being a distinctive attribute creating added functionality— of handheld mobile phones developed dramatically. For the first time, mobile devices were fast enough to support forms of online video and music streaming. And they got additional features, such as a photo and video camera, SMS texting, web browsing and a GPS-receiver.

*Camera Phone:* Take the feature of photography. After the traditional camera —using chemical processes to develop film— was overrun during the Age of the Digital Imaging by the digital camera, the digital photographic technology had exploded. At its core was the light-sensitive image sensors; the semiconductor-based Charge Coupled Devices (CCD) invented in 1969. Made in semiconductor technology (ie MOS technology) the device and its memory had replaced the 35 mm film, and by the 1980s the digital camera appeared on the market (eg the Nikon digital single-lens reflex camera QV-1000C, 1986).

This device became also the core of the phone camera. From the experimental resolution of the first CCD image sensors, emerged sensors with increasingly higher resolutions —from 0.35 Megapixel in 2000 growing to 1,3 Megapixels in 2004 and 3.2 Megapixel in 2006— and the phone camera sensor was soon physically small enough to be part of a mobile camera; the camera phone was born (Figure 440).

Such as the Kyocera Visual Phone VP-110 (resolution 80,000 pixels, 1997) followed by the Kyocera VP-210 (1999); the Sharp J-SH04 (resolution 110,000 pixels, 2000); the Samsung SCH-V200 (resolution 350,000 pixels, 2002), and the clamshell Sanyo SCP-5300 (resolution 300,000 pixels, 2002).

Over time the phone camera would mature with additional capabilities (auto-focus, multi-focus, wide angle, flash, etc.)

**Figure 440: Feature Mobile Phones (1): the Phone Camera.**

The digital camera feature: Kyocera VP-210 (top left, 1999), Samsung SCH-V200 (top right, 2000), Sanyo SCP-5300 (middle, 2002). Media player feature, Samsung SPH-M100 (bottom, 2000),

Source: Wikimedia Commons.

*Media Phone:* Other features were the feature of a portable media player: the functionality borrowed from dedicated devices such as the game changing Sony Walkman cassette player (1979), the MP3 media player (1998), and the Apple iPod portable media player (2001). This functionality became incorporated in the mobile phone such as the Samsung SPH-M100 (2000) (Figure 440, bottom).

*Text Messaging:* In order to add the new feature of sending short text messages, by 1993 the Short Message Service (SMS)-concept was implemented. It enabled the user to send a short text up to 160 characters long.

With it came the adapted numerical keyboards able to create the alphabetic signs. Nokia was the first handset manufacturer whose total GSM phone line in 1993 supported user-sending of SMS text messages (Figure 440, top). In 1997, it became the first manufacturer to produce a mobile phone with a full keyboard: the Nokia 9000i Communicator (Figure 452, bottom).

*E-mail:* From the early text messages send on mainframe computers creating the mail facility in the 1970s, the LAN mail facilities of the microcomputers, and the subsequent mail application programs

**Figure 441: Feature Mobile Phones (2): SMS texting.**

SMS-feature on different devices (top, late 1990s), GPS-feature on Beneton Esc! (upper middle, 1999), WAP-featured Ericsson R320 (lower middle, 2000) and Ericsson T39 Bluetooth (bottom, 2001).

Source: Wikimedia Commons.

for the personal computer, the e-mail feature was also incorporated on the mobile phone.

By the mid-1990s, Canada's phone company Research In Motion (RIM) was already working with partners on a messaging device that would work on a (proprietary) wireless data network. In 1999, RIM's Blackberry line of wireless-email solutions with a build-in mobile phone, started with the Model 850. By 2002 the Blackberry 5810 handheld (Figure 452) was introduced as 'wireless email solution with the convenience of a built-in phone' and targeted at the business segment of users.

*Connectivity Features:* And there were the features like the access to the navigation of the Global Position System (GPS), the WIFI- and Bluetooth connectivity, and mobile broad band access with a browser using the Wireless Application Protocol (WAP). Nokia and Ericsson were among the first to introduce the WAP feature on their model 7110 (1999), and the Hotline R250s Pro (1999) (Figure 442).

The WAP feature offered as a micro-browser, through a so-called WAP Gateway, the connection to the Internet. After opening the WAP-browser on the mobile device, the gateway gave him access to the selected website. And there was the website owner who decided what he wanted to share (and how he wanted to earn money). The WAP-browser became a hype as the next wave of internet access —next to the ADSL access through telephone landlines—, but failed to materialize on its expectations.

**Figure 442: Feature Mobile Phones (3): WAP browser.**

The WAP feature on the Nokia 7110 (left and middle, 1999), and WAP-featured Ericsson Hotline R250s Pro (right, 2000).

Source: Wikimedia Commons.

The initial excitement about WAP wasn't based on initial user response but on idealized projections of a science-fiction type future where we are all constantly connected; busy exchanging video and sound; handling e-mail and voice messages; and purchasing a variety of goods and services with marvellous, Flash Gordon-esque mobile communicators hardly bigger than a wristwatch.

**Figure 443: Feature Mobile Phones (4).**
Motorola RAZR V3 (top, 2003), Nokia Model Timeport 260 (bottom left, 2002), and Nokia 1100 (bottom right, 2003).
Source: Wikimedia Commons.

The more realistic expectations —sending photos and video with an SMS (aka Multi Media Service) and E-Commerce transactions— were implemented, but did not have a long time-span, as more open languages (eg HTML) took over.[616]

*The last of the Feature Phones*

With 3G Smart phone on the horizon, the 2G Feature phone was still popular in the early 2000s. The high-tech design attracted still the businessman. Its low-level pricing and broad availability in third countries made feature phones enjoyed unchallenged popularity.

In the mid-2000s, phone makers such as Motorola and Nokia enjoyed record sales of feature phones. Motorola, that sold 60 million StarTac's in 1996, and 130 million clamshell RAZR V3 (Figure 443, top) in 2004, was surpassed by NOKIA. At its peak in the early 2000s world leader Nokia supplied 40% of the world's mobile phones. Such as the Nokia 3310 that sold 126 million (2000), the candy bar NOKIA 1100 (Figure 443, bottom right) that sold 250 million units (2003). The Nokia Model Timeport 260, introduced in 2002, was capable of operating on the triband GSM and the GPRS network. The candy bar design had the SMS capability, and access to internet through a WAP-browser.

---

[616] Part of the problems was the need to reduce the webpages designed for the larger computer system to the small 4/8 line screen of the cell phone. WML was used as a technique to get content from an HTML Web site using WAP onto small-screen devices.

The Ericsson T39 (aka the Champion) GSM mobile phone (Figure 441, bottom), introduced in 2001, had the Bluetooth feature, next to Enhanced SMS, GPS, E-mail, and WAP features combined with GPRS technology of the Internet. It was the last mobile phone in the flip design, but packed with 'connectivity' features. Supporting a wide range of GSM frequencies, it could be used almost worldwide. These features of the GPS-era, made it herald the times to come.

## Personal Communication and Information Systems

The Cell Phone had grown from a simple 'mobile' phone —that was connected to a person instead of the 'fixed' phone being related to a place— into a multifunctional device with a broad functionality in the form of 'features.' And with those feature phones being supported by a development of the mobile 2G network improvements (eg 2.5 G, 2,75 G), other communication devices emerged. Such as the personal pager, the portable sound and video devices, personals assistants and information managers, as well as personal navigators that were widely published (Figure 444). It was the dawn of the *Great Mobile Telephone Race*.

### The Development of the Personal Pager

One of the off springs of wireless communication was a communication device known as 'pager.' Originating from the1960s —the Bellboy radio paging system from the Bell System for personal paging — these transistorized devices came widely used in the 1980s. By then they were used in a wide array of applications; from restaurant pager alerting the waiting customers that their table was ready, to the hospital pager alerting doctors and the emergency pager for alerting police and fire brigades. They were enabled by wide-area paging networks (Figure 445).

Figure 444: Publicity for the Personal Communication Devices.
Source: Wikimedia Commons.

In 1964 Motorola introduced the transistorized tone-only Pageboy I, a combination of the walkie-talkie concept and mobile radio technology. In 1974 it was succeeded by the Page boy II, sold by the companies of the Bell System (as the Bellboy Beeper), its huge success started a development trajectory of tone and voice paging in the 1970s. The Pageboy-series was succeeded by a range of improvements; such as Dimension IV (1977), the Bravo numeric pager (1986), the Bravo Express pager (1991), the NewsStream pager (1992).

> The Bravo pagers became worldwide best sellers. There were 3.2 million pager users worldwide at the beginning of the 1980s, and by 1994 that had risen to 61 million pagers.

Combined with the introduction of two-way pagers by RMI—the BlackBerry Interactive Pager in a clamshell design that incorporated a keyboard to allow recipients to respond— these devices reached the height of their popularity in the mid-1990s. In 1994, there were approximately 61 million pagers in use around the world, while in Japan, more than ten million pagers were active in 1996. In 1999 RIM introduced the Blackberry 850, a pager that could do email and was using an Intel 386 processor.

## The Development of Portable Audio and Video Players

After the battery powered portable cassette player —using the compact cassette— was introduced in the 1960s, and made popular by Sony's Walkman introduced in 1979, the trajectory of portable music came to life and enjoyed a dramatic rise (Figure 446).[617] The next versions of the

**Figure 445: Evolution of Pagers.**
Motorola Page Boy I (left, 1964), Motorola Page Boy II (middle left, 1974), Motorola Bravo Numeri Pager (middle right, 1984) and Motorola Scriptor LX2 (right, 1994),
Source: Wikimedi Commons. Zdnet.com

---

[617] In 1981 the next Model WM-2 sold one million units in nine months. By 2010, when production stopped, Sony had built about 200 million cassette-based Walkmans.

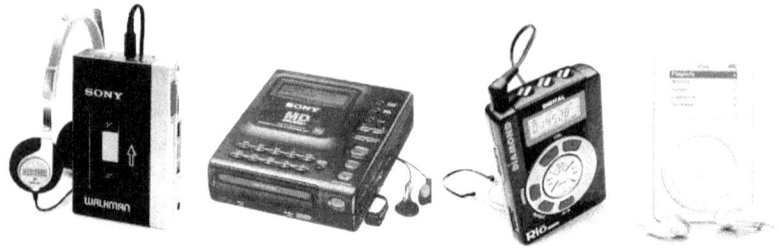

**Figure 446: Evolution of Portable Music.**

Sony Walkman TPS-L2 (left, 1979), Sony Minidisc player MZ1 (middle left, 1992), Diamond Multimedia's Rio PMP300 (middle right, 1998), and iPod Classic (right, 2001).

Source: Wikimedia Commons, Hifi Engine.com, Nippon.com

Walkman as well as the Compact disc players (eg the Sony Discman, 1984) emerged at a rapid pace. After the cassette as the music-carrier was replaced by solid state memory, the next generation of the digital MP (ie Media Protocol) [618] portable media players came on scene.

> The MPMan F-10 music player became the first mass produced solid state digital audio player (1998) together with the Rio PMP300 (1998). In 2001 Apple Computer Inc. introduced the iPod that was developed in less than a year by a team of designers. The electronics were based on a micro-controller and an audio chip. The digital music protocol was called MP3.

Next to portable music, also portable video was developed (Figure 448). The Japanese company Panasonic —already active in portable television (Figure 414)— introduced in 1978 a micro-television; the MTV-1 developed by the English designer Clive Sinclair. Sony introduced the Watchman FD-210 in 1982. The screen was a flat cathode ray tube. There were more than 65 models of the Watchman made before its discontinuation in 2000. In 1983 the CASIO Model TV-10 with a B/W LCD screen was introduced. Next, Epson and Casio introduced the first pocket colour TVs; Casio Model TV-300 and Epson ET-10 in 1983/1984. The era of the pocket-tv, however, came to an end in the early 1990s.

These portable audio and video devices, became rather popular over time. The brought entertainment on a personal level. The more when technological improvement (both in circuitry and protocols) and product design created highly attractive and affordable communication machines.

---

[618] In the 1990s the digital coding format for digital audio MPEG-1 was developed. A later version became the popular MP3 format.

**Figure 448: Evolution of Portable Video.**

Panasonic MTV-1 (left, 1978), Sony Watchman FD-210 (middle, 1982), and Casio TV-300 Colour (right, 1983).

Source: Wikimedia Commons. Sony.

## Development of the Personal Digital Assistant.

In the time of the paper versions of multifunctional agenda's (eg the Success Agenda), its electronic version appeared at the end of the twentieth century. The Personal Digital Assistant was in fact a pocket sized, handheld computer that had a wireless connectivity feature (after 2005). Next it had a range of applications programs: an appointment calendar, a to-do list, an address book for contacts, a calculator, and some sort of memo (or 'note') program. These applications were known as the PIM-applications.

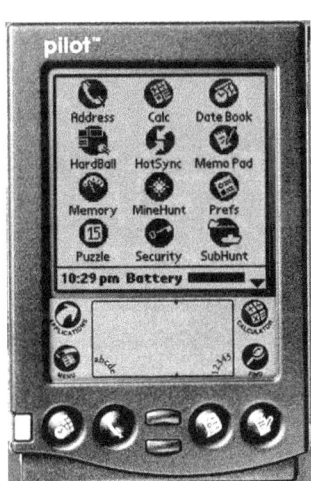

**Figure 447: Screenshot of the Palm PDA.**

Screen of the PalmPilot 1000 (1996). With cions to access productivity tools (eg Calculator, Clock, Notes, Calendar), games (eg Minehunt) and settings.

Source: Wikimedia Commons.

The development of the PDA-trajectory started in 1984 when the Psion Organiser was introduced (Figure 449). Over the year the Psion-series added functionality and changed in design, such as the Organizer II Model P (1988). IBM followed with the Simon in 1994 of which some 50,000 unit were made. Next Apple introduced the Apple Newton MessagePad in 1993 with a touch screen, of which that same year 80,000 units were sold.

The Palm Pilot 1000 (1996), using a Motorola 68328 microprocessor, became a mobile technology icon that sold for $299 (Figure 447). It came with a suit of Personal Information management (PIM) applications, offered a similar range of features. In 1999, Palm controlled 70% of the PDA market, with an over 5 million userbase. The PDAs were a huge success, as in the year 2000, nearly 12 million PDA units were sold worldwide.

The Palm Pilot first generation of Personal Digital Assistants (eg the PalmPilot 1000, Figure 449), released in 1996 used their own operating system PALM OS. The hardware of the Pilot 1000 debuted with 128k of memory and a monochrome black on green glass touch-screen display (Figure 447).

The PDA would lead to the development of the tablet computer. The technology advancing to increasing miniaturisation combined with growing functionality, made the thin slate design possible. In 1989 Grid Systems introduced the GridPad 1900, that would start the development trajectory of the tablets. By 2000 the project of the Microsoft Tablet PC specified the pen-enabled personal computer. Hundreds of such tablet personal computers have come onto the market since then.

**Figure 449: Evolution of the Personal Digital Assistant.**

The Psion Organiser (Model XP (top, 1988), the GRIDpad 1900 (upper middle, 1989), the Apple Newton MessagePad (lower middle,1993), and the Palm Pilot 1000 (bottom, 1996).

Source: Wikimedia Commons.

## Development of the Personal Navigator

The Global Positing System (GPS) —a satellite-based radionavigation system— had its origin in the US military development of a world-wide navigation system. Although early systems for electronic navigation had been developed by car manufacturers —such as the Driver Aid, Information & Routing system of General Motors Research (1966)— it took a while to become available as a consumer product. In 1983 the GPS facilities became available to the civilian sector[619] and by 1995 the GPS system, including 24 satellites, was fully operational. It spawned handheld and automobile navigation devises; the Personal Navigation Assistant (PNA) and the Automotive Navigation System that displayed the car's/user's location on an electronic map.

Various manufacturers began making GPS systems[620] for the commercial navigation market starting in the late 1980s, but between the reduced resolution available then and the initially high cost of the technology, they didn't exactly fly off store shelves.

Magellan Navigation Inc. introduced a handheld commercial GPS navigator, the NAV 1000, in 1989 for $3,000 (Figure 450, left). Over the years it expanded its consumer product line with a range of specialised

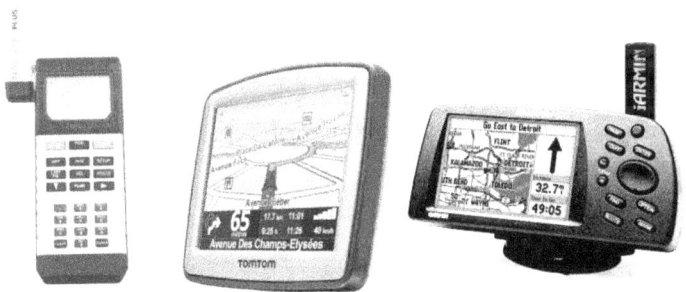

Figure 450: Evolution of the Personal Navigator.
Magellan NAV 1000 Plus (left, 1989), TomTom Navigator One (middle, 1991), and Garmin StreetPilot I (right, 1998).

Source: Wikimedia Commons. Retro-gps-info, TomTom

---

[619] In 1983, after Soviet interceptor aircraft shot down the civilian airliner KAL 007 of Korean Airlines that strayed into prohibited airspace because of navigational errors, killing all 269 people on board, US President Ronald Reagan announced that GPS would be made available for civilian uses once it was completed.
[620] The GPS receiver calculates its own four-dimensional position in spacetime based on data received from multiple GPS satellites. In the receiving device those data are translated to Cartesian coordinates, that are displayed on an electronic map.

handheld devices for automobile, outdoor and off-road navigation, and developed systems for truck fleet management and tracking.

The Swiss-domiciled company Garmin Ltd was founded in 1983 that created GPS-units for automotive applications. The first Garmin StreetPilot was introduced in 1998 for $400. In 2003 the 2000-series were introduced (Figure 450, right). By 2022, Garmin had sold 126 million units.

In 1991 the Dutch company Palmtop Software was founded, and started to design route planning software for mobile devices; EnRoute, Citymaps and Routeplanner. Next, they focussed on PDA-software. In 1991 they introduced their general-purpose devices; the TomTom Navigator One (Figure 450, middle) and by 2001 they released the first car satellite navigation software installed in automobiles. By 2004 the TomTom Go series were introduced. TomTom sold about 250,000 units of TomTom Go and this product represented 60% of the company's revenue for 2004. As of 2016, the company had sold nearly 80 million navigation devices worldwide.

## Development of the Personal Communicator

As the mobile phone had gotten increasing functionality and many additional features, it became rather 'communicative.' The more when its design-trajectory merged with the so-called Personal Digital Assistant.

IBM introduced the IBM Simon Personal Communicator in 1992 (Figure 451, left); it was a Personal Digital Assistant combined with cellular radio features. To replace the mechanical numerical keypad, it featured a black-and-white 160 x 293 LCD touchscreen measuring 4.5 inches by 1.4 inches that offered a virtual keypad in its Phone Screen (Figure 451, left). In addition, it featured 11 built-in programs, including a Calendar, To-do list, Calculator,

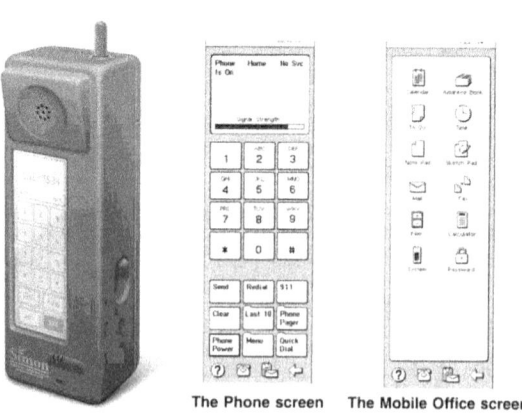

**Figure 451: The Screens of the IBM Simon.**
Source: IBM Simon's User Guide.

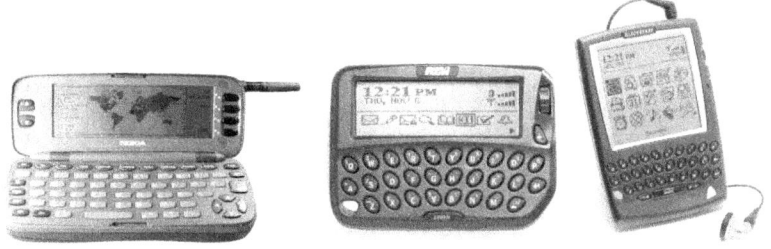

**Figure 452: Evolution of Personal Communicator.**
Nokia Communicator (left, 1996), Blackberry 850 (middle, 1999, and Blackberry 5810 (right, 2002).
Source: Wikimedia Commons.

Address book, Appointment scheduler, World time clock, electronic Note pad/Sketch pad in its Mobile Office Screen (Figure 451, left).[621]

The Simon was followed by the clamshell design Nokia 9000 Communicator (1996) (Figure 452, left). Next to being a cell phone, this device, powered by an Intel 24 MHz i386 CPU, had a web browser, email, SMS and fax facilities. And applications like Contacts, Notes, Calendar, Calculator, World time clock.

Also the Canadian BlackBerry Limited introduced the popular mobile phone Blackberry 850 in 1999 (Figure 452, middle). Its successor, the Blackberry 5810, was introduced in 2002 and aimed at the business market (Figure 452, right). It was both a handheld wireless Personal Data Assistant (PDA) and a mobile communication device. It offered features as push email, text messaging, WAP web browsing, internet faxing, mobile telephone, and many wireless informational services all accessible on a multi touch interface. But it came without an integrated microphone or speaker, you had to attach a headset to make calls.

---

[621] It was built by Mitsubishi Electric and commercially distributed by BellSouth Cellular, the device operated within a 15-state network, with approximately 50,000 units of Simon Personal Communicator sold at $899 over 6 months – between August 1994 and February 1995.

## From Tablet PDA to Pocket PC

The Personal Communicator had a spin off when in merged with the development of the Personal Computer that created the 'tablet computer.' Their origin laid in the table-like graphic input device that would reproduce the pen-made drawing onto the screen of a workstation.

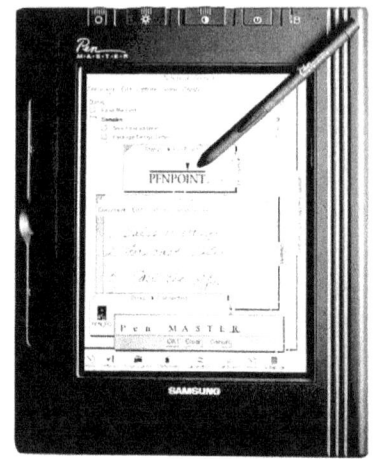

Early devices based on the INTEL 80286 processor — with the MS-DOS operating system— soon conquered the market (Figure 449d). The GRIDPad 1900 for DOS was released in 1989. It was a flat handheld computer with a touchscreen interface, based on an 8086 CPU and MS-DOS. It was soon followed by others models using more advanced micro-processors, larger memories and newly released operating systems like Microsoft Windows.

Samsung introduce the Penmaster in 1992 for $5,000 running Microsoft Pen Computing OS 1.0. As a competitor to the Apple Newton, AT&T introduced in 1993 the EO Personal Communicator 440/880, similar to a large personal digital assistant with wireless

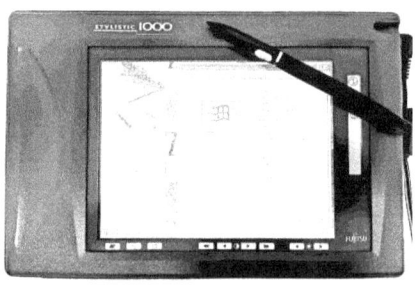

**Figure 453: The Tablet Computer.**
The Samsung Penmaster (top, 1992), EO 440/880 (middle, 1993), and Stylistic 1000 (bottom, 1994).
Source: Wikimedia Commons.

communications facilities (Figure 453, middle). In 1994, Fujitsu launched the Stylistic 500-tablet that ran on an Intel processor. This tablet came with Windows 95, which also featured on its improved version, the Stylistic 1000 (Figure 453, bottom).

In parallel to the trajectory of the PDAs, the development of the handheld Personal Computer took place.[622] The more when Microsoft introduced its Windows CE operating system in 1996 to be used in embedded systems. Such as the calculator manufacturer Hewlett-Packard who developed a line of handheld/palmtop PC's starting with the Palmtop PC HP 320 LX in 1996 (Figure 454, top).

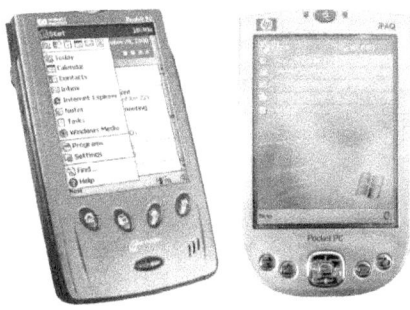

**Figure 454: Pocket PC.**
Palmtop PC HP 320 LX, (top 1996), HP Jonada 520 Pocket PC (bottom left, 1999), HP iPAG 4150 PDA (bottom right, 2000).

Source: Wikimedia Commons
6.6

This was the start of *pen-computing* as a pen like-device was used to operate the touchscreen. The tablet computer would functionally develop into the pocket PC; the palm sized class of 'Microsoft' Pocket PCs using the Windows Mobile OS.

    A range of companies started to design and manufacture these devices. Such as the Everex Freestyle (aka HTC Kangaroo), introduced in 1998. Other examples include Casio Cassiopeia E-10/E-11, Compaq Aero 1500/1520, Philips Nino and HP Jornada 420/430, 520/540 series (Figure 454).

---

[622] The labels PDA, Pocket PC, Palmtop PC and Handheld PC are mixed by the different sources (eg the manufacturers).

## The Technological Convergence of Mobile Personal Devices

In the preceding exploration into the development of a range of 'personal' communication and information devices was presented. Each of those devices was the result of independent —but often related— application development trajectories. Related by the different electronic technologies, but each unique in its application offering solutions to existing and latent needs in the market place.[623] Needs that all were part of the basic human need to communicate that constitutes such an important part of the Affairs of Man.

The Mobile Phone had come a long way. Sure, the bulky commutation devices like the suitcase-type analogue mobile telephones were cumbersome to handle. But for fast communication in emergency situations, they were quite usable. Compared to the

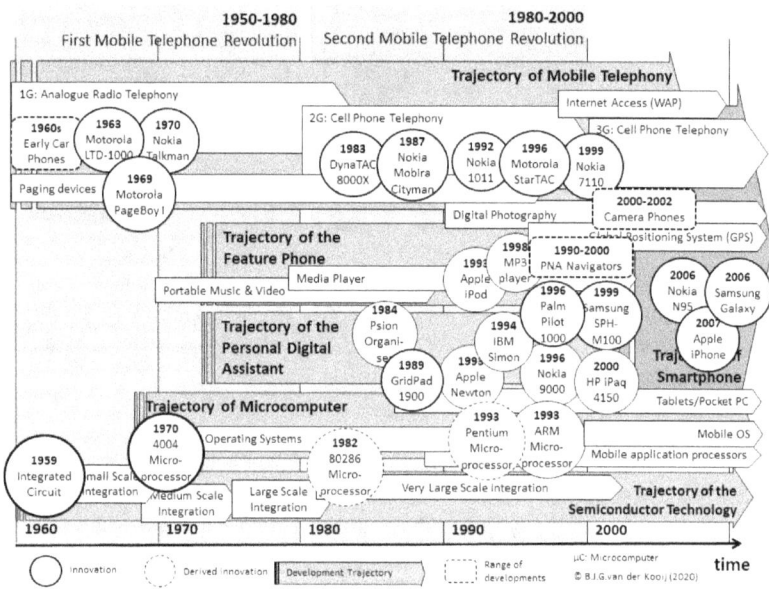

Figure 455: Overview Era of Personal Mobile Communication.

---

[623] *Existing Need:* When a human need is offered already a solution, that becomes improved by newer concepts, the market tends to absorb the new devices rapidly. *Latent Needs:* No (electronic) solution had been offered for an existing human need, and no-one is asking for it because nobody is aware of the need. Take the mobility need of navigation that was in the pre-digital times supported by printed maps. The introduction of the detailed map-books known as street guides (eg the famous Michelin Guide) was rapidly accepted. By then, not many people would have expressed their need for a device like a TomTom Go Navigator. But when technology made that feasible, it became rapidly absorbed. In the twenty first century, no car sold without a navigating facility.

fixed phones of the land-based systems, the mobility they offered was certainly a revolutionary improvement. And even more, when their performance improved over time, and when they became cheaper and physically smaller —from transportable to portable— they soon attracted new users (like the taxi-driver and business man).

The rapid progress of the electronic technologies (ie the semiconductor technologies) created the foundations on which all the application trajectories rested. The semiconductor technology for electronic integrated circuitry (IC's) designed for electronic transmission and receiving modules, as well as for electronic touch displays (both data presentation and data entry). Together with the wireless communication infrastructure of the cellular networks, the new phenomenon of mobile telephony penetrated society rapidly. Together they created the *Era of Personal Mobile Communication* (Figure 455).

In addition, there was the development of the computing machine. In a similar way, originating from bulky computing machines (ie Eniac) it was the micro-computer technology that laid the foundations for the Personal Computer. The development of this kind of computers laid the foundations for the basic division between 'hardware' (creating the physical device) and 'software' (creating the functionality).

The early information processing devices were replacing the paper-based information systems (eg Filofax, agenda, notebook) with their electronic counterparts (the Personal Digital Assistant and Personal Information Manager) during the *Era of Personal Information Systems*.

These application development trajectories became converged in mobile devices that would become known as 'smart phones.' In these mobile phones —that were part of the next phase of the Era of Mobile Communication— the before described development trajectories for Personal Communication and Personal Information came together (Figure 455) in the mobile telephone that became known as the 'smart phone.'

The manufacturing and service telecommunication industry, confronted with this explosion of novelty at the late-nineteenth century revolutionizing existing markets and creating new markets, struggled to adapt and survive. Spending enormous amounts on R&D-projects, looking for alliances to serve the exploding markets, many companies had grown to prominence. But even they could hardly cope with the technology fuelled disruption caused by the smartphone. It resulted in an industrial disruption that saw the failure of the old guard, and the rise of the new boys on the block.

## *The Smartphone*

For someone living in the 1970s in the days before the Personal Computer where the early transportable mobile phones could be seen used by emergency workers and professionals, it must have been hard, so not impossible, to predict that one day there would be a portable handheld mobile device that would dominate much of their lives; from their social life to their economic and financial life. A device they would use every day to communicate with others, to retrieve information in unlimited forms (eg music, video), to share experiences (ie by photos and videos), to execute economic activity (ie eCommerce) and financial transactions (ie eBanking). While there might have been difference in the adoption rates —young people embraced the social media through their mobile phones rapidly—, people became connected in unheard ways regardless of geography, spending hours per day with the device.[624] A device that brought the most dramatic events of world (eg the Arab Spring in 2011) on their little screen, that was used to swindle the unexperienced and gullible users of their savings, and that made them addicted to it (Figure 456).

Had the development of the mobile personal telephones been the outcome of preceding development trajectories during some decades, but the revolution of the smartphone took just a decade. Several different application development trajectories at the end of the twentieth century had created a massive range of intelligent connected devices for personal use. From the PIM- and PDA communicators to the Featured phones, these developments were paralleled by the development of the network infrastructure (Figure 455, top) as well as the development of the Silicon Engine (Figure 455, bottom). This was a development in a period of time that saw both the deregulation of the telecommunication industry, as well as the Mobile Telephone Crash that had created the overall context for a massive disruption.

**Figure 456: Smartphone Addiction.**
Source: Painting Erin Pollock (2015)

---

[624] With the rise of the smartphone, its use per day doubled between 2011 and 2013, from 1 hour 38 minutes to 3 hours 15 minutes.

## Inventions contributing to the Smartphone

The application trajectory of the development of the mobile telephone was paved with technology fuelled novelty. Novelty as result of the contributions of R&D teams in the large companies, as well as the those of their individual team members. Such as there were the following inventions:

*The Invention of Mobile Phone Protocols:* In order to be able to function as a mobile phone, the network infrastructure needs to build up —and maintain— the connection between the subscribing mobile caller and the called party.[625] The process of subscriber identification, radio channel selection, transmission techniques, applied in the late 1990s became labelled as the 3G standards. The development of the contributing 2G digital techniques, however, took place in the time before that.

*The Invention of WIFI:* In computer networks, in order to create wireless connectivity between Local Area Network (LAN) devices, next to the Ethernet-protocol for hardwired connectivity, a protocol for local interconnection by radio waves was developed (ie Wireless LAN). The protocol —known as the IEEE 802.11 standard— was in 1991 developed at the R&D centre of the NCR Corporation together with the AT&T Corporation and intended for use in cashier systems. The first wireless products were categorized under the name WaveLAN.

The Australian radio-astronomer John O'Sullivan with his colleagues Terence Percival, Graham Daniels, Diet Ostry, John Deane developed a wireless technique as a by-product of a Commonwealth Scientific and Industrial Research Organisation (CSIRO) research project. They applied on November 27th, 1992 for a patent that was granted as US Patent № 5,487,069 on January 23rd, 1996 for 'Wireless Wan,' That patent proved to be a key patent, and the patent loyalties (and dispute settlements) brought CSIRO $430 million.

The WIFI-feature came as an add-on PCI card to personal computers, a USB-key or a card to notebooks, or —as it became integrated in chips— as a System on Chip (SoC)[626] in a mobile phone (Figure 457).

---

[625] This is the wireless equivalent of building up a connection between land-based telephone. The techniques that were applied fall outside the scope of this case study.

[626] A system on a chip (SoC) is an integrated circuit (also known as a "chip") that integrates all or most components of a computer or other electronic system. These components almost always include a central processing unit (CPU), memory, input/output ports and secondary storage, often alongside other components such as radio modems (for Bluetooth/WIFI) and

*The Invention of Bluetooth:* Another form of interconnectivity found its origin in military research. In 1994 researchers of the Ericsson Mobile Terminal Division in Sweden were developing a wireless link between a headset and a cell phone. Together with Sony, Intel, IBM and Toshiba they created in 1998 the Bluetooth Special Interest group. And In 1999, Bluetooth specification 1.0 was released.

The Dutchman Jaap Haartsen, working in the R&D team was granted some five patents fundamental for the Bluetooth standard. Among which EP European Patents № 1,016,241 and № 1,016,242, both applications filed on September 15th, 1997, and both granted on March 14th, 2007.

**Figure 457: Wireless Connectivity Systems on a Chip (SoC).**
Wifi SoC (top), Bluetooth SoC (bottom).
Source: Wikimedia Commons.

The Bluetooth device became a hype. Several hundred million euros were invested in small start-ups between 1998 and 2001, particularly among the designers and manufacturers of Bluetooth chipsets and base stations. The hardware came as Add-on Card, and USB-modules.

*The Invention of the ARM Processor:* The Advanced RISC Machine (ARM) processor transformed computing by enabling maximum processing speed and simplified task management with low power consumption. The ARM processor emerged from a development at Acorn Computer

---

a graphics processing unit (GPU) – all on a single substrate or microchip. The SoC is the next step in integration of systems on a single chip.

Inc, hence the original name Acorn RISC Machine in 1985. Defining the *Reduced Instruction Set Computing* (RISC), reduced the chip complexity (to some 25,0000 transistors) and reduced its power consumption as well.

After in July 14th 1999, a patent application was filed for an 'instruction set for a computer,' GB Patent № 2,352,066 was granted on November 5th, 2003. For the ARM processor a patent application was filed on July 18th, 2000. The European Patent Office granted on February 9th, 2007 the European Patent № 1,206,737 for 'Setting condition values in a computer' to Sophy Wilson.

In 1990 the Advanced RISC Machines Ltd was formed as a joint venture between Acorn Computers (know how), Apple (capital), and VLSI Technology (tools). ARM Ltd only created and licensed its technology as intellectual property (IP), rather than manufacturing and selling its own physical CPUs, GPUs, SoCs or microcontrollers.

*The Invention of the SIM-card:* With the development of the credit card sized smart 'Integrated circuit cards'—a credit card with an embedded chip—came the Subscriber Identification Module (SIM). These cards were developed to facilitate the 2G GSM Network identification of the subscriber who was using the mobile phone.

The Frenchman Roland Moreno conceptualized and created a plastic card with a microchip in 1975. It took some eight years to gain widespread use and by 1983 France Telecom used it with its Télécarte pay phone payment cards. French banks used the cards for their debit cards; the Carte Blue. It was the birth of the smart card.

Next, early SIM cards were developed by in 1991 by Munich smart-card maker Giesecke & Devrient, and sold to the Finnish wireless network operator Radiolinja. From there on, its implementation was rapid, as well as its functionality (eg dual SIM) and size reduction (from the credit card size to the Mini-SIM, Micro-SIM and Nano-SIM, Figure 458). For those improvements different patents were issued. It resulted in the *SIM War*, part of the Smartphone War.

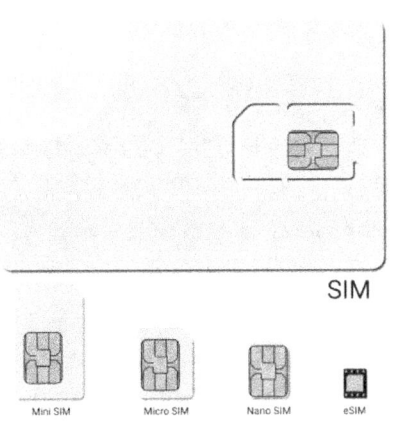

**Figure 458: Evolution SIM-card.**
Source: Wikimedia Commons.

The SIM War was about getting more subscribers. So, the SIM card became the device sprouting a new service; the prepaid SIM-card. Originally it was meant to expand the subscribers base from those with a low credit card rating, who could obtain a normal subscription (ie post-paid subscription). Later it became a market of its own as network companies created (eg Telecom Italia Mobile (TIM)) their own prepaid card service (eg the TIMcard, 1996). Many inventive engineers contributed to its development and patented their contributions.

Such as Keith Benson who worked on smart card improvements with an external card holder. He filed for a patent application on August 19th, 1996 that resulted in European Patent № 785,634 granted on July 23th, 1997 for a 'Mobile phone with SIM Card.' Andrew Wise, after founding the prepaid mobile phone service Banana Cellular in 1993, filed for a patent application on December 16th, 1996 that resulted in US Patent № 5,826,185 granted on October 20th, 1998 for a 'Cellular phone system wherein the air time use is predetermined.'

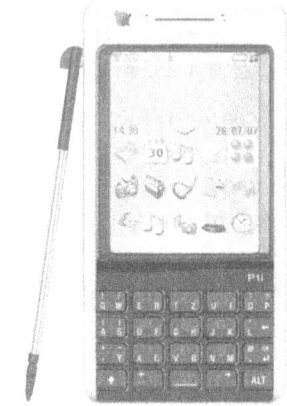
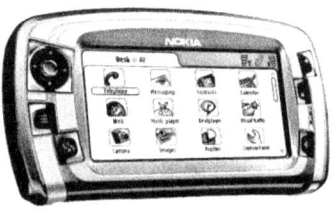

**Figure 459: Stylus operated Touch Screen Mobile Phones.**
Sony-Ericsson P1 (top, 2002), and Nokia 7710 (bottom, 2004).
Source: Wikimedia Commons.

*The Invention of the Touch Screen:* A major contribution to mobile telephones was the 'touch screen' replacing the traditional keyboards and combining the input device with the output device. The early touchscreen originated from many contributions. Such as the finger driven touch screen invented in 1965 by Eric A. Johnson who worked at the Royal Radar Establishment in Malvern, England. His first article, "Touch display—a novel input/output device for computers" describes his work and features a diagram of the design.

Johnson filed for a patent application and was granted GB № 117,222 in 1965, followed by US № 3,482,241 granted on December 12th, 1969 for 'Touch displays.'

He was not the only contributor to the development of the touchscreen technology.

George Samuel Hurst, working for the US company Elographics, developed a transparent implementation of the existing opaque touchpad technology. He applied for a patent on March 18th, 1974 and was granted US Patent № 3,911,215 on October 7th, 1975 (Figure 460).

With technology progressing, from the original stylus operated capacitive touchscreen[627] evolved the resistive touchscreen that featured two thinly-separated layers, which would make an electrical connection when pressed.

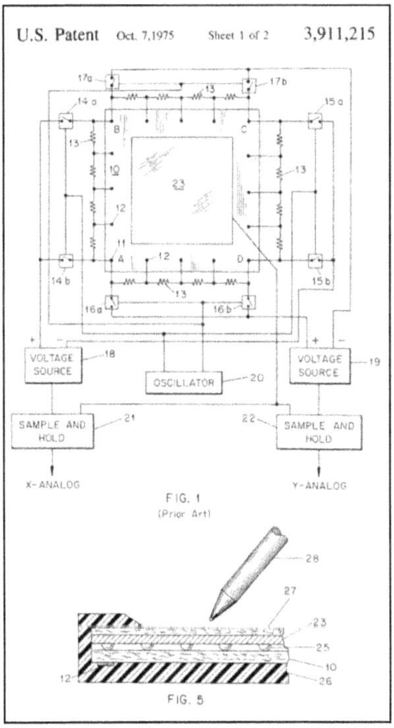

By the 1980s tech companies were starting to take notice of this new way to control computers. Hewlett-Packard was among the first to release a product —the HP 150 Personal Computer—that put touchscreens in the hands of everyday users. In the 1990s the touchscreen became used in hand held devices like IBM's Simon Personal Communicator, Apple's Newton and the Pilot PDA by Palm Computing introduced in 1996.

In 2002, Sony Ericsson introduced the Model P1. It was a Symbian based phone, packed with feature (from a 3.15 Mp Camera, FM-radio, Media Player to Bluetooth, WIFI, Web browser to 3D Games), that had both a keyboard and a pen-operated touchscreen. Nokia introduced in

**Figure 460: Touch Screen Patent.**
US Patent 3,911,215 (1975)
Source: USPTO

---

[627] A Stylus Pen, a pen-like device with a conductive point, used like a normal pen on the touch screen. The applied pressure is detected by the touched segment of the screen as the electrostatic field changes.

2004 the style operated feature phone Model 7710 with a TFT-captive screen (Figure 459).

*Design Feature Phones:* The use of touchscreen keyboard started in new development in the high-end phones; the design feature phones (Figure 461). In 2007, just before the release of the Apple iPhone the Korean company LG Electronics introduced the LG model KE850 (aka LG Prada), designed by the Italian luxury designer Prada. The design award winning had a captive touchscreen for display, and a slider keyboard for data-entry. In 18 months some 1 million units were sold. Also in 2007 Samsung introduced the B7620 Giorgio Armani as a luxurious slider phone, copy-catting the LG-Prada.

The invention and the improvements in the Touchscreen-technology proved essential in the further development of the mobile telephony.

*The Invention of the Virtual Keyboard:* From the times of the invention of the typewriter (ie 1867) up to the Personal Computer (ie 1980s), data-entry had used a mechanical keyboard (ie hardware). With the rise of the feature phones, the same keyboard with the physical keys in miniaturized form had frustrated the designers (and users) of the brick, flip, clamshell and slide mobile phones. However, with rise of the combined data-input and data-output facility of the touch screen, that barrier was lifted.; the virtual touch keyboard (aka soft keyboard) was the breakthrough.

The touchscreen as an input device was already explored since its conception. Texas Instruments had been granted on September 23rd, 1980 US Patent № 4,224,615 for a 'Method of using a liquid crystal display device as a data input device.' And at the Nokia Bell labs in the late 1980s engineers were working on solutions for "Simulated keyboards [that] can be displayed on the display element and, in response to the touching of the simulated keys, generate appropriate control signals." It resulted in US Patent № 4,725,694 granted on February 16th, 1988.

**Figure 461: Design Feature Phones.**

LG Prada Mobile Phone (top, 2007) and Samsung B7620 Giorgio Armani Mobile Phone (bottom).

Source: Wikimedia Commons.

By the turn of the century, the LCD-screen had got touch capabilities used a virtual numeric keypad to enter the desired numbers. It was used in small electronic devices like calculators. And then came the new development that took place at Apple.

The design of the Apple iPhone was focussed on using the features of the touchscreen, and getting rid of the mechanical keyboard/keypad that was so dominant in the Blueberry designs. Apple filed on September 16th, 2005 a patent application for 'vertical input device placement on a touch screen user interface, and was granted on February 16th, 2006 US Patent № 2006/0033724 for a 'Virtual Input Device Placement on Touch Screen User Interface' for activating virtual keys of a touch-screen virtual keyboard.

These are just a few of the more technical-based inventions that contributed to the development trajectory of the intelligent mobile phone. They illustrate how a multitude of technologies contributed to the device that became known as the 'smart phone.'

## Evolution of the Mobile Phone into Smart Phone

The *Great Mobile Telephone Race* was a period that had seen the evolution of the Mobile Phone in just two/three decades from a simple cell phone to a full-featured intelligent phone (Figure 455). That version of mobile telephony saw a meteoric rise; in the year 2000 some 400 million mobile phones were sold worldwide. A race that culminated in the device labelled as 'smart' phone[628] when 'Communication met Computing.' And with that technological development, the user-market exploded.

### Prominent Contributors to the Smart Phone

There were some major events in the technological development that brought forth the new concept of the smartphone. Again, many different devices had one way or the other intelligent features. But some devices showed the contours of a totally new communication device.

*Nokia N95 Smartphone:* In 2006 Nokia announced —and in 2007 released— its flagship and first 'smartphone' model N 95 (Figure 462, top). The first device that combined telephone capabilities with palmtop computer capabilities since IBM's Simon Personal Communicator in 1994. It had the dual slide design, a colour display and a 5 Mp VGA camera.

---

[628] The category of 'Smartphone' is not clearly defined. In general a smartphone refers to a multifunction mobile phone that has the ability to browse the Internet, play music and movies, equipped with a GPS chip, a touch screen, which can evolve over time using updates, and which has the ability to download and install new applications (aka apps).

Promoted as 'multimedia computer,' the connectivity features included WIFI, Bluetooth and GPS. Add to that additional feature such as ringtones and voice commands. The ARM11 processor had the Symbian OS at its core, and it operated on the UMTS network. But it still had a mechanical keyboard to enter data (Figure 462, top).

The N95 supported a range of audio formats (MP3, WMA, RealAudio, SP-MIDI, AAC+, eAAC+, MIDI, AMR, and M4A formats) and video formats (3GP, MPEG4, RealVideo, and, in newer firmware, Flash Video formats). In addition, the WIFI feature made it possible for the browser to access the World Wide Web.

*Apple's iPhone:* In 2007 Apple Computer Inc. introduced its iPhone ONE (Figure 462, middle). Steve Jobs announced it as a combination of three devices: a "widescreen iPod with touch controls"; a "revolutionary mobile phone"; and a "breakthrough Internet communicator".

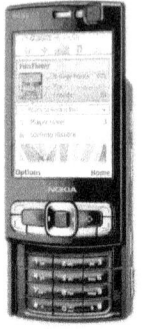

**Figure 462: The First 'Smart' Phones.**
The Nokia N95 (top, 2006) Apple iPhone ONE (middle, 2007), and Samsung Galaxy GT-17500 Galaxy (bottom, 2009).

Source: https://www.phonearena.com/, https://www.econstor.eu/ bitstream/ 10419/101414/1/794346243.pdf

The 2G mobile device was equipped with a 3.5" touchscreen, contained a 2 Mp camera, connectivity (WIFI, Bluetooth, GPS, USB). It was packed by applications programs (aka micro 'apps") like Phone, Contacts, Calendar, Photos, Stocks, Weather, Clock, Calculator, Notes, and a Settings apps. The iPhone's microcomputer circuitry was a 32-bit ARM microprocessor running Apple's proprietary iPhone OS operating system.

The iPhone featured e-mail, an iPOD music and video player app, a dedicated YouTube app and a Maps app powered by Google Maps. It had quad-band GSM cellular connectivity with GPRS and EDGE support for data transfer. Access to the Internet was through the SAFARI browser. The iPhone 3G version was announced in 2008. Its slate design set the norm for future smartphone designs (Figure 463). The subsequent iPhone were upgrade versions of the basic iPhone design; the iPhone 4 (2010), the iPhone 5 (2012) and iPhone 6 (2014).[629]

In addition, the iPhone One had the capability to download application programs (aka Aps) offering additional functionality, next to the functionality of the micro-apps included with its operating system (eg calculator, timer). The apps were developed by Apple itself, or by third parties developing software applications (eg access to the website Facebook). Apps could be acquired through Apple's online store. And…, it featured the virtual keyboard replacing the mechanical keyboards that had dominated so long the designs of the mobile phones.

Right from the start it was a sales success as it appealed to the general public. Supported by Apple's carefully orchestrated marketing and controlled media strategy helped heap fuel on consumer and media fascination. Thousands of people were reported to have waited outside Apple and AT&T retail stores days before the device's launch; many stores reported stock shortages within an hour of availability. The iPhone was, six most after Steve Jobs presented it, heralded by Times Magazine as the Invention of the Year. Within the year 6.1 million units were sold. In the next years it was succeeded by the iPhone 3G (2008) and the iPhone 3GS (2009).

*Samsung Galaxy Smartphone*: From a totally different background, the Samsung Galaxy Smartphone emerged (Figure 462, bottom). That new device was the result of the development of the Android Operating System, Google's involvement, and the de technical capabilities of the Korean company Samsung.

---

[629] Apple sold 5 million units of the iPhone 5 in its first week and 10 million units of the iPhone in the first week.

In June 2008 Samsung had introduced the M800 Instinct —being a mobile phone with a touch screen, with features as a camera phone, portable media player, text messenger, and a complete web browser and e-mail client— as competitor to the iPhone. A year later, in June 2009, Samsung Mobile released the Samsung GT-17500 Galaxy using the open-source Android operating system. The Quad-band mobile phone had an Amoled captive touchscreen, an ARM11 CPU with 128Mb Ram and 8 Gb storage. In addition, it had a 5 Mp camera with power Led Flash. For connectivity it had Bluetooth, Wife and GPS. That gave access to Google's services such as search, maps, YouTube and Gmail. Some considered it as the spitting image of the iPhone.

This was the first smartphone of a range Galaxy S series, and after the introduction of the Galaxy S (2010), it announced the Galaxy S2 (2011), Galaxy S3 (2012) and Galaxy S4 (2013). The Samsung Galaxy S2 became a huge commercial success, having sold over 40 million handsets (as of January 2013). It propelled Samsung from a second-rate electronics firm to a global tech player. It was from the phenomenal success of the S2 where the company had begun to build its own mobile empire with staggering sales figures.

The Galaxy S3 Samsung had already received about nine million orders of the Galaxy S3 prior to its official release. After its introduction the Galaxy S4 sold around 40 million handsets just in the first six months after its release. To date, it has sold about 80 million smartphones.

## Dominant Design for the Smart Phone

From the early brick-sized mobile phones introduced in the 1970s, followed by the mobile phones of the 1980s, to the slate design smartphone appearing in mid 2000s (Figure 463), the hardware design of the mobile telephone had followed an evolutionary path. Fuelled by a seemingly insatiable demand for mobile communication devices, technology driven by both network technologies and the digital electronic technologies of the Silicon Engine, in a whirlwind of phone models adding increasingly more features, better interfaces, longer battery life and elegant design, the mobile phone was to become a life changer in the Affairs of Man.

After 'Communication met Computing,' it was the slate design that became the Dominant Design[630] for the next hardware generations of

---

[630] The definition of dominant design is a specific path along an industry's design architecture, which establishes dominance among competing design paths. A dominant

**Figure 463: The Evolution of the Mobile Phone from brick phone to slate smart phone.**
Design from the 1G brick model (left) to the slate 3G model (right).

smartphones. Not so visible was the evolution of software of the mobile phones (ie the operating system) leading up to two dominant operating systems. Also the development of the mobile networking protocols leading up to 3G, as well as the Mobile Internet protocols, contributed to the revolutionary development that would disrupt the mobile phone business for ever. But the main catalyst was the replacement of the hardware keyboard with the virtual keyboard.

Since the end of the 1990s, most companies in the market embraced the early smart phone concept with keyboard; Nokia with its Communicator Line, Motorola with the MPX line, Ericsson with its T series, and later Samsung with its I series. These innovations were imitated by newcomers such as HTC, RIM, MiTAC International Corp. (under the brands Mio Technology and Navman) and Kyocera.

## Smartphone Patent Wars.

The iPhone became a hit, and was introduced in many countries, people lining up in front of the (physical) Apple stores,[631] sales were between 250,000-700,000 in the first weekend. In total some 6.1 million units were sold, the start of the subsequent iPhone-series that would sell in the

---

design usually takes the form of a new product (or novel set of features) synthesized from individual technological innovations introduced independently in prior product variants.
[631] In order to have more control over the retail sales of the Apple computer product, in 2001 Apple opened its first Apple Stores. The highly successful concept was complemented by 'online sales' from its website, copied from the mail-order sales concept.

millions. Times Magazine crowned it as the Invention of the Year (Figure 464). And with its rise in popularity came the unavoidable patent litigation as other manufacturers jumped on the intelligent phone patent bandwagon.[632]

The *Smartphone Patent Wars* started between Nokia, Motorola Mobility, Samsung, Google, Microsoft, HTC and Apple suing each other on patent infringements. It started in October 2009 with Nokia suing Apple for infringement of ten patents. Apple responded and sued Nokia over 13 patents. It was the beginning of hundreds of court cases. In 2010 Apple initiated a patent war against Android; the first of 21 cases brought to court, And in 2011 Apple sued Samsung for copying its design; it was one of the 41 cases brought to court that year. In less than seven years, the top ten smartphone manufacturer litigators amassed over 1,100 smartphone patent lawsuits as either plaintiff or defendant, including 142 such lawsuits involving Apple alone.

**Figure 464: Times Cover (2007).**

> Some of the patent cases were rather 'frivolous'; such as the 'Rounded corner' patent (US Patent № D670,286), the 'Slide-to-unlock' patents (eg US Patent № 8,046,721) and the 'Page-turn' patent (US Patent № D670,713).

The smartphone patents war not only were costly in terms of legal fees, the damages claimed ran into billion, and license fees to be paid could also run into millions.

---

[632] Samsung alone filed 2,179 smartphone patent applications in 2013 alone, which outnumbered the total patent applications filed by Apple in the same year.

*Apple Inc. v. Samsung Electronics Co. (2011/2012):* In two separate lawsuits, Apple accused Samsung of infringing on three utility patents (US Patent Nos. 7,469,381, 7,844,915, and 7,864,163) and four design patents (US Patent Nos. D504,889, D593,087, D618,677, and D604,305). Samsung accused Apple of infringing on US Patent Nos. 7,675,941, 7,447,516, 7,698,711, 7,577,460, and 7,456,893.

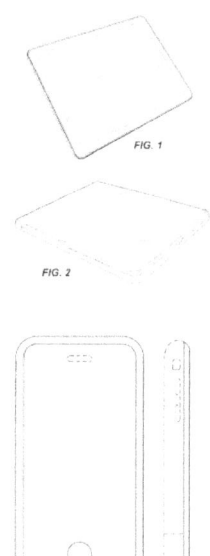

One 2005 design patent became "at the heart of the dispute is Design Patent 504,889", which consists of a one-sentence claim — "We claim the ornamental design for an electronic device, substantially as shown and described." — about the ornamental design of an electronic device, accompanied by nine figures depicting a thin rectangular cuboid with rounded corners. It listed fourteen inventors. The court case ran before Courts in different countries over numerous years.

On August 24th, 2012 the jury returned a verdict largely favourable to Apple. It found that Samsung had wilfully infringed on Apple's design and utility patents and had also diluted Apple's trade

**Figure 465: iPhone design figures from Apple Design Patents.**

Figures from US Design Patent № 504,889 (top), USD № 593,087 (middle,) and USD № 604,305 (bottom).

Source: USPTO. Wikimedia Commons.

dresses related to the iPhone. The jury awarded Apple $1.049 billion in damages and Samsung zero damages in its counter suit. Samsung appealed for a new trial. Apple had added a further amount of interest and damages totalling $707 million. That was lowered in a separate trial to $290 million.

In the second trial filed by Apple in February 2012, it was claiming $2 billion in damages. The trial began in early April and decision was delivered on May 2, 2014 and Samsung was instructed to pay US$119.6 million to Apple for smartphone patent violations. Samsung appealed the jury verdict to a three-judge panel of the United States Court of Appeals for the Federal Circuit in 2015, and won in February 2016, with the panel nullifying the jury verdic.t. Apple requested an en banc hearing from the full Federal Circuit, which ruled in favour of Apple by an 8-3 decision, restoring the $120 million award, in October 2016.

## The App Revolution

Another contribution to the popularity of the smartphone was its feature to download micro-applications with a specific functionality (eg games, ringtones, advanced calculators and calendars). Next to the build-in games (eg Solitaire and Brick on the iPhone) and built-in productivity tools (eg Calculator, Clock, Notes, Calendar), it became possible to download 'application software programs' (in short 'apps').

> As a result of failure of the Wireless Application Protocol, the operating systems of the mobile phones had to open up to third party development of apps. Unlike the webpages on the internet that were by that time loaded with text (ie Web 1.0), images, and other types of media, the (small) screen of the mobile phone needed new designs. Designs that were depending on the mobile platforms the handsets were designed for, and mobile software developers had to work with different programming environments, different tools, and different programming languages. In addition, the fierce competition in the mobile market in the 80s and 90s meant that mobile manufacturers sensibly guarded their secrets and didn't leave their platforms open for development. So, the eternal question of software operating systems —proprietary of open-source— became again an issue; now it was Apple versus Google.[633]

---

[633] *Open standards:* The introduction of the IBM Personal Computer in 1981, running PC-DOS operating system made by Microsoft, and later distributed as MS-DOS had been part of a similar situation. The open architecture of IBM-PC and release of its specifications, brought on an explosion of additional hardware made by third parties as well as clones of

Apple choose to control strictly the development of the micro-applications that could run on their iOS and iPhone, the Android consortium opened up to the app developers.

For both the Apple iOS and the Android OS, software developers — many already designing the web applications of the Internet for desktop applications— had to (re) design their smartphone apps. They were faced with a large variety of screen sizes, resolutions, and interfaces. And their applications became a large collection of apps derived from existing web-applications and of new functionality. In their totality the fast development of apps adapted and designed by a growing range of app developers, created the new functionality that popularized the smartphone. Such as the following examples (Figure 466, Figure 467).[634]

*Social Media Apps:* The line between mobile and web became blurred as mobile apps used existing social networks to create native communities and promote discovery, and web-based social networks took advantage of the mobile features and accessibility.

The most prominent social media websites —originating from the website Six Degrees (1997), Friendster (2001), LinkedIn (2003), Skype (2003), Facebook (2004), YouTube (2015), Yahoo 360 Degrees (2006), Twitter (2006) to Google+ (2011), Snapchat (2011)— were made accessible on the smartphone platforms. Each with millions of users, generating massive network traffic.

*Communication Apps:* Next to the text messaging features (eg SMS), voice messages developed as recorded messages, transmitted to the recipient's

**Figure 466: Icons given mobile access to the most popular Social Media apps (2020).**
Source: Cyberbullying Research Center.

---

the PC. And it contributed to the fall of the micro-computer industry that introduced their versions of the PC with their own proprietary operating system systems.
[634] This is just a limited overview to illustrate the development of the mobile telephone apps.

communications device (eg the pagers with voice capability). With the rise of the Internet infrastructure came the Voice Over Internet Protocol (VoIP). With it came the Instant Messaging (IM), real time text transmission aka as 'chatting' became the next step (Figure 467).

In the domain of communication apps, *Whatsapp* was the mobile phone app that combined instant messaging and voice over IP, added voice and video calls, creating a messaging service. Released exclusively for the iPhone as WhatsApp 2.0 in 2009, by early 2011, WhatsApp was one of the top 20 apps in Apple's US App Store, and by February 2013, WhatsApp had about 200 million active users. In that year another instant messaging service called *Telegram* started, offering end-to-end encrypted voice and video calls allowing 'secret chats.' In 2014, Telegram announced that it had reached 35 million monthly users and 15 million daily active users. The videotelephony and online chat service *Zoom*, capable of video conferencing, was launched in 2013 for use on mobile phones. At the end of the year in had over 3 million users.

*Gaming Apps:* Traditionally gaming was the driving forces for web-development. Games ventured from board games, Arcade games and PC-based games, to Web-games and to Mobile games.

Nokia introduced the game *Snake* in 1997 on the Nokia 6110. It started off the development of over 420 Snake-like games on iOS. Snake was a milestone moment for the mobile gaming industry, and much of what we see today can be linked back to the dawn of the cellular serpent navigating the small, black-and-white screen. The

**Figure 467: Examples of Social and Communication Mobile Apps.**
Screenshot of Friendster (left), Facebook (middle left), Whatsapp (middle right), and Zoom (right).
Source: Wikimedia Commons.

popular arcade video game of the eating icon *Pac-Man*, was made available on the iOS smartphone's screen in the mid-2000s. Also the game *Mario Bros*, figuring Mario, grew from an arcade game to a smartphone app as well as the classic game of *Pokemon* that was released as a mobile phone app. The Finish video game developer Rovio Entertainment, had developed in 2003 a puzzle game where a bird is flung at pigs using a slingshot. In 2009 the game *Angry Birds* was released for iOS. It reached No. 1 spot in the Apple App Store paid apps chart after six months, and remained charted for months after. In May 2012, Rovio announced that its game series Angry Birds had reached its one billionth download.

This is just a limited overview of the many apps that were developed for the smartphone. It created an App Ecosystem with a new and vibrant service industry of software developers.

## The Apple App Store

After its success with selling music for the iPod through the online website called iTunes, and the iTunes Music Store[635] in July 2008, Apple opened the App Store, a website from where the iPhone users (and the users of the iPod and iPad) could download micro-apps for Apple's iOS and iPad OS. The introduction of Apple's App Store for the iPhone and iPod Touch popularized manufacturer-hosted online distribution for third-

**Figure 468: Examples of Gaming Mobile Apps.**
Screenshot of Pac-Man (left), Mario (middle left), Angry Birds (middle right) and Pokemon (right).

Source: Wikimedia Commons

---

[635] The music store launched in 2003 but iTunes started life in 2001, when the first-generation iPod MP3 player —which stored 1,000 tunes— transformed the world of digital music. By April 2004, when the iTunes store arrived in the UK, Apple was able to report about 85 million songs had been legally acquired.

party applications focused on a single platform. The App Store became rapidly a billion-dollar enterprise, supporting the creativity and works of app developers worldwide. And it transformed the music, film and video industries.

From the originally 500 apps available, by June 2009, their number had grown into 50,000 apps, and in June 2010 it reached 225,000 apps. By then there had been 500 million downloads. In a decade it grew to 2.2 million apps by January 2017 with 130 billion total downloads. Apple pocketed in that year a revenue from the app store of $35,5 billion. Except the top-200 bestselling apps that grossed $82,500 a day, the average developer pocketed some $ 3,500 a day.

> The App Store made that any developer with a mobile business idea could create and share his app without any problem to reach its target audience, which was not possible before and unbelievable competing with the biggest developers. By 2012 there were over 200,000 developers publishing their apps. Most available for free, some had a free basic functionally and a pay upgrade version, others had a low-price tag for which Apple took a 30% commission.

App Store was received with the arms wide open from iOS users, and only one month after being launched, more than 60 million downloads had been made. Two months after its launch, 3.000 apps were available and 100 million downloads had been made already.

> The year 2009 for the App Store was a year of growth for the store, and thus of learning to improve its store. In April, 1.000.000.000 apps downloads were reached with a total of 35.000 apps. This data looks even stronger if you compare it with Android Market, who only had 2.300 published apps back then.

*The Android App Stores*

The Android Market launched in 2008 as a way for users to download apps and games for the new Android operating system. It started with a dozen apps, the store added support for paid apps in 2009 in the US and UK, and expanded to more international markets in 2010. Starting with 2,300 apps in March 2009, it grew to 250,000 available in July 2011. The number of total downloads reached 6 billion.

In 2012 the Google Play store came on scene. Google Play originated from three distinct categories: Android Market, Google Music (2011) and Google eBookstore (2010). By June 2012 some 600,000 apps were available. By 2017, Google Play featured more than 3.5 million Android applications. These were the App Stores that dominated the Android market. But there were also other companies that exploited the mobile apps. Such as the native platforms of the Amazon Appstore (2011), Blackberry World Appstore (2009), the Nokia Store (2009) and the Microsoft Store (2012).

## *The Mobile Telecom Industry*

The technological development trajectory from the mobile feature phone into the smartphone took place in the industrial context of the late twentieth century (Figure 455). A context where companies did spend massive resources on R&D-activities that resulted in an explosion of patenting, operated in a highly dynamic business environment. An environment with corporate developments of mergers and acquisitions. And an environment regulated by governmental institutions.

### **Research and Development**

The inventions (ie basic innovations) in the technological development trajectories —from vacuum tube to transistor, from integrated circuit to micro-computer—were more and more the result of teamwork. Teamwork in which different technical disciplines were working together in Research & Development project teams. Projects that became increasingly organized when the process of innovation itself became organized (ie the management of R&D).

The time of the lonely inventor working in a workshop in the basement (Alexander Graham Bell) or attic (Guillermo Marcon) of their parent's house, was gone. After early experiment with centrally organizing the creation of novelty (eg Thomas Edison in his invention factory in Menlo Park), and companies creating their Design Rooms (eg Henry Ford), large companies brought inventive people with a different professional backgrounds and expertise together in teams. Teams managed by project leaders, manning the R&D laboratories, working on specific task (eg TI's CAL-TECH project focusing on a miniature calculator). And with the rise of the large electric conglomerates (eg General Electric), came the rise of the centralised Research Laboratories (eg GE's Bell Laboratories) executing mostly their applied research (R) and engineering development (D). Often located close to academic institutions (eg MIT, Stanford) and closely cooperating with them. With the growing international markets many companies decentralized and specialized their R&D facilities. Universities were massively funded through military projects (eg guiding ballistic missiles). Or the government created Institutes with a mission (eg the Space R&D in NASA).

In the second half of the twentieth century, organisational structures were increasingly linked to specific application areas (eg the integration of electronic circuitry at Texas Instruments). R&D in large corporations was thus described as being more suitable for incremental R&D, whereas venture capital financed start-ups (eg INTEL) as being suitable for fast-growing and breakthrough R&D activities. Much of the development of the mobile phone took place in the late twentieth century, where R&D was recognized as being essential for the growth of telecom companies. Companies that reserved a large part of their revenues to finance that R&D (Figure 469).

*Ericsson R&D:* The major Swedish supplier for fixed telephone systems and selling them to other companies (OEM's), by 1990 combined their mobile telephone activity in a joint venture Ericsson GE Mobile Communication. Next to the research facility in Lund (Sweden), they opened a R&D lab in Raleigh, North Carlina (US). Ericsson's new strategy to be a major player in mobile telecommunication in the 1990s was both to dramatically increase in-house R&D, refocus activities, and also develop alliances to gain different types of competencies or access

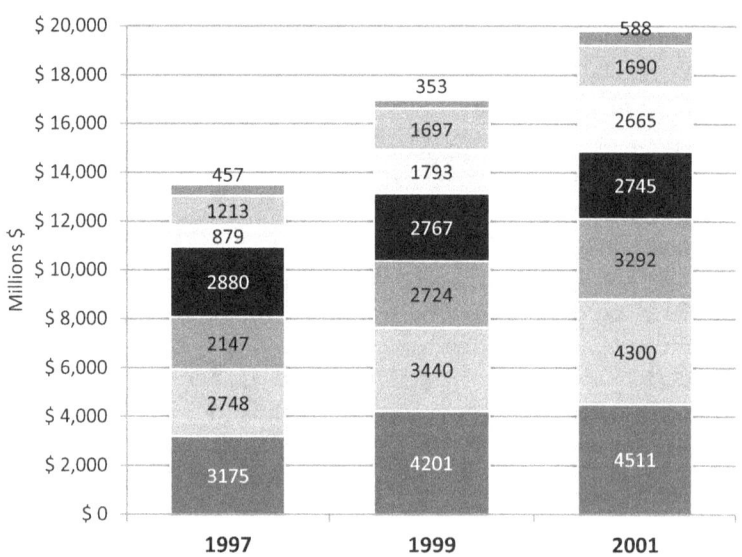

**Figure 469: R&D Expenditures for major Telecom companies (1967-2001).**

Source: Based on data from OECD Communications Outlook (2003), table 3.10.

to marketing. Spending between 15% (1997) -20% (2001) of the total revenues on R&D. An effort that resulted in a dramatic rise in patents granted from 181 (1997) to 775 (2001)

> Intellectual property rights came at the centre of the technologies that drove Ericsson into the 21st century. Ericsson was granted 33,000 patents and became the number-one holder of GSM/GPRS/EDGE, WCDMA/HSPA, and LTE essential patents.

*Nokia R&D:* By 1990 the former Pulp, Paper, Rubber, Electric Power and Bell System company had decided —at the end of the Cold War and the collapse of the Soviet Union— to make mobile telephony its core business. After the analogue radiotelephone business in the mid-1980s, Nokia first established its dominance in the business-to-business telecom equipment market in the early 1990s, and then expanded its reach to tap consumer markets worldwide in the latter half of the 1990s By the end of the 1990s it had transformed into a leading mobile telephone company.

An important part of this growth was their R&D strategy. First, Nokia increased its R&D expenditure dramatically. While in 1991, R&D spending as percentage of total sales represented approx. 5.5%, this share almost doubled to 9,7% (1996). Most spending was concentrated in the Mobile Phones and Networks business subsidiaries, with most in-house technological developments focussing on chipset and software platforms. In addition, Nokia adopted an adequate intellectual property rights protection (IPR), by registering patents, design, trademark and copyrights. Nokia has started to build its IPR portfolio since the early 1990s; from 47 (1997) to 291 (2001). By the end of 2006, it owned over 11,000 patent families, extending across all major mobile telecommunications standards and data applications, as such having one of the broadest patent portfolios in the mobile equipment industry.

From the early 1990s Nokia internationalised its R&D function, by setting up research centres abroad. By 1998, more than half of the company's R&D was conducted outside of Finland (though the main Nokia Research Centre remained in Finland). At the end of 2006, Nokia had eleven R&D centres in several foreign countries and employed approximately 21,000 people in their R&D-labs.

*Motorola R&D:* As a former supplier of military mobile communication devices, the Galvin Company (in 1947 renamed Motorola Inc.) was used to spend money on R&D. First in the solid-state technologies (ie transistors) later in IC's(ie microprocessors), defense electronic and mobile telephony (ie DynaTac), communication systems(ie Paging system, GPRS/GSM) and mobile phones (ie MicroTac).The result was an astronomical growth into a number one position (1998) in the mobile telecommunications market supported by a global network of R&D laboratories; from America (eg Chicago) to Israel, from Europe .... to India. At the end of the century, the company spend and increasing part of its revenues on R&D; from 9,2% (1997) to 14,3% (2001). The results were patented; from 1.058 (1997) to 778 (2001).

These are just a few examples of the massive funding of R&D-activities (Figure 469) that spiralled the mobile phone into a meteoric growth trajectory and brought their companies to the top of mobile phone markets.

## Mobile Phone Patents

All those R&D expenditures, spiralling their companies into the mobile phone business, resulted in increasing patent activity. Not only in America and Europe but also in the Far-East where Japanese and Korean companies had entered the market (Figure 470). In their totality the Mobile Phone

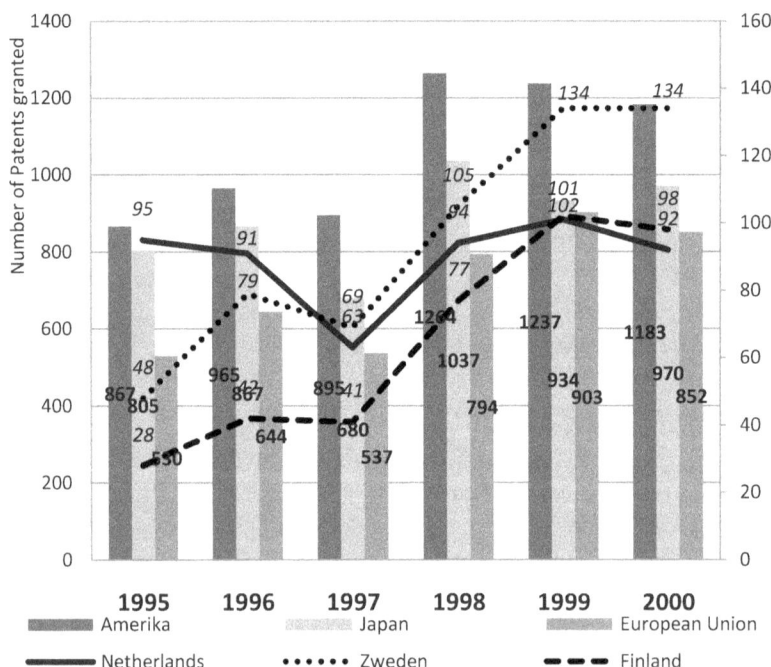

**Figure 470: Patent families granted by Country of Invention (1995-2000).**
Electric Communication technique patents granted by the USPTO filed for the European Patent Office (EPO) and Japan Patent Office (JP)
Source: OECD Communications Outlook 2003), Table 3.12

companies had seen a rise of granted USPTO patents from 6,392 (1997) to 11.526 (2001). The rise in patenting was notable for regions/continents like North America, Europe, Japan (Figure 470, bars), and for European countries (Figure 470, lines). With the development of patent positions came the conflicts of infringement on those patents by others. Due to the increasing complexity of the technology and functionality of the mobile phone, designer could hardly avoid infringing on other intellectual properties.[636]

The result were the patent disputes, with patent infringement accusations being fired regularly at companies like Apple, Samsung, Google, Research in Motion, Microsoft, Nokia, and Motorola. Many were settled out of court, other were subject to court rule. With the rise of the 'smartphone' it started the Smartphone Patent Wars.

> In October, 2009, Nokia sued Apple, claiming Apple used 10 of Nokia's patents related to Wi-Fi, phone calls and other Nokia cell phone features. Then Apple sued Nokia for copying the iPhone's "pattern and colour abstraction in a graphical user interface," phone connections to computers, touch screen menus, teleconferencing and power conservation. In June 2011, Apple agreed to pay an undisclosed sum as part of a settlement agreement and royalties for the use of parts of Nokia's patents.

## Merger and Acquisition in the Mobile Telecom Industry

The history of the mobile telephony industry is a history of mergers and acquisitions of companies. Depending of changing corporate strategies, divisions were split of and sold, or companies with complementary expertise were acquired. To give an impression of these dynamic development will paint a picture in large brushstrokes of the M&A activities in mobile telephony of the leading companies in the period 1980-2000.

*Ericsson M&A:* Ericsson Radio System (ERA) was created in 1982. In order to access lacking competencies in radio communication, small specialized firms were purchased. The first, *Magnetic*, specialized military radio and television equipment, was bought up in 1983. The second firm, *Radiosystem* —created in 1978 by three engineers from Magnetic to develop specific components of radio-base stations for the NMT standard— was bought up by Ericsson in 1988. In the meantime, analogue standards were made operational, research had started on digital technologies in order to improve the quality of the system. This implied new collaborations with universities. Ericsson entered on all

---

[636] Some 250,000 patents are commonly used in mobile technology. In the total patent universe, mobile technology patents account for one in six of all active patents.

three major standards, in the US, Japan and Europe (GSM). From 1990, the policy of Ericsson became firmly oriented towards high investments in research and development and a focus on mobile telecommunication systems. In 1991, the company strengthened its position in Europe's cellular phone market when it acquired 50 percent of *Orbitel Mobile Communications*, the manufacturing subsidiary of British concern Racal Telecom PLC. The following year it purchased a majority interest in *Fuba Telekom*, a German telecommunications concern.

*Nokia M&A:* In 1979 Nokia established a joint venture with the Swedish manufacturer of colour TVs *Salora Ltd* under the name Mobira Oy and by 1984 it purchased Salora Ltd. In 1990's, Nokia's top leadership decided to focus solely on the telecommunications market, and as a result, the company's data, power, television, tire, and cable units were sold off in the first few years of the decade. The British Company *Technophone*, active in the manufacturing of mobile pocket phones since 1984, was acquired in 1991 for £ 34 million by the Mobile Phones division, and propelled Nokia to become the world's second largest mobile manufacturer behind Motorola. In 1997 the Nokia Networks Division acquired the US-based Internet switches manufacturer Ipsilon Networks Inc for $120 Million. From 1999-2001 some fourteen companies were bought by the Nokia Telecom Divisions and it divested its Television, Radio Tuner, Loudspeaker and Monitor activities.

In 1998 Nokia partnered in Symbian Ltd, the software development and licensing consortium company, known for the Symbian operating system (OS). Together with Psion, Ericsson, Motorola, and Sony, the consortium exploited the convergence between personal digital assistants (PDAs) and mobile phones, trying to counter Microsoft's entry in the mobile devices market. Ten years later, in 2008, Nokia acquired all the shares for $264 million with the goal of developing an open-source and royalty-free mobile platform.

## Deregulation of the Telecom Industry

The monopolistic behaviour of the large telecom network providers had worried the regulating institutions in Europe and America in view of the rapid growth of the new (mobile) telecommunication technologies. As part of the worldwide trend to liberalization,[637] in the US, the monopoly of the AT&T was broken and the so-called Baby Bells were created. In Europe, the Nordic countries were opening up competition. It was followed by a new wave of liberalization all over Europe in the late 1990s.

---

[637] Governments all over the world, led by the United States, were opening up their telecom markets to competition. Public policy was inviting new entrants to jump in.

In the European Union, as part of the harmonization of the European market, common rules were needed to create a unified EU-wide telecoms market. The purpose was to set the rules for open access to the networks of the old monopolies so that the new entrants could offer services in competition —on equal terms— with the ex-monopolies. This policy was implemented in 1998.

In the US, the Telecommunication Act of 1996 —after decades of fierce regulation by the Federal Communications Commission (FCC) created after the Communications Act of 1934— had opened the telecom market under the motto 'let anyone enter any communication business.' The legislation's primary goal was deregulation of the converging broadcasting and telecommunications markets by open up local markets to competition by removing regulatory barriers to entry. And the effect was notable on state-level where the so-called Competitive Local Exchange carriers (CLECs) experienced a tremendous boom after the Act was passed. From 1996 to 2000 the number of CLECs rose from 30 to 711, and their revenue increased from less than $5 billion to $43 billion over the same period.

The great benefit of competition has been a rapid rollout of new technologies. Companies in the long-haul fiber-optic, cable-TV, satellite, local-telephone, wireless and other sectors of the industry did undertake massive capital expenditures to develop and upgrade networks. As a result, cable companies (through cable modems) and telephone companies (through digital subscriber lines, or DSL) both offered high-speed digital access to their subscribers in most areas of the country. With dramatic changes in basic technology, new products, and the liberal regulatory environment, it is not surprising that during the period from 1996 to 2002 the telecommunications industry[638] experienced significant volatility as existing firms and new ones began massively building networks over land, undersea and in the air.

## The Telecom Crash of 2000

At the end of the twentieth century, with the rise of the Internet, the dynamics of the telecom business had created an economic Telecom Bubble. The astronomic growth of the mobile business's traffic as a result from the explosion of mobile phone ownership and its increasing use due to the new facilities offered by the new network protocols, paralleled the Internet growth mania. Traditional players like AT&T in US, NTT in Japan and former national carriers in Europe —eg, France Telecom, Deutsche

---

[638] The telecommunication industry is the industry that enables tele-communication: the xchange of information over distance by electronic means.

Telecom, British Telecom— tried to become global, either by constructing new networks or via synergies with smaller telecoms.

Europeans in particular, assumed that the mythical demand for wired services would be followed by demand for mobile capacity (the 3G technology). Hundreds of competitors sprang up, borrowing heavily to finance their growth as the governments sold the new frequency bands in the air space in spectacular telecom auctions that raised billions of dollars.[639] Many of those competitors did not survive and filed bankruptcy. During the race to borrow and grow, the surviving telecommunications companies overbuilt their capacity.

The Telecom Bubble began deflating in 2000 and 2001, as the industry was straining under the weight of excess capacity and enormous debt. Stock prices tumbled up to 95%. A number of debt-laden telecom companies defaulted or went bankrupt and filed for court protection. WorldCom, XO Communications, Global Crossing LtD., and British NTL, were among the high-profile telecommunications business that collapsed.

> WorldCom was founded in 1983 as Long Distance Discount Service, Inc. After completing several mergers and acquisitions in the 1988-1994 period,[640] it became one of the largest long-distance providers in the US. WorldCom was forced to file for Chapter 11 bankruptcy protection in 2002 in the wake of a well-publicized accounting scandal.

For the communications industry, the year 2000 was a bad dream. Far and wide, established communications companies and aspiring start-ups were forced to alter course to survive in a cutthroat consumer and business market. If times were tough for some of the nation's largest communications carriers, the last half of the year was even worse for smaller entrants that had arisen during the bubble.

> In January of 2001 it was possible for a telecom firm to raise billions of dollars in the bond and debt markets and yet by April it was not possible to raise 5 cents those markets for a telecom business. The seemingly endless supply of funding had come to a screeching halt. Once investors noticed that little chance existed for some CLECs to raise the money they would need to continue

---

[639] Spectrum auctions for 3G in the United Kingdom in April 2000, raised £22.5 billion. In Germany, in August 2000, the auctions raised £30 billion. A 3G spectrum auction in the United States in 1999 had to be re-run when the winners defaulted on their bids of $4 billion. The re-auction netted 10% of the original sales prices.
[640] The $14 billion dollar acquisition of MFS by WorldCom kickstarted the great CLEC explosion.

expanding, an avalanche of selling began. Wall Street, venture capitalists and individual investors became spooked by a sudden stock market downturn (Figure 471), and began to look more favourably upon established, profitable companies.

The Telecom Crisis on the stock market was a hallmark in the development of the mobile phone as the whole mobile telecom industry was deeply affected. But not only the well-known mobile phone manufacturers suffered, and some even failed, also other segments[641] of the telecom industry were affected.

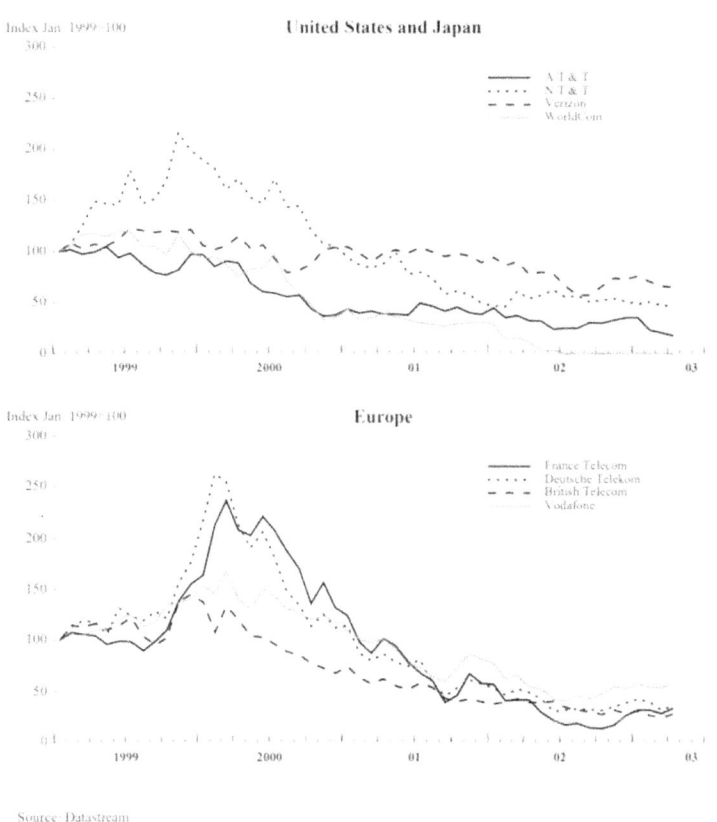

**Figure 471: Share price indices of selected Telecommunications Operators.**
Source: IV. AFTER THE TELECOMMUNICATIONS BUBBLE – OECD. https://www.oecd.org › eco › outlook.

---

[641] This segment of the mobile telecom industry manufactures fiber-optic networks, routers and other telecom equipment, satellites, wireless network systems.

When the speculative telecom bubble of the late 1990s reached its pinnacle late in the year 2000, the multinational telecommunications and data-networking equipment company Nortel was to become one of the most spectacular casualties. Nortel's market capitalization fell from C$398 billion in September 2000 to less than C$5 billion in August 2002, as Nortel's stock price plunged from C$124 to C$0.47. When Nortel's stock crashed, it took with it a wide swath of Canadian investors and pension funds and left 60,000 Nortel employees unemployed. Struggling the next decade, On January 14$^{th}$, 2009, Nortel filed for protection from creditors, that heralded the downfall of the company. In 2011, Sony, Apple, and Microsoft made a move challenging Google, by purchasing more than 6,000 Nortel patents. By 2014, A consortium including EMC, Ericsson, Microsoft, Research In Motion, and Sony bought the whole company for some $4.5 billion.

## The Mobile Phone Manufacturers

Originally the early transportable mobile phones and car phones (Figure 422) were considered as a variation of the fixed land line phones. So their manufacturers —such as the Scandinavian telephone manufacturer Ericsson— started manufacturing the mobile devices. Soon they were joined by other communication companies such as the US company Motorola, and start-ups like Finland's company Nokia. With the market for 2G mobile telephony exploding, other companies jumped on the bandwagon, such as the large Japanese companies Sony, Matsushita Communication (with the Panasonic brand) and Mitsubishi Melco (from 1999 to 2008). And the equipment companies[642] jumped on the bandwagon, adding their products to the massive mobile phone industry.

The pace of technological development of both the network technology and the phone technology was enormous and driven by the nearly insatiable market demand as the general public embraced the mobile phone. The result was that also the mobile phone industry and individual manufacturers saw a draconic growth. But many of those companies could not keep up with the dynamics and failed for a multitude of reasons.

*The Rise and Fall of Ericsson's mobile phone business*

After in 1981 the Nordic Mobile Telephony Network (NMT) was launched in Scandinavia, in 1987 Ericsson Radio Systems Division introduced the first mobile phone with the Model Hotline 900 Pocket. In 1989 it started a joint venture with General Electric: the Ericsson GE

---

[642] Such as the mobile network equipment suppliers like Cisco (1984), Northern Telecom Limited (1976) later Nortel Networks (1998).

Mobile Communication company that lasted till 1998. It was the start of a range of mobile phones that followed the network developments from the first generation NMT and AMPS, to the second-generation GSM. In 1994 mobile telephony made up 85% of the activities in the Ericsson business unit for Radio Systems and became renamed as the Ericsson Mobile Communications AB (ECS). Growth and volumes in ECS increased rapidly and during the early years all focus was on quickly ramping up production, which was met with success. During 1998 the company ran into supplier problems in the completion of their next flagship phone model, the Model T28. And their next mobile phone, the Model R380 ran into the Telecom Crisis. Notwithstanding, in terms of volume, 1999 was a record year. A total of 31 million mobile phones were sold, compared with 24 million in 1998.

> By 2000, the Telecom Crisis erupted after the burst of the Telecom Bubble that paralleled the DotCom Bubble of internet companies that had been developing in the preceding years. In a race for space in the radio spectrum bands reserved for 3G mobile telephony, with the government auctioning off the frequencies, many telecom operators had overstretched their financial capabilities. Many companies ran into problems, also Ericsson Mobile Communications was losing millions of SEK.

With a total of 104,000 employees worldwide, Ericsson laid off over 10 percent of its entire workforce. After 2001 further countermeasures to cut down on the losses resulted in outsourcing of production. Such as the joint venture with the Japanese Sony Corporation: *Sony-Ericsson Mobile Communications*. Sony was a marginal player in the mobile phone market with a market share of less than one percent, but Ericsson was at third position and was successful in mobile phone manufacturing industry. The venture introduced over the next decade a range of new handsets (eg digital camera phones), many produced by Sony Mobile India. In 2004, sales had reached 42 million phones and the profit for the year was SEK 3 billion. This healthy rising trend continued until 2007 when 103 million phones were sold and the profit was about SEK 15 billion. By 2008 the overall shipments of feature phone handsets dropped, the market introduction of the new smartphone Xperia-X1 failed. In a cost saving program, from a workforce of some 12,000 employees in 2008 they reduced it to 5,000 employees. In addition, they closed their R&D-activities all over the world.

In 2012 Sony acquired all the shares in the venture, and merged the operations in Japan under the name Sony Mobile Communications. The once Nr 3 in the mobile telephone market had failed to survive in the technology dominated mobile battleground.

*The Rise and Fall of Motorola*

Motorola started as a successful manufacturer of mobile car radios in the 1930s, moved into two-way radio communication mobile sets in the 1940s. After its entry into the microprocessor business in the 1970s that could hardly compete with INTEl, it went into other markets. Motorola was more successful in the market for embedded microprocessors, which became ubiquitous in automotive control units, industrial control systems, and such common items as kitchen appliances, pagers electronic game systems, routers, laser printers, and handheld personal digital assistants (PDAs). In this markets Motorola became a leading manufacturer. With the DynaTAC mobile phone (Figure 425), Motorola ventured into mobile phones in the 70's and by the 80's.

Motorola's mobile phone business became quite successful as it was at the head of the network developments. In 1991 Motorola released the world's first GSM cellular handset (Motorola 3200). In 1993 it manufactured the world's first mobile phone using the GSM 1800MHz band (Motorola M300), followed by the first tri-band mobile phone in 1999 (Motorola L7089). Together BT Cellnet and Motorola launched the world's first GPRS mobile data service in 2000, using the Motorola Timeport T-260. As a result, Motorola was the largest mobile phone manufacturer in the world before Nokia knocked them off their top spot in 1998.

> Motorola's problem was that it was a hardware technology company, but from the mid-2000s it was software driving the mobile phone business. It could not keep up with the development of the smartphones, failed in the satellite business supporting its wireless telephone system (aka Iridium) writing off 2.5 billion dollars on that investment, and had to refocus its business.

At its peak, Motorola had 150,000 employees. After Iridium's bankruptcy, Motorola started a string of worldwide layoffs. By the end of 2004, Motorola cut down its workforce to around 68,000 employees. After failing to deliver on its side project, the company wanted to focus on its core operations and launched Motorola Razr series. The 2G Razr V3, launched in 2004, sold around 120 million units worldwide, becoming the best-selling clamshell phone. But the sign was on the wall, as from 2007 to 2009, it was reported that Motorola lost 4.3 billion dollars.[643]

---

[643] Motorola was shielded from the customers by carrier operators, and could not catch-up speed with 3G as its partner carriers never felt the need of upgrading their eco-system.

## The Rise and Fall of Nokia

The Finish multinational Nokia Corp., that started in the 1860s as a pulp factory in the forestry business, moving into the rubber business (ie the Finish Rubber Works Ltd) and the electric cable industry in the 1920s (ie the Finnish Cable Works Ltd). The spirit of merging was in Nokia's DNA and in the following years they acquired a series of companies, including several of the main television manufacturers from Finland, Sweden, and Germany, turning into the third-largest TV maker company. Similar story with radio communicators, that they even manufactured for the army along with other products. In 1979 Nokia merged with Salora, under the name of Mobira Oy. Mobira developed analogue mobile phones for the 1G Nordic Mobile Telephony (NMT) standard (Figure 425). In 1984 Mobira was fully integrated in Nokia.

Nokia's transition to a primary focus on telecommunications began in the 1990s after it divested itself of all other businesses (eg Nokian Tyres, Nokian Footwear). In 1994, Nokia became the first manufacturer to launch a series of hand-held mobile telephones (the Nokia 2100 family) that could operate on any digital network around the world, making it a global player. By 1997, Nokia was supplying its phones to over 30 countries, more than any other hand-held manufacturer. By 1998, Nokia overtook Motorola to become the world's largest mobile phone brand. Between 1996 and 2001, Nokia's revenue increased from €6.5 billion to €31 billion. Nokia entered the 21st century still as the undisputed king of cell phones, having surpassed the 100th million manufactured phones in 1998. In that year alone, they had a sales revenue of $20 billion making $2.6 billion profit. By 2000 it employed over 55,000 people around 140 countries and had a market share of 30% in the mobile phone market, almost twice as large as its nearest competitor, Motorola.

In 2002 the company was restructured with four main divisions, ie Nokia Mobile Phones, Nokia Networks, Nokia Venture Organization, and Nokia Research Centre (NRC). In this year Nokia developed and marketed what can be labelled as Smartphones.[644] Nokia actually started to create the market that became disastrous for some (Motorola, RIM, and Nokia itself) and for the time being a goldmine for others (Apple and Samsung). Till 2007 Nokia dominated the smartphone market, mainly due to the Symbian operating system, and with the introduction of the N95 Nokia's market share jumped in two months from 33% to 36%.

---

[644] A core characteristic of smartphones is the convergence of telephony with computer technology in one handset.

Nokia's explosive growth in the telecom business was accompanied by a massive number of acquisitions; from multimedia companies to the digital mapping company Navteq. But on the market of the new smartphone devices, NOKIA lost market share. Still being a major player in the low-priced third world mobile phone market,[645] the hardware company[646] Nokia failed in the high-priced smartphone market that had become a war between operating systems (OS) with the rise of Apple's iOS and Samsung's Android OS. As upgrading efforts of the Symbian OS into the MeeGo OS failed, Nokia turned to Microsoft. The partnership with Microsoft and its Windows Phone OS, led to the introduction of the Lumia Windows phones. But they ultimo failed in the market. After massive losses and layoffs, on April 25th, 2014, Nokia sold its mobile phone business to Microsoft for $7 billion. In July 2014, Microsoft dismissed 12,500 former Nokia employees.[647]

Microsoft, however failed to enter the Mobil phone business in the competition with Android-based smartphones and Apple's iPhones. in 2015 Microsoft did write off $7.6 billion as a consequence of the Nokia acquisition and laid off 7,800 employees and a roughly $800 million restructuring charge, writing down the vast majority of the phone business purchase price. In 2016, Microsoft Mobile announced the sale of its feature phone business to the Finnish HMD Global and the Taiwanese FIH Mobile for $350 million.

*The Rise and Fall of RIM's Blackberry*

In Canada, BlackBerry began in 1984 under the name Research In Motion (RIM). Its specialty was building communications devices like pagers and modems. The company introduced its first line of mobile phones in 2000, the RIM-957 and its sibling the RIM-950, which came with internet functionality and push notifications for email. The introduction of the BlackBerry set the stage for future enterprise-oriented products from the company. The 2000s saw BlackBerry blossom into America's business community as preferred smartphone brand thanks to its focus on devices with workplace functionality and high-level security.

In 2002 the mobile phone BlackBerry 5810 —with a headphone-set— was introduced, followed by the BlackBerry 7230 (2003) with internet access. RIM rode high for the next six years. From 2005 to 2009, the

---

[645] Nokia's one billionth phone sold was a Nokia 1100 purchased in Nigeria in 2005.
[646] Nokia's core competence was in hardware design and engineering.
[647] In 2016 former executives created HDM Global and bought from Microsoft the Mobile Phone division back. From virtually zero sales in 2015-16 to 1.5 million Nokia smartphones sold in 2017, HMD Global revived Nokia. In Q3 of 2018, it shipped more than 4.8 million units, making it the 9th largest smartphone vendor worldwide.

BlackBerry brand dominated the business segment of the smartphone market in the US, with models on every carrier. For a businessman on the move, the iconic blackberry was like a status symbol he was getting addicted to (hence its nickname CrackBerry). For C-level executives circa 2008, carrying the BlackBerry Bold meant that you had arrived. Everything about this device was designed to feel luxurious, down to the leatherette back.

That dominant focus on the business market also contributed to its downfall when Apple's iOS and Google's Android OS fuelled the smartphone development. By 2008, the newly released Blackberry models (ie the Pearl, Curve and Bold-series) sustained unprecedented market share growth well into 2011.

> That was different for the Blackberry Storm smartphone, available at the end of 2008, that was to compete with iPhone. The Blackberry Storm sold 500,000 units in its first month and 1 million units by January 2009. However, Verizon had to replace almost all of the one million Storm smartphones sold in 2008 due to issues with the SurePress touch screen and claimed $500 million in losses.

As of December 1st, 2012, the company had 79 million BlackBerry users globally. But again the Android smartphones and the Apple iPhone (aka the Blackberry killer) exercised their powers, and caused a slowdown in BlackBerry sales. The stock market reacted with an 87% price drop in BlackBerry's stock price between 2010 and 2013. Reporting operational losses over the year 2013,[648] the company started to lay-off its employees, restructuring its top-management and looking for strategic alliances. In the year 2014 it lost $7,163 million. BlackBerry survived the darkest years in its history (2011 to 2016) because of geographic and technology peculiarity: its devices were extremely popular in the developing world.[649] But it did not turn Blackberry's downfall, and in September 2018, BlackBerry announced they would stop making phones which marks the end of the BlackBerry Smartphone Era.

*The Rise of Apple Inc.*

After its early beginnings in the Personal Computer business with the Apple II in the late 1970s, and the identity crisis in the early 1980s under John Scully with Steve Jobs leaving the company, Apple enjoyed strong growth. It created new products, including smart laser printers, Macintosh Portable, PowerBooks, the Newton PDA. When Jobs took Apple's reins

---

[648] On September 28th, 2013, media reports confirmed that BlackBerry lost US$1.049 billion during the second fiscal quarter of 2013.
[649] By February 2016, only 1.59 million (0.8%) of the 198.9 million smartphone users in the United States were running BlackBerry compared to 87.32 million (43.9%) on an iPhone.

once more in 1997, the hardware had caught up to his vision for all things digital. Next to the iPod and iPad, that vision resulted in the iPhone. All of these products moved Apple into a new business model of creating a tight ecosystem of hardware, software, and content. And it was the revolutionary, game changing iPhone that propelled Apple in the second decade of the twenty-first century. It was the combination of the iPhone and the Apple Store, that fuelled its meteoric rise.

> With the, iPod, iPhone and iPad, Apple did more than create great products, it created entirely new markets. With the iPod, it changed music listening; with the iPhone, it hijacked the phone industry; with the iPad, it resurrected tablets from the grave; and with the MacBook Air, it reshaped the laptop industry.

A rise that originated from a business model that enabled the firm to exercise unparalleled control over its multi-channel platform. A business model that relied on the integration of content (software, media, and apps) and hardware (laptops, phones, and tablets) to drive growth. A growth that was inspired by many other contributions.

> *Apple is a master at executing, but it steals ideas all of the time. Court documents show that many of Apple's design philosophies were inspired by old Sony products and other sources. The iPhone was definitely not the first touchscreen phone either, and the iPad would not exist without early tablets from Microsoft. Apple was just the first company to make touch products work in a more user-friendly way.*[650]

## The Rise of Samsung

The Korean company Samsung —started as grocery story in 1938— entered the electronics industry in 1969 with several electronics-focused divisions. Their first products were black-and-white television sets. During the 1970s the company began to export home electronics products overseas. At that time Samsung was already a major manufacturer in Korea. The late 1970s and early '80s witnessed the rapid expansion of Samsung's technology businesses. In the 1980s, Samsung began to explore the cell phone industry, and released its own mobile phone —the Samsung SC-1000— to the South Korean public in 1988. It was the start of a range of not too successful cell phones. It was not until 1995, some years after Samsung's initial cell phone launch, that it was decided Samsung needed a new business strategy for its future. As a result, the 2000s witnessed the birth of Samsung's Galaxy smartphone series, which quickly not only

---

[650] Source: Montgomerie, J., Roscoe, S.: Owning the consumer-Getting to the core of the Apple Business Model (2013). https://www.sciencedirect.com/science/article/pii/S015599821300032X

became the company's most-praised products but also were among the best-selling smartphones in the world.

During the first couple years of the iPhone, Samsung, RIM, and others released phones like the M800 "Instinct" and BlackBerry Storm, which kind of, sort of looked like the iPhone, but didn't include the full breadth of its functionality or elegance. Those products failed and the iPhone prospered. Samsung faced a choice: It could have gone the route of RIM or Nokia and retreated into more comfortable, familiar phone and software designs. Or it could have attempted to re-invent the wheel and create something completely different looking from the iPhone. Or it could choose to take the easiest route and compete with the iPhone.

Part of its success was resulting from its innovation-centred organizational culture that emphasizes employees' knowledge, skills, and abilities for innovating the company's technology products, such as smartphones and laptops. It was emphasized in Samsung's Mission Statement: *"We will devote our human resources and technology to create superior products and services, thereby contributing to a better global society."* Next to the human resources component, the technology component was mentioned and expanded in Samsung's Vision Statement: *'Inspire the world with our innovative technologies, products and design that enrich people's lives and contribute to social prosperity by creating a new future.'* And..., Samsung became a master in copying the iPhone concept.

> *So yes, Samsung definitely copied Apple, and yes, it got too excited in its quest to battle the success of the iPhone and other Apple products. But really, what choice did Samsung have? Apple has managed to mold entire industries into its image. It changed the rules of the game. To compete and prosper, every other company has been trying its best to come up with cool original ideas, but mostly, it seems like the only way to compete with Apple these days is to be just as good, but cheaper. So that's become the goal of Apple's competitors. Samsung is merely the best at it.'*[651]

## *Overview of the Early Days of the Cell Phone*

The preceding exploration into the mobile-phone-events illustrate the development trajectory of mobile phones from the cell phone and feature phone into the early smart phones.[652] A development trajectory over some

---

[651] Source: Van Camp, J.: Samsung copied Apple: Who cares? Everyone's doing it. (2012). https://www.digitaltrends.com/android/samsung-copied-apple-who-cares/
[652] We consider the *Cell Phone* as the wireless communication device with the basic communication functionality. The *Feature Phone* is the subsequent cell phone with added features extending the on board functionality (in terms of recreation, communication and information). Together they create the class of *Mobile phones*. The *Smart Phone* is the mobile

four decades that was fuelled by the IC-technologies capable to realise micro-computing systems applicable in a wide array of applications. Among them the circuitry needed for the mobile telephony.

The number of contributors to the cell phone during the Era of Mobile Telephony is too large to identify in detail. The more because much of the novelty creation was the work of R&D-teams working on specific projects in the R&D-departments of large corporations. Also the contributors working in the context of smaller companies are too numerous to identify in detail. So, we have to limit ourself to the detailing of the different development trajectories that constituted the Era of Mobile Telephony (Figure 472). A period that was characterized by two distinct phenomena.

*The Technology Push*: Fuelled by the semiconductor technology and its development of Integrated Circuits for mobile application processors

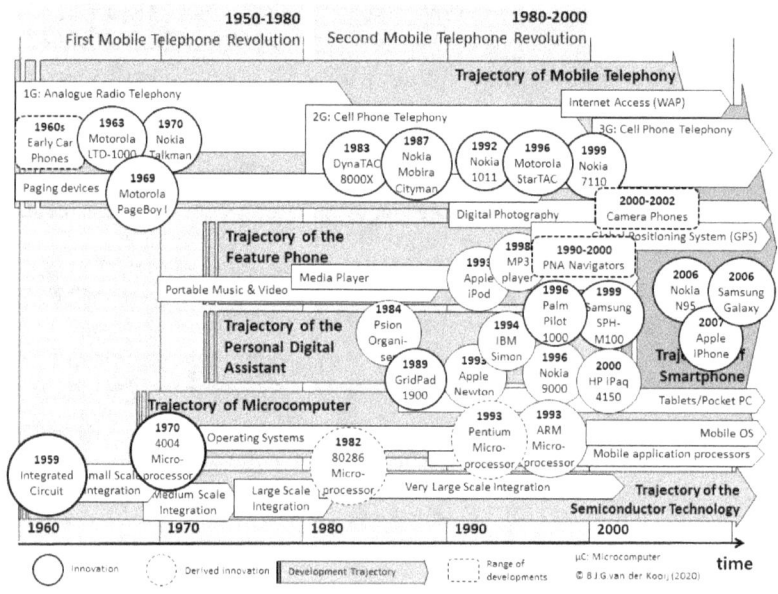

**Figure 472: Overview of the Developments in the Era of Mobile Telephony.**

The projects of development team in the Application Trajectory of the Mobile Phone in relation to the inventions of Semiconductor-Technology trajectory and the trajectory of the Micro-computer.

---

phone with extended computing capabilities. The 2012 Smartphone Patent Study defined smartphones as "hand-held computing devices that (a) have the ability to make phone calls over cellular networks and (b) can transfer data and run applications over mobile computing networks."

systems (Figure 472, bottom), combined with the application trajectory of Mobile Telephony with the different generations of networks and their corresponding cell phones (Figure 472, top), the mature 'Mobile Phone' came into existence. From the simple transportable phones emerged the hand-held feature phones adding functionality of media players and photography, SMS and internet access with the WAP-browser. The convergence with the development trajectory of the functionality of the Personal Digital Assistant laid the foundation for the next development of functionality.

In all these trajectories 'technology' was the driving factor; from the semiconductor technology creating the circuitry on the chips, to the network technology creating the equipment used for the transmission of the signals. But also the supporting technologies (eg the battery technology, touchscreen technology) contributed to its overall development.

*The Market Pull:* The handheld devices that emerged from the technology driven trajectory, fulfilled a basic human need to communicate. Communication being part of the social interaction. In this case the communication of the 'spoken word.' And then came, after the fixed phone of earlier times, this new mobile device. After the non-price sensitive early adapters (eg the user segment of the emergency professional, the businessman) had succumbed to its practicality, the masses discovered the advantages of the mobility of the handheld communication device. The more when the prices of the portable devices and the subscription fees dropped to a level affordable by the middle class. The rapid development of the mobile phone into the feature phone, and the continues release of new models, increased its attractiveness and usability. Like the ever-improving on-board photo-camera (that was actually killing the traditional camera market). And with the advancing network facilities expanding the mobility range with the roaming concept. Add to that the governmental regulation controlling the price-structures, and the worldwide mobile phone market exploded. The cell phone became a 'must-have' item.

*The Product Development:* The combination of technology push and market pull created a massive driving force for the development of the mobile phones as a new product the world had not seen before. A range of different development trajectories converged into the new development trajectory of the smartphone (Figure 472, middle). Companies owning the technology jumped on the smartphone bandwagon, telecom vendors having access to the market wanted their own versions. The result was a decade of impressive product development resulting in an evolutionary explosion of novelty. Both in design (Figure 473) as well as in functionality.

The outcome of this development was the revolutionary new version of a mobile telephone called the 'Smartphone' in the first decade of the twenty-first century. It had a huge impact on business, economy as well as society as a whole.

Figure 473: Evolution of the Design of the Mobile Telephone.

## *The Impact on Mobile Telephony*

The impact of the smartphone concept on mobile telephony was enormous. Not only in economic terms as the mobile telephone industry contributed to the economic developments of its time, but also on the user-side that became dominated by its use.

*Business Impact:* By the first decade of the twenty-first century, the mobile phone had already created an enormous industry. Originating from their countries of origin, the leading companies had spread over the world the handset production faculties as well as their R&D facilities. To serve all the telecom markets, the network providers had seen a worldwide development, starting in the developed countries and spreading out over the third world countries. Then came the disruption of the smartphone creating a new industry manufacturing devices (eg Apple, Samsung). The disruptive innovation caused the old industries that were unable to change, to fail (eg Ericsson, Motorola, Nokia, RIM). The same happened with the opportunistic new entrants, that had to leave the business after a fruitless effort (eg Microsoft, LG). Others struggled forth in an effort to copy the dominant design of the smartphone.

*Social Impact:* In 2008 some 12% of the global population owned a smartphone. In 2010, it was estimated that about 20% of the global population used smartphones, while by 2014 that number was estimated at 37%. The smartphone had an enormous impact on the Affairs of Man as it opened a new way of communication and instant access to information. The smartphone changed people's lifestyles and way of communicating from person to person. Distance and place were not a problem anymore to keep social contact between family members, friends and colleagues. That same characteristic also had a negative side in professional work situations where one had the feeling of being 'on call' all the time. For the younger part of the population the mobile phone would become an indispensable tool of their networking lifestyle (eg social networking). The smartphone enforced those effects as it, and the many apps that came with it, changed the way people communicate, do business, entertain, socially interact with others, and learn.

*Economic Impact:* On the economic side, the meteoric rise of mobile telephony had an economic impact on the western world economies. The mobile phone industry —both the manufacturers of phones and equipment, but also the service provider creating the network infrastructure— exploded, creating jobs by the thousands for both highly skilled engineers and industrial workers making the phones. The mobile economy contributed an estimated $2.4 trillion to the international economy in 2013, representing about 3.6% of global gross domestic product. This number includes more than 10 million jobs created, and contributed at least $336 billion to public funding in the United States alone. The creation of the "app economy" has also been a major boom for the economy at large. In fact, even as early as 2012, it was estimated that there were around 500,000 jobs created as a result of the app economy.

*Technological Impact:* The technological drive of the Silicon Engine had a disruptive effect outside the realm of technology itself. The touch screen and the virtual keyboard became used in a broad range of applications (eg tablet computing, public electronic displays). The Silicon Engine fuelled further technological development of the semiconductor industry. And the 'Internet of Information' became the 'Internet of Things.'

*Other Impacts.* The mobile technology revolution is unique in a way since it has achieved unparallel status as an agent of socio-economic change where landline telephones or other communication technology faltered. Especially rural areas of the developing countries (eg India, Africa) skipped the cable communications and embraced mobile telephony.

## Who invented the Mobile Phone?

In line with earlier efforts to answer this question for other developments, facing the same problems, trying to find an answer to this question is frustrated by the sheer number of contributions to the mobile phone that were made over time. When trying to address the classic question of 'who is the inventor,' one is faced with some problems of interpretation, as we already noted before in the case of the Radio Tube; from semantics to the subject, from moment in time to the consequences to the point of view (see chapter: Who invented the radio tube? for details).

In the case of the mobile telephone there is no obvious single inventor, nor a single patent that marks its invention. The mobile phone is the result of a Chain of subsequent Inventions, each with its own Cluster of Innovations (Scheme 2). In its totality it had an enormous impact created for each of the contributions (Scheme 1). The impact the invention had on the industry itself as shown by the patent wars. But also the impact in terms of business creation. In the case of the mobile telephony both are notable, and both were enormous.

When we try to reduce the enormity of developments contributing to the development of the 'Mobile Phone' (Figure 472), the pushing power of the Silicon Engine (ie both the trajectory of the semiconductor technology and the development trajectory of the microcomputer) is obvious (Figure 474, bottom). In addition, the development of the wireless infrastructure creating in a range of steps the Mobile Networks and their operators, paralleled the technological development (Figure 474, top). But there was

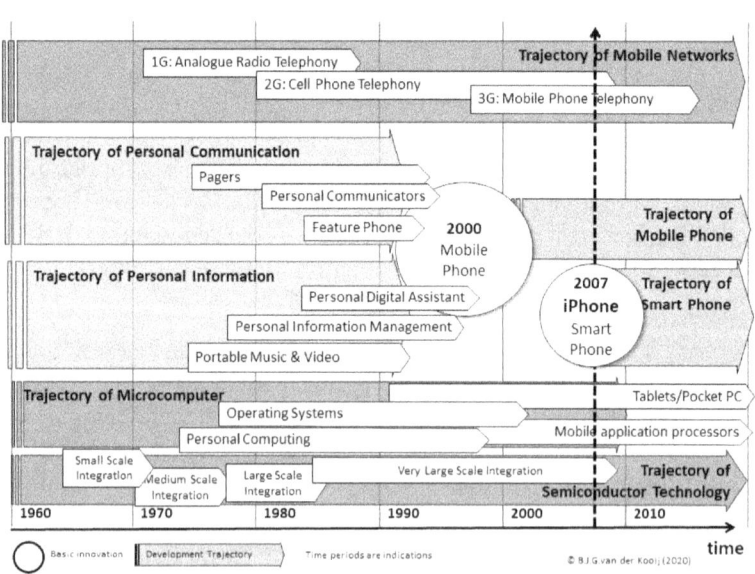

Figure 474: Overview of the Contributing trajectories to the Smartphone.

more as in related fields of Personal Communication (eg the Feature phone) and Personal Information (eg the PDAs and PIMs) a range of complementary developments took place (Figure 474, middle).

## Who invented the Smartphone?

The same question could be asked for the specific version of the mobile phone labelled as smartphone. Again, an individual inventor cannot be identified. When one considers its early conception, the IBM Simon was the early invention resulting from a R&D-effort to which many contributed (without much commercial impact though). Feature phones —such as the early Nokia 1100— were having a commercial impact but missed the computing facilities. But when one considers the computing facilities and the impact-factor (Scheme 1) the introduction of the Apple iPhone in 2007 stands out. This realisation of merging the mobile telephone concept and the computer concept certainly had a revolutionary character (Figure 475).

The iPhone was the result of two dominant trajectories merging together: the *Trajectory of Personal Communication* and the *Trajectory of Personal Information*. Each trajectory with its own contributing innovations having an impact of their own. Once introduced and gaining a massive success, the iPhone concept was followed by many derived innovations.

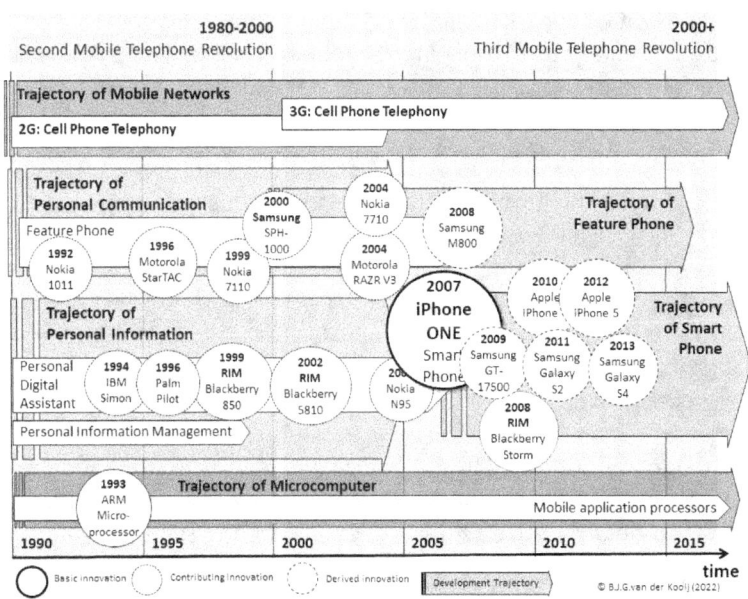

Figure 475: Overview Cluster of Innovations of the Smartphone.

## Periodization of the Mobile Phone Era

Looking at the development trajectory (Scheme 2) of the Mobile Telephony and seen in the perspective of the life technology cycle (Scheme 5), the Mobile Phone Era is marked by specific periods of events creating the basic innovations to which many thinkers and tinkerers contributed.

*Mobile Telephony 0.0* (1940-1950): The early days of wireless communication of the spoken word, saw the rise of mobile phone.[653] Originating from the wartime development of the 'walkie talkie,' it developed rapidly.

*Mobile Telephony 1.0* (1950-1980): This was the period in which the early (trans)portable mobile telephones were developed. After first applying the vacuum tube technology, followed by the transistor technology, early handsets operating on the 1G (analogue) networks were introduced for a limited market of early adaptors (Figure 455).

*Mobile Telephony 2.0* (1980-2000): With the introduction of the 2G networks, and the companion digital technologies, new standards for wireless communication by cell phone were established. The handset development saw a development into the Feature Phones. Paralleled by the development of the Personal Digital Assistant (PDA), the mobile phone took off and conquered the market (Figure 472).

*Mobile Telephony 3.0* (2000+): The next generation of mobile telephony standards (3G) solved technical and capacity problems adding bandwidth and transmission speed for new development. Thus creating room for the merger of Trajectories of the Personal Communication and Personal Information with the Trajectory of the Microcomputer creating the Smartphone-concept (Figure 474).

With the maturing Mobile Telephony came the next generations of networks. Such as 4G networks that supported amended mobile web access, IP telephony, gaming services, high-definition mobile TV, video conferencing, and 3D television. In 2009 a 4G network was deployed in Oslo, Norway and Stockholm, soon spreading over developed countries. By 2019 a new generation of networks were deployed; the 5G networks. Offering even more bandwidth, higher transfer rates, new frequencies and a larger capacity. It would open the door to the Internet of Things; physical devices that connect to the internet exchanging data with other devices and systems. Such as the devices of the 'Smart Home;' a home automation system that monitors and/or controls home attributes such as lighting, climate, entertainment systems, security systems, and appliances.

---

[653] The very first mobile phones were not really mobile phones at all. They were two-way radios that allowed people like taxi drivers and the emergency services to communicate.

## *The ICT Revolution (1950-2000)*

In the preceding we explored —next to the *First Calculating Revolution* (1850-1930), and the subsequent *First Computing Revolution* (1930-1980)— also a third revolution: the *First Communication Revolution* (1830-1920). Each representing a specific period of a specific technological development trajectory and the subsequent application development trajectories. Although spread out over a longer period of time, these revolutions were functionally intertwined and dominated by *mechanical technologies*. Next, the development of *electro-mechanical technology* would fuel the development calculating machines and computing machines, and the development of *wireless technology* would fuel the development of wireless communication in all its forms.

In our analysis of both the First Calculating Revolution and the First Computing Revolution, we observed how component, system and function are fundamentally interconnected. Calculating needs its mechanical calculating machines to perform arithmetic function, computing needs its computing machines to process data (ie information).[654] And there was a single component identifiable that was the basic engine: the Leibnitz wheel followed by the pinwheel.

Also in the analysis of the First Communication Revolution, we observed how component, system and function are fundamentally interconnected in the field of tele-communication. Telegraphy needs its electromagnetic relay, a Morse's receiver and the Morse code, and the cabled infrastructure. Telephony needs its microphone and earphone, Bell's telephone apparatus, and also a cabled infrastructure. Wireless communication became the coordinated system with a non-cabled infrastructure (ie the electromagnetic waves), a transmitter to prepare for the transport the information, and a receiver to present it.

So, the *Electro-Mechanical Technology* was a decisive factor in the development of the early communication, calculating and computing machines. The electro-mechanical technology with the dominant component called 'electromagnet' would create dedicated systems with a specific function to perform: from Morse's telegraphy of the coded word, to the relay-based calculating machine and computing machine complementing human mental capabilities. But, after it facilitated early

---

[654] Information is an item of intelligence incorporated in a collection of data; statements of facts about something or someone. Such as data originating from the environment and registered by the sensory facilities of living organisms. Data that is also processed (ie by association and combination), interpreted (ie by labelling) and transmitted as a meme. See: Van der Kooij, B.J.G.; Deep Origins of Innovation: Part II: Memes of Novelty.

development, the cabled infrastructure limited the exploding usage of telegraphy and telephony. And that same mechanical technology was the limiting factor in the further development of a more advanced functionality of the arithmetic machines. The machines that had 'hardwired'[655] functions that could also be realised by other technologies that emerged after the electro-mechanical technologies.

That limitation inherent to a specific technology was fundamentally interrupted when the *Electronic Technology* developed, and when 'Electronics'[656] enabled the creation of devices based on electronic circuits. Circuits as the specific configurations of active and passive electric elements, that would create their specific function; from the logic functions (ie AND, NOT and OR), to the analogue function (ie amplification).

## *The Rise of the ICT-Revolution*

Then came the second half of the twentieth century with the ICT-Revolution that was the successor of the CT & IT-Revolution (Figure 177) that occurred during the first half of the twentieth century. Again, it was about the evolution of the Information Technologies (IT) and the Communication Technologies (CT), but now technology was not limiting the development, but fuelling its further development.

After the First World War the First Electronic Revolution had produced the vacuum tube as a means to control electric currents in vacuum. After the Second World War the Second Electronic Revolution had produced the transistor as a means to control electric currents in semiconducting materials. The latter would become the core element of the Silicon Engine; with the digital technologies as the driving force of the ICT-Revolution (Figure 476).

*Second Electronic Revolution* (1950-1980): The inventions of the Silicon Engine (from transistor to Integrated Circuit, and from Microprocessor to Microcomputer), was the driving technological force for the ICT-Revolution. Over the years the technology matured rapidly with electronic circuit becoming more complex in functionality and smaller in size (Figure 476, bottom).

*Second Communication Revolution* (1950-2000): In the field of communication, the transmission of radio- and television signals created a massive

---

[655] Generally, "hardwired" implies a built-in immutable function. So, "hardwired logic" implies is a decision process that is always going to give the same result. It is the opposite of programmable logic, where the software program defines its result (ie output).
[656] The word 'electronics' is derived from electron mechanics, which means to study the behaviour of an electron under different conditions of an applied electric field.

novelty in Radio-broadcasting and Television broadcasting. Next to the evolution of the radio receivers and the television receiver, the broadcasting infrastructure was used by numerous radio- and television stations. A development that had a massive effect on society as news and entertainment was brought in the private home. The rise of mobile telephony added to that the two-way communication that made people connected all the time (Figure 476, middle).

*Second Computer Revolution* (1950-2000): In the field of information processing, the Silicon engine penetrated the computing devices. From the transistor and IC-logic circuits creating the Mini-computers, emerged ultimo the micro-computer based Personal Computers (Figure 476, top).

In the preceding content-analysis, we observed the events predominantly from a technological point of view, incorporating their impact. For example, in the form of the related business events. Defining technology as 'knowing how to make things,' we identified the major events (labelled as basic innovations) of the Information Technologies (IT) and Communication Technologies (CT), their contributing events and the derived events that created the Clusters of Innovation (Scheme 1, Scheme 2). Among those many events we tried to identify both the evolutionary as

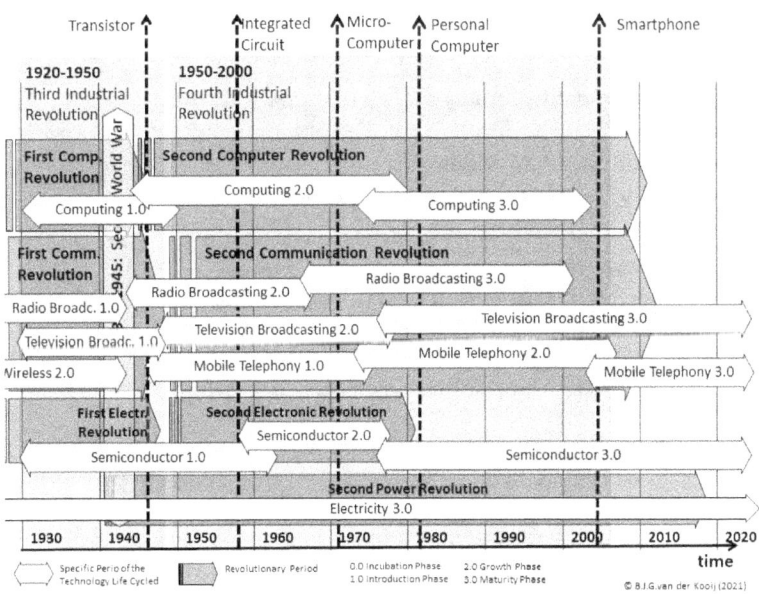

Figure 476: Overview of the contributors to the ICT-Revolution.

well as the revolutionary events (Scheme 3, Scheme 4). And to structure the events in the technological trajectories we applied the periodization of the Technology Life Cycle Concept (Scheme 5). After all these explorations and their explanations, we were able to place the periodized events in relation to each other in their specific time frames labelled as Industrial Revolutions, shaping up the Eras they covered.

Using the construct of the General Purpose Technology (GPT) enabled us to define the development phases of the specific technologies that constitute the GPT-ICT (Scheme 6). Each of these technologies had their own technological development trajectory in the 1950-2000 period ((Figure 476). This period encompassed two Industrial Revolutions; the Third Industrial Revolution (1920-1950) and the Fourth Industrial Revolution (1950-2000) both dominated by the industrialization of the manufacturing process mass-producing ICs, and the digitalization of society (eg television, mobile phone).

We observed how 'digitalization' (ie the use of digital circuits and systems in order to convert information into a digital format) heralded a new development that brought the Communication Technologies (CT) and Information Technologies (IT) together. Digitalization became of crucial importance in information processing and data transmission and was the basis of the GPT-ICT. It gave the impetus to a range of new technological and application developments (eg the Smartphone) that heralded a new age also labelled as the Digital Era heralding the Information Revolution.

Next to the earlier Industrial Revolutions with their 'industrialization,' the Information Revolution will transform the Affairs of Man fundamentally. From the advent of the personal computer in the late 1970s, to the Internet's reaching a critical mass in the early 1990s, and the adoption of such technology by the public in the two decades after 1990, it will have a massive impact on a society that is technology dominated. And, similar to the negative impacts of the Industrial Revolutions (eg working conditions, environmental damage), the Information Revolution will have its negative side-effects (eg Cyber Crime, Privacy issues).

In the timeframe of a century, the explosion of novelty resulting from the Electronic Revolutions, had influenced the Affair of Man dramatically. Both in his professional, public and social life the fruits of the creative efforts of many thinkers and tinkerers had influenced the lifestyle and worldviews of individuals. The collective impact of GPT-ICT was multi-fold: from its economic impact and social impact, to the expansion of man's intellectual capabilities and human development in general.

# Conclusions

In the preceding exploration into the Nature of Innovation with its Change and Novelty in relation to the Affairs of Man, we created in large brush stokes a picture of the innovative events that shaped the Twentieth Century. Events as part of technological trajectories as well as application trajectories, each with its own revolutions (Scheme 13). Some of those events were contributing to others to come, some derived from preceding events (Scheme 3). Some were part of an evolutionary trajectory, others part of a revolutionary trajectory (Scheme 4). Some events having a hardly notable impact, others having a dramatic impact on the Affairs of Man. Next to the events of a technical nature, we noted also the events of a social, political, scientific or economic nature (Scheme 10), creating the socio-political and the techno-economic contexts of their time (Scheme 11).

## *Change and Novelty of the ICT-Revolution*

The focus of our explorations is on Change and Novelty encompassing the concept of invention and innovation, and being the counterpart of Change and Destruction.[657] This case study focuses on the events in the second half of the twentieth century that shaped the Affairs of Man in the ICT-Revolutions. From the persons living before the First World War — used to horse traction—, to those living at the end of the century becoming hooked to their electronic gadgets like the cell phone—, society underwent the effects of Third Industrial Revolution (1920-1950) and the Fourth Industrial Revolution (1950-2000). Times characterized by the Roaring

---

[657] Change and Destruction is the conceptualization of periods in which the Madness of Times results in revolutionary and military warfarin with their destruction of property and lives. The First and Second World Wars are examples fitting in this concept.

Twenties as well the Great Depression, the Swinging Sixties followed by the Cold Seventies, and the Turbulent Eighties and Disruptive Nineties. Dynamic periods that created the specific context for Change and Novelty.

## The Historic Context for Change and Novelty

From Prehistoric Times, the *human existential needs* have ruled the Affairs of Man. For the personal affairs where the food, shelter, and safety needs determined his individual behaviour. To the social affairs where safety and procreation needs determined the social behaviour of bonding. And the totality brought forth egoistic, empathic, opportunistic behavioural patterns that configured (extended) families, tribes and clans banding together.

To fulfil those existential needs, the access to natural resources was essential (from drinking water to fruits and prey), and could become a source of territorial conflicts. Next to the friendly exchange of resources when they were available in abundance (ie trading), trying to get forced access to someone's resources (ie conquest) was another option. Or through colonization (forcefully) dominating others to make them part with the resources they had worked for (ie taxation, transferring wealth), that was also an option. As was using them as a submissive workforce (ie slavery). So, with growing populations the existential needs of the newcomers had to be fulfilled, causing increased trade and warfarin over the millennia.

### Human Need Fulfilment over Time

This basic pattern of fulfilling existential needs by bonding, trading, warfarin and transfer, ruled the Affairs of Man up to modern times.

*Ancient times:* Zooming in on the Ancient Times and their Affairs of Man, clans and tribes growing in numbers, started mutual cooperation when threatened by others (ie 'The enemy of my enemy is my friend'). That bonding took place on different level; knotting family ties by marriage, bonding tribal ties with religions, bonding armed forces by military agreements. And when the external threat became large enough, the internal unity increased creating 'polis' (eg Greek city states) and unions of military alliances (ie Aegean Coalitions) against foreign threats. So over the ages, tribes and states cooperated in harmony (ie trade), fought others (ie wars) or taxed them to the hilt (ie occupation). From the conquered peoples they took the wives to fulfil their hormonal needs (eg the Roman Rape of Sabine Women), their labour by slavery and their wealth transferring it to their home-base (eg Rome). And from their ranks emerged the rulers: from the democratic chosen to the hereditarily rulers and dictatorial rulers spear-heading a power-hungry elite. And when the subordinates felt cheated and mistreated, they revolted.

With those tribal unification into states, some social parties —from religious groups to family groups— were or became more dominant than others. Their leading individual —supported by the eminences of his clan— rose to power, unifying the state, creating aristocracies. And that state confronted other states in a similar way; trading, fighting and taxing others to get the resources they needed. A situation influenced by the climate creating both abundances and shortages of foods. And when those existential needs were not fulfilled, the subject peoples would revolt. As the Romans Rules understood when they created their Bread and Games policies and triumph marches showing the booty.[658]

*Historic Times:* Zooming in on more modern times we see the same pattern of need fulfilment repeated in the eighteenth and nineteenth century. Bonding social ties took place on the level of monarchies and aristocracies (eg Queen Victory being the grandmother of the European royalty, the Habsburg Rule by marriage). Nation formation ('The Family of Families') united former polities into nations (France) with shared (geographical, historic, language, of ethnic) bonds. In addition, trading policies changed from the protective, nation-centred Mercantilist ideology to the Free Trade ideology. Trade expanded physically from regional and national to international as a result of the improving transportation infrastructure (ie rail road expansion). Exploration, trade and Christianisation, flourished thanks to the steam powered shipping. Colonization of nearby regions (eg Ireland by English, Northern Italy by Austria) was succeeded by the expanding free trade movement to the colonial expansion in the Far East and Africa. It created new classes of workers; the civil servant and the bourgeoisie.

Industrialisation lay at the foundation of the Empires; first the British Empire with Brittainy as 'the workshop of the world', and the British Navy 'ruling the waves.' Followed by the German Empire rising from the remnants of the Holy Roman Empire flourishing during the Industrial Revolution into a geopolitical power that had to be taking into account by the other empires. The Empire creation as the driving force behind the colonial expansion went hand in hand with warfarin. Military confrontations that took place between the expanding empires themselves. Such as the Crimean War (1853-1856), a confrontation between the expansion aspirations of the Russian Empire, the collapsing Ottoman Empire, and the (trading) interests of the other Empires. And

---

[658] With the riches from trade, after the 1204-fall of Constantinople the Venetians brought also the booty of the Fourth Crusade home. That booty contained the statues of the Four Horses in gilded copper (ak Quadriga), that were exposed on St Marks Basilica in Venice. The same horses that ended up in Paris on the Arch de Triumph after being looted by Napoleon in 1779.

it saw military action in the overseas territories with the surmised people of the colonies (eg the Boer Wars in South Africa). Warfarin in which military technology (from canons to battleships and telegraphy) played an increasing role. In addition, the transfer of wealth from the colonies created individual richness never seen before; both the aristocracies of the 'Ancien Regime' as well as the upstarts of the 'Nouveau Riche.'

Next to the transfer of wealth (ie from the colonies), the transfer of know-how came on scene. Technology (ie 'knowing how to make things') created the power engines, and with the early industrialisation came new ways of manufacturing (eg the factory replacing the workshop). And, as that know-how was valuable to others, it geographically spread from is source of origin into other regions rapidly. Both the transfer of wealth and transfer of know-how had large social implications as it gave rise off the new class of industrial workers next to the agriculturally based peasantry. People who, like the other social new classes, wanted a place in their society or they would revolt (eg the American Revolution that was about 'No Taxation without Representation'). Their political efforts reflected in the Socialist ideology (ie Marx cs), confronting the Capitalist ideology with its non-intervention state policies that had succeeded the Laissez Faire policies. A policy that gave rise to the robber barons of the railroads and steel industry. But when people (the peasantry, industrial working class and the bourgeoisie) got hungry or were not recognised as a political participant in society, they revolted. For example, as happened during the 1848-Revolutions fuelled by the ideas of the French Revolution that spread around Europe in the mid-nineteenth century.

*Modern Times:* Reaching the twentieth century one can observe the same pattern of bonding, trade, war and transfer, but on a larger scale. Bonding came on an international level creating the geo-politics with nations sharing the same interest (acknowledging each other's 'Spheres of Interest'). Military bonding was created by alliance-formation between nations. Such as the French Revolution (1789-1799) being a catalyst for all the seven coalitions that fought Napoleon up to 1815 in the Coalition Wars. Trade became global, industrialisation overstepped the national borders, economies became intertwined. Earlier, in the 'Scramble for the Pacific', the Empires had already expanded their spheres of influence for commercial opportunities (eg Perry and his gunboat diplomacy showing up in the Bay of Tokyo in 1854). Now, colonization appeared in a new form in the late nineteenth century (the 'Scramble for Africa') accompanied by colonial warfarin (the Boer Wars). In addition, it saw the geo-political conflicts change into global warfarin resulting in the First World War and the Second World War.

Technology with its massive creation of novelty originating from the Third Industrial Revolution, reaching in many domains of application (ie mobility, communication, calculation) was transferred globally. The twentieth century saw massive revolutionary novelty in mobility (eg the Automobile Revolution, the Aeronautic Revolution) as well as in communication (eg the Mobile Telephone Revolution, Radio Revolution and Television Revolution) and information processing (eg the Calculator Revolution and the Computing Revolution). Novelty that was aften the spin-off of military funding.

Already from old times on, the transfer of wealth and labour was part of the Affairs of Man. The Romans brought the loot from their campaigns to Rome, and led the people of conquered regions into slavery working their estates on the Italian Peninsula. Wars meant booty, and that was transported to the homelands. In the times of imperialism and colonisation, the wealth from the colonies was also shipped to the homelands of the maritime empires creating individual wealth (the nouveau riche). Also Islamic chattel slavery had become a profitable trade for the Portuguese, Dutch and English. A development that continued during later times up to the twentieth-century's world wars that would see massive forced labour and prisoners of war as well as the plunder and pillages of war.[659]

Those processes of wealth and labour transfer were the walking companions of the warfarin for resources. Again the looting of valuables and the slaving of others was part of getting access to existential resources. From the looting of food (to fulfil the basic needs for nutrition) and the looting of women (to fulfil the existential need to procreate) by individuals, to the later collective military efforts to get access to natural resources (ie mineral, coal, oil, gas) and even pieces of Art.[660]

## Transfer of Science and Technology

Already from ancient times on, the transfer of knowhow from one generation to the other (ie the vertical transfer), was part of the Affairs of Man. A process to pass along not only the industrious experiences, practises and skills (eg stone, metal and wood working), but also the norms and values of groups (eg the codes of tribes, clans). This was part of the basic process of learning by imitation and the transfer of culture. In later time the

---

[659] Forced labour of people conquered or held in concentration camps (Japanese, Nazi-regime) to work supporting the war effort, was well organised (eg Birma Railroad).
[660] In recent times both the German Reich as well as the Japanese Empire —both lacking natural resources they needed for their expanding populations— had getting control over natural resources as a prime objective. Germany turned to the mineral rich Czech and Ukraine regions in the south of Central Europe. Japan turned to Manchuria in the north, as well as Borneo in the southern Indonesian Archipelago.

spreading of religions —the transfer of the religious ideology by the missionaries of the Christian and Muslim faith among the heathens, (ie the horizontal transfer)—, influenced cultures on a large scale.

Jumping to more recent centuries, the Scientific Revolution saw the understanding of the Nature of Matter by the natural philosophers rise. And the distribution of that knowledge by the arts and means of those times; such as the Art of Book-printing. Also the understanding of the Nature of Being and Society (ie the Enlightenment movements) that developed during the Renaissance Revolution saw the propagation of knowledge in printed form. Some of those books became sacred and ruled societies and individual lives; the holy scriptures of religions translated from Latin in native languages. Other books spread ideas disruptive to the intellectual establishment of their days: from Darwin's views on nature to Marx's views on society. Books that could be threatening existing power-structures trying to hold a grip on their societies.

> The subsequent book burnings,[661] censorship[662] or the damnation of the scholars that wrote them, could destroy the books but not the ideas. As they were anchored in the minds of people that sought to disseminate them (van der Kooij, 2021a).

## *The Invention of Innovation*

That was, painted in large brushstrokes, the historic evolution of the context in which the actual events of novelty blossomed during the twentieth century. Events we explored as they occurred in the *domain of electricity as medium of information*. Events with large variety reflected by their interpretation as invention, innovation or improvement based on the degree of their change. From a massive change with a fundamental character (ie invention); to change as the outcome of a new combination of knowledge/insight, and knowhow/skills (ie innovation); or a minor but recognizable change (ie. the improvement). We noted their differences in origin; from the technology-driven novelty (aka technology push) to the need-driven novelty (aka market pull). And we noted their impact; from those with a limited impact as their degree of change was limited (ie the evolutionary innovations) to those with massive impact (ie the disruptive innovations).

---

[661] Such as the destruction of the Library of Bagdad and Alexandria during the Middle Ages. The book burnings in Tudor England during the rise of the English Church, and in the Holy Roman Empire during the rise of Protestantism. Up to the Nazi book burning of recent times to cleanse German culture of Jewish and foreign influences.
[662] The Index Librorum Prohibitorum ("List of Prohibited Books") was a list of publications deemed heretical or contrary to morality by the Sacred Congregation of the Index (a former Dicastery of the Roman Curia), and Catholics were forbidden to read them

## The Different Dimensions of Novelty Creation

Our analytical approach (Scheme 1—Scheme 4) resulted in a wealth of events illustrating this heterogeneity in novelty. They also painted a picture of the diversity in novelty creation.

*Innovation as a New Combination:* Many of the presented events with a revolutionary character labelled as a 'basic innovation', had the character of a new combination resulting from different trajectories of contributing innovations (Scheme 2). The work of the thinkers (in modern times called the scientists) combined with the contributions of the tinkerers (in modern times called engineers) as the combination of Knowledge (ie Basic and Applied Science) and Knowhow (ie Engineering), was fundamental to the creation of novelty.

*Innovation as a Manageable Process:* Was the Second Industrial Revolution characterized by the creative efforts of the inventor-entrepreneur, the Third Industrial Revolution during the first half of the twentieth century saw the rise of 'organization.' First in the management of the manufacturing process where workshop-level manufacturing was replaced by the mass production techniques of conveyor belt manufacturing (eg the manufacturing of automobiles). From the artisan organizing his own work in its entirety (ie acquisition, production, sales), to the specialist rising from the working ranks to become a fulltime manager focussing on part of the total manufacturing processing (ie 'specialisation' in production management, sales management, etc.)

A similar development took place in the organization of Research and Development (R&D). The sole inventor was replaced by a multi-disciplinary team of specialists organized within Inventions Rooms, later known as R&D departments. And the organization of their work as a multi-phased process (ie the Management of R&D) improved their output of knowledge and knowhow. In the slipstream of the Management of R&D came the Management of Innovation, focussing on converting that knowledge and knowhow into commercial products. Both R&D-Management and Innovation Management became tools for companies to create Change and Novelty while adapting to changes in their business environment.

*Innovation as an Ecosystem:* The first half of the twentieth century saw the emergence of industrial regions with clusters of manufacturing industries. Such as the calculator and clock making industry-clusters of South-Germany and Switzerland, the car manufacturing industry cluster around the American city of Detroit. The second half of the twentieth century saw the rise of specific high-tech regions where technological innovations blossomed; such as Route 66 in the East US, and Silicon Valley in the West US. Regions with an ecosystem that facilitated innovation as initiated, implemented and embraced by individual people. With institutions like business-oriented universities educating management skills as well as engineering skills, and supporting the sprouting start-up companies. The availability of Venture Capital to finance them, combined with an educated workforce and visionary high-tech entrepreneurs. Those ecosystems could be local, regional and even national based (eg Japan) and were embedded in the bearers of their local, regional and national cultures.

*Innovation as a Culture:* With the recognition of the complex interaction between companies, R&D institutions, educational and finance institutions, came the recognition that the realization of Change and Novelty is embedded in the organisational culture. Cultures in organised entities shaped over time by the interaction with their business environment. Stable business environments had created the behemoths of the (electro)-mechanical industries —from the giants of the electric industry (eg AT&T) to the giants of the mobile telecom industry (eg Nokia)— exercising/managing a conserving business culture entangled in complex business alliances. On the other hand, the increasingly dynamic and disruptive business environment of the late twentieth century —fuelled by technological revolutions (eg the Electronic Revolutions)— had given birth to innovative companies embracing change and novelty as a business model (eg Intel, Microsoft, Apple).

That organizational culture of innovation could become complemented by the national culture of innovation. A national culture characterized for example by its governmental policies resulting from the political decision-making process.[663] Sometimes resulting in allocating resources into the creation of novelty (from tax-breaks and tax-incentives, to funding the universities and research institutions). Or in effort to copy Silicon Valley by facilitating the creation of Business Parks and Business Incubation Centres.[664] Resulting in the 'new economy.' In contrast with other, more conservative societies where conservative politics[665] allocated resources to extending the lifecycle of traditional industries (eg mining, railroads, textiles, shipbuilding); the 'old economy.'

---

[663] Politics —on a national level— is about advocacy of existing interests originating from the past. As innovation by definition is part of the 'unborn child of the future' with no interests in present time, it cannot compete in the advocacy process.
[664] In the 1980s many countries tried to implement the concept of an high-tech region (ie Silicon Valley). France created in the Provence near Nice the Sophia-Antipolis
[665] Britain missed out much of the modern high-tech developments in ICT as a consequence of the state monopoly of the Post Office.

These examples of the innovative characterizations reflected in the different scholarly views on innovation that emerged in the second half of the twentieth century.

*Organized Creation of Novelty*

In the nineteenth and twentieth century, the organized Creation of Insight (ie Research in modern parlance) enlarged the knowledge base of the specific domains of nature; from the study of Classical Physics to Solid State Physics. And it was the organized Creation of Skills (ie Development) that turned the ideas into reality by acts of novelty creation. From the individual Heroes of Invention toiling in their workshops, up to the R&D-teams with different expertise working in their laboratories. Research collectives that became well organized (ie the Management of R&D) and well-funded (ie R&D budgets).

> Although, the resulting novelty was well protected as Intellectual Property in patents, it would ultimo become part of the scientific knowledge base. And it was the transfer of Science and Technology that spurred the propagation of all that knowledge and knowhow, stimulating in turn the organized novelty creation in other R&D institutions.

Novelty creation was based on *Knowing* as in 'understanding things' (ie Knowledge and Insight) and as in 'how to make things' (ie Knowhow and Skills). And the transfer of both Science and Technology became the fuels of technological progress over recent times.

*Transfer of Technology*

Part of that transfer was the Transfer of Technology.[666] Already in the times of the First Industrial Revolution, the transfer of technology was an issue of protection, prevention and industrial espionage of the useful arts. One way to protect the transfer of accumulated knowledge and knowhow was to prevent the steeling, and restrict its exportation. Such as the French steeling the arts of steel- and glassmaking from the Brits in eighteenth century.[667] Already in the First Industrial Revolution, the British government as the 'owners' of the steam technology legally restricted

---

[666] Technology being defined as 'knowing how to make things', is the collection of techniques, skills, methods, and processes mastered by individuals and collectives.
[667] France had had considerable access to English technology in the 1760s through the movements of Gabriel Jars, John Wilkinson, and many others. Take Jars travelling England, who after his return to France, experimented with coke-making and coke-smelting and identified the Mont-Cenis-Le Creusot area (Burgundy) as one where the availability of coking coal would allow the use of coal in iron-making. And the iron-master Wilkinson who assisted the iron masters at Le Creusot.

efforts to transfer that technology to their former colony America.[668] Also the British arms technology and naval technology coming from the Workshop of the World that powered the British Navy, was of interest to many other nations. But the unavoidable spreading of industrialization over Europe in the nineteenth century made those restrictions an impossible task as it was also exported by British entrepreneurs and engineers themselves. They exported locomotives to European countries, setting up maintenance plants, and help creating manufacturing plants during the time that the First Industrial Revolution spread over Europe (Figure 10).

>Even in more recent times technology transfer was having great impact. The pre-war exodus of the scientists from Germany in the 1930s influenced the development of Physic Sciences in the US that produced the atomic bomb ending the Second World War in the Pacific. The after-war Japanese Economic Miracle was fuelled by the US-developed technologies. It was part of the continuous industrial revolution already reaching the East in earlier times.[669]

Another way to protect the intellectual property was with the legal system of patent laws. From the letters patent issued by English monarchy in the fourteenth century creating monopolies, followed by Venetian and Florence patents in the fifteenth century, emerged a rudimentary patent system during the eighteenth century. Embedded in the legal system, the Enlightenment views of intellectual property created the foundations of modern patent laws. Laws that were sometimes anchored in the Constitution (eg the Patent and Copyright Clause in the US-Constitution; Article 1, Section, Clause 18). Or they were created separately. With its national differences, after much criticism and debates, a fragmented system of Patent Laws emerged.[670]

>And with the Patent Laws came not only the system of patent licensing in which the intellectual property right was financially compensated,[671] but also the infringement disputes. Especially the

---

[668] The start of the American Industrial Revolution is often attributed to Samuel Slater who pirated technology and opened the first industrial mill in the United States in 1790 with a design that borrowed heavily from a British model.
[669] In the Pacific region, Japan had imported some three thousand foreign teachers, engineers and specialists at the beginning of the Meiji Restauration (1868-1889). They also send students to European and American Universities to become trained in technologies.
[670] Such as the US Patent Act (1790/1793), the French Patent Law (1791), the English Patent Law Amendment Act (1852) and the German Patent Act (1877).
[671] Already the inventors of the Electric Age (eg Thomas Edison) sold their patent rights for a fee to third parties. Charles Goodyear, inventor of the vulcanized rubber process, never manufactured or sold rubber products, but he licensed his rights to others. The sewing machine technology patent of John Bachfelder, who neither manufactured sewing machines nor licensed his patent, was sold it to Isaac Singer.

pioneering patents for new technological artifacts with their broad claims, created controversies and event Patent Wars.

The result was a process of technology transfer that spurred economic development as well as massive technology-induced novelty. And all this technology transfer needed two parties, each with different interests; one willing to share it, and the other eager and capable to accept it. This all in a cultural setting that determined the context for change: either stimulating or restricting the process.[672]

## The Novelty Potential of Electronic Technologies

This short conceptual approach of the transfer of technology illustrates how new technical systems evolved from preceding technical systems as the basic component(s) realized in a new technology changed. The distinction between component, system and function illustrates how that novelty emerged; from an invention at the component level (eg the electro-mechanical coil, the electro-magnetic wave creation and detection) to the innovation at system level (eg the telegraph and telephone systems, the wireless system) to the widespread use due to (cabled and wireless) infrastructures to realize the transmission function.

Especially the electronic technologies of the second half of the Twentieth Century proved to have a high novelty potential. Had the electric motor brought powered appliances, tools and machines, the new electronic technologies of the Silicon Engine brought a wealth of new opportunities. However, for many existing companies of the old IT and CT-industries (eg those that dominated the first half of the Twentieth Century) the adaptation and/or conversion to new technologies proved problematic. Many of the leading companies of the electromechanical IT and CT-industries serving their existing markets, proved unable to make the conversion to the new technology that fuelled novelty in a massive way.[673]

*Strategic Planning and Innovation Strategies*

With the rise of the large corporations came the need of strategic planning to cope with the dynamics of the business environment. Organization was not only for the management of current affairs (ie current business operations), also the future development of the companies should be influenced by directing the R&D activities. So, depending on the

---

[672] That the State of Californian became the Silicon Valley that was so essential to the development of semiconductor-technology, found it base in the laws from the time of the Gold Rush prohibiting restrictions on labour mobility.
[673] A classic example can be found in the Swiss Industry that failed to adapt the electronic technologies in their watches and faced competition by new (often Japanese) companies offering electronic watches eagerly absorbed by the market.

financial resources available, R&D-activities became directed to shape the future activities of their funding companies (eg Philips Nat Lab, Siemens Labs and Bell Laboratories).

*Strategic Thinking:* This corporate behaviour was part of the process of strategic thinking in which the strengths and weakness of the organization, as well as the opportunities and threats from its business environment were identified (aka the SWOT analysis). Combined with a vision of the future to come, the goal setting complemented that analysis in the creation of a corporate strategy. And part of that strategic thinking related to the domain of 'innovation strategies.'

Basically, there were different strategic routes for (existing) companies in terms of novelty creation. And depending on its formulated corporate strategy, the innovation strategy could be a mix of:

*Replacement Novelty by Horizontal Innovation:* The function of the existing system brought forth by the organization could be realised by a for the company new technology; eg existing mechanical systems could be *replaced* by electro-mechanical systems that performed the same function, in response to an existing and well defined market.[674]

> Such as the system using the electric lamp replacing the system of lighting with gas lamps, by still creating 'light.' The vacuum tube-based computing machines replacing the relay-based computer. Or the electronic watch replacing the mechanical watch, the electronic calculator replacing the mechanical calculator.

*Enhancement Novelty by Vertical Innovation:* Because the new electronic technologies had such a novelty potential, they made more advanced and complex functionality possible; eg the new technology created *enhanced systems* with *expanded functions.*

> The electric motor brought power to the individual machines creating electric powered handtools and more complex machinery. All powered through an infrastructure, made of electric cables. In communication, the early mobile telephone had limited capabilities. The digital capabilities of the 2G networks, however, offered opportunities in more recent times for cell phone companies to develop the Feature Phones with added functionality.

*Created Novelty by Lateral Innovation:* And finally, the new technology could create new systems that had not been in the marketplace before, and in the process creating a new market based on latent needs. Being it

---

[674] We consider a 'market' to be the collection of similar needs (of people); the communication market is the collective need of people who want to communicate.

analogue systems that would broadcast analogue signals over distance (eg music, speech), or being digital systems that transmitted digitally coded information over distance (ie telegraphy). And they could be the digital systems that could process information in its logic circuits (ie computing).

The Silicon Engine as a new technology, offered the possibility to create the personal information and communication devices (eg the PDA, PIM, Feature Phone). And the mergers of the trajectories of communication technology (CT) with information technology (IT) would see the birth of the Smartphone.

As we observed, these basic approaches to novelty creation can be found in the behaviour of the companies in the ICT-industries. Some became quite successful in a specific period of time. Others approaches led to Rise and Fall of large companies.

## From the IC&IT Revolution to the ICT-Revolution

And all this novelty was created in the context of its time. The electro-mechanical technologies were the fruits of mechanical thinking that was part of the technical culture in that period of time. Based on the mechanical skills originating from the skilled artisan (eg the fine-mechanical instrument maker), the mindset of the thinkers and tinkerers was 'mechanical' framed. It was part of the Era of Mechanics where the mechanical worldview dominated (Figure 8). The rise of electronic technologies and their ICT-technologies in the Digital Era, saw novelty that had not existed before and that was based on the different way of Logic thinking.

In the case of both the mechanic and electromechanical calculation machines, the technology was both enabling and limiting its further development. That changed when the different generations of electronic devices were invented. These basic innovations —the *General Purpose Engines* (GPEs)— of the vacuum tube (aka radio-tube), the transistor and the Integrated Circuit (IC) would constitute a new meta-technology called the *General Purpose Technology of Information and Communication Technologies* (GPT-ICT). A very fertile meta-technology that would follow a development trajectory characterized by continuous improvement, both in the technology itself (eg the vacuum tube technology, the semiconductor technology, the integrated circuit technology), as well in the ever expanding technical trajectories of the resulting artefacts (eg the information technology). But also a technology that would spawn off a range of complementary development trajectories; those of the computing applications (eg the minicomputer and the personal computer, computing devices for industrial control and office automation), as well as the tele-communication applications (eg the radio, the television, the mobile phone), creating technologies known as the *Information Technology (IC) and the Communication Technologies* (CT).

As a consequence, the basic innovations of the GPT-ICT would start new business cycles (Scheme 8) (eg the vacuum tube industry, the semiconductor industry), creating new technology-based industries and services (from the radio-broadcasting equipment to radio broadcasting stations), and fuel economic progress. Their novelty would, in a process of creative destruction, replace former technologies. On component level (eg the binary *flip-flop circuit* replacing the electro-mechanical relay) and on system level (eg the *transistor radio* replacing the tube-based radio) or by creating complementary systems (eg *mobile telephony* next to cabled telephony). And, not in the least, the GPT-ICT would also significantly influence the Affairs of Man; their individual lifestyle, their social interaction and their warfarin.

Just like the home-appliances sprouting from the GPT-Electricity contributed to early women's emancipation, the GPT-IT (also known as 'automation' contributed to societal development. Over time the novelty potential of the GPT-ICT spawned totally new application areas (eg Personal Computing, the Internet, Smart mobile telephony) that would change individual and social behaviour (eg Social Media). A development that started in the early twentieth century with the parallel developments creating the CT & IT Revolution (1920-1950). And continued when their developments merged in the Digital Era creating the ICT-Revolution (1950-2000).

## *Acts of Organization*

Novelty creation is not an incidental process, as it is always the result of the efforts of individual thinkers and tinkerers; sometimes a quite randomly activity (eg Alexander Bell working on the harmonic telegraphy and in the process inventing the telephone), sometimes a more or less organized (eg Thomas Edison and his staff working in his workshop in Menlo Park). [675] The early 'Heroes of Invention' would have been self-organizers; the inventor-entrepreneurs combined their technical experimenting with business creation and commercial activities (Scheme 12). With the growth of companies, they were followed by the organizations of people into formalized R&D departments, where the thinking and tinkering became an organised activity of scientist and engineers. What they have in common is

---

[675] The process of *organisation* can be defined as the act of putting activities into an orderly and logical order, or the act of taking an efficient and orderly approach to tasks.

that novelty creation takes place in an industrious context that emerged over time.

*The Development of Industrious Activities:* A key element in the Industrial Revolutions was the way artifacts were realized in an industrious context. They developed from the Nuclear Company into the Extended Company, followed by the Factory Company and the Manufacturing Company (Scheme 7). In the latter the division of labour was complemented by 'organisation;' the organization of the work-flow, the organisation of producing the interchangeable parts, and the organization of an (independent) sales force. Organization became a grouped-based process. And, as part of the total organisation, companies created their R&D department that became responsible for the novelty creation in an orderly, planned way.

*Organization as a Manageable Process:* The dominant element in the development of industrious activities was the Act of 'Organisation'. From self-organisation of the artisan to the grouped organisation of the manufacturing company. Organization executed by 'specialists' and their foremen (aka managers) who covered parts of the organization of the total manufacturing process. In modern parlance: the production managers organizing the workers in the production process, the marketing managers organizing the sales process, and the R&D managers organizing the R&D process.

## From Heroes of Invention to R&D Teams

Novelty creation had to be organized. First by the 'soloist' inventor himself, or by teams of (mostly) two individuals with complementary skills and interests; one possessing the technical creativity and technical drive, the other having the commercial creativity and commercial drive. Basically, it resulted in the following two forms of novelty creation.

*Individual Novelty Creation:* The individual Hero of Invention was self-organizing his creative work. He may have used the skills of artisans to physically create his prototype, but he was the man with the 'idea.' He conceptualized the new artifact, he tested the prototype, he realized the final product to be commercialized. The novelty creation depended on his personality, and his individual creative capabilities.

*Collective Novelty Creation:* With rise of the division of labour came the organization of novelty creation by dedicated groups of people; the R&D department with the R&D teams. Teams created out of the pool of researchers, each with his own speciality, and given a specific task. The novelty creation depended on the accumulation of the personalities, and their creative capabilities, as well as the company culture.

## Industrial Innovation

This collective novelty creation as a team effort became known as Industrial Innovation; a process with the phases of basic research, technology development, pre-development and development of solutions for example in the form of products, processes, services, business models or integrated combinations of those.

With the meteoric rise of the General Purpose Technologies ICT (GPT-ICT) originating from their General Purpose Engines (GPE), novelty creation in an industrial context rose to prominence in the Twentieth Century. The invention of the GPE Internal Combustion Engine had fuelled the novelty creation in the automobile industry. The invention of the GPEs of the electronic engines (vacuum tube, transistor and IC) fuelled the novelty creation in the ICT-industries. And that novelty creation was the result of the organisation of R&D in all its appearances.

> *The first approach for R&D organisation described in literature is referred to as "the strategy of hope" in which competent people were hired and provided with excellent surroundings, leaving them alone and hoping for the best. This organisational concept was followed by a project-based approach, introducing the quantification of costs and benefits of individual projects. Already more than two decades ago, organisational structures bringing together functional areas with required R&D disciplines aiming to break up the isolated structures of R&D departments were described in the third generation of R&D management.*[676]

The novelty creation became more and more recognized as a (linear) process that could be organized. Was 'innovation' originally the result of novelty creation (eg a product or manufacturing process), now innovation became considered as a linear process in itself that could be organized (ie the Management of Innovation).

The compliment of all the technological-driven innovation was organisational innovation. New forms of organisation (both the structure, the management as well as the corporate culture) that enabled companies to respond to environmental dynamics.

## *The Overall Impact of the GPT-ICT*

The business activities of all those contributors to the GPT-ICT had an enormous effect on the economies of the societies they were part of. Their companies created employment, improving the living standards of their workers. The income tax funded the budgets of their governments, as did

---

[676] Source: Bauer, W., Schimpf, S.: Understanding the history of industrial innovation: developments and milestones in key action fields of R&D management (p.4).

the corporate taxes of their companies. They even contributed to the geo-political conflicts by supplying the military with new tools of warfarin.

But there were other effects on society; such as the social and cultural effects. The radio and television stimulated home entertainment where news and culture were transmitted right into the homes. That 'broadcasting' in turn stimulated the rise of the radio- and television stations distributing news and culture (even propaganda). And the wireless technologies entered the field of mobile telephony connecting persons instead of fixed places.

> The Communication Technologies (CT) created the cell phone that evolved into the Feature Phone with its expanded communication functionality (eg SMS) followed by non-communication features (eg the Camera). The Information Technologies (IT) created the Personal Digital Assistant (PDA) and Personal Information Managers (PIM).

With the development along the technical and application trajectories, came the industrial development and the development of the service providers. Companies creating employment in the manufacturing and service-industries related to communication.

As those businesses operated in the context of their times —with the Madness of Times before major geo-political conflicts, the geo-political tensions around the World Wars, and the Spirit of Times during the Interbellum— in which the regional, national and international economies fluctuated. From times of economic growth to times of depressions. Fluctuations creating the long waves in the economies, as well as the business cycles.

## *Epilogue*

We started our scholarly endeavours with the basic reseach question "What is Innovation?" A scholarly topic —characterised by semantic heterogeneity— that expanded into a quest into the Nature of Innovation. Following the scholarly method of case studies as described in the preface, we explored the Context as well as the Content of novelty. After a first trial study (ie The invention of the Steam Engine), in some eight more detailed case studies labelled as the Invention Series we covered the development that became known as Industrial Revolutions. In order to avoid the semantic confusion, basis for our content research was the construct of the Basic Innovation as the core of the Innovation Cluster (Scheme 1). That Innovation Cluster being part of a complex development of contributing trajectories and derived trajectories (Scheme 2). Extended the construct to Basic Events enabled us to cover the Context (Scheme 3). A construct that covered both evolutionary as well as revolutionary developments (Scheme 4). And we added additonal concepts (eg the Life Cycle Concept).

In the second through sixth case study we explored profoudly the contributing events releated to electricity. In the seventh case study (the Invention of the Internal Combustion Engine) we explored a different development; the Power of Explosion. This eigthth case study (ie the Invention of the Silicon Engine) focusses on events that constituted the GPT-ICT; the events related to the Computer Revolutions as well as the Communication Revolutions. The forthcoming nineth case study (ie the Invention of the Internet Engine) will, by focussig of the Information Revolutions, complete our exploratory endeaovours in the Nature of Innovation

Together these case studies paint in large brushstrokes a picture of the development of Change and Novelty influencing the Affairs of Man fundamentally. Events that will be the foundation of the way how the Philosophers of Innovation looked at the real word to develop their Innovation Theories.

B.J.G. van der Kooij

-----

# References

Bergek, A., Jacobsson, S., Carlsson, B., Lindmark, S., & Rickne, A. (2008). Analyzing the functional dynamics of technological innovation systems: A scheme of analysis. *Research Policy, 37*(3), 407-429. doi:http://dx.doi.org/10.1016/j.respol.2007.12.003

Berlin, L. R. (2001). Robert Noyce and Fairchild Semiconductor, 1957-1968. *The Business History Review, 75*(1), 63-101. doi:10.2307/3116557

Bevir, M. (2012). *Governance: a very short introduction* (Vol. 333): Oxford University Press.

Bignell, J., & Fickers, A. (2008). *A European television history*: Wiley-Blackwell Oxford.

Clerk Maxwell, J. (1864). On Faraday's lines of force. *Transactions of the Cambridge Philosophical Society, 10*, 27.

Devezas, T. C. (2005). Evolutionary theory of technological change: State-of-the-art and new approaches. *Technological Forecasting and Social Change, 72*(9), 1137-1152. doi:http://dx.doi.org/10.1016/j.techfore.2004.10.006

Dosi, G. (1982). Technological paradigms and technological trajectories: A suggested interpretation of the determinants and directions of technical change. *Research Policy, 11*(3), 147-162. doi:http://dx.doi.org/10.1016/0048-7333(82)90016-6

Douglas, S. J. (1989). *Inventing American Broadcasting, 1899-1922*: Johns Hopkins University Press.

Duijn, J. J. v. (1983). *The long wave in economic life*: Allen & Unwin London.

Ellul, J., Wilkinson, J., & Merton, R. K. (1964). *The technological society* (Vol. 303): Vintage books New York.

Faraday, M. (1832). Experimental researches in electricity. *Philosophical transactions of the Royal Society of London, 122*, 125-162.

Fitzgerald, G. F. (1880). On the electromagnetic theory of the reflection and refraction of light. *Philosophical transactions of the Royal Society of London, 171*, 691-711.

Florida, R., & Kenney, M. (1988). Venture capital and high technology entrepreneurship. *Journal of Business Venturing, 3*(4), 301-319.

Garratt, G. R. (1994). *The early history of radio: from Faraday to Marconi* (Vol. 20): Iet.

Henderson, L. J. (1914). The Fitness of the Environment. *The American Journal of the Medical Sciences, 148*(3), 433.

Hobsbawm, E. (2010a). *Age of Empire: 1875-1914*: Hachette UK.

Hobsbawm, E. (2010b). *Age of Revolution: 1789-1848*: Hachette UK.

Jolly, W. P. (1972). *Marconi*. London: Constable.

Lipsey, R. G., Carlaw, K. I., & Bekar, C. T. (2005). *Economic transformations:*

*General purpose technologies and long-term economic growth*. Oxford: Oxford University Press.

Massey, H. S. W. (1952). Leslie John Comrie. 1893-1950. *Obituary Notices of Fellows of the Royal Society, 8*(21), 97-107.

Mazor, S. (1995). The history of the microcomputer-invention and evolution. *Proceedings of the IEEE, 83*(12), 1601-1608.

Mensch, G. (1979). *Stalemate in technology: innovations overcome the depression*: Ballinger Cambridge, Mass.

Merges, R. P., & Nelson, R. R. (1990). On the complex economics of patent scope. *Columbia law review, 90*(4), 839-916.

Morar, F.-S. (2014). Reinventing machines: The transmission history of the Leibniz calculator. *The British Journal for the History of Science, 48*, 1-24. doi:10.1017/S0007087414000429

Mouromtseff, I. (1950). Who is the true inventor? *Proceedings of the IRE, 38*(6), 609-611.

Pestrikov, V. M. (2019). *The invention of a tube audio amplifier*. Paper presented at the ITM Web of Conferences.

Raboy, M. (2016). *Marconi: The Man Who Networked the World*: Oxford University Press.

Riordan, M. (2007). From Bell Labs to Silicon Valley: A Saga of Semiconductor Technology Transfer, 1955-61. *The Electrochemical Society Interface, 16*(3), 36-41. doi:10.1149/2.f04073if

Riordan, M. (2007). The silicon dioxide solution. *IEEE Spectrum, 44*(12), 51-56.

Ross, I. M. (1998). The invention of the transistor. *Proceedings of the IEEE, 86*(1), 7-28.

Rutherford, E. (1897). A Magnetic Detector of Electrical Waves and Some of Its Applications. *Philosophical Transactions of the Royal Society of London. Series A, Containing Papers of a Mathematical or Physical Character, 189*, 1-24. doi:10.1098/rsta.1897.0001

Sarkar, T. K., & all, e. (2006). *History of wireless* (Vol. 177): John Wiley & Sons.

Schumpeter, J. A. (1939). *Business cycles; a theoretical, historical, and statistical analysis of the capitalist process (Fels)* (1st ed.). New York, London,: McGraw-Hill Book Company, inc.

Schumpeter, J. A., & Opie, R. (1934). *The theory of economic development; an inquiry into profits, capital, credit, interest, and the business cycle*. Cambridge, Mass.,: Harvard University Press.

Schweber, S. S. (1986). The empiricist temper regnant: Theoretical physics in the United States 1920-1950. *Historical Studies in the Physical and Biological Sciences, 17*(1), 55-98.

Shultz, S., Nelson, E., & Dunbar, R. I. M. (2012). Hominin cognitive evolution: identifying patterns and processes in the fossil and archaeological record. *Philosophical transactions of the Royal Society of London. Series B, Biological sciences, 367*(1599), 2130-2140. doi:10.1098/rstb.2012.0115

Simon, H. A. (1996). *The sciences of the artificial*: MIT press.

Thackray, A., & Myers, M. (2000). *Arnold O. Beckman: One hundred years of excellence* (Vol. 3): Chemical Heritage Foundation.

Tushman, M. L., Anderson, P. C., & O'Reilly, C. (1997). Technology cycles, innovation streams, and ambidextrous organizations: organization renewal through innovation streams and strategic change. *Managing strategic innovation and change, 34*(3), 3-23.

Tyne, G. F. J. (1977). *Saga of the vacuum tube*: Sams Indianapolis.

Usher, A. P. (1929). *A history of mechanical inventions*. New York: McGraw-Hill Book Company.

Usher, A. P. (2013). *A history of mechanical inventions (Revised Edition)* (Dover Edition (1988) ed.): Courier Corporation.

van der Kooij, B. J. G. (2016). How did the General Purpose Technology Electricity contribute to the Second Industrial Revolution (I): The Power Engines. Retrieved from Delft Repository website: https://repository.tudelft.nl/islandora/object/uuid%3A33eabc26-9220-41bc-8419-2f28087e8f32?collection=research

van der Kooij, B. J. G. (2017). How did the General Purpose Technology Electricity contribute to the Second Industrial Revolution (II): The Communication Engines. Retrieved from Delft Repository website: https://repository.tudelft.nl/islandora/object/uuid%3A009bdaa4-065f-4144-8b40-b2d86d02d0df?collection=research doi:10.13140/RG.2.2.26473.49768

van der Kooij, B. J. G. (2021a). *Deep Origins of Innovation (Part II) - Memes of Novelty*. Working Paper.

van der Kooij, B. J. G. (2021b). How did the General Purpose Technology 'Internal Combustion Engine'contribute to the Third Industrial Revolution: The Power Engines. *Available at SSRN 3852628*.

Williams, M. R. (1976). The difference engines. *The Computer Journal, 19*(1), 82-89. doi:10.1093/comjnl/19.1.82

Wolkonowicz, J. P. (1981). *The Philco Corporation: historical review and strategic analysis*. Massachusetts Institute of Technology,

Wood, J. (1992). *History of international broadcasting*: Iet.

# Acknowledgments

This scholarly work has been created by myself alone. Any errors made are mine. It is a work based on secondary research into 'technical/economic/scientific/political/social' phenomena investigated and described by others. Described from their point of view (as family-member, journalist, biographer, or scholar) published in the traditional way in books and articles in professional journals. According the fair use clause in the Copyright Acts, we took great care to acknowledge the sources, but are bound to fail in its multitude. Next to all the substantive contributions of the scholars acknowledged and mentioned extensively in the quotes and references, we were able —through the modern digital infrastructure called Internet— to get access to many primary sources. As provided on numerous websites and blogs owned and maintained by scholars, writers, collectors, museums large and small, and other institutions. They gave me the highly appreciated 'open access'[677] to their work, often too numerous to quote, but if the occasion arose acknowledged in a footnote for reference. Realising they observed our mutual subject of interest from their own perspectives, that diversity gave me the opportunity to look at the case from different angles. And, to finalize this enumeration of sources, I used for background orientation much of the information available on the digital encyclopaedia called Wikipedia. There we found the great treasury of information available as Wikimedia Commons. We are in debt to all those unknown people who contributed material that gave me the opportunity to illustrate my cases. If the original creator was known, we included their names. To the others we apologize.

# About the author

Drs.Ir.Ing. B. J. G. van der Kooij (b. 1947) in 1975 obtained his MBA (thesis: Innovation in SMEs) at the Interfaculteit Bedrijfkunde (nowadays part of the Rotterdam Erasmus University). In 1977 he obtained his MSEE (thesis: Technology Forecasting of Micro-electronics) at the Delft University of Technology. He started his career as assistant to the board of directors of Holec NV, a manufacturer of electrical power systems employing about 8,000 people at that time. His responsibilities were in the field of corporate strategy and innovation of Holec's electronic activities.

---

[677] Open access refers to online research outputs that are free of all restrictions on access (e.g. access tolls) and free of many restrictions on use (e.g. certain copyright and license restrictions). Our University of Technology in Delft aim is that TU Delft researchers publish each of their publications in Open Access.

Travelling extensively to Japan and California, he became well known as a Dutch guru on the topic of innovation and microelectronics.

In 1982–1986 he was a member of the Dutch Parliament (Tweede Kamer der Staten Generaal) and spokesman on the fields of economic, industrial, science, innovation, and aviation policy. He became the first member to introduce the personal computer in Parliament, but his work on topics like the TNO-Act, the Patent Act, the Chips-Act, and others went largely unnoticed.

After the 1986 elections and the massive loss for his party (VVD), he was dismissed from politics and became a part-time professor (Buitengewoon Hoogleraar) at the Eindhoven University of Technology. His endowed chair was the Management of Innovation. Additionally, in 1986 he started his own company, Ashmore Software BV, developer of software for professional tax applications on personal computers. After closing these activities in 2003, he became a real estate project developer, and in 2009 real estate consultant till his retirement in 2013.

Innovation being the focus of attention all his corporate, entrepreneurial, political, and scientific life, he wrote three books on the subject and published several articles. In his first book, he explored the technological dimension of innovation (the pervasive role of microelectronics). His second book focused on the management of innovation and the human role in the innovation process. And in his third book, he formulated "Laws of Innovation" based on the Dutch societal environment in the 1980s.

In 2012 he started studying the topic of innovation again. His focus is on the theory of innovation, and his aim is to develop a multidimensional model explaining innovation. For this he creates extensive and detailed case studies observing the inventions of the steam engine, the electromotive engines, the communication engines, the combustion engines and the computing engines. He studied their characteristics from a multidisciplinary perspective (economic, technical, and social). After a first fruitless effort to promote at the TU-Delft, in 2019 he was accepted at the Erasmus University by Prof. Dr. Huub Zwart as a PhD candidate. They agreed on his research proposal to investigate the Nature of Innovation by investigating through case studies what happened in the real world. Case studies that are presented in the Invention Series.

*INNOVARE NECESSE EST*

www.ingramcontent.com/pod-product-compliance
Lightning Source LLC
Chambersburg PA
CBHW052340220526
45465CB00003BA/882